Analog and Digital Signals and Systems

R.K. Rao Yarlagadda

Analog and Digital Signals and Systems

 Springer

R.K. Rao Yarlagadda
School of Electrical & Computer
    Engineering
Oklahoma State University
Stillwater OK 74078-6028
202 Engineering South
USA
rao.yarlagadda@okstate.edu

ISBN 978-1-4419-0033-3   e-ISBN 978-1-4419-0034-0
DOI 10.1007/978-1-4419-0034-0
Springer New York Dordrecht Heidelberg London

Library of Congress Control Number: 2009929744

Printed on acid-free paper

Springer is part of Springer Science+Business Media (www.springer.com)

This book is dedicated to my wife
Marceil, children, Tammy Bardwell, Ryan Yarlagadda
and Travis Yarlagadda and their families

Note to Instructors

The solutions manual can be located on the book's webpage http://www/
springer.com/engineering/cirucits+ %26+systems/bok/978-1-4419-0033-3

Preface

This book presents a systematic, comprehensive treatment of analog and discrete signal analysis and synthesis and an introduction to analog communication theory. This evolved from my 40 years of teaching at Oklahoma State University (OSU). It is based on three courses, Signal Analysis (a second semester junior level course), Active Filters (a first semester senior level course), and Digital signal processing (a second semester senior level course). I have taught these courses a number of times using this material along with existing texts. The references for the books and journals (over 160 references) are listed in the bibliography section. At the undergraduate level, most signal analysis courses do not require probability theory. Only, a very small portion of this topic is included here.

I emphasized the basics in the book with simple mathematics and the sophistication is minimal. Theorem-proof type of material is not emphasized. The book uses the following model:

1. Learn *basics*
2. Check the work using *bench marks*
3. Use *software* to see if the results are accurate

The book provides detailed *examples* (over 400) with applications. A three-number system is used consisting of chapter number – section number – example or problem number, thus allowing the student to quickly identify the related material in the appropriate section of the book. The book includes well *over 400 homework problems*. Problem numbers are identified using the above three-number system. Hints are provided wherever additional details may be needed and may not have been given in the main part of the text. A detailed *solution manual* will be available from the publisher for the instructors.

Summary of the Chapters

This book starts with an *introductory chapter* that includes most of the basic material that a junior in electrical engineering had in the beginning classes. For those who have forgotten, or have not seen the material recently, it gives enough background to follow the text. The topics in this chapter include singularity functions, periodic functions, and others. Chapter 2 deals with convolution and correlation of periodic and aperiodic functions. Chapter 3 deals with approximating a function by using a set of basis functions, referred to as the generalized Fourier series expansion. From these concepts, the three basic Fourier series expansions are derived. The discussion includes detailed discussion on the operational properties of the Fourier series and their convergence.

Chapter 4 deals with Fourier transform theory derived from the Fourier series. Fourier series and transforms are the bases to this text. Considerable material in the book is based on these topics. Chapter 5 deals with the relatives of the Fourier transforms, including Laplace, cosine and sine, Hartley and Hilbert transforms.

Chapter 6 deals with basic systems analysis that includes linear time-invariant systems, stability concepts, impulse response, transfer functions, linear and nonlinear systems, and very simple filter circuits and concepts. Chapter 7 starts with the Bode plots and later deals with approximations using classical analog Butterworth, Chebyshev, and Bessel filter functions. Design techniques, based on both amplitude and phase based, are discussed. Last part of this chapter deals with analysis and synthesis of active filter circuits. Examples of basic low-pass, high-pass, band-pass, band elimination, and delay line filters are included.

Chapter 8 builds a bridge to go from the continuous-time to discrete-time analysis by starting with sampling theory and the Fourier transform of the ideally sampled signals. Bulk of this chapter deals with discrete basis functions, discrete-time Fourier series, discrete-time Fourier transform (DTFT), and the discrete Fourier transform (DFT). Chapter 9 deals with fast implementations of the DFT, discrete convolution, and correlation. Second part of the chapter deals the z-transforms and their use in the design of discrete-data systems. Digital filter designs based on impulse invariance and bilinear transformations are presented. The chapter ends with digital filter realizations.

Chapter 10 presents an introduction to analog communication theory, which includes basic material on analog modulation, such as AM and FM, demodulation, and multiplexing. Pulse modulation methods are introduced.

Appendix A reviews the basics on matrices; Appendix B gives a brief introduction on MATLAB; and Appendix C gives a list of useful formulae. The book concludes with a list of references and Author and Subject indexes.

Suggested Course Content

Instructor is the final judge of what topics will best suit his or her class and in what depth. The suggestions given below are intended to serve as a guide only. The book permits flexibility in teaching analysis, synthesis of continuous-time and discrete-time systems, analog filters, digital signal processing, and an introduction to analog communications. The following table gives suggestions for courses.

	Topical Title	Related topics in chapters
One semester	$\begin{pmatrix} \text{Fundamentals of analog} \\ \text{signals and systems} \end{pmatrix}$	Chapters 1–4, 6
One semester	Systems and analog filters	Chapters 4, 5*, 6, 7
One semester	$\begin{pmatrix} \text{Introduction to digital} \\ \text{signal processing} \end{pmatrix}$	Chapters 4*, 6*, 8, 9
Two semesters	$\begin{pmatrix} \text{Signals and an introduction to} \\ \text{analog communications} \end{pmatrix}$	Chapters 1–4, 5*, 6, 8*, 10

*Partial coverage

Acknowledgements

The process of writing this book has taken me several years. I am indebted to all the students who have studied with me and taken classes from me. Education is a two-way street. The teachers learn from the students, as well as the students learn from the teachers. Writing a book is a learning process.

Dr. Jack Cartinhour went through the material in the early stages of the text and helped me in completing the solution manual. His suggestions made the text better. I am deeply indebted to him. Dr. George Scheets used an earlier version of this book in his signal analysis and communications theory class. Dr. Martin Hagan has reviewed a chapter. Their comments were incorporated into the manuscript. Beau Lacefield did most of the artwork in the manuscript. Vijay Venkataraman and Wen Fung Leong have gone through some of the chapters and their suggestions have been incorporated. In addition, Vijay and Wen have provided some of the MATLAB programs and artwork. I appreciated Vijay's help in formatting the final version of the manuscript.

An old adage of the uncertainty principle is, no matter how many times the author goes through the text, mistakes will remain. I sincerely appreciate all the support provided by Springer. Thanks to Alex Greene. He believed in me to complete this project. I appreciated the patience and support of Katie Chen. Thanks to Shanty Jaganathan and her associates of Integra-India. They have been helpful and gracious in the editorial process.

Dr. Keith Teague, Head, School of Electrical and Computer Engineering at Oklahoma State University has been very supportive of this project and I appreciated his encouragement.

Finally, the time spent on this book is the time taken away from my wife Marceil, children Tammy, Ryan and Travis and my grandchildren. Without my family's understanding, I could not have completed this book.

Oklahoma, USA   R.K. Rao Yarlagadda

Contents

List of Tables

Chapter 1
Basic Concepts in Signals

1.1 Introduction to the Book and Signals

The primary goal of this book is to introduce the reader on the basic principles of signals and to provide tools thereby to deal with the analysis of analog and digital signals, either obtained naturally or by sampling analog signals, study the concepts of various transforming techniques, filtering analog and digital signals, and finally introduce the concepts of communicating analog signals using simple modulation techniques. The basic material in this book can be found in several books. See references at the end of the book.

A signal is a pattern of some kind used to convey a message. Examples include smoke signals, a set of flags, traffic lights, speech, image, seismic signals, and many others. Smoke signals were used for conveying information that goes back before recorded history. Greeks and Romans used light beacons in the pre-Christian era. England employed a long chain of beacons to warn that Spanish Armada is approaching in the late sixteenth century. Around this time, the word signal came into use perceptible by sight, hearing, etc., conveying information. The present day signaling started with the invention of the Morse code in 1838. Since then, a variety of signals have been studied. These include the following inventions: Facsimile by Alexander Bain in 1843; telephone by Alexander Bell in 1876; wireless telegraph system by Gugliemo Marconi in 1897; transmission of speech signals via radio by Reginald Fessenden in 1905, invention and demonstration of television, the birth of television by Vladimir Zworykin in the 1920 s, and many others. In addition, the development of radar and television systems during World War II, proposition of satellite communication systems, demonstration of a laser in 1955, and the research and developments of many signal processing techniques and their use in communication systems. Since the early stages of communications, research has exploded into several areas connected directly, or indirectly, to signal analysis and communications. Signal analysis has taken a significant role in medicine, for example, monitoring the heart beat, blood pressure and temperature of a patient, and vital signs of patients. Others include the study of weather phenomenon, the geological formations below the surface and deep in the ground and under the ocean floors for oil and gas exploration, mapping the underground surface using seismometers, and others. Researchers have concluded that computers are powerful and necessary that they need to be an integral part of any communication system, thus generating significant research in digital signal processing, development of Internet, research on HDTV, mobile and cellular telephone systems, and others. Defense industry has been one of the major organizations in advancing research in signal processing, coding, and transmission of data. Several research areas have surfaced in signals that include processing of speech, image, radar, seismic, medical, and other signals.

1.1.1 Different Ways of Looking at a Signal

Consider a signal $x(t)$, a function representing a physical quantity, such as voltage, current, pressure, or any other variable with respect to a second variable t, such as time. The terms of interest are the time t and the signal $x(t)$. One of the main topics of

R.K.R. Yarlagadda, *Analog and Digital Signals and Systems*, DOI 10.1007/978-1-4419-0034-0_1,
© Springer Science+Business Media, LLC 2010

this book is the analysis of signals. Webster's dictionary defines the analysis as

1. Separation of a thing into the parts or elements of which it is composed.
2. An examination of a thing to determine its parts or elements.
3. A statement showing the results of such an examination.

There are other definitions. In the following the three parts are considered using simple examples. Consider the sinusoidal function and its expansion using *Euler's formula*:

$$x(t) = A_0 \cos(\omega_0 t + \theta_0)$$

$$= \left(\frac{A_0}{2} e^{j\theta_0}\right) e^{j\omega_0 t} + \left(\frac{A_0}{2} e^{-j\theta_0}\right) e^{-j\omega_0 t} \quad (1.1.1)$$

$$= \mathrm{Re}(A_0 e^{j\omega_0 t} e^{j\theta_0}).$$

In (1.1.1) A_0 is assumed to be positive and real and $A_0 e^{j\theta_0}$ is a complex number carrying the amplitude and phase angle of the sinusoidal function and is by definition the *phasor representation* of the given sinusoidal function. Some authors refer to this as *phasor transform* of the sinusoidal signal, as it transforms the time domain sinusoidal function to the complex frequency domain. A brief discussion on complex numbers is included later in Section 1.6. This signal can be described in another domain, i.e., such as the frequency domain. The amplitude is $(A_0/2)$ and the phase angles of $\pm\theta_0$ corresponding to the frequencies $\pm f_0 = \pm\omega_0/2\pi$ Hz. In reality, only positive frequencies are available, but Euler's formula in (1.1.1) dictates that both the positive and negative frequencies need to be identified as illustrated in Fig. 1.1.1a. This description is the two-sided amplitude and phase line spectra of $x(t)$. Amplitudes are always positive and are located at $f = \pm\omega_0/2\pi = \pm f_0$ Hz, symmetrically located around the zero frequency, i.e., with even symmetry. The phase spectrum consists of two angles $\theta = \pm\theta_0$ corresponding to the positive and negative frequencies, respectively, with odd symmetry. Since (t) is real, we can pictorially describe it by one- or two-sided *amplitude and phase line spectra* as shown in Fig. 1.1.1a,b,c,d. The following example illustrates the three steps.

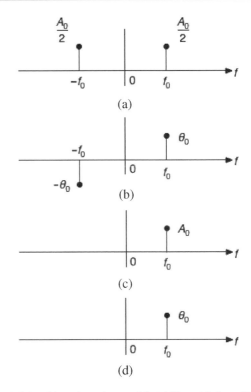

Fig. 1.1.1 $x(t) = A_0 \cos(\omega_0 t + \theta_0)$. (a) Two-sided amplitude spectrum, (b) two-sided phase spectrum, (c) one-sided amplitude spectrum, and (d) one-sided phase spectrum

Example 1.1.1 Express the following function in terms of a sum of cosine functions:

$$x(t) = -A_0 + A_1 \cos(\omega_1 t + \theta_1)$$
$$- A_2 \cos(\omega_2 t) - A_3 \sin(\omega_3 t + \theta_3), A_i > 0. \quad (1.1.2)$$

Solution: Using trigonometric relations to express each term in (1.1.2) in the form of $A_i \cos(\omega_i t + \theta_i)$ results in

$$x(t) = A_0 \cos((0)t - 180°)$$
$$+ A_1 \cos(\omega_1 t + \theta_1) + A_2 \cos(\omega_2 t - 180°) \quad (1.1.3)$$
$$+ A_3 \cos(\omega_3 t + \theta_3 + 90°).$$

In the first and the third terms either $180°$ or $-180°$ could be used, as the end result is the same. The two-sided line spectra of the function in (1.1.2) are shown in Fig. 1.1.2. How would one get the functions of the type shown in (1.1.3) for an arbitrary

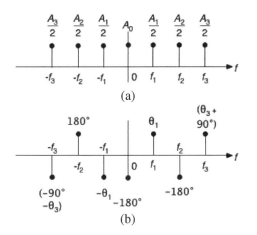

Fig. 1.1.2 (a) Two-sided amplitude spectra and (b) two-sided phase spectra

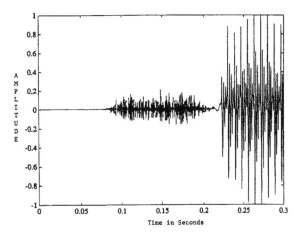

Fig. 1.1.3 Speech …sho in …show ._male 2000 Samples @ 8000 samples per second. Printed with the permission from Hassan et al. (1994)

function? The sign and cosine functions are the building blocks of the Fourier series in Chapter 3 and later the Fourier transforms in Chapter 4. The function $x(t)$ has four frequencies:

$$f_1(=0), f_2, f_3, f_4 \text{ with amplitudes } A_0, A_1, A_2, A_3$$
$$\text{and phases } -180^o, \theta_1, -180^o, \theta_3 + 90^o.$$

Figure 1.1.2 illustrates pictorially the discrete locations of the frequencies, their amplitudes, and phases. The signal in (1.1.2) can be described by using the time domain function or in terms of frequencies. In the figures, $\omega'(= 2\pi f)'$s in radians per second could have been used rather than f's in Hz. ∎

1.1.2 Continuous-Time and Discrete-Time Signals

A signal $x(t)$ is a continuous-time signal if t is a continuous variable. It can take on any value in the continuous interval (a, b). Continuous-time signal is an analog signal. If a function $y[n]$ is defined at discrete times, then it is a discrete-time signal, where n takes integer values. In Chapter 8 discrete-time signals will be studied by sampling the continuous signals at equal sampling intervals of t_s seconds and write $x(nt_s)$, where n an integer. This is expressed by

$$x[n] \equiv x(nt_s). \tag{1.1.4}$$

Example(s) 1.1.2 In this example several specific examples of interest are considered. In the first one, part of the time signal illustrating a male voice of speech in the sentence "…*Sho*w the rich lady out" is shown in Fig. 1.1.3. The speech signal is sampled at 8000 samples per second. There are three portions of the speech "/…/, /sh/, /o/" shown in the figure. The first part of the signal does not have any speech in it and the small amplitudes of the signal represent the noise in the tape recorder and/or in the room where the speech was recorded. It represents a random signal and can be described only by statistical means. Random signal analysis is not discussed in any detail in this book, as it requires knowledge of probability theory. The second part represents the phoneme "*sh*" that does not show any observable pattern. It is a time signal for a very short time and has finite energy. Power and energy signals are studied in Section 1.5. Third part of the figure represents the vowel "*o*," showing a structure of (almost) periodic pulses for a short time. In this book, aperiodic or non-periodic signals with finite energy and periodic signals with finite average power will be studied. One goal is to come up with a model for each portion of a signal that can be transmitted and reconstructed at the receiver.

Next three examples are from food industry. Small businesses are sprouting that use signal processing. For example, when we go to a grocery store we may like to buy a watermelon. It may not always be possible to judge the ripeness of the watermelon by outward characteristics such as external color, stem conditions, or just the way it looks. A sure way of looking at the quality is to cut the watermelon open and taste it before we buy it. This implies we break it first, which is destructive testing. Instead, we can use our grandmother's procedure in selecting a watermelon. She uses her knuckles to send a signal into the watermelon. From the audio response of the watermelon she decides whether it is good or not based on her prior experience. We can simulate this by putting the watermelon on a stand, use a small hammer like device, give a slight tap on the watermelon, and record the response. A simplistic model of this is shown in Fig. 1.1.4. The responses can be categorized by studying the outputs of tasty watermelons. For an interesting research work on this topic, see Stone et al (1996).

Image processing can be used to check for burned crusts, topping amount distribution, such as the location of pepperoni pizza slices, and others. For an interesting article on this subject, see Wagner (1983), which has several applications in the food industry.

The next two examples are from the surface seismic signal analysis. In the first one, we use a source in the form of dynamite sticks representing a source, dig a small hole, and blow them in the hole. The ground responds to this input and the response is recorded using a seismometer and a tape recorder. The analysis of the recorded waveform can provide information about the underground cavities and pockets of oil and other important measures.

Geologists drill holes into the ground and a small slice of the core sample is used to measure the oil content by looking at the percentage of the area with dark spots on the slice, which is image processing.

Another example of interest is measuring the distance from a ground station to an airplane. Send a signal with square wave pulses toward the airplane and when the signal hits it, a return signal is received at the ground station. A simple model is shown in Fig. 1.1.5. If we can measure the time between the time the signal left from the ground station and the time it returned, identified as T in the figure, we can determine the distance between the ground station and the target by the formula

$$x = 3(10^8) \ (m/s)$$
$$T(\text{signal round trip time in seconds})/2. \tag{1.1.5}$$

The constant $c = 3(10^8)$ m/s is the speed of light. *Radar and sonar* signal processing are two important areas of signal processing applications.

An exciting field of study is the biomedical area. We are well aware of a healthy heart that beats periodically, which can be seen from a record of *an electrocardiogram* (*ECG*). The ECG represents changes in the voltage potential due

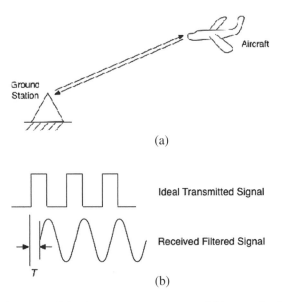

(a)

(b)

Fig. 1.1.5 (a) Radar range measurement and (b) transmitted and received filtered signals

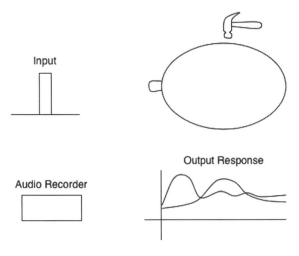

Fig. 1.1.4 Watermelon responses to a tap

to electrochemical processes in the heart cells. Inferences can be made about the health of the heart under observation from the ECG. Another important example is the *electroencephalogram* (*EEG*), which measures the electrical activity in brain. ■

Signal processing is an important area that interests every engineer. *Pattern recognition and classification* is almost on top of the list. See, for example, O'Shaughnessy (1987) and Tou and Gonzalez (1974). For example, how do we distinguish two phonemes, one is a vowel and the other one is a consonant. A rough measure of frequency of a waveform with zero average value is the number of zero crossings per unit time. We will study in much more detail the frequency content in a signal later in terms of Fourier transforms in Chapter 4. Vowel sounds have lower frequency content than the consonants. A simple procedure to measure frequency in a speech segment is by computing the number of zero crossings in that segment. To differentiate a vowel from a consonant, set a threshold level for the frequency content for vowels and consonants that differentiate between vowels and consonants. If the frequency content is higher than this threshold, then the phoneme is a consonant. Otherwise, it is a vowel. If we like to distinguish one vowel from another we may need more than one measure. Vocal tract can be modeled as an acoustic tube with resonances, called *formants*. Two formant frequencies can be used to distinguish two vowels, say /u/ and /a/. See *Problem 1.1.1*. Two formant frequencies may not be enough to distinguish all the phonemes, especially if the signal is corrupted by noise.

Consider a simple pattern classification problem with M prototype patterns $\underline{z}_1, \underline{z}_2, \ldots, \underline{z}_M$, where $\underline{z}_i$ is a vector representing an ith pattern. For simplicity we assume that each pattern can be represented by a pair of numbers, say $z_i = (z_{i1}, z_{i2}), i = 1, 2 \ldots M$ and classify an arbitrary pattern $\underline{x} = (x_1, x_2)$ to represent one of the prototype patterns. The *Euclidean distance* between a pattern $\underline{x}$ and the ith prototype pattern is defined by

$$D_i = \|\underline{x} - \underline{z}_i\| = \sqrt{(x_1 - z_{i1})^2 + (x_2 - z_{i2})^2}. \quad (1.1.6)$$

A simple classifier is a minimum distance classifier that computes the distance from a pattern $\underline{x}$ of the signal to be classified to the prototype of each class and assigns the unknown pattern to the class which it is closest to. That is, if $D_i < D_j$, for all $i \neq j$, then we make the decision that $\underline{x}$ belongs to the ith prototype pattern. Ties are rare and if there are, they are resolved arbitrarily. In the above discussion two measures are assumed for each pattern. More measures give a better separation between classes.

There are several issues that would interest a *biomedical signal processor*. These include removal of any noise present in the signals, such as 60-Hz interference picked up by the instruments, interference of the tools or meters that measure a parameter, and other signals that interfere with the desired signal. Finding the important facets in a signal, such as the frequency content, and many others is of interest. ■

1.1.3 Analog Versus Digital Signal Processing

Most signals are analog signals. Analog signal processing uses analog circuit elements, such as resistors, capacitors, inductors, and active components, such as operational amplifiers and non-linear devices. Since the inductors are made from magnetic material, they have inherent resistance and capacitance. This brings the quality of the components low. They tend to be bulky and their effectiveness is reduced. To alleviate this problem, active RC networks have been popular. Analog processing is a natural way to solve differential equations that describe physical systems, without having to resort to approximate solutions. Solutions are obtained in real time. In Chapter 10 we will see an example of analog encryption of a signal, wherein the analog speech is scrambled by the use of modulation techniques.

Digital signal processing makes use of a special purpose computer, which has three basic elements, namely adders, multipliers, and memory for storage. Digital signal processing consists of numerical computations and there is no guarantee that the processing can be done in real time. To encrypt a set of numbers, these need to be converted into another set of numbers in the digital encryption scheme, for example. The complete encrypted signal is needed before it can be decrypted. In addition, if the input and the output signals are analog, then an analog-to-digital converter (*A/D*), a *digital proces-sor*, and a *digital-to-analog converter* (*D/A*) are needed to implement analog processing by digital means. Special purpose processor with A/D and D/A converters can be expensive.

Digital approach has distinct advantages over analog approaches. Digital processor can be used to implement different versions of a system by chan-ging the software on the processor. It has flexibility and repeatability. In the analog case, the system has to be redesigned every time the specifications are changed. Design components may not be available and may have to live with the component values within some tolerance. Components suffer from parameter variations due to room temperature, humidity, supply voltages, and many other aspects, such as aging, component failure. In a particular situation, many of the above problems need to be investigated before a complete decision can be made. Future appears to be more and more digital. Many of the digital signal processing filter designs are based on using analog filter designs. Learning both analog and digital signal processing is desired.

Deterministic and random signals: Deterministic signals are specified for any given time. They can be modeled by a known function of time. There is no uncertainty with respect to any value at any time. For example, $x(t) = \sin(t)$ is a deterministic signal. A random signal $y(t) = x(t) + n(t)$ can take ran-dom values at any given time, as there is uncertainty about the noise signal $n(t)$. We can only describe such signals through statistical means, and the dis-cussion on this topic will be minimal.

1.1.4 Examples of Simple Functions

To begin the study we need to look into the concept of expressing a signal in terms of functions that can be generated in a laboratory. One such function is the sinusoidal function $x(t) = A_0 \cos(\omega_0 t + \theta_0)$ seen earlier in (1.1.1), where A_0, ω_0, and θ_0 are some constants. A digital signal can be defined as a sequence in the forms

$$x[n] = \begin{cases} a^n, n \geq 0 \\ 0, n < 0 \end{cases}, \quad (1.1.7)$$

$$\{x[n]\} = \{\dots, 0, 0, 0, 1, a, a^2, \dots, a^n, \dots\}, \quad (1.1.8)$$

$$\{x[n]\} = \{\dots, 1, a, a^2, \dots, a^n, \dots\} \text{ or } \{x[n]\}$$
$$\qquad\qquad\quad \uparrow$$
$$= \{1, a, a^2, \dots, a^n, \dots\} \qquad (1.1.9)$$
$$\quad \uparrow$$

In (1.1.8) reference points are not identified. In (1.1.9), the arrow below 1 is the 0 index term. The first term is assumed to be zero index term if there is no arrow and all the values of the sequence are zero for $n < 0$. We will come back to this in Chapter 8.

A signal $x(t)$ is a real signal if its value at some t is a real number. A complex signal $x(t)$ consists of two real signals, $x_1(t)$ and $x_2(t)$ such that $x(t) = x\_1(t) + jx_2(t)$, where $j = \sqrt{-1}$. The symbol j (or i) is used to represent the imaginary part.

Interesting functions: a. Π **Function:** The Π func-tion is centered at t_0 with a width of τ s shown in Fig. 1.1.6. It is not defined at $t = t_0 \pm \tau/2$ and is symbolically expressed by

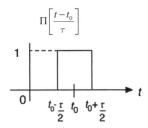

Fig. 1.1.6 A Π function

$$\Pi\left[\frac{t - t_0}{\tau}\right] \begin{cases} 1, & |t - t_0| < \tau/2 \\ 0, & \text{otherwise} \end{cases} \quad (1.1.10)$$

$$x(t)\Pi\left[\frac{t - t_0}{\tau}\right] = \begin{cases} x(t), & |t - t_0| < \tau/2 \\ 0, & \text{otherwise} \end{cases}$$

This Π function is a deterministic signal. It is even, as $\Pi[-t] = \Pi[t]$ and $\Pi[t_0 - t] = \Pi[t - t_0]$. Some use the symbol "rect" and $(1/\tau)\text{rect}((t - t_0)/\tau)$ is a rectangular pulse of width τ s centered at $t = t_0$ with a height $(1/\tau)$.

b. Λ **Function:** The triangular function shown in Fig. 1.1.7 is defined by

$$\Lambda\left[\frac{t - t_0}{\tau}\right] = \begin{cases} 1 - \frac{|t - t_0|}{\tau}, & |t - t_0| < \tau \\ 0, & \text{otherwise} \end{cases}. \quad (1.1.11)$$

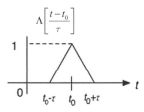

Fig. 1.1.7 A Λ function

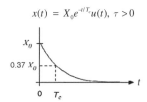

Fig. 1.1.9 Exponential decaying function

Rectangular function defined in (1.1.10) has a width of τ seconds, whereas the triangular function defined in (1.1.11) has a width of 2τ s. The symbol "tri" is also used for a triangular function and $\text{tri}((t - t_0)/\tau)$ describes the function in (1.1.11).

c. **Unit step function:** It is shown in Fig. 1.1.8 and is

$$u(t) = \begin{cases} 1, t > 0 \\ 0, t < 0 \end{cases} \text{(not defined at } t = 0). \quad (1.1.12)$$

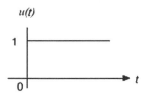

Fig. 1.1.8 Unit step function

The unit step function at $t = 0$ can be defined explicitly as 0 or 1 or $[u(0^+) + u(0^-)]/2 = .5$

d. **Exponential decaying function:** A simple such function is

$$x(t) = \begin{cases} X_0 e^{-t/T_c}, t \geq 0, T_c > 0 \\ 0, \quad \text{otherwise} \end{cases}. \quad (1.1.13)$$

See Fig. 1.1.9. It has a special significance, as $x(t)$ is the solution of a first-order differential equation. The constants X_0 and T_c can take different values and

$$\frac{x(T_c)}{X_0} = e^{-1} \approx .37 \text{ or } x(T_c) = .37 X_0. \quad (1.1.14)$$

$x(t)$ decreases to about 37% of its initial value in T_c s and is the *time constant*. It decreases to 2% in four time constants and $X(4T_c) \approx .018 X_0$. A measure associated with exponential functions is the *half life* T_h defined by

$$x(T_h) = \frac{1}{2} X_0, \; e^{-T_h/T_c}$$
$$= \frac{1}{2}, \; T_h = T_c \log_e(2) \cong .693 T_c. \quad (1.1.15)$$

e. **One-sided and two-sided exponentials:** These are described by

$$x_1(t) = \begin{cases} e^{-at}, t \geq 0 \\ 0, t < 0 \end{cases}, a > 0$$

$$x_2(t) = \begin{cases} 0, t \geq 0 \\ e^{at}, t < 0 \end{cases} a > 0, x_3(t) = e^{-a|t|}, a > 0.$$

$$(1.1.16)$$

$x_1(t)$ is the right-sided exponential, $x_2(t)$ is the left-sided exponential, and $x_3(t)$ is the two-sided exponential. These are sketched in Fig. 1.1.10. Using the unit step function, we have $x_2(t) = x_3(t)u(-t)$ and $x_1(t) = x_3(t)u(t)$.

Fig. 1.1.10 Exponential functions (a) $x_2(t) = e^{at}$ $u(-t), a > 0$,
(b) $x_1(t) = e^{-at}u(t), a > 0$,
and (c) $x_3(t) = e^{-a|t|}, a > 0$

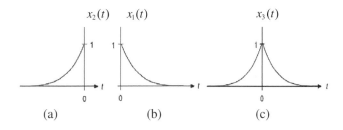

(a) (b) (c)

1.2 Useful Signal Operations

1.2.1 Time Shifting

Consider an arbitrary signal starting at $t = 0$ shown in Fig. 1.2.1a. It can be shifted to the right as shown in Fig. 1.2.1b. It starts at time $t = a > 0$, a delayed version of the one in Fig. 1.2.1a. Similarly it can be shifted to the left starting at time $-a$ shown in Fig. 1.2.1c. It is an advanced version of the one in Fig. 1.2.1a. We now have three functions: $x(t)$, $x(t - a)$, and $x(t + a)$ with $a > 0$. The delayed and advanced unit step functions are

$$u(t - a) = \begin{cases} 1, t > a \\ 0, t < a \end{cases},$$

$$u(t + a) = \begin{cases} 1, t > -a \\ 0, t < -a \end{cases}; (a > 0). \quad (1.2.1)$$

From (1.1.16), the right-sided delayed exponential decaying function is

$$x_1(t - \tau) = e^{-a(t-\tau)}u(t - \tau), \ a > 0, \ \tau > 0. \quad (1.2.2)$$

1.2.2 Time Scaling

The compression or expansion of a signal in time is known as time scaling. It is expanded in time if $a < 1$ and compressed in time if $a > 1$ in

$$\phi(t) = x(at), a > 0. \quad (1.2.3)$$

Example 1.2.1 Illustrate the rectangular pulse functions $\Pi[t], \Pi[2t]$, and $\Pi[t/2]$.

Solution: These are shown in Fig. 1.2.2 and are of widths 1, (1/2), and 2, respectively. The pulse $\Pi[2t]$ is a compressed version and the pulse $\Pi[t/2]$ is an expanded version of the pulse function $\Pi[t]$. ∎

1.2.3 Time Reversal

If $a = -1$ in (1.2.3), that is, $\phi(t) = x(-t)$, then the signal is *time reversed* (or *folded*).

Example 1.2.2 Let $x_1(t) = e^{-at}u(t)$. Give its time-reversed signal.

Solution: The time-reversed signal of $x_1(t)$ is $x_2(t) = e^{at}u(-t)$. ∎

1.2.4 Amplitude Shift

The amplitude shift of $x(t)$ by a constant K is $\phi(t) = K + x(t)$.

Combined operations: Some of the above signal operations can be combined into a general form. The signal $y(t) = x(at - t_0)$ may be described by one of the two ways, namely:

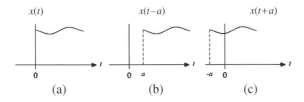

Fig. 1.2.1 (a) $x(t)$, (b) $x(t - a)$, (c) $x(t + a)$, $a > 0$

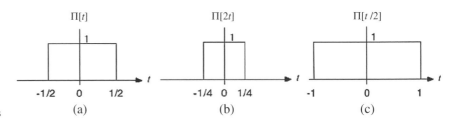

Fig. 1.2.2 Pulse functions

1. Time shift of t_0 followed by time scaling by a.
2. Time scaling by (a) followed by time shift of (t_0/a).

These can be visualized by the following:

1. $x(t) \xrightarrow[t \to t - t_0]{\text{shift}} v(t) = x(t - t_0) \xrightarrow[t \to at]{\text{scale}} y(t)$ (1.2.4)
$$= v(at) = x(at - t_0).$$

2. $x(t) \xrightarrow[t \to at]{\text{scale}} g(t) = x(at) \xrightarrow[t \to t - (t_0/a)]{\text{shift}} y(t)$ (1.2.5)
$$= g(t - (t_0/a)) = x(at - t_0).$$

Notation can be simplified by writing $y(t) = x(a(t - (t_0/a)))$. Noting that $\beta = at - t_0$ is a linear equation in terms of two constants a and t_0, it follows:

$$y(0) = x(-t_0) \text{ and } y(t_0/a) = x(0). \quad (1.2.6)$$

These two equations provide checks to verify the end result of the transformation. Following example illustrates some pitfalls in the order of time shifting and time scaling.

Example 1.2.3 Derive the expression for $y(t) = x(3t + 2)$ assuming $x(t) = \Pi[t/2]$.

Solution: Using (1.2.4) with $a = 3$ and $t_0 = -2$, we have

$$v(t) = x(t - t_0) = \Pi\left[\frac{t + 2}{2}\right],$$

$$y(t) = v(3t) = \Pi\left[\frac{3t + 2}{2}\right] = \Pi\left[\frac{t + (2/3)}{2/3}\right].$$

Using (1.2.5), we have

$$g(t) = x(at) = \Pi\left[\frac{3t}{2}\right],$$

$$y(t) = g(t - \frac{t_0}{a}) = g(t + \frac{2}{3})$$

$$= \Pi\left[\frac{3(t + (2/3))}{2}\right] = \Pi\left[\frac{t + (2/3)}{2/3}\right].$$

It is a rectangular pulse of unit amplitude centered at $t = -(2/3)$ with width $(2/3)$. We can check the

equations in (1.2.6) and $y(0) = x(3) = 0$ and $y(t_0/a) = y(3/2) = 0$. ∎

1.2.5 Simple Symmetries: Even and Odd Functions

Continuous-time even and odd functions satisfy

$x(t) = x(-t) \equiv x_e(t)$, an even function, $x(-t) = -x(t) \equiv x_0(t)$, an odd function. (1.2.7)

Examples of even and odd functions are shown in Fig. 1.2.3. The function $\cos(\omega_0 t)$ is an even function and $x_0(t) = \sin(\omega_0 t)$ is an odd function. An arbitrary real signal, $x(t)$, can be expressed in terms of its even and odd parts by

$$x(t) = x_e(t) + x_0(t), \quad x_e(t)$$
$$= [(x(t) + x(-t))/2], \quad x_0(t) \quad (1.2.8)$$
$$= [(x(t) - x(-t))/2].$$

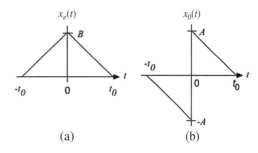

Fig. 1.2.3 (a) Even function and (b) odd function

1.2.6 Products of Even and Odd Functions

Let $x_e(t)$ and $y_e(t)$ be two even functions and $x_0(t)$ and $y_0(t)$ be two odd functions and arbitrary. Some general comments can be made about their products.

$x_e(-t)y_e(-t) = x_e(t)y_e(t)$, even function. (1.2.9)

$x_e(-t)y_0(-t) = -x_e(t)y_0(t)$, odd function. (1.2.10)

$x_0(-t)y_0(-t) = (-1)^2 x_0(t)y_0(t)$
$$= x_0(t)y_0(t), \text{even function.} \quad (1.2.11)$$

Fig. 1.2.4 (a) $x_1(t)$, (b) $x_{ie}(t)$ = $x_2(t)$ even part of $x_1(t)$, and (c) $x_{1o}(t) = x_3(t)$ odd part of $x_1(t)$

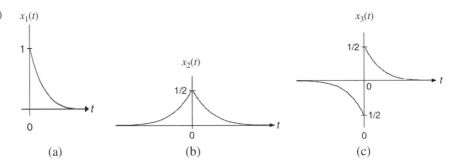

Note that the functions $\Pi[t]$, $\Lambda[t]$ and $\Pi[t]\cos(\omega_0 t)$ are even functions and $\Pi[t]\sin(\omega_0 t)$ is an odd function. The even and odd parts of the exponential pulse $x_1(t) = e^{-t}u(t)$ are shown in Fig. 1.2.4 and are

$$x_{1e}(t) = \frac{1}{2}(x_1(t) + x_1(-t)),$$
$$x_{1o}(t) = \frac{1}{2}(x_1(t) - x_1(-t)). \qquad (1.2.12)$$

1.2.7 Signum (or sgn) Function

The signum (or sgn) function is an odd function shown in Fig. 1.2.5:

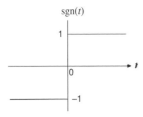

Fig. 1.2.5 Signum function sgn(t)

$$\mathrm{sgn}(t) = u(t) - u(-t) = 2u(t) - 1. \qquad (1.2.13)$$

$$\mathrm{sgn}(t) = \lim_{a \to 0}[e^{-at}u(t) - e^{at}u(-t)], a > 0. \qquad (1.2.14)$$

It is not defined at $t = 0$ and is chosen as 0.

1.2.8 Sinc and Sinc^2 Functions

The sinc and sinc^2 functions are defined in terms of an independent variable λ by

$$\mathrm{sinc}(\pi\lambda) = \frac{\sin(\pi\lambda)}{\pi\lambda}, \ \mathrm{sinc}^2(\pi\lambda) = \frac{\sin^2(\pi\lambda)}{(\pi\lambda)^2} \qquad (1.2.15)$$

Some authors use sinc(λ) for sinc($\pi\lambda$) in (1.2.15). Notation in (1.2.15) is common. Sinc ($\pi\lambda$) is indeterminate at $t = 0$. Using the L'Hospital's rule,

$$\lim_{\lambda \to 0}\frac{\sin(\pi\lambda)}{\pi\lambda} = \lim_{\lambda \to 0}\frac{\frac{d\sin(\pi\lambda)}{d\lambda}}{\frac{d(\pi\lambda)}{d\lambda}} = \lim_{\lambda \to 0}\frac{\pi\cos(\pi\lambda)}{\pi} = 1.$$

$$\qquad (1.2.16)$$

In addition, since $\sin(\pi\lambda)$ is equal to zero for $\lambda = \pm n$, n an integer, it follows that

$$\mathrm{sinc}(\pi\lambda) = 0, \quad \lambda = \pm n, \quad n \neq 0 \text{ and an integer.}$$
$$\qquad (1.2.17a)$$

Interestingly, the function $|\mathrm{sinc}(\pi\lambda)|$ is bounded by $|(1/\pi\lambda)|$ as $|\sin(\pi\lambda)| \leq 1$. The side lobes of $|\mathrm{sinc}(\pi\lambda)|$ are larger than the side lobes of $\mathrm{sinc}^2(\pi\lambda)$, which follows from the fact that the square of a fraction is less than the fraction we started with. Both the sinc and the sinc^2 functions are even. That is,

$$\mathrm{sinc}(-\pi\lambda) = \mathrm{sinc}(\pi\lambda) \text{ and } \mathrm{sinc}^2(-\pi\lambda) = \mathrm{sinc}^2(\pi\lambda).$$
$$\qquad (1.2.17b)$$

These functions can be evaluated easily by a calculator. For the sketch of a sinc function using MATLAB, see Fig. B.5.2 in Appendix B.

1.2.9 Sine Integral Function

The *sine integral function* is an odd function defined by (Spiegel, 1968)

$$Si(y) = \int_0^y \frac{\sin(\alpha)}{\alpha} \, d\alpha. \qquad (1.2.18a)$$

The values of this function can be computed numerically using the series expression

$$Si(y) = \frac{y}{(1)1!} - \frac{y^3}{(3)3!} + \frac{y^5}{(5)5!} - \frac{y^7}{(7)7!} + - \ldots \qquad (1.2.18b)$$

Some of its important properties are

$$Si(-y) = -Si(y), Si(0) = 0,$$
$$Si(\pi) \cong 2.0123, Si(\infty) = (\pi/2) . \qquad (1.2.18c)$$

Si function converges fast and only a few terms in (1.2.18b) are needed for a good approximation.

1.3 Derivatives and Integrals of Functions

It will be assumed that the reader is familiar with some of the basic properties associated with the derivative and integral operations. We should caution that derivatives of discontinuous functions do not exist in the conventional sense. To handle such cases, generalized functions are defined in the next section.

The three well-known formulas to approximate a derivative of a function, referred to as *forward difference*, *central difference*, and *backward difference*, are

$$x'(t) = \frac{dx(t)}{dt} : \frac{x(t+h) - x(h)}{h},$$
$$x'(t) : \frac{x(t+h) - x(t-h)}{2h},$$
$$x'(t) : \frac{x(t) - x(t-h)}{h}. \qquad (1.3.1)$$

MATLAB evaluations of the derivatives are given in Appendix B. If we have a function of two variables, then we have the possibility of taking the derivatives one or the other, leading to partial derivatives. Let $x(t, \alpha)$ be a function of two variables. The two partial derivatives of $x(t, \alpha)$ with respect to t, keeping α constant, and with respect to α, keeping t constant are, respectively, given by

$$\frac{\partial x(t, \alpha)}{\partial t} = \lim_{\Delta t \to 0} \frac{x(t + \Delta t, \alpha) - x(t, \alpha)}{\Delta t},$$
$$\frac{\partial x(t, \alpha)}{\partial \alpha} = \lim_{\Delta \alpha \to 0} \frac{x(t, \alpha + \Delta \alpha) - x(t, \alpha)}{\Delta \alpha}. \qquad (1.3.2)$$

Assuming the second (first) variable is not a function of the first (second) variable, the differential of $x(t, \alpha)$ is

$$dx = \frac{\partial x}{\partial t} dt + \frac{\partial x}{\partial \alpha} d\alpha.$$

The integral of a function over an interval is the area of the function over that interval.

Example 1.3.1 Compute the value of the integral of $x(t)$ shown in Fig. 1.3.1.

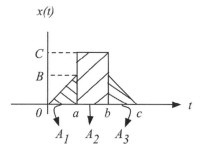

Fig. 1.3.1 Computation of Integral of $x(t)$ using areas

Solution: Divide the area into three parts as identified in the figure. The three parts are in the intervals $(0, a)$, (a, b), and (b, c), respectively. The areas of the two triangles are identified by A_1 and A_2 and the area of the rectangle by A_3. They can be individually computed and then add the three areas to get the total area. That is,

$$A_1 = \frac{1}{2}aB, A_2 = (b - a)C, A_3 = \frac{1}{2}(c - b)B,$$

$$A = A_1 + A_2 + A_3 = \int_0^c x(t) dt. \qquad \blacksquare$$

If the function is arbitrary and cannot be divided into simple functions like in the above example, we can approximate the integral by dividing the area into small rectangular strips and compute the area by adding the areas in each strip.

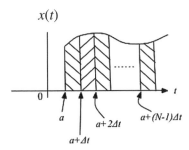

$x(t)$

a

$a+\Delta t$

$a+2\Delta t$

$a+(N-1)\Delta t$

Fig. 1.3.2 $x(t)$ and its approximation using its samples

Example 1.3.2 Consider the function $x(t)$ shown in Fig. 1.3.2. Find the integral of this function using the above approximation for $a < t < b$. Assume the values of the function are known as $x(a), x(a + \Delta t)$, $x(a + 2\Delta t), \ldots$, and $x((a + N - 1)\Delta t)$

Solution: Assuming Δt is small enough that we can approximate the area in terms of rectangular strips using the *rectangular integration formula* and

$$\int_a^b x(t)dt \approx \left[\sum_{n=0}^{N-1} x(a + n\Delta t)\right]\Delta t, \quad \Delta t = (b - a)/N.$$

(1.3.3a)

Note that $x(a)\Delta t$ gives the approximate area in the first strip. If the width of the strips gets smaller and the number of strips increases correspondingly, then the approximation gets better. In the limit, i.e., when $\Delta t \to 0$, approximation approaches the value of the integral. In computing the area of the kth rectangular strip, $x(a + (k - 1)\Delta t)$ was used to approximate the height of the pulse. Some other value of the function in the interval, such as the value of the function in the middle of the strip, could be used.

MATLAB evaluation of integrals is discussed in Appendix B. In Chapter 8 appropriate values for Δt will be considered in terms of the frequency content in the signal. Instead of rectangular integration formula, there are other formulas that are useful. One could assume that each strip is a trapezoid and using the *trapezoidal integration formula* the integral is approximated by

$$\int_a^b x(t)dt \approx [x(a) + 2x(a + \Delta t) + 2x(a + 2\Delta t) + \cdots$$

$$+ 2x(a + (N - 1)\Delta t) + x(a + N\Delta t)](\Delta t/2). \quad (1.3.3b)$$

If the time interval $\Delta t = (b - a)/N$ is sufficiently small, then the difference between the two formulas in (1.3.3a) and (1.3.3b) would be small and (1.3.3a) is adequate.   ∎

1.3.1 Integrals of Functions with Symmetries

The integrals of functions with even and odd symmetries around a symmetric interval are

$$\int_{-a}^a x_e(t)dt = 2\int_0^a x_e(t)dt \quad \text{and} \quad \int_{-a}^a x_0(t)dt = 0,$$

(1.3.4)

where a is an arbitrary positive number.

Example 1.3.3 Evaluate the integrals of the functions given below:

$$x_1(t) = \Pi[t/2a], \quad x_2(t) = tx_1(t). \quad (1.3.5)$$

Solution: $x_1(t)$ is a rectangular pulse with an even symmetry and $x_2(t)$ is an odd function with an odd symmetry. The integrals are

$$A_1 = \int_{-a}^a x_1(t)dt = 2\int_0^a dt = 2a,$$

$$A_2 = \int_{-a}^a x_2(t)dt = 0. \quad \blacksquare$$

1.3.2 Useful Functions from Unit Step Function

The ramp and the parabolic functions can be obtained by

$$x_r(t) = \int_0^t u(t)dt = tu(t) \quad \text{and}$$

$$x_p(t) = \int_0^t x_r(t)dt = (t^2/2)u(t). \quad (1.3.6)$$

Section 1.4 considers the derivatives of the unit step functions.

Now consider Leibniz's rule, interchange of derivative and integral, and interchange of integrals without proofs. For a summary, see Peebles (2001).

1.3.3 Leibniz's Rule

$$\text{Let } g(t) = \int_{\alpha(t)}^{\beta(t)} z(x,t)dx. \tag{1.3.7}$$

$\alpha(t)$ and $\beta(t)$ are assumed to be real differentiable functions of a real parameter t, and $z(x,t)$ and its derivative $dz(x,t)/dt$ are both continuous functions of x and t. The derivative of the integral with respect to t is Leibniz's rule Spiegel (1968) and is

$$\frac{dg(t)}{dt} = z[\beta(t),t]\frac{d\alpha(t)}{dt} - z[\alpha(t),t]\frac{d\beta(t)}{dt}$$
$$+ \int_{\alpha(t)}^{\beta(t)} \frac{\partial z(x,t)}{\partial t}dx. \tag{1.3.8}$$

1.3.4 Interchange of a Derivative and an Integral

When the limits in (1.3.7) are constants, say $\alpha(t) = a$ and $\beta(t) = b$, then the derivatives of these limits will be zero and (1.3.7) and (1.3.8) reduce to

$$g(t) = \int_a^b z(x,t)dx,$$

$$\frac{dg(t)}{dt} = \frac{d}{dt}\int_a^b z(x,t)dt = \int_a^b \frac{\partial z(x,t)}{\partial t}dx. \tag{1.3.9}$$

The derivative and the integral operations may be interchanged.

1.3.5 Interchange of Integrals

If any one of the following conditions,

$$\int_{-\infty}^{\infty} \int_{-\infty}^{\infty} |x(t,\alpha)|dtd\alpha < \infty, \quad \int_{-\infty}^{\infty} [\int_{-\infty}^{\infty} |x(t,\alpha)|d\alpha]dt < \infty,$$

$$\int_{-\infty}^{\infty} [\int_{-\infty}^{\infty} |x(t,\alpha)dt|]d\alpha < \infty, \tag{1.3.10}$$

is true, then *Fubini's theorem* (see Korn and Korn (1961) for a proof) states that

$$\int_{-\infty}^{\infty} \left[\int_{-\infty}^{\infty} x(t,\alpha)dt\right]d\alpha = \int_{-\infty}^{\infty} \left[\int_{-\infty}^{\infty} x(t,\alpha)dt\right]d\alpha$$

$$= \int_{-\infty}^{\infty} \int_{-\infty}^{\infty} x(t,\alpha)dtd\alpha \tag{1.3.11}$$

Signals generated in a lab are well behaved and they are valid.

In Chapter 3 on Fourier series, integrating a product of a simple function, say $h(t)$, with its nth derivative goes to zero; such a polynomial, and the other one is a sinusoidal function $g(t)$, such as $\sin(\omega_0 t)$ or $\cos(\omega_0 t)$ or $e^{j\omega_0 t}$ is applicable. The *generalized integration by parts formula* comes in handy.

$$\int h^{(n)}(t)g(t)dt = h^{(n-1)}(t)g(t) - h^{(n-2)}(t)g'(t)$$
$$+h^{(n-3)}(t)g''(t) - \cdots (-1)^n \int h(t)g^{(n)}(t)dt,$$
$$g^{(k)}(t) = \frac{d^k g(t)}{dt^k}, \quad \text{and} \quad h^{(k)}(t) = \frac{d^k h(t)}{dt^k}$$
$$\tag{1.3.12}$$

Using (1.3.12), the following equalities can be seen:

$$\int t\cos(t)dt = \cos(t) + t\sin(t);$$

$$\int t\sin(t)dt = \sin(t) - t\cos(t) \tag{1.3.13a}$$

$$\int t^2\cos(t)dt = 2t\cos(t) + (t^2 - 2)\sin(t);$$

$$\int t^2\sin(t)dt = 2t\sin(t) - (t^2 - 2)\cos(t). \tag{1.3.13b}$$

1.4 Singularity Functions

The *impulse function*, or the *Dirac delta function*, a singularity function, is defined by

$$\delta(t) = \begin{cases} 0, & t \neq 0 \\ \infty, & t = 0 \end{cases} \text{ with } \int_{-\infty}^{\infty} \delta(t)dt = 1. \quad (1.4.1)$$

$\delta(t)$ takes the value of infinity at $t = 0$ and is zero everywhere else. See Fig. 1.4.1b. Impulse function is a continuous function and the area under this function is equal to one. Note that a line has a zero area. Here, a *generalized* or a *distribution function* is defined that is nonzero only at one point and has a unit area. A delayed or an advanced impulse function can be defined by $\delta(t \mp t_0)$, where t_0 is assumed to be positive in the expressions. The ideal impulse function cannot be synthesized. It is useful in the limit. For example,

$$\delta(t) = \lim_{\varepsilon \to 0} \frac{1}{\varepsilon}\left(\Pi\left[\frac{t}{\varepsilon}\right]\right), \quad \delta(t) = \lim_{\varepsilon \to 0}\left(\frac{1}{\varepsilon}\right)\left(\Lambda\left[\frac{t}{\varepsilon}\right]\right).$$
$$(1.4.2)$$

Figure 1.4.1a illustrates the progression of rectangular pulses of unit area toward the delta function. As ε is reduced, the height increases and, in the limit, the function approaches infinity at $t = 0$ and the area of the rectangle is 1. There are other functions that approximate the impulse function in the limit. A nice definition is given in terms of an integral of a product of an impulse and a test function $\phi(t)$ by Korn and Korn (1961):

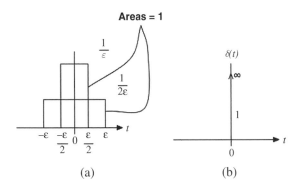

Areas = 1

(a)   (b)

Fig. 1.4.1 (a) Progression toward an impulse as $\varepsilon \to 0$ and (b) symbol for $\delta(t)$

$$\int_{t_1}^{t_2} \phi(t)\delta(t - t_0)dt$$

$$= \begin{cases} 0, & t_0 < t_1 \text{ or } t_2 < t_0 \\ (1/2)[\phi(t_0^+) + \phi(t_0^-)], & t_1 < t_0 < t_2 \\ (1/2)\phi(t_0^+), & t_0 = t_1 \\ (1/2)\phi(t_0^-), & t_0 = t_2 \end{cases}. \quad (1.4.3)$$

$\phi(t)$ is a *testing (or a test) function* of t and is assumed to be continuous and bounded in the neighborhood of $t = t_0$ and is zero outside a finite interval. That is $\phi(\pm\infty) = 0$. The integral in (1.4.3) is not an ordinary (Riemann) integral. In this sense, $\delta(t)$ is a generalized function. A simpler form of (1.4.3) is adequate. $\delta(t)$ has the property that

$$\int_{-\infty}^{\infty} \phi(t)\delta(t - t_0)dt = \phi(t_0), \quad (1.4.4a)$$

where $\phi(t)$ is a test function that is continuous at $t = t_0$. As a special case, consider $t_0 = 0$ in (1.4.4a) and $\phi(t) = 1$. Equation (1.4.4a) can be written as

$$\int_{-\infty}^{\infty} \delta(t)dt = \int_{0^-}^{0^+} \delta(t)dt = 1. \quad (1.4.4b)$$

That is, the area under the impulse function is 1. The integral in (1.4.4a) sifts the value of $\phi(t)$ at $t = t_0$ and $\delta(t)$ is called a *sifting function*. In summary, the impulse $\delta(t - t_0)$ has unit area (or weight) centered at the point $t = t_0$ and zero everywhere else. Since the $\delta(t - t_0)$ exists only at $t = t_0$, and $\phi(t)$ at $t = t_0$ is $\phi(t_0)$, we have an important result

$$\phi(t)\delta(t - t_0) = \phi(t_0)\delta(t - t_0). \quad (1.4.5)$$

There are many *limiting forms of impulse functions*. Some of these are given below.

Notes: In the limit, the following functions can be used to approximate $\delta(t)$:

$$\delta(t) = \lim_{\tau \to 0} x_i(t), \quad x_1(t) = \frac{2}{\tau}e^{-|t/\tau|},$$

$$x_2(t) = \frac{1}{\tau}\text{sinc}(t/\tau), \quad x_3(t) = \frac{1}{\tau}\text{sinc}^2(t/\tau)$$

$$x_4(t) = \frac{1}{\tau}e^{-\pi(t/\tau)^2}, \quad x_5(t) = \frac{\tau}{\pi(t^2 + \tau^2)}. \quad (1.4.6a)$$

$x_1(t)$ is a two-sided exponential function; $x_2(t)$ and $x_3(t)$ are sinc functions. Sinc function does not go to zero in the limit for all t. See the discussion by Papoulis (1962). $x_4(t)$ is a Gaussian function and $x_5(t)$ is a Lorentzian function. To prove these, approximate the impulse function in the limit, using (1.4.4a) and (1.4.4b).

Example 1.4.1 Show that the Lorentzian function $x_5(t)$ approaches an impulse function as $\tau \to 0$. Show the result by using the equations: *a.* (1.4.1) and *b.* (1.4.4a and b).

Solution: *a.* Clearly as $\tau \to 0$, $x_5(t) \to 0$ for t not equal to zero. As $\tau \to 0$, $x_5(t) \to \infty$. From tables,

$$\frac{1}{\pi}\int_{-\infty}^{\infty}\frac{\tau}{\tau^2 + t^2}dt = 1, \lim_{\tau \to 0}x_5(t) = \delta(t). \quad (1.4.6b)$$

b. First by (1.4.4a), with $t_0 = 0$ and $\phi(t) = 1$ in the neighborhood of $t = 0$ results in

$$\lim_{\tau \to 0}\int_{-\infty}^{\infty}\phi(t)x_5(t - t_0)dt = \lim_{\tau \to 0}\int_{-\infty}^{\infty}x_5(t)dt = 1.$$
$$(1.4.6c)$$

Using (1.4.4b) and the integral tables, it follows that

$$\int_{0^-}^{0^+}\frac{1}{\pi}\frac{\tau}{(t^2 + \tau^2)}dt = \frac{\tau}{\pi}\left[\frac{1}{\tau}\tan^{-1}\left(\frac{t}{\tau}\right)\right]_{0^-}^{0^+}$$

$$= \frac{1}{\pi}\left[\frac{\pi}{2} - \left(-\frac{\pi}{2}\right)\right] = 1. \qquad \blacksquare$$

1.4.1 Unit Impulse as the Limit of a Sequence

Another approach to the above is through a sequence. See the work of Lighthill (1958). Also see Baher, (1990). A *good function* $\phi(t)$ is

differentiable everywhere any number of times, and, in addition, the function and its derivative decrease at least as rapidly as $(1/t^n)$ as $t \to \infty$ for all n. The derivative of a good function is another good function and the sums and the products of two good functions are good functions. A sequence of good functions

$$\{x_n(t)\} \equiv \{x_1(t), x_2(t), \ldots, x_n(t)\} \quad (1.4.7)$$

is called regular, if for any good function $\phi(t)$, the following limit exists:

$$V_x(\phi) = \lim_{n \to \infty}\int_{-\infty}^{\infty}\{x_n(t)\}\phi(t)dt. \quad (1.4.8)$$

Example 1.4.2 Find the limit in (1.4.8) of the following sequence:

$$\{x_n(t)\} = \left\{e^{-(t^2/n^2)}\right\}. \quad (1.4.9)$$

Solution: The limit in (1.4.8) is

$$V_x(\phi) = \int_{-\infty}^{\infty}\phi(t)dt. \quad (1.4.10) \quad \blacksquare$$

Two regular sequences of good functions are considered *equivalent* if the limit in (1.4.8) is the same for the two sequences. For example, $e^{-(t^4/n^4)}$ and $e^{-(t^2/n^2)}$ are equivalent only in that sense. The function $V_x(\phi)$ defines a distribution $x(t)$ and the limit of the sequence

$$x(t) \equiv \lim_{n \to \infty}\{x_n(t)\}. \quad (1.4.11)$$

An impulse function can be defined in terms of a sequence of functions and write

$$\delta(t) \equiv \lim_{n \to \infty}\{x_n(t)\} \text{ if } \lim_{n \to \infty}\int_{-\infty}^{\infty}\phi(t)\{x_n(t)\}dt = \phi(0)$$
$$(1.4.12)$$

$$\Longrightarrow \int_{-\infty}^{\infty}\phi(t)\delta(t)dt = \phi(0). \quad (1.4.13)$$

Many sequences can be used to approximate an impulse.

Notes: A constant can be interpreted as a generalized function defined by the regular sequence $\{x_n(t)\}$ so that, for any good function $\phi(t)$

$$\lim_{n\to\infty} \int_{-\infty}^{\infty} \{x_n(t)\}\phi(t)dt = K \int_{-\infty}^{\infty} \phi(t)dt. \quad (1.4.14)$$

Using the function in (1.4.9), it follows that

$$\{x_n(t)\} \equiv Ke^{-t^2/n^2}, K = \lim_{n\to\infty}\{Ke^{-(t^2/n^2)}\}. \quad (1.4.15)$$

1.4.2 Step Function and the Impulse Function

Noting the area under the impulse function is one, it follows that

$$\int_{-\infty}^{t} \delta(\alpha)d\alpha = u(t). \quad (1.4.16)$$

Asymmetrical functions $(\delta_+(t)$ and $\delta_-(t))$ and $(u_+(t)$ and $u_-(t))$ can be defined and are

$$u_+(t) = \int_{-\infty}^{\infty} \delta_+(t)dt = \begin{cases} 1, & t > 0 \\ 0, & t \le 0 \end{cases},$$

$$u_-(t) = \int_{-\infty}^{\infty} \delta_-(t)dt = \begin{cases} 1, & t \ge 0 \\ 0, & t < 0 \end{cases} \quad (1.4.17)$$

$$u(t) = \begin{cases} 1, & t > 0 \\ (1/2), & t = 0 \\ 0, & t < 0 \end{cases} \quad (1.4.18)$$

These step functions differ in how the value of the function at $t = 0$ is assigned and are only of theoretical interest. Here the step function is assumed to be $u_-(t)$ and ignore the subscript. Derivative of the unit step function is an impulse function, which can be shown by using the generalized function concept.

Example 1.4.3 Using the property of the test function, $\lim_{t\to\pm\infty} \phi(t) = 0$, show that

$$\int_{-\infty}^{\infty} x'(t)\phi(t)dt = -\int_{\infty}^{\infty} x(t)\phi'(t)dt, \phi'(t) = \frac{d\phi}{dt}$$

$$(1.4.19)$$

Solution: Using the integration by parts, we have the result below and (1.4.19) follows:

$$\int_{-\infty}^{\infty} x'(t)\phi(t)dt = x(t)\phi(t)|_{-\infty}^{\infty} - \int_{-\infty}^{\infty} x(t)\phi\prime(t)dt.$$

$$(1.4.20) \quad \blacksquare$$

Notes: Let $g_1(t)$ and $g_2(t)$ be two generalized functions and

$$\int_{-\infty}^{\infty} \phi(t)g_1(t)dt = \int_{-\infty}^{\infty} \phi(t)g_2(t)dt \quad (1.4.21)$$

The two generalized functions are equal, i.e., $g_1(t) = g_2(t)$ only in the sense of (1.4.21). It is called the *equivalency property*.

Example 1.4.4 Show that the derivative of the unit step function is an impulse function using the equivalence property.

Solution: First

$$\int_{-\infty}^{\infty} u(t)\phi(t)dt = \int_{0}^{\infty} \phi(t)dt. \quad (1.4.22)$$

Noting that $\phi(\pm\infty) = 0$, it follows that

$$\int_{-\infty}^{\infty} u'(t)\phi(t)dt = -\int_{-\infty}^{\infty} u(t)\phi'(t)dt$$

$$= -\int_{0}^{\infty} \phi'(t)dt = -[\phi(\infty) - \phi(0)] = \phi(0), \quad (1.4.23)$$

$$\int_{-\infty}^{\infty} u'(t)\phi(t)dt = \int_{-\infty}^{\infty} \delta(t)\phi(t)dt \text{ and } u'(t)$$

$$= \frac{du(t)}{dt} = \delta(t). \quad (1.4.24)$$

The derivative of a parabolic function results in a ramp function, the derivative of a ramp function results in a unit step function, and the derivative of a unit step function results in an impulse function. All of these are true only in the generalized sense. What is the derivative of an impulse function? That is,

$$\delta'(t) = \frac{d\delta(t)}{dt}. \quad (1.4.25)$$

It is defined by the relation

$$\int_{-\infty}^{\infty} \delta'(t)\phi(t)dt = -\int_{-\infty}^{\infty} \delta(t)\phi'(t)dt = -\phi'(0).$$

(1.4.26)

Generalizing to higher-order derivatives of the impulse function results in

$$\int_{-\infty}^{\infty} \delta^{(n)}(t)\phi(t)dt = (-1)^n \phi^{(n)}(0).$$ (1.4.27)

Example 1.4.5 Evaluate the following integrals:

a. $A = \displaystyle\int_{-\infty}^{\infty} (t^2 + 2t + 1)\delta^{(2)}(t-1)dt, \delta^{(2)}(t) = \dfrac{d^2\delta(t)}{dt^2}$

(1.4.28)

b. $B = \displaystyle\int_{.5}^{2} [(t-1)^2\delta(t-1) + 5\delta(t+1) + 6t\delta(t)]dt.$

(1.4.29)

Solution: These follow

a. $A = \displaystyle\int_{-\infty}^{\infty} [(\alpha+1)^2 + 2(\alpha+1) + 1]\delta^{(2)}(\alpha)d\alpha$

$= (-1)^2 \dfrac{d^2}{d\alpha^2}[(\alpha+1)^2 + 2(\alpha+1) + 1]|_{\alpha=0} = 4$

b. $B = \displaystyle\int_{.5}^{2} [(t-1)^2\delta(t-1) + 5\delta(t+1)$

$+ 6t\delta(t)]dt = 0.$ (1.4.30)

When an impulse is outside the integration limits, then the integral is 0. In addition, it is assumed that the limits do not fall at the exact location of the impulses. ∎

$\delta(t)$ can be approximated using various functions in the limit. In Fig. 1.4.1 it is expressed in terms of a rectangular pulse function. That is,

$$\delta(t) = \lim_{\varepsilon \to 0} x_1(t), \ x_1(t) = \dfrac{1}{\varepsilon}\Pi\left[\dfrac{t}{\varepsilon}\right].$$ (1.4.31)

The generalized derivative of a pulse function is as follows:

$$\Pi[t/\varepsilon] = u(t + \varepsilon/2) - u(t - \varepsilon/2)$$ (1.4.32)

$$\dfrac{d\Pi\left[\frac{t}{\varepsilon}\right]}{dt} = \delta(t + \dfrac{\varepsilon}{2}) - \delta(t - \dfrac{\varepsilon}{2}), \delta'(t)$$

$$= \lim_{\varepsilon \to 0}\left[\delta(t + \dfrac{\varepsilon}{2}) - \delta(t - \dfrac{\varepsilon}{2})\right].$$ (1.4.33)

The derivative of the impulse function results in two impulses illustrated in Fig. 1.4.2. It is an odd function called a *doublet*. The square of an impulse function is not defined as

$$\lim_{\varepsilon \to 0}\dfrac{1}{\varepsilon^2}\left(\Pi\left[\dfrac{t}{\varepsilon}\right]\right)^2 = \lim_{\varepsilon \to 0}\left(\dfrac{1}{\varepsilon}\right) = \infty.$$ (1.4.34)

The impulse function is not square integrable. The square of a distribution is not defined Papoulis, (1962). Note that $\delta(t)$ is an even function and $\delta'(t)$ is an odd function.

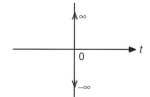

Fig. 1.4.2 Symbolic representation of $\delta'(t)$, a doublet

1.4.3 Functions of Generalized Functions

Using the equivalence property of the generalized function, the following is true:

$$g(t) = \delta(at - b) = \dfrac{\delta(t - b/a)}{|a|}, a \neq 0.$$ (1.4.35)

This can be seen from

$$\int_{-\infty}^{\infty} \phi(t)\delta(at - b)dt = \dfrac{1}{|a|}\int_{-\infty}^{\infty} \phi(y/a)\delta(y - b)dy$$

$$= \dfrac{1}{|a|}\phi(b/a).$$ (1.4.36)

From the equivalence property, the equality in (1.4.35) now follows. From (1.4.35), it follows that

$$\delta(\omega) = \dfrac{1}{2\pi}\delta(f) \text{ or } \delta(f) = 2\pi\delta(\omega).$$ (1.4.37)

Table 1.4.1 Properties of the impulse function

$$\int_{-\infty}^{\infty} \delta(t-t_0)\phi(t)dt = \phi(t_0).$$

$$\int_{-\infty}^{\infty} \delta(t)\phi(t-t_0)dt = \phi(-t_0).$$

$$\int_{-\infty}^{\infty} \phi(t)\delta(t)dt = \phi(0).$$

$$\int_{-\infty}^{\infty} \frac{d\delta(t)}{dt}\phi(t)dt = -\frac{d\phi(t)}{dt}\Big|_{t=0}.$$

$$\int_{-\infty}^{\infty} \delta^{(n)}(t)\phi(t)dt = (-1)^n\phi^{(n)}(0).$$

$$x(t)\delta(t) = x(0)\delta(t).$$

$$\int_{-\infty}^{\infty} \delta(t)\delta(t_0-t)dt = \delta(t_0).$$

$$\delta(at-b) = \frac{1}{|a|}\delta(t-\frac{b}{a}) \ ; \ a \neq 0.$$

$$\delta(j\omega) = \delta(\omega).$$

$$\delta(t) = \delta(-t).$$

1.4.4 Functions of Impulse Functions

In (1.4.35), $\delta(x(t))$ is considered with $x(t)$ being a linear function of time. Now consider other cases, where $x(t)$ is assumed to have *simple zeros*, i.e., no multiple zeros.

Example 1.4.6 Evaluate the following integral using (1.4.35) and the following cases for the limits. *a.* $x = 0, y = 4$ and *b.* $x = -10, y = 5$.

$$A = \int_{x}^{y} (t-1)(t+5)\delta(2t+5)dt.$$

Solution: First, changing the variables, $\alpha = 2t + 5$, i.e., $t = \frac{1}{2}(\alpha - 5)$, $dt = \frac{1}{2}d\alpha$, results in

$$a. \ t = 0 \Longrightarrow \alpha = 5, t = 4 \Longrightarrow \alpha = 13,$$

$$\Longrightarrow A = \int_{5}^{13} \left(\frac{1}{2}\alpha - \frac{7}{2}\right)\left(\frac{1}{2}\alpha + \frac{5}{2}\right)\delta(\alpha)\frac{1}{2}d\alpha = 0$$

($\alpha = 0$ is outside the range $5 < \alpha < 13$).

b. Similarly, changing the variables, $\alpha = 2t + 5$, i.e., $t = \frac{1}{2}(\alpha - 5)$, $dt = \frac{1}{2}d\alpha$, results in

$$t = -5 \Longrightarrow \alpha = -5, t = 0 \Longrightarrow \alpha = 5,$$

$$\Longrightarrow A = \int_{-5}^{5} \left(\frac{1}{2}\alpha - \frac{7}{2}\right)\left(\frac{1}{2}\alpha + \frac{5}{2}\right)\delta(\alpha)\left(\frac{1}{2}\right)d\alpha$$

$$= \left(\frac{1}{2}\alpha - \frac{7}{2}\right)\left(\frac{1}{2}\alpha + \frac{5}{2}\right)\left(\frac{1}{2}\right)\Big|_{\alpha=0} = -\frac{35}{8}. \qquad \blacksquare$$

Example 1.4.7 Using the equivalence property of the impulse functions, show that

$$\delta(t^2 - a^2) = \left(\frac{1}{|2a|}\right)(\delta(t+a) + \delta(t-a)), a \neq 0.$$

$$(1.4.38)$$

Solution: Since $t^2 - a^2 = (t-a)(t+a) = 0 \rightarrow t = \pm a \neq 0$ at $t = -a$, it follows that

$$\int_{-\infty}^{\infty} \delta(t^2 - a^2)\phi(t)dt = \int_{-\infty}^{0} \delta(t^2 - a^2)\phi(t)dt$$

$$+ \int_{0}^{\infty} \delta(t^2 - a^2)\phi(t)dt. \quad (1.4.39)$$

With $t = -\sqrt{(y+a^2)}$,

$$\int_{-\infty}^{0} \delta(t^2 - a^2)\phi(t)dt$$

$$= -\int_{\infty}^{-a^2} \delta(y)\phi\left(-\sqrt{y+a^2}\right)\left(\frac{1}{2\sqrt{y+a^2}}\right)dy$$

$$= \phi(-a)\left(\frac{1}{|2a|}\right).$$

Note that when $t = 0$, $y = -a^2$ and when $t = -\infty$, $y = \infty$. In a similar manner, we can evaluate the second integral in (1.4.39). Combining them, it follows that

$$\int_{-\infty}^{\infty} \frac{1}{|2a|}[\delta(t-a) + \delta(t+a)]\phi(t)dt = \frac{\phi(a)}{|2a|} + \frac{\phi(-a)}{|2a|}.$$

This can be generalized. If $x(t)$ has *simple roots* at $t = t_n$, then

$$\delta[x(t)] = \sum_{t_n} \frac{1}{|x'(t_n)|}\delta(t-t_n), x'(t_n) = \frac{dx(t)}{dt}\Big|_{t=t_n}$$

$$(1.4.40) \quad \blacksquare$$

Example 1.4.8 Give the expression for $\delta(\sin(t))$.

Solution: Since $\sin(t)|_{t=n\pi} = 0, d\sin(t)/dt = \cos(t)$, and $|\cos(n\pi)| = 1$, it follows that

$$\delta(\sin(t)) = \sum_{n=-\infty}^{\infty} \delta(t - n\pi). \quad (1.4.41) \quad \blacksquare$$

1.4.5 Functions of Step Functions

Example 1.4.9 Given $x(t) = t^2 - 1$, sketch the function $y(t) = u(x(t)) = u(t^2 - 1)$, where $u(t)$ is the unit step function.

Solution: Since $x(t) \leq 0$ for $-1 \leq t \leq 1$, it follows that

$$x(t) \leq 0 \text{ for } -1 \leq t \leq 1 \rightarrow y(t) = 0, \ -1 < t < 1,$$

$$x(t) > 0 \text{ for } -\infty < t < -1 \quad \text{and}$$
$$1 < t < \infty \rightarrow y(t) = 1, \quad t < -1, t > 1.$$

Fig. 1.4.3 shows the sketches for $x(t)$ and $y(t)$. $\blacksquare$

1.5 Signal Classification Based on Integrals

The *area of a signal* $x(t)$ is

$$\text{Area}[x(t)] = \int_{-\infty}^{\infty} x(t)dt. \quad (1.5.1)$$

If a signal $x(t)$ is said to be *absolutely integrable*, then

$$\text{Area}[|x(t)|] = \int_{-\infty}^{\infty} |x(t)|dt < \infty. \quad (1.5.2)$$

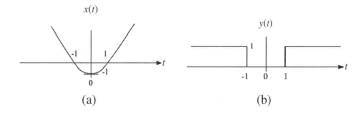

Fig. 1.4.3 (a) $x(t) = t^2 - 1$ and (b) $y(t) = u(t^2 - 1)$

(a)

(b)

If a signal is square integrable, i.e., a finite energy signal satisfies

$$\text{Area}[|x(t)|^2] = \int_{-\infty}^{\infty} |x(t)|^2 dt < \infty. \qquad (1.5.3)$$

Consider a resistor of value $R\ \Omega$. Ohm's law states that the voltage, $v(t)$, across this resistor is equal to R times the current $i(t)$ passing through the resistor and $v(t) = Ri(t)$. The instantaneous power delivered to the resistor is $p_R(t) = i^2(t)R$. The total energy delivered to the resistor is

$$E_R = \int_{-\infty}^{\infty} p_R(t)dt = R \int_{-\infty}^{\infty} i^2(t)dt = \frac{1}{R} \int_{-\infty}^{\infty} v^2(t)dt$$

If we normalize the resistor value to 1Ω, that is, $R = 1$, then we can consider the function, $x(t)$ as a generic (i.e., either voltage or current) function. The *energy* in $x(t)$ is defined by

$$E_x = \int_{-\infty}^{\infty} |x(t)|^2 dt. \qquad (1.5.4)$$

The normalized average power of a signal $x(t)$ is defined by

$$P_x = \lim_{T \to \infty} \frac{1}{T} \int_{-T/2}^{T/2} |x(t)|^2 dt. \qquad (1.5.5)$$

The signal $x(t)$ is an *energy signal* if $0 < E_x < \infty$, that is, E_x is finite and $P_x = 0$. The function $x(t)$ is a *power signal* if $0 < P_x < \infty$, i.e., P_x is finite and therefore E_x is infinite. If a signal does not satisfy one of these conditions, then it is *neither an energy signal nor a power signal*. The power and energy signals are mutually exclusive.

Example 1.5.1 Show that the function given below is an energy signal.

$$x(t) = Ae^{-\alpha t}u(t), \alpha > 0. \qquad (1.5.6)$$

Find its area, the absolute area, and its energy assuming $A \neq 0$ is finite.

Solution: The area, the absolute area, and the energy are, respectively, given by

$$\text{Area}[x(t)] = \int_{0}^{\infty} Ae^{-\alpha t} dt = -\frac{A}{\alpha} e^{-\alpha t}\Big|_{0}^{\infty} = \frac{A}{\alpha},$$

$$\text{Area}[|x(t)|] = |A|/\alpha, \quad E_x = \int_{-\infty}^{\infty} |x(t)|^2 dt = |A|^2$$

$$\int_{-\infty}^{\infty} \left[|e^{-\alpha t}|^2\right] u(t)dt = |A|^2 \int_{0}^{\infty} e^{-2\alpha t} dt$$

$$= \left(\frac{A}{-2\alpha}^2\right) e^{-2\alpha t}\Big|_{t=0}^{t=\infty} = \left(\frac{|A|^2}{2\alpha}\right).$$

From this it follows that E_x is finite. Clearly, $P_x = 0$ implying it is an energy signal. ∎

Example 1.5.2 Using $\alpha \to 0$ in the above example show that $x(t)$ is a power signal.

$$x(t) = Au(t), A \neq 0 \text{ and finite.} \qquad (1.5.7)$$

Solution: The energy contained in a step function E_x is infinite, whereas

$$P_x = \lim_{T \to \infty} \left[\left(\frac{1}{T}\right) \int_{-T/2}^{T/2} |x(t)|^2 dt\right]$$

$$= \lim_{T \to \infty} \left[\left(\frac{1}{T}\right) \int_{0}^{T/2} |A|^2 dt\right]$$

$$= \lim_{T \to \infty} \left[|A|^2 \left(\frac{T/2}{T}\right)\right] = \frac{|A|^2}{2}$$

is finite. It follows that $x(t)$ in (1.5.7) is a power signal. In determining whether a signal is a power or an energy signal, we can check either its energy or power. If E_x is finite, then $P_x = 0$. If P_x is finite, then E_x is infinite. We do not have to check both of them. If E_x is infinite, then we need to check the average power before making the decision on whether the signal is a power signal or neither a power nor an energy signal. ∎

Example 1.5.3 Show that $x(t)$ defined below is neither an energy nor a power signal.

$$x(t) = tu(t) \qquad (1.5.8)$$

Solution: The energy and the average power in this signal are, respectively, given by

$$E_x = \int\limits_0^\infty t^2 dt = \infty, P_x = \lim_{T\to\infty} \frac{1}{T} \int\limits_0^{T/2} t^2 dt$$

$$= \lim_{T_1\to\infty}\left[\left(\frac{1}{T}\right)\left(\frac{T^3}{3}\right)\right] = \infty.$$

Since both the energy and the average power go to infinity, the result follows. ∎

Example 1.5.4 Show that $x(t) = A$, a finite constant is a power signal.

Solution: The average power contained in $x(t)$ is given by

$$P_x = \lim_{T\to\infty}\left[\frac{1}{T}\int\limits_{-T/2}^{T/2} A^2 dt\right] = A^2 \Longrightarrow E = \infty.$$

It is a power signal. ∎

Notes: The signals $\delta(t)$ and $\delta'(t)$ are neither nor power signals since the squares of these functions are not defined. There are *two classes of power signals*. These are *periodic* and *random signals*. Random signals require some knowledge of probability theory, which is beyond the scope here. ∎

1.5.1 Effects of Operations on Signals

Example 1.5.5 In signal analysis, scaling and shifting of a function are quite common.
  a. Show that the functions $x(t)$ and $x(t - a)$ have the same areas and energies.
  b. Show

$$\text{Area}(x(at)) = (1/|a|)\text{Area}(x(t)), a \neq 0. \quad (1.5.9a)$$

$$\int\limits_{-\infty}^\infty |x(t-a)|^2 dt = \int\limits_{-\infty}^\infty |x(\beta)|^2 d\beta. \quad (1.5.9b)$$

Solution: a. Using a change of variable in the integral $\beta = t - a$ results in

$$\int\limits_{-\infty}^\infty x(t-a)dt = \int\limits_{-\infty}^\infty x(\beta)d\beta$$

b. Similarly, for $at = \beta$ results in

$$\int\limits_{-\infty}^\infty x(at)dt = \frac{1}{|a|}\int\limits_{-\infty}^\infty x(\beta)d\beta.$$

Equation (1.5.9b) can be shown in both cases and is left as an exercise. ∎

1.5.2 Periodic Functions

A function $x(t)$ is *periodic* or *T*-periodic if there is a number T for all time such that

$$x(t + T) = x(t). \quad (1.5.10)$$

It is common to use the actual period, such as T as a subscript on x and write

$$x_T(t) = x_T(t + T). \quad (1.5.11)$$

The smallest positive number T that satisfies (1.5.11) is called the *fundamental period* and it defines the duration of *one complete cycle*. The reciprocal of the fundamental period is the *fundamental frequency*. That is,

$$f_0 = 1/T \text{ Hz and the period } T = 1/f_0 s. \quad (1.5.12)$$

Clearly, if (1.5.11) is satisfied, then for all integers of n, $x_T(t + nT) = x_T(t)$. If there is *no T* that satisfies (1.5.10), then $x(t)$ is called an *aperiodic* or a *nonperiodic signal*. Note that the integral over any one period of a periodic function is the same. Earlier we were interested in finding the average power and the average energy in a signal. With periodic signals, we can make some simplifications of the integrals. Consider the normalized integral of a periodic signal with period T_1.

$$\lim_{T_1\to\infty}\frac{1}{T_1}\int\limits_{-\frac{T_1}{2}}^{\frac{T_1}{2}} y_T(t)dt = \lim_{N\to\infty}\frac{1}{2N+1}\sum_{k=-N}^N \frac{1}{T}\int\limits_{(2k-1)\frac{T}{2}}^{(2k+1)\frac{T}{2}} y_T(t)dt = \lim_{N\to\infty}\frac{2N+1}{2N+1}\frac{1}{T}\int\limits_{-\frac{T}{2}}^{\frac{T}{2}} y_T(t)dt = \frac{1}{T}\int\limits_{-\frac{T}{2}}^{\frac{T}{2}} y_T(t)dt. \quad (1.5.13)$$

Therefore, *any one period* can be used and written in symbolic *short hand notation* by

$$\lim_{T_1 \to \infty} \frac{1}{T_1} \int_{-T_1/2}^{T_1/2} y_T(t)dt = \frac{1}{T} \int_T y_T(t)dt. \quad (1.5.14)$$

The terms that are of interest in dealing with periodic functions are the *duty cycle* of an on–off signal (i.e., the ratio of on-time to the period), *average value* of the signal x_{ave}, the *average signal power* P_x, and the *root mean square (rms) value* x_{rms}. It is also called the *effective value* of the periodic function $y_T(t)$. These values are defined by

$$x_{ave} = \frac{1}{T} \int_T x_T(t)dt,$$

$$P_x = \frac{1}{T} \int_T |x_T|^2 dt, \quad x_{rms} = \sqrt{P_x}. \quad (1.5.15)$$

Since the average power in a periodic signal is finite, the energy is infinite, it follows that all *periodic signals are power signals*. In Chapter 3 periodic functions will be discussed in detail, where the average value of a periodic function can never exceed the rms value will be shown.

Example 1.5.6 Consider the function $x_T(t)$ with A, ω_0, and θ_0 being real constants given below. *a.* Show that it is a periodic function with period $T = (2\pi/\omega_0) = (1/f_0)$.

$$x_T(t) = A\cos(\omega_0 t + \theta_0). \quad (1.5.16)$$

b. Find x_{ave}, P_x, and x_{rms} for the above periodic signal.

Solution: *a.* Using tables it can be seen that $x_T(t)$ is a periodic function, i.e.,

$$x_T(t + T) = A\cos(\omega_0(t + (2\pi/\omega_0)) + \theta_0)$$
$$= A\cos(\omega_0 t + \theta_0)\cos(2\pi)$$
$$\quad - A\sin(\omega_0 t + \theta_0)\sin(2\pi)$$
$$= A\cos(\omega_0 t + \theta_0).$$

b. The average value of a sine or a cosine function is zero as their positive areas cancel out with their negative areas. The average power is independent of θ_0 and

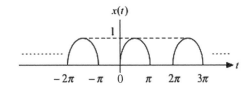

Fig 1.5.1 Half-rectified sine wave

$$P_x = \frac{1}{T} \int_T x_T^2(t)dt = \frac{A^2}{T} \int_T \cos^2(\omega_0 t + \theta_0)dt$$

$$= \frac{A^2}{T} \int_T \frac{1 + \cos(2(\omega_0 t + \theta_0))}{2} dt = \frac{A^2}{2}. \quad (1.5.17)$$

The rms value is

$$x_{rms} = (|A|/\sqrt{2}). \quad (1.5.18)$$

It is best to express sinusoidal functions in terms of Hertz to compute the period. The period is the inverse of the fundamental frequency. Interestingly, if the frequency $f_0 = \omega_0/2\pi$ is 1 MHz, then the signal completes 1 million cycles every second. ∎

Example 1.5.7 Consider the half-wave sinusoidal periodic function

$$x_{2\pi}(t) = \begin{cases} \sin(t), & 0 \leq t < \pi \\ 0, & \pi \leq t < 2\pi \end{cases}, \; x_{2\pi}(t) = x_{2\pi}(t + 2\pi)$$

$$(1.5.19)$$

shown in Fig. 1.5.1. Find its duty cycle, average, average signal power, and its rms value. Show that average value of the function is less than its rms value.

Solution: Clearly the duty cycle is (1/2), as the signal is on for half the time. The average, the power, and the corresponding root mean square values of $x_{2\pi}(t)$ are

$$x_{ave} = \frac{1}{2\pi} \int_0^\pi x_{2\pi}(t)dt = \frac{1}{2\pi} \int_0^\pi \sin(t)dt = \frac{1}{\pi} \quad (1.5.20a)$$

$$P_x = \frac{1}{2\pi} \int_0^\pi \sin^2(t)dt = \frac{1}{2\pi} \int_0^\pi \frac{1}{2}(1 - \sin(2t))dt$$

$$= \frac{1}{4\pi} \int_0^\pi (1 - \sin(2t))dt = \frac{1}{4} \quad (1.5.20b)$$

Fig. 1.5.2
$x_{T_1}(t) = \sin((2\pi/4)t)$,
$x_{T_2}(t) = \sin((2\pi/6)t)$

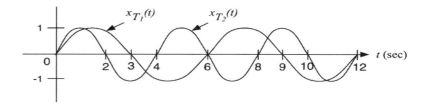

$$x_{\text{rms}} = \sqrt{P_x} = 1/2. \qquad (1.5.20c)$$

Note that the average value is less than the rms value, as $(1/\pi) \cong 0.3183 < (1/2) = 0.5$. ∎

Notes: The average value of a periodic function does not exceed the rms value, i.e., $x_{\text{ave}} \leq x_{\text{rms}}$. The rms value of the sinusoidal voltage supplied to the outlet of a US home is 120 V with a frequency of 60 Hz. The maximum value of the voltage at the outlet is $\sqrt{2}(120) = 169.71$ V ≈ 170 V. ∎

1.5.3 Sum of Two Periodic Functions

If $x_{T_1}(t)$ and $x_{T_2}(t)$ are two periodic functions with periods T_1 and T_2, respectively, then $x(t) = x_{T_1}(t) + x_{T_2}(t)$ is periodic with period T if

$$T = nT_1 = mT_2 \text{ or } [T_1/T_2] = [m/n]. \quad (1.5.21)$$

m and n are some integers and (T_1/T_2) is a rational number. The period of $x(t)$ is equal to the *least common multiple* (*LCM*) of T_1 and T_2. The LCM of two integers, m and n, is the smallest integer divisible by both m and n. If T_1/T_2 is an *irrational number*, it cannot be written in terms of a ratio of two integers and $x(t)$ is not periodic.

Example 1.5.8 Let $x_T(t) = a_1 \cos(\omega_0 t)$ and $y_T(t) = b_1 \sin(\omega_0 t)$, with $T = (2\pi/\omega_0)$. Show that $z_T(t + T) = z_T(t) = x_T(t) + y_T(t)$.

Solution: Since $y_T(t + T) = y_T(t)$ and $x_T(t + T) = x_T(t)$ implies $z_T(t)$ is periodic. ∎

Example 1.5.9 Let $x_{T_1}(t) = A_1 \sin(2\pi(1/4)t)$ and $x_{T_2}(t) = A_2 \sin(2\pi(1/6)t)$. Show that the period of $x_T(t) = x_{T_1} + x_{T_2}$ is 12. Sketch the function $x_T(t)$.

Solution: The period of $x_{T_1}(t)$ is $T_1 = 4$ s and the period of $x_{T_2}(t)$ is $T_2 = 6$ s. The ratio, $(T_1/T_2) = (4/6)$ is a rational number, and the

least common multiple of 4 and 6 is 12 and, therefore, $x_T(t) = x_{T_1}(t) + x_{T_2}(t)$ is periodic with period $T = 12$ s. This can be seen from the fact that in 12 s, $x_{T_1}(t)$ will have three full cycles, $x_{T_2}(t)$ will have two full cycles, and $x_T(t)$ will have one full cycle. Figure 1.5.2 gives sketches of $x_{T_1}(t)$ and $x_{T_2}(t)$. If each of the signals is shifted by different amounts, say if $y_T(t) = A_1 \sin(2\pi(1/4)t + \theta_1) + A_2 \sin(2\pi(1/6)t + \theta_2)$, then $y_T(t)$ is still periodic with period $T = 12$ s for any set of constants A_1, A_2 and angles θ_1 and θ_2.

This can be generalized and state that for any constants $X[0], h[k]$, and $\theta[k]$, the function

$$x_T(t) = X[0] + \sum_{k=1}^{\infty} h[k] \cos(k\omega_0 t + \theta[k]) \quad (1.5.22)$$

is periodic with period $T = \omega_0/2\pi$. Noting that $(\omega_0 T = 2\pi)$, we have

$$\cos(k\omega_0(t + T) + \theta[k]) = \cos(k\omega_0 t + \theta[k]) \cos(k\omega_0 T) - \sin(k\omega_0 t + \theta[k]) \sin(k\omega_0 T) = \cos(k\omega_0 t + \theta[k]).$$

The term for $k = 1$ is called the *fundamental* and the kth term is called the kth *harmonic*. The *dc term* is $X[0]$. The above can be generalized and state that the following function is periodic with period $T = 2\pi/\omega_0$ for any constants $X[0], A[k]$, and $B[k]$:

$$x_T(t) = X[0] + \sum_{k=1}^{\infty} A[k] \cos(k\omega_0 t)$$

$$+ \sum_{k=1}^{\infty} B[k] \sin(k\omega_0 t). \quad (1.5.23) \quad ∎$$

Example 1.5.10 Let $x_{T_1}(t) = \cos(4t)$, $x_{T_2}(t) = \cos(2\pi t)$, and $x(t) = x_{T_1}(t) + x_{T_2}(t)$. Show that $x(t)$ is not a periodic function.

Solution: The period of $x_{T_1}(t) = \cos(2\pi(2/\pi)t)$ is $T_1 = (\pi/2)$ and the period of $x_{T_2}(t)$ is $T_2 = 1$. The ratio $(T_1/T_2) = (\pi/2)$ is an irrational number and $x(t)$ is not periodic. ∎

For functions such as the one given in the above example, there is no repetition. These types of combination of periodic functions are referred as *quasi periodic or almost periodic*. For a study of almost periodic functions, see Chuanyi, (2003).

Example 1.5.11 Compute the average power in $x(t)$ given below for the two cases:

$$x(t) = C_1 \cos(\omega_1 t + \theta_1) + C_2 \cos(\omega_2 t + \theta_2)$$

a. $\omega_1 = n\omega_0 \neq \omega_2 = k\omega_0$, where n and k are integers

b. $\omega_1 = \omega_2 = \omega_0$. Assume $C_1, C_2, \theta_1,$ and θ_2 are arbitrary constants.

Solution: Without loosing any generality, assume $T = (2\pi/\omega_0)$.

$$x(t + \frac{2\pi}{\omega_0}) = C_1 \cos(n\omega_0(t + \frac{2\pi}{\omega_0}) + \theta_1)$$

$$+ C_2 \cos(k\omega_0(t + \frac{2\pi}{\omega_0}) + \theta_2)$$

$$= C_1 \cos(n\omega_0 t + \theta_1) + C_2 \cos(k\omega_0 t + \theta_2).$$

This indicates that $x(t) = x_T(t)$ is periodic for both cases with period $T = (\omega_0/2\pi)$.

a. $$P = \frac{1}{T_0} \int_{T_0} |x_T(t)|^2 dt = \frac{1}{T_0} \int_{T_0} [C_1 \cos(\omega_1 t + \theta_1)$$

$$+ C_2 \cos(\omega_2 t + \theta_2)]^2 dt$$

$$= \frac{1}{2T_0} \int_{T_0} [C_1^2 + C_2^2 + C_1^2 \cos(2(\omega_1 t + \theta_1))$$

$$+ C_2^2 \cos(2(\omega_2 t + \theta_2)] dt$$

$$+ \frac{1}{T_0} C_1 C_2 \int_{T_0} [\cos((\omega_1 + \omega_2)t + (\theta_1 + \theta_2))$$

$$+ \cos((\omega_2 - \omega_1)t + (\theta_2 - \theta_1))] dt = \frac{1}{2}[C_1^2 + C_2^2]$$

(1.5.24)

b. In the case of $\omega_2 = \omega_1$, the average power is

$$P = \frac{1}{2}[C_1^2 + C_2^2] + C_1 C_2 \cos(\theta_2 - \theta_1). \quad (1.5.25)$$

In the above equation $C_1 C_2 \cos(\theta_2 - \theta_1)$ is equal to zero only when $(\theta_2 - \theta_1) = \pm\pi/2$ and

$$x_T(t) = C_1 \cos(\omega_0 t + \theta_1) + C_2 \cos(\omega_0 t + \theta_1 \pm (\pi/2))$$

$$= C_1 \cos(\omega_0 t + \theta_1) \mp C_2 \sin(\omega_0 t + \theta_1). \quad \blacksquare$$

Notes: Consider

$$x(t) = C \cos(\omega_0 t + \theta) = C \cos(\theta) \cos(\omega_0 t)$$

$$- C \sin(\theta) \sin(\omega_0 t) = a \cos(\omega_0 t) + b \sin(\omega_0 t),$$

$$a = C \cos(\theta), b = -C \sin(\theta), \ C = \sqrt{a^2 + b^2},$$

$$\theta = \tan^{-1}(-b/a). \quad (1.5.26)$$

One should be careful in computing θ as

$$\tan^{-1}(-b/a) \neq \tan^{-1}(b/-a),$$

$$\tan^{-1}(-b/-a) \neq \tan^{-1}(b/a). \quad \blacksquare$$

Exponentially varying sinusoids: An example of such a function is

$$x(t) = Ae^{-at} \cos(\omega_0 t + \theta). \quad (1.5.27)$$

It becomes unbounded for $at < 0$. Our interest is for only positive t and such functions are referred as *causal signals*. If $x(t)$ is defined for all t, then its causal part is $y(t) = x(t)u(t)$. In the case of $x(t)$ in (1.5.27),

$$y(t) = x(t)u(t) = Ae^{-at} \cos(\omega_0 t + \theta)u(t). \quad (1.5.28)$$

If $a > 0$ ($a < 0$), $x(t)$ in (1.5.28) is an exponentially decaying (increasing) sinusoidal function. These functions can be sketched using the envelopes Ae^{-at} and $-Ae^{-at}$ as constraints and the function $\cos(\omega_0 t + \theta)$ between the envelopes.

Notes: Even for temporal signals, the analysis and design of noncausal systems is important. For example, the analysis of prerecorded data is applicable. $\blacksquare$

1.6 Complex Numbers, Periodic, and Symmetric Periodic Functions

A complex number $c_i = a_i + jb_i$, where $a_i = \text{Re}(c_i)$ is the real part and $b_i = \text{Im}(c_i)$ is the imaginary part. Similarly if $x(t)$ is a complex function, we can write it as

$$x(t) = \text{Re}(x(t)) + j\text{Im}(x(t)). \quad (1.6.1)$$

1.6.1 Complex Numbers

A complex number can be written in terms of its real and imaginary parts or in terms of its magnitude and phase. Consider the complex number

$$c_i = a_i + jb_i = r_i(\cos\theta_i + j\sin\theta_i) = r_i e^{j\theta_i}, \ r_i$$
$$= \sqrt{a_i^2 + b_i^2}, \theta_i = \arctan(c_i) = \tan^{-1}(b_i/a_i).$$

$$(1.6.2)$$

The representation $c_i = r_i e^{j\theta_i}$ is referred to as the polar form of the complex number, where r_i and θ_i are, respectively, called the amplitude (or the modulus) and the phase angle associated with the complex number. One needs to be careful in using the formula $\theta_i = \arctan(b_i/a_i)$, especially when the real part of a_i is negative.

Example 1.6.1 Sketch the following complex numbers as vectors on a complex plane.

$$c_1 = 1 + j1, c_2 = -1 + j1, c_3 = -1 - j1, c_4 = 1 - j1.$$

Solution: The complex numbers c_1, c_2, c_3, and c_4 can be represented on the complex plane as vectors, where the length of the vector is equal to r_i, and are illustrated in Fig. 1.6.1. They are located in the first, second, third, and fourth quadrants, respectively. Also,

$$|c_i| = r_i = 1/\sqrt{2}, i = 1, 2, 3, 4; \quad \theta_1 = \frac{\pi}{4} = 45°,$$
$$\theta_2 = \frac{3\pi}{4} = 135°, \theta_3 = \frac{5\pi}{4} = 225°, \theta_4 = \frac{7\pi}{4} = 315°$$

Noting that if we use the arctangent function, i.e., $\theta_i = \arctan(b_i/a_i)$, we have $\theta_1 = \theta_3$ and $\theta_2 = \theta_4$. Note the ambiguity in taking the ratio of positive (negative)/negative (positive) quantities. This ambiguity can be reconciled, for example, by noting that when the vector lies in the second quadrant, the angle must satisfy $90° \le \theta \le 180°$. The correct angle is

$$180° - \theta_2 = 180° - \arctan(b_2/a_2)$$
$$= 180 - 45° = 135°. \qquad \blacksquare$$

A general method for obtaining the phase angle of a complex number $c = a + jb$ is

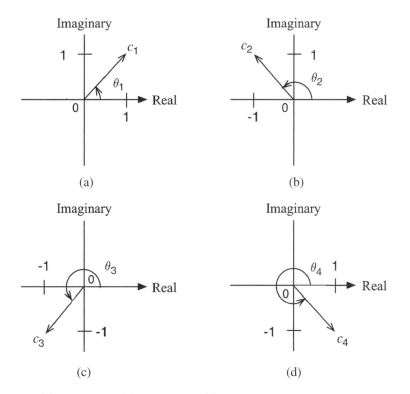

Fig. 1.6.1 (a) $c_1 = 1 + j1$, (b) $c_2 = -1 + j1$, (c) $c_3 = -1 - j1$, (d) $c_4 = 1 - j1$

$$\theta = \begin{cases} \arctan(b/a) & \text{if } a \geq 0 \\ \pm 180 + \arctan(b/a) & \text{if } a < 0 \end{cases} \quad (1.6.3)$$

When $a < 0$, select the appropriate one so that $|\theta| \leq 180^0$ or π radians. In terms of power series, in radians (1 rad $= \pi/180°$),

$$\arctan(y) = y - \frac{y^3}{3} + \frac{y^5}{5} - \frac{y^7}{7} + \cdots, \quad |y| < 1$$

$$\arctan(y) = \pm\frac{\pi}{2} - \frac{1}{y} + \frac{1}{3y^2} - \frac{1}{5y^5} + \cdots, \quad \begin{cases} +, & y \geq 1 \\ -, & y \leq -1 \end{cases}.$$
$$(1.6.4)$$

Convergence of the series is fast for most values of y and very few terms are needed to compute the arctangent function. The worst case is when $y = \pm 1$.

The conjugate, polar representation, sums, differences, multiplications, and divisions of complex numbers are given below, where we assume $c_i = a_i + jb_i$.

$$c_1^* = (a_1 - jb_1), c_i = r_i e^{j\theta_i}, c_1 \pm c_2$$
$$= (a_1 + a_2) \pm j(b_1 + b_2) \quad (1.6.5a)$$

$$c_1 c_2 = (a_1 + jb_1)(a_2 + jb_2) = (a_1 a_2 - b_1 b_2)$$
$$+ j(a_1 b_2 + b_1 a_2) = r_1 r_2 e^{j(\theta_1 + \theta_2)} \quad (1.6.5b)$$

$$\frac{c_1}{c_2} = \frac{a_1 + jb_1}{a_2 + jb_2} = \frac{a_1 + jb_1}{a_2 + jb_2}\frac{a_2 - jb_2}{a_2 - jb_2}$$
$$= \frac{(a_1 a_2 + b_1 b_2)}{a_2^2 + b_2^2} - j\frac{(a_1 b_2 - b_1 a_2)}{a_2^2 + b_2^2}, c_2 \neq 0$$

$$(c_1/c_2) = (r_1/r_2)e^{j(\theta_1 - \theta_2)}. \quad (1.6.5c)$$

$$\Longrightarrow \frac{c_1}{c_1^*} = \frac{(a_1^2 - b_1^2)}{(a_1^2 + b_1^2)} + j\frac{2a_1 b_1}{a_1^2 + b_1^2},$$

$$\frac{|c_1|}{|c_1^*|} = \frac{|r_1|}{|r_1|} = 1, \theta = 2\theta_1 = 2\arctan(b_1/a_1). \quad (1.6.6)$$

Consider the polar form of the complex number c and its natural log:

$$c = re^{j\theta}, \quad \ln(c) = \ln(re^{j\theta}) = \ln(r) + j\theta$$
$$r = |c|, \theta = \text{Im}(\ln(c)) \quad (1.6.7)$$

MATLAB can operate in terms of real or complex numbers. MATLAB commands for computing the amplitude and phase are

$$r = \text{abs}(c), \theta = \text{angle}(c) = \text{atan2}(\text{imag}(c), \text{real}(c)).$$
$$(1.6.8)$$

Notes: MATLAB atan(x) function computes the arctangent or inverse tangent of x. The function returns an angle in radians between $-(\pi/2)$ and $(\pi/2)$. MATLAB atan(x, y) computes the arctangent of (y/x). It returns an angle in radians between $-\pi$ and π and the signs on both x and y plays a role. See Appendix B for MATLAB. ■

In the last section we considered a sum of two sinusoids. *Euler's formula* can be used to express sinusoids in terms of complex exponentials and vice versa. These are

$$\cos(\theta) = [e^{j\theta} + e^{-j\theta}]/2, \quad \sin(\theta) = [e^{j\theta} - e^{-j\theta}]/2j.$$
$$(1.6.9a)$$

$$e^{j\theta} = \cos(\theta) + j\sin(\theta), \ e^{-j\theta} = \cos(\theta) - j\sin(\theta).$$
$$(1.6.9b)$$

Notes: MATLAB atan(x) function computes the arctangent or inverse tangent of x. The function returns an angle in radians between $-(\pi/2)$ and $(\pi/2)$. MATLAB atan(x, y) computes the arctangent of (y/x). It returns an angle in radians between $-\pi$ and π and the signs on both x and y plays a role. See Appendix B for MATLAB. ■

Example 1.6.2 Express the following functions in terms of single sinusoids. See (1.5.26).

$a.$ $x(t) = \cos(\omega_0 t) - \sqrt{2}\sin(\omega_0 t), \ b.\ x(t)$
$$= -\cos(\omega_0 t) + \sin(\omega_0 t). \quad (1.6.10)$$

Solution: $a.$ In this case, we have $a = 1, b = -\sqrt{2}$. From (1.6.10), we have

$$C = \sqrt{1 + (\sqrt{2})^2} = \sqrt{3}, \ \theta = \tan^{-1}(\sqrt{2}/1) \cong 54.73°,$$

$$x(t) = \sqrt{3}\cos(\omega_0 t + 54.73°)$$
$$= \sqrt{3}\cos(54.73°)\cos(\omega_0 t)$$
$$- \sqrt{3}\sin(54.73°)\sin(\omega_0 t).$$

$b.$ In this case $a = -1, b = 1$. From (1.6.10), it follows that

$$C = \sqrt{2}, \theta = \tan^{-1}[-1/-1] = -135° \rightarrow x(t)$$
$$= \sqrt{2}\cos(\omega_0 t - 135°).$$

Case *a*: a = 1; b = −1;
    [theta, C] = cart2pol(a, −b);
    theta_deg = (180/pi)*theta;
    C, theta_deg→   C = 1.7321, $\theta = 54.7356^0$
Case *b*: a = −1; b = −1;
    [theta, C] = cart2pol(a,-b);
    theta_deg = (180/pi)*theta;
    C, theta_deg→C = 1.4142, $\theta = -135°$.

De Moivre's theorem defines the power of a complex number c. For p a real number,

$$(c)^p = (re^{j\theta})^p = [r(\cos\theta + j\sin\theta)]^p$$
$$= r^p[\cos(p\theta) + j\sin(p\theta)].(c)^{1/n}$$
$$= r^{1/n}\left[\cos\frac{\theta + 2k\pi}{n} + j\sin\frac{\theta + 2k\pi}{n}\right],$$
$$k = 0, 1, 2, \ldots, n-1, n = \text{an integer}.$$

Approximation of the amplitude and the phase angle of a complex number: The magnitude of a complex number $c = a + jb$ is $|c| = \sqrt{(a^2 + b^2)} = r$. Representing c in terms of (a, b) is the rectangular coordinate representation and (r, θ) is the polar coordinate representation with $\theta = \arctan(b/a)$. Finding the square root of a number is not as simple as a multiplication. To save the computational time, the magnitude $|c|$ is usually approximated. For a simple algorithm see Problem 1.6.1.

Example 1.6.3 Consider the complex function below with the real variable ω. *a* . Derive the rectangular and polar form expressions for $X(j\omega)$:

$$X(j\omega) = \frac{a_1 - jb_1\omega}{a_2 + jb_2\omega}, a_i > 0, b_i > 0, i = 1, 2. \quad (1.6.12)$$

b. Simplify the expressions when $a_1 = a_2 = a$ and $b_1 = b_2 = b$.

Solution: *a.*
$$X(j\omega) = \frac{(a_1 - jb_1\omega)}{(a_2 + jb_2\omega)}\frac{(a_2 - jb_2\omega)}{(a_2 + jb_2)}$$
$$= \frac{(a_1a_2 - b_1b_2\omega^2)}{(a_2^2 + b_2^2\omega^2)} - j\frac{(a_1b_2 + a_2b_1)\omega}{(a_2^2 + b_2^2\omega^2)}.$$

b. In this part

$$|X(j\omega)| = \sqrt{\frac{a_1^2 + (b_1\omega)^2}{a_2^2 + (b_2\omega)^2}}, \theta(\omega)$$
$$= -\arctan\left(\frac{(a_1b_2 + a_2b_1)\omega}{a_1a_2 - b_1b_2\omega^2}\right) \Longrightarrow |X(j\omega)|_{a_1=a_2,b_1=b_2}$$
$$= |X(j\omega)| = \sqrt{\frac{a^2 + b^2\omega^2}{a^2 + b^2\omega^2}}$$
$$= 1, \theta(\omega) = -2\arctan(b\omega/a). \quad \blacksquare$$

In Chapter 7, systematic methods for sketching the magnitude and phase representations of complex rational functions of ω will be considered. MATLAB plots will be considered in Appendix B.

1.6.2 Complex Periodic Functions

For $A, \omega_0,$ and θ_0 constants, Euler's identity is

$$x(t) = Ae^{j(\omega_0 t + \theta_0)} = A\cos(\omega_0 t + \theta_0) + jA\sin(\omega_0 t + \theta_0).$$
$$(1.6.13)$$

Note that $x(t) = x(t + T) = x_T(t)$ with the period $T = 2\pi/\omega_0$ and $e^{-j2\pi} = 1$.

$$x_T(t + (2\pi/\omega_0)) = Ae^{-(j\omega_0(t+(2\pi/T)+\theta_0)}$$
$$= Ae^{-j(\omega_0 t + \theta_0)}e^{-j\omega_0(2\pi/T)} = Ae^{-j(\omega_0 t + \theta_0)} \quad (1.6.14)$$

In Chapter 3, periodic functions $x_T(t)$ will be approximated by the *complex Fourier series* expansion:

$$x_T(t) = \sum_{k=-\infty}^{\infty} X_s[k]e^{-jn\omega_0 t}. \quad (1.6.15)$$

1.6.3 Functions of Periodic Functions

If a function $x_T(t)$ is periodic, then

$$g(x_T(t + T)) = g_{T_1}(x_T(t)), T_1 \leq T. \quad (1.6.16)$$

That is, $g(t)$ is also periodic with the *same period T or smaller*. One problem of interest is given the period of $x_T(t)$ the period of the function $g(t)$ is to be computed.

Example 1.6.4 Find the fundamental periods of the following periodic functions:

a. $g_1(t) = |x_1(t)|,\ x_1(t) = \cos(\omega_0 t)$,

b. $g_2(t) = e^{jx_2(t)}, x_2(t) = \sin(\omega_0 t)$. (1.6.17)

Solution: *a.* Noting the absolute value $|\cos(\omega_0 t)|$, the fundamental period of $g_1(t)$ is $T/2 = (\pi/\omega_0)$.

b. In this case, $g_2(t)$ is periodic with the period $T = (2\pi/\omega_0)$ since

$$g_2(t + T) = e^{j\sin(\omega_0(t+T))} = e^{j\sin(\omega_0 t)}. \quad (1.6.18) \quad ∎$$

1.6.4 Periodic Functions with Additional Symmetries

Even and odd functions apply for both energy and periodic signals. $\cos(\omega_0 t)$ and $\sin(\omega_0 t)$ are periodic functions with period $(2\pi/\omega_0)$ and have even and odd symmetries, respectively. Interestingly sine and cosine functions have *four distinct parts* in one period. These are for $0 \le t < T/4, T/4 \le t < T/2$, $T/2 \le t < 3T/4$, and $3T/4 \le t < T$. Having information for one-fourth of the period of a sine wave provides the information about the other three parts. This gives a clue on the sizes of the transmitting and receiving antennas to transmit and receive the sine wave, which is discussed in Chapter 10.

Half-wave symmetric functions: A periodic function $x_T(t)$ is *half-wave symmetric* if

$$x_T(t) = -x_T(t - (T/2)). \quad (1.6.19)$$

Example 1.6.5 Show that if a function is half-wave symmetric, then

$$x_T(t) = -x_T(t + (T/2)). \quad (1.6.20)$$

Solution: Since $x_T(t)$ is periodic with period T, we can write

$$x_T\left(t - \frac{T}{2}\right) = x_T\left(t + T - \frac{T}{2}\right) = x_T\left(t + \frac{T}{2}\right). \quad ∎$$

Example 1.6.6 Show that the function $x_T(t)$ is a half-wave symmetric periodic function.

$$x_T(t) = \sum_{k=1}^{\infty}\{a[2k-1]\cos(2k-1)\omega_0 t$$
$$+ b[2k-1]\sin(2k-1)\omega_0 t\}. \quad (1.6.21)$$

Solution: Noting that $x_T(t)$ is an algebraic sum of sine and cosine terms with the same period $T = 2\pi/\omega_0$, it is a periodic function. Furthermore, since $\omega_0(T/2) = \pi$ and $(2k-1)$ is an odd integer, we can use $\cos[2k-1]\pi = -1$ and $\sin[2k-1]\pi = 0$. Using these and the following, we have $x(t + (T/2)) = -x(t)$.

$$\cos([2k-1]\omega_0(t + (T/2)))$$
$$= \cos[2k-1]\omega_0 t \cos([2k-1]\omega_0(T/2))$$
$$- \sin[2k-1]\omega_0 t \sin([2k-1]\omega_0(T/2))$$

$$\sin([2k-1]\omega_0(t + (T/2)))$$
$$= \sin[2k-1]\omega_0 t \cos([2k-1]\omega_0(T/2))$$
$$+ \cos[2k-1]\omega_0 t \sin([2k-1]\omega_0(T/2)) \quad ∎$$

Quarter-wave symmetric functions: If a periodic function $x_T(t)$ has half-wave symmetry and, in addition, is either even or odd function, then it is said to have even or odd quarter-wave symmetry. That is,

$$x_T(t) = \begin{cases} x_T(-t) = x_T(t) \\ x_T(t) = -x_T(t + \frac{T}{2}) \end{cases},$$
$$x_T(t + T) = x_T(t) : \text{even quarter} - \text{wave symmetry}. \quad (1.6.22)$$

$$x_T(t) = \begin{cases} x_T(-t) = -x_T(t) \\ x_T(t) = -x_T(t + \frac{T}{2}) \end{cases},$$
$$x_T(t + T) = x_T(t) : \text{odd quarter} - \text{wave symmetry} \quad (1.6.23)$$

Example 1.6.7 Show that the function below $x_T(t)$ has the even quarter-wave symmetry.

$$x_T(t) = \sum_{k=1}^{\infty} a[2k-1]\cos((2k-1)\omega_0 t). \quad (1.6.24a)$$

Solution: Since $x_T(t)$ is a sum of cosine terms with a zero phase, it is an even symmetric function. It has even quarter-wave symmetry since

$$\cos([2k-1]\omega_0(t+\frac{T}{2}))$$

$$= \cos([2k-1]\omega_0 t)\cos([2k-1]\omega_0\frac{T}{2})$$

$$- \sin([2k-1]\omega_0 t)\sin([2k-1]\omega_0\frac{T}{2})$$

$$= -\cos([2k-1]\omega_0 t) \qquad (1.6.24b) \quad \blacksquare$$

In a similar manner, it can be shown that the function below has odd quarter-wave symmetry. This is left as an exercise.

$$y_T(t) = \sum_{k=1}^{\infty} b[2k-1]\sin((2k-1)\omega_0 t). \qquad (1.6.25)$$

As examples, Fig. 1.6.2a and b has even and odd quarter-wave symmetries, respectively.

Hidden symmetries: The symmetries can be hidden within a constant as given by

$$x_T(t) = A + A\sin(\omega_0 t). \qquad (1.6.26)$$

It is neither even nor an odd function. On the other hand $(x_T(t) - A)$ is an odd function.

1.7 Examples of Probability Density Functions and their Moments

In this section, a brief introduction to the probability density function (PDF) $p(x)$ of a random variable $x(t)$, such as the amplitude of a noise signal, is presented. Noise, by its nature, is unpredictable. It is assumed that the amplitude can take any value in a continuous range $-\infty < a < x < b < \infty$. The level of the noise amplitude can only be described in terms of averages. Since noise is ever present,

probability of existence of the noise is always positive. That is $p(x) \geq 0$. Furthermore, the existence of noise is certain and therefore the integral of the probability density function must be 1. In summary,

$$p(x) \geq 0 \text{ and } \int_{-\infty}^{\infty} p(x)dt = 1. \qquad (1.7.1)$$

For a good discussion on probability theory, see Peebles (2001). Any nonnegative function with area 1 can serve as a probability density function. The nth moment of $p(x)$ is defined as

$$m_n = \int_{-\infty}^{\infty} x^n p(x)dx. \qquad (1.7.2)$$

The zero and the first moments are, respectively, defined by

$$m_0 = \int_{-\infty}^{\infty} p(x)dx = 1 \text{ and } m_1 = \int_{-\infty}^{\infty} xp(x)dx. \qquad (1.7.3)$$

The moments about the mean are called the central moments and are defined by

$$\mu_n = \int_{-\infty}^{\infty} (x - m_1)^n p(x)dx. \qquad (1.7.4a)$$

$$\Longrightarrow \mu_2(x) = \text{Variance of } x(t) = \sigma_x^2$$

$$= \int_{-\infty}^{\infty} (x - m_1)^2 p(x)dx. \qquad (1.7.4b)$$

The positive square root of the variance $\sigma_x = +\sqrt{\sigma_x^2}$ is called the *standard deviation*. It gives a measure of the spread of the probability density function. Now

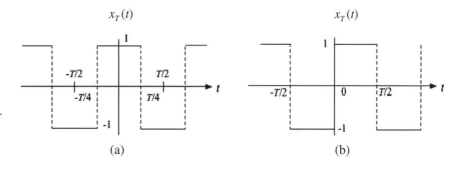

Fig. 1.6.2 (a) Even quarter-wave symmetric function and (b) odd quarter-wave symmetric function

$$\mu_2 = \int\limits_{-\infty}^{\infty} (x - m_0)^2 p(x)dx = \int\limits_{-\infty}^{\infty} x^2 p(x)dx - 2m_0$$

$$\int\limits_{-\infty}^{\infty} xp(x)dx + m_1^2 \int\limits_{-\infty}^{\infty} p(x)dx = m_2 - 2m_1^2 + m_1^2$$

$$= m_2 - m_1^2. \tag{1.7.5}$$

$$\sigma_x^2 = m_2 - m_1^2. \tag{1.7.6}$$

Mean and the variance are basic statistical values in the study of the probability density functions. In addition, $\sigma_x = +\sqrt{\sigma_x^2}$ measures its *effective width* or *duration*.

Example 1.7.1 Determine $m_0, m_1,$ and σ_x^2 for the function in Fig. 1.7.1.

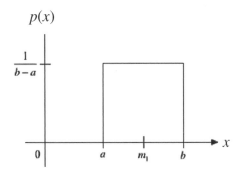

$$p(x)$$

Fig. 1.7.1 Uniform density function

$$p(x) = \frac{1}{b-a} \Pi\left[\frac{x - ((b+a)/2)}{b-a}\right] \text{ (uniform PDF)}.$$

$$\tag{1.7.7}$$

Solution: By inspection, we have the area under the function is 1. That is, $m_0 = 1$. Also

$$m_1 = \int\limits_{-\infty}^{\infty} xp(x)dx = \int\limits_{a}^{b} \frac{x}{b-a}dx$$

$$= \frac{b+a}{2}(\text{average value of } x). \tag{1.7.8}$$

$$m_2 = \text{Area}[t^2 x(t)] = \int\limits_{a}^{b} \frac{t^2}{b-a}dt$$

$$= \frac{1}{3}\frac{b^3 - a^3}{b-a} = \frac{1}{3}(b^2 + ab + a^2). \tag{1.7.9}$$

Variance :$\sigma_x^2 = m_2 - m_1^2 = (1/3)[b^2 + ab + a^2]$

$$- ((a+b)/2)^2 = \frac{(b-a)^2}{12}. \tag{1.7.10}$$

Standard deviation $= \sigma_x = (b-a)/2\sqrt{3}. \tag{1.7.11}$ ■
The function in Fig. 1.7.1 is the uniform density function, as the variable x is equally likely to take any value in the range $[a, b]$. It will be used in Chapter 10 to describe the error caused by quantization of samples.

Example 1.7.2 Consider the Gaussian probability density function shown in Fig. 1.7.2 and it is given in (1.7.12). Determine $m_0, m_1,$ and σ_x^2 of this PDF.

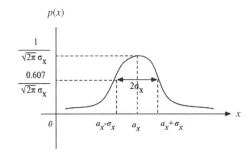

$$p(x)$$

Fig. 1.7.2 Gaussian density function

$$p(x) = \frac{1}{\sqrt{2\pi}(\sigma_x)} e^{-(x-a_x)^2/\sigma_x^2} \text{ (Gaussian PDF)}.$$

$$\tag{1.7.12}$$

Solution: From tables,

$$m_0 = \frac{1}{\sqrt{2\pi}\sigma_x} \int\limits_{-\infty}^{\infty} e^{-(t-a_x)^2/\sigma_x^2}dt = \frac{1}{\sqrt{2\pi}} \int\limits_{-\infty}^{\infty} e^{-y^2/2}dy = 1.$$

$$\tag{1.7.13}$$

That is, the area under the Gaussian function is equal to 1. The mean value is

$$m_1 = \int\limits_{-\infty}^{\infty} xp(x)dt = \frac{1}{\sqrt{2\pi}\sigma_x} \int\limits_{-\infty}^{\infty} xe^{-(x-a_x)^2/2\sigma_x^2}dx$$

$$\tag{1.7.14}$$

Using the change of variable $y = (x - a_x)/\sigma_x$, we have

$$m_1 = \frac{1}{\sqrt{2\pi}} \int_{-\infty}^{\infty} (\sigma_x y + a_x) e^{-y^2/2} dy = \frac{a_x}{\sqrt{2\pi}} \int_{-\infty}^{\infty} e^{-y^2/2} dy$$

$$+ \frac{\sigma_x}{\sqrt{2\pi}} \int_{-\infty}^{\infty} y e^{-y^2/2} dy = a_x. \tag{1.7.15}$$

These follow since the second one above on the right is an integral of an odd function of y over a symmetrical interval and it is zero. The first integral in (1.7.15) reduces to (1.7.13). To derive the variance σ_x^2, start with

$$m_0 = \frac{1}{\sqrt{2\pi}\sigma_x} \int_{-\infty}^{\infty} e^{-(x-a_x)^2/2\sigma_x^2} dt = 1 \text{ or}$$

$$\int_{-\infty}^{\infty} e^{-(x-a_x)^2/\sigma_x^2} dt = \sqrt{2\pi}\sigma_x.$$

Taking the derivative with respect to σ_x results in

$$\sqrt{2\pi} = \int_{-\infty}^{\infty} \frac{de^{-(x-a_x)^2/2\sigma_x^2}}{d\sigma_x} dx$$

$$= \int_{-\infty}^{\infty} \frac{(x-a_x)^2}{\sigma_x^3} e^{-(t-a_x)^2/2\sigma_x^2} dx.$$

$$\Rightarrow \sigma_x^2 = \frac{1}{\sqrt{2\pi}\sigma_x} \int_{-\infty}^{\infty} (x-a_x)^2 e^{(x-a_x)^2/2\sigma_x^2} dx. \tag{1.7.16}$$

Gaussian PDF (see Peebles (2001)) is one of the most important PDF, as most of the noise processes observed in practice are Gaussian. ∎

Example 1.7.3 Find m_0 and σ_x^2 for the Laplace PDF defined by

$$p(x) = (b/2)e^{-b|x|}, b > 0, -\infty < x < \infty. \tag{1.7.17}$$

Solution: From integral tables, the mean and the variance are

$$m_0 = \int_{-\infty}^{\infty} \frac{b}{2} e^{-b|x|} dx = b \int_{0}^{\infty} e^{-bx} dx = e^{-bx}|_{x=0}^{\infty} = 1,$$

$$\sigma_x^2 = \int_{-\infty}^{\infty} \frac{b}{2} x^2 e^{-|x|/b} dx = \frac{2}{b^2}. \tag{1.7.18} ∎$$

Notes: Noise is random and unpredictable. When it is added to the information bearing signal, the message signal is masked or even obliterated. Noise cannot be eliminated. A measure of corruption of the signal by noise is an important measure. It is the ratio of the average signal power to variance of the noise. It is

$$\text{Signal} - \text{to} - \text{noise ratio} = \text{SNR}$$

$$= \frac{\text{Average message signal power}}{\text{Variance of the noise}, \sigma_x^2}. \qquad ∎$$

1.8 Generation of Periodic Functions from Aperiodic Functions

Now we like to construct a periodic function from an *aperiodic function*, say $\varphi(t)$.

$$y_T(t) = \sum_{k=-\infty}^{\infty} \varphi(t + kT). \tag{1.8.1}$$

$\varphi(t)$ is the principal segment of the *periodic extension*. Clearly, $y_T(t)$ is a periodic function with period T s and is the periodic extension of $\varphi(t)$. If $\varphi(t)$ is not time limited to a T s interval (for example, $\varphi(t)$ is *nonzero* for $t > T$ and $t < 0$), then $\varphi(t)$ and $\varphi(t + T)$ terms will overlap and $\varphi(t)$ cannot be extracted from $y_T(t)$.

Example 1.8.1 Using the principal segment $\varphi(t) = \Lambda[t/\tau]$, sketch the periodic extensions for the following cases. *a.* $T \geq 2\tau$, *b.* $T < 2\tau$.

Solution: The periodic extension of the triangular function is

$$y_T(t) = \sum_{k=-\infty}^{\infty} \Lambda[\frac{t + kT}{\tau}]. \tag{1.8.2}$$

The function $\varphi(t)$ and its periodic extensions are sketched in Fig. 1.8.1a, b, and c. For simplicity, in the sketch for part *a*, $T = 2\tau$ is assumed.

a. The functions $\Lambda[(t + kT)/\tau]$ and $\Lambda[(t + (k + 1)T)/\tau]$ do not overlap and the function $\varphi(t)$ can be extracted

$$y_T(t)|_{k=0} = \Lambda[t/\tau] = \varphi(t), |t| \leq \tau. \tag{1.8.3}$$

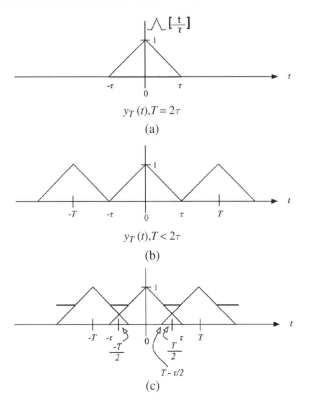

Fig. 1.8.1 (a) $\Lambda\left[\frac{t}{\tau}\right]$, (b) $y_T(t)$, $T = 2\tau$, (c) $y_T(t)$, $T<2\tau$

b.   For $T<2\tau$, $\Lambda[(t+kT)/\tau]$ and $\Lambda[(t+(k+1)T)/\tau]$ overlap (See Fig. 1.8.1b). Recovery of $\phi(t)$ from $y_T(t)$. is not possible.

It is interesting to note the area of one period of the periodic extension of $\varphi(t)$ equals to the area of the function $\varphi(t)$. This is a *consistency check*. See Ambardar (1995).

Example 1.8.2 Show that the area under one period of the periodic extension of $\varphi(t) = e^{-t/\tau}u(t)$ is equal to the area under $\varphi(t)$.

Solution: The periodic extension of $\varphi(t)$ with a period of T can be written as

$$y_T(t) = [e^{-(t/\tau)}u(t) + e^{-(t+T)/\tau}u(t+T)$$
$$+ e^{-(t+2T)/\tau}u(t+2T) + \cdots] + [e^{-(t-T)/\tau}u(t-T)$$
$$+ e^{-(t-2T)/\tau}u(t-2T) + \ldots] \qquad (1.8.4)$$

$$x_1(t) = e^{-t/\tau}[u(t) + e^{-T/\tau}u(t+\tau)$$
$$+ e^{-2T/\tau}u(t+2\tau) + \ldots], \tau > 0. \qquad (1.8.5)$$

$$\int_0^T x_1(t)dt = [1 + e^{-T/\tau} + e^{-2T/\tau} + \ldots]$$

$$\int_0^T e^{-t/\tau}dt = \frac{1}{1 - e^{-T/\tau}}\int_0^T e^{-t/\tau}dt = -\frac{1}{\tau}\left[\frac{e^{-T/\tau}-1}{1-e^{-T/\tau}}\right] = \frac{1}{\tau}.$$

$$\int_0^\infty \varphi(t)dt = \int_0^\infty e^{-t/\tau}u(t)dt = \frac{1}{\tau}. \qquad (1.8.6) \quad \blacksquare$$

1.9 Decibel

Decibel or dB is a logarithmic unit named after Alexander Graham Bell that is used to express power ratios. Given two powers P_1 and P_2 and their ratio (P_2/P_1), then

Table 1.9.1 Sound Power (loudness) Comparison

Threshold of audibility	0 dB
Whisper	15 dB
Average home	45 dB
Riveting machine (30' away)	100 dB
Threshold of hearing	120 dB
Jet plane	140 dB

$$\text{Power ratio in dB} = 10 \log_{10}[P_2/P_1]. \quad (1.9.1)$$

For computing the dB values from the amplitudes using MATLAB, see Section B.12 in Appendix B. The unit of bel is too large. For example, a human ear can detect audio power level differences of one-tenth of a bel or 1 dB. The loudness of a few typical activities are shown in table 1.9.1. Also, 1 dB is approximately equal to the attenuation of one mile of a standard telephone cable. Power ratio is expressed by

$$\text{Power ratio} = (P_2/P_1) = 10^{dB/10}. \quad (1.9.2)$$

Even though decibels were originally meant to be used with respect to power ratios, they can be used to express absolute values of power interpreted as a ratio of power P_2 to $P_1 = 1$ Watt, referred to as 1 dBW. That is, 1 Watt is a reference and

$$10 \log_{10}(P_2/P_1) = 10 \log_{10}(P_2 \text{ in W}/1 \text{ Watt}) \text{ dB W}. \quad (1.9.3)$$

Positive decibels imply $A = P_2/P_1 > 1$, zero decibels imply $A = 1$, and negative decibels imply $A < 1$. Decibels are used to express gains or losses in a system; gain is the output divided by input and loss is the input divided output. Instead of using 1 Watt as a reference, 1 mW can be used as a reference to compare small signal power levels, such as powers of radar echoes and the unit is dBm. Now

$$P_2/P_1|_{P_1=1\,mW} = 10 \log_{10}(P_2 \text{ in Watts}/1 \text{ mW}) \text{ dB m}. \quad (1.9.4)$$

As examples, a power of 1 Watt is 0 dBW, a power of 3 Watt is 4.77 dBW, and a power of 1 kW is 30 dBW. A power of 1 kilowatt is 0 dBm and a power of 10^{-10} mW is

$$10 \log_{10}(10^{-10} \text{ mW}/1 \text{ mW}) = -100 \text{ dBm}.$$

Table 1.9.2 Power ratios and their corresponding values in dB

dB	0	1	2	3	4	5	6	7	8	9
Power ratios	1	1.26	1.6	2	2.5	3.2	4	5	6.3	8

The logarithmic decibel function provides a greater resolution when the power ratio is small, indicating a good way to recognize very small differences in the power levels (Table 1.9.2).

The smaller the power ratio, that is less than 1, the larger the number of negative dBs required. As the ratio approaches zero, the negative dB increases without limit. Interestingly when you round of to the nearest whole decibel, the error in the power ratio is at most only 1 part in 7. The dB provides greater resolution when the power ratio is small. Small differences in the power levels are important in spectral analysis, analog and digital filter designs, control system designs, communication system designs, etc. A power ratio of 2 to 1 (1 to 2) is 3 dB (-3 dB). Bandwidths of 3 dB play a major role in filter designs. A ratio of 10^8 to is only 80 dB and 10^{-8} is -80 dB.

Power levels in radars have a large range. Dealing with large number of digits can be troublesome, as dropping a zero at the end of a large number makes the radar calculations wrong. The dB scale makes the numbers compressed. The power levels associated with seismic signals are low. One nice thing about dB scale is that we will be dealing with additions and subtractions rather than multiplications and divisions. Gain (or loss) is the term used for an increase (or decrease) in power level. For example, for an amplifier,

$$\text{Gain} = \frac{\text{Output signal power coming out of the amplifier}}{\text{Input signal power going into the amplifier}}. \quad (1.9.5)$$

If the output power is 100 times the input power, then

$$\text{Gain dB} = 10 \log(100) = 20 \text{ dB}. \quad (1.9.6)$$

If we use a wave guide or a cable, we have a loss in the power. That is,

$$\text{Loss} = \frac{\text{Input power to the device}}{\text{Output power of the device}}. \quad (1.9.7)$$

As an example, let an amplifier be connected to an antenna by a waveguide and the guide absorbs 20% of the power. If the ratio of the input power to the output power is 10 to 8, i.e., 1.25 indicating a loss of:

$$\text{Loss in dB} = 10 \ \log(1.25) \cong .9 \, \text{dB}$$
$$(\text{or, gain is}(-.9) \, \text{dB}).. \tag{1.9.8}$$

Since $\log(AB) = \log(A) + \log(B)$, we can simply find the power gain (or loss) in dB by adding or subtracting the numbers in dB. This is illustrated in the example below.

Example 1.9.1 Consider that the amplifier considered above is connected to an antenna by a waveguide. See Fig. 1.9.1. Assume the amplifier gain is $G_{\text{amplifier}} = 100$ and the waveguide absorbs about 20% of the power. Determine the total gain.

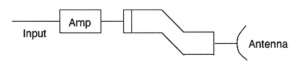

Fig. 1.9.1 Amplifier and a waveguide

Solution: The total gain from the input to the amplifier to the antenna is

$$G_{\text{Total}} = G_{\text{amplifier}} / G_{\text{wave guide}} = 100/1.25 = 80. \tag{1.9.9}$$

We can determine the gain or loss by simply subtracting gain from the loss. That is,

$$(G_{\text{Total}})_{\text{dB}} = (G_{\text{amplifier}})_{\text{dB}} - (G_{\text{wave guide}})_{\text{dB}}$$
$$\cong 20 - .97 = 19.03 \, \text{dB}. \tag{1.9.10}$$

$$\Longrightarrow \text{Power ratio} = P_r = 10^{(19.03/10)} \simeq 80. \ \blacksquare$$

Systems are designed part by part, arranged in a cascade, generally drawn symbolically by a block diagram shown in Fig. 1.9.2. We assume that the system identified by block k is not loading the system identified by block $(k-1)$. That is, there is no loading effect. Corresponding to this we can compute the

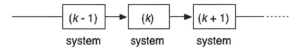

Fig. 1.9.2 A cascaded system

power gain or loss by adding or subtracting the appropriate quantities. It can be obtained by finding the transfer functions of each block and determine the total transfer function. Since the transfer functions are functions of frequency, the power gains or losses are functions of frequencies.

Notes: Filters keep approximately the same amplitude in the filter pass band and provide attenuation in the filter stop band. If $H(j\omega)$ is the transfer function of the filter, then we have attenuation (gain) at a frequency $\omega_1 = 2\pi f_1$, if $|H(j\omega_1)| \leq 1$ ($|H(j\omega_1)| > 1$). The corresponding attenuation and gain are as follows:

$$\alpha = -20 \ \log|H(j\omega_1)| \, \text{dB} \ > \ 0 \ (\text{Loss}),$$
$$A = 20 \ \log|H(j\omega_1)| \ (\text{Gain}). \tag{1.9.11}$$

Solving for $|H(j\omega)|$ from (1.9.11) results in

$$\text{Loss} : |H(j\omega_1)| = 1/(10^{.05})^{\alpha},$$
$$\text{Gain} : |H(j\omega_1)| = (10^{.05})^{A}. \tag{1.9.12}$$

Loss in terms of dB and the corresponding decrease in amplitudes are given below:

- $-1 \, \text{dB} \Longrightarrow$ approximately 10% decrease in $|H|$ from 1 to .891.
- $-2 \, \text{dB} \Longrightarrow$ approximately 20% decrease in $|H|$ from 1 to .794.
- $-3 \, \text{dB} \Longrightarrow$ approximately 30% decrease in $|H|$ from 1 to .708.
- $-6 \, \text{dB} \Longrightarrow$ approximately 50% decrease in $|H|$ from 1 to .501. $\blacksquare$

1.10 Summary

In this chapter some of the basics on signals are presented that a second semester junior in an electrical engineering program may have gone through. Some of the material, such as complex numbers, periodic functions, integrals, decibels, and others, are included to refresh the reader's memory. Specific principal topics that were included are

- Various types of continuous signals
- Useful signal operations involving time shifting, scaling, reversal, and amplitude shift

- Approximations and simplifications for integrals with symmetries
- Singularity functions that include impulse functions, step functions, etc., and functions that can be used to approximate impulses
- Signal classifications based on power and energy
- Periodic signals and special classes of periodic functions with symmetries
- Complex numbers and complex functions
- Energy signals and their moments
- Periodic extension of aperiodic functions
- Decibels

Problems

1.1.1 Peterson and Barney (1952) collected average formant frequencies for vowels spoken by adult male and female speakers. For example, first two average formant frequencies in Hertz for the two vowels $/i/$ and $/a/$ are given in the table below. For a particular subject the first two formant frequencies of one of the vowels given were determined and they are $F_1 = 500$ Hz and $F_2 = 1600$ Hz. Using the minimum distance classifier, and the below table, determine if this subject is a male or a female and what is the vowel $/i/$ or $/a/$?

		$|i|$	$|a|$
F_1	Male	270	730
	Female	310	850
F_2	Male	2290	1090
	Female	2790	1220

1.1.2 Sketch the following:

$a.\, x_1(t) = \Pi\left[\dfrac{t-1}{2}\right]\Pi\left[\dfrac{2-t}{2}\right], b.\, x_2(t) = \Pi\left[\dfrac{t-1}{2}\right].\Lambda\left[\dfrac{t}{2}\right],$

$c.\, x_3(t) = u(t-1).u(\beta - t) \text{ for } \beta = -5, 0, 1, 2$

1.1.3 Response of a first-order system is given by $y(t) = A(1 - e^{-t/2})u(t)$. What is the time constant of the system and compute the value of the time at the time $y(t)$ reaches 63.21% of the response? What do we call this time?

   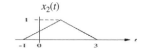

Fig. P1.2.1

1.2.1 Given the functions in Fig. P1.2.1, sketch

$a.\, x_i(t+1),$

$b.\, x_i(t-1),$

$c.\, x_i(2t-3),$

$d.\, x_i(-2t+1), i = 1, 2.$

1.2.2 Find the even and odd parts of the following functions:

$a.\, x_1(t) = e^{-t}u(t-1),$

$b.\, x_2(t) = e^t,$

$c.\, x_3(t) = [(t+1)/(t-1)].$

$d.\, x_4(t) = \cos(t) + \sin(t),$

$e.\, x_5(t) = \Pi[(t-1)/2].$

1.2.3 Sketch the functions $\text{sinc}(\pi\lambda)$ and $\text{sinc}^2(\pi\lambda)$ and give an approximate value of the magnitude of the first side lobe and their values in dB. Use MATLAB if you have the access. See Appendix B for information on MATLAB.

1.3.1 Approximate the following integral for the intervals shown by using $a.$ the rectangular integration formula and $b.$ the trapezoidal integration formula.

$$A = \int_0^\pi x(t)dt, \; x(t) = \sin(t), \; \text{intervals}: \left(0, \frac{\pi}{4}\right),$$

$$\left(\frac{\pi}{4}, \frac{\pi}{2}\right), \; \left(\frac{\pi}{2}, \frac{3\pi}{4}\right), \; \left(\frac{3\pi}{4}, \pi\right).$$

Use $x(t)$ at $t = k(\pi/4), k = 0, 1, 2, 3$ to approximate the area by assuming each strip is a rectangle or a trapezoid. Compare these values to the actual value of the integral.

1.3.2 Evaluate the following integral:

$$\int_0^{T/2} \left[\frac{1}{2} - \frac{t}{T}\right]\sin(\omega_0 t)dt.$$

1.3.3 Derive the expressions for the following partials:

$$G(j\omega) = 1/(\sigma + j\omega) = u(\omega) + jv(\omega),$$
$$\frac{\partial u(\sigma, \omega)}{\partial \sigma}, \frac{\partial v(\sigma, \omega)}{\partial \sigma}, \frac{\partial u(\sigma, \omega)}{\omega}, \text{ and } \frac{\partial v(\sigma, \omega)}{\partial \omega}.$$

1.4.1 Sketch $x_i(t)$ and $y_i(t) = u(x_i(t))$, $i = 1, 2, 3, 4, 5, 6$ for $-\pi/2 < t < 3\pi/2$ given

$a.\ x_1(t) = \sin(t), b.\ x_2(t) = \cos(t), c.\ x_3(t)$
$= \tan(t), d.\ x_4(t) = \sec(t), e.\ x_5(t)$
$= \cot(t), f.\ x_6(t) = \csc(t).$

1.4.2 Let $x(t) = \delta((t-1)(t-2))$. Express the function in terms of a sum of impulses.

1.4.3 Show the following functions are limiting forms of impulse functions:

$a.\ x_1(t) = (T/2)e^{-|t|T}, \quad b.\ x_2(t) = T(1 - (|t|/T)),$

$c.\ x_3(t) = \dfrac{T \sin Tt}{\pi \ tT}, d.\ x_4(t) = \dfrac{T}{\pi}\left(\dfrac{\sin Tt}{Tt}\right)^2$

$e.\ x_5(t) = Te^{-\pi t^2 T^2}, \ f.\ x_6(t) = \dfrac{1}{\pi}\dfrac{T}{t^2 + T^2}$

1.4.4 Show that $x_n(t)$ can be used as an impulse representation

$$\delta(t) = \lim_{n \to \infty} x_n(t), \ x_n(t) = .5\left[ne^{-n|t|}\right]$$

$$\left(\text{use } \int_{-\infty}^{\infty} x_n(t)dt = 1, \lim_{n \to \infty} x_n(t) = 0\right).$$

1.4.5 Show the following:

$a.\ \delta'(-t) = -\delta'(t), \quad b.\ t\delta'(t) = -\delta(t),$

$c.\ \int\limits_{-\infty}^{\infty} x(t)\dfrac{d\delta(t-t_0)}{dt}dt = -\dfrac{dx(t)}{dt}\Big|_{t=t_0},$

$d.\ \lim\limits_{\varepsilon \to 0}\dfrac{dx_\varepsilon(t)}{dt} = \dfrac{d\delta(t)}{dt}, \ x_\varepsilon(t) = \dfrac{1}{\pi}\dfrac{\varepsilon}{t^2 + \varepsilon^2}, \ \varepsilon > 0.$

1.4.6 Evaluate the following integrals using the properties of the impulse functions:

$a.\ \displaystyle\int\limits_0^5 (t^2 + 2t - 1)\delta(t-1)dt,$

$b.\ \displaystyle\int\limits_0^5 (t^2 + 2t - 1)\dfrac{d\delta(t-1)}{dt}dt,$

$c.\ \displaystyle\int\limits_{-\infty}^{\infty} e^{-t}u(t)\dfrac{\delta(t-1)}{dt}dt$

1.5.1a. Express an arbitrary real-valued signal in terms of its even and odd parts.

b. Express the unit step function $u(t)$ in terms of its even and odd parts and sketch the even and odd parts. Assume $u(0) = 1$ in sketching the function.

1.5.2 Classify each of the following functions as either an energy signal or a power signal or neither. If the functions are either energy or power signals, give the corresponding energy or power. Otherwise, explain why they are not.

$a.\ x_1(t) = \Lambda\left[\dfrac{t-1}{2}\right], \quad b.\ x_2(t) = e^{-a|t|}, a > 0,$

$c.\ x_3(t) = t\ e^{-t}u(t), \quad d.\ x_4(t) = \text{sinc}(\pi t),$

$e.\ x_5(t) = 1/[\pi(1+t^2)], \quad f.\ x_6(t) = e^{-\pi t^2},$

$g.\ x_7(t) = \delta'(t), \quad h.\ x_8(t) = \sin^2(t).$

1.5.3 Show that

$$\int\limits_{-\infty}^{\infty} x^2(t)dt = \int\limits_{-\infty}^{\infty} x_e^2(t)dt + \int\limits_{-\infty}^{\infty} x_0^2(t)dt$$

1.5.4 Given $x(t) = \cos(2\pi(1000)t)$, find the period of $x(at)$, $a > 0$. What can you say in the general case of a periodic function $x_T(t)$ and $x_T(at), a \neq 0$?

1.5.5 Find the mean, rms, and the peak values of the function

$$x(t) = A\cos(\omega_0 t + \phi_1) + B\cos(2\omega_0 t + \phi_2).$$

What is the effect of the phase angles on the peak value of this function?

1.6.1 The power series expansion of the square root function is Spiegel (1968)

$$\sqrt{(a^2 + b^2)} = a\sqrt{(1 + x^2)}$$
$$= a\left[1 + \frac{1}{2}x - \frac{1}{2(4)}x^2 + \frac{1.3}{(2)(4)6}x^3 - \cdots\right],$$
$$x = \frac{b}{a}, -1 < x \le 1.$$

Simplest approximation: $\sqrt{a^2 + b^2} \approx a + (b/2)$.

For $c = 1 + j1$, find $|c|$ by the approximation and the error between the direct computation and the simplified method.

1.6.2 Solve for the roots of the polynomial $1 + (-1)^n s^{2n} = 0$ in general terms. The roots may be complex. Give the roots on the left half-plane for $n = 1, 2, 3, 4, 5$. Plot these roots on a complex plane and comment on the magnitudes of these roots.

1.6.3 Give the even and odd parts of the function $x(t) = t$, $0 < t < T/2$ and zero elsewhere.

1.6.4 Give examples of half-wave, even, and odd quarter-wave symmetric functions.

1.6.5 Show that the function given in (1.6.25) has odd quarter-wave symmetry.

1.6.6 Consider the periodic function $x_T(t)$ given below. What can you say about the hidden symmetry in this periodic function?

$x_T(t) = (1 - (t/T)), 0 < t < T$ and $x_T(t + T) = x(t)$.

1.7.1 Determine the mean and the variances of the following two functions:

a. $x_L(t) = (b/2)e^{-b|t-a|}$, $-\infty < t < \infty, b > 0$: Laplace function.

b. $x_R(t) = (2/b)(t - a)e^{-(t-a)^2/b}u(t - a)$, $-\infty < t < \infty, b > 0$: Rayleigh function.

1.8.1 Assuming $T = \tau$, b. $T = \tau/2$, give the expressions for the periodic extension of the function $x_1(t) = \Pi[t/\tau]$.

1.9.1 Show that if you round off to the nearest whole decibel, the error in the power at most 1 in 7 by noting that the plot of decibel versus power ratio in the interval between 0 and 1 is approximately a straight line (Simpson, Hughes Air Craft Company, 1983).

1.9.2 Convert the following to power ratios and approximate them in dB:

Magnitude ratios : $1/\sqrt{10}$, $1/2$, $1/\sqrt{2}$, 1, $\sqrt{2}$, 2, $\sqrt{10}$, 5, 10, $100, 1000$.

1.9.3 Two radar signals $x_1(t)$ and $x_2(t)$ are assumed to have an average power of 3 dBm and -10 dBm, respectively. What are the corresponding absolute power levels?

Chapter 2
Convolution and Correlation

2.1 Introduction

In this chapter we will consider two signal analysis concepts, namely convolution and correlation. Signals under consideration are assumed to be real unless otherwise mentioned. *Convolution* operation is basic to linear systems analysis and in determining the probability density function of a sum of two independent random variables. Impulse functions were defined in terms of an integral (see (1.4.4a)) using a test function $\phi(t)$.

$$\int_{-\infty}^{\infty} \phi(t)\delta(t - t_0)dt = \phi(t_0). \quad (2.1.1)$$

This integral is the convolution of two functions, $\phi(t)$ and the impulse function $\delta(t)$ to be discussed shortly. In a later chapter we will see that the response of a linear time-invariant (LTI) system to an impulse input $\delta(t)$ is described by the convolution of the input signal and the impulse response of the system. Convolution operation lends itself to spectral analysis. There are two ways to present the discussion on convolution, first as a basic mathematical operation and second as a mathematical description of a response of a linear time-invariant system depending on the input and the description of the linear system. The later approach requires knowledge of systems along with Fourier series and transforms. This approach will be considered in Chapter 6. Although we will not be discussing random signals in any detail, convolution is applicable in dealing with random variables.

The process of *correlation* is useful in comparing two deterministic signals and it provides a measure of similarity between the first signal and a time-delayed version of the second signal (or the first signal). A simple way to look at correlation is to consider two signals: $x_1(t)$ and $x_2(t)$. One of these signals could be a delayed, or an advanced, version of the other. In this case we can write $x_2(t) = x_1(t + \tau), -\infty < \tau < \infty$. Multiplying point by point and adding all the products, $x_1(t)x_1(t + \tau)$ will give us a large number for $\tau = 0$, as the product is the square of the function. On the other hand if $\tau \neq 0$, then adding all these numbers will result in an equal or a lower value since a positive number times a negative number results in a negative number and the sum will be less than or equal to the peak value. In terms of continuous functions, this information can be obtained by the following integral, called the autocorrelation function of $x(t)$, as a function of τ *not t*.

$$R_{xx}(\tau) = \int_{-\infty}^{\infty} x(t)x(t + \tau)dt = \text{AC}\,[x(t)] \equiv R_x(\tau).$$

$$(2.1.2)$$

This gives a comparison of the function $x(t)$ with its shifted version $x(t + \tau)$. Autocorrelation (AC) provides a nice way to determine the spectral content of a random signal. To compare two different functions, we use the *cross-correlation* function defined by

$$R_{xh}(\tau) = x(\tau) * *h(\tau) = \int_{-\infty}^{\infty} x(t)h(t + \tau)dt. \quad (2.1.3)$$

R.K.R. Yarlagadda, *Analog and Digital Signals and Systems*, DOI 10.1007/978-1-4419-0034-0_2,
© Springer Science+Business Media, LLC 2010

Note the symbol (**) for correlation. Correlation is related to the convolution. As in autocorrelation, the cross-correlation in (2.1.3) is a function of τ, the time shift between the function $x(t)$, and the shifted version of the function $h(t)$.

2.1.1 Scalar Product and Norm

The scalar valued function $\langle x(t), y(t) \rangle$ of two signals $x(t)$ and $y(t)$ of the *same* class of signals, i.e., either energy or power signals, is defined by

$$\langle x(t),y(t)\rangle = \begin{cases} \int\limits_{-\infty}^{\infty} x(t)y^*(t)dt, \\ \text{energy signals.} \qquad (2.1.4a) \\ \lim\limits_{T\to\infty}\frac{1}{T}\int\limits_{-T/2}^{T/2} x(t)y^*(t)dt, \\ \text{power signals.} \qquad (2.1.4b) \end{cases}$$

Superscript (*) indicates complex conjugation. Our discussion will be limited to a subclass of power signals, namely periodic signals. In that case, assuming that both the time functions have the same period (2.1.4b) can be written in the symbolic form as follows:

$$\langle x_T(t)y_T(t)\rangle = \frac{1}{T}\int\limits_{T} x(t)y^*(t)dt. \qquad (2.1.4c)$$

Even though our interest is in real functions, for generality, we have used complex conjugates in the above equations. The norm of the function is defined by

$$\|x(t)\| = \langle x(t), x(t)\rangle^{1/2} = \begin{cases} E_x, & \text{energy signals} \\ P_x, & \text{power signals} \end{cases}. \qquad (2.1.5)$$

It gives the energy or power in the given energy or the power signal. The two functions, $x(t)$ and $y(t)$, are *orthogonal* if

$$\langle x(t), y(t)\rangle = 0. \qquad (2.1.6)$$

In that case,

$$\|x(t) + y(t)\|^2 = \|x(t)\|^2 + \|y(t)\|^2. \qquad (2.1.7)$$

If $x(t)$ and $y(t)$ are orthogonal, the energy and power contained in the energy or power signal $z(t) = x(t) + y(t)$ are respectively given by

$$E_z = E_x + E_y \text{ or } P_z = P_x + P_y. \qquad (2.1.8)$$

Some of the important properties of the norm are stated as follows:

1. $\|x(t)\| = 0$ if and only if $x(t) = 0$, \qquad (2.1.9a)

2. $\|x(t) + y(t)\| \leq \|x(t)\| + \|y(t)\|$,
   triangular inequality \qquad (2.1.9b)

3. $\|\alpha x(t)\| = |\alpha|\|x(t)\|$. \qquad (2.1.9c)

In (2.1.9c), α is some constant. One measure of *distance*, or *dissimilarity*, between $x(t)$ and $y(t)$ is $\|x(t) - y(t)\|$. A useful inequality is the *Schwarz's inequality* given by

$$|\langle x(t), y(t)\rangle| \leq \|x(t)\|\|y(t)\|. \qquad (2.1.9d)$$

The two sides are equal when $x(t)$ or $y(t)$ is zero or if $y(t) = \alpha x(t)$ where α is a scalar to be determined. This can be seen by noting that

$$\begin{aligned} \|x(t) + \alpha y(t)\|^2 &= \langle x(t) + \alpha y(t), x(t) + \alpha y(t)\rangle \\ &= \langle x(t)x(t)\rangle + \alpha^*\langle x(t), y(t)\rangle \\ &\quad + \alpha\langle x(t), y(t)\rangle^* + |\alpha|^2\langle y(t), y(t)\rangle \\ &= \|x(t)\|^2 + \alpha^*\langle x(t), y(t)\rangle \\ &\quad + \alpha\langle x(t), y(t)\rangle^* + |\alpha|^2\|y(t)\|^2. \end{aligned} \qquad (2.1.10)$$

Since α is arbitrary, select

$$\alpha = -\langle x(t), y(t)\rangle/\|y(t)\|^2. \qquad (2.1.11)$$

Substituting this in (2.1.10), the last two terms cancel out, resulting in

$$\|x(t) + \alpha y(t)\|^2 = \|x(t)\|^2 - \frac{\left|\langle x(t), y(t)\rangle^2\right|}{\|y(t)\|^2}$$

$$\Rightarrow \|x(t)\|^2 \|y(t)\|^2 - \left|\langle x(t), y(t)\rangle^2\right| \geq 0.$$

$$(2.1.12)$$

Equality exists in (2.1.9d) only if $x(t) + \alpha y(t) = 0$. Another possibility is the trivial case being either one of the functions or both are equal to zero. Ziemer and Tranter (2002) provide important applications on this important topic.

Correlations in terms of time averages: Cross-correlation and autocorrelation functions can be expressed in terms of the time average symbols and

$$R_{xh}(\tau) = \int\limits_{-\infty}^{\infty} x(t)h(t+\tau)dt = \langle x(t)h(t+\tau)\rangle, (2.1.13a)$$

$$R_{T,xh}(\tau) = \frac{1}{T} \int\limits_{-T/2}^{T/2} x_T(t)h_T(t+\tau)dt$$

$$(2.1.13b)$$

$$= \frac{1}{T}\int\limits_{T} x_T(t)h_T(t+\tau)dt = \langle x_T(t)h_T(t+\tau)\rangle.$$

In the early part of this chapter we will deal with convolution and correlation associated with aperiodic signals. In the later part we will concentrate on convolution and correlation with respect to both periodic and aperiodic signals. Most of the material in this chapter is fairly standard and can be seen in circuits and systems books. For example, see Ambardar (1995), Carlson (1975), Ziemer and Tranter (2002), Simpson and Houts (1971), Peebles (1980), and others.

2.2 Convolution

The convolution of two functions, $x_1(t)$ and $x_2(t)$, is defined by

$$y(t) = x_1(t) * x_2(t) = \int\limits_{-\infty}^{\infty} x_1(\alpha)x_2(t-\alpha)d\alpha$$

$$(2.2.1)$$

$$= \int\limits_{-\infty}^{\infty} x_2(\beta)x_1(t-\beta)d\beta = x_2(t) * x_1(t).$$

This definition describes a *higher algebra* and allows us to study the response of a linear time-invariant system in terms of a signal and a system response to be discussed in Chapter 6. It should be emphasized that the end result of the convolution operation is a function of time. Coming back to the sifting property of the impulse functions, consider the equation given in (2.1.1). Two special cases are of interest.

$$\phi(t) * \delta(t) = \int\limits_{-\infty}^{\infty} \phi(\alpha)\delta(t-\alpha)d\alpha$$

$$(2.2.2a)$$

$$= \int\limits_{-\infty}^{\infty} \phi(t-\beta)\delta(\beta)d\beta = \delta(t) * \phi(t),$$

$$\delta(t) * \delta(t) = \int\limits_{-\infty}^{\infty} \delta(\alpha)\delta(t-\alpha)d\alpha = \delta(t). \quad (2.2.2b)$$

2.2.1 Properties of the Convolution Integral

1. Convolution of two functions, $x_1(t)$ and $x_2(t)$, satisfies the *commutative property*,

$$y(t) = x_1(t) * x_2(t) = x_2(t) * x_1(t). \quad (2.2.3)$$

This equality can be shown by defining a new variable, $\beta = t - \alpha$, in the first integral in (2.2.1) and simplifying the equation.

2. Convolution operation satisfies the *distributive property*, i.e.,

$$x_1(t) * [x_2(t) + x_3(t)] = x_1(t) * x_2(t)$$
$$+ x_1(t) * x_3(t).$$

$$(2.2.4)$$

3. Convolution operation satisfies the *associative property*, i.e.,

$$x_1(t) * (x_2(t) * x_3(t)) = (x_1(t) * x_2(t)) * x_3(t). \quad (2.2.5)$$

The proofs of the last two properties follow from the definition.

4. The *derivative of the convolution* operation can be written in a simple form and

$$y'(t) = \frac{dy(t)}{dt} = \frac{d}{dt}\left[\int_{-\infty}^{\infty} x_1(\alpha)x_2(t-\alpha)d\alpha\right]$$

$$= \int_{-\infty}^{\infty} x_1(\alpha)\frac{dx_2(t-\alpha)}{dt}d\beta = x_1(t)*x_2'(t),$$

$$\Rightarrow \frac{dy(t)}{dt} = \frac{d}{dt}[x_1(t)*x_2(t)]$$

$$= \frac{dx_1(t)}{dt}*x_2(t) = x_1(t)*\frac{dx_2(t)}{dt}.$$

$$(2.2.6a)$$

Equation (2.2.6a) can be generalized for higher order derivatives. We can then write

$$x_1^{(m)}(t)*x_2(t) = \frac{d^m x_1(t)}{dt^m}*x_2(t) = \frac{d^m y(t)}{dt^m}$$

$$(2.2.6b)$$

$$= y^{(m)}(t)\left(\text{Note } x_i^{(m)}(t) = \frac{d^i x_i(t)}{dt^i}\right),$$

$$x_1^{(m)}(t)*x_2^{(n)}(t) = \frac{d^m x_1(t)}{dt^m}*\frac{d^n x_2(t)}{dt^n}$$

$$(2.2.6c)$$

$$= \frac{d^{m+n}y(t)}{dt^{m+n}} = y^{(m+n)}(t).$$

Since the impulse function is the *generalized derivative* of the unit step function $u(t)$ (see Section 1.4.2.), we have

$$y(t) = u(t)*h(t) \Rightarrow y'(t)$$

$$(2.2.7)$$

$$= u'(t)*h(t) = \delta(t)*h(t) = h(t).$$

5. Convolution is an integral operation and if we know the convolution of two functions and desire to compute its *running integral*, we can use

$$\int_{-\infty}^{t} y(\alpha)d\alpha = \int_{-\infty}^{t} [x_1(\alpha)*x_2(\alpha)]d\alpha$$

$$= \int_{-\infty}^{t}\left[\int_{-\infty}^{\infty} x_1(\beta)x_2(\alpha-\beta)d\beta\right]d\alpha,$$

$$= \int_{-\infty}^{\infty}\left[\int_{-\infty}^{t} x_2(\alpha-\beta)d\alpha\right]x_1(\beta)d\beta$$

$$= \int_{-\infty}^{\infty}\left[\int_{-\infty}^{t-\beta} x_2(\lambda)d\lambda\right]x_1(\beta)d\beta,$$

$$= \left[\int_{-\infty}^{t} x_2(\lambda)d\lambda\right]*x_1(t) = \left[\int_{-\infty}^{t} x_1(\beta)d\beta\right]*x_2(t).$$

$$(2.2.8)$$

Example 2.2.1 Find the convolution of a function $x(t)$ and the unit step function $u(t)$ and show it is a running integral of $x(t)$.

Solution: This can be seen from

$$x(t)*u(t) = \int_{-\infty}^{\infty} x(\beta)u(t-\beta)d\beta$$

$$= \int_{-\infty}^{t} x(\beta)d\beta, \quad \left[u(t-\beta) = \begin{cases} 1, & \beta < t \\ 0, & \beta > t \end{cases}\right].$$

$$(2.2.9) \quad \blacksquare$$

6. *Convolution of two delayed functions* $x_1(t-t_1)$ and $x_2(t-t_2)$ are related to the convolution of $x_1(t)$ and $x_2(t)$.

$$y(t) = x_1(t)*x_2(t) \Rightarrow x_1(t-t_1)*x_2(t-t_2)$$

$$= y(t-(t_1+t_2)).$$

$$(2.2.10)$$

This can be seen from

$$x_1(t-t_1)*x_2(t-t_2)$$

$$= \int_{-\infty}^{\infty} x_1(\alpha-t_1)x_2(t-\alpha-t_2)d\alpha$$

$$= \int_{-\infty}^{\infty} x_1(\beta)x_2([t-(t_1+t_2)]-\beta)d\beta$$

$$= y(t-(t_1+t_2)).$$

$$(2.2.11)$$

Example 2.2.2 Derive the expression for $y(t) = x_1(t)*x_2(t) = \delta(t-t_1)*\delta(t-t_2)$.

Solution: Using the integral expression, we have

$$\int_{-\infty}^{\infty} x_1(\alpha)x_2(t-\alpha)d\alpha = \int_{-\infty}^{\infty} \delta(\alpha - t_1)\delta(t - t_2 - \alpha)d\alpha$$

$$= \delta(t - t_2 - \alpha)|_{\alpha = t_1} = \delta(t - t_1 - t_2).$$

Noting $\delta(t) * \delta(t) = \delta(t)$ and using (2.2.11), we have
$\delta(t - t_1) * \delta(t - t_2) = \delta(t - t_1 - t_2)$. ∎

7. The *time scaling property* of the convolution operation is if $y(t) = x_1(t) * x_2(t)$, then

$$x_1(ct) * x_2(ct) = \int_{-\infty}^{\infty} x_1(c\beta)x_2(c(t - \beta))d\beta$$

$$= \frac{1}{|c|}y(ct), c \neq 0. \qquad (2.2.12)$$

Assuming $c < 0$ and using the change of variables $\alpha = c\beta$, and simplifying, we have

$$x_1(ct) * x_2(ct) = \frac{1}{c}\int_{\infty}^{-\infty} x_1(\alpha)x_2(ct - \alpha)dy$$

$$= \frac{1}{|c|}\int_{-\infty}^{\infty} x_1(\alpha)x_2(ct - \alpha)d\alpha = \frac{1}{|c|}y(ct).$$

A similar argument can be given in the case of $c > 0$. Scaling property applies only when both functions are scaled by the *same* constant $c \neq 0$. When $c = -1$, then

$$x_1(-t) * x_2(-t) = y(-t). \qquad (2.2.13)$$

This property simplifies the convolution if there are symmetries in the functions. In Chapter 1, even and odd functions were identified by subscripts e for even and 0 for odd (see (1.2.7)). From these

$$x_{ie}(-t) = x_{ie}(t), \text{ an even function;}$$
$$x_{i0}(-t) = -x_{i0}(t), \text{ an odd function} \qquad (2.2.14a)$$

$$x_{1e}(-t) * x_{2e}(-t) = y_{e_1}(t),$$

$$x_{10}(-t) * x_{20}(-t) = y_{e_2}(t), \text{ even functions}$$

$$(2.2.14b)$$

$$x_{1e}(-t) * x_{20}(-t) = y_{0_1}(t),$$

$$x_{10}(-t) * x_{2e}(-t) = y_{0_2}(t), \text{ odd functions.}$$

$$(2.2.14c)$$

8. The area of a signal was defined in Chapter 1 (see (1.5.1)) by

$$A[x_i(t)] = \int_{-\infty}^{\infty} x_i(\alpha)d\alpha. \qquad (2.2.15)$$

Area property of the convolution applies if the areas of the individual functions do *not change* with *a shift in time*. It is given by

$$A[y(t)] = A[x_1(t) * x_2(t)] = A[x_1(t)]A[x_2(t)].$$

$$(2.2.16)$$

This can be proved by

$$A[y(t)] = \int_{-\infty}^{\infty} y(\beta)d\beta = \int_{-\infty}^{\infty} [x_1(\beta) * x_2(\beta)]d\beta$$

$$= \int_{-\infty}^{\infty} \left[\int_{-\infty}^{\infty} x_1(\alpha)x_2(\beta - \alpha)d\alpha\right] d\beta$$

$$= \int_{-\infty}^{\infty} \left\{x_1(\alpha)\int_{-\infty}^{\infty} x_2(\beta - \alpha)d\beta]\right\}d\alpha$$

$$= A[x_2(t)]\int_{-\infty}^{\infty} x_1(\alpha)d\alpha = A[x_2(t)]A[x_1(t)].$$

9. Consider the signals $x_1(t)$ and $x_2(t)$ that are non-zero for the time intervals of t_{x1} and t_{x2}, respectively. That is, we have two time-limited signals, $x_1(t)$ and $x_2(t)$, with time widths t_{x_1} and t_{x_2}. Then, the time width t_y of the signal $y(t) = x_1(t) * x_2(t)$ is the sum of the time widths of the two convolved signals and $t_y = t_{x_1} + t_{x_2}$. This is referred to as the *time duration property of the convolution*. We will come back to some intricacies in this property, as there are some *exceptions* to this property.

Example 2.2.3 Derive the expression for the convolution $y(t) = x_1(t) * x_2(t)$, where $x_i(t)$, $i = 1, 2$ are as follows:

$$x_1(t) = 0.5\delta(t - 1) + 0.5\delta(t - 2),$$

$$x_2(t) = 0.3\delta(t + 1) + 0.7\delta(t - 3).$$

Solution: Convolution of these two functions is

$$y(t) = \int\limits_{-\infty}^{\infty} x_1(\alpha)x_2(t-\alpha)d\alpha$$

$$= \int\limits_{-\infty}^{\infty} [0.5\delta(\alpha-1) + 0.5\delta(\alpha-2)] \times$$

$$[0.3\delta(t-\alpha+1) + 0.7\delta(t-\alpha-3)]d\alpha$$

$$= \int\limits_{-\infty}^{\infty} (0.5)(0.3)\delta(\alpha-1)\delta(t-\alpha+1)d\alpha$$

$$+ \int\limits_{-\infty}^{\infty} (0.5)(0.7)\delta(\alpha-1)\delta(t-\alpha-3)d\alpha$$

$$+ \int\limits_{-\infty}^{\infty} (0.5)(0.3)\delta(\alpha-2)\delta(t-\alpha+1)d\alpha$$

$$+ \int\limits_{-\infty}^{\infty} (0.5)(0.7)\delta(\alpha-2)\delta(t-\alpha-3)d\alpha$$

$$= (0.15)\delta(t) + 0.35\delta(t-4)$$
$$+ 0.15\delta(t-1) + 0.35\delta(t-5). \qquad \blacksquare$$

Notes: If an impulse function is in the integrand of the form $\delta(at - b)$, then use (see (1.4.35), which is

$$\delta(at-b) = (1/|a|)\delta(t-(b/a)).$$

2.2.2 Existence of the Convolution Integral

Convolution of two functions exists if the convolution integral exists. Existence can be given only in terms of sufficient conditions. These are related to signal energy, area, and one sidedness. It is simple to give examples, where the convolution does not exist. Some of these are $a*a$, $a*u(t)$, $\cos(t)*u(t)$, $e^{at}*e^{at}$, $a > 0$. Convolution of energy signals and the same-sided signals always exist. In Chapter 4 we will be discussing Fourier transforms and the transforms make it convenient to find the convolution.

2.3 Interesting Examples

In the following, the basics of the convolution operation, along with using some of the above properties to simplify the evaluations are illustrated. A few comments are in order before the examples. First, the convolution $y(t) = x_1(t)* x_2(t)$ is an integral operation and can use either one of the integrals in (2.2.1). Note that $y(t)$, $-\infty < t < \infty$ is a time function. The expression for the convolution, say at $t = t_0$, will yield a zero value for those values of t_0 over which $x_1(\beta)$ and $x_2(t_0 - \beta)$ do not overlap. The area under the product $[x_1(\beta)x_2(t_0 - \beta)]$, i.e., the integral of this product gives the value of the convolution at $t = t_0$. Sketches of the function $x_1(\beta)$ and the time reversed and delayed function $x_2(t_0 - \beta)$ on the same figure would be helpful in identifying the limits of integration of the product $[x_1(\beta)x_2(t_0 - \beta)]$. As a check, the value of the convolution at end points of each range must match, except in the case of impulses and/or their derivatives in the integrand of the convolution integral. This is referred to as the *consistency check*. The following steps can be used to compute the convolution of two functions $x_1(t)$ and $x_2(t)$.

$$x_2(t) \xrightarrow{\text{New variable}} x_2(\beta) \xrightarrow{\text{Reverse}} x_2(-\beta) \xrightarrow{\text{Shift}} x_2(t-\beta)$$

$$x_1(t) \xrightarrow{\text{New variable}} x_1(\beta) \xrightarrow{\text{Multiply the two functions}} x_1(\beta)x_2(t-\beta).$$

$$\xrightarrow{\text{Integrate}} \int\limits_{-\infty}^{\infty} x_1(\beta)x_2(t-\beta)d\beta = y(t).$$

Example 2.3.1 Derive the expression for the convolution of the two pulse functions shown in Fig. 2.3.1 a,b. These are

$$x_1(t) = \frac{1}{a}\Pi\left[\frac{t-(a/2)}{a}\right] \text{ and }$$

$$x_2(t) = \frac{1}{b}\Pi\left[\frac{t-(b/2)}{b}\right], b \geq a > 0. \qquad (2.3.1)$$

Solution: First

$$y(t) = x_1(t) * x_2(t) = \int\limits_{-\infty}^{\infty} x_1(\beta)x_2(t-\beta)d\beta. \quad (2.3.2)$$

Figure 2.3.1c,d,e,f give the functions $x_1(\beta)$, $x_2(\beta)$, $x_2(-\beta)$, and $x_2(t-\beta)$, respectively. Note that the variable t is some value between $-\infty$ and ∞ on the β axis. Different cases are considered, and

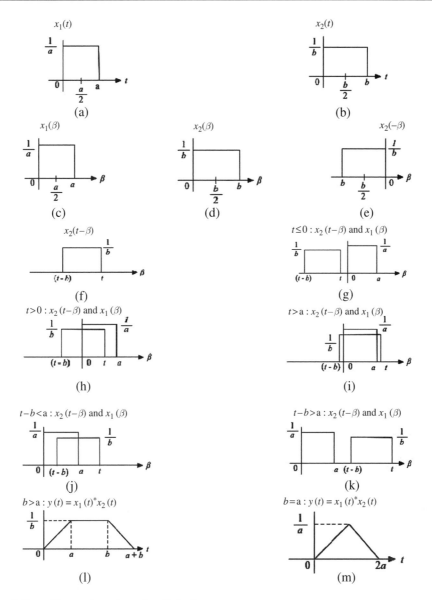

Fig. 2.3.1 Convolution of two rectangular pulses $(b \geq a)$

in each case, we keep the first function $x_1(\beta)$ stationary and move (or shift) the second function $x_2(-\beta)$, resulting in $x_2(t - \beta)$.

Case 1: $t \leq 0$. For this case the two functions are sketched in Fig. 2.3.1 g on the same figure. Noting that there is no overlap of these two functions, it follows that

$$y(t) = 0, t \leq 0. \tag{2.3.3}$$

Case 2: $0 < t \leq a$. The two functions are sketched for this case in Fig. 2.3.1 h. The two functions overlap and the convolution is

$$y(t) = \int_0^t x_1(\beta)x_2(t - \beta)d\beta = \frac{1}{ab}t, 0 < t \leq a. \tag{2.3.4}$$

Case 3: $a < t \leq b$. This corresponds to the complete overlap of the two functions and the functions are shown in Fig. 2.3.1i. The convolution integral and the area is

$$y(t) = \int\limits_{0}^{a} \frac{1}{ab} d\beta = \frac{1}{b}, a < t \leq b. \qquad (2.3.5)$$

Case 4: $0 < t - b \leq a < b$ or $b < t \leq (b + a)$. This corresponds to a partial overlap of the two functions and is shown in Fig. 2.3.1j. The convolution integral and the area is

$$y(t) = \int\limits_{t-b}^{a} \frac{1}{ab} d\beta = \frac{(a+b-t)}{ab}, b < t \leq (b+a). \qquad (2.3.6)$$

Case 5: $t - b > a$ or $t \geq a + b$. The two functions corresponding to this range are sketched in Fig. 2.3.1k and from the sketches we see that the two functions do not overlap and

$$y(t) = 0, \ t \geq (a + b). \qquad (2.3.7)$$

Summary:

$$y(t) = \begin{cases} 0, & t \leq 0 \\ \dfrac{t}{ab}, & 0 < t < a \\ \dfrac{1}{b}, & a \leq t < b \\ \dfrac{a+b-t}{ab}, & b \leq t < a+b \\ 0, & t \geq a+b \end{cases} \qquad (2.3.8)$$

This function is sketched in Fig. 2.3.1 l and m for the cases of $b > a$ and $b = a$. There are several interesting aspects in this example that should be noted. First, the two functions we started with have first-order discontinuous and the convolution is an integral operation, which is a smoothing operation. Convolution values at end points of each range must match (*consistency check*) as we do *not* have any *impulse functions or their derivatives* in the functions that are convolved. Some of these are discussed below.

The areas of the two pulses are each equal to 1 and the area of the trapezoid is given by

$$\begin{aligned} \text{Area}[y(t)] &= (1/2)a(1/b) + (b-a)(1/b) \\ &\quad + (1/2)a(1/b) = 1 \\ &= \text{Area}[x_1(t)]\text{Area}[x_2(t)]. \end{aligned} \qquad (2.3.9)$$

This shows that the area property is satisfied. Peebles (2001) shows the probability density function of the sum of the two independent random variables is also a probability density function. We should note that the probability density function is nonnegative and the area under this function is 1 (see Section 1.7). From the above discussion, it follows that the convolution of two rectangular pulses (these can be considered as uniform probability density functions) results in a nonnegative function and the area under this function is 1. The function $y(t)$ satisfies the conditions of a probability density function.

The time duration of $y(t)$, t_y is $t_y = t_{x_1} + t_{x_2}$ and

$$t_{x_1} = a, t_{x_2} = b \Rightarrow t_y = t_{x_1} + t_{x_2} = a + b. \qquad (2.3.10)$$

A special case is when $a = b$ and the function $y(t)$ given in Fig. 2.3.1m, a triangle, is

$$\Pi\left[\frac{t-a/2}{a}\right] * \Pi\left[\frac{t-a/2}{a}\right] = \Lambda\left[\frac{t-a}{a}\right]. \qquad (2.3.11) \quad \blacksquare$$

Example 2.3.2 Give the expressions for the convolution of the following functions:

$$x_1(t) = u(t) \text{ and } x_2(t) = \sin(\pi t)\Pi\left[\frac{t-1}{2}\right]. \qquad (2.3.12)$$

Solution: The convolution integral

$$y(t) = \int_{-\infty}^{\infty} x_2(\beta)x_1(t-\beta)d\alpha = \int_{0}^{2} \sin(\pi\beta)\,u(t-\beta)d\beta$$

$$= \int_{0}^{t} \sin(\pi\beta)d\beta = \begin{cases} 0, & t \le 0 \\ (1/\pi)(1-\cos(\pi t)), & 0 < t < 2, \\ 0, & t \ge 2 \end{cases}$$

$$y(t) = \int_{0}^{t} \sin(\pi\beta)d\beta$$

$$= \begin{cases} 0, & t \le 0 \\ (1/\pi)(1-\cos(\pi t)), & 0 < t < 2 . \\ 0, & t \ge 2 \end{cases} \quad (2.3.13)$$

The time duration of the unit step function is ∞ and the time duration of $x_2(t)$ is 2. The duration of the function $y(t)$ is 2, which illustrates a pathological case where the time duration property of the convolution is not satisfied.

The integral or the area of a sine or a cosine function over one period is equal to zero. The period of the function $\sin(\pi t)$ is equal to 2 and therefore

$$A[x_2(t)] = A\left[\sin(\pi t).\Pi\left[\frac{t-1}{2}\right]\right]$$

$$= 0 \Rightarrow A[y(t)] = (1/\pi)\int_{0}^{2}[1-\cos(\pi t)]dt$$

$$= 1/\pi \int_{0}^{2} dt = 2/\pi.$$

Noting that $A[x_1(t)] = A[u(t)] = \infty$ and $A[y(t)] = 2/\pi$, we can see that the area property of the convolution is not satisfied. See Ambardar (1995) for an additional discussion. ∎

Example 2.3.3 Derive the expression for the convolution of the following functions shown in Fig. 2.3.2a,b:

$$x(t) = \Pi\left[\frac{t}{2T}\right] \text{ and } h(t) = e^{-at}u(t), a > 0. \quad (2.3.14a)$$

Solution:

$$y(t) = x(t) * h(t) = \int_{-\infty}^{\infty} x(\beta)h(t-\beta)d\beta$$

$$= \int_{-\infty}^{\infty} h(\alpha)x(t-\alpha)d\alpha. \quad (2.3.14b)$$

In computing the convolution, we keep one of the functions at one location and the other function is time reversed and then shifted. In this example, since the function $h(t) = 0$ for $t < 0$, we have a *benchmark* to keep track of the movement of the function $h(t-\beta)$ as t varies. Therefore, the first integral in (2.3.14b) is simpler to use. The functions $x(\beta), h(\beta), h(-\beta)$, and $h(t-\beta)$ are shown in Fig. 2.3.2 c, d, e, and f respectively. As before, we will compute the convolution for different intervals of time.

Case 1: $t \le -T$: the two functions, $h(t-\beta)$ and $x(\beta)$, are sketched in Fig. 2.3.2 g. Clearly there is no overlap of the two functions and therefore the integral is zero. That is

$$y(t) = 0, t \le -T. \quad (2.3.15)$$

Case 2: $-T < t < T$: The two functions $h(t-\beta)$ and $x(\beta)$ are sketched in Fig. 2.3.2 h in the same figure for this interval. There is a partial overlap of the two functions in the interval $-T > t > T$. The convolution can be expressed by

$$y(t) = \int_{-\infty}^{\infty} x(\beta)h(t-\beta)d\beta = \int_{-T}^{t} (1)e^{-a(t-\beta)}d\beta$$

$$(2.3.16)$$

$$= e^{-at}\int_{-T}^{t} e^{a\beta}d\beta = \frac{1}{a}\left[1 - e^{-a(t+T)}\right], -T < t < T.$$

Case 3: $t > T$: From the sketch of the two functions in Fig. 2.3.2 h, the two functions overlap in this range $-T \le t \le T$ and the convolution integral is

$$y(t) = \int_{-T}^{T} e^{-a(t-\beta)}d\beta = \frac{1}{a}\left[e^{aT} - e^{-aT}\right]e^{-at}, \quad t > T.$$

$$(2.3.17)$$

Fig. 2.3.2 Convolution of a rectangular pulse with an exponentially decaying pulse

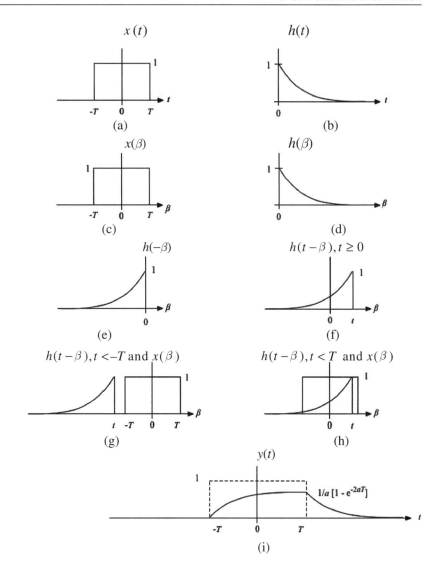

Summary:

$$y(t) = \begin{cases} 0, & t \leq -T \\ \dfrac{1}{a}\left[1 - e^{-a(t+T)}\right], & -T < t \leq T \\ \dfrac{1}{a}\left[e^{aT} - e^{-aT}\right]e^{-at}, & t > T \end{cases} \quad (2.3.18)$$

This function is sketched in Fig. 2.3.2i. Note $y(t)$ is smoother than either of the given functions used in the convolution. Computing the area of $y(t)$ is not as simple as finding the areas of the two functions, $x(t)$ and $h(t)$. Using the area property,

$$A[y(t)] = A[x(t)]A[h(t)] = (2T)(1/a). \quad (2.3.19) \quad \blacksquare$$

Notes: In computing the convolution, one of the sticky points is finding the integral of the product $[x(\beta)h(t-\beta)]$ in (2.3.14b), which requires finding the region of overlap of the two functions. Sketching both functions on the *same figure* allows for an easy determination of this overlap. The *delay property* is quite useful. For example, if $y(t) = x(t) * h(t)$ then it implies $y_1(t) = x(t-T) * h(t) = y(t-T)$. In Example 2.3.3, $x(t) = \Pi[t/2T] = u[t+T] - u[t-T]$. Therefore

$$y(t) = h(t) * x(t)$$
$$= h(t) * [u(t+T) - x(t-T)]$$
$$= h(t) * u(t+T) - h(t) * u(t-T). \qquad \blacksquare$$

Example 2.3.4 Determine the convolution $y(t) = x(t) * x(t)$ with $x(t) = e^{-at}u(t)$, $a > 0$.

Solution: The convolution is

$$y(t) = e^{-at}u(t) * e^{-at}u(t)$$

$$= \int_{-\infty}^{\infty} e^{-a\beta}e^{-a(t-\beta)}[u(\beta)u(t-\beta)]d\beta$$

$$= e^{-at}\int_{0}^{t} d\beta = te^{-at}u(t). \qquad (2.3.20)$$

In evaluating the integral, the following expression is used (see Fig. 2.3.3a):

$$[u(\beta)u(t-\beta)] = \begin{cases} 0, & \beta < 0 \text{ and } \beta > t \\ 1, & 0 < \beta > t \end{cases}. \qquad (2.3.21)$$

The functions $x(t)$ and $y(t)$ are shown in Fig. 2.3.3b,c. Note that the function $x(t)$ has a discontinuity at $t = 0$. The function $y(t)$, obtained by convolving two identically decaying signals, $x(t)$ and $x(t)$ is smoother than either one of the convolved signals. This is to be expected as the convolution operation is a smoothing operation. $\blacksquare$

Example 2.3.5 Derive the expression $y_i(t) = h(t) * x_i(t)$ for the following two cases:

$$a. x_1(t) = u(t), b. x_2(t) = \delta(t).$$

Solution: *a.* Since $u(t - \alpha) = 0, \alpha > t$, we have the running integral

$$y_1(t) = h(t) * u(t) = \int_{-\infty}^{\infty} h(\alpha)u(t-\alpha)d\alpha$$

$$= \int_{-\infty}^{t} h(\alpha)d\alpha. \qquad (2.3.22)$$

b. Noting that the impulse function is the *generalized derivative* of the unit step function, we can compute the convolution

$$y_2(t) = h(t) * \delta(t) = h(t) * \frac{du(t)}{dt} = y'_1(t) = h(t). \qquad (2.3.23) \quad \blacksquare$$

Example 2.3.6 Let $h(t) = e^{-at}u(t), a > 0$ *a.* Determine the running integral of $h(t)$.
b. Using (2.3.23), determine $y_2(t)$.

Solution:

$$a. \quad y_1(t) = \int_{-\infty}^{t} h(\beta)d\beta = \int_{-\infty}^{t} e^{-a\beta}u(\beta)d\beta$$

$$= \frac{1}{a}(1 - e^{-at})u(t), \qquad (2.3.24)$$

$$b. \; y_2(t) = \frac{dy_1(t)}{dt} = \frac{1}{a}\frac{d}{dt}(1 - e^{-at})u(t)$$

$$= \frac{1}{a}(1 - e^{-at})\frac{d}{dt}u(t) + \frac{1}{a}u(t)\frac{d(1-e^{-at})}{dt}$$

$$= (1/a)(1 - e^{-at})\delta(t) + e^{-at}u(t)$$

$$= (1/a)[\delta(t) - \delta(t)] + e^{-at}u(t)$$

$$= (1/a)e^{-at}u(t). \qquad (2.3.25) \quad \blacksquare$$

In a later chapter this result will be used in dealing with step and impulse inputs to an *RC* circuit with an impulse response $h(t) = e^{-at}u(t)$.

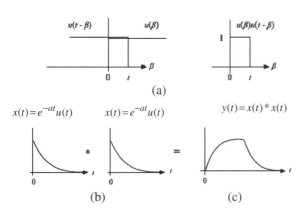

Fig. 2.3.3 Example 2.3.4

Example 2.3.7 Express the following integral in the form of $x(t) * p(t)$, $(p(t)$ is a pulse function:

$$y(t) = \int_{t-T/2}^{t+T/2} x(\alpha) d\alpha. \quad (2.3.26)$$

Solution:

$$y(t) = \int_{-\infty}^{t+(T/2)} x(\alpha) d\alpha - \int_{-\infty}^{t-(T/2)} x(\alpha) d\alpha$$

$$= x(t) * u(t + (T/2)) - x(t) * u(t - (T/2))$$

$$= x(t) * [u(t + (T/2)) - u(t - T/2))]$$

$$= x(t) * \Pi\left[\frac{t}{T}\right]. \quad (2.3.27)$$

The output is the convolution of $x(t)$ with a pulse width of T with unit amplitude and the process is a running average. ∎

Example 2.3.8 Find the derivative of the running average of the function in (2.3.27) and express the function $x(t)$ in terms of the derivative of $y(t)$.

Solution: McGillem and Cooper (1991) give an interesting solution for this problem.

$$y'(t) = x(t) * \left[\frac{du(t + (T/2))}{dt} - \frac{du(t - (T/2))}{dt}\right]$$

$$= x(t) * \delta\left(t + \frac{T}{2}\right) - x(t) * \delta\left(t - \frac{T}{2}\right)$$

$$= x(t + (T/2)) - x(t - (T/2))$$

$$\Rightarrow x(t) = y'(t - (T/2)) + x(t - T). \quad (2.3.28) ∎$$

Example 2.3.9 Derive the expressions *a.* $y_1(t) = u(t) * u(t)$, *b.* $y_2(t) = u(t) * u(-t)$.

Solution:

$$a.\ y_1(t) = u(t) * u(t) = \int_{-\infty}^{\infty} u(\alpha) u(t - \alpha) d\alpha$$

$$= \int_0^t (1) dt = \begin{cases} 0, & t > 0 \\ t, & t < 0 \end{cases} = tu(t), \quad (2.3.29)$$

$$b.\ y_2(t) = \int_{-\infty}^{\infty} u(\alpha) u(\alpha + t) d\alpha$$

$$= \begin{cases} \int_t^{\infty} u(\alpha) d\alpha \to \infty, & t \le 0 \\ \int_0^{\infty} u(\alpha + t) d\alpha \to \infty, & t > 0 \end{cases}. \quad (2.3.30)$$

It follows that $y_2(t) = \infty$, $-\infty < t < \infty$. In this case, convolution does not exist. ∎

2.4 Convolution and Moments

In the examples considered so far, except in the cases of impulses, convolution is found to be a smoothing operation. We like to quantify and compare the results of the convolution of nonimpulse functions to the Gaussian function. In Section 1.7.1, the moments associated with probability density functions were considered.

A useful result can be determined by considering the center of gravity convolution in terms of the centers of gravity of the factors in the convolution. First, the *moments* $m_n(x)$ of a waveform $x(t)$ and its *center of gravity* η are, respectively, defined as

$$m_n(x) = \int_{-\infty}^{\infty} t^n x(t) dt, \quad (2.4.1)$$

$$\eta = \frac{\int_{-\infty}^{\infty} t x(t) dt}{\int_{-\infty}^{\infty} x(t) dt} = \frac{m_1(x)}{m_0(x)}. \quad (2.4.2)$$

We note that we can define a term like the variance in Section 1.7.1 by

$$\sigma^2(x) = \frac{m_2(x)}{m_0(x)} - \eta^2. \quad (2.4.3)$$

Now consider the expressions for the convolution $y(t) = g(t) * h(t)$. First,

$$m_1(y) = \int_{-\infty}^{\infty} ty(t)dt = \int_{-\infty}^{\infty} \left[t \int_{-\infty}^{\infty} g(\lambda)h(t-\lambda)d\lambda \right] dt$$

$$= \int_{-\infty}^{\infty} g(\lambda) \left[\int_{-\infty}^{\infty} th(t-\lambda)dt \right] d\lambda.$$

Defining a new variable $\xi = t - \lambda$ on the right and rewriting the above equation results in

$$m_1(y) = \left[\int_{-\infty}^{\infty} g(\lambda) \int_{-\infty}^{\infty} (\xi+\lambda)h(\xi)d\xi \right] d\lambda$$

$$= \int_{-\infty}^{\infty} \lambda g(\lambda)d\lambda \int_{-\infty}^{\infty} h(\xi)d\xi + \int_{-\infty}^{\infty} \xi h(\xi)d\xi$$

$$\int_{-\infty}^{\infty} g(\lambda)d\lambda = m_1(g)m_0(h) + m_1(h)m_0(g). \qquad (2.4.4)$$

From the area property, it follows that $m_0(y) = m_0(g)m_0(h)$. The center of gravity is

$$\frac{m_1(y)}{m_0(y)} = \frac{m_1(g)}{m_0(g)} + \frac{m_1(h)}{m_0(h)} \Rightarrow \eta_y = \eta_g + \eta_h. \qquad (2.4.5)$$

Consider the expression for the squares of the spread of $y(t)$ in terms of the squares of the spreads of $g(t)$ and $h(t)$. The derivation is rather long and only results are presented.

$$\sigma_y^2 = \frac{m_2(y)}{m_0(y)} - \left(\frac{m_1(y)}{m_0(y)} \right)^2. \qquad (2.4.6)$$

Using the expressions for $m_0(y)$, $m_1(y)$ and $m_2(y)$ and simplifying the integrals results in

$$\sigma_y^2 = \sigma_g^2 + \sigma_h^2. \qquad (2.4.7)$$

That is, the variance of y is equal to the sum of the variances of the two factors. It also verifies that convolution is a broadening operation for pulses. Noting that if $g(t)$ and $h(t)$ are probability density functions then (2.4.7) is valid. In communications theory we are faced with a signal, say $g(t)$ is corrupted by a noise $n(t)$ with the variance, σ_n^2. The signal-to-noise ratio (SNR) is given by

$$\text{Signal-to-noise ratio} = \frac{\text{Average signal power}}{\text{Noise power, } \sigma_n^2}. \qquad (2.4.8)$$

Example 2.4.1 Verify the result is true in (2.4.7) using the functions

$$g(t) = h(t) = e^{-t} \text{ and } y(t) = g(t) * h(t).$$

Solution: Using integral tables, it can be shown that

$$m_0(g) = \int_0^{\infty} e^{-t}dt = 1, \, m_1(g) = \int_0^{\infty} te^{-t}dt = 1,$$

$$m_2(g) = \int_0^{\infty} t^2 e^{-t}dt = 2,$$

$$\eta_g = \frac{m_1(g)}{m_0(g)} = 1, \sigma_g^2 = \frac{m_2(g)}{m_0(g)} - \eta_g^2 = 1,$$

$$\sigma_h^2 = 1 \text{ (note } g(t) = h(t)),$$

$$y(t) = g(t) * h(t) = te^{-t}u(t)(\text{see Example 2.3.4}).$$

$$m_0(y) = \int_0^{\infty} te^{-t}dt = 1, m_1(y) = \int_0^{\infty} t^2 e^{-t}dt = 2,$$

$$m_2(y) = \int_0^{\infty} t^3 e^{-t}dt = 6,$$

$$\eta_y = \frac{m_1(y)}{m_0(y)} = 2, \sigma_y^2 = \frac{m_2(y)}{m_0(y)} - \eta_y^2 = 2 \Rightarrow$$

$$\sigma_y^2 = \sigma_g^2 + \sigma_h^2 = 1 + 1 = 2. \qquad \blacksquare$$

As an example, consider that we have signal $g(t) = A\cos(\omega_0 t)$ and is corrupted by a noise with a variance equal to σ_n^2. Then, the signal-to-noise ratio is

$$\text{SNR} = \frac{A^2/2}{\sigma_n^2}.$$

In Chapter 10, we will make use of this in quantization methods, wherein A and SNR are given and determine σ_n^2. This, in turn, provides the information on the size of the error that can be tolerated.

Notes: For readers interested in independent random variables, the probability density function of a sum of two independent random variables is the convolution of the density functions of the two factors of the convolution, and the variance of the sum of the two random variables equals the sum of their variances. For a detailed discussion on this, see Peebles (2001).■

2.4.1 Repeated Convolution and the Central Limit Theorem

Convolution operation is an integral operation, which is a smoothing operation. In Example 2.3.1, we have considered the special case of the convolution of two identical rectangular pulses and the convolution of these two pulses resulted in a triangular pulse (see Fig. 2.3.1m). The discontinuities in the functions being convolved are not there in the convolved signal. As more and more pulse functions convolve, the resultant functions become smoother and smoother. Repeated convolution begins to take on the bell-shaped Gaussian function. The generalized version of this phenomenon is called the *central limit theorem*. It is commonly presented in terms of probability density functions. In simple terms, it states that if we convolve N functions and one function does not *dominate* the others, then the convolution of the N functions approaches a Gaussian function as $N \rightarrow \infty$. In the general form of the central limit theorem, the means and variances of the individual functions that are convolved are related to the mean and the variance of the Gaussian function (see Peebles (2001)).

Given $x_i(t)$, $i = 1, 2, ..., N$, the convolution of these functions is

$$y(t) = x_1(t) * x_2(t) * ... * x_N(t). \quad (2.4.9)$$

The function $y(t)$ can be approximated using $(m_0)_N$, the sum of the individual means of the functions, and σ_N^2 the sum of the individual variances by

$$y(t) \approx \frac{1}{\sqrt{2\pi\sigma_N^2}} e^{-(t-(m_0)_N)^2/2\sigma_N^2}. \quad (2.4.10)$$

Example 2.4.2 Illustrate the effects of convolution and compare $y(t)$ to a Gaussian function by considering the convolution

$$y(t) = x_1(t) * x_2(t),$$

$$x_i(t) = \frac{1}{a}\Pi\left[\frac{t - a/2}{a}\right], i = 1, 2. \quad (2.4.11)$$

Solution: $y(t)$ is a triangular function (see Example 2.3.1) given by

$$y(t) = \frac{1}{a}\Lambda\left[\frac{t - a}{a}\right]. \quad (2.4.12)$$

The mean values of the two rectangular pulses are $a/2$ (see Section 1.7). The mean value of $y(t)$ is $2(a/2) = a$. The variance of each of the rectangular pulses is

$$\sigma_i^2 = m_2 - m_1^2 = a^2/12, i = 1, 2. \quad (2.4.13a)$$

The variance is given by $\sigma_y^2 = \sigma_1^2 + \sigma_2^2 = a^2/6$. The Gaussian approximation is

$$(y(t))|_{N=2} \approx \frac{1}{\sqrt{\pi(a^2/3)}} e^{-(t-a)^2/(a^2/3)}. \quad (2.4.13b)$$

This Gaussian and the triangle functions are symmetric around a. They are sketched in Fig. 2.4.1. Even with $N = 2$, we have a good approximation. ■

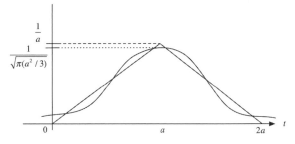

Fig. 2.4.1 Triangle function $y(t)$ in (2.4.12) and the Gaussian function in (2.4.13b)

Example 2.4.3 In Example 2.4.1 we considered two identically exponentially decaying functions: $x_1(t) = e^{-t}u(t) = x_2(t)$. The convolution of these two functions is given by $y_2(t) = te^{-t}u(t)$. Approximate this function using the Gaussian function.

Solution: The Gaussian function approximations of $y_n(t)$, considering $n = 2$ and for n large, are, respectively, given below. Note that $m_0(y) = 2$.

$$y_2(t) \approx \frac{1}{\sqrt{2\pi(2)}} e^{-\left((t-2)^2/2(2)\right)},$$

$$\quad (2.4.14)$$

$$y_n(t) \approx \frac{1}{\sqrt{2\pi(n)}} e^{-\left((t-n)^2/2(n)\right)}.$$

For sketches of these functions for various values of n, see Ambardar (1995). ■

2.4.2 Deconvolution

In this chapter, we have defined the convolution $y(t) = h(t) * x(t)$ as a mathematical operation. If $x(t)$ needs to be recovered from $y(t)$, we use a process called the *deconvolution* defined by

$$x(t) = y(t) * h_{inv}(t) = x(t) * h(t) * h_{inv}(t)$$
$$= x(t) * [h(t) * h_{inv}(t)], \Rightarrow h(t) * h_{inv}(t)$$
$$= \delta(t) \text{ and } x(t) * \delta(t) = x(t). \qquad (2.4.15)$$

It is a difficult problem to find $h_{inv}(t)$, which may not even exist.

Table 2.4.1 Properties of aperiodic convolution

Definition:

$$y(t) = x_1(t)^* x_2(t) = \int_{-\infty}^{\infty} x_1(\alpha) x_2(t - \alpha) d\alpha = \int_{-\infty}^{\infty} x_2(\alpha) x_1(t - \alpha) d\alpha.$$

Amplitude scaling:

$$\alpha x_1(t)^* \beta x_2(t) = \alpha\beta(x(t)^* h(t)), \alpha \text{ and } \beta \text{ are constants.}$$

Commutative:

$$x_1(t)^* x_2(t) = x_2(t)^* x_1(t).$$

Distributive:

$$x_1(t)^* [x_2(t) + x_3(t)] = x_1(t)^* x_2(t) + x_1(t)^* x_3(t).$$

Associative:

$$x_1(t)^* [x_2(t)^* x_3(t)] = [x_1(t)^* x_2(t)]^* x_3(t).$$

Delay:

$$x_1(t - t_1)^* x_2(t - t_2) = x_1(t - t_2)^* x_2(t - t_1) = y(t - (t_1 + t_2)).$$

Impulse response:

$$x(t)^* \delta(t) = x(t).$$

Derivatives:

$$x_1(t)^* x_2'(t) = x_1'(t)^* x_2(t) = y'(t), \quad x_1^{(m)}(t)^* x_2^{(n)}(t) = y^{(m+n)}(t).$$

Step response:

$$y(t) = x(t)^* u(t) = \int_{-\infty}^{t} x(\alpha) d\alpha, \quad y'(t) = x(t)^* \delta(t) = x(t).$$

Area:

$$A[x_1(t)^* x_2(t)] = A[y(t)], \text{ where } A[x(t)] = \int_{-\infty}^{\infty} x(t) dt.$$

Duration:

$$t_{x_1} + t_{x_2} = t_y.$$

Symmetry:

$$x_{1e}(t)^* x_{2e}(t) = y_e(t), \quad x_{1e}(t)^* x_{20}(t) = y_0(t), \quad x_{10}(t)^* x_{20}(t) = y_e(t).$$

Time scaling:

$$x_1(ct)^* x_2(ct) = \tfrac{1}{|c|} y(ct), \ c \neq 0.$$

2.5 Convolution Involving Periodic and Aperiodic Functions

2.5.1 Convolution of a Periodic Function with an Aperiodic Function

Let $h(t)$ be an aperiodic function and $x_T(t)$ be a periodic function with a period T. We desire to find the convolution of these two functions. That is, find $y(t) = x_T(t) * h(t)$.

Example 2.5.1 Derive the expressions for the convolution of the following two functions: $\delta_T(t)$ and $h(t)$ assuming $T = 1.5$ and $T = 2$ and sketch the results for the two cases.

$$\delta_T(t) = \sum_{k=-\infty}^{\infty} \delta(t - nT), \quad h(t) = \Lambda[t]. \quad (2.5.1)$$

Derive the expressions for the convolution of these two functions assuming $T = 1.5$ and $T = 2$ and sketch the results of the convolution for the two cases.

$$y(t) = h(t) * \delta_T(t) = h(t) * \sum_{k=-\infty}^{\infty} \delta(t - kT)$$

$$= \sum_{k=-\infty}^{\infty} h(t) * \delta(t - kT). \quad (2.5.2)$$

Noting that $h(t) * \delta(t - kT) = h(t - kT)$, it follows that

$$y(t) = \sum_{k=-\infty}^{\infty} h(t - kT) = y_T(t). \quad (2.5.3)$$

Figure 2.5.1a,b gives the sketches of the functions $\delta_T(t)$ and $h(t)$. The sketches for the convolution are shown in Fig. 2.5.1c,d. In the first case, there were no overlaps, whereas in the second case there are overlaps. ∎

Example 2.5.2 Derive an expression for the convolution $y(t) = h(t) * x_T(t)$,

$$x_T(t) = \cos(\omega_0 t + \theta) \quad \text{and} \quad h(t) = e^{-at} u(t). \quad (2.5.4)$$

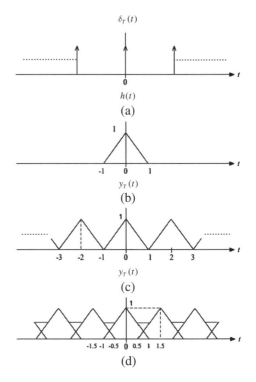

Fig. 2.5.1 (a) Periodic impulse sequence, (b) $\Lambda[t]$, (c) $y_T(t)$, $T = 2$, and (d) $y_T(t)$, $T = 2$

Solution: $y(t) = h(t) * x_T(t) = \int\limits_{0}^{\infty} e^{-a\beta} \cos(\omega_0(t - \beta) + \theta) d\beta$

$$= \int\limits_{0}^{\infty} e^{-a\beta} [\cos(\omega_0 t + \theta) \cos(\omega_0 \beta)$$

$$+ \sin(\omega_0 t + \theta) \sin(\omega_0 \beta)] d\beta$$

$$= \left[\int\limits_{0}^{\infty} e^{-a\beta} \cos(\omega_0 \beta) d\beta \right] \cos(\omega_0 t + \theta)$$

$$+ \left[\int\limits_{0}^{\infty} e^{-a\beta} \sin(\omega_0 \beta) d\beta \right] \sin(\omega_0 t + \theta). \quad (2.5.5)$$

Using the identities given below (see (2.5.7 a, b, and c.)), (2.5.5) can be simplified.

$$y(t) = [a/(a^2 + \omega_0^2)] \cos(\omega_0 t + \theta)$$

$$+ [\omega_0/(a^2 + \omega_0^2)] \sin(\omega_0 t + \theta)$$

$$= \frac{1}{\sqrt{a^2 + \omega_0^2}} \cos(\omega_0 t + \theta - \tan^{-1}(\omega_0/a))$$

$$(2.5.6)$$

$$\equiv y_T(t),$$

$$\int_0^\infty e^{-a\beta} \sin(\omega_0 \beta)d\beta = \frac{e^{-a\beta}}{a^2 + \omega_0^2}[-a\sin(\omega_0\beta)$$

$$-\omega_0 \cos(\omega_0\beta)]\Big|_0^\infty = \frac{\omega_0}{a^2 + \omega_0^2},$$

(2.5.7a)

$$\int_0^\infty e^{-a\beta} \cos(\omega_0 \beta)d\beta = \frac{e^{-a\beta}}{a^2 + \omega_0^2}[-a\cos(\omega_0\beta)$$

$$+\omega_0 \sin(\omega_0\beta)]\Big|_0^\infty = \frac{a}{a^2 + \omega_0^2},$$

(2.5.7b)

$$\alpha \cos(\omega_0 t + \theta) + \beta \sin(\omega_0 t + \theta) = c \cos(\omega_0 t + \phi),$$

$$c = \sqrt{\alpha^2 + \beta^2}, \phi = [-\tan^{-1}(\beta/\alpha) + \theta].$$ (2.5.7c)

The functions $y(t) = y_T(t)$ and $x_T(t)$ are sinusoids at the same frequency ω_0. The amplitude and the phase of $y(t)$ are *different* compared to that of $x_T(t)$. ∎

The derivation given above can be generalized for a periodic function

$$x_T(t) = X_s[0] + \sum_{k=0}^\infty c[k]\cos(k\omega_0 t + \theta[k]), \omega_0 = 2\pi/T,$$

(2.5.8a)

$$y(t) = x_T(t) * h(t) = X_s[0] * h(t) + \sum_{k=0}^\infty c[k][\cos(k\omega_0 t$$

$$+ \theta[k]) * h(t)], \omega_0 = 2\pi/T.$$ (2.5.8b)

2.5.2 *Convolution of Two Periodic Functions*

In Section 1.5 energy and power signals were considered. The energy in a periodic function is infinity and its average power is finite. One period of a periodic function has all its information. In the same vein, the *average convolution* is a useful measure of periodic convolution. Such averaging process is called *periodic* or *cyclic convolution*. The convolution of two periodic functions with *different* periods is very difficult and is limited here to the convolution of two periodic functions, each with *the same period*.

The periodic convolution of two periodic functions, $x_T(t)$ and $h_T(t)$, is defined by

$$y_T(t) = x_T(t) \otimes h_T(t) = \frac{1}{T}\int_{t_0}^{t_0+T} x_T(\alpha)h_T(t-\alpha)d\alpha$$

$$= \frac{1}{T}\int_T x_T(\alpha)h_T(t-\alpha)d\alpha = \frac{1}{T}\int_T x_T(t-\alpha)h_T(\alpha)d\alpha$$

$$= h_T(t) \otimes x_T(t).$$ (2.5.9a)

Note that the symbol $\otimes$ used for the periodic convolution and the constant (T) in the denominator in (2.5.9a) indicates that it is an average periodic convolution. $y_T(t)$ is periodic since

$$h_T(t + T - \alpha) = h_T(t - \alpha) \text{ and}$$

$$x_T(t + T - \alpha) = x_T(t - \alpha).$$ (2.5.9b)

Also, periodic convolution is *commutative*. Many of the aperiodic convolution properties discussed earlier are applicable for periodic convolution with some modifications. The expression for the periodic convolution can be obtained by considering aperiodic convolution for one period of each of the two functions.

Consider the periodic functions in the form

$$x_T(t) = \sum_{n=-\infty}^\infty x(t - nT) \text{ and } h_T(t) = \sum_{n=-\infty}^\infty h(t - nT),$$

(2.5.10a)

$$x(t) = \begin{cases} x_T(t), & t_0 \le t < t_0 + T \\ 0, & \text{otherwise} \end{cases},$$

$$h(t) = \begin{cases} h_T(t), & t_0 \le t < t_0 + T \\ 0, & \text{otherwise}. \end{cases}$$ (2.5.10b)

Note that the time-limited functions, $x(t)$ and $h(t)$, are defined from the periodic functions $x_T(t)$ and $h_T(t)$. Using (2.5.10b) the periodic convolution is

$$y_T(t) = \frac{1}{T}\int_T x_T(\alpha)h_T(t - \alpha)d\alpha$$

$$= \frac{1}{T}\int_T x_T(\alpha)\sum_{n=-\infty}^\infty h(t - \alpha - nT)d\alpha$$

$$= \frac{1}{T} \sum_{n=-\infty}^{\infty} \int_T x(\alpha)h(t - \alpha - nT)d\alpha$$

$$= \frac{1}{T} \sum_{n=-\infty}^{\infty} x(t) * h(t - nT),$$

(2.5.11a)

$$y_T(t) = x_T(t) \otimes h_T(t)$$

$$= \frac{1}{T} \sum_{n=-\infty}^{\infty} y(t - nT), y(t) = x(t) * h(t). \quad (2.5.11b)$$

That is, $y_T(t)$ can be determined by considering one period of each of the two functions and finding the aperiodic convolution.

Example 2.5.3 *a.* Determine and sketch the aperiodic convolution $y(t) = h(t) * x(t)$.

$$x(t) = \frac{1}{2} \Pi \left[\frac{t - 1}{2} \right], \quad h(t) = \frac{1}{3} \Pi \left[\frac{t - 1.5}{3} \right]. \quad (2.5.12)$$

b. Determine and sketch the periodic convolution $y_T(t) = x_T(t) \otimes h_T(t)$ for periods $T = 6$ and 4.

$$x_T(t) = \sum_{k=-\infty}^{\infty} x(t - kT) \text{ and } h_T(t) = \sum_{k=-\infty}^{\infty} h(t - kT).$$

(2.5.13)

Solution: *a.* From (2.5.13), the results for the aperiodic convolution can be derived. The sketches of the two functions and the result of the convolution are shown in Fig. 2.5.2a. The periodic convolutions for the two different periods are shown in Fig. 2.5.2b,c. There are no overlaps of the functions from one period to the next in Fig. 2.5.2b, whereas in Fig. 2.5.2c, the pulses overlap. ∎

Convolution of *almost periodic* or *random signals*, $x(t)$ and $h(t)$, is defined by

$$y(t) = \lim_{T \to \infty} \frac{1}{T} \int_{-T/2}^{T/2} x(\alpha)h(t - \alpha)d\alpha. \quad (2.5.14)$$

This reduces to the periodic convolution if $x(t)$ and $h(t)$ are periodic with the same period.

2.6 Correlation

Equation (2.1.3) gives the cross-correlation of $x(t)$ and $h(t)$ as the integral of the product of two functions, one displaced by the other by τ between the interval $a < t < b$ and is given by

$$R_{xh}(\tau) = x(\tau) ** h(\tau) = \int_a^b x(t)h(t + \tau)dt = \langle x(t)h(t + \tau) \rangle.$$

Cross-correlation function gives the *similarity* between the two functions: $x(t)$ and $h(t + \tau)$. Many a times the second function $h(t)$ may be a corrupted version of $x(t)$, such as $h(t) = x(t) + n(t)$, where $n(t)$ is a noise signal. In the case of $x(t) = h(t)$, cross-correlation reduces to autocorrelation. In this case, at $\tau = 0$, the autocorrelation integral gives the highest value at $\tau = 0$. Comparison of two functions appears in many identification situations. For example, to identify an individual based upon his speech pattern, we can store his speech

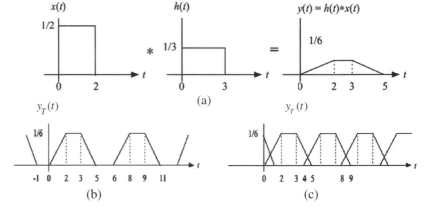

Fig. 2.5.2 Example 2.5.1
(a) Aperiodic convolution;
(b) periodic convolution
$T = 6$; (c) periodic
convolution, $T = 4$

segment in a computer. When he enters, say a secure area, we can request him to speak and compute the cross-correlation between the stored and the recorded. Then decide on the individual's identification based on the cross-correlation function. Generally, an individual is identified if the peak of the cross-correlation is close to the possible peak auto-correlation value. Allowance is necessary since the speech is a function of the individual's physical and mental status of the day the test is made. Quantitative measures on the cross-correlation will be considered a bit later.

The order of the subscripts on the cross-correlation function $R_{xh}(\tau)$ is important and will get to it shortly. In the case of $x(t) = h(t)$, we have the autocorrelation and the function is referred to as $R_x(\tau)$ with a *single subscript*. The cross- and autocorrelation functions are functions of τ and not t. Correlation is applicable to periodic, aperiodic, and random signals. In the case of periodic functions, we assume that both are periodic with *the same period*.

Cross-correlation: Aperiodic:

$$R_{xh}(\tau) = \int_{-\infty}^{\infty} x(t)h(t+\tau)dt \quad (2.6.1a)$$

Cross-correlation: Periodic:

$$R_{T,xh}(\tau) = \frac{1}{T}\int_T x_T(t)h_T(t+\tau)dt \quad (2.6.1b)$$

Autocorrelation: Aperiodic:

$$R_x(\tau) = \int_{-\infty}^{\infty} x(t)x(t+\tau)dt \quad (2.6.1c)$$

Autocorrelation: Periodic:

$$R_{T,x}(\tau) = \frac{1}{T}\int_T x_T(t)x_T(t+\tau)dt. \quad (2.6.1d)$$

Notes: Cross- and autocorrelations of periodic functions and random signals are referred to as *average periodic cross- and autocorrelation functions*. In the case of random or noise signals, the average cross-correlation function is defined by

$$R_{a,xh}(\tau) = \lim_{T\to\infty} \frac{1}{T}\int_{-T/2}^{T/2} x(t)h(t+\tau)dt. \quad (2.6.1e)$$

For periodic functions, (2.6.1e) reduces to (2.6.1b).

2.6.1 Basic Properties of Cross-Correlation Functions

Folding relationship between the two cross-correlation functions is

$$R_{xh}(\tau) = R_{hx}(-\tau), \quad (2.6.2)$$

$$\Rightarrow R_{xh}(\tau) = \int_{-\infty}^{\infty} x(t)h(t+\tau)dt$$

$$= \int_{-\infty}^{\infty} x(\alpha-\tau)h(\alpha)d\alpha = R_{hx}(-\tau). \quad (2.6.3)$$

2.6.2 Cross-Correlation and Convolution

The cross-correlation function is related to the convolution. From (2.6.3) we have

$$R_{xh}(\tau) = x(\tau) ** h(\tau) = x(-\tau) * h(\tau), \quad (2.6.4a)$$

$$R_{hx}(\tau) = h(\tau) ** x(\tau) = h(-\tau) * x(\tau). \quad (2.6.4b)$$

Equation (2.6.4a) can be seen by first rewriting the first integral in (2.6.3) using a new variable $t = -\alpha$, and then simplifying it. That is,

$$R_{xh}(\tau) = \int_{-\infty}^{\infty} x(t)h(t+\tau)dt = \int_{-\infty}^{\infty} x(-\alpha)h(\tau-\alpha)d\alpha$$

$$= x(-\tau) * h(\tau). \quad (2.6.4c)$$

Equation (2.6.4b) can be similarly shown. Noting the explicit relation between correlation and convolution, many of the convolution properties are applicable to the correlation. To compute the cross–

correlation, $R_{xh}(\tau)$, one can use either of the integral in (2.6.3) or the integral in (2.6.4c). $R_{xh}(\tau)$ is not always equal to $R_{hx}(\tau)$. In case, if one of the functions is *symmetric*, say $x(t) = x(-t)$, then

$$R_{xh}(\tau) = x(-\tau) * h(\tau) = x(\tau) * h(\tau). \qquad (2.6.5)$$

Example 2.6.1 illustrates the use of this property. In particular, the area and duration properties for convolution also apply to the correlation. We should note that the correlations *are functions of τ* and *not t*, where τ is the time shift between $x(t)$ and $h(t+\tau)$. In the case of energy signals, the energies in the real signals, $g(t)$ and $h(t)$, are

$$R_g(0) = \int\limits_{-\infty}^{\infty} g^2(t)dt = E_g, \quad R_h(0) = \int\limits_{-\infty}^{\infty} h^2(t)dt = E_h.$$

$$(2.6.6) \quad \blacksquare$$

2.6.3 Bounds on the Cross-Correlation Functions

Consider the integral

$$\int\limits_{-\infty}^{\infty} [x(t) \pm h(t+\tau)]^2 dt = \int\limits_{-\infty}^{\infty} x^2(t)dt$$

$$+ \int\limits_{-\infty}^{\infty} h^2(t+\tau)dt \pm 2 \int\limits_{-\infty}^{\infty} x(t)h(t+\tau)dt$$

$$= R_x(0) + R_h(0) \pm 2R_{xh}(\tau) \geq 0. \qquad (2.6.7a)$$

This follows since the integrand in (2.6.7a) is non-negative and

$$|R_{xh}(\tau)| \leq (R_x(0) + R_h(0))/2. \qquad (2.6.7b)$$

An interesting bound can be derived using the *Schwarz's inequality*. See (2.1.9d).

$$\langle x(t)h(t+\tau)^2 = \left[\int\limits_{-\infty}^{\infty} x(t)h(t+\tau)dt \right]^2$$

$$\leq \left(\int\limits_{-\infty}^{\infty} |x(t)|^2 dt \right) \left(\int\limits_{-\infty}^{\infty} |h(t+\tau)|^2 dt \right), \quad (2.6.8)$$

$$\Rightarrow |R_{xh}(\tau)|^2 \leq \left(\int\limits_{-\infty}^{\infty} x^2(t)dt \right) \left(\int\limits_{-\infty}^{\infty} h^2(t)dt \right)$$

$$= R_x(0)R_h(0), \qquad (2.6.9a)$$

$$|R_{xh}(\tau)| \leq \sqrt{R_x(0)R_h(0)}. \qquad (2.6.9b)$$

Equation (2.6.9b) represents a tighter bound compared to the one in (2.6.7b), as the geometric mean cannot exceed the arithmetic mean. That is,

$$\sqrt{R_x(0)R_h(0)} \leq (R_x(0) + R_h(0))/2. \qquad (2.6.9c)$$

Another way to prove (2.6.9b) is as follows. Start with the inequality below. Expand the function and identify the auto- and cross-correlation terms.

$$\int_{-\infty}^{\infty} [x(t) + \alpha h(t+\tau)]^2 dt \geq 0. \qquad (2.6.10)$$

Write the resulting equation in a quadratic form in terms of α. In order for the equation in (2.6.10) to be true, the roots of the quadratic equation have to be real and equal or the roots have to be complex conjugates. The proof is left as a homework problem.

Example 2.6.1 Determine the cross-correlation of the functions given in Fig. 2.3.2.

$$x(t) = \Pi\left[\frac{t}{2T}\right], \; h(t) = e^{-at}u(t), a > 0. \quad (2.6.11a)$$

Solution: Example 2.3.3 dealt with computing the convolution of these two functions. The cross-correlation functions are as follows:

$$R_{hx}(\tau) = \int\limits_{-\infty}^{\infty} h(t)x(t+\tau)dt = h(-\tau) * x(\tau),$$

$$R_{xh}(\tau) = \int\limits_{-\infty}^{\infty} x(t)h(t+\tau)dt = x(-\tau) * h(\tau).$$

$$(2.6.11b)$$

Note that we have $x(-\tau) = x(\tau)$, and therefore the cross-correlation $R_{xh}(\tau) = x(\tau) * h(\tau)$ is the convolution determined before (see (2.3.18).), except the cross-correlation is a function of τ rather than t. It is given below. The two cross-correlation functions are sketched in Fig. 2.6.1a,b. Note $R_{hx}(\tau) = R_{xh}(-\tau)$

Fig. 2.6.1 Cross-correlations (a) $R_{xh}(\tau)$, (b) $R_{hx}(\tau)$ $(R_{xh}(T) = \frac{1}{a}[1 - e^{-2aT}] = R_{hx}(-T))$

(a)   (b)

$$R_{xh}(\tau) = \begin{cases} 0, & \tau \leq -T \\ \frac{1}{a}\left[1 - e^{-a(\tau+T)}\right], & -T < \tau \leq T \\ \frac{1}{a}\left[e^{aT} - e^{-aT}\right]e^{-a\tau}, & \tau > T \end{cases}$$

(2.6.11c) ∎

2.6.4 Quantitative Measures of Cross-Correlation

The amplitudes of $R_{xh}(\tau)$ (and $R_{hx}(\tau)$) vary. It is appropriate to consider the *normalized correlation coefficient* (or correlation coefficient) of two energy signals defined by

$$\rho_{xh}(\tau) = \frac{R_{xh}(\tau)}{\sqrt{\left[\int_{-\infty}^{\infty} x^2(t)dt\right]\left[\int_{-\infty}^{\infty} h^2(t)dt\right]}} = \frac{R_{xh}(\tau)}{\sqrt{E_x E_h}},$$

(2.6.12a)

$$\Rightarrow |\rho_{xh}(\tau)| \leq 1.$$

(2.6.12b)

Equation (2.6.12b) can be shown as follows. From (2.1.13a) and using the Schwarz's inequality (see (2.1.9d)), we have

$$R_{xh}(\tau) = \langle x(t)h(t+\tau)\rangle \leq \|x(t)\|\|h(t+\tau)\| = \sqrt{E_x E_h}$$

It should be noted that the case of $x(t) = h(t)$, the correlation coefficient reduces to

$$\rho_{xx}(\tau) = \frac{R_x(\tau)}{R_x(0)}.$$

(2.6.13)

Correlation measures are very useful in statistical analysis. See Yates and Goodman (1999), Cooper and McGillem (1999) and others.

The significance of $\rho_{xh}(\tau)$ can be seen by considering some extreme cases. When $x(t) = \alpha h(t)$, $\alpha > 0$, we have the correlation coefficient $\rho_{xh}(\tau) = 1$. In the case of $x(t) = \alpha h(t)$, $\alpha < 0$ and $\rho_{xh}(\tau) = -1$. In communication theory, we will be interested in signals that are corrupted by *noise*, usually identified by $n(t)$, which can be defined only in *statistical terms*. In the following, we will consider the analysis without going through statistical analysis. *Noise signal $n(t)$ is assumed to have a *zero average* value. That is,

$$\lim_{T\to\infty} \frac{1}{T} \int_{-T/2}^{T/2} n(t)dt = 0.$$

(2.6.14)

Cross-correlation function can be used to compare two signals. The signals $x(t)$ and $h(t)$ are *uncorrelated* if the average cross-correlation satisfies the relation

$$R_{a,xh}(\tau) = \lim_{T\to\infty} \frac{1}{T} \int_{T/2}^{T/2} x(t)h(t+\tau)dt$$

$$= \left[\lim_{T\to\infty} \frac{1}{T} \int_{T/2}^{T/2} x(t)dt\right]\left[\lim_{T\to\infty} \frac{1}{T} \int_{T/2}^{T/2} h(t)dt\right].$$

(2.6.15)

Example 2.6.2 If the signals $x(t)$ and a zero average noise signal $n(t)$ are uncorrelated, then show

$$\lim_{T\to\infty} \frac{1}{T} \int_{-T/2}^{T/2} x(t)n(t-\tau)dt = 0 \text{ for all } \tau.$$

(2.6.16)

Fig. 2.6.2 Correlation
detector

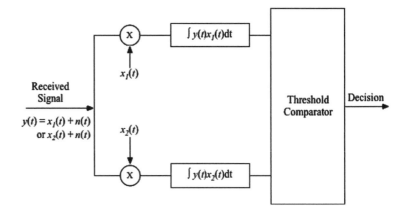

Solution: Using (2.6.14) and (2.6.15), we have

$$\lim_{T\to\infty} \frac{1}{T} \int_{-T/2}^{T/2} x(t)n(t+\tau)dt$$

$$= \left[\lim_{T\to\infty} \frac{1}{T} \int_{T/2}^{T/2} x(t)dt \right] \left[\lim_{T\to\infty} \frac{1}{T} \int_{T/2}^{T/2} n(t+\tau)dt \right] = 0.$$

(2.6.17)

Cross-correlation function can be used to estimate the *delay* caused by a system. Suppose we know that a finite duration signal $x(t)$ is passed through an ideal transmission line resulting in the output function $y(t) = x(t - t_0)$. The delay t_0 caused by the transmission line is unknown and can be estimated using the cross-correlation function $R_{xy}(\tau)$. At $\tau = t_0$, $R_{xy}(t_0)$ gives a maximum value. Then determine τ corresponding to the maximum value of $R_{xy}(\tau)$. ∎

Example 2.6.3 Consider the transmitted signals $x_1(t)$ and $x_2(t)$ in the interval $0 < t < T_s$ and zero otherwise. Use the cross-correlation function to determine which signal was transmitted out of the two. They are assumed to be mutually *orthogonal* (see Section 2.1.1) over the interval and satisfy

$$\int_0^{T_s} x_i(t)x_j(t)dt = \begin{cases} E_{x_i} = E_x, & i=j \\ 0, & i \neq j, i = 1,2 \end{cases}.$$

(2.6.18)

E_x is the energy contained in each signal. The two signals to be transmitted are assumed to be available at the receiver. A simple receiver is the *binary*

correlation detector (or receiver) shown in Fig. 2.6.2. The received signals $y_i(t)$ are assumed to be of the form in (2.6.19). Decide which signal has been transmitted using the cross-correlation function.

$$y_i(t) = x_i(t) + \text{noise}, \ i = 1 \text{ or } 2. \quad (2.6.19)$$

Solution: Let the transmitted signal be $x_1(t)$. Using the top path in Fig. 2.6.2, we have

$$\int_0^{T_s} [x_1(t) + n(t)]x_1(t)dt = A \int_0^{T_s} x_1^2(t)dt. \quad (2.6.20)$$

Using the bottom path, with the transmitted signal equal to $x_1(t)$, we have

$$\int_0^{T_s} [x_1(t)+n(t)]x_2(t)dt = \int_0^{T_s} [x_1(t)x_2(t)+x_1(t)n(t)]dt$$

$$= \int_0^{T_s} x_1(t)n(t)dt = B. \quad (2.6.21)$$

Since the noise signal has no relation to $x_1(t)$, B will be near zero and $A \gg B$, implying $x_1(t)$ was transmitted. If $x_2(t)$ was transmitted, the roles are reversed and $B \gg A$. The *correlation method of detection* is based on the following:

1. If $A > B \Rightarrow$ transmitted signal is $x_1(t)$.
2. If $B > A \Rightarrow$ transmitted signal is $x_2(t)$.
3. If $B = A \Rightarrow$ no decision can be made as noise swamped the transmitted signal. ∎

Example 2.6.4 Derive the expressions for the cross-correlation $R_{xh}(\tau)$ and $R_{hx}(\tau)$ assuming $x(t) = e^{-t}u(t),\ h(t) = e^{-2t}u(t)$.

Solution: Using the expression in (2.6.3), we have

$$R_{xh}(\tau) = \int_{-\infty}^{\infty} x(\alpha - \tau)h(\alpha)d\alpha$$

$$= \int_{-\infty}^{\infty} e^{-(\alpha - \tau)}e^{-2\alpha}[u(\alpha - \tau)u(\alpha)]d\alpha \quad (2.6.22a)$$

Consider the following and then the corresponding correlations:

$$\tau \geq 0:\quad [u(\alpha)u(\alpha - \tau)] = \begin{cases} 1, & \alpha \geq \tau \\ 0, & \text{Otherwise} \end{cases},$$

$$\tau < 0:\quad [u(\alpha)u(\alpha - \tau)] = \begin{cases} 1, & \alpha \geq 0 \\ 0, & \text{Otherwise} \end{cases},$$

$$\tau \geq 0: R_{xh}(\tau) = e^{\tau}\int_{\tau}^{\infty} e^{-3\alpha}d\alpha$$

$$= e^{\tau}\frac{1}{-3}e^{-3\alpha}\Big|_{\tau}^{\infty} = \frac{1}{3}e^{-2\tau}u(\tau),\quad (2.6.22b)$$

$$\tau < 0: R_{xh}(\tau) = e^{\tau}\int_{0}^{\infty} e^{-3\alpha}d\alpha = \frac{1}{(-3)}e^{\tau}e^{-3\alpha}\Big|_{0}^{\infty} = \frac{e^{\tau}}{3}.$$

$$(2.6.22c)$$

$R_{xh}(\tau)$ is shown in Fig. 2.6.3. Note that $R_{hx}(\tau) = R_{xh}(-\tau)$. ∎

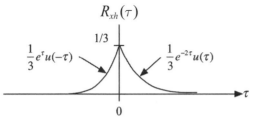

Fig. 2.6.3 $R_{xh}(\tau)$

Example 2.6.5 Derive the cross–correlation $R_{xh}(\tau)$ for the following functions:

$$x(t) = \Pi[t - .5],\ h(t) = t\Pi\left[\frac{t-1}{2}\right],\ R_{xh}(\tau)$$

$$= \int_{-\infty}^{\infty} x(t)h(t + \tau)dt. \quad (2.6.23)$$

Solution: See Fig. 2.6.4c for $h(t + \tau)$ for an arbitrary τ. The function $h(t + \tau)$ starts at $t = -\tau$ and ends at $t = 2 - \tau$. As τ varies from $-\infty$ to ∞, there are five possible regions we need to consider. These are sketched in Fig. 2.6.4 d,e,f,g,h. In each of these cases both the functions are sketched in the same figure, which allows us to find the regions of overlap. The regions of overlap are listed in Table 2.6.1.

Case 1: $\tau \geq -1$: See Fig. 2.6.4d. There is no overlap between $x(t)$ and $h(t + \tau)$ and

$$R_{xh}(\tau) = 0, -\tau > 1 \text{ or } \tau \leq -1. \quad (2.6.24)$$

Case 2: $0 < -\tau \leq 1$ or $-1 < \tau \leq 0$: See Fig. 2.6.4e. Using Table 2.6.1

$$R_{xh}(\tau) = \int_{-\tau}^{1} (t + \tau)dt = \frac{t^2}{2} + \tau t\Big|_{t=-\tau}^{t=1}$$

$$= \frac{(\tau + 1)^2}{2}, -1 \leq \tau < 0. \quad (2.6.25)$$

Case 3: $0 < \tau \leq 1$: See Fig. 2.6.4 f. Using Table 2.6.1, we have

$$R_{xh}(\tau) = \int_{0}^{1} (t + \tau)dt = \frac{t^2}{2} + \tau t\Big|_{t=0}^{t=1} = \frac{1 + 2\tau}{2}, 0 < \tau \leq 1.$$

$$(2.6.26)$$

Case 4: $1 < \tau \leq 2$: See Fig. 2.6.3 g. Using Table 2.6.1 we have

$$R_{xh}(\tau) = \int_{0}^{2-\tau} (t + \tau)dt = \frac{t^2}{2} + \tau t\Big|_{t=0}^{t=2-\tau}$$

$$= \frac{4 - \tau^2}{2}, 1 < \tau \leq 2. \quad (2.6.27)$$

Case 5: $2 < \tau$: See Fig. 2.6.4 h. There is no overlap and

Fig. 2.6.4 (a) $x(t)$, (b) $h(t)$,
(c) $h(t + \tau)$,
(d) $x(t)$ and $h(t + \tau)$,
$-\tau > 1$(or $\tau \leq -1$),
(e) $x(t)$ and $h(t+\tau), -1 < \tau \leq 0$,
(f) $x(t)$ and $h(t+\tau), 0 < \tau \leq 1$,
(g) $x(t)$ and $h(t+\tau), 1 < \tau < 2$,
(h) $x(t)$ and $h(t+\tau), \tau > 2$,
(i) $R_{xh}(\tau)$

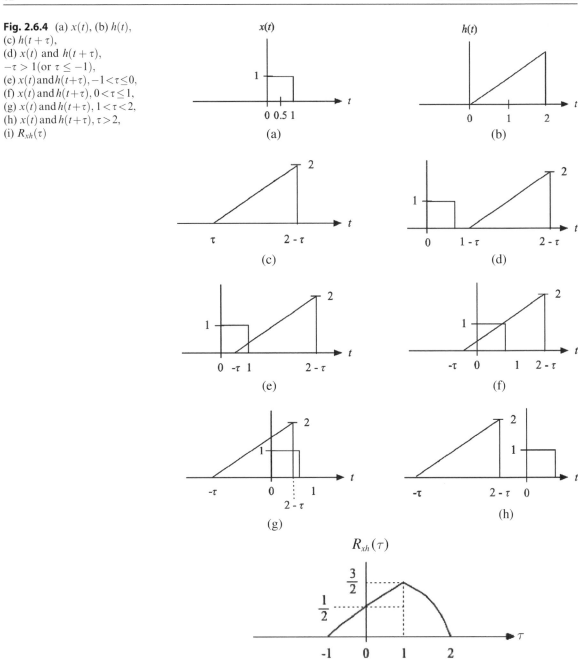

Table 2.6.1	Example 2.6.4	
Case	τ	Range of overlap/ integration range
1	$\tau \leq -1$	No over lap
2	$-1 < \tau \leq 0$	$-\tau < t < 1$
3	$0 < \tau \leq 1$	$0 < t < 1$
4	$1 < \tau \leq 2$	$1 < t < 2 - \tau$
5	$\tau > 2$	No over lap

$$R_{xh}(\tau) = 0, \tau > 2. \qquad (2.6.28)$$

See Fig. 2.6.4i for the cross-correlation $R_{xh}(\tau)$ sketch. There are no impulses in either of the two functions and therefore the cross-correlation function is continuous. ∎

2.7 Autocorrelation Functions of Energy Signals

Autocorrelation function describes *the similarity* or *coherence* between the given function $x(t)$ and its delayed or its advanced version $x(t \pm \tau)$. It is an even function. The autocorrelation (AC) function of an aperiodic signal $x(t)$ was defined by

$$AC\{x(t)\} = R_x(\tau) = \int_{-\infty}^{\infty} x(t)x(t+\tau)dt$$

$$= \int_{-\infty}^{\infty} x(t)x(t-\tau)dt = \langle x(t)x(t-\tau) \rangle. \quad (2.7.1)$$

$$R_x(-\tau) = R_x(\tau).\text{(even function)} \quad (2.7.2)$$

Second, the maximum value of the autocorrelation function occurs at $\tau = 0$. That is

$$|R_x(\tau)| \le R_x(0). \quad (2.7.3)$$

The proof of the symmetry property in (2.7.2) can be shown by changing the variable $\beta = t + \tau$ and simplifying the integral. The proof on the upper bound on the autocorrelation function is shown below by noting the integral with a nonnegative integrand is nonnegative.

$$\int_{-\infty}^{\infty} [x(t) \pm x(t-\tau)][x(t) \pm x(t-\tau)]dt \ge 0. \quad (2.7.4)$$

$$\int_{-\infty}^{\infty} x^2(t)dt + \int_{-\infty}^{\infty} x^2(t-\tau)dt \pm 2\int_{-\infty}^{\infty} x(t)x(t-\tau)dt$$

$$= 2[\int_{-\infty}^{\infty} x^2(t)dt \pm \int_{-\infty}^{\infty} x(t)x(t-\tau)dt] \ge 0$$

$$\Rightarrow R_x(0) = \int_{-\infty}^{\infty} x^2(t)dt \ge |R_x(\tau)|$$

$$= \left| \int_{-\infty}^{\infty} x(t)x(t-\tau)dt \right|. \quad (2.7.5)$$

Third,

$$E_x = R_x(0) = \int_{-\infty}^{\infty} x^2(t)dt.\text{(energy in } x(t)\text{).} \quad (2.7.6)$$

In addition, if $y(t) = x(t \pm \alpha)$, then

$$R_x(\tau) = R_y(\tau). \quad (2.7.7)$$

This can be seen first for $\tau > 0$ from

$$R_y(\tau) = \int_{-\infty}^{\infty} y(t)y(t-\tau)dt = \int_{-\infty}^{\infty} x(t-\alpha)x(t-\alpha-\tau)dt$$

$$= \int_{-\infty}^{\infty} x(\beta)x(\beta-\tau)d\beta = R_x(\tau). \quad (2.7.8)$$

Change of a variable $\beta = (t - \alpha)$ was made in the above integral and then simplified. Since the autocorrelation function is even, the result follows for $\tau < 0$.

Example 2.7.1 Find the AC of $x(t) = e^{-at}u(t), a > 0$ by first computing the AC for $\tau > 0$ and then use the symmetry property to find the other half of the autocorrelation function.

Solution: First,

$$u(t)u(t-\tau) = u(t-\tau) = \begin{cases} 1, & t > \tau \\ 0, & \text{otherwise} \end{cases}, \quad (2.7.9)$$

$$\tau > 0 : R_x(\tau) = \int_{-\infty}^{\infty} x(t)x(t-\tau)dt$$

$$= \int_{-\infty}^{\infty} e^{-at}u(t)e^{-a(t-\tau)}u(t-\tau)dt$$

$$= e^{a\tau}\int_{\tau}^{\infty} e^{-2at}dt = \frac{e^{-a\tau}}{2a}.$$

Using the symmetry property of the AC, we have

$$R_x(\tau) = (1/2a)e^{-a|\tau|}. \quad (2.7.10)$$

The energy contained in the exponentially decaying pulse is $E = R_x(0) = (1/2a)$. The autocorrelation function is sketched in Fig. 2.7.1.   ∎

Example 2.7.2 Consider the function $x(t) = \Pi[t - 1/2]$. Determine its autocorrelation function and its energy using this function.

Solution: The AC function for $\tau \ge 0$ is

$$R_x(\tau) = \int_{-\infty}^{\infty} x(t)x(t-\tau)dt = \int_{-\infty}^{\infty} \Pi[t-.5]\Pi[t-\tau-.5]dt.$$

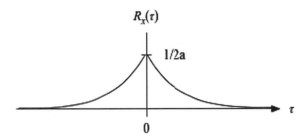

Fig. 2.7.1 Example 2.7.1

The function $\Pi[t-1/2]$ is a rectangular pulse centered at $t=1/2$ with a width of 1, and $\Pi[t-(\tau+(1/2))]$ is a rectangular pulse centered at $(\tau+0.5)$ with a width of 1. See Fig. (2.7.2a) for the case $0<\tau<1$. In the case of $\tau\geq1$, there is no overlap indicating that $R_x(\tau)=0, \tau\geq1$.

$$R_x(\tau)=\int_{\tau}^{1} dt=(1-\tau), 0\leq\tau<1.$$

Using the symmetry property, we have

$$R_x(\tau)=R_x(-\tau)=\begin{Bmatrix}(1-|\tau|), 0\leq|\tau|\leq1\\ 0, \qquad \text{Otherwise}\end{Bmatrix}=\Lambda[\tau].$$

$$(2.7.11a)$$

This is sketched in Fig. 2.7.2b indicating that there is correlation for $|\tau|<1$ and no correlation for $|\tau|\geq1$. The peak value of the autocorrelation is when $\tau=0$ and is $R_x(0)=1$. The energy contained in the unit rectangular pulse is equal to 1 and by using the autocorrelation function, i.e., $R_x(0)=1$, the same by both the methods. Noting that the autocorrelation function of a given function and its delayed or advanced version are the same, the

AC function is much easier to compute using this property. The AC of the pulse function $\Pi[t-.5]$ can be computed by ignoring the delay. That is, $AC\{\Pi[t-.5]\}=AC\{\Pi[t]\}$. Interestingly,

$$AC\left\{\Pi\left[\frac{t}{T}\right]\right\}=T\Lambda\left[\frac{\tau}{T}\right]. \qquad (2.7.11b)$$

The AC function of a rectangular pulse of width T is a triangular pulse of width $2T$ and its amplitude at $t=0$ is T. We can verify the last part by noting

$$AC\left\{\Pi\left[\frac{t}{T}\right]\right\}|_{\tau=0}=T\Lambda\left[\frac{\tau}{T}\right]|_{\tau=0}=T. \qquad \blacksquare$$

Note $x(t)\Pi\left[\frac{t}{T}\right]$ extracts $x(t)$ for the time $-T/2<t<T/2$. That is,

$$x(t)\Pi\left[\frac{t}{T}\right]=\begin{cases}x(t), -T/2<t<T/2\\ 0, \quad \text{otherwise}\end{cases}. \qquad (2.7.12)$$

Example 2.7.3 Find the autocorrelation of the function $y(t)=\cos(\omega_0 t)\Pi[t/T]$.

Solution:

$$R_y(\tau)=\int_{-\infty}^{\infty}\Pi\left[\frac{t}{T}\right]\Pi\left[\frac{t-\tau}{T}\right]\cos(\omega_0 t)\cos(\omega_0(t-\tau))dt$$

$$=\frac{\cos(\omega_0\tau)}{2}\int_{-\infty}^{\infty}\Pi\left[\frac{t}{T}\right]\Pi\left[\frac{t-\tau}{T}\right]dt$$

$$+\frac{1}{2}\int_{-\infty}^{\infty}\Pi\left[\frac{t}{T}\right]\Pi\left[\frac{t-\tau}{T}\right]\cos(2\omega_0 t-\tau)dt.$$

(a)   (b)

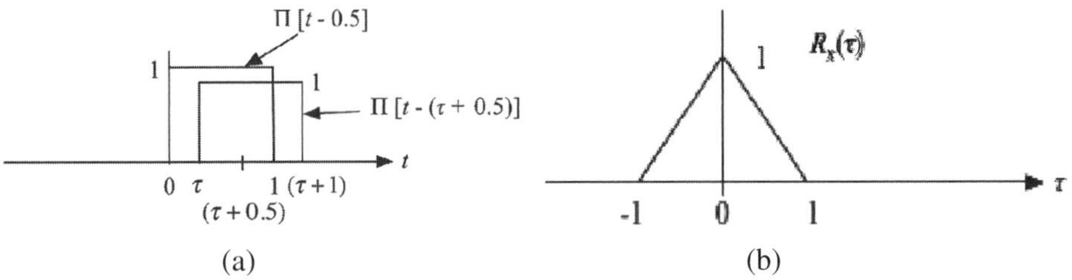

Fig. 2.7.2 Example 2.7.2 Autocorrelation of a rectangular pulse

$$= \begin{cases} (1/2)T\Lambda\left[\frac{\tau}{T}\right]\cos(\omega_0\tau) + B, |\tau| \leq T \\ 0, |\tau| > T \end{cases}. \quad (2.7.13)$$

Now consider the evaluation of B. For $\tau \geq 0$,

$$B = \frac{1}{2}\int_{-\infty}^{\infty} \Pi\left[\frac{t}{T}\right]\Pi\left[\frac{t-\tau}{T}\right]\cos(2\omega_0 t - \tau)dt$$

$$= \frac{1}{2}\int_{\tau}^{T/2} \cos(2\omega_0 t - \omega_0\tau)dt$$

$$= \frac{1}{4\omega_0}[\sin(\omega_0 T - \omega_0\tau) - \sin(\omega_0\tau)]. \quad (2.7.14)$$

If ω_0 is large, $R_y(\tau)$ in (2.7.13) can be *approximated* by the first term and

$$R_Y(\tau) \simeq \frac{1}{2}T\Lambda\left[\frac{t}{T}\right]\cos(\omega_0\tau). \quad (2.7.15)$$

The envelope of the autocorrelation function in (2.7.16) is a triangular function, which follows since the correlation of the two identical rectangular functions is a triangular function. Noting that the cosine function oscillates between ± 1, the envelope of the autocorrelation function in (2.7.15) is shown in Fig. 2.7.3. ∎

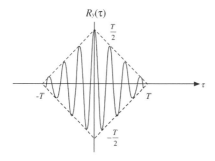

Fig. 2.7.3 Sketch of $R_y(\tau)$

Notes: Conditions for the *existence* of an *aperiodic autocorrelation* are similar to those of convolution (see Section 2.2.3). But there are a few exceptions. For example, the autocorrelation of the unit step function does not exist.

2.8 Cross- and Autocorrelation of Periodic Functions

The cross- and the autocorrelation functions of periodic functions of $x_T(t)$ and $h_T(t)$ are

$$R_{T,xh}(\tau) = \frac{1}{T}\int_T x_T(t)h_T(t+\tau)dt = \langle x_T(t)h_T(t+\tau)\rangle,$$
$$(2.8.1a)$$

$$R_{T,x}(\tau) = \frac{1}{T}\int_T x_T(t)x_T(t+\tau)dt = \langle x_T(t)x_T(t+\tau)\rangle.$$
$$(2.8.1b)$$

Note that the periods of the functions, $x_T(t)$ and $h_T(t)$, are assumed to be the *same* and the constant $(1/T)$ before the integrals in (2.8.1a and b). If they have different periods, computation of (2.8.1a) is difficult and these cases will not be discussed here. Many of the cross-correlation and AC function properties derived earlier for the aperiodic case apply for the periodic functions with some modifications. Note that

$$R_{T,xh}(\tau) = R_{T,hx}(-\tau),$$
$$R_{T,x}(0) + R_{T,h}(0) \geq 2|R_{T,xh}(\tau)|. \quad (2.8.2)$$

In Section 2.5.1, aperiodic convolution was used to find periodic convolution. The same type of analysis can be used to determine periodic cross-correlations using aperiodic cross-correlations. Furthermore, as discussed before, correlation is related to convolution. First define two finite duration functions, $x(t)$ and $h(t)$, over the interval $t_0 \leq t < t_0 + T$. Assume that they are zero outside this interval. Now create two periodic functions:

$$x_T(t) = \sum_{n=-\infty}^{\infty} x(t - nT), \quad h_T(t) = \sum_{n=-\infty}^{\infty} h(t - nT).$$
$$(2.8.3a)$$

The periodic cross-correlation function is defined by

$$R_{T,xh}(\tau) = \frac{1}{T} \int_{t_0}^{t_0+T} x_T(t) h_T(t+\tau) dt, \ h_T(t+\tau)$$

$$= \sum_{n=-\infty}^{\infty} h(t+\tau-nT). \qquad (2.8.3b)$$

The expression for periodic convolution is given in terms of aperiodic convolution and

$$R_{T,xh}(\tau) = \frac{1}{T} \sum_{n=-\infty}^{\infty} R_{xh}(\tau-nT),$$

$$R_{xh}(\tau-nT) = \int_{t_0}^{t_0+T} x(t) h(t+\tau-nT). \qquad (2.8.3c)$$

The details of the derivation are left as an exercise. Copies of $R_{xh}(\tau)$ will overlap if the width of $R_{xh}(\tau)$ is wider than T.

Example 2.8.1 Give the lower bound on the period T so that there are no overlaps in the cross-correlation of the functions $x_T(t)$ and $h_T(t)$ given below. See Example 2.6.5.

$$x(t) = \Pi[t-.5], h(t) = t\Pi\left[\frac{t-1}{2}\right],$$

$$x_T(t) = \sum_{n=-\infty}^{\infty} x(t+nT), h_T(t) = \sum_{n=-\infty}^{\infty} h(t+nT).$$

Solution: If the period T is larger than 3, then there are no overlaps in the periodic cross-correlation function. In that case, one period of the cross-correlation function can be obtained from the aperiodic cross-correlation in that example and dividing it by the period T. If the period is less than 3, then there will be overlaps. ∎

Example 2.8.2 Consider the periodic functions

$$x_{T,1}(t) = X_s[0], \quad x_{T,2}(t) = c[k]\cos(k\omega_0 t + \theta[k]). \qquad (2.8.4)$$

a. Find the AC functions for the functions in (2.8.4). b. Find the cross-correlation of the two functions.

Solution:

$$a. \ R_{T,x_{T,1}}(\tau) = \frac{1}{T} \int_T X_s^2[0] dt = X_s^2[0], \qquad (2.8.5a)$$

$$R_{T,x_{T,2}}(\tau) = \frac{1}{T} \int_T x_{T,2}(t) x_{T,2}(t+\tau) dt$$

$$= \frac{c^2[k]}{2T} \int_T \cos(k\omega_0\tau) dt$$

$$+ \frac{c^2[k]}{2T} \int_T \cos(k\omega_0(2t+\tau) + 2\theta[k]) dt$$

$$= \frac{c^2[k]\cos(k\omega_0\tau)}{2T} \int_0^T dt = \frac{c^2[k]\cos(k\omega_0\tau)}{2}.$$

$$(2.8.5b)$$

Note that the integral of a cosine function over any integer number of periods is zero.

b. The cross-correlation of a constant and a cosine function over one period is zero. Also note that the two functions are orthogonal. That is $\langle x_{T_1}(t), x_{T_2}(t) \rangle = 0$. ∎

Example 2.8.3 Find the AC of $x_T(t)$ given below with $k \neq m$, k and m are integers.

$$x_T(t) = x_{T,1}(t) + x_{T,2}(t), x_{T,1}(t)$$

$$= c[k]\cos(k\omega_0 t + \theta[k]), x_{T,2}(t)$$

$$= c[m]\cos(m\omega_0 t + \theta[m]).$$

Solution: The periodic autocorrelations are determined as follows:

$$R_{T,x}(\tau) = \frac{1}{T} \int_T [x_{T,1}(t) + x_{T,2}(t)][x_{T,1}(t+\tau)$$

$$+ x_{T,2}(t+\tau)] dt \ = \frac{1}{T} \int_T x_{T,1}(t) x_{T,1}(t+\tau) dt$$

$$+ \frac{1}{T} \int_T x_{T,2}(t) x_{T,2}(t+\tau) dt$$

$$+ \frac{1}{T} \int_T x_{T,1}(t) x_{T,2}(t+\tau) dt$$

$$+ \frac{1}{T} \int_T x_{T,2}(t) x_{T,1}(t+\tau) dt. \qquad (2.8.6)$$

Note

$$\frac{1}{T}\int_T x_{T,1}(t)x_{T,2}(t+\tau)dt$$

$$=\frac{1}{2T}\int_T c[k]c[m]\cos[(k+m)\omega_0 t+m\omega_0\tau+\theta[k]$$

$$+\theta[m])dt+\frac{1}{2T}\int_T h[k]h[m]\cos[(k-m)\omega_0 t-m\omega_0\tau)$$

$$+(\theta[k]-\theta[m])]dt=0.$$

Similarly the fourth term in (2.8.6) goes to zero. From the last example,

$$R_{T,x}(\tau)=\frac{c^2[k]}{2}\cos(k\omega_0\tau)+\frac{c^2[m]}{2}\cos(m\omega_0\tau), k\neq m.$$

$$(2.8.7) \quad \blacksquare$$

These results can be generalized using the last two examples and the autocorrelation of a periodic function $x_T(t)$ is given as follows:

$$x_T(t)=X_s[0]+\sum_{k=1}^{\infty}c[k]\cos(k\omega_0 t+\theta[k]), \quad (2.8.8)$$

$$\Rightarrow R_{T,x}(\tau)=X_s^2[0]+\frac{1}{2}\sum_{k=1}^{\infty}c^2[k]\cos(k\omega_0\tau),\omega_0=2\pi/T.$$

$$(2.8.9)$$

AC function of a periodic function is also a periodic function with the same period. It is independent of $\theta[k]$. It does not have the phase information contained in (2.8.8). In the next chapter, (2.8.8) will be derived for an arbitrary periodic function and will be referred to as the harmonic form of Fourier series of a periodic function $x_T(t)$. $\blacksquare$

Notes: The AC function of a constant $X_s[0]$ is $X_s^2[0]$. The AC of the sinusoid $c[k]\cos(k\omega_0 t+\theta[k])$ is $(c^2[k]/2)\cos(k\omega_0\tau)$. That is, it *loses* the phase information in the function in the sinusoid. The power contained in the periodic function $x_T(t)$ in (2.8.8) can be computed from the autocorrelation function evaluated at $\tau=0$. That is,

$$P=X_s^2[0]+\frac{1}{2}\sum_{k=1}^{\infty}c^2[k]. \quad (2.8.10)$$

The difference between the total power and the dc power is the variance and is given by

$$\text{Variance}=\frac{1}{2}\sum_{k=1}^{\infty}c^2[k]. \quad (2.8.11) \quad \blacksquare$$

Example 2.8.4 Consider the corrupted signal $y(t)=x(t)+n(t)$, where $n(t)$ is assumed to be noise. Assuming the signal $x(t)$ and noise $n(t)$ are uncorrelated, derive an expression for the autocorrelation function of $y(t)$.

Solution:

$$R_{yy}(\tau)=\lim_{T\to\infty}\frac{1}{T}\int_{-T/2}^{T/2}y(t)y(t+\tau)dt$$

$$=\lim_{T\to\infty}\frac{1}{T}\int_{-T/2}^{T/2}[x(t)+n(t)][x(t+\tau)n(t+\tau)]dt,$$

$$=\lim_{T\to\infty}\int_{-T/2}^{T/2}x(t)x(t+\tau)dt+\lim_{T\to\infty}\int_{-T/2}^{T/2}x(t)n(t+\tau)dt$$

$$+\lim_{T\to\infty}\int_{-T/2}^{T/2}n(t)x(t+\tau)dt+\lim_{T\to\infty}\int_{-T/2}^{T/2}x(t)x(t+\tau)dt.$$

$$(2.8.12)$$

Noting that the signal and the noise are *uncorrelated*, i.e., $R_{xn}(\tau)=R_{nx}(\tau)=0$, we have

$$R_{yy}(\tau)=R_{xx}(\tau)+R_{nn}(\tau). \quad (2.8.13) \quad \blacksquare$$

The average power contained in the signal and the noise is given by

$$P_y=P_x+P_n=R_x(0)+R_n(0)=R_x(0)+\sigma_n^2.$$

$$(2.8.14)$$

The signal-to-noise ratio (SNR), P_x/P_n, can be computed. It is normally identified in terms of decibels. See Section 1.9.

2.9 Summary

We have introduced the basics associated with the two important signal analysis concepts: convolution and correlation. Specific principal topics that were included are

- Convolution integral: its computations and its properties
- Moments associated with functions
- Central limit theorem
- Periodic convolutions
- Auto- and cross-correlations
- Examples of correlations involving noise without going into probability theory
- Quantitative measures of cross-correlation functions and the correlation coefficient
- Auto- and cross-correlation functions of energy and periodic signals
- Signal-to-noise ratios

Problems

2.1.1 Consider the following functions defined over $0 < t < 1$. Using (2.1.3), identify the two functions that give the maximum cross-correlation at $\tau = 0$.
$$x_1(t) = e^{-t}, \ x_2(t) = \sin(t), x_3(t) = (1/t).$$

2.2.1 Prove the commutative, distributive, and the associate properties of the convolution.

2.2.2 Find the convolution $y(t) = h(t) * x(t)$ for the following functions:

a. $x(t) = .5\delta(t-1) + .5\delta(t-2),$
   $h(t) = .5\delta(t-2) + .5\delta(t-3)$

b. $x(t) = (t-1)\Pi[t-1], \quad h(t) = x(t),$

c. $x(t) = (1 - t^2), \ -1 \le t \le 1,$
   $h(t) = \Pi[t],$

d. $x(t) = e^{-at}u(t), \quad h(t) = e^{-bt}u(t)$
   for cases : $1.a > 0, \ b > 0, \quad 2.a = 0, b > 0$

e. $x(t) = \Pi[t/2], \quad h(t) = \Pi[t - .5] - \Pi[t - 1.5]$

f. $x(t) = \delta(t - 1), \quad h(t) = e^{-t}u(t)$

g. $x(t) = \cos(\pi t)\Pi[t], \quad h(t) = e^{-t}u(t).$

2.2.3 Use the area property of convolution to find the integrals of $y(t)$ in Problem 2.2.2.

2.3.1 a. Derive the expression for the convolution of two pulse functions given by $x(t) = \Pi[t - 1]$ and $h[t] = \Pi[t - 2]$. Compute this directly first and then verify your result by using the delay property of convolution.
b. Verify the time duration property of the convolution using the above problems.

2.3.2 Determine the area of $y(t)$ in (2.3.18) using the area property of the convolution.

2.4.1 Approximate the function $y(t)$ in Example 2.3.1 using the Gaussian function.

2.4.2 Use the derivative property of the convolution to derive the convolution of the two functions given below using the results in Example 2.5.2.

$$x_T(t) = \sin(\omega_0 t), h(t) = e^{-at}u(t), a > 0.$$

2.4.3 Use the delay property of the convolution to determine

$$x(t) = e^{-at}u(t) * u(t - 1).$$

2.5.1 Derive the expressions for the periodic convolution of the two periodic functions

$$x(t) = \sum_{n=-\infty}^{\infty} \delta(t - nT), \quad h(t) = \sum_{n=-\infty}^{\infty} \Pi\left[\frac{t - nT}{T/2}\right].$$

2.6.1 Find the cross-correlation of the functions $x(t)$ and $h(t)$ given in (2.6.11a) by directly deriving the result and verify the result using the results in Example 2.6.1.

2.6.2 Show the bounds given in (2.6.7a and b) and (2.6.9b) are valid. Use (2.6.11a).

2.6.3 Show (2.6.9b) using (2.6.10).

2.7.1 Find the autocorrelations of the following functions:

a. $x_1(t) = \Pi[t - .5] - \Pi[t - 1.5],$
b. $x_2(t) = u(t - .5) - u(t + .5), \quad c. x_3(t) = t\Pi[t].$

Compute the energies contained in the functions directly and then verify the results using the autocorrelation functions derived in the first part.

2.7.2 Verify the result in (2.7.3) using the results in Example 2.7.1.

2.7.3 Show the identity

$$AC[x(t - t_0)] = AC[x(t)].$$

2.7.4 Derive the AC function step by step for the function $x(t) = \cos(\omega_0 t)\Pi[t/T]$.

Use the integral formula by assuming $\omega_0 = \pi$ and $T = 4$. Verify the results in Example 2.7.3 using the information provided in this problem. Give the appropriate bounds.

2.7.5 Show that the autocorrelations of the function $x_2(t) = e^{at}u(t)$ for $a \geq 0$ do not exist.

2.8.1 *a*. Derive the time-average periodic autocorrelation function $R_{x, T}(\tau)$ for the following periodic function using the integral formula.

$$x_T(t) = A_1 \cos(\omega_0 t + \theta_1) + A_2 \cos(2\omega_0 t + \theta_2).$$

b. Verify the result using (2.8.8) and (2.8.9).

c. Compute the average power contained in the function directly and by evaluating the autocorrelation function at $\tau = 0$. Sketch the function $x(t)$ by assuming the values $A_1 = 5, A_2 = 2, \theta_1 = 20^0, \theta_2 = 120^0$. Sketch the autocorrelation function using

these constants. Suppose we are interested in determining the period T from these two sketches, which function is better, the given function or its autocorrelation? Why?

2.8.2 Let $y_T(t) = A + x_T(t)$, $A -$ constant. Repeat the last problem, except for the plots.

2.8.3 *a*. Show that the following functions are orthogonal over a period: $x_T(t) = \cos(\omega_0 t + \theta)$, $y(t) = A$ *b*. Show the functions $x(t) = \Pi[t]$, $y[t] = t$ are orthogonal.

2.8.4 Consider the signal $z(t) = x(t) + y(t)$. Show that the AC of this function is given by

$$R_z(\tau) = R_x(\tau) + R_y(\tau) + R_{xy}(\tau) + R_{yx}(\tau).$$

Simplify the expression for $R_z(\tau)$ by assuming that $x(t)$ is orthogonal to $y(t)$ for all τ.

2.8.5 Complete the details in deriving the periodic cross-correlation function in terms of the aperiodic convolution leading up to Equation (2.8.3c).

2.8.6 Show (2.8.3c) using (2.6.5).

Chapter 3
Fourier Series

3.1 Introduction

In this chapter we will consider approximating a function by a linear combination of *basis functions*, which are simple functions that can be generated in a laboratory. *Joseph Fourier* (1768–1830) developed the mathematical theory of heat conduction using a set of trigonometric (sine and cosine) series of the form we now call Fourier series (Fourier, J.B.J., 1955 (A. Freeman, translation)). He established that an arbitrary mathematical function can be represented by its Fourier series. This idea was new and startling and met with vigorous opposition from some of the leading mathematicians at the time, see Hawking (2005). Fourier series and the Fourier transform are basics to mathematics and science, especially to the theory of communications. For example, a phoneme in a speech signal is smooth and wavy. A linear combination of a few sinusoidal functions would approximate a segment of speech within some error tolerance. Suppose we like to build a structure that allows us to climb from the first floor to the second floor of a building. We can have a staircase approximating a ramp function using a linear combination of pulse functions. The amplitudes and the width of the pulses can be determined based on the error between the ramp and the staircase. Apart from the staircase problem, this type of analysis is important in electrical engineering, for example, when converting an analog signal to a discrete signal.

The term "well-behaved" function, $x(t)$ defined in the interval, $(t_0, t_0 + T)$ is given in terms of the following *Dirichlet* conditions:

1. The function $x(t)$ must be *single valued* within the given interval of T seconds.

2. The function $x(t)$ can have at most a *finite number* of *discontinuities* and a *finite number* of *maxima and minima* in the time interval.

3. The function $x(t)$ must be *absolutely integrable* on the interval, i.e.,

$$\int_{t_0}^{t_0+T} |x(t)|dt = \text{finite} < \infty.$$

Fortunately, all signals that we will be interested in satisfy these properties. The functions that do not satisfy the Dirichlet conditions are only of theoretical interest. Dirichlet gave an example that does not satisfy the conditions mentioned above and is

$$x_{2\pi}(t) = \begin{cases} 1, & t\text{-rational} \\ 0, & t\text{-irrational} \end{cases}.$$

Our goal is to express a well-behaved function $x(t)$ by an approximate function $x_a(t)$ in terms of an independent set of functions $\{\phi_k(t)\}$ and a set of constants $c[k]$ in the form

$$x_a(t) = \sum_{k=-N}^{N} c[k]\phi_k(t). \tag{3.1.1}$$

The subscript a on x in (3.1.1) denotes that it is an approximation of the function $x(t)$. Without loosing any generality we can assume that the limits on the sum $\pm N \to \pm\infty$. We will be interested in a finite N that satisfies some constraints on the error signal, i.e., the difference between the given signal and its approximation. The entries in the expansion are assumed to have the following properties:

R.K.R. Yarlagadda, *Analog and Digital Signals and Systems*, DOI 10.1007/978-1-4419-0034-0_3,
© Springer Science+Business Media, LLC 2010

1. The constants $c[k]$ are assumed to be some constants and k is an integer.
2. The set $\{\phi_k(t), k = -N, -(N-1), \ldots -1, 0, 1, \ldots, (N-1), N\}$ is a linearly independent set. That is, $\phi_k(t)$ cannot be obtained as a linear combination of the other $\phi_n(t), n \neq k$. Such a set is called *a basis function set* and the members of this set are called *basis functions*. The basis functions can be real or complex.
3. Finally, we like to consider a basis set that is *independent* of $x(t)$.

These properties are based on common sense. The first one allows for a level adjustment. The second property allows for the use of a set of independent basis functions. The third property allows for a general analysis. Later on we will see that some basis functions may be more attractive than others for a particular application. Fourier used the sine and cosine functions as basis functions. The most important aspect of generalized Fourier series expansion is that it allows an arbitrary function, defined over a finite interval, and may have discontinuities to be represented as a sum of basis functions, such as sine and cosine functions instead of using Taylor's series:

$$x(t) = x(a) + x'(a)(x-a) + \frac{x''(a)(x-a)^2}{2!} + \cdots$$
$$+ \frac{x^{(n)}(a)(t-a)^n}{n!} + \cdots; \; x^{(n)}(a) = \frac{d^n x(t)}{dt^n}\Big|_{t=a}$$
$$(3.1.2)$$

This is an approximation of the function $x(t)$ based upon the value of the given function at a point $t = a$ and the values of the derivatives of the function at that point. The Taylor series gives a strict prediction of $x(t)$ at a finite distance from $x(t)|_{t=a}$, whereas the Fourier series gives information of the function *over the entire range* $t_0 \leq t < t_0 + T$. Another striking difference is the coefficients in the Taylor series are based upon the derivatives of the function at $t = a$ and the Fourier series coefficients are obtained by integration. Furthermore we can use (3.1.2) only if we know all the derivatives of the function at $t = t_0$. If not, we have to resort to other methods, such as approximating them. The material in this chapter is fairly standard and can be found in

most of the standard circuits and systems text books. For example, see Ambardar (1995), Haykin and Van Veen (1999), Carlson (1998), Hsu (1967), and many others. Also, see Carslaw (1950), Jeffrey, (1956), Tolstov (1962), and Zygmund (1955). In Chapter 8 we will approximate a function by making use of samples of a signal in combination with some interesting interpolation functions.

The presentation starts with the generalized Fourier series and later the Fourier series as a member of the generalized class of series. The basis functions are *independent*. The expansion gets *easier* if the basis functions are orthogonal.

3.2 Orthogonal Basis Functions

The set of basis functions $\{\phi_k(t)\}$ is an *orthogonal basis set* if the functions satisfy

$$\int_{t_0}^{t_0+T} \phi_k(\alpha)\phi_m^*(\alpha)d\alpha = \begin{cases} E_k & k = m \\ 0 & k \neq m \end{cases}. \quad (3.2.1)$$

The superscript (*) on $\phi_m(t)$ indicates complex conjugation. If $\phi_m(t)$ is real, then $\phi_m^*(t) = \phi_m(t)$. The symbol E_k is used to denote the energy in the basis function, $\phi_k(t)$ in the given time interval and E_k is real. That is,

$$E_k > 0 \; (\text{assuming } \phi_k(t) \neq 0). \quad (3.2.2)$$

When $k \neq m$ in (3.2.1), the integral is zero, which is the orthogonality property of the basis functions. If $E_k = 1$ in (3.2.1), then the basis set is an orthonormal set. Orthonormality is not critical in our expansion, as we can create an orthonormal set by normalizing an orthogonal set, i.e., by replacing $\phi_k(t)$ by $[\phi_k(t)/\sqrt{E_k}]$. Therefore, we will concentrate on using orthogonal basis sets instead of orthonormal basis sets.

Example 3.2.1 Show that the set $\{\phi_1(t), \phi_2(t)\}$ given below is an independent set:

$$\phi_1(t) = \Pi[t - 0.5] \text{ and } \phi_2(t) = \Pi\left[\frac{t-1}{2}\right]. \quad (3.2.3)$$

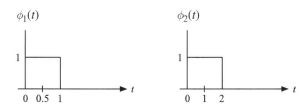

Fig. 3.2.1 Pulse functions $\phi_i(t)$, $i = 1, 2$

Solution: The members of the set are sketched in Fig. 3.2.1 and the set is an independent set since any one of the members cannot be expressed in terms of the others. Since they overlap, and the time width of the second member is longer than the first member, they are not orthogonal. This can be seen from the integral

$$\int_0^2 \phi_1(\alpha)\phi_2(\alpha)d\alpha = \int_0^1 \Pi[\alpha - .5]\Pi[(\alpha - 1)/2]d\alpha$$

$$= 1 \neq 0. \qquad (3.2.4) \blacksquare$$

Example 3.2.2 Consider the pulse functions given below and show that they form an orthogonal basis set. Find the value of A that makes the set an orthonormal set.

$$\phi_1(t) = A\Pi\left[\frac{t - (T/6)}{T/3}\right],$$

$$\phi_2(t) = A\Pi\left[\frac{t - (T/2)}{T/3}\right],$$

$$\phi_3(t) = A\Pi\left[\frac{t - (5T/6)}{T/3}\right]. \qquad (3.2.5)$$

Solution: The pulse functions are shown in Fig. 3.2.2. Clearly, each pulse function exists in a different interval and therefore they are orthogonal. Since all the pulses are of the same width and the same height, we can write

$$E_1 = E_2 = E_3 = \int_0^{T/3} A^2 d\alpha = (A^2 T)/3. \qquad (3.2.6)$$

The functions are orthonormal if $A = \sqrt{\frac{3}{T}}$ and $E_i = 1$, $i = 1, 2, 3$. $\blacksquare$

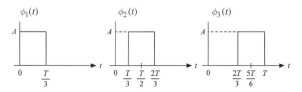

Fig. 3.2.2 Pulse functions $\phi_i(t), i = 1, 2, 3$

Example 3.2.3 Show that the following set is an orthogonal basis set over the time interval $t_0 \leq t < t_0 + T$ and give the values of E_k:

$$\left\{\phi_k(t) = e^{jk\omega_0 t}, k = 0, \pm 1, \pm 2, \ldots\right\}. \qquad (3.2.7a)$$

Solution: Note that $\phi_k(t) = \phi_k(t + T)$ with the period $T = 2\pi/\omega_0$ since

$$e^{jk\omega_0(t+T)} = e^{jk\omega_0 t}e^{jk\omega_0 T} = e^{jk\omega_0 t},$$

$$\omega_0 T = 2\pi, \quad f_0 = 2\pi/\omega_0 = 1/T. \qquad (3.2.7b)$$

If a function $x_T(t) = x_T(t + T)$, then a short hand notation (see (1.5.14)) is

$$\int_{t_0}^{t_0+T} x_T(\alpha)d\alpha = \int_T x_T(\alpha)d\alpha. \qquad (3.2.8)$$

The integral on the right is over any period. Using the orthogonality property, we have

$$k \neq m : \int_T \phi_k(\alpha)\phi_m^*(\alpha)d\alpha = \int_T e^{jk\omega_0\alpha}e^{-jm\omega_0\varepsilon}dt$$

$$= \int_0^T e^{j(k-m)\omega_0\alpha}d\alpha = e^{j(k-m)\omega_0\alpha}\left(\frac{1}{j(k-m)\omega_0}\right)\Big|_0^T$$

$$= \frac{1 - 1}{(k-m)\omega_0} = 0. \qquad (3.2.9a) \blacksquare$$

$$k = m : \int_T e^{jk\omega_0\alpha}e^{-jm\omega_0\alpha}d\alpha$$

$$\qquad (3.2.9b)$$

$$= \int_0^T d\alpha = T, \quad E_k = T = E.$$

Example 3.2.4 Test the orthogonality over the interval (t_0, t_1) of the set of functions

$$\{\phi_k(t), k = 0, 1, 2, \ldots\} = \{1, t, t^2, \ldots\}. \qquad (3.2.10)$$

Solution: First

$$\int_{t_0}^{t_1} \phi_1(\alpha)\phi_2(\alpha)d\alpha = \int_{t_0}^{t_1} \alpha d\alpha = \frac{1}{2}[t_1^2 - t_0^2].$$

The functions $\phi_1(t)$ and $\phi_2(t)$ are orthogonal if $t_1 = -t_0$ or if $t_1 = t_0$. Now consider the two functions $\phi_0(t) = 1$ and $\phi_2(t) = t^2$. With these, we have

$$\int_{t_0}^{t_1} \phi_0(\alpha)\phi_2(\alpha)d\alpha = \int_{t_0}^{t_1} \alpha^2 d\alpha = \frac{1}{3}(t_1^3 - t_0^3).$$

The functions $\phi_0(t)$ and $\phi_2(t)$ are not orthogonal over any interval, except in the trivial case of $t_0 = t_1$. The set in (3.2.10) is not a good set to represent all signals. ∎

3.2.1 Gram–Schmidt Orthogonalization

Consider a linear set of independent real functions, $\phi_1(t), \phi_2(t), \ldots, \phi_k(t), \ldots$, defined on the interval $[a, b]$. Now define a new set of functions $\{\varphi_1(t), \varphi_2(t), \ldots, \varphi_k(t), \ldots\}$ by

$$\varphi_1(t) = \phi_1(t),$$

$$\varphi_2(t) = \phi_2(t) - \frac{\int_a^b \varphi_1(\alpha)\phi_2(\alpha)d\alpha}{\int_a^b \varphi_1^2(\alpha)d\alpha} \varphi_1(t)$$

$$\varphi_3(t) = \phi_3(t) - \frac{\int_a^b \varphi_1(\alpha)\phi_3(\alpha)d\alpha}{\int_a^b \varphi_1^2(\alpha)d\alpha}$$

$$\varphi_1(t) - \frac{\int_a^b \varphi_2(\alpha)\phi_3(\alpha)d\alpha}{\int_a^b \varphi_2^2(\alpha)d\alpha} \varphi_2(t), \ldots \quad (3.2.11)$$

This process of generating an orthogonal set of functions starting with an independent set is called the *Gram–Schmidt orthogonalization process*.

Example 3.2.5 Use the Gram–Schmidt process to generate an orthogonal basis set, $\{P_n(t), n = 0, 1, 2, 3, \ldots\}$, in the interval $-1 \le t \le 1$ using (3.2.11) in Example 3.2.4.

Solution: First two are

$$P_0(t) = \phi_0(t) = 1.$$

$$P_1(t) = \phi_1(t) - \frac{\int_{-1}^{1} P_0(\alpha)\phi_1(\alpha)d\alpha}{\int_{-1}^{1} P_0^2(\alpha)d\alpha}$$

$$P_0(t) = t - \frac{\int_{-1}^{1} (1)\alpha d\alpha}{\int_{-1}^{1} d\alpha} P_0(t) = t.$$

Similarly we can determine $P_2(t) = t^2 - (1/3)$. We can multiply these polynomials by a constant since multiplying a polynomial in the set by a constant does not change the orthogonality of the polynomials. The above process generates the *Legendre polynomials* within a constant. The first five Legendre polynomials are listed below:

$$L_0(t) = 1,$$
$$L_1(t) = t,$$
$$L_2(t) = (1/2)(3t^2 - 1),$$
$$L_3(t) = (1/2)(5t^3 - 3t),$$
$$L_4(t) = (1/8)(35t^4 - 30t^2 + 3).$$

Note the constant factors between $P_i(t)$ and $L_i(t)$. These polynomials can be generated by *Rodrigue's formula* Spiegel (1968):

$$L_k(t) = \frac{1}{2^k k!} \frac{d(t^2 - 1)^k}{dt^k}, k = 0, 1, 2, 3, \ldots. \quad (3.2.12a)$$

The polynomials generated by this process are referred to as *special Legendre polynomials*. Note the subscript k is used as an index, which is different from p used in the L_p measures. They satisfy the orthogonality property

$$\int_{-1}^{1} L_m(\alpha)L_k(\alpha)d\alpha = \begin{cases} 0, & m \ne k \\ E_k = \frac{2}{(2k+1)}, & m = k \end{cases}.$$

$$(3.2.12b) \quad ∎$$

Example 3.2.6 Show the set of periodic functions given below is an orthogonal basis set over one period and compute the energy in each of the basis functions in one period:

$$\{1, \cos(\omega_0 t), \cos(2\omega_0 t), \ldots, \cos(k\omega_0 t), \ldots,$$
$$\sin(\omega_0 t), \sin(2\omega_0 t), \ldots, \sin(k\omega_0 t), \ldots\}. \qquad (3.2.13)$$

Solution: The members of the set are periodic with period $T = 2\pi/\omega_0$ and we need to show (3.2.1) using the members of the given set:

$$\int_T (1)d\alpha = T, \quad \int_T (1)\cos(k\omega_0\alpha)d\alpha = 0, \quad \int_T (1)\sin(k\omega_0\alpha)d\alpha = 0, \quad k = 1, 2, \ldots. \qquad (3.2.14a)$$

Using trigonometric identities, we have

$$\int_T \cos(k\omega_0\alpha)\cos(m\omega_0\alpha)d\alpha = \begin{cases} \int_T \cos^2(k\omega_0\alpha)d\alpha = \frac{1}{2}\int_T d\alpha + \frac{1}{2}\int_T \cos(2k\omega_0\alpha)d\alpha = \frac{T}{2}, \; k = m \\ \frac{1}{2}\int_T \cos((k+m)\omega_0\alpha)d\alpha + \frac{1}{2}\int_T \cos((k-m)\omega_0\alpha)d\alpha = 0, \; k \neq m \end{cases}.$$

$$\int_T \sin(k\omega_0\alpha)\sin(m\omega_0\alpha)d\alpha = \frac{1}{2}\int_T \cos(k-m)\omega_0\alpha d\alpha - \frac{1}{2}\int_T \cos(k+m)\omega_0\alpha d\alpha = \begin{cases} \frac{T}{2}, k = m \\ 0, k \neq m \end{cases}.$$

$$\int_T \sin(k\omega_0\alpha)\cos(m\omega_0\alpha)d\alpha = \frac{1}{2}\int_T \sin((k+m)\omega_0\alpha)d\alpha + \frac{1}{2}\int_T \sin((k-m)\omega_0\alpha)d\alpha = 0,$$
$$(3.2.14b)$$

for all k and m.

These prove that the set in (3.2.13) is an orthogonal set. The *energies* contained in the members of the basis set *in one period* are as follows:

$$(E)_1 = T, (E)_{\text{sine or a cosine function}}$$
$$= T/2. \qquad (3.2.15) \; \blacksquare$$

The set is an orthogonal set and not orthonormal set. There are many other basis sets.

3.3 Approximation Measures

We are interested in approximating a given function $x(t)$ over an interval $(t_0, t_0 + T)$ by $x_a(t)$ using a set of *orthogonal basis functions*. How do we measure the approximation and then how good is the approximation? It can be measured by the error $[x(t) - x_a(t)]$. Figure 3.3.1 illustrates an example where $x(t)$ is the given function and its approximation is $x_a(t)$. The hatched area represents the error.

Since the functions can be complex and a positive error is just as bad as a negative error, and to make it general, we would like to consider the magnitude of

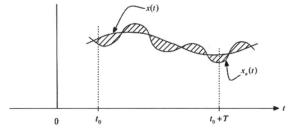

Fig. 3.3.1 $x(t)-$ Given function, $x_a(t)-$ Function approximating $x(t)$

the error. In addition, if the error measure is a number, we can compare and evaluate a particular approximation with respect to a number of basis sets. These goals can be achieved by considering the integral of the pth power of the magnitude of the error function, i.e.,

$$\int_{t_0}^{t_0+T} |x(t) - x_a(t)|^p dt, 1 \leq p. \qquad (3.3.1)$$

Notes: There is a good deal of interest in the area of inverse problems, such as deconvolution of signals based on $L_p, 1 \geq p$ measures. For a review see Tarantola (1987) and Hassan et al. (1994). Statisticians have investigated the L_p measures based on probabilistic behavior of signals. Our discussion here does *not* involve any details of statistical analysis. For readers interested in statistical details of these measures, p is selected based upon the *kurtosis* defined as the fourth moment normalized by the square of the variance of a probability density function Money (1982). In Section 1.7 we have considered three important density functions, uniform, Gaussian, and Laplacian. These can be used in selecting the value of p. Kurtosis values are k = 1.8 (Uniform, 3 (Gaussian), and 6 (Laplacian). The constant p is selected by

$$p = \left(\frac{9}{k^2 + 1}\right), 1 \le k \le \infty,$$

$$\left\{\begin{array}{c} k > 3.8 \text{ use } L_1 \\ 2.2 < k < 3.8 \text{ use } L_2 \\ k < 2.2 \text{ use } L_\infty \end{array}\right\}. \qquad (3.3.2)$$

L_p measures are used for speech, seismic, radar, and other signal coding. For example, for vowel sounds, L_1 is preferable and for nonvowel sounds, L_2 is preferable, see Lansford and Yarlagadda (1988). Since seismic signals have spiky noise, L_1 seems to work well, see Yarlagadda et al. (1985). See Schroeder and Yarlagadda (1989) on spectral estimation using L_1 norm. An old adage is if you have a fork in the road and have a choice to select either L_1 or L_2 measure, L_2 tells you to go in the middle of the two roads, not one of the two possible paths, whereas L_1 suggests taking one or the other paths. L_2 measure thinks like a machine, whereas L_1 measure thinks like a human, see Problem 3.2.2 at the end of the chapter. For most applications, L_2 the least-squares error measure is adequate and simple to use.

In Section 1.7.1, the Gaussian probability density function was introduced, see (1.7.12). Writing it terms of the error ε with mean 0 and variance σ_ε^2, the density function is

$$f_\varepsilon(\varepsilon) = \frac{1}{\sqrt{2\pi\sigma_\varepsilon^2}} e^{-\varepsilon^2/2\sigma_\varepsilon^2}. \qquad (3.3.3)$$

Minimizing the error (see ε^2 in (3.3.a)) corresponds to maximizing the probability, thus

providing a mathematical basis for the least-squares approach.   ∎

The *mean-squared error (MSE)* is defined by considering $(2N + 1)$ terms, we have

$$\text{MSE} = \frac{1}{T} \int_{t_0}^{t_0+T} |x(t) - x_a(t)|^2 dt, \ x_a(t)$$

$$= \sum_{k=-N}^{N} c[k]\phi_k(t) \equiv S_{2N+1}. \qquad (3.3.4a)$$

The interval T will be the same for different approximations in a particular situation and the normalization constant can be omitted and compare the approximations by using an orthogonal basis set $\{\phi_k(t)\}$ and the *integral-squared error (ISE)*, i.e.,

$$\text{ISE} = \int_{t_0}^{t_0+T} |x(t) - x_a(t)|^2 dt$$

$$= \int_{t_0}^{t_0+T} \left| x(t) - \sum_{k=-N}^{N} c[k]\phi_k(t) \right|^2 dt \equiv \varepsilon_{2N+1}. \qquad (3.3.4b)$$

The subscript on $\varepsilon, 2N + 1$ corresponds to the number of terms in the approximation and some of the coefficients $c[k]$ may be zero and N could go to infinity. It is convenient to consider odd number of terms. Since $c[k]$'s are unknowns, there is no loss in generality. The constants $c[k]$ can be determined from the basis set $\{\phi_k(t), k = -N, \ldots, -2, -1, 0, 1, 2, \ldots, N\}$ by minimizing the ISE.

We will consider two ways of computing the constants that minimize the integral-squared error. The first one is based upon taking the *partials* of the ISE with respect to $c[k]$, equating the partials to zero, and then solving for them. The second one is based on using *perfect squares* by rewriting the ISE in terms of two parts. First term is independent of $c[k]$ and the second is a sum of perfect square terms involving $c[k]$. Equating the perfect square terms to zero and solving for $c[k]$ give the desired result.

3.3.1 Computation of c[k] Based on Partials

First,

$$
\varepsilon_{2N+1} = \int_{t_0}^{t_0+T} \left| x(t) - \sum_{k=-N}^{N} c[k]\phi_k(t) \right|^2 dt
$$

$$
= \int_{t_0}^{t_0+T} |x(t) - S_{2N+1}|^2 dt
$$

$$
= \int_{t_0}^{t_0+T} \left[x(t) - \sum_{m=-N}^{N} c[m]\phi_m(t) \right]
$$

$$
\times \left[x^*(t) - \sum_{k=-N}^{N} c^*[k]\phi_k^*(t) \right] dt. \quad (3.3.5)
$$

Note the two different variables k and m in the above summations. This allows us to keep track of the terms in the summation products. Multiplying the product terms and since the integral of a sum is equal to the sum of the integrals, we have

$$
\varepsilon_{2N+1} = \int_{t_0}^{t_0+T} x(t)x^*(t)dt - \sum_{m=-N}^{N} c[m] \int_{t_0}^{t_0+T} x^*(t)\phi_m(t)dt
$$

$$
- \sum_{k=N}^{N} c^*[k] \int_{t_0}^{t_0+T} x(t)\phi_k^*(t)dt
$$

$$
+ \sum_{k=-N}^{N}\sum_{m=-N}^{N} c[m]c^*[k] \int_{t_0}^{t_0+T} \phi_m(t)\phi_k^*(t)dt.
$$

$$
(3.3.6)
$$

Take the partial derivatives of ε_{2N+1} with respect to $c[m]$ and equate them to zero:

$$
\frac{\partial \varepsilon_{2N+1}}{\partial c[m]} = \int_{t_0}^{t_0+T} [-x^*(t)\phi_m(t)]dt
$$

$$
+ \sum_{k=-N}^{N} c*[k] \int_{t_0}^{t_0+T} \phi_m(t)\phi_k^*(t)dt = 0. \quad (3.3.7)
$$

Orthogonality of the basis functions implies

$$
\int_{t_0}^{t_0+T} \phi_k(t)\phi_m^*(t)dt = \begin{cases} E_k > 0 \text{ and real, } k=m \\ 0, \quad \text{otherwise} \end{cases}. \quad (3.3.8)
$$

Using this in (3.3.6) results in

$$
\int_{t_0}^{t_0+T} x^*(t)\phi_k(t)dt + c*[k](E_k) = 0.
$$

The coefficients $c[k]$ are the *generalized Fourier series coefficients* giving an *explicit formula* given below to compute $c[k]$ given $x(t)$ *and* the orthogonal basis set $\{\varphi_k(t)\}$:

$$
c[k] = \frac{1}{E_k} \int_{t_0}^{t_0+T} x(t)\phi_k^*(t)dt. \quad (3.3.9)
$$

3.3.2 Computation of c[k] Using the Method of Perfect Squares

Example 3.3.1 Consider the second-order polynomial $y(t) = t^2 + bt$. Determine the minimum value of $y(t)$ by having a part of the expression that is a perfect square.

Solution: By adding and subtracting the term $(b/2)^2$ to $y(t)$, we have

$$
y(t) = t^2 + bt = t^2 + bt + \left(\frac{b}{2}\right)^2 - \left(\frac{b}{2}\right)^2
$$

$$
= \left(t + \frac{b}{2}\right)^2 - \left(\frac{b}{2}\right)^2.
$$

This function takes a minimum value for $t = -b/2$ and $y(t) = -(b/2)^2$. ∎

This idea can be used in minimizing the integral-squared error, See Ziemer and Tranter (2002). *Add and subtract* the following term to ε_{2N+1} in (3.3.6):

$$
\sum_{k=-N}^{N} \frac{1}{E_k} \left| \int_{t_0}^{t_0+T} x(t)\phi_k^*(t)dt \right|^2.
$$

$$
\varepsilon_{(2N+1)} = \int_{t_0}^{t_0+T} |x(t)|^2 dt + \left\{ -\sum_{k=-N}^{N} c[k] \int_{t_0}^{t_0+T} x^*(t)\phi_k(t)dt \right.
$$

$$
- \sum_{k=-N}^{N} c*[k] \int_{t_0}^{t_0+T} x(t)\phi_k^*(t)dt
$$

$$+ \sum_{k=-\infty}^{\infty} \frac{1}{E_k} \left| \int_{t_0}^{t_0+T} x(t)\phi_k^*(t)dt \right|^2 + \sum_{k=-\infty}^{\infty} E_k(c[k]c*[k]) \}$$

$$- \sum_{k=-\infty}^{\infty} \frac{1}{E_k} \left| \int_{t_0}^{t_0+T} x(t)\phi_k^*(t)dt \right|^2. \qquad (3.3.10)$$

The terms inside the brackets {.} can be expressed as a sum of squares and

$$\varepsilon_{2N+1} = \int_{t_0}^{t_0+T} |x(t)|^2 dt + \sum_{k=-N}^{N}$$

$$\left| \sqrt{E_k}c[k] - \frac{1}{\sqrt{E_k}} \int_{t_0}^{t_0+T} x(t)\phi_k^*(t)dt \right|^2$$

$$- \sum_{k=-N}^{N} \frac{1}{E_k} \left| \int_{t_0}^{t_0+T} x(t)\phi_k^*(t)dt \right|^2. \qquad (3.3.11)$$

To show that (3.3.11) reduces to (3.3.10), expand the middle term on the right and then cancel the equal terms that have opposite signs. The parameters $c[k]$ are not in the first and the third terms in (3.3.11). It is only included in the middle term, which is a sum of absolute values. Therefore, minimization of ε_{2N+1} is achieved when the middle term is zero and $c[k]$ is given by (3.3.9).

3.3.3 Parseval's Theorem

The minimum ISE with $(2N+1)$ terms is (see (3.3.5))

$$\varepsilon_{2N+1} = \int_{t_0}^{t_0+T} |x(t)|^2 dt - \sum_{k=-N}^{N} E_k|c[k]|^2$$

$$= \frac{1}{T} \int_{t_0}^{t_0+T} |x(t) - S_{2N+1}|^2 dt \geq 0. \qquad (3.3.12)$$

As N increases, the partial the quantity (ε_{2N+1}) can only decrease. Therefore, as N increases, the partial sums of the F-series give a closer and closer

approximation to the function $x(t)$, only in the sense that the approximation gives a smaller mean square error. In the limit, for any complete set of basis functions defined below:

$$\lim_{N \to \infty} \varepsilon_{2N+1} = 0. \qquad (3.3.13)$$

From (3.3.12), we have *Parseval's equation, or formula, or identity* given by

$$\int_{t_0}^{t_0+T} |x(t)|^2 dt = \sum_{k=-\infty}^{\infty} E_k|c[k]|^2. \qquad (3.3.14)$$

Summary: Given a time-limited function $x(t)$, $t_0 \leq t < t_0 + T$ and a set of orthogonal basis functions $\{\phi_k(t), k = 0, \pm1, \pm2, \ldots\}$, the function $x(t)$ is approximated by

$$x_a(t) \simeq \sum_{k=-\infty}^{\infty} c[k]\phi_k(t),$$

$$c[k] = \frac{1}{E_k} \int_{t_0}^{t_0+T} x(t)\phi_k^*(t)dt \qquad (3.3.15)$$

and the integral-squared error is equal to zero. The function $x_a(t)$ is an *approximation* of $x(t)$ in the identified interval. Only *in the sense that the integral-squared error goes to zero*, we write the equality of the given function to the approximate function by

$$x(t) = x_a(t) = \sum_{k=-\infty}^{\infty} c[k]\phi_k(t). \qquad (3.3.16)$$

It should be emphasized that $x(t)$ and $x_a(t)$ are *not equal in the true sense*. Differences between these two functions will be considered a bit later. In simple terms, the coefficients $c[k]$ of the generalized Fourier series are

$$c[k] = \left[\frac{1}{\substack{E_k, \text{Energy in } \phi_k(t) \\ \text{in the interval } (t_0, t_0 + T)}} \right] \int_{t_0}^{t_0+T} x(t) \left[\substack{\text{(conjugate of the} \\ \text{basis function, } \phi_k^*(t))} \right] dt. \qquad (3.3.17a)$$

The error in (3.3.12) with $(2N+1)$ coefficients in the F-series expansion is equal to

$$E_{2N+1} = (\text{Energy in the given } T \text{ second interval of the function})$$

$$- \left[\sum_{k=-N}^{N} (\text{Squared magnitude of the } k\text{th coefficient}) (\text{Energy in the } k^{\text{th}} \text{basis function}) \right]$$

$$= \int_{t_0}^{t_0+T} |x_T(t)|^2 dt - \sum_{k=-N}^{N} |c[k]|^2 E_k = \text{ISE}. \quad (3.3.17b)$$

Later, the convergence of the approximated signal $x_a(t)$ to $x(t)$ in terms of the number of coefficients in the approximation will be considered. The signal $x(t)$ is assumed to satisfy the *Dirichlet* conditions. The value of the integral in (3.3.17b) is unique and the generalized Fourier series is *unique* for a given set of basis functions. Let us illustrate the above by a detailed example, see Simpson and Houts (1971). ∎

Example 3.3.2 Find the generalized Fourier series expansion of the function $x(t)$ given below in Fig. 3.3.2 using the three orthogonal basis functions in (3.2.5). Then compute the mean-squared error using the direct method and Parseval's theorem.

Solution: The expression for the time function $x(t)$ and the three basis functions are

$$x(t) = \frac{3}{G} t \, \Pi \left[\frac{t - \frac{T}{2}}{T} \right],$$

$$\phi_1(t) = A\Pi \left[\frac{t - \frac{T}{6}}{T/3} \right],$$

$$\phi_2(t) = A\Pi \left[\frac{t - \frac{T}{2}}{T/3} \right], \text{ and}$$

$$\phi_3(t) = A\Pi \left[\frac{t - \frac{5T}{6}}{T/3} \right]. \quad (3.3.18)$$

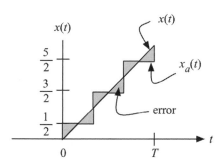

$x(t)$

$\frac{5}{2}$

$\frac{3}{2}$

$\frac{1}{2}$

$x(t)$

$x_a(t)$

error

0 T t

Fig. 3.3.2 $x(t)$, $x_a(t)$ and the error between these two functions

Earlier we have shown that $\phi_i(t), i = 1, 2, 3$ form an orthogonal basis set with $E_k = A^2 T/3, k = 1, 2, 3$, see Example 3.2.2. Noting that the time interval of the given function is $0 \le t < T$ and from (3.3.9), the coefficients are as follows:

$$c[1] = \frac{1}{E_1} \int_{t_0=0}^{t_0+T=T} x(t)\phi_1^*(t)dt$$

$$= \frac{1}{A^2(T/3)} \int_0^T \frac{3}{T} t \phi_1(t)dt$$

$$= \frac{1}{A^2(T/3)} \frac{3}{T} \int_0^{T/3} t.A dt$$

$$= \frac{1}{A^2} \left(\frac{3}{T}\right) \left(\frac{3}{T}\right) A \left(\frac{t^2}{2}\right) \Big|_{t=0}^{t=\frac{T}{3}} = \frac{1}{2A}$$

$$c[2] = \frac{1}{A^2\left(\frac{T}{3}\right)} \int_{T/3}^{2T/3} \left(\frac{3}{T}t\right) A dt = \frac{3}{2A}, c[3] = \frac{5}{2A}.$$

$$x_a(t) = c[1]\phi_1(t) + c[2]\phi_2(t) + c[3]\phi_3(t)$$

$$= \frac{1}{2}\Pi\left[\frac{t - T/6}{T/3}\right] + \frac{3}{2}\Pi\left[\frac{t - T/2}{T/3}\right]$$

$$+ \frac{5}{2}\Pi\left[\frac{t - 5t/6}{T/3}\right]. \quad (3.3.19)$$

The functions $x(t)$ and $x_a(t)$ are sketched in Fig. 3.3.2. The error is shown by the hatched marks and it has six equal parts. Clearly,

$$\text{ISE} = 6\int_0^{T/6} [x(t) - x_a(t)]^2 dt = 6\int_0^{T/6} \left(\frac{3}{T}t - \frac{1}{2}\right)^2 dt$$

$$= 6\int_0^{T/6} \left(\frac{9t^2}{T^2} + \frac{1}{4} - \frac{3t}{T}\right) dt$$

$$= 6\left[\frac{1}{4}t - \frac{3}{T}\frac{t^2}{2} + \frac{9}{T^2}\frac{t^3}{3}\right]\Big|_0^{T/6} = \frac{T}{12} \quad (3.3.20a)$$

This procedure is complicated and unnecessary since some of this work has already been done in deriving the generalized Fourier series coefficients. Using Parseval's theorem and (3.3.17b) results in the same value as in (3.3.20a) and is shown below.

$$\text{ISE} = \int_0^T \left(\frac{3}{T}t\right)^2 dt - A^2 \left(\frac{T}{3}\right)\left[\left(\frac{1}{2A}\right)^2 + \left(\frac{3}{2A}\right)^2 + \left(\frac{5}{2A}\right)^2\right]$$

$$= 3T - \frac{35}{12}T = \frac{T}{12}. \qquad (3.3.20b) \ \blacksquare$$

3.4 Fourier Series

The generalized Fourier series developed earlier can be used to study both the complex and the trigonometric Fourier series of a periodic function $x_T(t)$ with period of T seconds. The functions $e^{jn\omega_0 t}$, $\sin(n\omega_0 t)$, and $\cos(n\omega_0 t)$ are *nice* functions in the sense that all the derivatives of these functions exist. Approximation of a given function by the *Fourier series (F-series)* gives a smooth function even when the function being approximated has discontinuities.

3.4.1 Complex Fourier Series

The complex Fourier series makes use of the basis functions $\{e^{jk\omega_0 t}, k = 0, \pm 1, \pm 2, \dots\}$ given in (3.2.7a) with $\omega_0 = 2\pi/T$. The basis functions are periodic and the energy in one period is $E_k = T$ (see (3.2.9b)). The complex Fourier series expansion of a periodic function $x_T(t)$ that is either real or complex with the period T is identified by

$$x_a(t) = \sum_{k=-\infty}^{\infty} X_s[k]e^{jk\omega_0 t},$$

$$X_s[k] = \frac{1}{T}\int_T x_T(t)e^{-jk\omega_0 t} dt. \qquad (3.4.1a)$$

See (3.3.15) with $X_s[k] = c[k]$ and $\phi_k(t) = e^{jk\omega_0 t}$. $x_a(t)$ is a periodic function and we are approximating the given function $x_T(t)$ in an interval $(t_0, t_0 + T)$. Some authors do *not* show the subscript T on x

and should be evident from the context. The equality is only true in the sense that the integral-squared error between the periodic function and the F-series in the given interval is zero. The given function and the corresponding Fourier series coefficients are identified by the symbolic notation:

$$x_T(t) \xleftrightarrow{\text{FS},T} X_s[k] \qquad (3.4.1b)$$

The subscript s on X is used to denote that the coefficients are the complex F-series coefficients. Note the difference in the sign of the exponents in the sum and the integral expressions in (3.4.1a). Fourier coefficients are computed by an integral and the integral has a unique value. That is, *F-series expansion is unique*. The complex F-series can be used to approximate an aperiodic function in a time interval $(t_0, t_0 + T)$, where t_0 is arbitrary and T is the interval of the function that is under consideration. The approximation, in terms of periodic basis functions, will be valid only in the given time interval and, outside this interval, it is not valid. Complex Fourier series is applicable to both real and complex functions. When the function $x_T(t)$ is *real*, the F-series coefficients $X_s[k]$ and $X_s[-k]$ are related.

$$X_s[-k] = \frac{1}{T}\int_T x_T(t)e^{jk\omega_0 t} dt$$

$$= \left\{\frac{1}{T}\int_T \left[x_T(t)e^{-jk\omega_0 t}\right] dt\right\}^* = X_s^*[k]. \qquad (3.4.1c)$$

In Section 1.3.1, we have seen that the computation of the integrals can be simplified for even and odd functions. An arbitrary real periodic function $x_T(t)$ can be written in terms of its even and odd parts, $x_{Te}(t)$ and $x_{T0}(t)$, respectively, see (1.2.8).

$$x_T(t) = x_{Te}(t) + x_{T0}(t),$$
$$x_{Te}(t) = .5[x_T(t) + x_T(-t)],$$
$$x_{T0}(t) = .5[x_T(t) - x_T(-t)]. \qquad (3.4.2)$$

Example 3.4.1 If $x_T(t)$ is real and even, show the complex F-series coefficients are real and even. If $x_T(t)$ is real and odd, show the complex F-series coefficients are imaginary.

Solution: These can be seen from the following:

$$X_s[k] = \frac{1}{T}\int_T x_T(t)e^{-jk\omega_0 t}$$

$$dt = \frac{1}{T}\int_T x_T(t)\cos(k\omega_0 t)dt$$

$$-\frac{j}{T}\int_T x_T(t)\sin(k\omega_0 t)dt.$$

If $x_T(t)$ is real and even and since $\sin(n\omega_0 t)$ is odd, the integrand in the second integral in the above equation is odd and therefore it is zero and $X_s[k]$ is real and even. That is,

$$X_s[k] = \frac{1}{T}\int_T x_T(t)\cos(k\omega_0 t)dt \Rightarrow X_s[-k]$$

$$= X_s[k]. \quad (3.4.3)$$

In the case of $x_T(t)$ being odd, the product $x_T(t)\cos(n\omega_0 t)$ is odd and $X_s[-k] = -X_s[k]$.

$$X_s[k] = \frac{-j}{T}\int_T x_T(t)\sin(k\omega_0 t)dt. \quad (3.4.4) \blacksquare$$

Summary:

$$x_{Te}(t) \overset{FS,T}{\longleftrightarrow} X_s[k] = X_s[-k], x_{T0}(t) \overset{FS,T}{\longleftrightarrow} X_s[k]$$

$$= -X_s[-k]. \quad (3.4.5)$$

Example 3.4.2 Find the complex F-series of the pulse sequence shown in Fig. 3.4.1.

Solution: Using the expression in (3.4.1a) and using sinc functions defined in (1.2.15), we have the following. Note $\omega_0 = 2\pi f_0$, $f_0 = (1/T)$.

$$X_s[k] = \frac{1}{T}\int_T x_T(t)e^{-jk\omega_0 t}dt = \frac{1}{T}\int_{-\tau/2}^{\tau/2} Ae^{-jk\omega_0 t}$$

$$dt = \frac{A}{T}\left(\frac{1}{-jk\omega_0}\right)e^{-jk\omega_0 t}\Big|_{t=-\tau/2}^{t=\tau/2}.$$

$$= \frac{A\tau}{T}\frac{\left(e^{jk\omega_0\tau/2} - e^{-jk\omega_0 T/2}\right)}{2jk\omega_0\frac{\tau}{2}} = \frac{A\tau}{T}\frac{\sin(k\omega_0\frac{\tau}{2})}{k\omega_0\frac{\tau}{2}}$$

$$= \frac{A\tau}{T}\frac{\sin(\pi k f_0\tau)}{\pi k f_0\tau} = \frac{A\tau}{T}\sin c\left(\frac{k\pi\tau}{T}\right) \quad (3.4.6)$$

The complex F-series coefficients are real as $x_T(t)$ is an even function and

$$x_T(t) = \sum_{k=-\infty}^{\infty} X_s[k]e^{jk\omega_0 t} = \sum_{k=-\infty}^{\infty}\frac{A\tau}{T}\text{sinc}\left(\frac{k\pi\tau}{T}\right)e^{jk\omega_0 t}$$

$$(3.4.7)$$

The complex Fourier coefficients are inversely proportional to k in this example, i.e.,

$$X_s[k] = \frac{A\tau}{T}\text{sinc}\left(\frac{k\pi\tau}{T}\right) = \frac{A\tau}{T}\frac{\sin(k\pi\tau/T)}{(k\pi\tau/T)} \propto 1/k.$$

$$(3.4.8)$$

Note that $|\sin(k\pi\tau/T)| \leq 1$. Equation (3.4.8) provides a measure of the rate of decay of the F-series coefficients. $\blacksquare$

We are interested in reconstructing a pulse waveform using the complex Fourier series, i.e., using exponential (later sine and cosine) functions. How many terms are needed to keep in the series to get a good approximation? This will be answered shortly.

Example 3.4.3 Consider the saw-tooth waveform shown in Fig. 3.4.2 expressed by

$$x_{2\pi}(t) = t, -\pi < t < \pi, x_{2\pi}(t) = x_{2\pi}(t+2\pi), T = 2\pi.$$

$$(3.4.9)$$

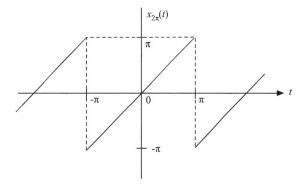

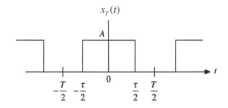

Fig. 3.4.1 Periodic pulse sequence $x_T(t)$

Fig. 3.4.2 Saw-tooth waveform

Solution: Since $\omega_0 = 2\pi/T$, we have $\omega_0 = 1$ and $x_{2\pi}(t)$ is an odd function, we have

$$X_s[0] = 0. \qquad (3.4.10)$$

Now use the integration by parts given below to compute $X_s[k], k \neq 0$:

$$\int y e^{ay} dy = \frac{e^{ay}}{a}\left(y - \frac{1}{a}\right). \qquad (3.4.11)$$

The coefficients are given by

$$X_s[k] = \frac{1}{2\pi}\int_{-\pi}^{\pi} t e^{-jkt} dt = \frac{e^{-jkt}}{-2\pi(jk)}(t + \frac{1}{jk})\big|_{t=-\pi}^{t=\pi}.$$

$$= -\frac{e^{-jk\pi}}{2\pi(jk)}\left[\pi + \frac{1}{jk}\right] + \frac{e^{jk\pi}}{2\pi(jk)}\left[-\pi + \frac{1}{jk}\right]$$

$$= -\frac{(-1)^k}{jk}, k \neq 0. \qquad (3.4.12)$$

Since $x_{2\pi}(t)$ is an odd function, the coefficients are pure imaginary and

$$x_{2\pi}(t) = \cdots - \frac{e^{j2t}}{2j} + \frac{e^{jt}}{j} + 0 + \frac{e^{-jt}}{j} - \frac{e^{-j2t}}{2j} - \cdots. \qquad (3.4.13a)$$

Euler's equation can be used and write (3.4.13a) in the form

$$x_{2\pi}(t) = 2\left[\frac{\sin(t)}{1} - \frac{\sin(2t)}{2} + \frac{\sin(3t)}{3} + \cdots\right]. \qquad (3.4.13b)$$

The F-series in (3.4.13b) contains only sine terms since $x_{2\pi}(t)$ is an odd function. It is more appealing since the given function is real. ∎

Example 3.4.4 Derive the complex F-series given the periodic impulse sequence

$$x_T(t) = \sum_{k=-\infty}^{\infty} A\delta(t - kT). \qquad (3.4.14)$$

Solution: The complex F-series coefficients are

$$X_s[k] = \frac{1}{T}\int_T x(t) e^{-jk\omega_0 t} dt$$

$$= \frac{1}{T}\int_{-T/2}^{T/2} A\delta(t) dt = \frac{A}{T}. \qquad (3.4.15a)$$

$$x_T(t) = \frac{A}{T}\sum_{k=-\infty}^{\infty} e^{jk\omega_0 t} = \frac{2A}{T}\sum_{k=0}^{\infty} \cos(k\omega_0 t). \qquad (3.4.15b)$$

That is, the F-series coefficients $X_s[k]$ of a periodic impulse sequence are *independent* of k. Since the impulse sequence is an even function, it follows that the coefficients have a dc term and cosine terms and no sine terms. Independence of k in this example will be used in determining the convergence of F-series coefficients a bit later. ∎

Example 3.4.5 Find the complex F-series of the trigonometric function $x_T(t) = \sin^2 \omega_0 t + 2\cos\omega_0 t$.

Solution: Using Euler's equation, we have

$$x_T(t) = \left[\frac{e^{j\omega_0 t} - e^{-j\omega_0 t}}{2j}\right]^2 + 2\left[\frac{e^{j\omega_0 t} + e^{-j\omega_0 t}}{2}\right]$$

$$= -\frac{1}{4}e^{-j2\omega_0 t} + e^{-j\omega_0 t} + \frac{1}{2} + e^{j\omega_0 t} - \frac{1}{4}e^{j2\omega_0 t}$$

$$= \sum_{k=-\infty}^{\infty} X_s[k] e^{jk\omega_0 t}$$

$$\Longrightarrow X_s[-2] = -.25, X_s[-1] = 1, X_s[0] = 1/2,$$
$$X_s[1] = 1, X_s[2] = -.25, X_s(k) = 0, |k| > 2.$$

Since the function is real and even, the above validates $X_s[-k] = X_s[k]$. The F-series coefficients are *unique*. Using trigonometric identity, it follows that

$$x_T(t) = (1/2) + 2\cos(\omega_0 t)$$
$$- (1/2)\sin(2\omega_0 t). \qquad ∎$$

The complex F-series expansion is applicable for both real and complex periodic functions and the complex F-series leads to Fourier transforms. Most functions, in reality, are real functions and the trigonometric F-series are more desirable.

3.4.2 Trigonometric Fourier Series

The set of basis functions for the trigonometric Fourier series with period $T = 2\pi/\omega_0$ is

$$\{1, \cos(k\omega_0 t), \sin(k\omega_0 t), k = 1, 2, 3, \ldots\}. \quad (3.4.16a)$$

Equation (3.2.15) gives the energy contents in one period of these functions and are

$$(E)_1 = T, (E) \text{ sine or a cosine function}$$
$$= T/2. \quad (3.4.16b)$$

In the trigonometric F-series, $X_s[0]$ for the dc term, $a[k]$ for the coefficients for the cosine terms, and $b[k]$ for the coefficients of the sine terms will be used. With (3.4.16b) and (3.3.17a), the following *trigonometric F-series* result:

$$x_T(t) = X_s[0] + \sum_{k=1}^{\infty} \{a[k]\cos(k\omega_0 t) + b[k]\sin(k\omega_0 t)\},$$
$$(3.4.17a)$$

$$\begin{cases} X_s[0] = \frac{1}{T}\int_T x_T(t)dt \\\\ a[k] = \frac{2}{T}\int_T x_T(t)\cos(k\omega_0 t)dt. \quad (3.4.17b) \\\\ b[k] = \frac{2}{T}\int_T x_T(t)\sin(k\omega_0 t)dt \end{cases}$$

Note the *factor 2* before the integrals in computing $a[k]$ and $b[k]$. Also, note $x_T(t)$ instead of $x(t)$ in (3.3.17a). If $x_T(t)$ is an arbitrary real periodic function, then it can be written in terms of its even and odd parts and they can be expressed in terms of $x_T(t)$:

$$x_T(t) = x_{Te}(t) + x_{T0}(t),$$
$$\Longrightarrow x_{Te}(t) = (x_T(t) + x_T(-t))/2,$$
$$x_{T0}(t) = (x_T(t) - x_T(-t))/2.$$

Since constant and the cosine terms are even and the sine terms are odd, the trigonometric F-series can be written for the even and odd periodic functions by

$$x_{Te}(t) = X_s[0] + \sum_{k=1}^{\infty} a[k]\cos(k\omega_0 t);$$

$$x_{T0}(t) = \sum_{k=1}^{\infty} b[k]\sin(k\omega_0 t). \quad (3.4.18)$$

3.4.3 Complex F-series and the Trigonometric F-series Coefficients-Relations

By using Euler's formula and comparing the results in (3.4.1a) and (3.4.17), we have

$$X_s[k] = \frac{1}{T}\int_T x(t)e^{-jk\omega_0 t}dt$$

$$= \frac{1}{T}\int_T x(t)\cos(k\omega_0 t)dt - \frac{j}{T}\int_T x(t)\sin(k\omega_0 t)dt$$

$$= \frac{a[k]}{2} - j\frac{b[k]}{2}. \quad (3.4.19a)$$

Since the cosine and sine functions are even and odd functions, we can see that

$$X_s[-k] = \frac{a[k]}{2} + j\frac{b[k]}{2}. \quad (3.4.19b)$$

$$\Rightarrow a[k] = X_s[k] + X_s[-k] \text{ and}$$
$$b[k] = j(X_s[k] - X_s[-k]). \quad (3.4.20)$$

3.4.4 Harmonic Form of Trigonometric Fourier Series

The trigonometric Fourier series given in (3.4.17a) can be written in the *harmonic (or compact form)* by

$$x_T(t) = X_s[0] + \sum_{k=1}^{\infty} d[k]\cos(k\omega_0 t + \theta[k]). \quad (3.4.21)$$

$$= X_s[0] + \sum_{k=1}^{\infty} d[k]\cos(\theta[k])\cos(k\omega_0 t)$$

$$+ \sum_{k=1}^{\infty} d[k](-\sin(\theta[k]))\sin(k\omega_0 t). \quad (3.4.22a)$$

Equating the terms in the trigonometric F-series in (3.4.17) to the above, we have

$$a[k] = d[k]\cos(\theta[k]),$$
$$b[k] = -d[k]\sin(\theta[k]);$$
$$d[k] = \sqrt{a^2[k] + b^2[k]},$$
$$\theta[k] = -\tan^{-1}\left(\frac{b[k]}{a[k]}\right). \quad (3.4.22b)$$

The harmonic form given in (3.4.21) contains the *dc* component $X_s[0]$, the *fundamental* $d[1]\cos(\omega_0 t + \theta[1])$, and $d[k]\cos(k\omega_0 t + \theta[k])$, the kth *harmonic component*. The trigonometric and harmonic F-series expansions are for real functions.

3.4.5 Parseval's Theorem Revisited

From (3.3.14), by noting $E_k = T$ and $X_s[k] = c[k]$, Parseval's theorem is expressed by

$$P = \frac{1}{T}\int_{t_0}^{t_0+T} |x_T(t)|^2 dt = \sum_{k=-\infty}^{\infty} |X_s[k]|^2. \quad (3.4.23)$$

Since $|X_s[k]| = |X_s[-k]|$, the average power in $x_T(t)$, in terms of the trigonometric F-series coefficients $X_s[0], a[k]$, and $b[k]$ and the harmonic coefficients $d[k]$, is

$$P = \sum_{k=-\infty}^{\infty} |X_s[k]|^2 = |X_s[0]|^2 + \frac{1}{4}\sum_{k=1}^{\infty} |a[k] + jb[k]|^2$$

$$+ \frac{1}{4}\sum_{k=1}^{\infty} |a[k] - jb[k]|^2$$

$$= |X_s[0]|^2 + \frac{1}{2}\sum_{k=1}^{\infty} (a^2[k] + b^2[k])$$

$$= |X_s[0]|^2 + \frac{1}{2}\sum_{k=1}^{\infty} d^2[k]. \quad (3.4.24a)$$

Table 3.4.1 Summary of the three Fourier series representations

Complex Fourier series: $x_T(t)$ – a complex or real periodic function

$$x_T(t) = \sum_{k=-\infty}^{\infty} X_s[k]e^{jk\omega_0 t}; \qquad \omega_0 = 2\pi/T.$$

$$X_s[k] = \frac{1}{T}\int_T x_T(t)e^{-jk\omega_0 t}dt.$$

Trigonometric Fourier series: $x_T(t)$ – a real periodic function

$$x_T(t) = X_s[0] + \sum_{k=1}^{\infty} (a[k]\cos(k\omega_0 t) + b[k]\sin(k\omega_0 t)).$$

$$X_s[0] = \frac{1}{T}\int_T x_T(t)dt; \quad a[k] = \frac{2}{T}\int_T x_T(t)\cos(k\omega_o t)dt; \quad b[k] = \frac{2}{T}\int_T x_T(t)\sin(k\omega_0 t)dt.$$

Harmonic (compact form) of Fourier series: $x_T(t)$ – a real periodic function

$$x_T(t) = X_s[0] + \sum_{k=1}^{\infty} d[k]\cos(k\omega_0 t + \theta[k]).$$

$$f_0 = \frac{\omega_0}{2\pi} = \frac{1}{T} = \text{Fundamental frequency}, \quad kf_0 = k^{th} \text{ harmonic frequency},$$

$$d[k] = \text{Amplitude of the } k^{th} \text{ harmonic}, \quad \theta[k] = \text{Phase angle of the } k^{th} \text{ harmonic}.$$

Conversion formulae:

$$X_s[k] = |X_s[k]|e^{j\theta[k]}.$$

$$X_s[k] = \frac{1}{2}(a[k] - jb[k]), \quad X_s[-k] = \frac{1}{2}(a[k] + jb[k]) = X^*[k].$$

$$a[k] = (X_s[k] + X_s[-k]) = 2\text{Re}\{X_s[k]\}, \quad b[k] = j(X_s[k] - X_s[-k]) = -2\text{Im}\{X[k]\}.$$

$$d(k) = \sqrt{a^2[k] + b^2[k]}, \quad \theta[k] = -\tan^{-1}\left(\frac{b[k]}{a[k]}\right).$$

$$a[k] = d[k]\cos(\theta[k]), \quad b[k] = -d[k]\sin(\theta[k]).$$

Parseval's theorem:

$$P = \sum_{k=-\infty}^{\infty} |X_s[k]|^2 = |X_s[0]|^2 + \frac{1}{2}\sum_{k=1}^{\infty} (a^2[k] + b^2[k]) = |X_s[0]|^2 + \frac{1}{2}\sum_{k=1}^{\infty} d^2[k].$$

$$\Longrightarrow \sum_{k=-n}^{n} |X_s[k]|^2 \leq P = \int_{t_0}^{t_0+T} |x_T(t)|^2 dt$$

(assumed to be finite). $\qquad$ (3.4.24b)

This bound is called *Bessel's inequality* and each coefficient magnitude must have a limit as $n \to \infty$, which is true for all n. Thus

$$\lim_{k\to\infty} |X_s[k]| = 0, \quad \lim_{k\to\infty} |a[k]| = 0, \text{ and}$$

$$\lim_{k\to\infty} |b[k]| = 0. \qquad (3.4.24c)$$

3.4.6 Advantages and Disadvantages of the Three Forms of Fourier Series

Complex F-series: Advantages: The *complex F-series* is applicable for real and complex functions, although most of the functions we deal with are real functions. The series can be extended conceptually to the aperiodic case leading to the Fourier transform discussed in Chapter 4. Making use of Euler's identities, the complex F-series of a real function can be written in terms of trigonometric F-series.

Disadvantages: The complex F-series is non-intuitive, as complex functions are used to approximate real functions. Since the complex F-series coefficients of a function are generally complex functions, it is hard to visualize a real function as a sum of complex exponential functions. Besides, to sketch the F-series approximation of the periodic function, sinusoidal functions need to be used. The complex F-series introduces negative frequencies since k in $X_s[k]$ can be positive or negative.

Trigonometric F-series: Advantages: The trigonometric series contains a constant and a set of sinusoids, cosine, and sine functions. It is the most used series in the literature. The constant and the cosine terms represent the even part of the periodic function and the sine terms represent the odd part of the periodic function.

Disadvantages: Since there are two Fourier series coefficients $a[k]$ and $b[k]$ for each frequency, the magnitude and phase spectra, at each frequency, need to be computed using both the coefficients. The trigonometric F-series are for real functions.

Harmonic form: Advantages: The harmonic form is compact and it has only positive frequencies and is simple to generate the magnitude and phase spectra, which are used to visualize the spectral behavior of the signal. We can see that if we like to transmit a segment of a signal using the F-series, we only need to send the amplitudes and phases of the harmonic form at the appropriate frequencies.

Disadvantages: It is used for real functions. The amplitudes and the phases of the harmonic coefficients are determined from one of the other two forms.

Notes: The F-series can be used to approximate a function over an arbitrary interval $t_0 \leq t < t_0 + T$, where t_0 and T are arbitrary. The approximation is *valid* only in that interval. Outside this interval, the series results in a periodic function. $\qquad\blacksquare$

3.5 Fourier Series of Functions with Simple Symmetries

Equation (3.4.18) gives explicit formulas for the F-series of even and odd functions.

Example 3.5.1 Find the trigonometric Fourier series of the function

$$x_T(t) = \begin{cases} -1, & -(T/2) < t < 0 \\ 1, & 0 < t < (T/2) \end{cases},$$

$$x_T(t) = x_T(t+T). \qquad (3.5.1)$$

Solution: This function is an odd function and the F-series contains only sine functions.

$$b[k] = \frac{2}{T} \int_{-T/2}^{T/2} x_T(t) \sin(k\omega_0 t) dt$$

$$= \frac{4}{T} \int_0^{T/2} x_T(t) \sin(k\omega_0 t) dt = \frac{4}{T} \int_0^{T/2} \sin(k\omega_0 t) dt$$

$$= \frac{4}{T} \left[\frac{1-\cos(k\omega_0(T/2))}{k\omega_0} \right] = \frac{4}{k(2\pi)}[1-\cos(k\pi)]$$

$$= \begin{cases} \dfrac{4}{k\pi}, & k-\text{odd} \\ 0, & k-\text{even} \end{cases}, \quad \omega_0 = \frac{2\pi}{T}.$$

$$x_T(t) = \sum_{k=1}^{\infty} b[k] \sin(k\omega_0 t) = \frac{4}{\pi}[\sin(\omega_0 t)$$

$$+ \frac{1}{3}\sin(3\omega_0 t) + \frac{1}{5}\sin(5\omega_0 t) + \ldots . \quad (3.5.2) \quad \blacksquare$$

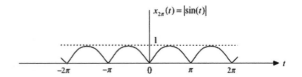

Fig. 3.5.1 Example 3.5.2

The decay rate of the nonzero F-series coefficients is inversely proportional to k and

$$b[k] \infty (1/k). \quad (3.5.3)$$

3.5.1 Simplification of the Fourier Series Coefficient Integral

A change of variable $\lambda = \omega_0 t$ in the computation of the Fourier series coefficients can be used to *standardize* the integral. As $t \rightarrow \pm T/2$, we have $\lambda \rightarrow \pm \pi$ and $dt = d\lambda/\omega_0$ and

$$X_s[k] = \frac{1}{T} \int_T x_T(t) e^{-jk\omega_0 t} dt$$

$$= \frac{1}{2\pi} \int_{2\pi} x_T(\lambda/\omega_0) e^{-jk\lambda} d\lambda. \quad (3.5.4a)$$

The complex Fourier coefficients are expressed in terms of an integral over a period 2π since $x_T(\lambda/\omega_0)$ is periodic with period 2π. Equation (3.5.4a) standardizes the F-series coefficients. Similarly, we can express the trigonometric Fourier series:

$$X_s[0] = \frac{1}{2\pi} \int_{2\pi} x_{2\pi}(\lambda/\omega_0) d\lambda,$$

$$a[k] = \frac{1}{\pi} \int_{2\pi} x_{2\pi}(\lambda/\omega_0) \cos(k\lambda) d\lambda,$$

$$b[k] = \frac{1}{\pi} \int_{2\pi} x_{2\pi}(\lambda/\omega_0) \sin(k\lambda) d\lambda. \quad (3.5.4b)$$

Example 3.5.2 Find the complex and the trigonometric Fourier series of the full-wave rectified function given by $x_{2\pi}(t) = |\sin(t)|$. That is $x_T(t) = |\sin(\omega_0 t)|$, $\omega_0 = 2\pi/T$, and $T/2 = \pi$ (the fundamental period), see Fig. 3.5.1.

Solution: The complex F-series coefficients are

$$X_s[k] = \frac{1}{T} \int_{-T/2}^{T/2} |\sin(\omega_0 t)| e^{-jk\omega_0 t} dt. \quad (3.5.5a)$$

$$= \frac{1}{T} \int_{-T/2}^{T/2} |\sin(\omega_0 t)| \cos(k\omega_0 t)$$

$$- j\frac{1}{T} \int_{-T/2}^{T/2} |\sin(\omega_0 t)| \sin(k\omega_0 t) dt. \quad (3.5.5b)$$

$$= \frac{2}{T} \int_0^{T/2} \sin(\omega_0 t) \cos(k\omega_0 t) dt. \quad (3.5.5c)$$

The integrand in the first integral in (3.5.5b) is an even function and therefore the integral can be computed by integrating for half the period and multiplying it by 2. The integrand in the second integral is an odd function and therefore the integral is zero. Using the change of variable $\alpha = \omega_0 t$ in (3.5.5c), we have $d\alpha = \omega_0 dt$ and $\alpha = \pm\pi$. Now

$$X_s[k] = \frac{1}{\pi} \int_0^{\pi} \sin(\alpha) \cos(k\alpha) d\alpha$$

$$= \frac{1}{2\pi} \int_0^{\pi} [\sin((1-k)\alpha) + \sin((1+k)\alpha)] d\alpha$$

$$= \frac{1}{2\pi} \left[\frac{1 - \cos((1-k)\pi)}{(1-k)} + \frac{1 - \cos((1+k)\pi)}{(1+k)} \right], k \neq 1.$$

$$(3.5.6)$$

For $k = \pm 1$, one of the terms in the above equation becomes indeterminate and the other term goes to zero. As $k \rightarrow 1$, using L'Hospital's rule results in

$$X_s[k]|_{k=1} = \lim_{k \rightarrow 1} \left[\frac{1 - \cos((1-k)\pi)}{1-k} + \frac{1 - \cos((1+k)\pi)}{(1+k)} \right]$$

$$= \lim_{k \rightarrow 1} \left[\frac{-(\pi)\sin((1-k)\pi)}{-1} + 0 \right] = 0.$$

From (3.4.1c), we have $X_s[-1] = 0$. Since $\cos[(1 \pm k)\pi] = 1$, the odd coefficients are zero. The complex F-series coefficients and the complex F-series expansion of the full-wave rectified signal are as follows:

$$X_s[k] = \begin{cases} 0, & k\text{-odd} \\ \dfrac{2}{(1-k^2)\pi}, & k\text{-even} \end{cases}. \quad (3.5.7)$$

$$x_T(t) = \frac{2}{\pi} \frac{1}{\pi}$$

$$\sum_{k \neq 0, k \text{ even}}^{\infty} \frac{2}{(k-1)(k+1)} \left[e^{jk\omega_0 t} + e^{-jk\omega_0 t} \right]. \quad (3.5.8)$$

By using Euler's formula, the trigonometric F-series are given by

$$x_T(t) = \frac{2}{\pi} - \frac{4}{\pi}$$

$$\sum_{k \text{ even}, k \neq 0}^{\infty} \frac{1}{(k-1)(k+1)} \cos(k\omega_0 t) = \frac{2}{\pi} + \frac{4}{\pi}$$

$$\sum_{n=1}^{\infty} \frac{1}{1 - 4n^2} \cos(2n\omega_0 t). \quad (3.5.9)$$

There are some interesting aspects to note. First, $x_T(t)$ is periodic with the *fundamental* period $T/2$. This follows from Fig. 3.5.1 and from (3.5.9), as the fundamental frequency is $f_0 = 2\omega_0/2\pi = 2/T$. So, the period of $T/2 = \pi$ could have been used and go through the expansion. Note the decay rate is

$$|X_s[k]| \infty (1/k^2). \quad (3.5.10) \quad \blacksquare$$

3.6 Operational Properties of Fourier Series

3.6.1 Principle of Superposition

Computing the F-series coefficients is an integral operation. The integral of a sum is equal to the sum of the integrals. Therefore, if α and β are some constants, we have

$$x_T(t) \overset{\text{FS},T}{\longleftrightarrow} X_s[k] \text{ and } y_T(t) \overset{\text{FS},T}{\longleftrightarrow} Y_s[k]$$

$$\Longrightarrow \alpha x_T(t) + \beta y_T(t) \overset{\text{FS},T}{\longleftrightarrow}$$

$$\alpha X_s[k] + \beta Y_s[k]. \quad (3.6.1)$$

Fourier series coefficients are added with appropriate multipliers to get the Fourier series coefficients of the sum of two periodic functions.

3.6.2 Time Shift

The time shift of a signal $x_T(t)$ by τ seconds corresponds to $y_T(t) = x_T(t \pm \tau)$, $\tau > 0$, where the positive sign is for an advance and a negative sign for a delay. The F-series $y_T(t)$ is

$$y_T(t) = \sum_{k=-\infty}^{\infty} Y_s[k] e^{jk\omega_0 t},$$

$$Y_s[k] = \frac{1}{T} \int_T y_T(t) e^{-jk\omega_0 t} dt = \frac{1}{T} \int_T x_T(t \pm \tau) e^{-jk\omega_0 t} dt.$$

Using the change of variable $\beta = t \pm \tau$, we have $t = \beta \mp \tau$ and

$$Y_s[k] = \frac{1}{T} \int_T x_T(\beta) e^{-jk\omega_0 (\beta \mp \tau)} d\beta$$

$$= \left[\frac{1}{T} \int_T x_T(\beta) e^{-jk\omega_0 \beta} d\beta \right] e^{\pm jk\omega_0 \tau}$$

$$= X_s[k] e^{\pm jk\omega_0 \tau}. \quad (3.6.2)$$

Since the period remains the same, there is no need to consider in (3.6.2). In summary,

$$x_T(t+\tau) \overset{\text{FS},T}{\longleftrightarrow} X_s[k] e^{jk\omega_0 \tau}, x_T(t-\tau) \overset{\text{FS},T}{\longleftrightarrow}$$

$$X_s[k] e^{-jk\omega_0 \tau}, \tau > 0. \quad (3.6.3)$$

The magnitudes of the F-series coefficients of a periodic signal and its delayed (or advanced) version are the same. The delay (or advance) τ appears explicitly in the phase angle $(k\omega_0 \tau)$ for the advance and $(-k\omega_0 \tau)$ for the delay. In case of trigonometric F-series, for $\tau > 0$ (for $\tau < 0$, replace τ by $-\tau$ in the following.), we have

$$x_T(t-\tau) = X_s[0] + \sum_{k=1}^{\infty} [a[k]\cos(k\omega_0(t-\tau))$$

$$+ b[k]\sin(k\omega_0(t-\tau))] = X_s[0]$$

$$+ \sum_{k=1}^{\infty} [a[k]\cos(k\omega_0\tau) - b[k]\sin(k\omega_0\tau)]$$

$$\cos(k\omega_0 t) + \sum_{k=1}^{\infty} [a[k]\sin(k\omega_0\tau)$$

$$+ b[k]\cos(k\omega_0\tau)]\sin(k\omega_0 t). \qquad (3.6.4)$$

3.6.3 Time and Frequency Scaling

Given a function $x(t)$, its time-scaled version is given by $x(at), a > 0$. If $a < 1$, the signal is expanded and if $a > 1$, the signal is compressed (see (1.2.3)). The F-series of the scaled signal is obtained by replacing $t \to at$ in the complex F-series and

$$x_T(at) = \sum_{k=-\infty}^{\infty} X_s[k]e^{jk\omega_0(at)}. \qquad (3.6.5)$$

The *frequency locations* moved from $k\omega_0$ to $k(a\omega_0)$ and the F-series coefficients are $X_s[k]$. The time-scaled signal changes the period from T to (T/a). Harmonics are now located at $k\omega_0 a = k(2\pi)f_0 a = k(2\pi)(a/T)$. The more compressed the time function is, farther apart its harmonics are. Harmonics of the compressed signal are located at frequencies kf_0a and $a > 1$. If $a = -1$, then a time-reversed or a folded function results and

$$x_T(-t) = \sum_{k=-\infty}^{\infty} X_s[k]e^{-jk\omega_0 t} = \sum_{k=-\infty}^{\infty} \left[X_s^*[k]e^{jk\omega_0 t}\right]^*.$$

$$x_T(-t) \overset{FS,T}{\longleftrightarrow} X^*[k], |X_s[k]| = |X_s^*[k]|. \qquad (3.6.6)$$

Example 3.6.1 Using Example 3.5.2, find the trigonometric F-series of the full-wave rectified signal $x_{2\pi}(t) = |\sin(t)|$ assuming the period is 2π.

Solution: Substituting $T = 2\pi$, i.e., $\omega_0 = 1$ in (3.5.9), we have

$$x_{2\pi}(t) = \frac{2}{\pi} - \frac{4}{\pi}$$
$$\left[\frac{\cos(2t)}{1(3)} + \frac{\cos(4t)}{3(5)} + \frac{\cos(6t)}{5(7)} + \cdots\right]. \qquad (3.6.7) \blacksquare$$

Example 3.6.2 Find the trigonometric F-series of the signal below using the full-wave rectified signal series:

$$y_{2\pi}(t) = \begin{cases} \sin(t), & 0 < t < \pi \\ 0, & \pi < t < 2\pi \end{cases}. \qquad (3.6.8)$$

Solution: The full-wave rectified function $x_{2\pi}(t)$ was shown in Fig. 3.5.1 with period $T = 2\pi$. The half-wave rectified signal is shown in Fig. 3.6.1 and

$$y_{2\pi}(t) = [x_{2\pi}(t) + \sin(t)]/2. \qquad (3.6.9)$$

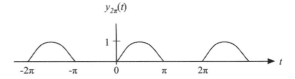

Fig. 3.6.1 $y_{2\pi}(t)$ Half-wave rectified signal

Using the linearity property of the Fourier series, we have

$$y_{2\pi}(t) = \frac{1}{\pi} + \frac{1}{2}\sin(t) - \frac{2}{\pi}$$
$$\left[\frac{\cos(2t)}{1(3)} + \frac{\cos(4t)}{3(5)} + \frac{\cos(6t)}{5(7)} + \cdots\right]. \qquad (3.6.10)$$

In Chapter 1, we studied the one-sided and the two-sided line spectra by expressing each term in terms of cosines and sines. Noting that $\cos(\alpha \pm 90^0) = \mp \sin(\alpha)$ and $\cos(\alpha \pm 180^0) = -\cos(\alpha)$, we can write

$$y_{2\pi}(t) = \frac{1}{\pi} - \frac{1}{2}\cos(t + 90^0) + \frac{2}{\pi}\left[\frac{\cos(2t - 180^0)}{1(3)} + \frac{\cos(4t - 180^0)}{3(5)} + \frac{\cos(6t - 180^0)}{5(7)} + \cdots\right]. \qquad (3.6.11)$$

$$x_{2\pi}(t) = \frac{2}{\pi} + \frac{4}{\pi}\left[\frac{\cos(2t - 180^0)}{1(3)} + \frac{\cos(4t - 180^0)}{3(5)} + \frac{\cos(6t - 180^0)}{5(7)} + \cdots\right]. \qquad (3.6.12)$$

These are the harmonic forms of the trigonometric Fourier series for the two given functions. The two-sided amplitude line spectra associated with these two functions are sketched in Fig. 3.6.2. The only difference between the two functions $x_{2\pi}(t)$ and $y_{2\pi}(t)$ is the component at the frequency $\omega_0 = 1$ or $f_0 = 1/2\pi$. If we can remove or *filter out* this frequency component from $y_{2\pi}(t)$, we can obtain the function $x_{2\pi}(t)$ illustrating one of the remarkable insights into the description of signals provided by the Fourier series. Filter design will be discussed in later chapters. ∎

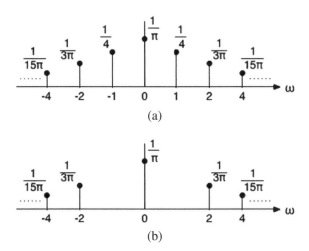

Fig. 3.6.2 Two-sided amplitude line spectra (a)$x_{2\pi}(t)$ and (b) $y_{2\pi}(t)$

3.6.4 Fourier Series Using Derivatives

All the derivatives of the F-series exist since all the derivatives of the functions $e^{jn\omega_0 t}$, $\sin(n\omega_0 t)$, and $\cos(n\omega_0 t)$ exist. In that sense, F-series is a *nice function*. The complex F-series of a periodic function and its derivative are

$$x_T(t) = \sum_{k=-\infty}^{\infty} X_s[k]e^{jk\omega_0 t},$$

$$x_T'(t) = \frac{dx_T(t)}{dt} = \sum_{\substack{k=-\infty \\ k \neq 0}}^{\infty}[X_s[k]j(k\omega_0)]e^{jk\omega_0 t}$$

The dc term in the F-series of the derivative goes to zero and the other coefficients are multiplied by $(jk\omega_0)$, $k \neq 0$. The spectral components of $x_T'(t)$ have significantly higher frequency content compared to the spectral components of $x_T(t)$. Note the F-series coefficients of $x_T(t)$ are multiplied by $jk\omega_0$ to obtain the F-series coefficients of $x_T'(t)$. Derivative operation *enhances* the *details* in the signal. We can state that

$$\frac{d^n x_T(t)}{dt^n} \xrightarrow{FS,T} (jk\omega_0)^n X_s[k], k \neq 0. \qquad (3.6.13)$$

In a similar manner, the derivatives of the trigonometric F-series are given by

$$x_T(t) = X_s[0] + \sum_{k=0}^{\infty} a[k]\cos(k\omega_0 t)$$

$$+ \sum_{k=0}^{\infty} b[k]\sin(k\omega_0 t).$$

$$\frac{dx_T(t)}{dt} = \sum_{k=1}^{\infty}(-k\omega_0)a[k]\sin(k\omega_0 t)$$

$$+ \sum_{k=1}^{\infty}(k\omega_0)b[k]\cos(k\omega_0 t).$$

$$= \sum_{k=1}^{\infty} a_1[k]\cos(k\omega_0 t)$$

$$+ \sum_{k=1}^{\infty} b_1[k]\sin(k\omega_0 t); \quad a_1[k] = b[k](k\omega_0),$$

$$b_1[k] = a[k](-k\omega_0)$$

$$a[k] = -\frac{b_1[k]}{k\omega_0}, \quad b[k] = \frac{a_1[k]}{k\omega_0}, k \neq 0,$$

$$X_s[0] = \frac{1}{T}\int_T x_T(t)dt. \qquad (3.6.14)$$

In (3.6.14), the subscript "1" on a and b indicates the trigonometric F-series are for the first derivative of the periodic function. The dc component *needs to be computed directly* from the given periodic function. This approach of finding the F-series is the *derivative method of finding F-series*.

The derivative property allows simplifies the computing the F-series coefficients. Most of the signals we deal with are pulses that do not have derivatives in the conventional sense. The derivatives of such functions can only be considered in the sense of generalized functions discussed in Section 1.4, resulting in impulse functions in the derivatives. See, for example, the derivative of the rectangular pulse in (1.4.33). Fourier series coefficients are determined by use of integrals. Integrals involving impulses are trivial to compute and, therefore, computing the Fourier series coefficients by the derivative method makes it simple.

Example 3.6.3 Find the trigonometric F-series of the trapezoidal waveform shown in Fig. 3.6.3a using the derivative method.

Solution: The first two derivatives of the wave form are shown in Figs. 3.6.3b and c. The second derivative of the function $x_T(t)$ is

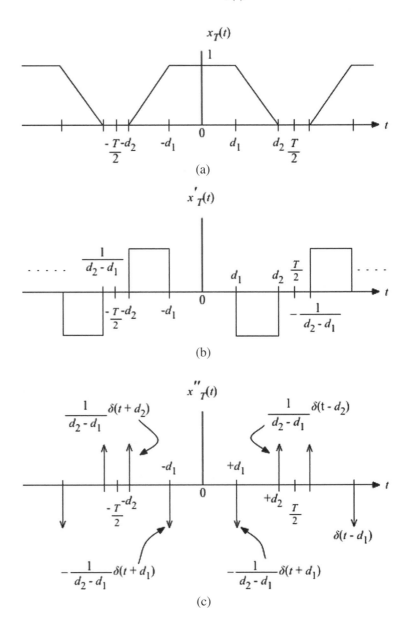

Fig. 3.6.3 (a) Trapezoidal wave-form $x_T(t)$, (b) $x'_T(t)$, and (c) $x''_T(t)$

$$x_T''(t) = \frac{d^2 x_T(t)}{dt^2} = \frac{1}{(d_2 - d_1)}$$

$$[\delta(t + d_2) - \delta(t + d_1) - \delta(t - d_1) + \delta(t - d_2)],$$

$$d_2 \geq d_1, -T/2 < t < T/2, x_T(t + T) = x_T(t).$$

$$(3.6.15)$$

Since the second derivative has an even symmetry, we can use some simplifications:

$$b_2[k] = 0, \quad a_2[k] = \frac{2}{T} \int_{-T/2}^{T/2} x_T''(t) \cos(k\omega_0 t) dt$$

$$= \frac{4}{(d_2 - d_1)T} \int_0^{T/2} [-\delta(t - d_1) + \delta(1 - d_2)]$$

$$\cos(k\omega_0 t) dt,$$

$$a_2[k] = \frac{4}{T(d_2 - d_1)} [-\cos(k\omega_0 d_1) + \cos(k\omega_0 d_2)].$$

Noting that $a_2[k] = -(k\omega_0)^2 a[k]$, we have the dc term (to be computed directly) and

$$X_s[0] = \frac{1}{T} \int_{-d_2}^{d_2} x_T(t) dt = \frac{1}{T}(d_1 + d_2).$$

$$a[k] = \frac{4}{T(d_2 - d_1)(k\omega_0)^2} [\cos(k\omega_0 d_1) - \cos(k\omega_0 d_2)].$$

The trigonometric F-series are

$$x_T(t) = \frac{d_1 + d_2}{T} + \frac{4}{\omega_0^2 T(d_2 - d_1)}$$

$$\sum_{k=1}^{\infty} \frac{1}{k^2} [\cos(k\omega_0 d_1) - \cos(k\omega_0 d_2)] \cos(k\omega_0 t).$$

$$(3.6.16a)$$

$$x_T(t)|_{d_1=0} = \frac{d_2}{T} + \frac{4}{d_2 T(\omega_0^2)}$$

$$\sum_{k=1}^{\infty} \frac{1}{k^2} [1 - \cos(k\omega_0 d_2)] \cos(k\omega_0 t). \quad (3.6.16b)$$

Note that when $d_1 = 0$, we have a triangular pulse wave form. The F-series coefficients for the two cases

considered above decay proportional to $(1/k^2)$. In the case of $d_2 = d_1$, we have a rectangular pulse waveform. Using L'Hospital's rule, the coefficients in (3.6.16a) can be simplified. Assuming $d_2 = d_1 + \varepsilon$,

$$a[k] = \lim_{\varepsilon \to 0} \left\{ -\frac{4}{T\varepsilon(k\omega_0)^2} [\cos(k\omega_0(d_1+\varepsilon)) - \cos(k\omega_0 d_1)] \right\}$$

$$= \lim_{\varepsilon \to 0} \frac{4}{T(k\omega_0)^2} \frac{(k\omega_0)\sin(k\omega_0(d_1+\varepsilon))}{1} = \frac{4\sin(k\omega_0 d_1)}{kT\omega_0}$$

$$x_T(t)\Big|_{d_1=d_2} = \frac{2d_1}{T} + \sum_{k=1}^{\infty} \frac{4}{k\omega_0 T} \frac{\sin(k\omega_0 d_1)}{\cos(k\omega_0 t)} \cos(k\omega_0 t).$$

$$(3.6.16c)$$

From (3.6.16b and c), the F-series coefficients of the trapezoidal and the rectangular pulse sequences decay at a rate proportional to $(1/k^2)$ and $(1/k)$, respectively.

The derivative of an even (odd) function is an odd (even) function. See Fig. 3.6.3 a,b,c and the F-series to verify the above assertion and the *chain rule* below:

$$\frac{dy}{dt} = \frac{dy}{d\alpha} \frac{d\alpha}{dt} \to y(t) = \frac{dx_e(t)}{dt},$$

$$y(-t) = \frac{dx_e(-t)}{dt} = \frac{dx_e(t)}{dt}(-1) = -y(t).$$

$$(3.6.17) \blacksquare$$

3.6.5 Bounds and Rates of Fourier Series Convergence by the Derivative Method

The Fourier series coefficients $X_s[k]$ of a periodic signal $x_T(t)$ *usually* decay at a rate inversely proportional to k^n, where k is the harmonic index. An exception is the periodic impulse sequence. In Example 3.4.4, we have seen that the periodic impulse sequence has the complex F-series given by $X_s[k] = (A/T)$, i.e., the coefficients are *independent* of k. Higher the value of n in k^n is, the faster the high-frequency component decays. An estimated value of n can be determined without actually computing the Fourier series coefficient $X_s[k]$ using the derivative properties of $x_T(t)$. In (3.6.13) we have seen that the Fourier series coefficients of the nth derivative of a periodic function are related to the Fourier coefficients of the function multiplied by $(jk\omega_0)^n$. If we differentiate an arbitrary periodic function $x_T(t)$ n times before the *first* set of impulses appear, then the F-series coefficients have the property that $X_s[k] \propto (1/|k\omega_0|)^n$. That is, the decay rate of the coefficients is proportional to $1/(|k|)^n$. The decay rate is a good indicator for *large k*, as some of the early coefficients may be even zero.

Since the complex F-series and trigonometric F-series are related, the decay rate of the trigonometric F-series coefficients is tied to the decay rate of the complex F-series coefficients. The derivative property of the F-series provides bounds on the F-series coefficients, referred to as the *spectral bounds*. For $k \neq 0$,

$$|X_s(k)| = \left| \frac{1}{jk\omega_0} \right|^n \frac{1}{T} \left| \int_T x_T^{(n)}(t) e^{-jk\omega_0 t} dt \right|$$

$$\leq \frac{1}{T|k|^n |\omega_0|^n} \int_T \left| x_T^{(n)}(t) \right| dt. \qquad (3.6.18a)$$

In deriving the right side of the above equation, the fundamental theorem of calculus is used and

$$\left| \int y(t) dt \right| \leq \int |y(t)| dt \left(\text{note } \left| e^{-jk\omega_0 t} \right| = 1 \right).$$

For example, the F-series coefficients in Example 3.4.2 show their decay rate is proportional to $(1/k)$, see (3.4.8). The function $x_T(t)$ has discontinuities at $t = \pm\tau/2$ in one period of the time function. Correspondingly, the bound on the F-series coefficient is

$$|X_s[k]| < B/k, B \text{ a constant.} \qquad (3.6.18b)$$

The *number of nonzero bounds* that can be determined equals the number of times the function can be differentiated before *derivatives of impulses* occur in the derivatives, see Ambardar (1995), Morrison, (1994), and others.

Example 3.6.4 Find all the nonzero spectral bounds for the rectangular pulse $x(t) = \Pi[t]$, $x_T(t) = x_T(t+T)$, $T = 2$.

Solution: The function, its first, and second derivatives are shown in Fig. 3.6.4. Noting that $\omega_0 = (2\pi/2) = \pi$, we have

$n = 0$ bound :

$$|X_s[k]| \leq \frac{1}{2(\omega_0)^0} \int_{-1/2}^{1/2} |x_T(t)| dt = \frac{1}{2}. \qquad (3.6.19a)$$

$n = 1$ bound :

$$|X_s[k]| \leq \frac{1}{2|\omega_0|^1} \int_{-1}^{1} \left| \frac{dx_2(t)}{dt} \right| dt = \frac{1}{2\pi} \int_{-1}^{1}$$

$$\left[\delta(t + \frac{1}{2}) + \delta(t - \frac{1}{2}) \right] dt = \frac{1}{\pi}. \qquad (3.6.19b)$$

The bounds above $n = 1$ are *not* defined. ∎

Bounds on the trigonometric Fourier series coefficients: If $x_T(t)$ has discontinuities, then its trigonometric F-series coefficients satisfy *for large k*

$$|a[k]| < \frac{K_{a1}}{k} \text{ and } |b[k]| < \frac{K_{b1}}{k}$$

$(K_{a1}$ and K_{b1} are some constants). $\qquad (3.6.20)$

If $x_T(t)$ is continuous but $x_T'(t)$ is discontinuous, then for large k

$$|a[k]| < \frac{K_{a2}}{k^2} \text{ and } |b[k]| < \frac{K_{b2}}{k^2} (K_{a2} \text{ and } K_{b2}$$

are some constants). $\qquad (3.6.21)$

Convergence is a function of the continuity of the highest derivative of $x_T(t)$. The convergence rates of the coefficients $a[k]$ and $b[k]$ may be different.

Example 3.6.5 Consider the function and its F-series coefficients given below. Comment on the convergence rates.

$$x_{2\pi}(t) = e^t, -\pi < t < \pi, x_{2\pi}(t+2\pi) = x_{2\pi}(t). \quad (3.6.22a)$$

$$x_{2\pi}(t) = \frac{2\sinh(\pi)}{\pi} \left[\frac{1}{2} + \sum_{k=1}^{\infty} \frac{(-1)^k}{(1+k^2)} \right]$$

$$(\cos(kt) - k\sin(kt)). \qquad (3.6.22b)$$

Solution: The sine series coefficients $b[k]$ converge like (K_b/k), whereas the cosine series coefficients $a[k]$ converge like (K_a/k^2), implying that the Fourier series, as a whole, converges like (K/k); K's

Fig. 3.6.4 (a) Periodic pulse waveform, (b) periodic impulse sequence, and (c) periodic doublet sequence

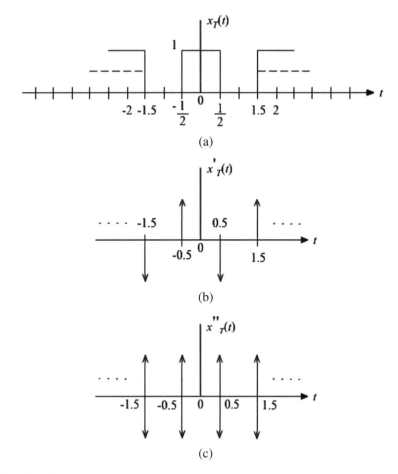

are constants. The dc term and the first few harmonics contain bulk of the power for low-frequency signals. ∎

3.6.6 Integral of a Function and Its Fourier Series

Consider the integral of a periodic function and its F-series. The integral of a periodic function with a dc component cannot be periodic as the integral of a constant is a ramp. In the case of $X_s[0] = 0$, we can derive the complex F-series of an integral of a function:

$$
\begin{aligned}
y_T(t) &= \int_0^t x_T(\alpha)d\alpha = \int_0^t \left[\sum_{k=-\infty,k\neq 0}^{\infty} X_s[k]e^{jk\omega_0\alpha} \right] d\alpha \\
&= \sum_{k=-\infty}^{\infty} Y_s[k]e^{jk\omega_0 t} \\
&= \sum_{k=-\infty,k\neq 0}^{\infty} \{(1/jk\omega_0)X_s[k]\}e^{jk\omega_0 t}] + \text{constant}.
\end{aligned}
$$

(3.6.23a)

$$
Y_s[k] = \begin{cases} [X_s[k]/jk\omega_0] \ , k \neq 0 \\ \text{constant}, \ k = 0 \end{cases}.
$$

(3.6.23b)

Note the division by $k\omega_0$ above indicating the F-series of the integral of a function $x_T(t)$ converges faster than the F-series convergence of $x_T(t)$. Integration is a smoothing operation. Therefore, the integrated signal has much smaller high-frequency content than $x_T(t)$. Without the dc term $X_s[0]$, integration and differentiation can be thought of as inverse operations.

3.6.7 Modulation in Time

Consider the F-series

$$
\begin{aligned}
x_T(t)e^{\pm j\alpha t} &= \sum_{k=-\infty}^{\infty} X_s[k]e^{jk\omega_0 t}e^{\pm j\alpha t} \\
&= \sum_{k=-\infty}^{\infty} X_s[k]e^{j(k\omega_0 \pm \alpha)t}.
\end{aligned}
$$

(3.6.24)

Multiplying a function by $e^{\pm j\alpha t}$ shifts the frequencies from $k\omega_0$ to $k\omega_0 \pm \alpha$. This is *modulation*. Multiplying $x_T(t)$ by $\cos(\alpha t)$ and using Euler's formula results in

$$y_1(t) = x_T(t)\cos(\alpha t) = \frac{1}{2}\sum_{k=-\infty}^{\infty} X_s[k]e^{j(k\omega_0+\alpha)t}$$

$$+ \frac{1}{2}\sum_{k=-\infty}^{\infty} X_s[k]e^{j(k\omega_0-\alpha)t}. \qquad (3.6.25)$$

Example 3.6.6 Consider the periodic signal $x_T(t) = \cos(\omega_0 t)$ modulated by the same function. The resulting function is $y_T(t) = \cos^2(\omega_0 t)$. Determine the frequency shifts.

Solution:

$$y_T(t) = .5(1 + \cos(2\omega_0 t))$$
$$= .25e^{j2\omega_0 t} + 0e^{j\omega_0 t} + .5 + 0e^{-j\omega_0 t} + .25e^{-j2\omega_0 t}.$$

Modulation shifted the frequencies from $\pm\omega_0$ to $\pm 2\omega_0, \pm\omega_0, 0$, with one of the frequencies having zero amplitude. ∎

3.6.8 Multiplication in Time

Let $x_T(t)$ and $w_T(t)$ be two periodic functions with the same period T. The product $y_T(t) = x_T(t)w_T(t)$ is also periodic with period T. The relation between the F-series coefficients is derived below. First, let the F-series expansions of these are as follows:

$$x_T(t) = \sum_{n=-\infty}^{\infty} X_s[n]e^{jn\omega_0 t}, \ w_T(t) = \sum_{m=-\infty}^{\infty} W_s[m]e^{jm\omega_0 t},$$

$$y_T(t) = \sum_{k=-\infty}^{\infty} Y_s[k]e^{jk\omega_0 t}. \qquad (3.6.26a)$$

Following shows that the complex F-series coefficients of the product $y_T(t) = x_T(t)w_T(t)$ are

$$Y_s[k] = \sum_{n=-\infty}^{\infty} X_s[n]W_s[k-n] = \sum_{n=-\infty}^{\infty} W_s[n]X_s[k-n]\ .$$
$$\qquad (3.6.26b)$$

First, use the change of variable $m = k - n$ in F-series expansion of $w_T(t)$, substitute this expression in the F-series expansion of $y_T(t)$, and then simplify the expression $y_T(t)$:

$$w_T(t) = \sum_{k=-\infty}^{\infty} W_s[k-n]e^{j(k-n)\omega_0 t}.$$

$$y_T(t) = \sum_{k=-\infty}^{\infty} X_s[n]e^{jn\omega_0 t}\left[\sum_{n=-\infty}^{\infty} W_s[k-n]e^{j(k-n)\omega_0 t}\right]$$

$$= \sum_{k=-\infty}^{\infty}\sum_{n=-\infty}^{\infty} X_s[n]W_s[k-n]e^{jn\omega_0 t}e^{j(k-n)\omega_0 t}$$

$$= \sum_{k=-\infty}^{\infty}\left[\sum_{n=-\infty}^{\infty} X_s[n]W_s[k-n]\right]e^{jk\omega_0 t}$$

$$= \sum_{k=-\infty}^{\infty} Y_s[k]e^{jk\omega_0 t}. \qquad (3.6.27)$$

Equation (3.6.26b) now follows from (3.6.27). The expression $Y_s[k]$ in terms of $X_s[k]$ and $W_s[k]$ in (3.6.26b) is a *discrete convolution*.

Periodic time convolution: If $x_T(t)$ and $w_T(t)$ are two periodic functions and their F-series are (3.6.26a), then (see Section 2.5)

$$y_T(t) = \frac{1}{T}\int_T x_T(t-\tau)w_T(\tau)dt$$

$$= \frac{1}{T}\int_T x_T(t)w_T(t-\tau)d\tau$$

$$= \sum_{k=-\infty}^{\infty} X_s[k]W_s[k]e^{jk\omega_0 t}. \qquad (3.6.28a)$$

$$y_T(t) = x_T(t) \otimes w_T(t) = w_T(t) \otimes x_T(t). \qquad (3.6.28b)$$

The Fourier series expansion of the periodic convolution in (3.6.28a) can be determined by using the F-series for the two functions as shown below:

$$y_T(t) = \frac{1}{T}\int_T x_T(t-\tau)w_T(\tau)dt$$

$$= \frac{1}{T}\int_T \sum_{k=-\infty}^{\infty} X_s[k]e^{jk\omega_0(t-\tau)}w_T(\tau)d\tau$$

$$= \sum_{k=-\infty}^{\infty} X_s[k]\left[\frac{1}{T}\int_T w_T(\tau)e^{-jk\omega_0\tau}d\tau\right]e^{jk\omega_0 t}$$

$$= \sum_{k=-\infty}^{\infty} X_s[k]W_s[k]e^{jk\omega_0 t}\ . \qquad (3.6.29a)$$

$$y_T(t) = x_T(t) \oplus w_T(t) \overset{FS,T}{\longleftrightarrow} X_s[k]W_s[k]$$
$$= Y_s[k]. \qquad (3.6.29b)$$

3.6.9 Frequency Modulation

The dual of time-domain modulation is *frequency modulation*. Using the superposition and the delay properties of the F-series, we can determine the F-series coefficients of the function $y_T(t) = x_T(t + \alpha) + x_T(t - \alpha)$. It follows that

$$y_T(t) = \sum_{k=-\infty}^{\infty} X_s[k] \left[e^{jk\omega_0\alpha} + e^{-jk\omega_0\alpha} \right] e^{jk\omega_0 t}$$

$$= \sum_{k=-\infty}^{\infty} Y_s[k] e^{jk\omega_0 t}. \qquad (3.6.30a)$$

$$\Rightarrow Y_s[k] = 2 X_s[k] \cos(k\omega_0\alpha). \qquad (3.6.30b)$$

3.6.10 Central Ordinate Theorems

The following results from the F-series at $t = 0$ and the F-series coefficient at $k = 0$:

$$x_T[0] = \sum_{k=-\infty}^{\infty} X_s[k], \quad X_s[0] = \frac{1}{T} \int_T x_T(t) dt. \quad (3.6.31)$$

3.6.11 Plancherel's Relation (or Theorem)

Let $x_T(t)$ and $y_T(t)$ be two periodic functions. Then, *Plancherel's relation* is

$$\frac{1}{T} \int_T x_T(t) y_T^*(t) dt = \sum_{k=-\infty}^{\infty} X_s[k] Y_s^*[k]. \quad (3.6.32)$$

Note, for generality, the expression in (3.6.32) is given for complex functions and the superscript (*) corresponds to the conjugation. The above relation can be derived by substituting the F-series coefficients for the two time functions:

$$\frac{1}{T} \int_T x_T(t) y_T^*(t) dt = \frac{1}{T} \int_T x_T(t) \left[\sum_{k=-\infty}^{\infty} Y_s^*[k] e^{-jk\omega_0 t} \right] dt$$

$$= \sum_{k=-\infty}^{\infty} Y_s^*[k] \left[\frac{1}{T} \int_T x_T(t) e^{-jk\omega_0 t} dt \right]$$

$$= \sum_{k=-\infty}^{\infty} X_s[k] Y_s^*[k]. \qquad (3.6.33)$$

If $y(t) = x(t)$ (i.e., $Y_s^*[k] = X_s^*[k]$), then the average power in a complex or a real periodic function with period T is

$$P_x = \frac{1}{T} \int_T |x_T(t)|^2 dt$$

$$= \sum_{k=-\infty}^{\infty} |X_s[k]|^2 \text{ (Parseval's formula).} \quad (3.6.34)$$

3.6.12 Power Spectral Analysis

The *power density spectrum* of the periodic signal $x_T(t)$ is defined by

$$S_x[k] = |X_s[k]|^2 \left(x_T(t) = \sum_{k=-\infty}^{\infty} X_s[k] e^{-jk\omega_0 t} \right) (3.6.35)$$

The average power contained in $x_T(t)$ is then given by

$$P_x = \sum_{k=-\infty}^{\infty} |X_s[k]|^2 = \sum_{k=-\infty}^{\infty} S_x[k]. \quad (3.6.36)$$

Notes: In Chapter 1, it was pointed that periodic and random signals are power signals. Although we will not be going through any discussion on random signals or processes, as it is beyond our scope, the average power contained in a random process is expressed in terms of power spectral density, which is real, even, and nonnegative function of frequency (f), identified by $S_x(f)$. The average power in the process is expressed by

$$P_x = \int_{-\infty}^{\infty} S_x(f) df = \frac{1}{2\pi} \int_{-\infty}^{\infty} S_x(\omega) d\omega. \quad (3.6.37)$$

It is interesting to tie the average power P_x in (3.6.36) and (3.6.37). This is achieved by using impulse functions (note $\delta(f) = 2\pi\delta(\omega)$, see (1.4.37)

$$S_x(\omega) = 2\pi \sum_{k=-\infty}^{\infty} |X_s[k]|^2 \delta(\omega - k\omega_0). \quad (3.6.38)$$

Using this expression results in (3.6.37)

$$\frac{1}{2\pi} \int_{-\infty}^{\infty} S_x(\omega)d\omega$$

$$= \frac{1}{2\pi} \int_{-\infty}^{\infty} \left[2\pi \sum_{k=-\infty}^{\infty} |X_s[k]|^2 \delta(\omega - k\omega_0) \right] d\omega$$

$$= \sum_{k=-\infty}^{\infty} |X_s(k)|^2 = P_x. \quad (3.6.39)$$

3.7 Convergence of the Fourier Series and the Gibbs Phenomenon

In Chapter 1, the average value, the average power, and the root mean-squared (rms) values were defined (see (1.5.15)). It was pointed out that the average value of a periodic function $X_s[0]$ can never exceed the rms value $\sqrt{P_x}$. Using (3.4.23), we have

$$P_x = |X_s(0)|^2 + \sum_{k=-\infty, k\neq 0}^{\infty} |X_s[k]|^2$$

$$\Rightarrow P_x \geq |X_s[0]|^2 \text{ or } \sqrt{P_x} \geq |X_s[0]|. \quad (3.7.1)$$

Furthermore, from (3.4.24a), the mean-squared value of a periodic function is equal to the sum of the mean-squared values of its dc component and its harmonics.

Example 3.7.1 Consider the periodic function given in (3.5.1) with $T = 2\pi$. It is discontinuous at $t = k\pi$ and its F-series is given below (it follows from (3.5.2)). Identify how fast the F-series coefficients decrease.

$$x_{2\pi}(t) = \frac{4}{\pi} \left[\sin(t) + \frac{1}{3}\sin(3t) + \frac{1}{5}\sin(5t) + \ldots \right]$$

$$= \sum_{k=1}^{\infty} \frac{4}{(2k-1)\pi} \sin((2k-1)t). \quad (3.7.2)$$

Solution: The nonzero F-series coefficients are proportional to $(1/k)$. ∎

Notes: Before considering the convergence of F-series let us briefly summarize the theoretical constraints on the periodic function $x_T(t)$ of interest and its F-series existence. It is assumed that $x_T(t)$ is square integrable. That is,

$$\int_T |x_T(t)|^2 dt < \infty. \quad (3.7.3)$$

Second, the periodic function is assumed to satisfy the Dirichlet conditions, see Section 3.1. All physically realizable functions satisfy these conditions and therefore, we will not be dealing with these. ∎

3.7.1 Fourier's Theorem

Dirichlet proved first that Fourier series approximation converges to $x_T(t)$ at every point $x_T(t)$ is continuous and to $[x_T(t^+) + x_T(t^-)]/2$, the *half-value*, wherever the function $x_T(t)$ is discontinuous, i.e., the F-series converges to the average value of the function. This result is called Fourier's theorem.

Example 3.7.2 Let $x_T(t)$ has a discontinuity at $t = t_0$ as shown in Fig. 3.7.1. The Fourier series approximation of $x_T(t)$ with $(2n + 1)$ terms is assumed to be

$$x_{T,2n+1}(t) = \sum_{k=-n}^{n} X_s[k]e^{jk\omega_0 t}. \quad (3.7.4)$$

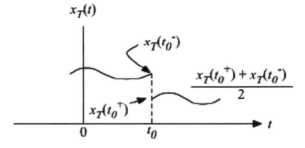

Fig. 3.7.1 $x_T(t)$ with a discontinuity

The mean-squared error at the discontinuity between $x_T(t)$ and its $(2n + 1)$ term F-series is

$$\varepsilon^2 = \left[\{x_{T,2n+1}(t) - x_T(t_0^-)\}^2 + \{x_{T,2n+1}(t) - x_T(t_0^+)\}^2 \right].$$

$$(3.7.5)$$

The time entries in the above equation t_0^- and t_0^+ are the values of t before and after the discontinuity at $t = t_0$. Find the minimum value of the mean-squared error, with respect to $x_{T(2n+1)}(t)$ by taking the partial derivative of ε^2 with respect to $x_{T,2n+1}(t)$, equating the result to zero, and then solving for $x_{T,2n+1}(t)$ at $t = t_0$.

Solution: Taking the partial derivative and equating it to zero at $t = t_0$, we have

$$\frac{\partial \varepsilon^2}{\partial x_{T,2n+1}} = 2[x_{T,2n+1}(t) - x_T(t_0^-)]$$

$$+ 2[x_{T,2n+1}(t) - x_T(t_0^+)]|_{t=t_0} = 0. \quad (3.7.6)$$

$$\Rightarrow x_{T,2n+1}(t)|_{t=t_0} = [x_T(t_0^+) + x_T(t_0^-)]/2. \quad (3.7.7)$$

That is, the F-series approximation gives the average value of the function before and after the discontinuity. This value is referred to as the *half-value* of $x_T(t)$ at the discontinuity $t = t_0$. ■

3.7.2 Gibbs Phenomenon

From Fourier's theorem and the following discussion we will see that at both sides of a discontinuity the finite F-series approximation exhibits ripples before and after the discontinuity. This behavior is called the *Gibbs phenomenon* (or *effect*). Historically, Albert Michalson observed the phenomenon and reported to Josiah Gibbs, a theoretical physicist. Gibbs investigated this behavior of oscillations with overshoots and undershoots before and after the discontinuity associated with Fourier series. Equality of the function to its F-series is only in the sense the integral-squared error between the two goes to zero when infinite number of terms is included in the F-series approximation. Fourier's theorem points out that at a point of discontinuity the series converges to the average value or the half-value given in (3.7.7).

Example 3.7.3 Illustrate the convergence of the F-series expansion to the function given in Example 3.7.1 by using the first few terms in the series.

Solution: Consider the approximations by considering the first few terms. Let

$$s_1(t) = \frac{4}{\pi}\sin(t), \; s_3(t) = \frac{4}{\pi}[\sin(t) + \frac{1}{3}\sin(3t)],$$

$$s_5(t) = \frac{4}{\pi}[\sin(t) + \frac{1}{3}\sin(3t) + \frac{1}{5}\sin(5t)],\ldots. \quad (3.7.8)$$

The functions $s_1(t), s_3(t)$, and $s_{(2k-1)}(t)$, k large are sketched in Fig. 3.7.2 for one period. They are odd periodic functions. Fourier series approximation gives the value of 0 at $t = 0$, the average value (or half-value) of the function equals to $(1 - 1)/2 = 0$. First, consider only the positive values of $t, 0 < t < \pi$. The maximum value of $s_1(t)$ is equal to $(4/\pi) = 1.2732$. This function crosses the value of 1 when $(s_1(t) - 1) = 0$ for positive t. The roots of this equation are $t = .9033$ and $t = 2.2383$ located symmetrically around the middle $t = 1.5708$. More number of terms we consider, the better the approximation of the given function is, and in the limit, the integral-squared error goes to zero.

Summary on $s_{2k-1}(t)(t > 0)$:

- The function rises rapidly as t goes from 0.
- It overshoots the value of 1 and oscillates about the line $x(t) = 1$ with increasing frequency and decreasing amplitudes.
- Although the magnitude of the peak overshoots and undershoots before and after the discontinuity at $t = 0$ diminish as k increases, there is a *lower bound of* 9% on the overshoots or undershoots even as $k \to \infty$. Furthermore, the F-series converges to every point of $x_T(t)$ that is continuous with rare exceptions. It is possible that the Fourier series of a continuous function to be divergent at some point. Kolmogoroff Zygmund (1955) has given a function whose Fourier series is everywhere divergent.
- At the point of discontinuity in $x_T(t)$, the series converges to the half-value of the function, i.e., the average value of the function before and after the discontinuity.
- Since $s_k(t)$ is a periodic odd function, the sketches follow for $-\pi < t < 0$. ■

Fig. 3.7.2 (a)$s_1(t)$, (b)$s_3(t)$, and (c)$s_{2k-1}(t)$, k large

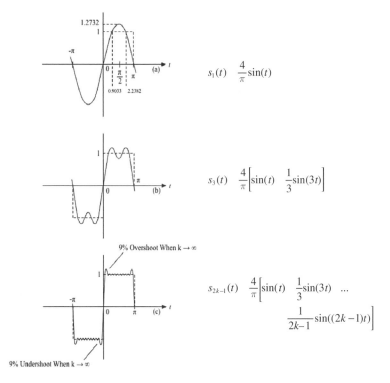

$$s_1(t) \quad \frac{4}{\pi}\sin(t)$$

$$s_3(t) \quad \frac{4}{\pi}\left[\sin(t) \quad \frac{1}{3}\sin(3t)\right]$$

$$s_{2k-1}(t) \quad \frac{4}{\pi}\left[\sin(t) \quad \frac{1}{3}\sin(3t) \quad \cdots \right.$$
$$\left. \frac{1}{2k-1}\sin((2k-1)t)\right]$$

Notes: Gibbs published his results on the phenomenon (or effect) in *Nature Magazine* in 1899. Fourier did *not* discuss the convergence of F-series in his paper. If N is small, the value of the overshoot may be different. The Gibbs *effect occurs* only for waveforms with jump discontinuities, see Carslaw (1950). ∎

Example 3.7.4 Find a series expansion for π using the results in Example 3.7.1.

Solution: At $t = (1/2)\pi$ the function is continuous and therefore the series converges to the actual value of the function. Substituting $t = \pi/2$ in (3.7.2), and simplifying, the expression for π is

$$\frac{4}{\pi}\left[1 - \frac{1}{3} + \frac{1}{5} - \frac{1}{7} + - \cdots\right] = 1$$

$$\Longrightarrow \pi = 4\left(1 - \frac{1}{3} + \frac{1}{5} - \frac{1}{7} + - \cdots\right). \qquad (3.7.9)$$

The series converge at a rate of $4(1/(2k - 1))$ corresponding to the $(2k-1)$th term. ∎

Notes: Beckmann (1971), in his book on *A History of π*, gives many interesting aspects associated with respect to the constant π. ∎

Example 3.7.5 Find a series expansion for π^2 using the function in Example 3.7.3.

Solution: The average power is

$$P = \frac{1}{2\pi}\int\limits_{-\pi}^{\pi}(1)^2 dt = 1 = \frac{1}{2}\sum_{k=1}^{\infty}b^2[2k-1])$$

$$= \frac{1}{2}\left(\frac{16}{\pi^2}\right)\left[1 + \frac{1}{9} + \frac{1}{25} + \cdots\right].$$

$$\Longrightarrow \pi^2 = 8\left(1 + \frac{1}{9} + \frac{1}{25} + \cdots\right). \qquad (3.7.10) \ \blacksquare$$

How many terms in the F-series are needed to have a"good"approximation of the given function? The answer can only be given for a particular application. The integral-squared error (ISE) between the periodic function $x_T(t)$ and its Fourier series approximation is

$$\text{ISE} = \int_T |x_T(t) - x_a(t)|^2 dt = \int_T x_T^2(t) dt - T \left[\sum_{k=-\infty}^{\infty} |X[k]|^2 \right]$$

$$= \int_T x_T^2(t) dt - \left[TX_s^2[0] + \frac{T}{2} \sum_{k=1}^{\infty} (a^2[k] + b^2[k]) \right] = 0.$$

(3.7.11)

The ISE goes to zero *only if* an infinite number of terms are used in the expansion, which is impractical. A goal is to approximate the function $x_T(t)$ using N trigonometric F-series with a bound on the ISE. First,

$$x_{T,N}(t) = X_s[0] + \sum_{k=1}^{N} a[k] \cos(k\omega_0 t)$$

$$+ \sum_{k=1}^{N} b[k] \sin(k\omega_0 t).$$

(3.7.12)

Find the smallest integer value of N that results within certain percentage of ISE. The ISE, keeping only the dc term and N harmonics, is

$$(\text{ISE})_N = \int_T x_T^2(t) dt - T \left\{ X_s^2[0] + \frac{1}{2} \sum_{k=1}^{N} (a^2[k] + b^2[k]) \right\}.$$

(3.7.13)

Example 3.7.6 Consider the half-wave rectified periodic function in Example 3.6.2 with

$$y_{2\pi}(t) = \begin{cases} \sin(t), & 0 \le t < \pi \\ 0, & \pi \le t < 2\pi \end{cases},$$

$$y(t + 2\pi) = y(t).$$

(3.7.14)

The trigonometric F-series of this periodic function was given by (see (3.6.10))

$$y_{2\pi}(t) = \frac{1}{\pi} + \frac{1}{2} \sin(t)$$

$$- \frac{2}{\pi} \left[\frac{\cos(2t)}{1(3)} + \frac{\cos(4t)}{3(5)} + \frac{\cos(6t)}{5(7)} + \cdots \right].$$

(3.7.15)

Find the smallest N in (3.7.13) for the two cases: *a.*10% or less *b.* 2% or less, see Gibson (1993).

Solution: First

$$\int_{2\pi} y_{2\pi}^2(t) dt = \frac{1}{2} \int_0^{\pi} (1 - \cos(2t)) dt = \frac{\pi}{2} = 1.571.$$

$N = 1$:

$$(\text{ISE})_1 = \int_{2\pi} y_{2\pi}^2(t) dt - 2\pi \left(X_s^2[0] + \frac{1}{2}(a^2[1] + b^2[1]) \right)$$

$$= 1.571 - 2\pi[(1/\pi)^2 + (1/2)(1/2)^2]$$

$$= .149.$$

(3.7.16)

The percentage error is $(.149/1.571) = 9.5\%$. This implies that $N = 1$ satisfies the requirements for Part *a*. Continuing this procedure, we have

$$N = 2: (\text{ISE})_2 = 1.571 - 2\pi[(1/\pi)^2$$

$$+ (1/2)(1/2)^2 + (1/2)(4/9\pi^2)] = .0075.$$

(3.7.17)

In this case the percentage integral-squared error is $(.0075/1.571) = .5\%$ and $N = 2$ satisfies the requirement for part *b*. The coefficients die out like (K/k^2), where K is a constant. Very few terms are needed to approximate the half-wave rectified signal. Unfortunately, there is *no* general formula to determine the number of harmonics needed for a given set of specifications. ∎

The ISE can be reduced by increasing N. It goes to 0 when $N \to \infty$. On the other hand, the peak 9% overshoots and undershoots before and after the discontinuities discussed earlier cannot be reduced even if infinite number of terms is included in the F-series expansion. There is another measure, average error, which can be used in judging an approximation, which is not very attractive as the positive errors may cancel out with the negative errors. Squaring the error function accentuates the larger errors. Least-squares error measure gives a convenient and a simple way to calculate the parameters in the approximation. Overshoots and undershoots can be reduced by smoothing.

3.7.3 Spectral Window Smoothing

The ripples generated by the Fourier series approximation of a discontinuous function are due to the abrupt change of the function before and after the discontinuity. The use of a taper, instead of a discontinuity at a transition, yields a smoother reconstruction from the basis functions. The windowed signal is defined by the periodic convolution (see Section 2.5) by

$$y_T(t) = x_T(t) \otimes w_T(t)$$
$$= \frac{1}{T} \int_T x_T(\alpha) w_T(t - \alpha) d\alpha. \qquad (3.7.18a)$$

Considering $(2N+1)$ complex F-series coefficients of $y_T(t)$ (see (3.6.29a)), we have

$$y_{T,N}(t) = \sum_{k=-N}^{N} W_s[k] X_s[k] e^{jk\omega_0 t}. \qquad (3.7.18b)$$

The sequence $W_s[k]$ is a window and its weights (or coefficients) typically decrease with increasing $|k|$. The rectangular and hamming window sequences are

$$W_{s,R}[k] = 1, \quad -N \leq k \leq N. \qquad (3.7.19)$$
$$W_{s,H}[k] = .54 + .46 \cos(k\pi/N),$$
$$- N \leq k \leq N. \qquad (3.7.20)$$

The use of special windows reduces or even eliminates the overshoots and undershoots in the approximated signal before and after a discontinuity, see Ambardar (1995).

Example 3.7.7 Consider the trigonometric F-series in (3.7.2). Give the expression using the rectangular window with $N = 7$. Illustrate the window smoothing by first sketching the F-series and then the Hamming windowed series.

Solution: The trigonometric F-series approximation is

$$(x_{2\pi}(t))|_{N=7} = \frac{4}{\pi} \left[\sin(t) + \frac{1}{3}\sin(3t) + \frac{1}{5}\sin(5t) + \frac{1}{7}\sin(7t) \right].$$
$$(3.7.21)$$

Note the odd harmonic terms are all zero. The 15 tapered Hamming window coefficients and the Hamming windowed function $y_{2\pi}(t)$ are, respectively, given by

$$W_{s,H}[k] = \left\{ \begin{array}{l} .0800, .1256, .2532, .4376, .6424, .8268, .9544, 1, \\ .9544, .8268, .6424, .4376, .2532, .1256, .0800 \end{array} \right\}.$$

$$(y_{2\pi}(t))|_{N=7} = \frac{4}{\pi} \left[(.9544) \sin(t) + \frac{1}{3}(.6424) \sin(3t) \right.$$
$$\left. + \frac{1}{5}(.2532)\sin(5t) + \frac{1}{7}(.0800)\sin(7t) \right]$$
$$(3.7.22)$$

The two functions $x_{2\pi}(t)$ (identified as $x(t)$ on the top figure) and $y_{2\pi}(t)$ (identified as $y(t)$ on the bottom figure) are plotted in Fig. 3.7.3 using MATLAB. Note the overshoots and undershoots before and after the discontinuities in each case. The later case has hardly any ripples. The slope in the transition region is much higher in the case of the rectangular window compared to the Hamming window case.   ∎

3.8 Fourier Series Expansion of Periodic Functions with Special Symmetries

In Section 1.6, periodic functions with half-wave and quarter-wave symmetries were considered. Computation of the F-series for these cases is considered next.

3.8.1 Half-Wave Symmetry

Figure 3.8.1 illustrates a periodic function with period T and with *half-wave symmetry* (or *rotation symmetry*). Such functions satisfy (see (1.6.19) and (1.6.20))

$$x_T(t) = -x_T\left(t \pm \frac{T}{2}\right). \qquad (3.8.1)$$

Periodic functions with half-wave symmetry have odd harmonics, i.e., $X_s[k] = 0, k$ even, which can be seen from the following. First

$$X_s[k] = \frac{1}{T} \int_T x_T(t) e^{-jk\omega_0 t} dt$$

Fig. 3.7.3 Window smoothing

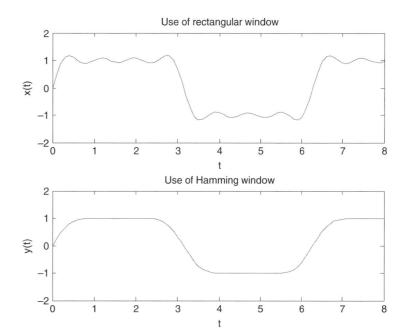

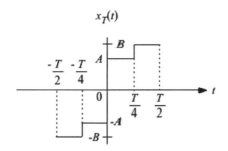

Fig. 3.8.1 A half-wave symmetric function

$$= \frac{1}{T} \int_{-T/2}^{0} x_T(t)e^{-jk\omega_0 t}dt + \frac{1}{T} \int_{0}^{T/2} x_T(t)e^{-jk\omega_0 t}dt. \quad (3.8.2)$$

Consider the change of variable from t to $\alpha - (T/2)$ in the first integral on the right in (3.8.2), which results in

$$X_s[k] = \frac{1}{T} \int_{0}^{T/2} x_T(\alpha - \frac{T}{2})e^{-jk\omega_0(\alpha - \frac{T}{2})}d\alpha$$

$$+ \frac{1}{T} \int_{0}^{T/2} x_T(t)e^{-jk\omega_0 t}dt. \quad (3.8.3)$$

The limits follow from $t = 0 \Longrightarrow \alpha = T/2$ and $t = T/2 \Longrightarrow \alpha = 0$ and

$$\frac{1}{T} \int_{0}^{T/2} x_T(\alpha - \frac{T}{2})e^{-jk\omega_0 \alpha}e^{jk\omega_0 T/2}d\alpha$$

$$= \left[\frac{1}{T} \int_{0}^{T/2} x_T(\alpha - \frac{T}{2})e^{-jk\omega_0 \alpha}d\alpha \right] e^{jk\pi}$$

$$= (-1)^k \frac{1}{T} \int_{0}^{T} x_T(t - \frac{T}{2})e^{-jk\omega_0 t}dt. \quad (3.8.4)$$

Using (3.8.4) in (3.8.3), we have

$$X_s[k] = \frac{1}{T} \int_{0}^{T/2} \left[x(t) + (-1)^k x\left(t - \frac{T}{2}\right) \right] e^{-jk\omega_0 t}dt.$$

$$(3.8.5)$$

From the half-wave symmetry property in (3.8.1) we see that $X_s[k] = 0$, k even, thus establishing the half-wave symmetric functions that contain only odd harmonics. The complex F-series of these functions and the corresponding trigonometric F-series are

$$x_T(t) = \sum_{\substack{k=-\infty \\ k\neq 0, k-\text{odd}}}^{\infty} X_s[k]e^{jk\omega_0 t}, X_s[k] = \frac{2}{T} \int_{0}^{T/2} x_T(t)e^{-jk\omega_0 t}dt.$$

$$(3.8.6)$$

$$x_T(t) = \sum_{k=1}^{\infty} \{a[2k-1]\cos((2k-1)\omega_0 t) + b[2k-1]$$
$$\sin((2k-1)\omega_0 t)\},$$

$$a[2k-1] = \frac{4}{T}\int_0^{T/2} x_T(t)\cos((2k-1)\omega_0 t)dt$$

$$b[2k-1] = \frac{4}{T}\int_0^{T/2} x_T(t)\sin((2k-1)\omega_0 t)dt$$

$$(3.8.7)$$

3.8.2 Quarter-Wave Symmetry

If $x_T(t)$ is a periodic function with half-wave symmetry and, in addition, is either even or an odd function, then $x(t)$ is said to have *even* or *odd quarter-wavesymmetry*, respectively, see Section 1.6.4. They satisfy the following properties:

Even quarter-wave symmetry : $x_T(t) =$
$$x_T(-t) \text{ and } x_T(t) = -x_T(t \pm T/2). \quad (3.8.8a)$$

Odd quarter-wave symmetry : $x_T(t) =$
$$-x_T(-t) \text{ and } x_T(t) = -x_T(t \pm T/2). \quad (3.8.8b)$$

3.8.3 Even Quarter-Wave Symmetry

Since the function must be a half-wave symmetric to be a quarter-wave symmetric, it follows that $X_s[0] = 0$ and $a[2k] = 0$. In addition, $x(t)$ is even and $b[k] = 0$. Therefore,

$$X_s[0] = 0, b[k] = 0, a[2k-1]$$
$$= \frac{4}{T}\int_0^{T/2} x_T(t)\cos((2k-1)\omega_0 t)dt, a[2k] = 0.$$

$$(3.8.9a)$$
$$a[2k-1] = \frac{4}{T}\int_0^{T/4} x_T(t)\cos((2k-1)\omega_0 t)dt$$

$$+\frac{4}{T}\int_{T/4}^{-T/2} x_T(t)\cos((2k-1)\omega_0 t)dt. \quad (3.8.9b)$$

$$= \frac{4}{T}\int_0^{T/4} x_T(t)\cos((2k-1)\omega_0 t)dt$$

$$+\frac{4}{T}\int_{-T/4}^{0} x_T(t)\cos[(2k-1)\omega_0 t]dt$$

Trigonometric and complex F-series for the even quarter-wave symmetric function:

$$x_T(t) = \sum_{n=1}^{\infty} a[2k-1]\cos[(2k-1)\omega_0 t],$$

$$a[2k-1] = \frac{8}{T}\int_0^{T/4} x_T(t)\cos((2k-1)\omega_0 t)dt.$$

$$(3.8.10)$$

$$= \sum_{k=1}^{\infty}(1/2)a[2k-1]e^{j(2k-1)\omega_0 t}$$

$$+\sum_{k=1}^{\infty}(1/2)a[2k-1]e^{-j(2k-1)\omega_0 t}$$

$$= \sum_{k=-\infty}^{\infty} X_s[k]e^{jk\omega_0 t}$$

$$\Longrightarrow X_s[2k-1] = X_s[-(2k-1)]$$
$$= a[2k-1]/2. \quad (3.8.11)$$

3.8.4 Odd Quarter-Wave Symmetry

The F-series for this case are as follows. Derivation is left as an exercise.

$$x_T(t) = \sum_{k=1}^{\infty} b[2k-1]\sin[(2k-1)\omega_0 t],$$

$$b[2k-1] = \frac{8}{T}\int_0^{T/4} x_T(t)\sin[(2k-1)\omega_0 t]dt. \quad (3.8.12)$$

$$x_T(t) = \sum_{k=1}^{\infty} X[2k-1]e^{j(2k-1)\omega_0 t}$$

$$+\sum_{k=1}^{\infty} X[-(2k-1)]e^{-j(2k-1)\omega_0 t}. \quad (3.8.13)$$

Compare this with the F-series expansion and equate the corresponding coefficients.

$$x_T(t) = \sum_{n=-\infty, n\neq 0}^{\infty} X_s[n]e^{jn\omega_0 t}$$

$$\Rightarrow X[2k-1] = \frac{b[2k-1]}{2j}, X[-(2k-1)]$$

$$= -\frac{b[2k-1]}{2j}. \qquad (3.8.14)$$

3.8.5 Hidden Symmetry

Example 3.8.1 Symmetry of a periodic function can be obscured by a constant. Consider the periodic saw-tooth waveform, $x_{2\pi}(t) = (1 - (t/2\pi))$, $0 < t < 2\pi$ in Fig. 3.8.2. It does not have any obvious symmetry. Find the F-series of the function $x_{2\pi}(t)$ by noting $(x_{2\pi}(t) - (1/2))$ is an odd function.

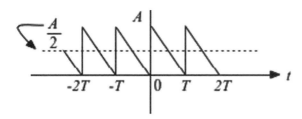

Fig. 3.8.2 $x_T(t)$ with hidden symmetry, $T = 2\pi$

Solution: From the odd symmetry,

$$\left(x_{2\pi}(t) - \frac{1}{2}\right) = \sum_{k=1}^{\infty} b[k]\sin(k\omega_0 t)$$

$$\Rightarrow x_{2\pi}(t) = \frac{1}{2} + \sum_{k=1}^{\infty} b[k]\sin(k\omega_0 t). \qquad (3.8.15) \blacksquare$$

3.9 Half-Range Series Expansions

Consider an aperiodic function $x(t)$ over the interval $(0, T/2)$ and zero everywhere else. Even and odd functions can be generated in the interval $-T/2 < t < T/2$ by

$$x_e(t) = x(t) + x(-t), x_0(t) = x(t) - x(-t). \qquad (3.9.1)$$

Even and odd periodic extensions (see Section 1.8.1.) of these are

$$x_{eT}(t) = \sum_{k=-\infty}^{\infty} x_e(t + kT) \text{ and}$$

$$x_{0T}(t) = \sum_{k=-\infty}^{\infty} x_0(t + kT). \qquad (3.9.2)$$

The trigonometric F-series of the even periodic function has dc and cosine terms and the odd periodic function has only sine terms. The two periodic functions $x_{eT}(t)$ and $x_{0T}(t)$ can be expressed by the following with $\omega_0 = 2\pi/T$ (see (3.4.18)):

$$x_{eT}(t) = X_s[0] + \sum_{k=1}^{\infty} a[k]\cos(k\omega_0 t). \qquad (3.9.3a)$$

$$X_s[0] = \frac{2}{T}\int_0^{T/2} x(t)dt,$$

$$a[k] = \frac{2}{(T/2)}\int_0^{T/2} x(t)\cos(k\omega_0 t)dt. \qquad (3.9.3b)$$

$$x_{0T}(t) = \sum_{k=1}^{\infty} b[k]\sin(k\omega_0 t),$$

$$b[k] = \frac{4}{T}\int_0^{T/2} x(t)\sin(k\omega_0 t)dt. \qquad (3.9.4)$$

The functions $x_{eT}(t)$ and $x_{0T}(t)$ in (3.9.3a and b) and (3.9.4) represent the *same function* in the interval $(0, T/2)$. Outside this interval (3.9.3a) represents an even periodic function and (3.9.4) represents an odd periodic function. These expansions are called the *half-range Fourier series expansions* of the aperiodic function $x(t)$.

Example 3.9.1 Given the aperiodic function $x(t) = \sin(t), 0 < t < \pi$ and 0 otherwise, expand this function in terms of a cosine series expansion and the sine series expansion in the *interval* $0 < t < \pi$. Give the even and odd periodic extensions of $x(t)$.

Solution: It is simple to see that $x_{eT}(t) = |\sin(t)|$ and the odd periodic extension $x_{0T}(t) = \sin(t)$. In the interval $0 < t < \pi$, the two functions $x_{eT}(t)$ and $x_{0T}(t)$ are equal. Since $|\sin(t)|$ and $\sin(t)$ are continuous functions, the F-series converges and (see (3.6.7))

$$x_{eT}(t) = |\sin(t)|$$

$$= \frac{2}{\pi} - \frac{4}{\pi}\left[\frac{\cos(2t)}{(1)(3)} + \frac{\cos(4t)}{(3)(5)} + \cdots\right], 0 < t < \pi .$$

$$(3.9.5)$$

$$x_{0T}(t) = \sin(t), 0 < t < \pi. \qquad (3.9.6) \blacksquare$$

Example 3.9.2 Expand the function given below in terms of a cosine series expansion in the interval $0 < t < \pi$. Give the even periodic extensions of $x(t)$.

$$x(t) = \begin{cases} 0, & 0 < t < \pi/2 \\ 1, & \pi/2 < t < \pi \end{cases}. \qquad (3.9.7)$$

Solution: Cosine series need a symmetric function defined by $x_e(t) = x(t) + x(-t)$. The F-series can now be obtained by considering the interval $-\pi < t < \pi$, i.e., the period $T = 2\pi$. Noting that $\omega_0 = 2\pi/T = 1$, the trigonometric F-series can be computed. Since it is even, it follows that $b[k] = 0$. The coefficients $X_s[0]$ and $a[k]$ are, respectively, given by

$$X[0] = \frac{1}{2\pi}\int_{-\pi}^{\pi} x_{eT}(t)dt = \frac{2}{\pi}\int_{\pi/4}^{\pi/2} dt$$

$$= \frac{2}{\pi}[(\pi/2) - (\pi/4)]dt = \frac{1}{2}.$$

$$a[k] = \frac{2}{\pi}\int_{0}^{\pi} x_{eT}(t)\cos(kt)dt$$

$$= \frac{2}{\pi}\int_{\pi/2}^{\pi} \cos(kt)dt = -\frac{2}{k\pi}\sin(k\pi/2).$$

$$x_{eT}(t) = \frac{1}{2} - \frac{2}{\pi}\left[\cos(t) - \frac{1}{3}\cos(3t) + \frac{1}{5}\cos(5t) - \cdots\right],$$
$$0 < t < \pi.$$

This gives an approximation of $x(t)$ in terms of cosine series in the time interval $0 < t < \pi$. In a similar manner, the odd periodic extension of the function can be determined. Since the function is odd, it follows that $X_s[0] = 0$ and $a[k] = 0, k = 0, 1, 2, \ldots$. The coefficients $b[k]$ and the corresponding odd period extension are given by

$$b[k] = \frac{2}{\pi}\int_{\pi/2}^{\pi} \sin(kt)dt$$

$$= -\frac{2}{k\pi}[\cos(k\pi) - \cos((1/2)k\pi)].$$

$$x_0(t) = \frac{2}{\pi}\left[\sin(t) + \frac{1}{3}\sin(3t) + \frac{1}{5}\sin(5t) + \cdots\right]$$
$$- \frac{2}{\pi}\left[\sin(2t) + \frac{1}{3}\sin(6t) + \frac{1}{5}\sin(10t) + \cdots\right].$$

$$0 < t < \pi. \qquad \blacksquare$$

3.10 Fourier Series Tables

Refer tables 3.10.1 and 3.10.2 for Fourier Series.

3.11 Summary

In this chapter we have introduced some of the basis functions and their use in approximating a given function in an interval. The important set of basis functions are the sine and the cosine functions and the periodic exponential function leading to the discussion on Fourier series. This chapter dealt with Fourier series, their properties and the computations of the Fourier series coefficients, in general, and in cases of special symmetries in the given function. Convergence of the coefficients and the number of coefficients required for a given set of specifications are discussed. Approximation measures are discussed in terms of basis functions. Specific principal topics that were included are

- Various basis functions and error measures of a function and their approximations
- Basics of complex and trigonometric Fourier series and the relationships between the trigonometric and the complex F-series
- Computation and simplification of the F-series coefficients of periodic functions that have simple and special symmetries
- Operational properties of the Fourier series that include simple methods that allow for simplification in the computation of the F-series

Table 3.10.1 Symmetries of real periodic functions and their Fourier-series coefficients

Type of symmetry	Constraints Periodic, $x_T(t) = x_T(t + T)$; $\omega_0 = 2\pi/T$.	Trignometric Fourier-series	Fourier series coefficients
Even	$x_T(t) = x_T(-t)$	$x_T(t) = X_S[0] + \sum\limits_{k=1}^{\infty} a[k]\cos(k\omega_0 t)$	$X_S[0] = \frac{2}{T}\int\limits_0^{T/2} x_T(t)dt$
			$a[k] = \frac{4}{T}\int\limits_0^{T/2} x_T(t)\cos(k\omega_0 t)dt$
Odd	$x_T(t) = -x_T(-t)$	$x_T(t) = \sum\limits_{k=1}^{\infty} b[k]\sin(k\omega_0 t)$	$b[k] = \frac{4}{T}\int\limits_0^{T/2} x_T(t)\sin(k\omega_0 t)dt$
Half-wave	$x_T(t) = -x_T(t + T/2)$	$x(t) = \sum\limits_{k=1}^{\infty} a[2k-1]\cos[2k-1]\omega_0 t$	$a[2k-1] = \frac{4}{T}\int\limits_0^{T/2} x_T(t)\cos((2k-1)\omega_0 t)dt$
		$+ \sum\limits_{k=1}^{\infty} b[2k-1]\sin(2k-1)\omega_0 t$	$b[2k-1] = \frac{4}{T}\int\limits_0^{T/2} x_T(t)\sin((2k-1)\omega_0 t)dt$
Even quarter-wave	$x_T(t) = -x_T(-t)$, $x_T(t) = -x_T(t + T/2)$	$x_T(t) = \sum\limits_{k=1}^{\infty} a[2k-1]\cos(2k-1)\omega_0 t$	$a[2k-1] = \frac{8}{T}\int\limits_0^{T/4} x_T(t)\cos((2k-1)\omega_0 t)dt$
Odd quarter-wave	$x_T(t) = -x_T(-t)$, $x_T(t) = -x_T(t + T/2)$	$x_T(t) = \sum\limits_{k=1}^{\infty} b[2k-1]\sin(2k-1)\omega_0 t$	$b[2k-1] = \frac{8}{T}\int\limits_0^{T/4} x_T(t)\sin((2k-1)\omega_0 t)dt$

There are extensive tables in literature that list Fourier series of functions. See Abramowitz and Stegun (1964), Gradshteyn and Ryzhik (1980), and others. To standardize the tables, we will assume the period is $T = 2\pi$. Spiegel (1968) has several other interesting periodic functions and the corresponding Fourier series.

Table 3.10.2 Periodic functions and their Trigonometric Fourier Series

Periodic Function $x_{2\pi}(t) = x_{2\pi}(t + 2\pi)$	Trigonometric Fourier-series
$x_{2\pi}(t) = \begin{cases} 1, & 0 < t < \pi \\ -1, & -\pi < t < 0 \end{cases}$	$\frac{4}{\pi}\left[\frac{\sin(t)}{1} + \frac{\sin(3t)}{3} + \frac{\sin(5t)}{5} + \cdots\right]$
$x_{2\pi}(t) = \lvert t\rvert = \begin{cases} t, & 0 < t < \pi \\ -t, & -\pi < t < 0 \end{cases}$	$\frac{\pi}{2} - \frac{4}{\pi}\left[\frac{\cos(t)}{1^2} + \frac{\cos(3t)}{3^2} + \frac{\cos(5t)}{5^2}\right]$
$x_{2\pi}(t) = t, -\pi < t < \pi$	$2\left[\frac{\sin(t)}{1} - \frac{\sin(2t)}{2} + \frac{\sin(3t)}{3} + \cdots\right]$
$x_{2\pi}(t) = t^2, -\pi < t < \pi$	$\frac{\pi^2}{3} - 4\left[\frac{\cos(t)}{1} - \frac{\cos(2t)}{2^2} + \frac{\cos(3t)}{3^2} - \cdots\right]$
$x_{2\pi}(t) = \lvert\sin(t)\rvert, -\pi < t < \pi$	$\frac{2}{\pi} - \frac{4}{\pi}\left[\frac{\cos(2t)}{(1)(3)} + \frac{\cos(4t)}{(3)(5)} + \frac{\cos(6t)}{(5)(7)} + \cdots\right]$
$x_{2\pi}(t) = \begin{cases} \sin(t), & 0 < t < \pi \\ 0, & \pi < t < 2\pi \end{cases}$	$\frac{1}{\pi} + \frac{1}{2}\sin(t) - \frac{2}{\pi}\left[\frac{\cos(2t)}{(1)(3)} + \frac{\cos(4t)}{(3)(5)} + \frac{\cos(6t)}{(5)(7)} + \cdots\right]$
$x_{2\pi}(t) = \begin{cases} \cos(t), & 0 < t < \pi \\ -\cos(t), & -\pi < t < 0 \end{cases}$	$\frac{8}{\pi}\left[\frac{\sin(2t)}{(1)(3)} + \frac{2\sin(4t)}{(3)(5)} + \frac{3\sin(6t)}{(5)(7)} + \cdots\right]$
$x_{2\pi}(t) = e^t, -\pi < t < \pi$	$\frac{2\sinh\pi}{\pi}\left[\frac{1}{2} + \sum\limits_{k=1}^{\infty} \frac{(-1)^k}{(1+k^2)}\left(\cos(kt) - k\sin(kt)\right)\right]$

- Bounds and convergence of the F-series to the given function
- Half-range expansions

Problems

3.1.1 The set of functions $\phi_i(t), i = 1, 2, 3, 4$ shown in Fig. P3.1.1 are a set of *Walsh functions*. Show that they are orthogonal in the interval $[0, 1]$.

3.1.2 Consider the set $\{\phi_1(t), \phi_2(t)\}$ with $\phi_1(t) = 1$ and $\phi_2(t) = c(1 - 2t)$. Is this an orthogonal set in the interval $[0, 1]$? If so, compute the value of c that makes the functions $\phi_1(t)$ and $\phi_2(t)$ become an orthonormal basis set.

3.1.3 The function $x(t) = \sin(t)$ is approximated by $x(t) = c_1\phi_1(t) + c_2\phi_2(t)$. Use the results in Problem 3.1.2 and find the constants c_1 and c_2 so that the

mean-squared error is minimized between the given function and the approximated function.

3.1.4 Show the set $\phi_k(t) = \sqrt{(2/T)}\sin(k\pi/T)t$, $k = 1, 2, 3, \ldots$ is an orthogonal set in the interval $(0, T)$.

3.1.5 Use the first five Legendre polynomials to approximate the following function:

$$x(t) = \begin{cases} 0, -1 < t < 0 \\ 1, \quad 0 < t < 1 \end{cases}.$$

3.2.1 Use the Walsh functions given in Problem 3.1.1 to approximate the following function and find the mean-squared error between the given function and approximation:

$$x(t) = \begin{cases} t, \quad 0 < t < 1 \\ 0, \quad \text{otherwise} \end{cases}.$$

3.2.2 Consider the equations given in matrix form given below. There is no value of α that satisfies the set of equations given below. The system is called an *overdetermined system of equations*:

$$\begin{bmatrix} 1 \\ 2 \end{bmatrix} = \begin{bmatrix} 1 \\ 1 \end{bmatrix}\alpha.$$

The L_p error between the two sides of the above equation is $E_p = |(1 - \alpha)|^p + |(2 - \alpha)|^p$. If $p = 2$, we call that as a least-squares error or L_2 error and for $p = 1$, we call that as the L_1 error. Find the value of α that minimizes the L_1 and L_2 errors. The least-squares error can be computed by taking the partial of E_2 with respect to α, equating it to zero and solving for α, see Section A.8.1. A simple way to solve the L_1 error problem is solve each equation and find the value of α out of the two solutions that gives the minimum L_1 error. For iterative L_p solutions, see Yarlagadda, Bednar and Watt (1986).

3.3.1 Determine the complex Fourier series of the function

$$x_T(t) = \cos(2\pi f_0 t - 1) + \sin(2\pi(2f_0)t - 2).$$

3.3.2 *a.* Determine the period of the function

$$x_T(t) = \sum_{k=-\infty}^{\infty} \frac{1}{(a + j\pi kb)} e^{j(3\pi kt/2)}.$$

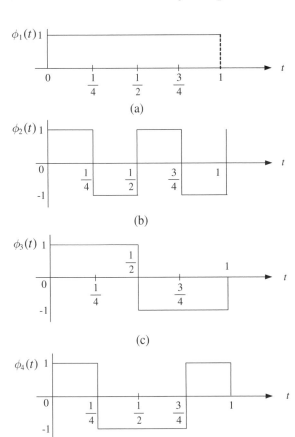

(a)

(b)

(c)

(d)

Fig. P3.1.1 Walsh functions

b. What is the average value of the periodic function $x_T(t)$?

c. Determine the amplitude and phase values of the third harmonic.

3.3.3 Expand the following periodic functions with $x_i(t) = x_i(t + 2\pi)$ in trigonometric F-series:

a. $x_1(t) = e^{at}, -\pi < t < \pi, a$ is some constant

b. $x_2(t) = \cosh at, -\pi \le t \le \pi$.

c. $x_3(t) = t(\pi - t), 0 < t < \pi$,

d. $x_4(t) = \begin{cases} \cos(t), & 0 < t < \pi \\ -\cos(t), & -\pi < t < 0 \end{cases}$.

3.3.4 Find the F-series of the function $x_1(t) = At^2 + Bt + C, -\pi < t < \pi$.

3.3.5 Show that the equation in (3.3.11) reduces to the equation in (3.3.10).

3.4.1 Use the derivative method to determine the trigonometric Fourier series of the periodic function $x_{2\pi}(t) = t, 0 \le t < 2\pi$.

3.4.2 Find the trigonometric F-series of the function shown below using a. derivative method. b. What can you say about the convergence of its F-series coefficients?

$$x_T(t) = \begin{cases} 2\cos[\pi t/2], & -1 < t < 1 \\ 0, & 1 < |t| < T/2 \end{cases}.$$

c. What can you say about the convergence of the trigonometric F-series coefficients?

3.4.3 Use the generalized derivatives of the function given in Problem 3.4.2 and see how fast the series converge without actually computing the Fourier series and verify the results obtained in that problem.

3.4.4 Show that $|X_s[k]| = |X_s[-k]|$ for a real periodic function $x_T(t)$. Give a complex periodic function where this is not true.

3.5.1 The function and its trigonometric F-series expansion are given by

$$x_{2\pi}(t) = |\sin(t)|, 0 \le t < 2\pi,$$

$$|\sin(t)| = \frac{2}{\pi} - \frac{4}{\pi}$$

$$\left[\frac{\cos(2t)}{3} + \frac{\cos(4t)}{15} + \frac{\cos(6t)}{35} + \cdots \right].$$

Derive an expression for π using the F-series. See Example 3.7.4. What is the value of the F-series at $t = 0$? Compare this to the actual value of the function at that location.

3.5.2 Using the Fourier series expansion, determine the sum A identified below

$$x(t) = \cosh at = \frac{2}{\pi} \sinh(a\pi)$$

$$\left[\frac{1}{2a} + \sum_{k=1}^{\infty} (-1)^k \frac{a}{k^2 + a^2} \cos(kt) \right], -\pi \le t \le \pi.$$

$$A = \frac{1}{2a} + \sum_{k=1}^{\infty} (-1)^k \frac{1}{k^2 + a^2}$$

3.5.3 Find the F-series expansion of the function $y(t) = \sinh(at)$ using the above results.

3.6.1 Consider the full-wave rectified function $x(t) = |\sin(t)|, -\pi < t < \pi$. Estimate the rms value of the full-wave rectified signal by using the first four nonzero terms in the Fourier series representation of the function. Calculate the percentage of error in the estimation.

3.6.2 Consider the triangular wave function given below and derive the trigonometric F-series of this function.

$$x_T(t) = \begin{cases} 1 + \frac{4t}{T}, & -\frac{T}{2} < t \le 0 \\ 1 - \frac{4t}{T}, & 0 \le t < \frac{T}{2} \end{cases}, x_T(t + T) = x_T(t).$$

3.6.3 Show that

$$\frac{1}{T} \int_T x_T(t) x_T(t - \tau) dt =$$

$$\frac{1}{T} \int_T x_{Te}(t) x_{Te}(t - \tau) dt + \frac{1}{T} \int_T x_{T0}(t) x_{T0}(t - \tau) dt.$$

3.6.4 Show the integral of $x_T(t)$ with a nonzero average value is non-periodic.

3.6.5 Show the derivative of an even function is an odd function and the derivative of an odd function is an even function. Verify this property using the trigonometric F-series.

3.7.1 Using the F-series in Table 3.10.2, determine

$$A = \sum_{k=1}^{\infty} [1/(2k-1)^2].$$

3.7.2 Verify the Fourier series of the periodic function $x_{2\pi}(t) = t$, $-\pi < t < \pi$ are

$$x_{2\pi}(t) = 2 \sum_{k=1}^{\infty} \left[(-1)^{k-1}/k \right] \sin(kt).$$

What value does this function converges to at $t = 0, (\pi/2), \pi$?

3.7.3 Give the convergence rate of the F-series coefficients of the periodic functions without actually using the F-series.

$$a. x_1(t) = t - 1, 0 < t < 1, \; x_1(t + T) = x_1(t),$$

$$T = 1, \quad b. x_{2\pi}(t) = \begin{cases} \sin(t), 0 < t < \pi \\ 0, \pi < t < 2\pi \end{cases}.$$

3.8.1 Derive the expressions for the coefficients $b[k]$ in (3.8.13).

3.8.2 Consider the periodic function $x_{2\pi}(t)$ below and identify any symmetry this has.

$$x_{2\pi}(t) = \begin{cases} 1, & 0 < t < \pi \\ -1, & -\pi < t < 0 \end{cases}, x_{2\pi}(t) = x_{2\pi}(t + 2\pi).$$

a. Use that to derive the trigonometric F-series of this function. Give the corresponding complex F-series. *b.* Use the derivative method to derive the trigonometric F-series.

3.8.3 Consider the periodic triangular wave function below. Does this function have any symmetry? Derive the trigonometric F-series of this function.

$$g_{2\pi}(t) = \begin{cases} t - (\pi/2), & 0 < t < \pi \\ -t - (\pi/2), & -\pi < t < 0 \end{cases},$$

$$g_{2\pi}(t) = g_{2\pi}(t + 2\pi).$$

3.8.4 Give the complex F-series of the full-wave rectified signal $x_{2\pi}(t) = |\cos(t)|$. Use the Fourier series of the results on the full-wave rectified sine wave in Table 3.10.2.

3.9.1 Verify the trigonometric Fourier series given in Table 3.10.2 for the following periodic functions in the range $-\pi < t < \pi$.

$$a. \; x_{2\pi}(t) = t^2, \; b. \; x_{2\pi}(t) = e^t, c. \; x_{2\pi}(t) = \cos(at).$$

3.9.2 Use the results in Table 3.10.2 and the derivative method to derive the F-series of the periodic function $x_{2\pi}(t) = \sin(at)$, $-\pi < t < \pi$.

3.9.3 Derive an expression for the F-series of the function $x_{2\pi}(t - \pi)$, $-\pi < t < \pi$ given

$$x_{2\pi}(t) = X_s[0] + \sum_{k=1}^{\infty} a[k] \cos(k\omega_0 t)$$

$$+ \sum_{k=1}^{\infty} b[k] \sin(k\omega_0 t), -\pi < t < \pi.$$

3.9.4 Consider the functions $x(t) = \cos(t)$ and $y(t) = \sin(t)$ over the interval $0 < t < \pi$ and 0 otherwise. Expand these functions using the Fourier sine series and cosine series by directly going through the procedure discussed in Section 3.9. Can you think of a simpler method knowing the results given in Example 3.9.1?

Chapter 4
Fourier Transform Analysis

4.1 Introduction

In Chapter 3 we have discussed the frequency representation of a periodic signal. Fourier series expansions of periodic signals give us a basic understanding how to deal with signals in general. Since most signals we deal with are aperiodic energy signals, we will study these in terms of their Fourier transforms in this chapter. Fourier transforms can be derived from the Fourier series by considering the period of the periodic function going to infinity. Fourier transform theory is basic in the study of signal analysis, communication theory, and, in general, the design of systems. Fourier transforms are more general than Fourier series in some sense. Even periodic signals can be described using Fourier transforms. Most of the material in this chapter is standard (see Carlson, (1975), Lathi, (1983), Papoulis, (1962), Morrison, (1994), Ziemer and Tranter, (2002), Haykin and Van Veen, (1999), Simpson and Houts, (1971), Baher, (1990), Poulariskas and Seely, (1991), Hsu, (1967, 1993), Roberts, (2007), and others).

4.2 Fourier Series to Fourier Integral

Consider a periodic signal $x_T(t)$ with period T and its complex Fourier series

$$x_T(t) = \sum_{k=-\infty}^{\infty} X_s[k]e^{jk\omega_0 t},$$

$$X_s[k] = \frac{1}{T} \int_{-T/2}^{T/2} x_T(t)e^{-jk\omega_0 t}dt, \quad \omega_0 = 2\pi/T. \tag{4.2.1}$$

The frequency $f_0 = \omega_0/2\pi$ is the fundamental frequency of the signal, which is the inverse of the period of the signal $f_0 = (1/T)$. The Fourier series coefficients are complex in general. To make the analysis simple we assume the signal under consideration is real and the amplitude of the Fourier coefficients is given by $|X_s[k]|$. Figure 4.2.1b gives the sketch of the amplitude line spectra of the complex Fourier series of a periodic function shown in Fig. 4.2.1a. The frequencies are located at $k\omega_0 = 2\pi(kf_0)$, $k = 0, \pm 1, \pm 2, \dots$ and the frequency interval between the adjacent line spectra is $\omega_0 = 2\pi f_0 = 2\pi/T$. In this example, we assumed $\tau/T = 1/5$. As $T = 2\pi/\omega_0 \to \infty$, ω_0 goes to zero and the spectral lines merge. To quantify this, let

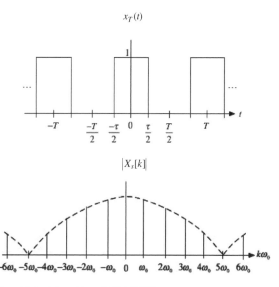

Fig. 4.2.1 (a) $x_T(t)$ and (b) $|X_s(k)|$

R.K.R. Yarlagadda, *Analog and Digital Signals and Systems*, DOI 10.1007/978-1-4419-0034-0_4,
© Springer Science+Business Media, LLC 2010

us extract one period of the periodic signal by defining

$$x(t) = \begin{cases} x_T(t), & -\frac{T}{2} < t < \frac{T}{2} \\ 0, & \text{Otherwise} \end{cases}. \quad (4.2.2)$$

We can consider the function in (4.2.2) as a periodic signal with period equal to ∞. In the expression for the Fourier series coefficients in (4.2.1), k and ω_0 appear as product $(k\omega_0) = (k(2\pi)/T)$. As $T \to \infty$, the expression for the Fourier series coefficients in (4.2.1) results in a value equal to zero, which does *not* provide any spectral information of the signal. To avoid this problem, define

$$X(k\omega_0) = TX_s[k] = \int_{-T/2}^{T/2} x(t)e^{-jk\omega_0 t}dt. \quad (4.2.3)$$

Note that k is an integer and it can take any integer value from $-\infty$ to ∞. Furthermore, as $T \to \infty$, $\omega_0 = (2\pi/T)$ becomes an incremental value, $k\omega_0$ becomes a continuous variable ω on the frequency axis, and $X(k\omega_0)$ becomes $X(\omega)$. From this, we have the *analysis equation* for our single pulse:

$$X(j\omega) = \lim_{T\to\infty}(TX_s[k]) = \lim_{T\to\infty}\int_{-T/2}^{T/2} x(t)e^{-jk\omega_0 t}dt$$

$$= \int_{-\infty}^{\infty} x(t)e^{-j\omega t}dt. \quad (4.2.4)$$

Now consider the *synthesis equation* in terms of the Fourier series in the forms

$$x_T(t) = \sum_{k=-\infty}^{\infty} X_s[k]e^{jk\omega_0 t}, \quad -T/2 < t < T/2, \quad (4.2.5)$$

$$x_T(t) = \frac{1}{T}\sum_{k=-\infty}^{\infty} X(k\omega_0)e^{jk\omega_0 t}. \quad (4.2.6)$$

One can see that as $T \to \infty$, $k\omega_0 \to \omega$, a continuous variable, and $\omega_0 = 2\pi/T \to d\omega$, an incremental value and the summation becomes an integral. These result in

$$x(t) = \lim_{T\to\infty}\left[\frac{1}{2\pi}\sum_{k=-\infty}^{\infty} X(k\omega_0)e^{j(k\omega_0)t}\omega_0\right]$$

$$= \frac{1}{2\pi}\int_{-\infty}^{\infty} X(j\omega)e^{j\omega t}d\omega. \quad (4.2.7)$$

We now have the Fourier transform of the time function $x(t)$, $X(j\omega)$, and its inverse Fourier transform. Some authors use $X(j\omega)$ instead of $X(\omega)$ for the Fourier transform, indicating the transform is a function of complex variable $(j\omega)$. The pair of functions $x(t)$ and $X(j\omega)$ is referred to as a Fourier transform pair:

$$X(j\omega) = F[x(t)] = \int_{-\infty}^{\infty} x(t)e^{-j\omega t}dt, \quad (4.2.8a)$$

$$x(t) \cong \tilde{x}(t) = F^{-1}[X(j\omega)] = \frac{1}{2\pi}\int_{-\infty}^{\infty} X(j\omega)e^{j\omega t}d\omega,$$

$$(4.2.8b)$$

$$x(t) \overset{\text{FT}}{\longleftrightarrow} X(j\omega). \quad (4.2.8c)$$

The transform and its inverse transform can be written in terms of frequency variable f in Hertz instead of $\omega = 2\pi f$:

$$X(jf) = \int_{-\infty}^{\infty} x(t)e^{-j2\pi ft}dt, \quad x(t) = \int_{-\infty}^{\infty} X(jf)e^{j2\pi ft}df.$$

$$(4.2.8d)$$

Equation (4.2.8c) shows that the transform and its inverse have the same general form, one has the time function and an exponential term with negative exponent and the other has the transform and an exponential term with a positive exponent in the corresponding integrands. F-transforms are applicable for both real and complex functions. Most practical signals are real signals. The transforms are generally complex. Integrals in the transform and its inverse are with respect to a *real variable*. The relations between the Laplace transforms, considered in Chapter 5, and the Fourier transforms become evident with this form. The form in (4.2.8d) is adopted by the engineers in the communications area. The transforms are computed by integration and the inverse transforms are determined by using transform tables. Since the Fourier transform is derived from the Fourier series, we can now say that

$$x(t) \xrightarrow{FT} X(j\omega) \xrightarrow{\text{Inverse FT}} \tilde{x}(t) \simeq x(t).$$

The inverse transform of $X(j\omega), F^{-1}[X(j\omega)]$ identified by $\tilde{x}(t)$ is an *approximation* of $x(t)$ and it may not be the same as the function $x(t)$. We will have more on this shortly.

Fourier transform is applicable to signals that obey the Dirichlet conditions (see Section 3.1), with the exception now that $x(t)$ must be absolutely integrable over all time, which is a *sufficient* but *not a necessary* condition. Periodic functions violate the last condition of absolute integrability over all time and will be considered in a later section. There are many functions that do not have Fourier transforms. For example, the Fourier transform of the function e^{at}, $a>0$ is not defined. The functions that can be generated in a laboratory have Fourier transforms. Existence of the Fourier transforms will not be discussed any further. In the synthesis equation, the inverse transform given by (4.2.8b) is an approximation of the function $x(t)$. The function $x(t)$ and its approximation $\tilde{x}(t)$ are equal in the sense that the error $e(t) = [x(t) - \tilde{x}(t)]$ is *not equal to zero* for all t and may differ significantly from zero at a discrete set of points t, but

$$\int_{-\infty}^{\infty} |e(t)|^2 dt = 0. \qquad (4.2.9)$$

In the sense that the integral squared error is zero, the equality $x(t) = \tilde{x}(t)$ between the function and its approximation is valid. Physically realizable signals have F-transforms and when they are inverted, they provide the original function. Physically realizable signals do not have any jump discontinuities. That is $\tilde{x}(t) = x(t)$. If the function $x(t)$ has jump discontinuities, then the reconstructed function $\tilde{x}(t)$ exhibits Gibbs phenomenon. The reconstructed function converges to the *half-point* at the discontinuity and will have overshoots and undershoots before and after the discontinuity (see Section 3.7.1).

Example 4.2.1 Determine $F\{x(t)\} = F\{A\Pi[t/\tau]\}$ and $F\{y(t)\} = F\{x(t - (\tau/2))\}$ using the F-series of

$$x_T(t) = x(t), \ |t|<\tau/2, \ x_T(t) = \sum_{n=-\infty}^{\infty} x(t + nT).$$

$$x_T(t) = A\Pi[t/\tau], \ |\tau|<T/2, \ x_T(t) = x_T(t + T),$$
$$y_T(t) = x_T(t - (\tau/2)). \qquad (4.2.10)$$

Solution: The complex Fourier series coefficients of $x_T(t)$ were given in Example 3.4.2. The Fourier series coefficients and their amplitudes of the two functions are given by

$$X_s[k] = A(\tau/T) \frac{\sin(k\omega_0\tau/2)}{(k\omega_0\tau/2)},$$

$$Y_s[k] = X_s[k]e^{-jk\omega_0(\tau/2)}, \ \omega_0 = 2\pi/T. \qquad (4.2.11)$$

$$|Y_s[k]| = |X_s[k]| = \left|\frac{A\tau}{T}\right| \left|\frac{\sin(k\omega_0(\tau/2))}{(k\omega_0\tau/2)}\right|.$$

By using the complex Fourier series in (4.2.4), the transforms of the two pulses $x(t)$ and $y(t)$ are given below:

$$y(t) = A\Pi[(t - \tau/2)/\tau], \qquad (4.2.12)$$

$$Y(j\omega) = \lim_{T\to\infty} TY_s[k] = A\tau \frac{\sin(\omega\tau/2)}{(\omega\tau/2)} e^{-j\omega(\tau/2)}, \qquad (4.2.13a)$$

$$X(j\omega) = \lim_{T\to\infty} TX_s[k] = A\tau \frac{\sin(\omega\tau/2)}{(\omega\tau/2)}. \qquad (4.2.13b)$$

Obviously it is simpler to compute the transform directly. For example,

$$Y(j\omega) = \int_{-\infty}^{\infty} x(t)e^{-j\omega t} dt = \int_{0}^{\tau} Ae^{-j\omega t} dt = \frac{A}{-(j\omega)} e^{-j\omega t}\Big|_0^\tau$$

$$= A \frac{1 - e^{-j\omega\tau}}{j\omega} = A\tau \frac{e^{j\omega(\tau/2)} - e^{-j\omega(\tau/2)}}{2j\omega\tau/2} e^{-j\omega\tau/2}$$

$$= A\tau \frac{\sin(\omega\tau/2)}{(\omega\tau/2)} e^{-j\omega\tau/2}. \qquad (4.2.13c)$$

The transform of the pulse $x(t)$ is

$$X(j\omega) = A\tau \frac{\sin(\omega\tau/2)}{(\omega\tau/2)}. \qquad (4.2.14)$$

We should note that $y(t)$ is a delayed version of $x(t)$ and the delay explicitly appears in the phase spectra. See the difference between (4.2.13c) and (4.2.14). ∎

Since the F-transform is derived from the F-series, many of the properties for the F-series can be modified to derive the transform properties with some exceptions. Let a *time limited function* transform $x(t) \overset{\text{FT}}{\longleftrightarrow} X(j\omega)$ be defined over the interval $t_0 < t < t_0 + T$ and zero everywhere else. This implies

$$x_T(t) = \sum_{n=-\infty}^{\infty} x(t + nT) = \sum_{k=-\infty}^{\infty} X_s[k] e^{jk\omega_0 t}$$
$$\Rightarrow X_s[k] = (1/T) X(j\omega)|_{\omega=k\omega_0}. \qquad (4.2.15)$$

If $x(t)$ is not time limited to a T second interval, then the function $x(t)$ cannot be extracted from $x_T(t)$ and (4.2.15) is not valid.

4.2.1 Amplitude and Phase Spectra

Let the Fourier transform of a real signal $x(t)$ be given by $X(j\omega)$. It is usually complex and can be written as either in terms of the rectangular or the polar form:

$$X(j\omega) = R(\omega) + jI(\omega) = A(\omega) e^{j\theta(\omega)}. \qquad (4.2.16a)$$

The functions $R(\omega)$ and $I(\omega)$ are the *real* and the *imaginary* parts of the spectrum. In the polar form, the magnitude or the *amplitude* and the *phase spectra* are given by

$$A(\omega) = |X(j\omega)| = \sqrt{R^2(\omega) + I^2(\omega)},$$
$$\theta(\omega) = \tan^{-1}[I(\omega)/R(\omega)].$$
$$(4.2.16b)$$

When $x(t)$ is *real*, the transform satisfies the properties that $R(\omega)$ and $I(\omega)$ are *even* and *odd* functions of ω, respectively. That is,

$$R(-\omega) = R(\omega) \quad \text{and} \quad I(-\omega) = -I(\omega). \qquad (4.2.17)$$

These can be easily verified using Euler's identity and

$$X(j\omega) = \int_{-\infty}^{\infty} x(t) e^{-j\omega t} dt = \int_{-\infty}^{\infty} x(t) \cos(\omega t) dt$$
$$- j \int_{-\infty}^{\infty} x(t) \sin(\omega t) dt = R(\omega) + jI(\omega),$$

$$R(\omega) = \int_{-\infty}^{\infty} x(t) \cos(\omega t) dt,$$

$$I(\omega) = - \int_{-\infty}^{\infty} x(t) \sin(\omega t) dt. \qquad (4.2.18a)$$

Since $\cos(\omega t)$ is even and $\sin(\omega t)$ is odd, the equalities in (4.2.18a) follow. As a consequence, for any *real signal* $x(t)$, we have

$$X(-j\omega) = R(-\omega) + jI(-\omega) = R(\omega) = jI(\omega) = X^*(j\omega). \qquad (4.2.18b)$$

From (4.2.16b) and (4.2.17), the amplitude spectrum $|X(j\omega)|$ of a real signal is even and the phase spectrum $\theta(\omega)$ is odd. That is,

$$|X(-j\omega)| = |X(j\omega)|, \quad \theta(-\omega) = -\theta(\omega). \qquad (4.2.19)$$

Interesting transform relations in terms of the even and odd parts of a real function: If $x(t) = x_e(t) + x_0(t)$, a real function, then the following is true:

$$x(t) = x_e(t) + x_0(t) \overset{\text{FT}}{\longleftrightarrow} R(\omega) + jI(\omega), \qquad (4.2.20)$$

$$x_e(t) = [x(t) + x(-t)]/2 \overset{\text{FT}}{\longleftrightarrow} R(\omega) \text{ and}$$
$$x_0(t) = [x(t) - x(-t)]/2 \overset{\text{FT}}{\longleftrightarrow} jI(\omega). \qquad (4.2.21)$$

These can be seen from

$$X(j\omega) = \int_{-\infty}^{\infty} x(t) e^{-j\omega t} dt$$
$$= \int_{-\infty}^{\infty} [x_e(t) + x_0(t)][\cos(\omega t) - j\sin(\omega t)] dt$$
$$= \int_{-\infty}^{\infty} x_e(t) \cos(\omega t) dt - j \int_{-\infty}^{\infty} x_0(t) \sin(\omega t) dt$$
$$- j \int_{-\infty}^{\infty} x_e(t) \sin(\omega t) dt + \int_{-\infty}^{\infty} x_0(t) \cos(\omega t) dt$$
$$= \int_{-\infty}^{\infty} x_e(t) \cos(\omega t) dt - j \int_{-\infty}^{\infty} x_0(t) \sin(\omega t) dt$$
$$= R(\omega) + jI(\omega) \qquad (4.2.22a)$$

$$R(\omega) = \int\limits_{-\infty}^{\infty} x_e(t)\cos(\omega t)dt,$$

$$I(\omega) = -\int\limits_{-\infty}^{\infty} x_0(t)\sin(\omega t)dt. \quad (4.2.22b)$$

Note that integral of an odd function over a symmetric interval is zero. The F-transform of a *real and even function* is *real and even* and the F-transform of a *real and odd function* is *pure imaginary*. The transform of a real function $x(t)$ can be expressed in terms of a real integral:

$$x(t) = \frac{1}{2\pi}\int\limits_{-\infty}^{\infty} X(j\omega)e^{j\omega t}d\omega$$

$$= \frac{1}{2\pi}\int\limits_{-\infty}^{\infty} |X(j\omega)|e^{j\theta(\omega)}e^{j\omega t}d\omega$$

$$= \frac{1}{2\pi}\int\limits_{-\infty}^{\infty} |X(j\omega)|e^{j(\omega t+\theta(\omega))}d\omega$$

$$= \frac{1}{\pi}\int\limits_{0}^{\infty} |X(j\omega)|\cos(\omega t + \theta(\omega))d\omega. \quad (4.2.23)$$

Note $|X(j\omega)|$ and $\sin(-\omega t + \theta(-\omega))$ are even and odd functions, respectively.

Example 4.2.2 Rectangular (or a *gating pulse*) is given by $x(t) = \Pi[(t - t_0)/\tau]$.

 a. Give the expression for the transform using (4.2.13a).

 b. Compute the amplitude and the phase spectra associated with the gating pulse function.

 c. Sketch the magnitude and phase spectra of this function assuming $t_0 = \tau/2$.

Solution: *a.* From (4.2.13a), we have

$$X(j\omega) = \tau\frac{\sin(\omega\tau/2)}{(\omega\tau/2)}e^{-j\omega t_0} = \tau\text{sinc}(\omega\frac{\tau}{2})e^{-j\omega t_0}$$

$$= \tau\text{sinc}(2\pi f\frac{\tau}{2})e^{-j\omega t_0},$$

$$\Pi\left[\frac{t - t_0}{\tau}\right] \xrightarrow{\text{FT}} \tau\,\text{sinc}(\omega\tau/2)e^{j\omega t_0} = \tau\,\text{sinc}(\pi f\tau)e^{-j2\pi f t_0}.$$

$$(4.2.24a)$$

 b. The magnitude and the phase spectra are, respectively, given by

$$|X(j\omega)| = \tau\left|\frac{\sin(\omega(\tau/2))}{\omega(\tau/2)}\right|\left|e^{-j\omega t_0}\right| = \tau|\text{sinc}(\omega\tau/2)|,$$

$$(4.2.24b)$$

$$\theta(\omega) = \begin{cases} -(\omega t_0), & \text{sinc}(\omega\tau/2) > 0 \\ -(\omega t_0) \pm \pi, & \text{sinc}(\omega\tau/2) < 0 \end{cases}. \quad (4.2.24c)$$

The time function $x(t)$, the magnitude spectrum $|X(j\omega)|$, and the phase spectrum are sketched in Fig. 4.2.2 assuming $t_0 = \tau/2$. At $\omega = k2\pi/\tau$,

$$\left|X(j\frac{k2\pi}{\tau})\right| = \begin{cases} \tau, & k = 0 \\ 0, & k \neq 0 \text{ and } k, \text{ an integer} \end{cases}.$$

$$(4.2.25a)$$

The discontinuity in the phase spectrum at $\omega = 2\pi/\tau$ can be seen from

$$\theta\left(\frac{2\pi^-}{\tau}\right) = -2\pi\left(\frac{1}{\tau}\right)\left(\frac{\tau}{2}\right) = -\pi, \theta\left(\frac{2\pi^+}{\tau}\right)$$

$$= (-\pi + \pi) = 0. \quad (4.2.25b)$$

We have added π in determining the phase angle in determining $\theta(2\pi^+/\tau)$ taking into consideration that the sinc function is negative in the range $0 < (2\pi/\tau) < \omega < 2(2\pi/\tau)$, i.e., in the first side lobe. In sketching the plots, appropriate multiples of (2π)

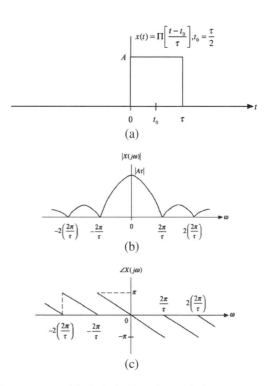

Fig. 4.2.2 (a) $x(t)$, (b) $|X(j\omega)|$, and (c) $\angle X(j\omega)$

have been subtracted to make the phase spectrum compact and the phase angle is bounded between $-\pi$ and π. Noting

$$|X(j\omega)| = \tau \left| \frac{\sin(\omega(\tau/2))}{\omega(\tau/2)} \right| \propto \frac{1}{|\omega|}, \qquad (4.2.25c)$$

it can be seen that the envelope of the magnitude spectrum gets smaller for larger frequencies. The exact frequency representation of the square pulse should include all frequencies in the reconstruction of this pulse, which is impractical. Therefore, keep only the frequencies that are significant and the range or the width of those significant frequencies is referred to as the "*bandwidth*". Keeping the desired frequencies is achieved by using a filter. Filters will be discussed in later chapters. There are several interpretations of bandwidth. For the present, the following explanation of the bandwidth is adequate and will quantify these measures at a later time.   ∎

4.2.2 Bandwidth-Simplistic Ideas

1. The width of the band of positive frequencies passed by a filter of an electrical system.
2. The width of the positive band of frequencies by the central lobe of the spectrum.
3. The band of frequencies that have most of the signal power.
4. The bandwidth includes the positive frequency range lying between two points at which the power is reduced to half that of the maximum. This width is referred to as the half-power bandwidth or the 3 dB bandwidth.

Note that only *positive frequencies* are used in defining the *bandwidth*. With bandwidth in mind, let us look at $|X(j\omega)|$ in Example 4.2.2, where τ is assumed to be the width of the pulse. Any signal that is nonzero for a finite period of time is referred to as a *time-limited* signal and the signal given in Example 4.2.2 is a time-limited signal. The main lobe width of the magnitude spectrum for positive frequencies is $(1/\tau)$ Hz. The spectrum is *not frequency limited*, as the spectrum occupies the entire frequency range, except that it is zero at $\omega = k(2\pi/\tau)$, $k \neq 0$, k – integer. The amplitude

spectrum $|X(j\omega)|$ gives the value $|X(j\omega_i)|$ at the frequency $\omega_i = 2\pi f_i$. Noting that most of the energy is in the main lobe, a standard assumption of bandwidth is generally assumed to be equal to k times half of the main lobe width $(2\pi/\tau)$, i.e., $(k(2\pi/\tau))$ rad/s or (k/τ) Hz. An interesting formula that ties the time and frequency widths is

(Time width) (Frequency width or bandwidth)

   = Constant

$$(4.2.26)$$

Clearly as τ decreases (increases), the main lobe width increases (decreases) and we say that bandwidth is inversely proportional to the time width. For most applications, $k = 1$ is assumed. Bandwidth is generally given in terms of Hz rather than rad/s.

4.3 Fourier Transform Theorems, Part 1

We will consider first a set of theorems or properties associated with the energy function $x(t)$ and its F-transform $X(j\omega)$. Transforms are applicable for both the real and complex functions. In Chapter 3, Parseval's theorem was given and it can be generalized to include energy signals and is referred to as *generalized Parseval's theorem, Plancheral's theorem* or *Rayleigh's energy theorem*.

4.3.1 Rayleigh's Energy Theorem

The energy in a complex or a real signal is

$$E_x = \int_{-\infty}^{\infty} |x(t)|^2 dt = \frac{1}{2\pi} \int_{-\infty}^{\infty} |X(j\omega)|^2 d\omega$$

$$= \int_{-\infty}^{\infty} |X(jf)|^2 df. \qquad (4.3.1)$$

This is proved in general terms first.

   Given $x(t) \overset{FT}{\longleftrightarrow} X(j\omega)$ and $y(t) \overset{FT}{\longleftrightarrow} Y(j\omega)$, then

$$E_{xy} = \int_{-\infty}^{\infty} x(t) y^*(t) dt = \frac{1}{2\pi} \int_{-\infty}^{\infty} X(j\omega) Y^*(j\omega) d\omega.$$

$$(4.3.2)$$

First,

$$E_{xy} = \int\limits_{-\infty}^{\infty} y^*(t)x(t)dt$$

$$= \int\limits_{-\infty}^{\infty} y^*(t)[\frac{1}{2\pi} \int\limits_{-\infty}^{\infty} X(j\omega)e^{j\omega t}d\omega]dt$$

$$= \frac{1}{2\pi} \int\limits_{-\infty}^{\infty} X(j\omega)[\int\limits_{-\infty}^{\infty} y^*(t)e^{j\omega t}dt]d\omega$$

$$= \frac{1}{2\pi} \int\limits_{-\infty}^{\infty} X(j\omega)[\int\limits_{-\infty}^{\infty} y(t)e^{-j\omega t}dt]^* d\omega$$

$$= \frac{1}{2\pi} \int\limits_{-\infty}^{\infty} X(j\omega) Y^*(j\omega)d\omega.$$

When $x(t) = y(t)$, the proof of Rayleigh's energy theorem in (4.3.1) follows. The transform of the function or the function itself can be used to find the energy.

Example 4.3.1 Compute the energy in the pulse E_1 using Rayleigh's energy theorem, the transform pair, and the identity Spiegel (1968) given below:

$$E_1 = \frac{1}{2\pi} \int\limits_{-\infty}^{\infty} \text{sinc}^2(\omega\tau/2)d\omega \left(\int\limits_{-\infty}^{\infty} \frac{\sin^2(p\alpha)}{\alpha^2} d\alpha = \pi p \right).$$

$$(4.3.3)$$

$$\Pi\left[\frac{t}{\tau}\right] \overset{\text{FT}}{\longleftrightarrow} \tau \, \text{sinc}(\omega\tau/2).$$

Solution: Using the energy theorem,

$$E_1 = \frac{1}{2\pi} \int\limits_{-\infty}^{\infty} \text{sinc}^2(\omega\tau/2)d\omega = \left(\frac{1}{\tau}\right)^2 \int\limits_{-\infty}^{\infty} \Pi^2\left(\frac{t}{\tau}\right)dt$$

$$= \left(\frac{1}{\tau}\right)^2 (\tau) = \frac{1}{\tau} = \frac{1}{\tau^2} \int\limits_{-\tau/2}^{\tau/2} dt. \qquad (4.3.4)$$

Bandwidth of a rectangular pulse of width τ is usually taken $(1/\tau)$ Hz corresponding to the first zero crossing point of the spectrum. Energy contained in the main lobe of the sinc function can be computed by numerical methods, such as the

rectangular formula or the trapezoidal formula discussed in Chapter 1. ∎

Example 4.3.2 Compute the energy in the main lobe of the sinc function and compare with the total energy in the function using the following:

$$E_{\text{Main lobe}} = \frac{1}{2\pi} \int\limits_{-2\pi/\tau}^{2\pi/\tau} \text{sinc}^2(\omega\tau/2)d\omega, \qquad (4.3.5)$$

$$= \frac{1}{\tau} \int\limits_{-1}^{1} \text{sinc}^2(\pi\alpha)d\alpha. \qquad (4.3.6)$$

Solution: To obtain (4.3.6) from (4.3.5), change of variable, $\pi\alpha = \omega\tau/2$, is used and the limits are from $\omega = \pm 2\pi/\tau$ to $\alpha = \pm 1$. Using the rectangular method of integration, the energies in the main lobe and in the pulse E_1 (see (4.3.4)) are

$$E_{\text{Main lobe}} \approx 0.924/\tau, \; E_1 = (1/\tau). \qquad (4.3.7)$$

The ratio of the energy in the main lobe, $E_{\text{Main lobe}}$, of the spectrum to the total energy in the pulse is 92.4%. That is, the main lobe has over 90% of the total energy in the pulse function. Therefore, a bandwidth of $(1/\tau)$ Hz is a reasonable estimate of the pulse function. ∎

4.3.2 Superposition Theorem

The Fourier transform of a linear combination of functions $F[x_i(t)] = X_i(j\omega)$, $i = 1, 2, ..., n$ with constants a_i, $i = 1, 2..., n$ is

$$F\left[\sum_{i=1}^{n} a_i x_i(t)\right] = \sum_{i=1}^{n} a_i X_i(j\omega),$$

$$\sum_{i=1}^{n} a_i x_i(t) \overset{\text{FT}}{\longleftrightarrow} \sum_{i=1}^{n} a_i X_i(j\omega). \qquad (4.3.8)$$

Since the integral of a sum is equal to the sum of the integrals, the proof follows. This theorem is useful in computing transforms of a function expressible as a sum of simple functions with known transforms. The F-transform of the function $x^*(t)$ is related to the transform of $x(t)$. This can be seen from

$$\int_{-\infty}^{\infty} x^*(t)e^{-j\omega t}dt = \left[\int_{-\infty}^{\infty} x(t)e^{j\omega t}dt\right]^* = X^*(-j\omega),$$

$$\Rightarrow x^*(t) \overset{\text{FT}}{\longleftrightarrow} X^*(-j\omega). \qquad (4.3.9)$$

4.3.3 Time Delay Theorem

The F-transform of a delayed function is given by

$$F[x(t - t_d)] = e^{-j\omega t_d} X(j\omega). \qquad (4.3.10)$$

This can be shown directly by using the change of variable $\alpha = t - t_d$ in the transform integral and

$$F[x(t - t_d)] = \int_{-\infty}^{\infty} x(t - t_d)e^{-j\omega t}dt$$

$$= \left[\int_{-\infty}^{\infty} x(\alpha)e^{-j\omega\alpha}d\alpha\right]e^{-j\omega t_d} = X(j\omega)e^{-j\omega t_d}$$

$$\Rightarrow |F[x(t \mp t_d)]| = |X(j\omega)e^{\pm j\omega t_d}| = |X(j\omega)| = |F[x(t)]|.$$
$$(4.3.11)$$

Superposition and delay theorems are useful in finding the Fourier transform pairs:

$$x(t - \tau) + x(t + \tau) \overset{\text{FT}}{\longleftrightarrow} 2X(j\omega)\cos(\omega\tau),$$
$$x(t - \tau) - x(t + \tau) \overset{\text{FT}}{\longleftrightarrow} -2jX(j\omega)\sin(\omega\tau). \qquad (4.3.12)$$

Example 4.3.3 Using the superposition and the delay theorem, compute the F-transform of the function shown in Fig. 4.3.1.

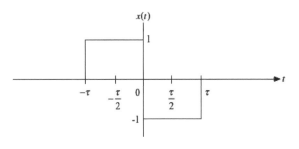

Fig. 4.3.1 Example 4.3.3

Solution: $x(t)$ can be expressed as a sum of two rectangular pulses and is

$$x(t) = \Pi\left[\frac{t + (\tau/2)}{\tau}\right] - \Pi\left[\frac{t - (\tau/2)}{\tau}\right]. \qquad (4.3.13a)$$

Using the superposition and delay theorems, we have

$$X(j\omega) = F[x(t)] = F\left\{\Pi\left[\frac{t + (\tau/2)}{\tau}\right] - \Pi\left[\frac{t - (\tau/2)}{\tau}\right]\right\}$$

$$= \left[e^{j\omega\tau/2} - e^{-j\omega\tau/2}\right]F\left\{\Pi\left[\frac{t}{\tau}\right]\right\}$$

$$= 2j\tau\sin(\omega\tau/2)\frac{\sin(\omega\tau/2)}{(\omega\tau/2)}. \qquad (4.3.13b)$$

Note that $x(t)$ is an *odd function* and therefore the transform is *pure imaginary*. ∎

Notes: In the above example a time-limited function, i.e., $x(t) = 0$ for $|t| > \tau$, was used and its transform is not frequency limited, as its spectrum occupies the entire frequency range. A signal $x(t)$ and its transform $X(j\omega)$ *cannot be both* time and frequency limited. We will come back to this later. ∎

4.3.4 Scale Change Theorem

The scale change theorem states that

$$F[x(at)] = \int_{-\infty}^{\infty} x(at)e^{-j\omega t}dt = \frac{1}{|a|}X\left(j\frac{\omega}{a}\right), a \neq 0.$$
$$(4.3.14)$$

This can be shown by considering the two possibilities, $a < 0$ and $a > 0$. For $a < 0$, by using the change of variable $\beta = at$ in (4.3.14) in the integral expression, we have

$$F[x(t)] = \int_{\infty}^{-\infty} x(\beta)e^{-j(\omega/a)\beta}\left(\frac{1}{a}\right)d\beta$$

$$= \left(-\frac{1}{a}\right)\int_{-\infty}^{\infty} x(\beta)e^{-j(\omega/a)\beta}d\beta$$

$$= \frac{1}{|a|}\int_{-\infty}^{\infty} x(\beta)e^{-j(\frac{\omega}{a})\beta}d\beta = \frac{1}{|a|}X\left(j\frac{\omega}{a}\right). \qquad (4.3.15)$$

When a is a negative number, $-a = |a|$. For $a > 0$, the proof similarly follows. The scale change theorem states that *timescale contraction (expansion)* corresponds to the *frequency-scale expansion (contraction)*.

Example 4.3.4 Use the scale change theorem to find the F-transforms of the following:

$$x_1(t) = \Pi\left[\frac{t}{(\tau/2)}\right] \quad \text{and} \quad x_2(t) = \Pi\left[\frac{t}{2\tau}\right]. \quad (4.3.16a)$$

Solution: Consider the transform pair (see (4.2.24a)) with $t_0 = 0$:

$$x(t) = \Pi[t/\tau] \overset{\text{FT}}{\longleftrightarrow} \tau \operatorname{sinc}(\omega\tau/2) = X(j\omega). \quad (4.3.16b)$$

Using the result in (4.3.14), we have

$$x_1(t) = x(2t) = \Pi\left[\frac{t}{\tau/2}\right] \overset{\text{FT}}{\longleftrightarrow} \frac{\tau}{2}\frac{\sin(\omega(\tau/4))}{\omega(\tau/4)}$$

$$= \frac{\tau}{2}\operatorname{sinc}(\omega(\tau/4)) = X_1(j\omega), \quad (4.3.16c)$$

$$x_2(t) = x\left(\frac{t}{2}\right) = \Pi\left[\frac{t}{2\tau}\right] \overset{\text{FT}}{\longleftrightarrow} (2\tau)\frac{\sin(\omega\tau)}{\omega\tau}$$

$$= (2\tau)\operatorname{sinc}(\omega\tau) = X_2(j\omega). \quad (4.3.16d)$$

The two functions and their amplitude spectra are sketched in Figs. 4.3.2a–d. Comparing the magnitude spectra, the main lobe width of $|X_1(j\omega)|$ is twice that of the main lobe width of $|X(j\omega)|$, whereas the main lobe width of $|X_2(j\omega)|$ is half the main lobe width of $|X(j\omega)|$. Consider Figs. 4.2.2 and 4.3.2. The main lobe width times its height in each of the cases are equal and $\tau(2\pi/\tau) = (\tau/2)(2\pi(2/\tau)) = (2\tau)(\pi/\tau) = 2\pi$. The

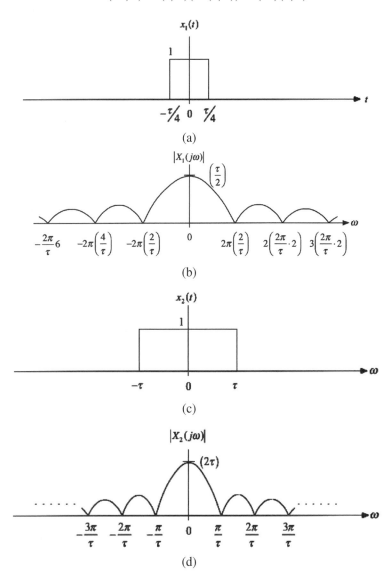

Fig. 4.3.2 (a) $x_1(t)$, (b) $|X_1(j\omega)|$, (c) $x_2(t)$, and (d) $|X_2(j\omega)|$

pulse amplitudes are all assumed to be equal to 1 for simplicity. For any a

$$\left| F[x(t)] \right| = \left| F\left\{ \Pi\left[\frac{t-a}{\tau}\right] \right\} \right| = \left| F\left\{ \Pi\left[\frac{t}{\tau}\right] \right\} \right|. \quad \blacksquare$$

Time reversal theorem: A special case of the scale change theorem is time reversal and

$$F[x(-t)] = X(-j\omega). \quad (4.3.17a)$$

This follows from the scale change theorem by using $a = -1$ in (4.3.14). We note that

$$x(t)\overset{FT}{\longleftrightarrow}X(j\omega), \quad x_1(t) = x(-t)\overset{FT}{\longleftrightarrow}X(-j\omega) = X_1(j\omega),$$

$$|X_1(j\omega)| = |X(-j\omega)| = |X(j\omega)|,$$
$$\angle X_1(\omega) = \angle X(-j\omega) = -\angle X(j\omega). \quad (4.3.17b)$$

Example 4.3.5 Find the F-transform of the following functions:

a. $x_1(t) = e^{-\alpha t}u(t)$,   b. $x_2(t) = e^{\alpha t}u(-t)$,

   $x_3(t) = e^{-\alpha|t|}, \quad \alpha > 0. \quad (4.3.18a)$

Solution: *a.* Using the F-transform integral results in

$$X_1(j\omega) = \int_{-\infty}^{\infty} e^{-\alpha t}u(t)e^{-j\omega t}\,dt = \int_{0}^{\infty} e^{-(\alpha+j\omega)t}\,dt$$

$$= \left[\frac{e^{-(\alpha+j\omega)t}}{-(\alpha+j\omega)}\right]\Big|_{t=0}^{t=\infty} = \frac{1}{(\alpha+j\omega)}. \quad (4.3.18b)$$

b. Using the time reversal theorem and the last part results in

$$X_2(j\omega) = X_1(-j\omega) = [1/(\alpha-j\omega)]. \quad (4.3.18c)$$

c. Noting that $x_3(t) = e^{-\alpha t}u(t) + e^{\alpha t}u(-t)$, the F-transform can be computed using the superposition theorem and the results in the last two parts. That is,

$$X_3(j\omega) = F[e^{-\alpha t}u(t)] + F[e^{\alpha t}u(-t)]$$
$$= \frac{1}{(\alpha+j\omega)} + \frac{1}{(\alpha-j\omega)} = \frac{2\alpha}{\alpha^2+(\omega)^2}. \quad (4.3.18d)$$

$\blacksquare$

Summary:

$$e^{-\alpha t}u(t)\overset{FT}{\longleftrightarrow}\frac{1}{(a+j\omega)}, \quad \alpha > 0, \quad (4.3.19a)$$

$$e^{\alpha t}u(-t)\overset{FT}{\longleftrightarrow}\frac{1}{\alpha-j\omega}, \quad \alpha > 0, \quad (4.3.19b)$$

$$e^{-\alpha|t|}\overset{FT}{\longleftrightarrow}\frac{2\alpha}{\alpha^2+\omega^2}, a > 0. \quad (4.3.20)$$

The time and frequency functions are not limited in time and frequency, respectively. $\blacksquare$

4.3.5 Symmetry or Duality Theorem

$$x(t)\overset{FT}{\longleftrightarrow}X(j\omega) \Rightarrow X(t)\overset{FT}{\longleftrightarrow}2\pi x(-\omega). \quad (4.3.21)$$

Starting with the expression for $2\pi x(t)$ and changing the variable from t to $-t$, we have

$$2\pi x(t) = \int_{-\infty}^{\infty} X(j\omega)e^{j\omega t}\,d\omega \rightarrow 2\pi x(-t)$$

$$= \int_{-\infty}^{\infty} X(j\omega)e^{-j\omega t}\,d\omega. \quad (4.3.22)$$

Interchanging t and $j\omega$ in (4.3.22) results in

$$2\pi x(-j\omega) = \int_{-\infty}^{\infty} X(t)e^{-j\omega t}\,dt. \quad (4.3.23)$$

This proves the result in (4.3.21). In terms of f (4.3.21) can be written as follows:

$$x(t)\overset{FT}{\longleftrightarrow}X(jf), \quad X(t)\overset{FT}{\longleftrightarrow}x(-jf). \quad (4.3.24)$$

A consequence of the symmetry property is if an F-transform table is available with N entries, then this property allows for doubling the size of the table.

Example 4.3.6 Using the duality theorem, show that

$$y(t) = \frac{1}{a^2+t^2}\overset{FT}{\longleftrightarrow}\frac{\pi}{a}e^{-a|\omega|} = Y(j\omega). \quad (4.3.25)$$

Solution: Using (4.3.20) and the duality property of the F-transforms, we have

$$\frac{1}{2a}e^{-a|t|}\overset{FT}{\longleftrightarrow}\frac{1}{a^2+\omega^2},$$

$$a > 0 \rightarrow \frac{1}{a^2+t^2}\overset{FT}{\longleftrightarrow}\frac{1}{2a}(2\pi)e^{-a|-j\omega|} = \frac{\pi}{a}e^{-a|\omega|}.$$

One can appreciate the simplicity of using the duality theorem compared to finding the transform directly in terms of difficult integrals given below:

$$Y(j\omega) = \int_{-\infty}^{\infty} \frac{1}{a^2 + t^2} e^{-j\omega t} d\omega = \int_{-\infty}^{\infty} \frac{1}{a^2 + t^2} \cos(\omega t) d\omega$$

$$- j \int_{-\infty}^{\infty} \frac{1}{a^2 + t^2} \sin(\omega t) d\omega. \qquad \blacksquare$$

Example 4.3.7 Determine the F-transform of $x(t)$ using the transform of the rectangular pulse given below and the duality theorem:

$$x(t) = \frac{\sin(at)}{\pi t}, \quad \Pi\left[\frac{t}{\tau}\right] \xleftrightarrow{\text{FT}} \tau \frac{\sin(\omega\tau/2)}{(\omega\tau/2)}.$$

Solution: Using the duality theorem and noting that Π-function is even, it follows

$$\tau \frac{\sin(t\tau/2)}{(t\tau/2)} \xleftrightarrow{\text{FT}} 2\pi\Pi\left[\frac{-j\omega}{\tau}\right] = 2\pi\Pi\left[\frac{\omega}{\tau}\right], \quad (4.3.26)$$

$$\frac{\sin(at)}{\pi t} \xleftrightarrow{\text{FT}} \Pi\left[\frac{\omega}{2a}\right]. \qquad (4.3.27)$$

Note $(\tau/2) = a$ in (4.3.26). For later use, let $a = 2\pi B$. Using this in (4.3.27) results in

$$y(t) = \frac{\sin(2\pi B t)}{(2\pi B t)} \xleftrightarrow{\text{FT}} \frac{1}{2B}\Pi\left[\frac{\omega}{2\pi(2B)}\right] = Y(j\omega). \quad (4.3.28)$$

Time domain *sinc pulses* are *not time limited* but are *band limited*. The sinc pulse and its transform in (4.3.28) are sketched in Fig. 4.3.3a,b, respectively. $\blacksquare$

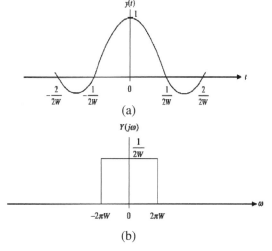

(a)

(b)

Fig. 4.3.3 (a) $y(t) = \frac{\sin(2\pi W t)}{(2\pi W t)}$ and (b) $Y(j\omega) = \frac{1}{2W}\Pi\left[\frac{\omega}{2\pi(2W)}\right]$

4.3.6 Fourier Central Ordinate Theorems

The value of the given function at $t = 0$ and its transform value at $\omega = 0$ are given by

$$X(0) = \int_{-\infty}^{\infty} x(t) dt, \qquad x(0) = \frac{1}{2\pi} \int_{-\infty}^{\infty} X(j\omega) d\omega.$$

$$(4.3.29)$$

Equation (4.3.29) points out that if we know the transform of a function, we can compute the integral of this function for all time by evaluating the spectrum at $\omega = 0$. In a similar manner the integral of the spectrum for all frequencies is given by $(2\pi)x(0)$.

Example 4.3.8 Use the transform pair in (4.3.28) to illustrate the ordinate theorems in (4.3.29) using the identity Spiegel (1968)

$$\int_{-\infty}^{\infty} \frac{\sin(p\alpha)}{\alpha} d\alpha = \begin{cases} \pi, & p > 0 \\ 0, & p = 0 \\ -\pi, & p < 0 \end{cases}. \qquad (4.3.30)$$

Solution: The integrals of the sinc function and the area of the pulse are

$$A_1 = \int_{-\infty}^{\infty} \frac{\sin(2\pi W t)}{(2\pi W t)} dt = \frac{1}{2\pi W} \int_{-\infty}^{\infty} \frac{\sin(2\pi W t)}{t} dt$$

$$= \frac{\pi}{2\pi W} = \frac{1}{2W}. \qquad (4.3.31a)$$

$$A_2 = \frac{1}{2W}\Pi\left[\frac{\omega}{2\pi(2W)}\right]\Big|_{\omega=0} = \frac{1}{2W} \Rightarrow A_2 = A_1.$$

$$(4.3.31b)$$

In a similar manner,

$$B_1 = \frac{\sin(2\pi W t)}{(2\pi W t)}\Big|_{t=0} = 1,$$

$$B_2 = \frac{1}{2\pi} \int_{-\infty}^{\infty} Y(j\omega) d\omega = \frac{1}{2\pi(2W)} 2\pi(2W)$$

$$= 1 \Rightarrow B_1 = B_2. \qquad (4.3.32)$$

$\blacksquare$

4.4 Fourier Transform Theorems, Part 2

Impulse functions are used in finding the transforms of periodic functions below.

Example 4.4.1 Find the Fourier transform of the impulse function in time domain and the inverse transform of the impulse function in the frequency domain.

Solution: Clearly

$$F[\delta(t - t_0)] = \int_{-\infty}^{\infty} \delta(t - t_0)e^{-j\omega t}dt = e^{-j\omega t}|_{t=t_0} = e^{-j\omega t_0}.$$

(4.4.1)

That is, an *impulse function* contains *all frequencies* with the same amplitude. That is $|F[\delta(t - t_0)]| = 1$. The inverse transform

$$F^{-1}[\delta(\omega - \omega_0)] = \frac{1}{2\pi} \int_{-\infty}^{\infty} \delta(\omega - \omega_0)e^{j\omega t}d\omega = \frac{1}{2\pi}e^{j\omega_0 t},$$

(4.4.2)

$$\Rightarrow F^{-1}[\delta(\omega)] = 1/2\pi, \quad F[1] = 2\pi\delta(\omega). \quad (4.4.3)$$

A constant contains only the single frequency at $\omega = 0$ (or $f = 0$). We refer to a constant as a *dc signal*. Symbolically we can express

$$\delta(t - t_0) \overset{FT}{\longleftrightarrow} e^{-j\omega t_0}, \quad e^{j\omega_0 t} \overset{FT}{\longleftrightarrow} 2\pi\delta(\omega - \omega_0). \quad (4.4.4)$$

The result on the right in the above equation follows by using the duality theorem. ∎

4.4.1 Frequency Translation Theorem

Multiplying a time domain function $x(t)$ by $e^{-j\omega_c t}$ shifts all frequencies in the signal $x(t)$ by ω_c. In general, the following transform pair is true:

$$x(t)e^{\pm j\omega_c t} \overset{FT}{\longleftrightarrow} X(j(\omega \mp \omega_c)). \quad (4.4.5)$$

Note

$$F^{-1}[X(j(\omega - \omega_c))] = \frac{1}{2\pi} \int_{-\infty}^{\infty} X(j(\omega - \omega_c))e^{j\omega t}d\omega$$

$$= \left[\frac{1}{2\pi} \int_{-\infty}^{\infty} X(ja)e^{jat}da\right] e^{j\omega_c t} = x(t)e^{j\omega_c t}.$$

(4.4.6)

This provides a way to modify a time function to shift its frequencies. The scale change and the frequency translation theorems can be combined.

Example 4.4.2 Show the following:

$$x(at)e^{j\omega_c t} \overset{FT}{\longleftrightarrow} \frac{1}{|a|} X\left(\frac{j(\omega - \omega_c)}{a}\right). \quad (4.4.7a)$$

Solution: Using the scale change theorem results in

$$x(at) \overset{FT}{\longleftrightarrow} \frac{1}{|a|} X\left(\frac{j\omega}{a}\right). \quad (4.4.7b)$$

Using the frequency translation theorem, i.e., multiplying the function by $e^{j\omega_c t}$ causes a shift in frequency. That is, replace ω by $\omega - \omega_c$ and the result in (4.4.7a) follows. ∎

4.4.2 Modulation Theorem

The frequency translation theorem directly leads to the modulation theorem. Given $F[x(t)] = X(j(\omega))$ and $y(t) = x(t)\cos(\omega_c t + \theta)$, the modulation theorem results in

$$Y(j\omega) = F[x(t)\cos(\omega_c t + \theta)]$$

$$= F\left[\frac{1}{2}(x(t)e^{j\theta})e^{j\omega_c t} + \frac{1}{2}(x(t)e^{-j\theta})e^{-j\omega_c t}\right]$$

$$= \frac{1}{2}e^{-j\theta}X(j(\omega + \omega_c)) + \frac{1}{2}e^{j\theta}X(j(\omega - \omega_c)).$$

(4.4.8)

In simple words, multiplying a signal by a sinusoid translates the spectrum of a signal around $\omega = 0$ to the locations around ω_c and $-\omega_c$. If the spectrum of the signal $x(t)$ is frequency (or band) limited to ω_0, i.e., $|X(j\omega)| = 0, |\omega| > \omega_0$, then

$$|Y(j\omega)| = 0 \text{ for } |\omega| > |\omega_c + \omega_0| \text{ and } |\omega| < |\omega_c - \omega_0|.$$

(4.4.9)

Figure 4.4.1 gives sketches of the signals and their spectra. The signal $x(t)$ is assumed to cross the time axis. There is no real significance in the shape of the spectrum. Since $x(t)$ is real, it has even magnitude and odd phase spectrum. The signal is band limited to $f_0 = \omega_0/2\pi$ Hz. The modulated signal $y(t)$ shown in Fig. 4.4.1b assumes $\theta = 0$ in (4.4.8). The positive and negative envelopes of the modulated signal are shown by the dotted lines. Note the envelopes cross the axis wherever the

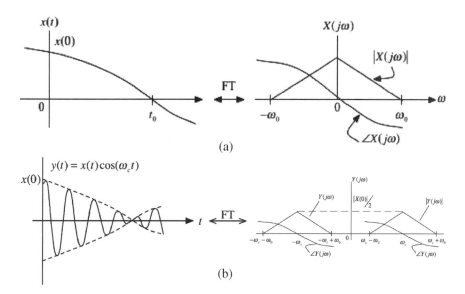

Fig. 4.4.1 (a) $x(t)$ and $|X(j\omega)|$, (b) $y(t)$ and $|Y(j\omega)|$

function $x(t) = 0$. The magnitude and phase spectra of the modulated signal are shown in Fig. 4.4.1b. The bandwidth of the modulated signal is *twice* the bandwidth of the message signal. Note the factor $(1/2)$ in both terms in (4.4.8). ∎

Example 4.4.3 Find the $F[x(t)\cos(\omega_c t)]$ and $F[x(t)\sin(\omega_c t)]$ in terms of $F[x(t)]$.

Solution: Clearly when $\theta = 0$ and $\theta = -\pi/2$ in (4.4.8), the F-transform pairs are

$$x(t)\cos(\omega_c t) \xrightarrow{\text{FT}} \frac{1}{2}X(j(\omega - \omega_c)) + \frac{1}{2}X(j(\omega + \omega_c)),$$

(4.4.10a)

$$x(t)\sin(\omega_c t) \xrightarrow{\text{FT}} \frac{1}{2j}X(j(\omega - \omega_c)) - \frac{1}{2j}X(j(\omega + \omega_c)).$$

(4.4.10b)

Modulation theorem provides a powerful tool for finding the Fourier transforms of functions that are seen (or *windowed*) through a function $x(t)$. For example,

$$x(t) = \Pi\left[\frac{t}{\tau}\right] \rightarrow y(t) = \Pi\left[\frac{t}{\tau}\right]\cos(\omega_c t)$$

$$= \begin{cases} \cos(\omega_c t), & |t| < \frac{\tau}{2} \\ 0, & \text{otherwise} \end{cases}.$$

The signal $y(t)$ is being seen through a rectangular (window) function $x(t)$. Outside this window, no signal is available. The study of windowed signals is an important topic for signal processors and it is humorously called as window carpentry. We will come back to this topic later. ∎

4.4.3 Fourier Transforms of Periodic and Some Special Functions

Modulation theorem gives a back door way to find the Fourier transforms of periodic functions, such as sine and cosine functions. A sufficient condition for the existence of $F[x(t)]$ is

$$\int_{-\infty}^{\infty} |x(t)|dt < \infty \text{ (absolute integrability condition)}.$$

Clearly the sine, cosine, unit step, and many other functions violate this condition. Use of the generalized functions allows for the derivation of the F-transforms of these functions.

Example 4.4.4 Use the transform $F[1] = 2\pi\delta(\omega)$ and the modulation theorem to find the Fourier transforms of $x(t) = \cos(\omega_0 t)$ and $y(t) = \sin(\omega_0 t)$.

Solution: Using (4.4.8), we have the transforms. These are given below in terms of ω and f. In the latter case, we have used $\delta(\omega) = \delta(f)/2\pi$:

$$x(t)=\cos(\omega_0 t)\overset{FT}{\longleftrightarrow}X(j\omega)=\pi\delta(\omega+\omega_0)+\pi\delta(\omega-\omega_0),$$

$$y(t)=\sin(\omega_0 t)\overset{FT}{\longleftrightarrow}Y(j\omega)=j\pi\delta(\omega+\omega_0)-j\pi\delta(\omega-\omega_0),$$
$$(4.4.11a)$$

$$\cos(2\pi f_0 t)\overset{FT}{\longleftrightarrow}\tfrac{1}{2}\delta(f+f_0)+\tfrac{1}{2}\delta(f-f_0),$$
$$\sin(2\pi f_0 t)\overset{FT}{\longleftrightarrow}\tfrac{j}{2}\delta(f+f_0)-\tfrac{j}{2}\delta(f-f_0).$$
$$(4.4.11b)$$

The spectra of the cosine and the sine functions are shown in Figs. 4.4.2. The spectra of these are located at $\omega = \pm\omega_0 = \pm 2\pi f_0$ with the same magnitude, but the phases are different. In reality, we do not have any negative frequencies. Euler's formula illustrates that $\sin(\omega_c t)$ and $\cos(\omega_c t)$ are not the same functions, even though they have the same frequencies. Noting that the real part of the transform of a real signal is even and the phase spectra is odd, the negative frequency component does not give any additional information regarding what frequency is present. The average power represented by the negative frequency component simply adds to the average power of the positive frequency component resulting in the total average power at that frequency. In the case of an arbitrary signal resolved into in-phase and quadrature-phase components, the negative frequency terms do contribute additional information. A cosine wave reaches its positive peak 90° before a sine wave does. By convention the cosine wave is called the in-phase or i (or I) component and the sine wave is called the quadrature-phase or the q (or Q) component. ∎

Notes: A narrowband band-pass signal with a slowly changing envelope $R(t)$ and phase $\phi(t)$ has the forms

$$x(t) = R(t)\cos(\omega_c t + \phi(t)), R(t) \geq 0, \qquad (4.4.12a)$$

$$x(t) = x_i(t)\cos(\omega_c t) - x_q(t)\sin(\omega_c t),$$

$$x_i(t) \overset{\Delta}{=} R(t)\cos(\phi(t)), x_q(t) = R(t)\sin(\phi(t)).$$
$$(4.4.12b)$$

Equation (4.4.12a) gives the envelope-and-phase description and (4.4.12b) gives the in-phase and quadrature-carrier description. The components $x_i(t)$ and $x_q(t)$ are the *in-phase* and *quadrature-phase* components.

Now consider a windowed cosine function and see the effects of that window. ∎

Example 4.4.5 Find the Fourier transform of the cosinusoidal pulse function. Plot the functions $X(j\omega)$ and $Y(j\omega)$ and identify the important parameters:

$$y(t) = x(t)\cos(\omega_0 t), \quad x(t) = \Pi\left[\frac{t}{\tau}\right]\overset{FT}{\longleftrightarrow}\tau\sin c(\omega\tau/2).$$
$$(4.4.13a)$$

Solution: The transform of $y(t)$ is

$$Y(j\omega) = \frac{\tau}{2}\frac{\sin[(\omega - \omega_0)(\tau/2)]}{[(\omega - \omega_0)(\tau/2)]} + \frac{\tau}{2}\frac{\sin[(\omega + \omega_0)(\tau/2)]}{[(\omega + \omega_0)(\tau/2)]}.$$
$$(4.4.13b)$$

The functions $x(t)$, $X(j\omega)$, $y(t)$, and $Y(j\omega)$ are sketched in Fig. 4.4.3a–d, respectively. Noting that $X(j\omega)$ and $Y(j\omega)$ are *real functions*, the main lobe

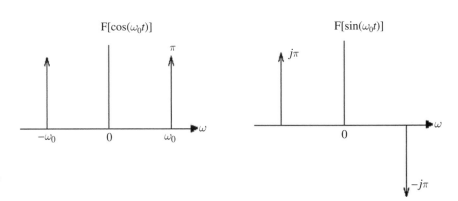

Fig. 4.4.2 Transform of the cosine and sine functions

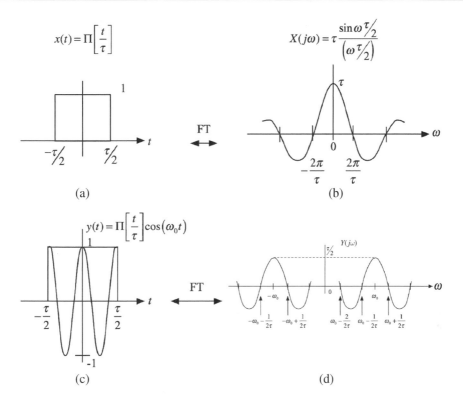

Fig. 4.4.3 (a) and (b) Pulse function and its transform; (c) and (d) windowed cosine function and its transform

width of $X(j\omega)$ corresponding to the positive frequencies is $(2\pi/\tau)$. The function $Y(j\omega)$ has two main lobes centered at $\omega = \pm\omega_0 = \pm2\pi f_0$. Again considering only positive frequencies, the main lobe width of $Y(j\omega)$ is twice that of $X(j\omega)$ equal to $(4\pi/\tau)$. That is, the process of modulation doubles the bandwidth. As in $X(j\omega)$, we have side lobes in $Y(j\omega)$ that decay as we go away from the center frequency. The peak of the main lobe in $X(j\omega)$ is τ, whereas the peaks of the main lobes of $Y(j\omega)$ are equal to $(\tau/2)$. Clearly, if we are interested in finding the frequency $\omega_0 = 2\pi f_0$, the steps could include the following:

1. Find the transform.
2. Find the peak value of the spectrum and its location.

In a practical problem, we may have several frequencies. Finding the locations of these frequencies and their amplitudes is of interest. This problem is usually referred to as *spectral estimation*. The spectrum of a cosine function consists of two impulses located at $\omega = \pm\omega_0$. The spectrum of the windowed cosine function contains two sinc functions. We generally assume that $\omega_0 \gg (2\pi/\tau)$ and therefore the overlap of the two sinc functions at dc is

assumed to be negligible. Rectangular window modified the impulse spectra of the signal to a spectra consisting of sinc functions. Windowing a function results in *spectral leakage*. ∎

Fourier transforms of arbitrary periodic functions: In Chapter 3, we derived that if $x_T(t)$ is a periodic function with period T and $x_T(t)$ can be expressed into its F-series,

$$x_T(t) = \sum_{k=-\infty}^{\infty} X_s[k]e^{jk\omega_0 t},$$

$$X_s[k] = \frac{1}{T}\int_T x(t)e^{-jk\omega_0 t}dt, \quad \omega_0 = \frac{2\pi}{T}. \quad (4.4.14)$$

where $X_s[k]'$s in (4.4.14) are generally complex. The transform can be derived by noting that

$$F[e^{\pm jk\omega_0 t}] = 2\pi\delta(\omega \mp n\omega_0), \quad (4.4.15)$$

$$F[x_T(t)] = \sum_{k=-\infty}^{\infty} X_s[k]F[e^{jk\omega_0 t}]$$

$$= \sum_{k=-\infty}^{\infty} X_s[k](2\pi)\delta(\omega - k\omega_0). \quad (4.4.16)$$

Example 4.4.6 Find the transform $F[\delta_T(t)[= F[\sum_{k=-\infty}^{\infty} \delta(t - kT)]$.

Solution: The F-series of the function $\delta_T(t)$ is given by (see (3.4.15b))

$$\delta_T(t) = \frac{1}{T} \sum_{k=-\infty}^{\infty} e^{jk\omega_0 t}, \quad \omega_0 = 2\pi/T.$$

Using the linearity and frequency shift properties of the Fourier transforms, we have

$$\delta_T(t) = \sum_{n=-\infty}^{\infty} \delta(t - nT) \overset{FT}{\longleftrightarrow} \frac{2\pi}{T}$$

$$\sum_{k=-\infty}^{\infty} \delta(\omega - k\omega_0) = \frac{2\pi}{T} \delta_{\omega_0}(\omega). \quad (4.4.17)$$

We have an interesting result: the Fourier transform of a *periodic impulse sequence* $\delta_T(t)$ with period T is also a *periodic impulse sequence* $(2\pi/T)\delta_{\omega_0}(\omega)$ with period ω_0. They are sketched in Fig. 4.4.4. ∎

One question should come to our mind, that is, are there other functions and their transforms have the same general form? The answer is yes.

Example 4.4.7 Show the Gaussian pulse transform pair as follows:

$$e^{-\alpha t^2} \overset{FT}{\longleftrightarrow} \sqrt{\frac{\pi}{\alpha}} e^{-\frac{(\omega)^2}{4\alpha}}, \quad \alpha > 0. \quad (4.4.18)$$

Solution:

$$X(j\omega) = \int_{-\infty}^{\infty} x(t)e^{-j\omega t}dt = \int_{-\infty}^{\infty} e^{-\alpha t^2} e^{-j\omega t}dt$$

$$= \int_{-\infty}^{\infty} e^{-\alpha(y)}dt, \quad y = t^2 + j\frac{\omega t}{\alpha}.$$

Now add and subtract the term (ω^2/α) to the term y in the exponent inside the integral

$$y = t^2 + j\frac{\omega t}{\alpha} = t^2 + j\frac{\omega t}{\alpha} + \frac{\omega^2}{4\alpha^2} - \frac{\omega^2}{4\alpha^2}$$

$$= (t + \frac{j\omega}{2\alpha})^2 + \frac{(\omega)^2}{4\alpha^2} \Rightarrow X(j\omega)$$

$$= e^{-\frac{\omega^2}{4\alpha}} \int_{-\infty}^{\infty} e^{-\alpha(t + j\frac{\omega}{2\alpha})^2} dt.$$

By the change of variable, we have $r = \sqrt{\alpha}(t + \frac{j\omega}{2\alpha})$, $dt = dr/\sqrt{\alpha}$, $t \Rightarrow \pm\infty$, $r \to \pm\infty$, and

$$X(j\omega) = e^{-\frac{\omega^2}{4\alpha}} \left[\frac{1}{\sqrt{\alpha}} \int_{-\infty}^{\infty} e^{-r^2} dr \right] = e^{-\frac{\omega^2}{4\alpha}} \sqrt{\frac{\pi}{\alpha}}. \quad (4.4.19)$$

Integral tables are used in (4.4.19). The transform pair in (4.4.18) now follows, that is, the *Fourier transform of a Gaussian function* is also a *Gaussian function*. Both time and frequency functions are *not limited in time and in frequency*, respectively. ∎

The following pairs are valid and can be verified using Fourier transform theorems.

$$\cos(\alpha t^2) \overset{FT}{\longleftrightarrow} \sqrt{\frac{\pi}{\alpha}} \cos\left[\frac{(\omega^2 - \alpha\pi)}{4\alpha} \right],$$

$$\sin(\alpha t^2) \overset{FT}{\longleftrightarrow} -\sqrt{\frac{\pi}{\alpha}} \sin\left[\frac{(\omega^2 - \alpha\pi)}{4\alpha} \right], \quad (4.4.20a)$$

$$|t|^{-1/2} \overset{FT}{\longleftrightarrow} \sqrt{2\pi}|\omega|^{-1/2}. \quad (4.4.20b)$$

For a catalog of Fourier transform pairs, see Abromowitz and Stegun (1964) and Poularikis (1996).

4.4.4 Time Differentiation Theorem

If $F[x(t)] = X(j\omega)$ and $x(t)$ is differentiable for all time and vanishes as $t \to \pm\infty$, then

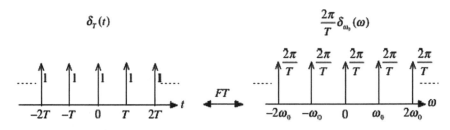

Fig. 4.4.4 Periodic Impulse Sequence and its transform

$$F\left[\frac{dx(t)}{dt}\right] = F[x'(t)] = (j\omega) X(j\omega). \quad (4.4.21)$$

Using integration by parts, we have

$$F\left[\frac{d}{dt}x(t)\right] = \int_{-\infty}^{\infty} x'(t)e^{-j\omega t}dt = x(t)e^{-j\omega t}\Big|_{t=-\infty}^{t=\infty} + j\omega$$

$$\int_{-\infty}^{\infty} x(t)e^{-j\omega t}dt = j\omega\, X(j\omega).$$

Differentiation of a function in time corresponds to multiplication of its transform by $(j\omega)$, provided that the function $x(t) \to 0$ as $t \to \pm\infty$. If $x(t)$ has a finite number of discontinuities, then $x'(t)$ contains impulses. Then, (4.4.21) can be generalized and

$$F\left[\frac{d^n x(t)}{dt^n}\right] = (j\omega)^n X(j\omega), \quad n = 1, 2, \dots. \quad (4.4.22)$$

The above does *not* provide a proof of the existence of the Fourier transform of the nth derivative of the function. It merely shows that if the transform exists, then it can be computed by the above formula. This theorem is useful if the transforms of derivatives of functions can be found easier than finding the transforms of functions. For example,

$$\frac{dx(t)}{dt} \overset{FT}{\longleftrightarrow} j\omega X(j\omega) \Rightarrow x(t) \overset{FT}{\longleftrightarrow} \frac{1}{j\omega}F[x'(t)].$$

Use of this approach in finding transforms is referred to as the *derivative method*.

Example 4.4.8 Find the Fourier transform of the triangular function

$$x(t) = \Lambda\left[\frac{t}{\tau}\right] = \begin{cases} (1 - |\frac{t}{\tau}|), & |t| \le \tau \\ 0, & \text{Otherwise} \end{cases} \quad (4.4.23)$$

using the derivative method. Sketch the transform of the triangular function and compare the transform of the rectangular or Π function with the transform of the Λ function.

Solution: $x(t)$, $x'(t)$, and $x''(t)$ are sketched in Fig. 4.4.5 a–c. Clearly,

$$x''(t) = \frac{1}{\tau}\delta(t+\tau) - \frac{2}{\tau}\delta(t) + \frac{1}{\tau}\delta(t-\tau). \quad (4.4.24)$$

Using the derivative theorem and solving for $X(j\omega)$, we have

$$x''(t) \overset{FT}{\longrightarrow} (j\omega)^2 X(j\omega) = \frac{1}{\tau}e^{j\omega\tau} - \frac{2}{\tau} + \frac{1}{\tau}e^{-j\omega\tau},$$

$$(j\omega)^2 X(j\omega) = \frac{V}{\tau}\left[\frac{e^{j\omega\tau/2} - e^{-j\omega\tau/2}}{2j}\right]^2 (-4)$$

$$= -\frac{4}{\tau}\sin^2(\omega\tau/2),$$

$$x(t) = V\lambda\left[\frac{t}{\tau}\right] \overset{FT}{\longleftrightarrow} V\tau\frac{\sin^2(\omega\tau/2)}{(\omega\tau/2)^2}$$

$$= V\tau\,\text{sinc}^2(\omega\tau/2) = X(j\omega). \quad (4.4.25) \ \blacksquare$$

Equation (4.4.25) gives the spectrum of the triangular function of width (2τ) s. The rectangular function of width τ s and its transform were given earlier by

$$V\Pi\left[\frac{t}{\tau}\right] \overset{FT}{\longleftrightarrow} V\tau\frac{\sin(\omega\tau/2)}{(\omega\tau/2)} = V\tau\,\text{sinc}(\omega\tau/2). \quad (4.4.26)$$

The time width of the rectangular window function in (4.4.26) is τ s, whereas the time width of the triangular window function in (4.4.25) is 2τ s. Note the square of the sinc^2 function in the spectrum of the triangle function and the sinc function in the spectrum of the rectangular window. Since

$$|\text{sinc}(\omega\tau/2)|^2 \le |\text{sinc}(\omega\tau/2)|,$$

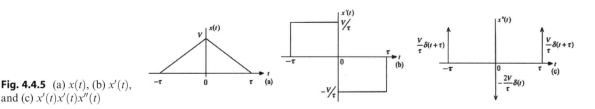

Fig. 4.4.5 (a) $x(t)$, (b) $x'(t)$, and (c) $x'(t)x'(t)x''(t)$

the spectral amplitudes of the triangular function have lower side lobe levels compared to the spectral amplitudes of the rectangular pulses. Since the square of a fraction is smaller than the fraction itself, the side lobes in the transform of the triangular function are much smaller than the side lobes in the transform of the rectangular function. There is *less leakage* in the side lobes for the triangular (window) pulse function compared to the rectangular (window) function. See Fig. B.4.1 for a sketch of the sinc function.

High-frequency decay rate: In the Fourier series discussion, the decay rate of the F-series coefficients $X_s[k]$ was determined using the derivatives of the periodic function (see Section 3.6.5). Similarly, the F-transforms of pulse functions decay rate can be determined without actually finding the transform of the function. Given a pulse function $x(t)$, find the successive derivatives, $x^{(n)}(t)$, of the function until the first set of impulses appear in the nth derivative, then the decay rate is proportional to $(1/\omega^n)$. In Example 4.4.7 the triangular pulse was considered and, in this case, the second derivative exhibits impulses indicating that the high-frequency decay rate of the transform is $(1/\omega^2)$ (see (4.4.25)). Similarly the first derivative of the rectangular pulse function and the exponential decaying function $e^{-at}u(t)$, $a>0$ exhibit impulses indicating that the high-frequency decay rate of these transforms is $(1/|\omega|)$.

4.4.5 Times-t Property: Frequency Differentiation Theorem

If $X(j\omega) = F[x(t)]$ and if the derivative of the transform exists, then

$$F[(-jt)x(t)] = \frac{dX(j\omega)}{d\omega}. \qquad (4.4.27)$$

This can be shown by

$$\frac{dX(j\omega)}{d\omega} = \frac{d}{d\omega}\int_{-\infty}^{\infty} x(t)e^{-j\omega t}dt = \int_{-\infty}^{\infty} x(t)\frac{d(e^{-j\omega t})}{d\omega}dt$$

$$= \int_{-\infty}^{\infty} [(-jt)x(t)]e^{-j\omega t}dt = F[-jtx(t)].$$

The similarities between the time and frequency differentiation theorems illustrate the duality properties with the F-transform pairs.

Example 4.4.9 Show the following relationship using the times-t property:

$$te^{-at}u(t) \overset{FT}{\longleftrightarrow} \frac{1}{(a+j\omega)^2}, \ a>0. \qquad (4.4.28)$$

Solution: Noting the times-t property given above with $x(t) = e^{-at}u(t)$, we have

$$F[te^{-at}u(t)] = j\frac{dX(j\omega)}{d\omega} = j\frac{d(1/(a+j\omega))}{d\omega} = \frac{1}{(a+j\omega)^2}.$$

This can be generalized to obtain the following and the proof is left as an exercise:

$$\frac{t^{n-1}}{(n-1)!}e^{-at}u(t) \overset{FT}{\longleftrightarrow} \frac{1}{(a+j\omega)^n}, \ a>0. \qquad (4.4.29) \ \blacksquare$$

Example 4.4.10 Noting that

$$e^{-(a\pm jb)t}u(t) \overset{FT}{\longleftrightarrow} \frac{1}{a+j(\omega\pm b)}, \ a>0, \qquad (4.4.30a)$$

show the following is true:

$$x(t) = e^{-at}\sin(bt)u(t) \overset{FT}{\longleftrightarrow} \frac{b}{(a+j\omega)^2+b^2} = X(j\omega),$$
$$(4.4.30b)$$

$$y(t) = e^{-at}\cos(bt)u(t) \overset{FT}{\longleftrightarrow} \frac{(a+j\omega)}{(a+j\omega)^2+b^2} = Y(j\omega).$$
$$(4.4.30c)$$

Solution: These can be shown by first expressing the sine and cosine functions by Euler's formulas, taking the transforms and then combining the complex–conjugate terms. $\blacksquare$

Example 4.4.11 Using $\lim_{a\to 0} e^{-at}u(t) = u(t)$, $a>0$, find $F[u(t)]$:

$$F[u(t)] = \lim_{a\to 0}\frac{1}{a+j\omega}.$$

Solution: Noting that the *limiting process* is on the *complex function*, we need to take the limits on the real and the imaginary parts of the complex function *separately*. That is,

$$\lim_{a \to 0}\left[\frac{1}{a+j\omega}\right] = \lim_{a \to 0}\left[\frac{a}{a^2+\omega^2}\right] + j\lim_{a \to 0}\left[\frac{-\omega}{a^2+\omega^2}\right].$$

$$(4.4.31)$$

The second term in the above, i.e., the Lorentzian function, approaches an impulse function. That is,

$$\lim_{a \to 0}\left[\frac{a}{a^2+\omega^2}\right] = \pi\delta(\omega).$$ (4.4.32)

Using this result in (4.3.30),

$$\lim_{a \to 0}\left[\frac{1}{a+j\omega}\right] = \pi\delta(\omega) + \frac{1}{j\omega},$$ (4.4.33)

$$\Rightarrow U(j\omega) = F[u(t)] = \pi\delta(\omega) + \frac{1}{j\omega},$$ (4.4.34a)

$$|U(j\omega)| = \pi\delta(\omega) + |1/\omega|, \quad \angle U(j\omega) = \begin{cases} -\pi/2, & \omega>0 \\ \pi/2, & \omega<0 \end{cases}.$$

Note that the amplitude is an even function and the phase angle function is an odd function, as the

unit step function is real. These are illustrated in Fig. 4.4.6.

Interestingly the spectrum of the delayed unit step $u(t-1)$ is

$$F[u(t-1)] = [\pi\delta(\omega) + \frac{1}{j\omega}]e^{-j\omega},$$

$$|F[u(t-1)]| = \left[\pi\delta(\omega) + \left|\frac{1}{\omega}\right|\right], \quad \angle F[u(t-1)]$$

$$= \begin{cases} -\omega - \pi/2, & \omega>0 \\ -\omega + \pi/2, & \omega<0 \end{cases}.$$ (4.4.34b)

Since delay of a function depends on the phase angle, it follows that $|F[u(t-1)]| = |F[u(t)]|$. The phase spectrum of the delayed unit step function is sketched in Fig. 4.4.7. The Fourier transform of the unit function has two parts. The first part corresponds to the transform of the average value of the unit step function and the other part is the transform of the signum function. That is,

$$F[u(t)] = F[(1/2) + (1/2)\,\mathrm{sgn}(t)]$$
$$= F[1/2] + (1/2)F[\mathrm{sgn}(t)] = \pi\delta(\omega) + (1/j\omega).$$

∎

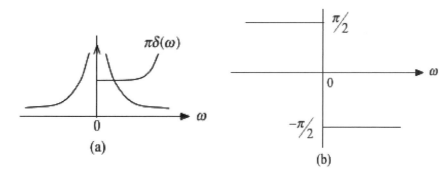

(a)   (b)

Fig. 4.4.6 (a) Magnitude and (b) phase spectra of the unit step function

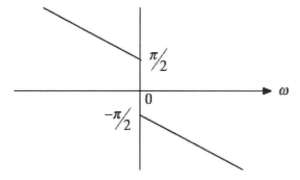

Fig. 4.4.7 Phase spectrum of $u(t-1)$

Notes: If we had ignored that we had to take the limit on the real and the imaginary parts of the complex function in (4.4.31) separately and take the limit on the *complex function as a whole*, the result would be *wrong*. That is,

$$\lim_{a \to 0} \frac{1}{a + j\omega} = \frac{1}{j\omega} \neq F[u(t)].$$

This indicates the transform is imaginary and the time function must be odd. This cannot be true since the unit step function is not an odd function and $F[u(t)] \neq (1/j\omega)$. ∎

The *sgn* (or *signum*) *function* is used in communications and control theory and can be expressed in terms of the unit step function. The sgn function and its transform are as follows:

$$\text{sgn}(t) = 2u(t) - 1 = \begin{cases} 1, & t > 0 \\ 0, & t = 0. \\ -1, & t < 0 \end{cases} \quad (4.4.35)$$

$$F[2u(t)] - F[1] = 2\pi\delta(\omega) + (2/j\omega) - 2\pi\delta(\omega)$$
$$= (2/j\omega). \quad (4.4.36)$$

The times-t property and the transform of the unit step function can be used to determine the *Fourier transform of the ramp function* and is given as

$$tu(t) \overset{\text{FT}}{\longleftrightarrow} j\pi\delta'(\omega) - (1/\omega^2). \quad (4.4.37)$$

Noting that $|t| = 2t\,u(t) - t$, we have the following transform pair:

$$|t| \overset{\text{FT}}{\longleftrightarrow} -(2/\omega^2). \quad (4.4.38)$$

Example 4.4.12 Find the Fourier transform of the function $x(t) = (1/t)$ using the duality theorem and the Fourier transform of the signum function.

Solution: Using the duality theorem, we have

$$x(t) \overset{\text{FT}}{\longleftrightarrow} X(j\omega) \xrightarrow{\text{Duality theorem}} X(t) \overset{\text{FT}}{\longleftrightarrow} 2\pi x(-j\omega),$$

$$\text{sgn}(t) \overset{\text{FT}}{\longleftrightarrow} \frac{2}{j\omega}, \quad \frac{1}{jt} \overset{\text{FT}}{\longleftrightarrow} (1)\pi\,\text{sgn}(-\omega) = -\pi\,\text{sgn}(\omega).$$

We can write sgn $(-j\omega) = \text{sgn}(-\omega) = -\text{sgn}(\omega)$. It follows that

$$(1/t) \overset{\text{FT}}{\longleftrightarrow} -j\pi\,\text{sgn}(\omega) = j\pi - j2\pi u(\omega). \quad (4.4.39a)$$

This can be generalized and

$$\frac{1}{t^n} \overset{\text{FT}}{\longleftrightarrow} -\frac{(-j\omega)^{n-1}}{(n-1)!} j\pi\,\text{sgn}(\omega). \quad (4.4.39b)$$
∎

4.4.6 Initial Value Theorem

The initial value theorem is applicable for the right-sided signals, i.e., the functions of the form $y(t) = x(t)u(t)$ and is stated below without proof:

$$y(0^+) = \lim_{\omega \to \infty} j\omega Y(j\omega). \quad (4.4.40)$$

Example 4.4.13 The unit step function is not defined at $t = 0$, whereas $u(0^+) = 1$, which is well defined. Verify the initial value theorem for the unit step function by noting $\delta(\omega) = 0, \omega \neq 0$, and $\omega\delta(\omega) = 0$.

Solution:

$$u(0^+) = \lim_{\omega \to \infty} \{j\omega F[u(t)]\}$$

$$= \lim_{\omega \to \infty} \left[j\omega \left(\pi\delta(\omega) + \frac{1}{j\omega} \right) \right] = 1. \quad ∎$$

4.4.7 Integration Theorem

It states that

$$y(t) = \int_{-\infty}^{t} x(\alpha)d\alpha \overset{\text{FT}}{\longleftrightarrow} \frac{X(j\omega)}{j\omega} + \pi X(0)\delta(\omega) = Y(j\omega).$$

$$(4.4.41)$$

This is true only if $X(0)$, i.e., the area under $x(t)$, is finite. If the area under $x(t)$ is zero, then the second term on the right in (4.4.41) disappears. Note that if $X(0) = 0$, *integration and differentiation* operations are *inverse operations*. Integration operation is a smoothing operation. Integral of a function has lower frequency content than the function that is integrated. On the other hand, since $x'(t) = j\omega X(j\omega)$, differentiation accentuates the higher frequencies. Integration theorem is not applicable if $X(0)$ is infinity. This theorem will be proved in Section 4.5.

Example 4.4.14 Find the Fourier transform of $u(t)$ using the integration theorem and

$$\delta(t)\overset{FT}{\longleftrightarrow}1, \ u(t) = \int\limits_{-\infty}^{t}\delta(\alpha)d\alpha.$$

Solution: By the integration theorem

$$F[u(t)] = F\left[\int\limits_{-\infty}^{t}\delta(\alpha)d\alpha\right] = \frac{1}{j\omega} + \pi\delta(\omega).$$

∎

Example 4.4.15 Use the Fourier transform of the function $x(t) = \cos(\omega_0 t)$, $\omega_0 \neq 0$, and the integration theorem to find the Fourier transform of the function $\sin(\omega_0 t)$.

Solution: First for $\omega_0 \neq 0$, from (4.4.11a), we have

$$x(t) = \cos(\omega_0 t)\overset{FT}{\longleftrightarrow}\pi\delta(\omega + \omega_0) + \pi\delta(\omega - \omega_0)$$
$$= X(j\omega), \ X(0) = 0$$

and

$$y(t) = \int\limits_{-\infty}^{t}x(\alpha)d\alpha = \int\limits_{-\infty}^{t}\cos(\omega_0\alpha)d\alpha$$
$$= (1/\omega_0)\ \sin(\omega_0 t). \quad (4.4.42a)$$

See the comment below in regard to the evaluation of the limit at $-\infty$ in the above integral. The integration property gives us

$$\omega_0 y(t) = \omega_0 \int\limits_{-\infty}^{t}\cos(\omega_0\alpha)d\alpha\overset{FT}{\longleftrightarrow}$$

$$\frac{\omega_0[\pi\delta(\omega + \omega_0) + \pi\delta(\omega - \omega_0)]}{j\omega} + \omega_0\pi X(0)\delta(\omega).$$

With $X(0) = 0$ and $\delta(\omega \pm \omega_0)/\omega = \mp\delta(\omega \pm \omega_0)/\omega_0$, we have result as in (4.4.11a).

$$\int\limits_{-\infty}^{t}\cos(\omega_0\alpha)d\alpha = \sin(\omega_0 t)\overset{FT}{\longleftrightarrow}j\pi\delta(\omega + \omega_0)$$

$$- j\pi\delta(\omega - \omega_0). \quad (4.4.42b) \ ∎$$

Notes: Papoulis (1962) discusses the concepts of generalized limits. For example

$$\lim_{t\to\infty} e^{-j\omega t} = 0. \quad (4.4.43a)$$

The limit does *not* exist as an *ordinary limit* and is a *generalized limit* in the sense of *distributions*. Using Euler's formula and the limit in (4.4.43a), computation of the integral in (4.4.42a) follows. *Switched functions* are very useful in system theory. In computing the derivatives of such functions, one needs to be careful. For example,

$$\frac{d[\cos(t)u(t)]}{dt} = \frac{d[\cos(t)]}{dt}u(t) + \cos(t)\frac{d[u(t)]}{dt}$$
$$= -\sin(t)u(t) + \delta(t), \quad (4.4.43b)$$
$$\frac{d[\sin(t)u(t)]}{dt} = \cos(t)u(t) + \sin(t)\delta(t) = \cos(t)u(t).$$

$$(4.4.43c)$$

To find the transforms of such functions we can make use of modulation theorem. Derivative theorem can be used to find transforms of many functions such as $x(t) = e^{-\alpha t}u(t)$, $\alpha > 0$. We should keep in mind that if the pulse is not time limited, we need to add a frequency domain delta function, whose weight is equal to 2π times the average of the pulse over the entire time axis to the transform result of the successive differentiation. See the discussion on finding the transform of a unit step function. ∎

4.5 Convolution and Correlation

Chapter 2 considered convolution and correlation. Here we will consider the transforms of the signals that are convolved and correlated.

4.5.1 Convolution in Time

Convolution of two time functions $x_1(t)$ and $x_2(t)$ is defined by

$$y(t) = x_1(t) * x_2(t) = \int\limits_{-\infty}^{\infty}x_1(\alpha)x_2(t - \alpha)d\alpha$$

$$= \int\limits_{-\infty}^{\infty}x_2(\beta)x_1(t - \beta)d\beta = x_2(t) * x_1(t). \quad (4.5.1)$$

Assuming that $x_i(t) \overset{FT}{\longleftrightarrow} X_i(j\omega)$, $i = 1, 2$, the *convolution theorem* is given by

$$x_1(t) * x_2(t) \overset{FT}{\longleftrightarrow} X_1(j\omega)X_2(j\omega). \quad (4.5.2)$$

This can be proven by using the transform pair $F[x_2(t - \alpha)] = X_2(j\omega)e^{-j\omega\alpha}$ in (4.5.1) and the resulting integral is the inverse transform of $[X_1(j\omega)X_2(j\omega)]$. That is,

$$y(t) = x_1(t) * x_2(t)$$

$$= \int_{-\infty}^{\infty} x_1(\alpha) \left[\frac{1}{2\pi} \int_{-\infty}^{\infty} X_2(j\omega)e^{j\omega(t-\alpha)} d\omega \right] d\alpha$$

$$= \frac{1}{2\pi} \int_{-\infty}^{\infty} X_2(j\omega) \left[\int_{-\infty}^{\infty} x_1(\alpha)e^{-j\omega\alpha} d\alpha \right] e^{j\omega t} d\omega$$

$$= \frac{1}{2\pi} \int_{-\infty}^{\infty} [X_2(j\omega)X_1(j\omega)]e^{j\omega t} d\omega.j \quad (4.5.3)$$

Convolution theorem follows from the above equation. It gives a method for computing the convolution of two aperiodic functions via Fourier transforms. This method is the *transform method* of computing the convolution. The *direct method* is by the use of the convolution integral and all the operations are in the time domain. The transform method involves the following steps:

a. Find $F[x_1(t)] = X_1(j\omega)$ and $F[x_2(t)] = X_2(j\omega)$.
b. Determine $Y(j\omega) = X_1(j\omega)X_2(j\omega)$.
c. Find the inverse transform of the function $Y(j\omega)$ to obtain $y(t)$.

There are several problems with the transform method of computing the convolution. First, the given function may not have analytical expressions for the transforms. Even if does, we may not be able to find the

inverse transform of $Y(j\omega)$. Second, in most applications, the function may not be given in an analytical or equation form and may be given in the form of a plot or a set of data and we have to resort to digital means to find the values for $y(t)$. We will consider the discrete Fourier transforms in Chapters 8 and 9.

Example 4.5.1 Determine the function $y(t) = \Pi[t - .5] * \Pi[t - .5]$ by using the transforms.

Solution: Using the transforms of the pulse functions, we have

$$F\left[\Pi\left[t - \frac{1}{2} \right] \right] = \frac{\sin(\omega/2)}{(\omega/2)} e^{-j\omega/2},$$

$$Y(j\omega) = \left[\frac{\sin^2(\omega/2)}{(\omega/2)^2} \right] e^{-j\omega}. \quad (4.5.4)$$

Using (4.4.25) and the time delay theorem, we have a triangle or a tent function given by

$$y(t) = \Lambda[t - 1]. \quad (4.5.5)$$

The given time functions and the result of the convolution are shown in Fig. 4.5.1. ∎

Example 4.5.2 Consider the two delayed functions $x_1(t - t_1)$ and $x_2(t - t_2)$. *a.* Assuming $y(t) = x_1(t)* x_2(t)$ is known, show the following is true by using the transform method:

$$z(t) = x_1(t - t_1) * x_2(t - t_2) = y(t - (t_1 + t_2)). \quad (4.5.6)$$

b. Using the results in (4.5.6), determine the convolution of the two impulse functions

$$y(t) = \delta(t - t_1) * \delta(t - t_2).$$

Solution: *a.* Using the convolution and time delay theorems, we have

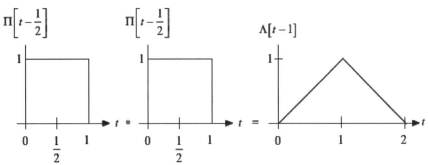

Fig. 4.5.1 Convolution of two square pulses

$$Z(j\omega) = F[z(t)] = F[x_1(t - t_1)]F[x_2(t - t_2)]$$
$$= F[x_1(t)]e^{-j\omega t_1} F[x_2(t)]e^{-j\omega t_2}$$
$$= X_1(j\omega)X_2(j\omega)e^{-j\omega(t_1 + t_2)} = Y(j\omega)e^{-j\omega(t_1 + t_2)}.$$

The inverse transform of $Z(j\omega)$ is given by $y(t - (t_1 + t_2))$.

b. Noting that $F[\delta(t - t_i)] = e^{-j\omega t_i}$, $i = 1, 2$, we have

$$Y(j\omega) = e^{-j\omega(t_1 + t_2)}, F^{-1}\left[e^{-j\omega(t_1 + t_2)}\right] = \delta(t - (t_1 + t_2)).$$

That is, the inverse transform is a delayed impulse function and

$$y(t) = \delta(t - t_1) * \delta(t - t_2) = \delta(t - (t_1 + t_2)). \quad (4.5.7)$$

∎

Example 4.5.3 Show by using the transform method

$$y(t) = x(t) * \delta(t) = x(t). \quad (4.5.8)$$

Solution: We have

$$F[x(t) * \delta(t)] = F[x(t)]F[\delta(t)]$$
$$= F[x(t)] \text{ and } y(t) = x(t). \quad ∎$$

Example 4.5.4 Determine the convolution $y(t) = e^{-at}u(t) * u(t), a > 0$ by a. the direct method and b. by the transform method.

Solution: a. By the direct method,

$$y(t) = \int_{-\infty}^{\infty} e^{-a(t - \beta)}[u(t - \beta)u(\beta)]d\beta = e^{-at}\int_0^t e^{a\beta}d\beta$$

$$= e^{-at}\frac{1}{a}e^{a\beta}\Big|_{\beta=0}^{\beta=t} = \frac{1}{a}(1 - e^{-at})u(t). \quad (4.5.9a)$$

b. By the transform method,

$$Y(j\omega) = F[e^{-at}u(t)]F[u(t)] = \frac{1}{(a + j\omega)}\left[\pi\delta(\omega) + \frac{1}{j\omega}\right]$$

$$= \frac{\pi\delta(\omega)}{a} + \frac{1}{j\omega(a + j\omega)},$$

$$Y(j\omega) = \frac{1}{a}\left[\pi\delta(\omega) + \frac{1}{j\omega}\right] - \frac{1}{a}\frac{1}{a + j\omega} \longrightarrow$$

$$y(t) = \frac{1}{a}(1 - e^{-at})u(t), \quad a > 0. \quad (4.5.9b)$$

In (4.5.9b) we have made use of partial fraction expansion and the transforms of the unit step function and the exponential decaying function. ∎

Example 4.5.5 Determine the convolution $y(t) = x_1(t) * x_2(t)$ in each case below using the transforms. a. The two Gaussian functions and their transforms are given by

$$x_i(t) = \frac{1}{\sigma_i\sqrt{2\pi}}e^{-t^2/2\sigma_i^2} \xrightarrow{FT} e^{-(\omega\sigma_i)^2/2} = X_i(j\omega), \; i = 1, 2..$$

$$(4.5.10a)$$

b. The two sinc functions and their transforms are given by

$$x_i(t) = \tau_i\frac{\sin(t\tau_i/2)}{(t\tau_i/2)} \xrightarrow{FT} 2\pi\Pi\left[\frac{\omega}{\tau_i}\right] = X_i(j\omega),$$

$$i = 1, 2, \quad \tau_1 < \tau_2. \quad (4.5.10b)$$

c. $\quad x_i(t) = 1/j\pi t \xrightarrow{FT} \text{sgn}(\omega), \; i = 1, 2..$

Solution: a. Noting that the transform of a Gaussian pulse is a Gaussian pulse, the product of the two Gaussian pulses is a Gaussian pulse:

$$y(t) = x_1(t) * x_2(t) \xrightarrow{FT} e^{-(\omega\sigma_1)^2/2} e^{-(\omega\sigma_2)^2/2}$$

$$= e^{-\omega^2(\sigma_1^2 + \sigma_2^2)/2} = Y(j\omega). \quad (4.5.10c)$$

The inverse transform of this function is again a Gaussian pulse with

$$y(t) = \frac{1}{\sigma\sqrt{2\pi}}e^{-t^2/2\sigma^2}, \quad \sigma^2 = \sigma_1^2 + \sigma_2^2. \quad (4.5.10d)$$

b. The rectangular pulses in the Fourier domain overlap. The product of the two rectangular pulses is a rectangular pulse and its inverse transform is a sinc pulse. The details are left as an exercise.

c. The convolution of the two functions and its transform are given by

$$y(t) = \frac{1}{j\pi t} * \frac{1}{j\pi t} \xrightarrow{FT} \text{sgn}^2(\omega) = 1, \; F^{-1}[1] = \delta(t).$$

$$(4.5.10e)$$

∎

Notes: The transform method is simpler if the transforms of the individual functions and the inverse transform of the convolution are known. In Chapter 2 we discussed the duration property associated with convolution and pointed out that there are exceptions. Part c of the above example illustrates an exception. Convolutions of some functions do not exist. For example, $y(t) = u(t) * u(t)$ does not exist since its transform has a term that is a square of an impulse function, which is not defined. ∎

4.5.2 Proof of the Integration Theorem

In Section 4.4.6 the integration theorem is stated (see (4.4.42)) and is

$$y(t) = \int_{-\infty}^{t} x(\alpha)d\alpha \overset{FT}{\longleftrightarrow} \frac{X(j\omega)}{j\omega} + \pi X(0)\delta(\omega) = Y(j\omega).$$

(4.5.11)

Since $u(t - \alpha) = 0$ for $\alpha > t$, we can write the above running integral as a convolution:

$$\int_{-\infty}^{t} x(\alpha)d\alpha = \int_{-\infty}^{\infty} x(\alpha)u(t - \alpha)d\alpha$$

$$= x(t) * u(t) \overset{FT}{\longleftrightarrow} X(j\omega)F[u(t)],$$

$$Y(j\omega) = X(j\omega)\left[\pi\delta(\omega) + \frac{1}{j\omega}\right]$$

$$= \pi X(0)\delta(\omega) + \frac{1}{j\omega}X(j\omega).$$

This proves the integration theorem.

Example 4.5.6 Find the inverse transform of the function $X(j\omega)$ given below for two cases: $a.$ $a \neq b$, $a>0$, $b>0$ and $b.$ $a = b>0$

$$X(j\omega) = 1/[(a + j\omega)(b + j\omega)].$$ (4.5.12a)

Solution: $a.$ This can be solved by first noting that convolution in time domain corresponds to the multiplication in the frequency domain. Therefore,

$$x(t) = F^{-1}\left[\frac{1}{a+j\omega}\right] * F^{-1}\left[\frac{1}{b+j\omega}\right]$$

$$= e^{-at}u(t) * e^{-bt}u(t) = \int_{0}^{t} e^{-a\alpha}u(\alpha)e^{-b(t-\alpha)}u(t-\alpha)d\alpha$$

$$= \int_{0}^{t} e^{-a\tau}e^{-b(t-\tau)}d\tau = e^{-bt}\int_{0}^{t} e^{(b-a)\tau}d\tau$$

$$= e^{-bt}\frac{1}{(b-a)}\left[e^{(b-a)\tau}\right]_{\tau=0}^{\tau=t} = \frac{e^{-at}-e^{-bt}}{(b-a)}u(t).$$

(4.5.12b)

Second, by using partial fraction expansion, we have

$$X(j\omega) = \frac{1}{(b-a)(a+j\omega)} - \frac{1}{(b-a)(b+j\omega)}, a \neq b$$

(4.5.12c)

$$\Rightarrow x(t) = F^{-1}[X(j\omega)] = \frac{1}{(b-a)}e^{-at}u(t)$$

$$- \frac{1}{(b-a)}e^{-bt}u(t).$$ (4.5.12d)

This coincides with the solution in (4.5.12b).

$b.$ When $a = b$, the transform function has a double pole. By the convolution method,

$$x(t) = e^{-at}u(t) * e^{-at}u(t) = \int_{0}^{t} e^{-a\tau}e^{-a(t-\tau)}d\tau$$

$$= e^{-at}\int_{0}^{t} d\tau = te^{-at}u(t).$$ (4.5.12e)

Since the function has a double pole, we can find its inverse transform by using times-t property of the Fourier transforms or from tables. Now

$$\frac{1}{(-j)}\frac{d}{d\omega}\left[\frac{1}{(a+j\omega)}\right] = \frac{1}{(a+j\omega)^2} = X(j\omega),$$

$$x(t) = F^{-1}\left\{\frac{1}{(a+j\omega)^2}\right\} = F^{-1}\left\{\frac{1}{(-j)}\frac{d}{d\omega}\left[\frac{1}{(a+j\omega)}\right]\right\}$$

$$= (-jt)F^{-1}\left\{\frac{1}{(-j)}\frac{1}{a+j\omega}\right\} = te^{-at}u(t).$$

This coincides with the result obtained in (4.5.12e). ■

Example 4.5.7 Find $y(t)$ for the function below by using *a.* the derivative theorem and *b.* the long division:

$$Y(j\omega) = \frac{j\omega}{(a+j\omega)}, \quad a>0. \qquad (4.5.13)$$

Solution: *a.* Using the derivative theorem, we have

$$y(t) = \frac{d[F^{-1}(1/(a+j\omega))]}{dt} = \frac{d(e^{-at}u(t))}{dt}$$

$$= e^{-at}\frac{du(t)}{dt} + u(t)\frac{d(e^{-at})}{dt}$$

$$= e^{-at}\delta(t) - ae^{-at}u(t) = \delta(t) - ae^{-at}u(t). \,(4.5.14)$$

b. Also, dividing $j\omega$ by $(a+j\omega)$ by long division and using the superposition property of the

Fourier transforms gives the same result as in (4.5.14). That is,

$$\frac{j\omega}{a+j\omega} = 1 - \frac{a}{a+j\omega} \overset{FT}{\longleftrightarrow} \delta(t) - ae^{-at}u(t). \qquad ■$$

4.5.3 Multiplication Theorem (Convolution in Frequency)

The dual to the time convolution theorem is the convolution in frequency theorem. It is given below and can be shown directly by using a proof similar to the time domain convolution theorem. An alternate way of showing is by using the symmetry theorem

$$y(t) = x_1(t)x_2(t) \overset{FT}{\longleftrightarrow} \frac{1}{2\pi}X_1(j\omega) * X_2(j\omega). \qquad (4.5.15)$$

Summary: Convolution in time and in frequency:

Convolution in time: $[x(t) * x_2(t)] \overset{FT}{\longleftrightarrow} [X_1(j\omega)X_2(j\omega)]$: Multiplication in frequency

Multiplicaion in time: $[x_1(t)x_2(t)] \overset{FT}{\longleftrightarrow} \frac{1}{2\pi}[X_1(j\omega) * X_2(j\omega)]$: Convolution in frequency

---

Example 4.5.8 Consider the time function and its transform

$$x(t) \overset{FT}{\longleftrightarrow} X(j\omega) = \Pi\left[\frac{\omega}{W}\right]. \qquad (4.5.16)$$

Find the Fourier transform of the function $y(t) = x^2(t)$ and its bandwidth by assuming the bandwidth of $x(t)$ is $(W/2)$ rad/s.

Solution: Example 2.3.1 considered the time domain convolution of two rectangular pulse functions. Using these results, we have

$$Y(j\omega) = X(j\omega) * X(j\omega) = \Pi\left[\frac{\omega}{W}\right] * \Pi\left[\frac{\omega}{W}\right] = W\Lambda\left[\frac{\omega}{W}\right].$$
$$(4.5.17) \quad ■$$

Notes: It is instructive to review the properties of the convolution of the transform functions in (4.5.17). The bandwidth of the pulse function $\Pi[\omega/W]$ is $W/2$, whereas the bandwidth of the

function $\Lambda[\omega/W]$ is (W). Note the duration property of the convolution is satisfied since the width of the triangular pulse is twice that of the rectangular pulse. From the area property of the convolution, we have using the *time averages*

$$A\left\{\Pi\left[\frac{\omega}{W}\right]\right\}A\left\{\Pi\left[\frac{\omega}{W}\right]\right\} = W^2.$$

Coming back to the bandwidths, if $x_1(t)$ and $x_2(t)$ have bandwidths of B_1 and B_2 Hz, respectively, then the bandwidth of $y(t) = x_1(t)x_2(t)$ is equal to $(B_1 + B_2)$ Hz. *Multiplication of two time functions increases the bandwidth* of the resulting time function. The above property is dual to the time width property of the convolution. We have seen some pathological cases where the time width property of the convolution does not hold. What about in the frequency domain? Obviously, the same is true in the frequency domain for pathological cases. For practical signals, the above discussion applies. We will come back to this topic at a later time, as it pertains to the important topic of nonlinear systems and the bandwidth requirements of such systems.

Fourier transform computation of windowed periodic functions: Stanley et al. (1984) present a nice approach in finding the transforms of windowed time-limited trigonometric functions using the multiplication theorem, which is presented below.

Let $g_T(t)$ be a periodic function with period T and $F[g_T(t)] = G(j\omega)$. Let $p(t)$ be a pulse function with $P(j\omega) = F[p(t)]$ and a function $F[w(t)] = W(j\omega)$ is defined by

$$w(t) = p(t)g_T(t) \xrightarrow{\text{FT}} P(j\omega) * G(j\omega). \quad (4.5.18)$$

We like to find $F[w(t)]$ using $F[e^{jk\omega_0 t}] = 2\pi\delta(\omega - k\omega_0)$ and

$$g_T(t) = \sum_{k=-\infty}^{\infty} G_s[k]e^{jk\omega_0 t},$$

$$G_s[k] = \frac{1}{T}\int_T g_T(t)e^{-jk\omega_0 t}dt,$$

$$\omega_0 = 2\pi/T,$$

$$G(j\omega) = F[g_T(t)]$$

$$= 2\pi\sum_{k=-\infty}^{\infty} G_s[k]\delta(\omega - k\omega_0). \quad (4.5.19)$$

The transform of the function $w(t)$ can be obtained by the convolution of the two transform functions $P(j\omega) = F[p(t)]$ and $G(j\omega)$. That is,

$$W(j\omega) = P(j\omega) * \left[2\pi\sum_{k=-\infty}^{\infty} G_s[k]\delta(\omega - k\omega_0)\right]$$

$$= 2\pi\sum_{k=-\infty}^{\infty} G_s[k]\{P(j\omega) * \delta(\omega - k\omega_0)\}$$

$$= 2\pi\sum_{k=-\infty}^{\infty} G_s[k]P(j(\omega - k\omega_0)). \quad (4.5.20)$$

Example 4.5.9 Find the Fourier transform of the Hamming window function given below using the above method:

$$w_H(t) = \begin{cases} .54 + .46\cos(2\pi t/T), & |t| \le T/2 \\ 0, & \text{Otherwise} \end{cases}.$$

$$(4.5.21)$$

Solution: Define a periodic function using the window function in (4.5.21) by

$$g_T(t) = .54 + .46\ \cos(2\pi t/T), \quad g_T(t+T) = g_T(t),$$

$$(4.5.22a)$$

$$\Rightarrow w_H(t) = g_T(t)\Pi\left[\frac{t}{T}\right]. \quad (4.5.22b)$$

This function $g_T(t)$ contains a constant and a cosine function. Its Fourier transform is

$$G(j\omega) = .54(2\pi)\delta(\omega) + .23(2\pi)\delta(\omega - \omega_0)$$
$$+ .23(2\pi)\delta(\omega + \omega_0), \omega_0 = 2\pi/T, \quad (4.5.23)$$

$$W_H(j\omega) = F\left[g_T(t)\Pi\left[\frac{t}{T}\right]\right] = G(j\omega)$$

$$* T\frac{\sin(\omega T/2)}{(\omega T/2)}.54(2\pi)\left[\delta(\omega) * \frac{T\sin(\omega T/2)}{(\omega T/2)}\right]$$

$$+ .23(2\pi)\left[\delta(\omega - \omega_0) * \frac{T\sin(\omega T/2)}{(\omega T/2)}\right]$$

$$+ .23(2\pi)\left[\delta(\omega + \omega_0) * \frac{T\sin(\omega T/2)}{(\omega T/2)}\right].$$

With

$$\delta(\omega \mp \omega_0) * Y(j\omega) = \frac{1}{2\pi}\int_{-\infty}^{\infty}\delta(\alpha \mp \omega_0)Y(j(\omega - \alpha))d\alpha$$

$$= \frac{1}{2\pi}Y(j(\omega \mp \omega_0)),$$

we have

$$W_H(j\omega) = .54\frac{T\sin(\omega T/2)}{(\omega T/2)} + .23\frac{T\sin((\omega - \omega_0)T/2)}{((\omega - \omega_0)T/2)}$$

$$+ .23\frac{T\sin((\omega + \omega_0)T/2)}{((\omega + \omega_0)T/2)}.$$

$$(4.5.24)$$

Noting $\sin(\omega(T/2) \mp \omega_0(T/2)) = \sin(\omega(T/2) \mp \pi) = -\sin(\omega T/2)$ and using this in (4.5.24) results in

$$w_H(j\omega) = .54T\frac{\sin(\omega T/2)}{(\omega T/2)}$$

$$- .23T\sin(\omega T/2)\left[\frac{1}{(\omega T/2) - \pi} + \frac{1}{(\omega T/2) + \pi}\right].$$

In terms of f, we have

$$W_H(j\omega) = \frac{T\sin(\pi fT)}{(\pi fT)}\left[\frac{.54 - .08(fT)^2}{1 - (fT)^2}\right], \quad \omega = 2\pi f.$$

$$(4.5.25) \blacksquare$$

4.5.4 Energy Spectral Density

From Rayleigh's energy theorem, the energy contained in an energy signal $F[x(t)] = X(j\omega)$ can be computed either by the time domain function or by the frequency domain function and the energy contained in the signal is

$$
E = \int_{-\infty}^{\infty} |x(t)|^2 dt = \frac{1}{2\pi} \int_{-\infty}^{\infty} |X(j\omega)|^2 d\omega
$$

$$
= \int_{-\infty}^{\infty} |X(jf)|^2 df, \quad G(f) = |X(jf)|^2 = |X(j\omega)|^2/2\pi.
$$

$$(4.5.26)$$

Note that $G(f) = |X(jf)|^2 = |X(j\omega)|^2/2\pi$ is the *energy spectral density*.

Example 4.5.10 *a*. Derive the energy spectral density of the function

$$
x(t) = e^{-at}u(t) \overset{\text{FT}}{\longleftrightarrow} [1/(a+j\omega)] = X(j\omega), \quad a > 0
$$

b. Illustrate the validity of Rayleigh's energy theorem.

c. Select the frequency band $-W < \omega < W$ so that 95% of the total energy is in this band.

Solution: *a*. The energy spectral density is given by

$$
|X(j\omega)|^2 = 1/[a^2 + \omega^2]. \qquad (4.5.27a)
$$

b. By using the time domain function, the energy contained in the function is

$$
E_{\text{Total}} = \int_{-\infty}^{\infty} |x(t)|^2 dt = \int_{0}^{\infty} e^{-2at} dt = \frac{1}{2a}. \quad (4.5.27b)
$$

We can make use of the frequency function to determine the energy as well and is

$$
E_{\text{Total}} = \frac{1}{2\pi} \int_{-\infty}^{\infty} |X(j\omega)|^2 d\omega = \frac{1}{2\pi a} \tan^{-1}\left(\frac{\omega}{a}\right)\Big|_{-\infty}^{\infty} = \frac{1}{2a}.
$$

$$(4.5.27c)$$

The above two equations validate Rayleigh's energy theorem.

c. $$
E_{.95} = \frac{.95}{2a} = \frac{1}{2\pi} \int_{-W}^{W} \frac{d\omega}{a^2 + \omega^2} d\omega
$$

$$
= \frac{1}{a\pi} \tan^{-1}\left(\frac{W}{a}\right) \to a \tan(.95(\pi/2)) = W,
$$

$$
W = 2\pi F, \quad F \approx (2.022a) \text{ Hz}. \qquad (4.5.27d) \quad \blacksquare
$$

Example 4.5.11 Consider the pulse function $x(t) = \Pi[t/\tau]$. Find the percentage of energy contained in the frequency range $-W < \omega < W$, $W = 2\pi f_c$.

Solution: The spectrum and the energy spectral densities are, respectively, given by

$$
X(j\omega) = \tau \frac{\sin((\omega/2)\tau)}{((\omega/2)\tau)},
$$

$$
\frac{1}{2\pi} |X(j\omega)|^2 = \frac{1}{2\pi} \frac{\tau^2 \sin^2((\omega/2)\tau)}{((\omega/2)\tau)^2}. \qquad (4.5.28)
$$

The total energy and the energy contained in the frequency range $-f_c < f < f_c$ of the pulse are

$$
E_{\text{Total}} = (1)^2 \tau = \tau, \quad E_{f_c} = \int_{-f_c}^{f_c} \tau^2 \frac{\sin^2(\pi f \tau)}{(\pi f \tau)^2} df.
$$

$$(4.5.29)$$

Using the change of variable $\beta = f\tau, df = d\beta/\tau$, and $f = \pm f_c \to \beta = \pm f_c \tau$, the energy contained in the frequency range $-f_c < f < f_c$ can be computed and the ratio of this to the total energy contained in the pulse. These follow

$$
E_{f_c} = 2\tau \int_{0}^{f_c\tau} \frac{\sin^2(\pi\beta)}{(\pi\beta)^2} d\beta,
$$

$$
\frac{E_{f_c}}{E_{\text{Total}}} = 2 \int_{0}^{f_c\tau} \frac{\sin^2(\pi\beta)}{(\pi\beta)^2} d\beta. \qquad (4.5.30a)
$$

We can only compute this integral numerically. In the case of $f_c\tau = 1$, we have

$$
(E_{f_c}/E_{\text{Total}}) \approx .9028. \qquad (4.5.30b)
$$

That is, approximately 90% of the energy is contained in the spectral *main lobe* of the signal. If we include the side lobes, more energy will be included and $E_{f_c \to \infty} = E_{\text{Total}}$. Ninety percent of the energy is reasonably sufficient to represent a rectangular pulse. ∎

An *interesting formula* can be derived to find the energy of a *causal signal* $x(t)$, i.e., $x(t) = 0$, $t < 0$ in terms of its real and imaginary parts of its transform $X(j\omega) = R(\omega) + jI(\omega)$. The energy is given by Papoulis (1977) as follows:

$$\int_0^\infty x^2(t)dt = \frac{2}{\pi}\int_0^\infty R^2(\omega)d\omega = \frac{2}{\pi}\int_0^\infty I^2(\omega)d\omega. \quad (4.5.31)$$

This can be shown by noting $x(t) = 2x_e(t)u(t) = 2x_0(t)u(t)$ and then using the transforms of real and imaginary parts. Details are left as an exercise.

4.6 Autocorrelation and Cross-Correlation

In this section we will see that the inverse Fourier transform of the energy spectral density discussed in the last section is the autocorrelation (AC) function defined in Chapter 2 (see (2.7.1)). The AC function of a real function $x(t)$ is

$$R_{xx}(\tau) = \int_{-\infty}^\infty x(t)x(t+\tau)dt$$

$$= \int_{-\infty}^\infty x(t)x(t-\tau)dt = R_x(\tau). \quad (4.6.1)$$

Note the single subscript in the case of autocorrelation and a double subscript in the case of cross-correlation below. The cross-correlations (see (2.6.3)) are

$$R_{xh}(\tau) = \int_{-\infty}^\infty x(t)h(t+\tau)dt = \int_{-\infty}^\infty x(\alpha - \tau)h(\alpha)d\alpha,$$

$$(4.6.2a)$$

$$R_{hx}(\tau) = \int_{-\infty}^\infty h(t)x(t+\tau)dt = \int_{-\infty}^\infty h(\beta - \tau)x(\beta)d\beta,$$

$$(4.6.2b)$$

$$R_{hx}(\tau) = R_{xh}(-\tau). \quad (4.6.3)$$

Cross-correlation reduces to the autocorrelation when $h(t) = x(t)$. The Fourier transform of the AC function is *the energy spectral density* and is

$$G_x(\omega) = \int_{-\infty}^\infty R_x(\tau)e^{-j\omega\tau}d\tau,$$

$$R_x(\tau) = \frac{1}{2\pi}\int_{-\infty}^\infty G_x(\omega)e^{j\omega\tau}d\omega = F^{-1}[G_x(\omega)],$$

$$R_x(\tau) \xleftrightarrow{\text{FT}} G_x(\omega). \quad (4.6.4)$$

Note that the autocorrelation function is the integral of the product of two functions, the function and its shifted version. It is a function of τ, which is the shift between the given function and its shifted version. The Fourier transform pair relationship in (4.6.4) is referred to as the *Wiener–Khintchine theorem*. Also, we should note that a function $x(t)$ and its delayed or advanced version $x(t \pm t_0)$ have the *same* autocorrelation function and therefore they have the same energy spectral densities. That is,

$$R_y(\tau) = \int_{-\infty}^\infty x(t \pm t_0)x(t \pm t_0 + \tau)dt$$

$$= \int_{-\infty}^\infty x(\alpha)x(\alpha + \tau)dt = R_x(\tau) \quad (4.6.5)$$

$$G_y(\omega) = G_x(\omega). \quad (4.6.6)$$

Correlations were expressed in terms of convolution (see (2.6.4a and b)). Now,

$$R_x(\tau) = x(\tau) * x(-\tau). \quad (4.6.7)$$

Using the convolution theorem and $F[x(-\tau)] = X(-j\omega) = X^*(j\omega)$, it follows that

$$F[R_x(\tau)] = F[x(\tau) * x(-\tau)] = F[x(\tau)]F[x(-\tau)]$$
$$= |X(j\omega)|^2 = G_x(\omega). \quad (4.6.8)$$

Example 4.6.1 Show that the energy spectral densities of $x(t)$ and $x(t \mp t_0)$ are the same and therefore the autocorrelation functions of the two functions are identical (see (4.6.5) and (4.6.6)).

Solution: Noting $x(t \mp t_0) \overset{FT}{\longleftrightarrow} e^{\mp j\omega t_0} X(j\omega)$, we have

$$e^{\mp j\omega t_0} X(j\omega) e^{\pm j\omega t_0} X^*(j\omega) = |X(j\omega)|^2 = G_x(\omega). \quad (4.6.9)$$

The corresponding autocorrelation function is given by

$$R_x(\tau) = F^{-1}[G_x(\omega)] = F^{-1}\left[|X(j\omega)|^2\right]. \quad (4.6.10)$$

■

This example illustrates that the autocorrelation function $R_x(\tau)$ does not have the *phase* information contained in the function $x(t)$. This can be seen from the fact that t_0 is not in either of the expressions $R_x(\tau)$ or $G_x(\omega)$. The autocorrelation function is even and its spectrum, the energy spectral density, is real and even and

$$E_x = R_x(\tau)|_{\tau=0} = \int_{-\infty}^{\infty} x(t)x(t+\tau)dt|_{\tau=0} = \int_{-\infty}^{\infty} |x(t)|^2 dt.$$

$$(4.6.11a)$$

This gives the energy in the signal. Using the Wiener–Khintchine theorem, we have

$$E_x = R_x(\tau)|_{\tau=0} = \frac{1}{2\pi} \int_{-\infty}^{\infty} |X(j\omega)|^2 d\omega. \quad (4.6.11b)$$

Example 4.6.2 Consider the pulse function and its transform given by

$$x(t) = e^{-at}u(t) \overset{FT}{\longleftrightarrow} 1/(a+j\omega) = X(j\omega), \quad a>0.$$

$$(4.6.12a)$$

  a. Give the expression for the energy spectral density and its inverse transform, the corresponding autocorrelation function.
  b. Compute the energy in $x(t)$ using its AC function and its energy spectral density.

Solution: *a.* From Example 2.7.1 and (2.7.10), we have

$$\phi_x(\tau) = \frac{1}{2a} e^{-a|\tau|}.$$

The energy spectral density and its inverse transform are (see (4.3.20))

$$|X(j\omega)|^2 = \frac{1}{(a+j\omega)(a-j\omega)} = \frac{1}{a^2+\omega^2},$$

$$F^{-1}\left[\frac{1}{a^2+\omega^2}\right] = \frac{1}{2a} e^{-a|\tau|} = R_x(\tau).$$

$$(4.6.12b)$$

  b. By using the AC function and the energy spectral density, the energy in $x(t)$ is $E_x = R_x(\tau)|_{\tau=0} = 1/2a$,

$$\frac{1}{\pi} \int_0^{\infty} \frac{1}{a^2+\omega^2} d\omega = \frac{1}{a\pi} \tan^{-1}\left(\frac{\omega}{a}\right)\Big|_0^{\infty} = \frac{1}{2a} = E_x.$$

$$(4.6.12c) \quad ■$$

The *cross-correlation theorem* for aperiodic signals is

$$R_{hx}(\tau) \overset{FT}{\longleftrightarrow} H^*(j\omega)X(j\omega). \quad (4.6.13)$$

It can be shown by

$$R_{hx}(\tau) = \int_{-\infty}^{\infty} h(t)x(t+\tau)dt$$

$$= \frac{1}{2\pi} \int_{-\infty}^{\infty} h(t)\left[\int_{-\infty}^{\infty} X(j\omega)e^{j\omega(t+\tau)}d\omega\right]dt$$

$$= \frac{1}{2\pi} \int_{-\infty}^{\infty} X(j\omega)\left[\int_{-\infty}^{\infty} h(t)e^{j\omega t}dt\right]e^{j\omega\tau}d\omega$$

$$= \frac{1}{2\pi} \int_{-\infty}^{\infty} H^*(j\omega)X(j\omega)e^{j\omega\tau}d\omega.$$

Comparing these, (4.6.13) follows. For $\tau = 0$,

$$R_{hx}(0) = \int_{-\infty}^{\infty} h(t)x(t)dt = \frac{1}{2\pi} \int_{-\infty}^{\infty} H^*(j\omega)X(j\omega)d\omega.$$

$$(4.6.14)$$

Equation (4.6.14) is a generalized version of *Parseval's theorem*. The cross-correlation theorem reduces to the case of autocorrelation by replacing $H(j\omega)$ by $X(j\omega)$ in (4.6.13).

4.6.1 Power Spectral Density

Earlier we have studied the power signals that include periodic signals and random signals. We will not be discussing random signals in this book in any detail. The autocorrelation of a periodic signal $x_T(t)$ is given by

$$R_{T,x}(\tau) = \frac{1}{T} \int_T x_T(t) x_T(t+\tau) dt. \qquad (4.6.15a)$$

Note that the above integral is over any one period. We have seen in Chapter 2 that the autocorrelation function of a periodic function is also a periodic function with the same period. The Fourier transform of the autocorrelation function is called the *power spectral density (PSD)* and its inverse is the autocorrelation function. The autocorrelation function and the corresponding spectral density function form a Fourier transform pair

$$S_x(\omega) = F\big[R_{x,T}(\tau)\big], \quad R_{x,T}(\tau) = F^{-1}[S_x(\omega)],$$
$$R_{x,T}(\tau) \overset{\text{FT}}{\longleftrightarrow} S_x(\omega). \qquad (4.6.15b)$$

This relation is referred to as the *Wiener–Khintchine theorem* for periodic signals. The power spectral density of a periodic signal can be determined from

$$P = \frac{1}{2\pi} \int\limits_{-\infty}^{\infty} S_x(\omega) d\omega = R_{x,T}(0)$$
$$= \frac{1}{T} \int_T x(t) x(t+\tau) d\tau\big|_{\tau=0}. \qquad (4.6.15c)$$

Formal proof of the general Wiener–Khintchine theorem is beyond the scope here (see Ziemer and Tranter, 2002 and Peebles, 2001). In the following we will assume that $S_x(\omega)$ is given by the transform of the periodic autocorrelation function.

Notes: Power signals include periodic and random signals. The autocorrelation function of a periodic function is periodic and the *power spectral density* contains impulses. ∎

Example 4.6.3 Consider the harmonic form of Fourier series of a periodic function

$$x_T(t) = X_s[0] + \sum_{k=1}^{N} d[k] \cos(k\omega_0 t + \theta[k]), \quad \omega_0 = 2\pi/T.$$
$$(4.6.16)$$

Find its auto correlation.

Solution: The autocorrelation function is given by (see (2.8.9))

$$R_{T,x}(\tau) = X_s^2[0] + \frac{1}{2} \sum_{k=1}^{N} d^2[k] \cos(k\omega_0 \tau). \qquad (4.6.17)$$

The phase terms $\theta[k]$s are not in the autocorrelation function. The PSD is

$$S_x(\omega) = 2\pi X_s^2[0] \delta(\omega)$$
$$+ \frac{\pi}{2} \sum_{k=1}^{N} d^2[k] \{\delta(\omega - k\omega_0) + \delta(\omega + k\omega_0)\}.$$
$$(4.6.18)$$

The AC function and the PSD do not have any phase information and the frequencies are located at $\omega = k\omega_0, k = 1, 2, ..., N$. The average power can be computed from the AC function or from the power spectral density. From (4.6.17), we have the average power

$$P_x = R_{T,x}(0) = X_s^2[0] + \frac{1}{2} \sum_{k=1}^{N} d^2[k]. \qquad (4.6.19)$$

Using the power spectral density, we have

$$P_x = \frac{1}{2\pi} \int\limits_{\infty}^{\infty} S_x(\omega) d\omega$$
$$= \frac{2\pi}{2\pi} \int\limits_{-\infty}^{\infty} \left\{ X_s^2[0] \delta(\omega) + \frac{1}{4} \sum_{k=1}^{N} d^2[k].\{\delta(\omega - k\omega_0) \right.$$
$$\left. + \delta(\omega + k\omega_0)\} \right\} d\omega.$$

Since the integrand contains only impulses, the integral can be evaluated by inspection and is $R_{T,x}(0)$ in (4.6.19). ∎

Notes: In the case of energy signals the square of the magnitude spectrum gives the *energy spectral* *density (ESD)*, and the energy content is obtained by integrating the ESD. In the case of periodic signals, the spectrum contains impulses and the *square* of *an impulse function* is *not defined*. The AC function and the Wiener–Khintchine theorem are used to compute the *power spectral density (PSD)* of the periodic signal and its integral is the average power contained in the power signal.

Most signals are corrupted by noise. Autocorrelation function "cleans" the signal and it provides a better insight into the essential qualities of the signal. Note that the AC function does not have the phase information in the signal. Since the AC of a periodic function is also a periodic with the same period, it can be determined by identifying the peaks in the autocorrelation function, thereby identifying the fundamental frequency. Finding the pitch period of a vowel in a noisy speech signal by using the autocorrelation function is very effective on a short-time basis (see Rabiner and Schafer, 1978). ∎

4.7 Bandwidth of a Signal

In Section 4.2.2, bandwidth (BW) of a signal was discussed in simple terms. One definition is the range of positive frequencies in which *most* of the signal energy or power is contained. This is vague since the word "most" can be interpreted differently. We will consider this here in more detail.

A signal is considered time limited if the signal is zero outside an interval. For example, a pulse function $\Pi[(t - t_0)/\tau]$ is nonzero for $|t - t_0| < \tau/2$ and zero outside this range. It is nonzero for τ s. The signals considered in this book are real signals and their spectral amplitudes are even and the spectral phase angles are odd. A signal $x(t)$ is said to be band limited to B Hz if

$$|X(j\omega)| = 0, \quad \omega = |2\pi f| > W = 2\pi B. \quad (4.7.1)$$

Since it is band limited to B Hz, B is defined as the bandwidth of the signal. In this case the signal occupies only low frequencies, i.e., it is a low-frequency signal or sometimes referred to as a *low-pass signal*. Note that the bandwidth is defined using *only positive frequencies*. Band-pass signals are common in communication theory. Band-pass spectrum can be defined as follows:

$$|X(j\omega)| = \begin{cases} 0, & |\omega| < \omega_0 - (W/2) \\ H_0, & |\omega - \omega_0| < (W/2) \\ H_0, & |\omega + \omega_0| < (W/2) \\ 0, & |\omega| > \omega_0 + (W/2) \end{cases}. \quad (4.7.2)$$

It is an *ideal band-pass signal*. Most practical signals are not band limited. There are functions that are neither time limited nor band limited. For example, consider the double exponential function given below and its transform derived earlier

$$x(t) = e^{-a|t|} \xleftrightarrow{\text{FT}} 2a/[a^2 + (\omega^2)] = X(j\omega), \ a > 0. \quad (4.7.3)$$

The question we need to answer is, what is a meaningful definition of the *time width* of an arbitrary *non-time-limited* signal? How about a meaningful definition of the *frequency width* of an arbitrary *non-band-limited* signal? In Section 4.2.2, we have seen that a shorter time width signal corresponds to a broader spectrum. For example, the Fourier transform pair of a rectangular pulse function is given by

$$x(t) = \Pi\left[\frac{t}{\tau}\right] \xleftrightarrow{\text{FT}} \tau \frac{\sin(\omega\tau/2)}{(\omega\tau/2)} = X(j\omega). \quad (4.7.4)$$

Most of the energy is contained in the main lobe of the spectrum, which occupies the frequency band between the two zeros of $X(j\omega)$ located at $\omega = \pm 2\pi/\tau$. The energy content of this pulse is quantified in terms of the frequency content in Example 4.5.12. The side lobes contain a small portion of the energy. The time width of the pulse is obviously equal to τ s and the frequency width is approximately $1/\tau$ Hz, considering only positive frequencies. Increasing (decreasing) the time width reduces (increases) the frequency width. At least, from this example, we see that the two widths are inversely proportional to each other. In the following we will consider a few standard definitions of time and frequency widths and they give some meaning.

4.7.1 Measures Based on Areas of the Time and Frequency Functions

Using the ordinate theorems discussed earlier, we have

$$X(0) = \int_{-\infty}^{\infty} x(t)dt, \ x(0) = \frac{1}{2\pi}\int_{-\infty}^{\infty} X(j\omega)d\omega. \quad (4.7.5)$$

The dc value of a signal is zero if it is an odd function. Interestingly,

$$x(t) = d^n y(t)/dt^n \xrightarrow{\text{FT}} (j\omega)^n Y(j\omega) = X(j\omega). \quad (4.7.6)$$

It is zero at $\omega = 0$ provided $Y(j\omega)$ has *no* poles at the origin that can be canceled by $(j\omega)^n$. If the time function has a discontinuity at the origin, then $x(0)$ is obtained from the integral in (4.7.5) and is the average value or the *half-value* at the discontinuity. Let the time width and the frequency widths, respectively, be defined by

$$t_w = \frac{\int_{-\infty}^{\infty} x(t)dt}{x(0)}, \quad \omega_w = \frac{\int_{-\infty}^{\infty} X(j\omega)d\omega}{X(0)}. \quad (4.7.7)$$

Then the product

$$t_w \omega_w = 2\pi. \quad (4.7.8)$$

That is, the product of time and frequency widths is a constant. Some authors use the frequency f in Hz rather than $\omega = 2\pi f$ in rad/s in the time–bandwidth product. These simple measures have drawbacks illustrated below.

Example 4.7.1 Consider the function given by $x(t) = \Pi[t - .5] - \Pi[t + .5]$. It is an odd function. If the above definition is used, the time width is zero, even though the actual width of this function is 2 s. ∎

Example 4.7.2 Consider the pair

$$\Pi[t] \xrightarrow{\text{FT}} \frac{\sin(\omega/2)}{(\omega/2)}.$$

The area under the pulse function is 1. The area under the sinc function is 2π and the time–bandwidth product is 2π. ∎

Example 4.7.3 Consider the Fourier transform pair corresponding to the exponential decaying function to find the values of the functions a. $X(0)$ and b. $F^{-1}[X(j\omega)]|_{t=0}$:

$$x(t) = e^{-at}u(t) \xrightarrow{\text{FT}} \frac{1}{a + j\omega} = X(j\omega), \ a > 0.$$

Solution: a. $X(0) = \int_{0}^{\infty} e^{-at}dt = \frac{1}{a}$.

a. $x(0) = \frac{1}{2\pi}\int_{-\infty}^{\infty} X(j\omega)e^{j\omega t}d\omega|_{t=0}$

$$= \frac{1}{2\pi}\int_{-\infty}^{\infty} \frac{a}{a^2 + (\omega)^2}d\omega + \frac{j}{2\pi}\int_{-\infty}^{\infty} \frac{(-\omega)}{a^2 + (\omega)^2}d\omega.$$

The integrand in the second integral is an odd function and therefore it is zero:

$$x(0) = \frac{1}{2\pi}\int_{-\infty}^{\infty} \frac{a}{a^2 + \omega^2}d\omega = \frac{1}{2\pi}\tan^{-1}(\omega/a)|_{-\infty}^{\infty} = \frac{1}{2}.$$

Note the exponential time-decaying function $x(t)$ is discontinuous at $t = 0$ and the above result verifies that the inverse transform converges to the *half-value*, i.e., the average value of the function before and after the discontinuity. ∎

4.7.2 Measures Based on Moments

The time width T_w of a real non-time-limited function is defined by

$$(T_w)^2 = \frac{1}{\|x\|^2}\int_{-\infty}^{\infty} (t - \bar{t})^2 x^2(t)dt, \ \|x\|^2 = \int_{-\infty}^{\infty} x^2(t)dt, \quad (4.7.9)$$

$$E_x = \|x\|^2 = \int_{-\infty}^{\infty} x^2(t)dt = \frac{1}{2\pi}\int_{-\infty}^{\infty} |X(j\omega)|^2 d\omega < \infty. \quad (4.7.10)$$

The *center of gravity* of the area of the function is

$$\bar{t} = \frac{1}{E_x}\int_{-\infty}^{\infty} tx^2(t)dt. \quad (4.7.11)$$

T_w is a measure of the signal spread about $\bar{t}$ and is the signal *dispersion in time*.

Notes: These measures can be seen noting $p(t) = [x^2(t)/E_x] > 0$ is a valid probability density function, as is nonnegative for all t and the area under it is 1. In statistical terminology $\bar{t}$ is the mean and $(T_W)^2$ is the variance (see Peebles, 2001). ∎

Example 4.7.4 Consider the exponential decaying function $x(t) = e^{-at}u(t)$, $a > 0$. Find the center of gravity and the time dispersion T_w.

Solution: The energy contained in the pulse is $E = (1/2a)$. The center of gravity is

$$\bar{t} = \frac{\int_0^\infty te^{-2at}dt}{(1/2a)} = 2a\frac{-e^{-2at}[1+2at]}{(2a)^2}\Big|_0^\infty = \frac{1}{2a}.$$

$$(4.7.12)$$

In addition, using the following integral formulas, T_w is as follows:

$$\int t^2 e^{\beta t}dt = e^{\beta t}\left[\frac{t^2}{\beta} - \frac{2t}{\beta^2} + \frac{2}{\beta^3}\right],$$

$$\int te^{\beta t}dt = e^{\beta t}\left[\frac{t}{\beta} - \frac{1}{\beta^2}\right], \text{ and } \int e^{\beta t}dt = e^{\beta t}\frac{1}{a},$$

$$(T_w)^2 = 2a\int_0^\infty (t-(1/2a))^2 e^{-2at}dt$$

$$= (2a)\int_0^\infty [t^2 - (1/a)t + (1/2a)^2]e^{-2at}dt = (1/2a)^2$$

$$\Rightarrow T_w = (1/2a). \qquad (4.7.13)$$

The frequency width, a frequency measure, $W_w = 2\pi F_w$, can be defined by

$$W_w^2 = \frac{1}{\|X\|^2}\int_{-\infty}^\infty (\omega - \bar{\omega})^2|X(\omega)|^2 d\omega$$

$$\left[\text{Note}: \bar{\omega} = \frac{1}{E_x(2\pi)}\int_{-\infty}^\infty \omega|X(j\omega)|^2 d\omega = 0\right], (4.7.14a)$$

$$\|X\|^2 = \int_{-\infty}^\infty |X(j\omega)|^2 d\omega = 2\pi\|x\|^2$$

(Rayleigh's energy theorem). $\qquad$ (4.7.14b)

This follows since the integrand in the above equation is odd and the integral of an odd function over a symmetric interval is zero. ∎

A bound on the time–bandwidth product $T_w W_w = T_w(2\pi F_w)$ is derived using Rayleigh's energy theorem and Schwarz's inequality (see Section 2.1.). The inequality is briefly reviewed below.

Schwarz's inequality: The inequality is (see (2.1.9d))

$$\|\langle x(t) + y(t)\rangle\| \leq \|x(t)\|\|y(t)\|$$

$$\Rightarrow \int_a^b |x(t)y(t)|^2 dt \leq \int_a^b |x(t)|^2 dt\int_a^b |y(t)|^2 dt.$$

$$(4.7.15)$$

4.7.3 Uncertainty Principle in Fourier Analysis

The uncertainty principle in spectral analysis states that if the integrals in (4.7.9) and (4.7.14a) are *finite* and

$$\lim_{t\to\infty}\sqrt{t}x(t) = 0, \qquad (4.7.16)$$

then

$$T_w W_w \geq \frac{1}{2} \text{ or } T_w F_w \geq \frac{1}{2(2\pi)}, \quad W_w = 2\pi F_w.$$

$$(4.7.17)$$

Using the expressions for T_w and W_w from (4.7.9) and (4.7.14a) results in

$$(T_w W_w)^2 = \frac{1}{\|x\|^2\|X\|^2}\left[\int_{-\infty}^\infty (t-\bar{t})^2 x^2(t)dt\right]$$

$$\times\left[\int_{-\infty}^\infty \omega^2|X(j\omega)|^2 d\omega\right],$$

$$(4.7.18)$$

$$\|X\|^2 = \int_{-\infty}^\infty |X(j\omega)|^2 d\omega, \|x\|^2 = \int_{-\infty}^\infty |x^2(t)|dt,$$

$$\|X\|^2 = 2\pi\|x\|^2. \qquad (4.7.19)$$

Noting the Fourier transform derivative theorem, i.e., $F[x'(t)] = (j\omega)X(j\omega)$ and using Rayleigh's energy theorem results in

$$\int_{-\infty}^{\infty} |x'(t)|^2 dt = \frac{1}{2\pi} \int_{-\infty}^{\infty} \omega^2 |X(j\omega)|^2 d\omega. \quad (4.7.20)$$

Therefore

$$(T_w W_w)^2$$

$$= \frac{1}{\|x\|^2 \|X\|^2} \int_{-\infty}^{\infty} (t - \bar{t})^2 x^2(t) dt \left[\int_{-\infty}^{\infty} \omega^2 |X(j\omega)|^2 d\omega \right]$$

$$= \frac{1}{\|x\|^2 \|X\|^2} \int_{-\infty}^{\infty} (t - \bar{t})^2 x^2(t) dt (2\pi) \int_{-\infty}^{\infty} [x'(t)]^2 dt$$

$$= \frac{1}{\|x\|^4} \int_{-\infty}^{\infty} (t - \bar{t})^2 x^2(t) dt \int_{-\infty}^{\infty} [x'(t)]^2 dt \quad (4.7.21)$$

Using Schwarz's inequality in (4.7.15) results in

$$\left[\int_{-\infty}^{\infty} (t - \bar{t})^2 x^2(t) dt \right] \left[\int_{-\infty}^{\infty} [x'(t)]^2 dt \right]$$

$$\geq \left| \int_{-\infty}^{\infty} (t - \bar{t}) x(t) x'(t) dt \right|^2. \quad (4.7.22)$$

Considering the right-hand side of the above equation and integrating it by parts, we have $\int u dv = uv - \int v du$, with $u = (t - \bar{t})$, $du/dt = 1$, $dv/dt = x(t)x'(t)$, and $v = (1/2)x^2(t)$

$$\int_{-\infty}^{\infty} (t - \bar{t}) x(t) \frac{dx}{dt} dt = \frac{1}{2}(t - \bar{t}) x^2(t) \Big|_{-\infty}^{\infty} - \frac{1}{2} \int_{-\infty}^{\infty} x^2(t) dt.$$

Assuming that $\lim_{t \to \pm\infty} (t - \bar{t}) x^2(t) = 0$, it follows that

$$\int_{-\infty}^{\infty} (t - \bar{t}) x(t) x'(t) dt = -\frac{1}{2} \int_{-\infty}^{\infty} x^2(t) dt = -\frac{1}{2}\|x\|^2. \quad (4.7.23)$$

Using this and (4.7.21) in (4.7.20), we have

$$(T_w W_w)^2 \geq \frac{\|x\|^4}{4\|x\|^4} = \frac{1}{4} \quad \text{or} \quad T_w W_w \geq \frac{1}{2} \quad \text{or}$$

$$T_w F_w \geq \frac{1}{2(2\pi)}, \quad W_w = 2\pi F_w. \quad (4.7.24)$$

Example 4.7.5 Illustrate the uncertainty principle using the Gaussian transform pair

$$x(t) = e^{-at^2} \xleftrightarrow{\text{FT}} \sqrt{\frac{\pi}{\alpha}} e^{-(\omega)^2/4\alpha} = X(j\omega), \quad a > 0. \quad (4.7.25a)$$

Solution: First, differentiate both sides of the following equation with respect to α:

$$\int_{-\infty}^{\infty} e^{-\alpha t^2} dt = \sqrt{\frac{\pi}{\alpha}} \Rightarrow \int_{-\infty}^{\infty} \frac{de^{-\alpha t^2}}{d\alpha} dt = \int_{-\infty}^{\infty} (-t^2) e^{-\alpha t^2} dt$$

$$= \sqrt{\pi} \frac{d(\alpha^{-1/2})}{d\alpha} = -\sqrt{\pi} \frac{\alpha^{-1.5}}{2}$$

$$= -\frac{1}{2\alpha} \sqrt{\frac{\pi}{\alpha}}. \quad (4.7.25b)$$

Canceling the negative signs in (4.7.25b) results in

$$\int_{-\infty}^{\infty} t^2 e^{-\alpha t^2} dt = \frac{1}{2\alpha} \sqrt{\frac{\pi}{\alpha}}. \quad (4.7.25c)$$

From tables,

$$\|x\|^2 = \int_{-\infty}^{\infty} e^{-2at^2} dt = \sqrt{\frac{\pi}{2a}}. \quad (4.7.26)$$

Now let $\alpha = 2a$ in (4.7.25c), which results in

$$\int_{-\infty}^{\infty} t^2 e^{-\alpha t^2} dt = \int_{-\infty}^{\infty} t^2 e^{-2at^2} dt = \frac{1}{4a} \sqrt{\frac{\pi}{2a}}.$$

Noting that the Gaussian pulse in this example is even and the integrand in (4.7.11) is odd, it follows that $\bar{t} = 0$. The time width can be computed from (4.7.9) and

$$(T_w)^2 = \frac{1}{\|x\|^2} \int_{-\infty}^{\infty} t^2 e^{-2at} dt = (1/\sqrt{\frac{\pi}{2a}}) \frac{1}{4a} \sqrt{\frac{\pi}{2a}} = \frac{1}{4a}. \quad (4.7.27)$$

Noting that $\|X\|^2 = 2\pi\|x\|^2 = 2\pi\sqrt{\pi/2a}$, it follows that

$$(W_w)^2 = \frac{1}{\|X\|^2} \int_{-\infty}^{\infty} \omega^2 |X(j\omega)|^2 d\omega$$

$$= \frac{1}{\|x\|^2 (2\pi)} \int_{-\infty}^{\infty} \omega^2 \frac{\pi}{a} e^{-\frac{2(\omega)^2}{4a}} d\omega$$

$$= \frac{\sqrt{2a}}{(2\pi)\sqrt{\pi}} (\frac{\pi}{a}) \int_{-\infty}^{\infty} \omega^2 e^{-\frac{\omega^2}{2a}} d\omega$$

$$= \frac{\sqrt{2a}}{(2\pi)\sqrt{\pi}a} a \sqrt{\pi(2a)} = a \Rightarrow (T_w W_w)^2$$

$$= \frac{1}{4a} a = \frac{1}{4}. \tag{4.7.28}$$

The time–bandwidth product of a Gaussian pulse is obtained by using $W_w = 2\pi F_w$ and

$$T_w F_w = 1/(2(2\pi)). \tag{4.7.29}$$

This shows the equality in (4.7.24) in the Gaussian case. See Hsu (1967) for additional examples. ∎

4.8 Moments and the Fourier Transform

The nth moment m_n of $x(t)$ is defined by (see Section 1.7.)

$$m_n = \int_{-\infty}^{\infty} t^n x(t) dt, \quad n = 0, 1, 2, \dots . \tag{4.8.1}$$

The moment theorem relates the derivatives of the transform of a function at $\omega = 0$:

$$(-j)^n m_n = \frac{d^n X(j\omega)}{d\omega^n} |_{\omega=0}, \quad n = 0, 1, 2, \dots , \tag{4.8.2}$$

$$m_0 = \int_{-\infty}^{\infty} x(t) dt = X(j\omega)|_{\omega=0} \text{ (Ordinate theorem)},$$

$$\frac{dX(j\omega)}{d\omega}|_{\omega=0} = \int_{-\infty}^{\infty} x(t) \frac{de^{-j\omega t}}{d\omega} dt|_{\omega=0}$$

$$= j \int_{-\infty}^{\infty} t x(t) e^{-j\omega t} dt|_{\omega=0} = -j m_1.$$

Repeating this process and evaluating the derivatives at $\omega = 0$ proves the result in (4.8.2). Now we

will derive the transform in terms of m_i by using the power series expansion

$$e^{-j\omega t} = \sum_{n=0}^{\infty} \frac{1}{n!} (-j\omega t)^n. \tag{4.8.3}$$

Substituting this in the transform and using (4.8.1) and (4.8.2) result in

$$X(j\omega) = F[x(t)] = \int_{-\infty}^{\infty} x(t) e^{-j\omega t} dt$$

$$= \int_{-\infty}^{\infty} \sum_{n=0}^{\infty} \frac{(-j\omega)^n t^n}{n!} x(t) dt$$

$$= \sum_{n=0}^{\infty} \frac{(-j\omega)^n}{n!} \int_{-\infty}^{\infty} t^n x(t) dt$$

$$= \sum_{n=0}^{\infty} \left[\frac{d^n X(\omega)}{d\omega^n} \right]|_{\omega=0} \frac{\omega^n}{n!}. \tag{4.8.4}$$

This holds *only if* the integral of the terms in the above equation is valid. From (4.8.2)

$$X(j\omega) = \sum_{n=0}^{\infty} (-j)^n m_n [\omega^n / n!]. \tag{4.8.5}$$

Although the moment theorem is given in terms of a series expansion, it can be used to compute the transforms of functions (see Papoulis, 1962).

Example 4.8.1 Use the moment theorem and the following identity to show that Fourier transform of the Gaussian pulse $x(t) = e^{-at^2}$ is also a Gaussian pulse.

$$\int_{-\infty}^{\infty} e^{-at^2} dt = \sqrt{\frac{\pi}{a}} \Rightarrow \int_{-\infty}^{\infty} t^2 e^{-at^2} dt = \frac{1}{2} \sqrt{\frac{\pi}{a^3}}, \quad a > 0. \tag{4.8.6a}$$

Solution: Equation (4.8.6a) on the right can be generalized and

$$\int_{-\infty}^{\infty} t^{2n} e^{-at^2} dt = \frac{1.3\dots(2n-1)}{2^n} \sqrt{\frac{\pi}{a^{2n+1}}}$$

$$= \frac{1.3\dots(2n-1)}{(2a)^n} \sqrt{\frac{\pi}{a}} = m_{2n}. \tag{4.8.6b}$$

This gives *even moments* and the odd moments are zero since $x(t)$ is even:

$$X(j\omega) = \sum_{n=0}^{\infty} (-j\omega)^{2n} \frac{m_{2n}}{(2n)!}$$

$$= \sum_{n=0}^{\infty} (-1)^n (\omega)^{2n} \frac{1}{(2n)!} \frac{1.3...(2n-1)}{(2a)^n} \sqrt{\frac{\pi}{a}}.$$

This can be expressed in a compact form by noting

$$\frac{(1)(3)(5)...(2n-1)}{(1/2^n)(2n)!} = \frac{(1)(3)(5)...(2n-1)}{(1/2^n)(1)(2)(3)...(2n-1)(2n)}$$

$$= \frac{1}{(1/2^n)(2)(4)...(2n)}$$

$$= \frac{1}{(1)(2)...(n)} = \frac{1}{n!},$$

$$X(j\omega) = \sqrt{\frac{\pi}{a}} \sum_{n=0}^{\infty} \frac{(-1)^2 \omega^{2n}}{(4a)^n n!}$$

$$= \sqrt{\pi/a} e^{-\omega^2/4a} \Rightarrow e^{-at^2} \overset{FT}{\longleftrightarrow} \sqrt{\pi/a} e^{-\omega^2/4a}. \quad \blacksquare$$

Example 4.8.2 Find the first two moments of $x(t) = ae^{-at}u(t)$, $a>0$.

Solution: Noting

$$F[x(t)] = F[ae^{-at}u(t)] = \frac{a}{[a+j\omega]} = X(j\omega), \quad a>0,$$

it follows that

$$m_0 = X(0) = 1, m_1 = j\frac{dX(j\omega)}{d\omega}\Big|_{\omega=0}$$

$$= j\frac{-ja}{(a+j\omega)^2}\Big|_{\omega=0} = \frac{a}{a^2} = \frac{1}{a}. \quad \blacksquare$$

4.9 Bounds on the Fourier Transform

In Chapter 3, we have learned that the derivative of a periodic function plays a role on the bounds on its F-series coefficients. We can use the Fourier time differentiation theorem to find the bounds on the transform. First

$$\frac{d^n x(t)}{dt^n} \overset{FT}{\longleftrightarrow} (j\omega)^n X(j\omega). \quad (4.9.1)$$

Spectral bounds: The following bounds are valid *if* the appropriate *derivatives exist*:

$$|X(j\omega)| \leq \begin{cases} \int_{-\infty}^{\infty} |x(t)|dt \\ \frac{1}{|\omega|} \int_{-\infty}^{\infty} \left|\frac{dx(t)}{dt}\right|dt \\ \frac{1}{\omega^2} \int_{-\infty}^{\infty} \left|\frac{d^2 x(t)}{dt^2}\right|dt \end{cases} \quad (4.9.2)$$

First bound:

$$|X(j\omega)| = \left| \int_{-\infty}^{\infty} x(t)e^{-j\omega t}dt \right| \leq \int_{-\infty}^{\infty} |x(t)e^{-j\omega t}|dt$$

$$= \int_{-\infty}^{\infty} |x(t)|dt.$$

Second bound: With $x\prime(t) \overset{FT}{\longleftrightarrow} j\omega X(j\omega)$, we have

$$(j\omega)X(j\omega) = \int_{-\infty}^{\infty} \frac{dx(t)}{dt} e^{-j\omega t}dt \rightarrow |X(j\omega)|$$

$$\leq \frac{1}{|\omega|} \int_{-\infty}^{\infty} \left|\frac{dx(t)}{dt}\right|dt.$$

In a similar manner, we can prove the third bound in (4.9.2) *if* it exists.

Example 4.9.1 Find the first two bounds in (4.9.2) using the Fourier transform pair

$$x(t) = \Pi[t] \overset{FT}{\longleftrightarrow} [\sin(\omega/2)/(\omega/2)] = X(j\omega).$$

Solution: Since the sinc function is bounded by one, we have

$$|X(j\omega)| = \left|\frac{\sin(\omega/2)}{(\omega/2)}\right| \leq \int_{-\infty}^{\infty} |x(t)|dt = 1.$$

The derivative of the unit pulse function has two impulses. That is,

$$\frac{dx(t)}{dt} = \frac{d\Pi[t]}{dt} = \delta(t+(1/2)) - \delta(t-(1/2))$$

$$\Rightarrow \left|\frac{\sin(\omega/2)}{(\omega/2)}\right| \leq \frac{1}{|\omega|} \int_{-\infty}^{\infty} \left|\frac{dx(t)}{dt}\right|dt = \frac{1}{|\omega|} \int_{-\infty}^{\infty} |\delta(t+(1/2))$$

$$- \delta(t-(1/2))|dt = \frac{2}{|\omega|}. \quad \blacksquare$$

Higher bounds in this case are *not* defined since the area of the function $|\delta'(t)|$ is infinite.

4.10 Poisson's Summation Formula

Let $x(t)$ be an arbitrary function with its transform $X(j\omega)$ and let $y_T(t)$ be

$$y_T(t) = \sum_{n=-\infty}^{\infty} x(t+nT) = \frac{1}{T}\sum_{k=-\infty}^{\infty} X_s[k]e^{jk\omega_0 t},$$

$$\omega_0 = 2\pi/T. \tag{4.10.1}$$

This can be seen by first noting that

$$x(t) = x(t)*\delta(t) \text{ and } \delta_T(t) = \sum_{m=-\infty}^{\infty}\delta(t+nT)$$

$$= \sum_{n=-\infty}^{\infty}\delta(t-nT).$$

$$\sum_{n=-\infty}^{\infty} x(t+nT) = x(t) * \sum_{n=-\infty}^{\infty}\delta(t-nT)$$

$$= \sum_{n=-\infty}^{\infty}\int_{-\infty}^{\infty} x(\beta)\delta(t-nT-\beta)d\beta$$

$$= \sum_{m=-\infty}^{\infty} x(t-mT) = y_T(\text{t}) .$$

Noting $F[\delta_T(t)] = (2\pi/T)\sum_{k=-\infty}^{\infty}\delta(\omega-k\omega_0)$ and the convolution theorem, we have

$$F[x(t)*\delta_T(t)] = X(j\omega)\frac{2\pi}{T}\sum_{k=-\infty}^{\infty}\delta(\omega-k\omega_0)$$

$$= \frac{2\pi}{T}\sum_{k=-\infty}^{\infty} X(jk\omega_0)\delta(\omega-k\omega_0). \tag{4.10.2}$$

$$\sum_{n=-\infty}^{\infty} x(t+nT) = \frac{2\pi}{T}\sum_{k=-\infty}^{\infty} X(jk\omega_0)F^{-1}[\delta(\omega-k\omega_0]$$

$$= \frac{1}{T}\sum_{k=-\infty}^{\infty} X(k\omega_0)e^{jk\omega_0 t}. \tag{4.10.3}$$

Comparing this equation with (4.10.1), we have $X_s[k] = X(k\omega_0)$. Using $t=0$ in (4.10.3) results in *Poisson's summation formula*:

$$\sum_{n=-\infty}^{\infty} x(nT) = \frac{1}{T}\sum_{k=-\infty}^{\infty} X(k\omega_0) = \frac{1}{T}\sum_{k=-\infty}^{\infty} X_s[k]. \tag{4.10.4}$$

Notes: Since $x(t)$ is arbitrary, its shifted copies may overlap and $x(t)$ may not be recoverable from $y_T(t)$. The transform pair derived in (4.4.17) can be proved using the above results. ∎

Example 4.10.1 Use Poisson's sum formula to derive an expression for

$$A = \sum_{n=-\infty}^{\infty} e^{-a|n|}. \left(\text{Note}: e^{-a|t|}\overset{\text{FT}}{\longleftrightarrow}[2a/(a^2+\omega^2)], a>0\right).$$

Solution: With $T=1$, i.e., $\omega_0 = 2\pi$, results in

$$A = \sum_{n=-\infty}^{\infty} e^{-a|n|} = \sum_{k=-\infty}^{\infty}\frac{2a}{a^2+(2\pi k)^2}$$

$$= \left(2a/(2\pi)^2\right)\sum_{k=-\infty}^{\infty}\frac{1}{(a/2\pi)^2+(k)^2}. \tag{4.10.5}$$

∎

Example 4.10.2 With $a=2\pi$ and the closed-form expression for the geometric series (given below), derive an expression for

$$B = \sum_{k=-\infty}^{\infty}\frac{1}{1+k^2} \left(\text{Assume } \pi\sum_{n=-\infty}^{\infty} e^{-|n2\pi|}\right.$$

$$\left.= 2\pi\left[\frac{1}{2}+\sum_{n=1}^{\infty}(e^{-2\pi})^n\right]\right).$$

Solution:

$$\sum_{n=1}^{\infty}(e^{-2\pi})^n = \frac{e^{-2\pi}}{1-e^{-2\pi}} \Rightarrow B = \sum_{k=-\infty}^{\infty}\frac{1}{1+k^2}$$

$$= 2\pi\left[\frac{1}{2}+\frac{e^{-2\pi}}{1-e^{-2\pi}}\right] = \pi\frac{1+e^{-2\pi}}{1-e^{-2\pi}}. \quad ∎$$

4.11 Interesting Examples and a Short Fourier Transform Table

Example 4.11.1 Use the transform pair $F[u(t)] = \pi\delta(\omega)+(1/j\omega)$ and the Fourier modulation theorem to find the transform of the switched function $x(t) = u(t)\cos(\omega_c t)$.

Solution:

$$u(t)\cos(\omega_c t)\overset{\text{FT}}{\longleftrightarrow}\frac{j\omega}{\omega_c^2-\omega^2}+\frac{\pi}{2}[\delta(\omega-\omega_c)$$

$$+\delta(\omega+\omega_c)]. \tag{4.11.1} ∎$$

Example 4.11.2 Determine the transform of the *LORAN pulse* defined by

$$p(t) = A[t/T]^2 e^{-2(t/T)} \sin(2\pi f_0 t) u(t). \quad (4.11.2)$$

Solution: LORAN system is used for navigational purposes (Shenoi, 1995). The carrier frequency of the Loran transmission is $f_0 = 100$ kHz and the envelope of the pulse is determined by the time constant $T = 65$ s. Find the transform of the Loran pulse using the following transform pair and the modulation theorem:

$$(-jt)^k x(t) \overset{\text{FT}}{\longleftrightarrow} d^k X(j\omega)/d\omega^k. \quad (4.11.3)$$

$$m(t) = A\left[\frac{t}{T}\right]^2 e^{-2(t/T)} u(t) \overset{\text{FT}}{\longleftrightarrow} \frac{2A}{T^2}\left[\frac{1}{\left(\frac{2}{T}+j\omega\right)^3}\right]$$

$$= \frac{AT}{4}\left[\frac{1}{1+j\omega T/2}\right]^3 = M(j\omega), \quad (4.11.4)$$

$$m(t)\sin(\omega_0 t) \overset{\text{FT}}{\longleftrightarrow} \frac{1}{2j} M(j(\omega-\omega_0))$$
$$\quad (4.11.5) \ \blacksquare$$
$$- \frac{1}{2j} M(j(\omega+\omega_0)).$$

Note from (4.11.4) that $|M(j\omega)|$ is proportional to $1/|\omega|^3$ for large ω.

4.11.1 Raised-Cosine Pulse Function

Earlier the transforms of the rectangular and the sinc pulses (see (4.2.14) and (4.3.27)) were given by

$$x(t) = \Pi\left[\frac{t}{\tau}\right] \overset{\text{FT}}{\longleftrightarrow} \tau\frac{\sin(\omega\tau/2)}{(\omega\tau/2)} = X(j\omega), \quad (4.11.6)$$

$$y_0(t) = \frac{\sin(2\pi Bt)}{(2\pi Bt)} \equiv \frac{\sin(wt)}{wt} \overset{\text{FT}}{\longleftrightarrow} \frac{1}{2B}\Pi\left[\frac{\omega}{2\pi(2B)}\right]$$

$$= \frac{w}{\pi}\Pi\left[\frac{\omega}{2w}\right] = Y_0(j\omega). \quad (4.11.7)$$

For simplicity, $w = 2\pi B$ is used in (4.11.7). In (4.11.6), $x(t)$ is a time-limited rectangular pulse function with its transform, a sinc pulse that is not frequency limited. In (4.11.7), $y_0(t)$ is a sinc pulse in time that is not time limited and its transform is frequency limited to B Hz. In addition to the band-limited property of the sinc pulse, it has some interesting properties. For example,

$$y_0(0) = 1 \quad \text{and} \quad y_0(t)\big|_{t=n/2B} = 0, \quad n = \pm1, \pm2, \dots . \quad (4.11.8)$$

In Section 8.2, it will be shown that a set of sinc pulses is a generalized Fourier series basis set and can be used to represent a signal that is band limited to B Hz from its sample values separated by $t_s = 1/2B = \pi/w$ s. Note $y_0(nt_s) = 0$, $n \neq 0$. The sinc pulses are attractive to use in transmitting data. Unfortunately, systems cannot be built that produce sinc pulses, as they are not time limited. Are there other pulses that have zeros at uniformly spaced time intervals and their transforms decrease toward zero gradually, rather than abruptly? One such family of pulses has this property and the spectra of these pulses have the so-called raised cosine spectra. The pulse and its transform $F[p(t)] = P(j\omega)$ (a real function) are given by

$$p(t) = \frac{1}{t_s}\left(\frac{\sin(wt)}{wt}\right)\left[\frac{\cos(\beta wt)}{1-(2\beta wt/\pi)^2}\right], \quad (4.11.9a)$$

$$P(j\omega) = \begin{cases} 1, & 0 \leq |\omega| \leq (1-\beta)w \\ \frac{1}{2}\left[1 - \sin\left(\frac{\pi}{2\beta w}[|\omega|-w]\right)\right], & (1-\beta)w \leq |\omega| \leq (1+\beta)w \\ 0, & |\omega| > (1+\beta)w \end{cases}$$

$$\quad (4.11.9b)$$

The spectral bandwidth of the raised cosine function depends on the *roll-off factor* β, $0 \leq \beta \leq 1$, as the spectrum roll-off is a function of β. The pulse function in (4.11.9a) has a sinc function as a multiplicative part, thus allowing for the function $p(t)$ to become zero at $t = nt_s = n\pi/w$, $n \neq 0$. Plots of $P(j\omega)$ and $p(t)$ are sketched in Fig. 4.11.1 for three roll-off factors, namely $\beta = 0$, .5, 1. When $\beta = 0$, the transform is a rectangular pulse function and the bandwidth is $(1/2t_s)$ Hz or w rad/s. For $0 < \beta < 1$, the bandwidth of the pulse is $((1 + \beta)/2t_s)$ Hz. The parameter $\beta/2t_s$ represents the additional **or** *excess bandwidth*, which is a fraction of the *minimum bandwidth* $(1/2t_s)$ Hz. When $\beta = 1$, it has a bandwidth of $(1/t_s)$ Hz. The corresponding transform is

$$P(j\omega) = \begin{cases} \dfrac{1}{2}\left(1 + \cos\dfrac{\pi\omega}{2w}\right), & |\omega| \leq 2w \\ 0, & \text{Otherwise} \end{cases} \quad (4.11.10)$$

It has the full cosine roll-off characteristic. Raising the roll-off factor increases the bandwidth of the pulse and the corresponding pulse is easier to generate. The raised cosine pulse decays at a rate inversely proportional to the cube of $|t|$ (see (4.11.9a)). For a detailed discussion on using these pulses in data transmission, see Lathi (1983).

4.12 Tables of Fourier Transforms Properties and Pairs

4.13 Summary

In this chapter we have related the Fourier series to the Fourier transforms by considering the period of a function going to infinity. Basic properties associated with Fourier transforms have been introduced. Several interesting pulse functions and their transforms are included. Specific topics that were included in this chapter are given below:

- Fourier transforms from Fourier series
- Various measures of bandwidth of a signal
- Basic Fourier transforms theorems; examples
- Fourier transforms of periodic functions using impulse functions
- Time domain and frequency domain convolutions; convolution theorem is discussed
- Autocorrelation and cross-correlation; energy and power spectral densities
- Moments of functions; various measures dealing with bandwidths
- Bounds on the transform
- Special topics including Poisson summation formula and some special pulse functions
- Tables of some of the Fourier transform theorems and Fourier transform pairs

Problems

4.1.1 We derived the trigonometric Fourier series of a trapezoidal waveform (see (3.6.16a)). Derive the Fourier transform of this waveform using the F-series coefficients.

4.2.1 Consider the Fourier transforms pair

$$A\Pi\left[\frac{t}{\tau}\right] \overset{\text{FT}}{\longleftrightarrow} A\tau\frac{\sin(\omega\tau/2)}{(\omega\tau/2)}$$

a. Assuming the main lobe of the transform can be approximated by a triangle, find the energy contained in the main lobe. Estimate the percentage of energies in the main lobe versus the pulse.

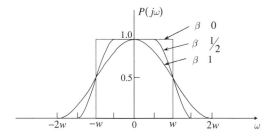
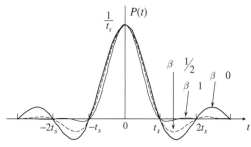

Fig. 4.11.1 $P(j\omega)$ and $p(t)$

Table 4.12.1 Fourier transform properties

$F[x_i(t)] = X_i(j\omega)$ or $F[x_i(t)] = X_i(jf)$

Linearity:

$$x(t) = \sum_{i=1}^{N} a_i x_i(t) \xleftrightarrow{\text{FT}} \sum_{i=1}^{N} a_i X_i(j\omega); \quad a_i\text{'s are constants.}$$

Time delay or shift:

$$x(t - t_0) \xleftrightarrow{\text{FT}} X(j\omega)e^{-j\omega t_0}; \quad t_0 \text{ is a constant.}$$

Frequency shift:

$$x(t)e^{\pm j\omega t_0} \xleftrightarrow{\text{FT}} X(j(\omega \mp \omega_0)); \quad \omega_0 \text{ is a constant.}$$

Time scaling:

$$x(at) \xleftrightarrow{\text{FT}} \frac{1}{|a|} X\left(\frac{j\omega}{a}\right); \quad a \neq 0.$$

Time reversal:

$$x(-t) \xleftrightarrow{\text{FT}} X(-j\omega).$$

Time scale and frequency shift:

$$x(at)e^{\pm j\omega_0 t} \xleftrightarrow{\text{FT}} \frac{1}{|a|} X\left[\frac{j(\omega \mp \omega_0)}{a}\right].$$

Duality or symmetry:

$$X(t) \xleftrightarrow{\text{FT}} 2\pi x(-j\omega); \quad X(t) \xleftrightarrow{\text{FT:}f} x(-jf).$$

Differentiation in time:

$$\frac{d^n x(t)}{dt} \xleftrightarrow{\text{FT}} (j\omega)^n X(j\omega).$$

Times-t or differentiation in frequency:

$$(-jt)^n x(t) \xleftrightarrow{\text{FT}} \frac{d^n X(j\omega)}{d\omega^n}; \quad (-j2\pi t)^n x(t) \xleftrightarrow{\text{FT:}f} \frac{d^n X(jf)}{df^n}.$$

Integration in time:

$$\int_{-\infty}^{t} x(\alpha)d\alpha \xleftrightarrow{\text{FT}} \pi X(0)\delta(\omega) + \frac{X(j\omega)}{j\omega};$$

$$\int_{-\infty}^{t} x(\beta)d\beta \xleftrightarrow{\text{FT:}f} \frac{1}{2}\delta(f)X(0) + \frac{X(jf)}{j2\pi f}.$$

Integration in frequency:

$$\pi x(0)\delta(t) - \frac{x(t)}{jt} \xleftrightarrow{\text{FT}} \int_{-\infty}^{\omega} X(j\beta)d\beta;$$

$$\frac{x(0)}{2}\delta(t) - \frac{x(t)}{j2\pi t} \xleftrightarrow{\text{FT:}f} \int_{-\infty}^{f} X(j\alpha)d\alpha.$$

Conjugation:

$$x*(t) \xleftrightarrow{\text{FT}} X*(-j\omega).$$

Table 4.12.1 (continued)

Even and odd parts of a real function:

$$x(t) = x_e(t) + x_o(t) \xleftrightarrow{\text{FT}} X(j\omega) = Re[X(j\omega)] + jIm[X(j\omega)].$$

$$x_e(t) \xleftrightarrow{\text{FT}} Re[X(j\omega)]; \quad x_o(t) \xleftrightarrow{\text{FT}} jIm[X(j\omega)]$$

Convolution in time:

$$x(t) = x_1(t) * x_2(t) = \xleftrightarrow{\text{FT}} X_1(j\omega)X_2(j\omega).$$

Convolution in frequency:

$$x(t) = x_1(t)x_2(t) \xleftrightarrow{\text{FT}} \frac{1}{2\pi} X_1(j\omega) * X_2(j\omega).$$

$$x_1(t)x_2(t) \xleftrightarrow{\text{FT}} X_1(jf) * X_2(jf).$$

Cross correlation:

$$R_{hx}(\tau) = \int_{-\infty}^{\infty} h(t)x(t+\tau)dt \xleftrightarrow{\text{FT}} H*(j\omega)X(j\omega).$$

Autocorrelation: aperiodic and periodic:

$$R_x(\tau) = \int_{-\infty}^{\infty} x(t)x(t+\tau)dt \xleftrightarrow{\text{FT}} |X(j\omega)|^2 = G_x(\omega).$$

$$R_{T,x}(\tau) = \frac{1}{T}\int_{T} x_T(t)x_T(t+\tau)dt \xleftrightarrow{\text{FT}} S_x(\omega).$$

Generalized Parseval's theorem:

$$\int_{-\infty}^{\infty} h^*(t)x(t)dt = \frac{1}{2\pi} \int_{-\infty}^{\infty} H*(j\omega)X(j\omega)d\omega.$$

Rayleigh's energy theorem:

$$\int_{-\infty}^{\infty} |x(t)|^2 dt = \frac{1}{2\pi} \int_{-\infty}^{\infty} |X(j\omega)|^2 d\omega;$$
$$\int_{-\infty}^{\infty} |x(t)|^2 dt = \int_{-\infty}^{\infty} |X(jf)|^2 df.$$

Modulation:

$$x(t)\cos(\omega_0 t) \xleftrightarrow{\text{FT}} \frac{1}{2} X(j(\omega + \omega_0)) + \frac{1}{2} X(j(\omega - \omega_0)).$$

$$x(t)\sin(\omega_0 t) \xleftrightarrow{\text{FT}} \frac{1}{2j} X(j(\omega - \omega_0)) - \frac{1}{2j} X(j(\omega + \omega_0)).$$

Initial value theorem:

$$x(0^+) = \lim_{\omega \to \infty} [j\omega X(j\omega).$$

Central ordinate theorems:

$$x(0) = \frac{1}{2\pi} \int_{-\infty}^{\infty} X(j\omega)d\omega; \quad X(0) = \int_{-\infty}^{\infty} x(t)dt.$$

Table 4.12.2 Fourier Transform Pairs

$$x(t)\overset{\text{FT}}{\longleftrightarrow}X(j\omega)$$

Rectangular and triangular pulses:

$$\Pi\left[\tfrac{t}{a}\right]\overset{\text{FT}}{\longleftrightarrow}a\,\frac{\sin(a\omega/2)}{a\omega/2}\,;\quad \Lambda\left[\tfrac{t}{\tau}\right]\overset{\text{FT}}{\longleftrightarrow}\tau\,\frac{\sin^2(\omega\tau/2)}{(\omega\tau/2)^2}\,.$$

Sinc pulse:

$$\frac{\sin(2\pi Wt)}{(2\pi Wt)}\overset{\text{FT}}{\longleftrightarrow}\frac{1}{2\,W}\,\Pi\left[\frac{\omega}{2\pi(2\,W)}\right].$$

Impulse functions:

$$\delta(t-t_0)\overset{\text{FT}}{\longleftrightarrow}e^{-j\omega t_0}\,;\quad \frac{d^n\delta(t)}{dt^n}\overset{\text{FT}}{\longleftrightarrow}(j\omega)^n\,;\quad \delta^{(n)}(t)\overset{\text{FT}}{\longleftrightarrow}(j\omega)^n$$

Constants in time and frequency:

$$1\overset{\text{FT}}{\longleftrightarrow}(2\pi)\delta(\omega)\,;\quad 1\overset{\text{FT:}f}{\longleftrightarrow}\delta(f)\,;\quad \delta(t)\overset{\text{FT}}{\longleftrightarrow}1.$$

Exponential functions in time and frequency:

$$e^{-at}u(t)\overset{\text{FT}}{\longleftrightarrow}\frac{1}{a+j\omega}\,,a>0;\ e^{at}u(-t)\overset{\text{FT}}{\longleftrightarrow}\frac{1}{a-j\omega}\,,a>0.$$

$$e^{-a|t|}\overset{\text{FT}}{\longleftrightarrow}\frac{2a}{a^2+\omega^2}\,,a>0;\qquad \frac{t^{n-1}}{(n-1)!}e^{-at}u(t)\overset{\text{FT}}{\longleftrightarrow}\frac{1}{(a+j\omega)^n}\,.$$

$$\frac{1}{a^2+t^2}\overset{\text{FT}}{\longleftrightarrow}\frac{\pi}{a}e^{-a|\omega|}\,.$$

Signum function:

$$sgn(t)\overset{\text{FT}}{\longleftrightarrow}\frac{2}{j\omega}\,;\quad t^n sgn(t)\overset{\text{FT}}{\longleftrightarrow}(-j)^{n+1}\frac{2(n!)}{\omega^{n+1}}\,.$$

Unit step function:

$$u(t)\overset{\text{FT}}{\longleftrightarrow}\pi\delta(\omega)+\frac{1}{j\omega}\,;\quad u(t)\overset{\text{FT:}f}{\longleftrightarrow}\frac{1}{2}\delta(f)+\frac{1}{j2\pi f}\,.$$

Functions involving t:

$$t\overset{\text{FT}}{\longleftrightarrow}j2\pi\delta'(\omega)\,;\quad t^n\overset{\text{FT}}{\longleftrightarrow}j^n 2\pi\delta^{(n)}(\omega).$$

$$tu(t)\overset{\text{FT}}{\longleftrightarrow}j\pi\delta'(\omega)-\frac{1}{\omega^2}\,;\quad |t|\overset{\text{FT}}{\longleftrightarrow}\frac{-2}{\omega^2}\,.$$

$$\frac{1}{t}\overset{\text{FT}}{\longleftrightarrow}\pi j-2\pi ju(\omega).$$

Gaussian function:

$$\frac{1}{\sqrt{2\pi}\sigma}e^{-t^2/2\sigma^2}\overset{\text{FT}}{\longleftrightarrow}e^{-(\omega\sigma)^2/2}\,.$$

Periodic function:

$$x_T(t)=\sum_{k=-\infty}^{\infty}X_s[k]e^{-jk\omega_0 t}\overset{\text{FT}}{\longleftrightarrow}2\pi\sum_{k=-\infty}^{\infty}X_s[k]\delta(\omega-k\omega_0)\,;\ \omega_0=2\pi/T.$$

$$e^{j\omega_0 t}\overset{\text{FT}}{\longleftrightarrow}2\pi\delta(\omega-\omega_0).$$

Cosine and sine functions:

$$\cos(\omega_0 t)\overset{\text{FT}}{\longleftrightarrow}\pi\delta(\omega+\omega_0)+\pi\delta(\omega-\omega_0).$$

$$\sin(\omega_0 t)\overset{\text{FT}}{\longleftrightarrow}j\pi\delta(\omega+\omega_0)-j\pi\delta(\omega-\omega_0).$$

Table 4.12.2 (continued)

Periodic impulse sequence:

$$\delta_T(t) = \sum_{n=-\infty}^{\infty} \delta(t - nT) \xleftrightarrow{\text{FT}} \omega_0 \sum_{k=-\infty}^{\infty} \delta(\omega - k\omega_0) = \omega_0 \delta_{\omega_0}(\omega); \quad \omega_0 = 2\pi/T.$$

Exponentially decaying functions involving t and sinusoidal functions:

$$\frac{t^{n-1}}{(n-1)!} e^{-at} u(t) \xleftrightarrow{\text{FT}} \frac{1}{(j\omega+a)^n}, \quad a > 0, n1.$$

$$e^{-at} \cos(bt) u(t) \xleftrightarrow{\text{FT}} \frac{(a+j\omega)}{(a+j\omega)^2+b^2}, \quad a > 0.$$

$$e^{-at} \sin(bt) u(t) \xleftrightarrow{\text{FT}} \frac{b}{(a+j\omega)^2+b^2}, \quad a > 0.$$

Cosinusoidal pulse:

$$\cos(\omega_0 t) \Pi\left[\frac{t}{T_1}\right] \xleftrightarrow{\text{FT}} T_1 2\left[\frac{\sin(\omega-\omega_0)T_1/2}{(\omega-\omega_0)T_1/2}\right] + T_1 2\left[\frac{\sin(\omega+\omega_0)T_1/2}{(\omega+\omega_0)T_1/2}\right].$$

Switched functions:

$$\sin(\omega_0 t) u(t) \xleftrightarrow{\text{FT}} \frac{\omega_0}{\omega_0^2-\omega^2} + \frac{\pi}{2j}[\delta(\omega - \omega_0) - \delta(\omega + \omega_0)].$$

$$\cos(\omega_0 t) u(t) \xleftrightarrow{\text{FT}} \frac{j\omega}{\omega_0^2-\omega^2} + \frac{\pi}{2}[\delta(\omega - \omega_0) + \delta(\omega + \omega_0)].$$

b. Approximate the energy contained in the main lobe by dividing the main lobe into 10 rectangles and compute the percentage of energy in the main lobe.

4.2.2 Consider the function $x(t) = e^{-at}u(t)$, $a > 0$. Derive the transforms of the even and odd parts of this function. Compare this result by deriving these results from $X(j\omega)$.

4.2.3 Are the following functions Fourier transformable? If not, give the reasons:

a. $x_1(t) = tu(-t)$,

b. $x_2(t) = e^{\alpha t}u(t)$,

c. $x_3(t) = 1/(t - 2)$.

4.3.1 Use the Fourier central ordinate theorems to evaluate the following integrals:

a. $A = \displaystyle\int_{-\infty}^{\infty} e^{-at^2}\, dt,$

b. $B = \displaystyle\int_{0}^{\infty} te^{-at}\, dt, \quad a > 0,$

c. $C = \displaystyle\int_{-\infty}^{\infty} \frac{t^n}{(n-1)!} e^{-at} u(t)\, dt, \quad a > 0.$

4.3.2 Find the Fourier transforms of the following functions:

a. $x_1(t) = u(1 - |t|)$,

b. $x_2(t) = \dfrac{t}{2}\Pi\left[\dfrac{t-1}{2}\right]$,

c. $x_3(t) = -t\Pi\left[t + \dfrac{1}{2}\right]$.

d. Use the results in part *b.* to obtain the transform in part *c.*

4.3.3 Find the Fourier transform of the function given below either by using the transform integral or by using the derivative method discussed in Section 4.4, see Example 4.4.7.

$$x(t) = \begin{cases} 4, & -1 \leq t \leq 1 \\ -2t + 6, & 1 < t < 3 \\ 2t + 6, & -3 \leq t < -1 \end{cases}$$

4.3.4 Find the transforms of the following functions using the times-t property:

a. $x(t) = te^{-\alpha t}u(t)$,

b. $r(t) = tu(t)$,

c. $y(t) = te^{-\alpha|t|}, \quad \alpha > 0$.

4.3.5 Use the derivative method to find the Fourier transforms of the following functions:

a. $x(t) = e^{-t}\Pi[t - .5]$,

b. $x(t) = \cos(\pi t)$, $-.5 < t < .5$; 0, otherwise .

4.3.6 *a.* Find the Fourier transform of the function $y(t) = 1/[\alpha^2 + t^2]$ by noting that

$$x(t) = \frac{1}{\alpha^2 + (2\pi t)^2} \overset{FT}{\longleftrightarrow} \frac{1}{2\alpha} e^{-\alpha|\omega/2\pi|} = X(j\omega).$$

b. Verify the Fourier transform derived in Part *a* by considering the transform pair of the double exponential and the duality theorem of the transforms.

4.3.7 Use the derivative theorem to find the transform of the Gaussian pulse.

4.3.8 Let $x(t) = (1 - t/T)$, $0 < t < T$ and 0 otherwise. Determine $X(j\omega) = F[x(t)]$ by using the following methods: *a.* the derivative method and *b.* direct method.

4.3.9 Use Rayleigh's energy theorem, or some other procedure, evaluate the following integrals. Assume the constants identified by *a* in the integrals equal to 1:

a. $\displaystyle\int_{-\infty}^{\infty} \frac{2a}{a^2 + \omega^2} d\omega$,

b. $\displaystyle\int_{-\infty}^{\infty} e^{-at^2} dt, a > 0$,

c. $\displaystyle\int_{-\infty}^{\infty} \frac{\sin^4(\omega\tau/2)}{(\omega\tau/2)^4} d\omega$,

d. $\displaystyle\int_{-\infty}^{\infty} \left[\frac{\sin^2(t)}{t^2}\right]\left[\frac{\sin(t)}{t}\right] dt$.

4.3.10 Validate the Fourier ordinate theorems using the transform pair in (4.4.25).

4.3.11 Derive the F-transform of the function $x(t) = -u(t - 1)$. Give the phase angle of the transform at the frequencies $\omega = 0^+$ and 0^-.

4.3.12 Show the equation in (4.3.9) by first writing the complex function in terms of its real and imaginary parts $x(t) = x_R(t) + jx_I(t)$ and then using the superposition theorem.

4.4.1 Use derivative theorem to find the transforms for the following functions:

a. $x_1(t) = \text{sgn}(t) = 2u(t) - 1$,

b. $x_2(t) = e^{-2t}\Pi[t - .5]$.

4.4.2 Find $X_i(j\omega)$ for the following functions by the methods identified in each part: *a.* $x_1(t) = t^2\Pi[t - 1]$ by the direct method, *b.* $x_2(t) = \Pi[t/T] \sin(\pi t/T)$ by the multiplication and the frequency convolution theorems, and *c.* $x(t) = u(-t)$ by the time reversal theorem.

4.4.3 Find the Fourier transform of the function $x_3(t) = x(t - a) + x(2t)$ in terms of the transform of $x(t)$ using some of the Fourier transforms theorems.

4.4.4 Find the inverse transform of the function $X(j\omega) = u(\omega)$, the unit step function in the frequency domain (use the duality property and the time reversal property).

4.4.5 Use the timescale property and the transform pair below to find $X(j\omega)$:

$$e^{-\pi t^2} \overset{FT}{\longleftrightarrow} e^{-\pi f^2}, \quad x(t) = \frac{1}{\sqrt{2\pi}\sigma} e^{-\pi t^2/2\sigma^2} \overset{FT}{\longleftrightarrow} X(j\omega).$$

4.4.6 Determine Fourier transforms of the following functions:

a. $x_1(t) = \dfrac{\cos(bt)}{a^2 + t^2}$,

b. $x_2(t) = \dfrac{\sin(bt)}{a^2 + t^2}$,

c. $x_3(t) = \dfrac{t}{1 - jt}$,

d. $x_4(t) = e^{-2t}u(t - 1)$.

4.4.7 Use the time or frequency differentiation theorem to find $X_i(j\omega)$ for the functions

a. $x_1(t) = te^{-at}u(t)$, *b.* $x_2(t) = \left(1 - \left[\dfrac{2t}{\tau}\right]^2\right)\Pi\left[\dfrac{t}{\tau}\right]$.

4.4.8 Find the initial values of the following functions using the transforms of the function and the initial value theorem:

a. $u(t)$, *b.* $e^{-t}u(t)$.

4.4.9 Use the differentiation theorem to find the transform of the function $x(t) = t\Pi(t/T)$.

4.4.10 Find the Fourier transform of the function $x(t) = t^n \text{sgn}(t)$ and then use the fact that $y(t) = |t| = t \text{ sgn}(t)$ to find $Y(j\omega) = F[y(t)]$.

4.5.1 Determine the transforms of the following functions:

a. $x(t) = e^{-\alpha|t|} * \Pi[t]$, $\alpha > 0$,
b. $x_2(t) = \delta(2t - \alpha) * \delta(3t - \beta)$, find $X_i(j\omega)$.

4.5.2 Show that if $x(t)$ is band limited to f_c Hz then the following equality is true:

$$x(t) * [\sin(\alpha t)/\pi t] = x(t) \text{ for all } \alpha > 2\pi f_c.$$

4.5.3 Find $Y(j\omega)$ assuming $F[y(t)[= F[x(t+T) + x(t-T)] = Y(j\omega)$, $F[x(t)] = X(j\omega)$.

4.5.4 Solve the differential equation $y'(t) + ay(t) = x(t) + x'(t)$ using transforms.

4.5.5 Show the relation in (4.5.31).

4.5.6 Find $F[x(t)] = F[|\cos(\omega_0 t)|\Pi[t/(T/2)]]$.

4.5.7 Sketch the function $x(t) = \Pi[t/T] * [\delta(t) - \delta(t-T)]$ and derive its transform.

4.6.1 Find $x(t) = F^{-1}[X(j\omega)]$, given

$X(j\omega) = e^{-\omega}, -1 < \omega < 0; -e^{\omega}, 0 < \omega < 1; 0, \text{otherwise}.$

Use the central ordinate theorems to find $x(0)$.

4.6.2 Find the inverse transforms of the following using the Fourier transform properties:

a. $X_1(j\omega) = u(1 - |\omega|)$,   b. $X_2(j\omega) = \dfrac{8}{(a + j\omega)^3}$,

c. $X_3(j\omega) = \dfrac{a}{(a + j\omega)}\left[\pi\delta(\omega) + \dfrac{1}{j\omega}\right]$,

d. $X_4(j\omega) = e^{-\beta\omega}u(\omega)$,   e. $X_5(j\omega) = \dfrac{\cos(\omega/2)e^{-j\omega/2}}{1 + j\omega}$,

f. $X_6(j\omega) = \Pi\left[\dfrac{\omega}{2W}\right]e^{-j\omega t_0}$,

g. $|X_7(j\omega)| = \Pi\left[\dfrac{\omega}{2W}\right]$, $\angle X_7(j\omega) = \begin{cases} \pi/2, \omega < 0 \\ -\pi/2, \omega > 0 \end{cases}$.

4.6.3 Find the inverse transforms of the following functions by using the partial fraction expansion. Use the results in Part a to find the partial fraction expansion in Part b:

a. $X_1(j\omega) = \dfrac{(2 + j\omega)}{(1 + j\omega)(3 + j\omega)}$,

b. $X_2(j\omega) = \dfrac{1}{(1 + j\omega)}\left[\dfrac{(2 + j\omega)}{(1 + j\omega)(3 + j\omega)}\right]$.

4.6.4 Find the inverse transform of the function $X(j\omega) = e^{-|\omega|}$ by using the direct method and by using the duality theorem.

4.6.5 Find $F^{-1}[\cos(\omega)]$, $|\omega| < \pi/2$ and 0 otherwise.

4.6.6 Find the energy spectral density for the function in Problem 4.6.5 and compute the energy contained in the pulse and verify the results by computing the energy directly.

4.6.7 a. Find the power spectral density of the periodic function $x(t) = \cos(t) + \sin(t)$.
b. Compute the average power contained in this function directly and then by making use of the power spectral density to find the average power.

4.7.1 By making use of the convolution theorem in the frequency domain show that

$$\Pi\left[\dfrac{t}{T}\right]\cos\left(\dfrac{\pi t}{T}\right) \overset{\text{FT}}{\longleftrightarrow} \dfrac{2T\cos(\pi fT)}{\pi\left[1 - (2fT)^2\right]}.$$

4.7.2 Find the transforms of the functions

a. $x_1(t) = \delta'(t)$, b. $x_2(t) = tu(t)$, c. $x_3(t) = |t|$,
d. $x_4(t) = \sin^2(t)$.

4.7.3 Use the appropriate properties to find the transforms of the following:

a. $x_1(t) = \text{sgn}(t)$ (Derivative theorem),
b. $x_2(t) = \sin(\omega_0 t)$ (Transform of the cosine).

4.7.4 Verify the pulse width–bandwidth product bounds for the following functions:

a. $x_1(t) = e^{-t^2/2\tau^2}$,   b. $x_2(t) = e^{-\alpha|t|}$

4.7.5 Find the inverse transforms of the following functions with $H_0 > 0$ and $t_0 > 0$. The subscripts on the transform functions indicate the type of filter under consideration: Lp – Lowpass, Hp – High pass, Bp – Band pass, Be – Band elimination.

a. $H_{Lp}(j\omega) = H_0 \Pi \left[\frac{\omega}{2W} \right] e^{-j\omega t_0}$,

b. $H_{Hp}(j\omega) = H_0 \left[1 - \Pi \left[\frac{\omega}{2W} \right] \right] e^{-j\omega t_0}$,

c. $H_{Bp}(j\omega) = H_0 \left[\Pi \left[\frac{\omega - \omega_0}{W} \right] + \Pi \left[\frac{\omega + \omega_0}{W} \right] \right] e^{-j\omega t_0}$,

d. $H_{Be}(j\omega) = H_0 \left[1 - \left[\Pi \left[\frac{\omega - \omega_0}{W} \right] + \Pi \left[\frac{\omega + \omega_0}{W} \right] \right] \right] e^{-j\omega t_0}$.

4.7.6 Use the functions $g_1(t) = \Lambda[t]$ and $g_2(t) = e^{-t}u(t)$ to verify Schwarz's inequality.

4.8.1 Using the Fourier transform of a Gaussian pulse, show that

$$\int_0^\infty e^{-at^2} dt = \frac{1}{2} \sqrt{\frac{\pi}{a}}, \quad a > 0.$$

4.8.2 Find the variance of the exponential decaying pulse in Example 4.8.2.

4.9.1 Determine the transforms of the following functions. Use the spectral bounds (see (4.9.2)) to verify the central ordinate theorems using these:

a. $x_1(t) = \Lambda[t/T]$,   b. $x_2(t) = e^{-a|t|}$, $a > 0$

4.9.2 Comment on the spectral bounds on the periodic function $x(t) = \cos(\omega_0 t)$.

4.10.1 Use the Gaussian pulse function in (4.4.18) and give an expression for

$$\sum_{n=-\infty}^{\infty} x(t+n) \, (\text{see} (4.10.3)).$$

4.11.1 Show that the pulse function in (4.11.9a) has the transform given in (4.11.9b). The proof is lengthy. However there are few shortcuts we can make use of. First, the transform is even and therefore we can use the real integrals. Since the integrand has products of cosine functions, use the trigonometric formulas to simplify the integrands.

4.12.1 Using the times-t property to derive $X_n(j\omega)$ for $n = 1, 2, 3, 4$:

$$x_1(t) \overset{FT}{\longleftrightarrow} e^{-at}u(t), x_n(t) = \frac{t^{n-1}}{(n-1)!} e^{-at}u(t) \overset{FT}{\longleftrightarrow} \frac{1}{(a+j\omega)^n}$$
$$= X_n(j\omega), \ a > 0.$$

4.12.2 Use the central ordinate theorems to evaluate the integral

$$\int_0^\infty \frac{t^{n-1}}{(n-1)!} e^{-at} dt, \quad a > 0.$$

Chapter 5
Relatives of Fourier Transforms

5.1 Introduction

There are various ways of introducing a student to different forms of transforms. We chose the approximation of a function by using Fourier series first and then came up with the Fourier transforms. Fourier cosine and sine series were considered using the Fourier series. The next step is to study some of the other transforms that are related to the Fourier transforms. These include cosine, sine, Laplace, discrete, and fast Fourier transforms. Discrete and fast Fourier transforms will be included in Chapters 8 and 9. In many of the undergraduate engineering curricula, Laplace transforms are introduced first and then the Fourier transforms. Fourier transforms are considered more theoretical. The development of the sine and cosine transforms parallel to the Fourier cosine and sine series discussed in Chapter 3. For a good review on many of these topics, see the handbook by Poularikas (1996). We can consider the Laplace transform as an independent transform or a modified version of the Fourier transform. One problem with Fourier transforms is that the signal under consideration must be absolutely integrable. (the periodic functions are exceptions). Therefore, the transformation to the Fourier domain is limited to energy signals or to finite power signals that are convergent in the limit. Fourier and Laplace transforms have been widely used in engineering; Fourier transforms in the signal and communications area and the Laplace transform in the circuits, systems, and control area. Neither one is a generalization of the other. Both transforms have their own merits.

As we have learned earlier in studying Fourier transforms, they describe a function (signal) in a different domain and they are used in almost every aspect of system analysis and design. We will study two transforms that are useful in signals and communications. These are Hilbert transforms (see the Chapter by Hahn) and Hartley transforms (see the Chapter by Olejniczak) in the handbook by Poularikas (1996). There are many other transforms that can be studied. Many of these are given in this handbook. Data compression is one of the important applications of transform coding. Retrieval of a signal from a large database makes use of data compression. A simple example that illustrates the idea of data compression is the compression of a single sinusoid. Here we need only its amplitude, phase, and the corresponding frequency. Extending this thought, pack a *time signal* into a small set of coefficients in *another domain*, such as the Fourier domain. That is, keep the Fourier series coefficients with large amplitudes and discard the coefficients with small amplitudes. We considered this approach using Fourier series. Obviously discarding some coefficients will amount to loosing some information, the details in the signal, and the original signal cannot be reconstructed *exactly*. The process is irreversible in the ideal sense. In most applications the approximation is good within some tolerance.

Digital coding provides for simple manipulations of data allowing for data compression. To learn these concepts we need to have some statistical knowledge of the data. The signals that can be described in terms of Markov models are best suited for discrete sine and cosine transforms. Interested readers should consult books in this area. For a good discussion on this topic, see the chapter by Yip in the handbook by Poularikas (1996). We will not be discussing statistical methods in any detail in this book.

R.K.R. Yarlagadda, *Analog and Digital Signals and Systems*, DOI 10.1007/978-1-4419-0034-0_5, © Springer Science+Business Media, LLC 2010

A good portion of this chapter is on Laplace transforms, named after Marquis Pierre Simon De Laplace for his work, see Hawking (2005). Most of the material is standard and can be found in any of the undergraduate texts on systems and control, see, for example, Haykin and Van Veen (1999), Close (1966), Poularikas and Seely (1991), and others.

5.2 Fourier Cosine and Sine Transforms

In the last chapter we considered the Fourier transform and its inverse as

$$X(j\omega) = \int_{-\infty}^{\infty} x(t)e^{-j\omega t}dt = \int_{-\infty}^{\infty} x(t)\cos(\omega t)dt + j(-1)$$

$$\int_{-\infty}^{\infty} x(t)\sin(\omega t)dt, \ X(j\omega) = R(\omega) + jI(\omega)$$

$$x(t) = \frac{1}{2\pi}\int_{-\infty}^{\infty} X(\omega)e^{j\omega t}d\omega; \ x(t) \xleftrightarrow{\text{FT}} X(j\omega).$$

$$(5.2.1)$$

The real and the imaginary parts of the transform are, respectively, given by

$$R(\omega) = \int_{-\infty}^{\infty} x(t)\cos(\omega t)dt, I(\omega)$$

$$= -\int_{-\infty}^{\infty} x(t)\sin(\omega t)dt \qquad (5.2.2)$$

If $X(j\omega) = R(\omega)$, i.e., $I(\omega) = 0$, then the integrand in the integral to compute $I(\omega)$ must be odd, which implies $x(t)$ is even. These result in

$$R(\omega) = 2\int_{0}^{\infty} x(t)\cos(\omega t)dt,$$

$$x(t) = \frac{1}{\pi}\int_{0}^{\infty} R(\omega)\cos(\omega t)d\omega. \qquad (5.2.3)$$

Note that we are using $x(t)$ for $t > 0$. Suppose we have been given the function $x(t)$ for $t > 0$ only. Consider an even function by defining $x(-t) = x(t)$ and

$$F\left[(x(t) + x(-t)\right] = \int_{0}^{\infty} x(t)e^{-j\omega t}dt + \int_{0}^{\infty} x(t)e^{j\omega t}dt$$

$$= 2\int_{0}^{\infty} x(t)\cos(\omega t)dt \qquad (5.2.4)$$

We define the *Fourier cosine transform (FCT)* and *its inverse* by

$$X_c(\omega) = \int_{0}^{\infty} x(t)\cos(\omega t)dt = X_c(-\omega),$$

$$x(t) = \frac{2}{\pi}\int_{0}^{\infty} X_c(\omega)\cos(\omega t)d\omega. \qquad (5.2.5)$$

From (5.2.5 a and b), we have

$$x(t) + x(-t) = \frac{1}{2\pi}\int_{-\infty}^{\infty} 2X_c(\omega)e^{j\omega t}d\omega$$

$$= \frac{2}{\pi}\int_{0}^{\infty} X_c(\omega)\cos(\omega t)d\omega. \qquad (5.2.6)$$

Now considering only for positive time, we have the Fourier cosine transform pair

$$X_c(\omega) = \int_{0}^{\infty} x(t)\cos(\omega t)dt,$$

$$x(t) = \frac{2}{\pi}\int_{0}^{\infty} X_c(\omega)\cos(\omega t)d\omega, \qquad (5.2.7a)$$

$$x(t) \xleftrightarrow{\text{FCT}} X_c(\omega). \qquad (5.2.7b)$$

The subscript c on X indicates it is a cosine transform. The function $x(t)$ is defined only for the time interval $0 < t < \infty$. The inverse transform converges to the function $x(t)$ wherever the function is continuous and it converges to half-value, i.e., the average of the values of the function before and after the discontinuity wherever the function is discontinuous. We can see that the Fourier cosine transform is related to the F-transform using a trigonometric identity and an even extension of the function defined over the interval $0 \leq t < \infty$,

$$\cos(\omega t) = \frac{1}{2}[e^{j\omega t} + e^{-j\omega t}],$$

$$x_e(t) = x(|t|) = \begin{cases} x(t), & t \geq 0 \\ x(-t), & t < 0 \end{cases}.$$

The Fourier transform of the function $x(|t|)$ is

$$F[x_e(t)] = \int_{-\infty}^{\infty} x(|t|)e^{-j\omega t}\,dt = \int_0^{\infty} x(t)e^{-j\omega t}\,dt$$

$$+ \int_0^{\infty} x(t)e^{j\omega t}\,dt = 2\int_0^{\infty} x(t)\cos(\omega t)\,dt$$

$$= 2F_c[x(t)] \Rightarrow F_c[x(t)] = \frac{1}{2}F[x(|t|)]. \quad (5.2.8)$$

This gives the relation between the cosine transform and the Fourier transform.

In a similar manner to the Fourier cosine transform, we can define the *Fourier sine transform (FST)* and its *inverse* as

$$X_s(\omega) = \int_0^{\infty} x(t)\sin(\omega t)\,d\omega,$$

$$x(t) = \frac{2}{\pi}\int_0^{\infty} X_s(\omega)\sin(\omega t)\,d\omega$$

$$x(t) \overset{\text{FST}}{\longleftrightarrow} X_s(\omega). \quad (5.2.9)$$

Just as in the case of cosine transforms, Fourier sine transforms are related to the Fourier transforms. This can be seen using the trigonometric identity and the odd extension of the function defined over the interval $0 \le t < \infty$,

$$\sin(\omega t) = \frac{1}{2j}[e^{j\omega t} - e^{-j\omega t}],$$

$$x_0(t) = \begin{cases} x(t), & t \ge 0 \\ -x(-t), & t < 0 \end{cases}.$$

Then the Fourier transform of the function $x_0(t)$ is given by

$$F[x_0(t)] = \int_{-\infty}^{\infty} x_o(t)e^{-j\omega t}\,dt = -\int_0^{\infty} x(t)e^{j\omega t}\,dt$$

$$+ \int_0^{\infty} x(t)e^{-j\omega t}\,dt = -2j\int_0^{\infty} x(t)\sin(\omega t)\,dt$$

$$= -2jF_s[x(t)] \rightarrow F_s[x(t)] = -\frac{1}{2j}F[x_0(t)]. \quad (5.2.10)$$

This gives the relation between the sine transforms and the Fourier transforms.

Example 5.2.1 Find the Fourier cosine and sine transforms of $x(t) = e^{-at}u(t)$, $a > 0$.

Solution: Let

$$A = \int_0^{\infty} e^{-at}\cos(\omega t)\,dt, B = \int_0^{\infty} e^{-at}\sin(\omega t)\,dt.$$

The first integral can be evaluated by using the integration by parts

$$X_c(\omega) = A = \int_0^{\infty} e^{-at}\cos(\omega t)\,dt = -\frac{e^{-at}\cos(\omega t)}{a}\Big|_0^{\infty}$$

$$-\frac{\omega}{a}\int_0^{\infty} e^{-at}\sin(\omega t)\,dt = \frac{1}{a} - \frac{\omega}{a}B.$$

In a similar manner

$$B = \int_0^{\infty} e^{-at}\sin(\omega t)\,dt = -\frac{e^{-at}\sin(\omega t)}{a}\Big|_0^{\infty}$$

$$+\frac{\omega}{a}\int_0^{\infty} e^{-at}\cos(\omega t)\,dt = \frac{\omega}{a}A.$$

Solving for A and B, we have the cosine and sine transforms of the function and

$$x(t) = e^{-at}u(t) \overset{\text{FT}}{\longleftrightarrow} \frac{a}{a^2 + \omega^2} = X_c(\omega),$$

$$x(t) = e^{-at}u(t) \overset{\text{FT}}{\longleftrightarrow} \frac{\omega}{a^2 + \omega^2} = X_s(\omega), a > 0.$$

$$(5.2.11) \quad \blacksquare$$

Example 5.2.2 Find the FCT of the Gaussian pulse function $x(t) = e^{-at^2}$.

Solution: Noting that $x(t)$ is an even function, the following follows from the F-transform of a Gaussian function:

$$X_c(\omega) = \int_0^{\infty} e^{-at^2}\cos(\omega t)\,dt = \frac{1}{2}\sqrt{\frac{\pi}{a}}e^{-\omega^2/4a}.$$

$$(5.2.12) \quad \blacksquare$$

Example 5.2.3 Find the Fourier cosine and sine transforms of the pulse function

$$x(t) = \Pi\left[\frac{t - (a/2)}{a}\right].$$

Solution:

$$X_c(\omega) = \int_0^a \cos(\omega t)dt = \frac{1}{\omega}\sin(\omega a),$$

$$X_s(\omega) = \int_0^a \sin(\omega)d\omega = \frac{1 - \cos(\omega a)}{\omega}. \quad (5.2.13) \blacksquare$$

Noting the correspondence between the Fourier transforms and the Fourier cosine and sine transforms, we can see that the cosine and sine transforms can be obtained from the Fourier transforms, but we need to be careful about the ranges. We can make use of tables of trigonometric functions and integrals to compute these transforms. See Yip's chapter in Poularikis (1996) for tables on the cosine and sine transforms.

Example 5.2.4 Find the Fourier cosine transforms of the following functions:

$$a.\ x_1(t) = \begin{cases} t/a, & 0 < t < a \\ (2a - t)/a, & a < t < 2a, \\ 0, & t > 2a \end{cases} \quad (5.2.14a)$$

$$b.\ x_2(t) = \frac{\sin(at)}{t}, a > 0, \quad c.\ x_3(t) = e^{-bt}\sin(at).$$

Use integral tables wherever appropriate.

Solution:

$$a.\ X_c(\omega) = \int_0^a \frac{t}{a}\cos(\omega t)dt + \int_a^{2a} \frac{2a - t}{a}\cos(\omega t)dt$$

$$= \frac{1}{a\omega^2}[2\cos(a\omega) - \cos(2a\omega - 1)].\ (5.2.14b)$$

$$b.\ X_{2c}(\omega) = \int_0^\infty \frac{\sin(at)}{t}\cos(\omega t)dt = \begin{cases} \pi/2, \omega < a \\ \pi/4, \omega = a, \\ 0, \omega > a \end{cases}$$

$$(5.2.14c)$$

$$c.\ X_{3c}(\omega) = \int_0^\infty e^{-bt}\sin(at)\cos(\omega t)dt$$

$$= \frac{1}{2}\left[\frac{a + \omega}{b^2 + (a + \omega)^2} + \frac{a - \omega}{b^2 + (a - \omega)^2}\right].$$

$$(5.2.14d)$$

The last integral can be seen by using the trigonometric formula $\sin(at)\cos(\omega t) = (1/2)[\sin(a + \omega)t + \sin(a - \omega)t]$ and using the results in the last chapter. $\blacksquare$

Example 5.2.5 Find the FSTs of the functions given in the last example.

Solution:

$$a.\ X_{1s}(\omega) = \int_0^a (t/a)\sin(\omega t)dt$$

$$+ \int_a^{2a} [(2a - t)/a]\sin(\omega t)dt$$

$$= \frac{1}{a\omega^2}[2\sin(a\omega) - \sin(2a\omega)],$$

$$b.\ X_{2s}(\omega) = \int_0^\infty \frac{\sin(at)}{t}\sin(\omega t)dt = (1/2)\ln\left|\frac{\omega + a}{\omega - a}\right|.$$

C. Using the F-transform of the exponential decaying function and the results in the last example,

$$c.\ X_{3s}(\omega) = \int_0^\infty e^{-bt}\sin(at)e^{-j\omega t}dt$$

$$= \int_0^\infty e^{-bt}\sin(at)u(t)[\cos(\omega t) - j\sin(\omega t)]dt$$

$$= \frac{a}{(b + j\omega)^2 + a^2}.$$

Making use of the last example results in the sine transform. $\blacksquare$

Cosine and sine transforms properties can be derived using Fourier transform properties. Computation of these transforms can be simplified by using those properties. Without going through the proofs of these properties, we will consider some examples. Discrete version of the Fourier cosine transform is popular in the digital image processing.

Example 5.2.6 Show the Fourier cosine and sine transforms pairs given below:

$$a.\ x_1(t) = \frac{1}{t^2 + a^2} \xleftrightarrow{\text{FCT}} \frac{\pi}{2a}e^{-a\omega} = X_{1c}(\omega), \quad (5.2.15a)$$

$$b.\ x(t) = \frac{t}{t^2 + a^2} \overset{\text{FST}}{\longleftrightarrow} \frac{\pi}{2} e^{-a\omega} = X_{2s}(\omega). \quad (5.2.15b)$$

Solution: These can be shown using the integral tables and are left as exercises. ■

5.3 Hartley Transform

Just like the cosine and sine transforms, the Hartley transform is an integral transformation that maps a *real-valued time function* into a *real-valued frequency function*. For a detailed discussion on this topic, see Bracewell (1986). It makes use of the kernel $\text{cas}(\nu t) = \cos(\nu t) + \sin(\nu t)$ and allows for decomposition of a function into two independent sets of sinusoidal components represented in terms of positive and negative frequency components. The signal we are dealing will be a real-valued signal.

The Hartley transform of $x(t)$ is defined by

$$X_H(\omega) = \int_{-\infty}^{\infty} x(t)\text{cas}(\omega t)dt. \quad (5.3.1)$$

The integral kernel, *cosine and sine* referred to as the **cas** function, is

$$\begin{aligned}\text{cas}(\omega t) &= \cos(\omega t) + \sin(\omega t)\\ \text{cas}(\omega t) &= \sqrt{2}\sin(\omega t + (\pi/4)),\\ \text{cas}(\omega t) &= \sqrt{2}\cos(\omega t - (\pi/4)). \quad (5.3.2)\end{aligned}$$

The existence of the Hartley transform of $x(t)$ is equivalent to the existence of the Fourier transform. The inverse Hartley transform is

$$\begin{aligned}x(t) &= \frac{1}{2\pi}\int_{-\infty}^{\infty} X_H(\omega)\text{cas}(\omega t)d\omega \text{ or } x(t)\\ &= \int_{-\infty}^{\infty} X_H(2\pi ft)\text{cas}(2\pi ft)df.\end{aligned} \quad (5.3.3)$$

Symbolically we will identify the Hartley transform pair by

$$x(t) \overset{\text{Hart}}{\longleftrightarrow} X_H(\omega) \text{ or } x(t) \overset{\text{Hart}}{\longleftrightarrow} X_H(f).$$

The second integral on the right in (5.3.3) provides an interesting property since the Hartley transform and its inverse have the same exact form, usually referred to as the *self-inverse property*, except that the forward transform has the time and the inverse has the frequency functions in the integral. The self-inverse property is given in Fig. 5.3.1.

$$x(t) \xrightarrow{\text{Hartley tranform}} X_H(f) \xrightarrow{\text{Hartley transform}} x(t)$$

Fig. 5.3.1 Self-inverse property of the Hartley transforms

The even and the odd parts of the Hartley transform can be expressed by

$$\begin{aligned}X_H(\omega) &= \int_{-\infty}^{\infty} x(t)\text{cas}(\omega t)dt\\ &= \int_{-\infty}^{\infty} x(t)\cos(\omega t)dt + \int_{-\infty}^{\infty} x(t)\sin(\omega t)dt\\ &= X_{He}(\omega) + X_{Ho}(\omega), \quad (5.3.4)\end{aligned}$$

$$X_{He}(-\omega) = X_{He}(\omega) \text{ and } X_{Ho}(-\omega) = -X_{Ho}(\omega).$$

The *Hartley transform* has a very simple relationship with *the Fourier cosine and sine transforms*. They are defined for real functions. These follow from the fact that

$$\begin{aligned}X_H(\omega) &= X_c(\omega) + X_s(\omega),\\ X_c(\omega) &= X_{He}(\omega), \quad (5.3.5)\\ X_s(\omega) &= X_{H0}(\omega).\end{aligned}$$

The real and the imaginary parts of the Fourier transform are related to the even and odd parts of the Hartley transform as

$$\begin{aligned}X(j\omega) &= \int_{-\infty}^{\infty} x(t)\cos(\omega t)dt - j\int_{-\infty}^{\infty} x(t)\sin(\omega t)dt\\ &= \text{Re}[X(j\omega)] + j\,\text{Im}[X(j\omega)], \quad (5.3.6)\end{aligned}$$

$$\begin{aligned}\text{Re}[X(j\omega)] = X_{He}(\omega) \text{ and } \text{Im}[X(j\omega)]\\ = -X_{H0}(\omega). \quad (5.3.7)\end{aligned}$$

Noting the relationship between the Hartley and the Fourier transforms many of the properties discussed for Fourier transforms can be modified to describe the properties of Hartley transforms. See Olejniczak's chapter in Poularikis, ed. (1996).

Example 5.3.1 Find the Hartley transforms of the following functions, where $\alpha > 0$:

a. $x_1(t) = \Pi\left[\dfrac{t}{\tau}\right]$,

b. $x_2(t) = e^{-at}u(t)$,

c. $x_3(t) = e^{-\alpha|t|}$,

d. $x_4(t) = e^{-\alpha t^2}$,

e. $x_4(t) = e^{-\alpha t}\sin(\omega_0 t)$,

f. $x_5(t) = e^{-\alpha t}\cos(\omega_0 t)$.   (5.3.8a)

Solution: *a.* Since the time function is a real symmetric function it follows that the Hartley transform is equal to the Fourier transform and

$$\Pi\left[\frac{t}{\tau}\right] \overset{\text{Hart}}{\longleftrightarrow} \tau\frac{\sin(\omega\tau/2)}{(\omega\tau/2)}. \qquad (5.3.8b)$$

b. The Hartley transform of this function follows from the Fourier transform.

$$e^{-at}u(t) \overset{\text{FT}}{\longleftrightarrow} \frac{1}{\alpha + j\omega} = \frac{\alpha}{\alpha^2 + \omega^2} - j\frac{\omega}{\alpha^2 + \omega^2}, \alpha > 0$$
$$X_H(\omega) = \text{Re}\{F[e^{-at}u(t)]\}$$
$$- \text{Im}\{F[e^{-at}u(t)]\} = (\alpha + \omega)/(\alpha^2 + \omega^2). \quad (5.3.8c)$$

c. and *d.* The transforms follow from the Fourier transform as the time functions are even functions and they coincide with their F-transforms:

$$e^{-\alpha|t|} \overset{\text{Hart}}{\longleftrightarrow} \frac{2\alpha}{\alpha^2 + \omega^2}, \quad e^{\alpha t^2} \overset{\text{Hart}}{\longleftrightarrow} \sqrt{\frac{\pi}{\alpha}}e^{-\omega^2/4\alpha}. \quad (5.3.8d)$$

e. and *f.* We can make use of the Fourier transforms and use the real and the imaginary parts to compute the even and odd parts of the Hartley transforms of these functions. The results are given below and are left as homework problems:

$$e^{-\alpha t}\sin(\omega_0 t)u(t) \overset{\text{FT}}{\longleftrightarrow} \frac{\omega_0}{(\alpha + j\omega)^2 + \omega_0^2},$$

$$e^{-\alpha t}\cos(\omega_0 t)u(t) \overset{\text{FT}}{\longleftrightarrow} \frac{\alpha + j\omega}{(\alpha + j\omega)^2 + \omega_0^2},$$

$$e^{-\alpha t}\sin(\omega_0 t)u(t) \overset{\text{Hart}}{\longleftrightarrow} \frac{\omega_0(\alpha^2 + \omega_0^2 - \omega^2) + 2(\alpha\omega)}{(\alpha^2 + \omega_0^2 - \omega^2) + 2(\alpha\omega)^2}$$
$$(5.3.8e)$$

$$e^{-\alpha t}\cos(\omega_0 t)u(t) \overset{\text{Hart}}{\longleftrightarrow} \frac{(\alpha - \omega)(\alpha^2 + \omega_0^2 - \omega^2) + 2\omega(\alpha + \omega)}{(\alpha^2 + \omega_0^2 - \omega^2) + 2(\alpha\omega)^2}.$$
$$(5.3.8f)$$

∎

Example 5.3.2 Show that the following transforms of the power signals are valid:

a. $x_1(t) = \delta(t) \overset{\text{Hart}}{\longleftrightarrow} 1$ (5.3.9a)

b. $1 \overset{\text{Hart}}{\longleftrightarrow} 2\pi\delta(\omega)$ (5.3.9b)

c. $u(t) \overset{\text{Hart}}{\longleftrightarrow} \pi\delta(\omega) + (1/\omega)$ (5.3.9c)

d. $\text{sgn}(t) \overset{\text{Hart}}{\longleftrightarrow} \dfrac{2}{\omega}$ (5.3.9d)

e. $\cos(\omega_0 t) \overset{\text{Hart}}{\longleftrightarrow} \pi[\delta(\omega - \omega_0) + \delta(\omega + \omega_0)]$ (5.3.9e)

f. $\sin(\omega_0 t) \overset{\text{Hart}}{\longleftrightarrow} \pi[\delta(\omega - \omega_0) - \delta(\omega + \omega_0)]$ (5.3.9f)

g. $e^{\pm j\omega_0 t} \overset{\text{Hart}}{\longleftrightarrow} 2\pi\delta(\omega \mp \omega_0)$.   (5.3.9g)

Solution: Parts *a, b,* and *d* follow directly from the Fourier transforms of these functions, as they are real and even. Part *c.* follows from

$$u(t) \overset{\text{FT}}{\longleftrightarrow} \pi\delta(\omega) + \frac{1}{j\omega},$$
$$X_H(\omega) = \text{Re}\{F[u(t)]\} - \text{Im}\{F[u(t)]\}$$
$$= \pi\delta(\omega) + \frac{2}{\omega}.$$

The other parts are left as exercises.   ∎

There are several advantages and disadvantages of the Hartley transform. Hartley transform avoids the complex integration. It is its own inverse. When dealing with real signals the Hartley transform provides a simple and efficient approach in dealing with Fourier spectrum. Some of the disadvantages include that the Fourier amplitude and phase information are not readily available in the Hartley transform.

5.4 Laplace Transforms

As mentioned in the last chapter, the Fourier transform is limited to finite energy signals and to finite power signals. The condition for the existence of the Fourier transform of a signal is that it must be absolutely integrable. Although the power signals are not absolutely integrable, they have Fourier transforms, in the limit, by using impulse functions. The signals that are not absolutely convergent can be made convergent by introducing a convergence factor $e^{-\sigma t}$, where σ is a real number, into the basis function. The range of σ that ensures the existence of the Laplace transform for a particular function defines the *region of convergence (ROC)*. The Laplace transform exists only if $x(t)e^{-\sigma t}$ is absolutely integrable. As an example, the Laplace transform of $x(t) = e^{3t}u(t)$ exists only if $\sigma > 3$. That is, replace $e^{-j\omega t}$ by $e^{-(\sigma+j\omega)t}$ in the Fourier integral. In other words we consider the Fourier transform of the function $(x(t)e^{-\sigma t})$. Using the convergence factor, the forward transform and its inverse are defined by

$$X'(j\omega) = \int_{-\infty}^{\infty} (x(t)e^{-\sigma t})e^{-j\omega t}dt,$$

$$x(t)e^{-\sigma t} = \frac{1}{2\pi}\int_{-\infty}^{\infty} X'(j\omega)e^{j\omega t}d\omega. \qquad (5.4.1)$$

We can combine the exponential terms in both expressions resulting in the pair

$$X'(j\omega) = \int_{-\infty}^{\infty} x(t)e^{-(\sigma+j\omega)t}dt,$$

$$x(t) = \frac{1}{2\pi}\int_{-\infty}^{\infty} X'(j\omega)e^{(\sigma+j\omega)t}d\omega. \qquad (5.4.2)$$

The transform and its inverse have the term $(\sigma + j\omega)$ in their integrands. We can use a complex variable $s = \sigma + j\omega$ in (5.4.2). It has the same units as ω measured in rad/s. The forward *two-sided (or bilateral Laplace) transform and its inverse* are, respectively, defined and their symbolic representation by

$$X_{II}(s) = \int_{-\infty}^{\infty} x(t)e^{-st}dt,$$

$$x(t) = \frac{1}{2\pi j}\int_{\sigma-j\omega}^{\sigma+j\omega} X_{II}(s)e^{st}ds. \qquad (5.4.3a)$$

$$x(t) \overset{LT_{II}}{\longleftrightarrow} X_{II}(s). \qquad (5.4.3b)$$

Note the subscript on X indicating it is a bilateral (a two-sided transform). The most useful one in our study is the one-sided or the *unilateral Laplace transform* and is

$$X_I(s) = \int_{0^-}^{\infty} x(t)e^{-st}dt \equiv X(s). \qquad (5.4.4)$$

If there is no subscript on X as in (5.4.4), the transform is the unilateral transform.

Example 5.4.1 Consider the signal $x(t) = e^{-at}u(t)$, a – real and positive. Find the bilateral Laplace transform of $x(t)$ and compare it with the corresponding Fourier transform.

Solution: The Laplace transform of $x(t)$ is given by

$$X_{II}(s) = \int_0^{\infty} e^{-at}e^{-st}dt = \int_0^{\infty} e^{-(s+a)t}dt$$

$$= \frac{-1}{(s+a)}e^{-(s+a)t}\Big|_0^{\infty}. \qquad (5.4.5a)$$

Since $\lim_{t\to\infty}\left[e^{-(s+a)t}\right] = \begin{cases} 0, & \sigma > -a \\ \infty, & \sigma < -a \end{cases}$, $\qquad (5.4.5b)$

the Laplace transform of the function is therefore

$$X_{II}(s) = \frac{1}{(s+a)}, \sigma = \mathrm{Re}(s) > -a. \qquad (5.4.5c)$$

Thus the ROC of this function is $\mathrm{Re}(s) = \mathrm{Re}(\sigma + j\omega) = \sigma < -a$ in the complex plane. The complex plane is usually referred to as the *s-plane*. The horizontal and the vertical axes are referred to as the σ – axis and the $j\omega$ – axis, respectively. Next question we want to ask ourselves is what happens when $\sigma = -a$? Clearly $X(s)$ is a well-behaved function *except* at $s = -a$, that is when $\omega = 0$. This simply means that the contour of integration is as shown in Fig. 5.4.1. See the half moon around the point $s = -a$ in this figure. The region of convergence can include the line $\sigma = -a$, *except* in the near vicinity of the point $s = -a$. The ROC includes the $j\omega$-axis, and, therefore, the Laplace transform of the function in this example is equivalent to the Fourier

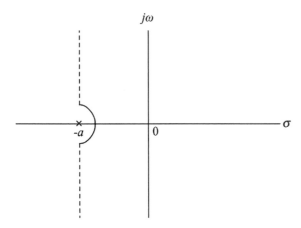

Fig. 5.4.1 $> s = \sigma + j\omega$-plane, contour of integration, Example 5.4.1

transform and can be obtained from the Laplace transform by simply substituting $s = j\omega$ in the Laplace transform. The example we have considered above exists for positive time. ∎

A two-sided signal can be written in two parts – one for positive time (*causal*) including $t = 0$ and the other part for negative time (*anti-causal*). The two-sided function $x(t)$ can be separated into two parts $x_1(t)$ and $x_2(t)$, representing the causal and the anti-causal part, respectively. See Fig. 5.4.2, where we have the following:

$$x(t) : \text{two–sided},$$

$$x_1(t) = x(t)u(t) : \text{causal part}$$

$$x_2(t) = x(t)u(-t) : \text{anti–causalpart},$$

$$x_2(-t) = x(-t)u(t) : \text{invertedanti–causalpart}$$

$x_1(t)$ and $x_2(-t)$ are now causal signals. The bilateral or two-sided transform is given by

$$X_{II}(s) = \int_{-\infty}^{\infty} x(t)e^{-st}dt = \int_{-\infty}^{0} x_2(t)e^{-st}dt$$

$$+ \int_{0}^{\infty} x_1(t)e^{-st}dt = X_{2,II}(s) + X_{1,II}(s).$$

$$(5.4.6a)$$

Note that $X_{1,II}(s)$ is the one-sided Laplace transform of the causal part and $X_{2,II}(s)$ is the Laplace transform of the anti-causal part. The anti-causal part can be rewritten as

$$X_{2,II}(s) = \int_{-\infty}^{0} x_2(t)e^{-st}dt = \int_{0}^{\infty} x_2(-\alpha)e^{st}d\alpha$$

$$= \int_{0}^{\infty} x_2(-\alpha)e^{st}d\alpha$$

$$\Rightarrow X_{2,II}(-s) = \int_{0}^{\infty} x_2(-t)e^{-st}dt. \qquad (5.4.6b)$$

That is, we have expressed the two-sided Laplace transform in terms of two one-sided Laplace transforms. Changing the sign of s in the function $X_{2,II}(-s)$ gives $X_{2,II}(s)$.

Notes:

1. Express
   $x(t) = [x(t)u(t)] + [x(t)u(-t)] = x_1(t) + x_2(t)$.
2. Use the causal signals $x_1(t)$ and $x_2(-t)$ and find their one-sided transforms.
3. The two-sided Laplace transform of the signal is given by $X(s) = X_1(s) + X_2(-s)$. ∎

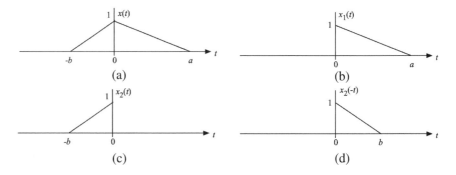

Fig. 5.4.2 (a) Two-sided signal, (b) causal, (c) anti-causal, and (d) inverted anti-causal parts

Example 5.4.2 Find the two-sided Laplace transform of the function

$$x(t) = \left\{ \begin{array}{ll} e^{bt}, & b>0, t<0 \\ e^{-at}, & a>0, t \geq 0 \end{array} \right\} \qquad (5.4.7)$$

$$= [e^{bt}u(-t)] + [e^{-at}u(t)] = x_2(t) + x_1(t)$$

Solution: From Example 5.4.1, we have

$$x_1(t) = e^{-at}u(t) \overset{LT_{II}}{\longleftrightarrow} \frac{1}{(s+a)}$$
$$= X_1(s), \operatorname{Re}(s) = \sigma > -a. \qquad (5.4.8a)$$

$$x_2(-t) = e^{-bt}u(t) \overset{LT_{II}}{\longleftrightarrow} \frac{1}{(s+b)}$$
$$= X_2(s), \sigma = \operatorname{Re}(s) > -b,$$

$$x_2(t) = e^{bt}u(-t) \overset{LT_{II}}{\longleftrightarrow} X_2(-s)$$
$$= -\frac{1}{s-b}, \sigma = \operatorname{Re}(s) < b. \qquad (5.4.8b)$$

Combining the two equations in (5.4.8a) and (5.4.8b), we have

$$x(t) \overset{LT_{II}}{\longleftrightarrow} -\frac{1}{(s-b)} + \frac{1}{(s+a)}, -a < \sigma < b. \qquad (5.4.9)$$

Note the constraints on σ. The transform does *not* exist *if* $(-a) > b$. This will be tied to the region of convergence shortly. From this example the *causal signal* results in the transform that has poles on the *left half* of the s-plane and the *anti-causal* signal results in the transform that has poles on the *right half* of the s-plane. ∎

5.4.1 Region of Convergence (ROC)

We defined the two-sided Laplace transform of a signal $x(t)$ by assuming that the function $(x(t)e^{-\sigma t})$ is absolutely convergent. This implies that there exists a pair of constants α and β and M a real positive number, such that

$$x(t) \leq \left\{ \begin{array}{ll} Me^{-\alpha t}, & t>0 \\ Me^{\beta t}, & t<0 \end{array} \right. . \qquad (5.4.10a)$$

To see these constraints, we can separate the bilateral Laplace transform into two integrals, one for positive time and the other for negative time. That is,

$$X_{II}(s) = L_{II}[x(t)] = \int_{-\infty}^{\infty} x(t)e^{-st} dt$$
$$= \int_{-\infty}^{0} x(t)e^{-st} dt + \int_{0}^{\infty} x(t)e^{-st} dt. \qquad (5.4.10b)$$

The integrals in the above equation must be absolutely integrable in order for the transform to exist. Using (5.4.10a), we have

$$X_{II}(s) \leq \int_{-\infty}^{0} Me^{\beta t}e^{-\sigma t} dt + \int_{0}^{\infty} Me^{-\alpha t}e^{-\sigma t} dt$$
$$= \int_{-\infty}^{0} Me^{(\beta - \sigma)t} dt + \int_{0}^{\infty} Me^{-(\alpha + \sigma)t} dt.$$
$$(5.4.10c)$$

Noting that limits on the integration, we can state that the transform exists if

$$(\beta - \sigma) > 0, \quad (\alpha + \sigma) > 0 \rightarrow -\alpha < \sigma < \beta. \qquad (5.4.11)$$

This defines the ROC and is illustrated by the dark area in Fig. 5.4.3.

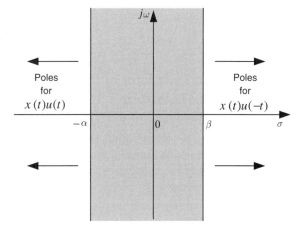

Fig. 5.4.3 Region of convergence for bilateral Laplace transform

Now we will relate this to a rational Laplace transform function $X_{II}(s)$ in terms of its poles and zeros. We say that $X_{II}(s)$ has a *pole* at $s = -p_i$ if $X(s)|_{s=-p_i} = \infty$ and has a *zero* at $s = -z_i$

if $X(s)|_{s=-z_i} = 0$. The poles of the function $X(s)$ must lie to the right of the line $s = \beta + j\omega$ for $t < 0$, whereas the poles corresponding to the positive time, i.e., $x(t)$, $t > 0$ must lie to the left of the line $s = -\alpha + j\omega$ in the complex s-plane.

Example 5.4.3 Find the two-sided Laplace transform of the following function directly:

$$x(t) = e^{-a|t|} = e^{-at}u(t) + e^{at}u(-t), a > 0. \quad (5.4.12)$$

Solution: The transform of this function is

$$X(s) = \int_{-\infty}^{0} e^{(a-s)t}e^{-st}dt + \int_{0}^{\infty} e^{-(a+s)t}dt$$
$$= \frac{1}{(a-s)} + \frac{1}{(s+a)} = \frac{2a}{a^2-s^2}, -a < \sigma < a.$$
$$(5.4.13) \blacksquare$$

The region of convergence is indicated on the right of (5.4.13). The Laplace transform does not converge at the pole locations and in Fig. 5.4.1 we have shown that the region of convergence goes around the poles (see the *half moons around the poles*). The region of convergence includes the $j\omega$-axis and the function evaluated on the $j\omega$-axis gives the Fourier transform of the function. That is, $X(j\omega) = X_{II}(s)|_{s=j\omega}$.

5.4.2 Inverse Transform of Two-Sided Laplace Transform

One needs to be careful in finding the inverse transforms of a two-sided Laplace transform, as the positive and negative time portions must be handled separately. If the region of convergence is not specified, then there is some ambiguity.

Example 5.4.4 Consider the time functions

$$x(t) = e^{-at}u(t), \quad y(t) = -e^{-at}u(-t). \quad (5.4.14)$$

Solution: The two-sided Laplace transforms of these functions are given by

$$X_{II}(s) = \int_{0}^{\infty} e^{-at}e^{-st}dt = \frac{1}{s+a},$$
$$Y_{II}(s) = \int_{-\infty}^{0} -e^{-at}e^{-st}dt = \frac{1}{s+a}, \operatorname{Re}(s) < -a.$$
$$(5.4.15)$$

We can see that the *transforms* are the *same*, except that they have *different regions of convergence*. These indicate the ambiguity in identifying the time function that the transform came from *if* the region of convergence of a two-sided Laplace transform is *not* given. In this case it is not possible to compute the corresponding time function. However, in practice, the ambiguity can be resolved on the basis of physical considerations, as the time functions increase without limit as t approaches either $+\infty$ or $-\infty$. This gives a way to select a time function among many possibilities.   ∎

Procedure to find the two-sided inverse LT of a rational function:

1. Expand the given rational transform function by using partial fraction expansion.
2. The terms in the partial fraction expansion that come from the *left half-plane poles* will result in time functions that exist only for $t \geq 0$.
3. The terms in the partial fraction expansion that come from *the right half-plane poles* will result in the time functions that exist only for $t < 0$.

Example 5.4.5 Find the inverse transform of the $X_{II}(s) = 1/[(s-1)(s+2)]$.

Solution: The partial fraction expansion of this function is

$$X_{II}(s) = \frac{A}{(s-1)} + \frac{B}{(s+2)},$$
$$A = \frac{1}{(s+2)}\Big|_{s=1} = \frac{1}{3},$$
$$B = \frac{1}{(s-1)}\Big|_{s=-2} = -\frac{1}{3} \quad (5.4.16)$$

$$\Rightarrow X_{II}(s) = \frac{1/3}{(s-1)} + \frac{-1/3}{(s+2)}, \quad (5.4.17)$$

$$\Rightarrow x(t) = (1/3)e^{t}u(t) - (1/3)e^{-2t}u(t),$$
$$-2 < \operatorname{Re}(s) < 1. \quad (5.4.18)$$

The region of convergence can be obtained by noting that the first term has a pole at $s = 1$ and the region of convergence for this term is $\sigma = \text{Re}(s) < 1$. The second term has a pole at $s = -2$ and the region of convergence for this term is $\sigma > -2$. The intersection of these two regions of convergence is given by $-2 < \text{Re}(s) < 1$. ∎

The two-sided Laplace transform of a function can be determined by decomposing the two-sided function into two one-sided functions and then transforming each one. In the case of finding the inverse Laplace transform, we first separate the transform into two parts, one with poles on the right half-plane and other with poles in the left half-plane and the imaginary axis. Then determine the two time functions. Most of our discussion on inverse transforms covers rational functions.

5.4.3 Region of Convergence (ROC) of Rational Functions – Properties

1. The ROC does not contain any poles of the function.
2. If $x(t) = 0$, except in a finite interval, then the ROC is the entire s-plane except possibly $s = 0$ and $s = \infty$.
3. If $x(t)$ is right-sided, then the ROC is right-sided, i.e., $\sigma = \text{Re}(s) > -a$, where $(-a)$ is the real part of the *left-most pole*.
4. If $x(t)$ is left-sided, then the ROC is left-sided, i.e., $\sigma = \text{Re}(s) < b$, where b is the real part of the *right-most pole*.
5. If $x(t)$ is two-sided function, i.e., the sum of left- and right-sided functions, then the ROC is either a strip defined by $-a < \text{Re}(s) < b$ or the individual regions of convergence will not overlap and, in that case, the ROC is the null set.

5.5 Basic Two-Sided Laplace Transform Theorems

Now consider some of the important two-sided Laplace transform theorems that are given below without proofs for most. Assume the region of convergence of $x_i(t)$ is R_{x_i}.

5.5.1 Linearity

The Laplace transform of a sum is the sum of the Laplace transforms and can be stated as

$$x(t) = \sum_{i=1}^{2} a_i x_i(t) \xrightarrow{LT_{II}} \sum_{i=1}^{2} a_i X_i(s) \tag{5.5.1}$$

$$= X_{II}(s), \text{ROC: at least } R_{x_1} \cap R_{x_2}$$

Example 5.5.1 Let $x(t) = x_1(t) + x_2(t), x_1(t) = e^{-t}u(t), x_2(t) = e^t u(t)$. Find the Laplace transform of the function $x(t)$ and identify the region of convergence.

Solution:

$$X_{II}(s) = \frac{1}{s+1} + \frac{1}{s-1}.$$

The ROC of $x_1(t)$ is $\text{Re}(s) > -1$ and the ROC of $x_2(t)$ is $\text{Re}(s) < 1$. The ROC of $x(t)$ is the intersection of the two and is given by $-1 < \text{Re}(s) < 1$. ∎

5.5.2 Time Shift

$$L_{II}[x(t - t_0)] = e^{-st_0} X_{II}(s). \tag{5.5.2}$$

The region of convergence is the same for both the original and its shifted version.

5.5.3 Shift in s

$$L_{II}[e^{-at} x(t)] = X_{II}(s + a). \tag{5.5.3}$$

Since the poles will be shifted to the left by a, the *ROC* will be shifted to the left by a.

5.5.4 Time Scaling

$$L_{II}[x(at)] = \frac{1}{|a|} X_{II}(s/a), (\text{ROC})_{\text{new}} = (\text{ROC})_{\text{old}}/a. \tag{5.5.4}$$

Time scaling makes the ROC scaled as well.

5.5.5 Time Reversal

$$L_{II}[x(-t)] = X_{II}(-s),$$
$$(\text{ROC})_{\text{new}} = -(\text{ROC})_{\text{old}}.$$
(5.5.5)

Note that the right half-plane poles become left half-plane poles and vice versa.

5.5.6 Differentiation in Time

$$L_{II}\left[\frac{dx(t)}{dt}\right] = sX_{II}(s), \text{ROC}_{\text{new}} \supset \text{ROC}_{\text{old}}.$$
(5.5.6)

The ROC will not change unless there is a pole–zero cancellation in the product $(sX_{II}(s))$.

5.5.7 Integration

$$L_{II}\left[\int_{-\infty}^{t} x(\alpha)d\alpha\right] = \frac{1}{s}X_{II}(s).$$
(5.5.7)

Noting the term $(1/s)$ in the transform and $\text{ROC}_{\text{new}} = \text{ROC}_{\text{old}}\text{I}\{\text{Re}(s) > 0\}$.

5.5.8 Convolution

In Chapter 2 we defined the convolution of two functions by

$$y(t) = \int_{-\infty}^{\infty} x_1(\alpha)x_2(t - \alpha)d\alpha$$
$$= \int_{-\infty}^{\infty} x_2(\alpha)x_1(t - \alpha)d\alpha.$$
(5.5.8a)

The transform is

$$Y_{II}(s) = \int_{-\infty}^{\infty}\left[\int_{-\infty}^{\infty} x_1(\alpha)x_2(t - \alpha)d\alpha\right]e^{-st}dt$$
$$= \int_{-\infty}^{\infty} x_1(\alpha)\left[\int_{-\infty}^{\infty} x_2(t - \alpha)e^{-st}dt\right]d\alpha$$
$$= \int_{-\infty}^{\infty} x_1(\alpha)X_{II,2}(e^{-s\alpha})d\alpha$$
$$= X_{II,2}(s)\int_{-\infty}^{\infty} x_1(\alpha)e^{-s\alpha}d\alpha = X_{II,2}(s)X_{II,1}(s).$$
(5.5.8b)

The ROC satisfies $\text{ROC}_{\text{new}} \supset (\text{ROC})_1 \cap (\text{ROC})_2$. The ROC of the convolution may be larger. When two transforms are multiplied, there is a possibility of pole cancellations.

5.6 One-Sided Laplace Transform

So far we have been discussing the bilateral or two-sided transform. A special form of the bilateral transform is the one-sided or unilateral or simply Laplace transform, which was defined in (5.4.4). It was pointed out that the bilateral transform can be computed by using the one-sided Laplace transform. In real-life systems, there is no negative time. However, bilateral transforms provide a structure that we can work with, as the bilateral Laplace transforms relate to the Fourier transforms. We can make the discussion simpler by considering the unilateral Laplace transform. The unilateral transform is fundamental in circuits, systems, and control, where we are interested in the response of a system with initial conditions. In a later chapter we will describe a linear time-invariant system by constant coefficient differential equations. The unilateral Laplace transform provides a powerful tool in the analysis and design of systems. As mentioned earlier we will use the notation $X(s) = L\{x(t)\}$ and $x(t)$ is a causal signal. Furthermore, the *ROC* is the *right half s-plane* for the *unilateral Laplace transforms*. For simplicity, we generally do *not* explicitly identify the region of convergence. Unless otherwise mentioned, we will assume that the transform functions are unilateral and will not be mentioned explicitly.

The unilateral Laplace transform of a signal $x(t)$ is defined earlier and is repeated below:

$$X(s) = L[x(t)] = \int_{0^-}^{\infty} x(t)e^{-st}dt.$$
(5.6.1)

Symbolic relation:

$$x(t) \xleftrightarrow{\text{LT}} X(s).$$
(5.6.2)

Notes: In defining the transform integral in (5.6.1), we have used the lower limit of 0^-. This allows us to include signals such as the unit impulse function $\delta(t)$. From now on we will use the lower limit on the integral as zero except in special cases and assume that the limit is 0^-. In cases where there is some ambiguity we will explicitly identify the lower limit on the integral as 0^-. Some texts use the limits of integration on the Laplace integral as 0^+ to infinity. This implies that origin is excluded. This approach is impractical in the theoretical study of linear systems. The Laplace transform of a function exists if $(x(t)e^{-\sigma t})$ is absolutely integrable. We can select the range of σ that ensures the convergence and this is referred to as the region of convergence.This is one of the nice aspects of the Laplace transform. For example, the L-transform of $e^{at}u(t), a > 0$ *exists only if* $\sigma > a$ and we can select such a range. Noting that the Laplace transform is an integral operation, the transform is unique.

Example 5.6.1 Find the Laplace transforms of the functions by using the definition

$$a.\ x_1(t) = \delta(t - t_0), t_0 > 0,$$

$$b.\ x_2(t) = u(t), \qquad (5.6.3)$$

$$c.\ x_3(t) = e^{-at}u(t), a > 0.$$

Solution: *a*. The Laplace transform of the impulse function is given by

$$X_1(s) = \int_{0^-}^{\infty} \delta(t - t_0)e^{-st}dt = e^{-s\,t_0},$$
$$\qquad (5.6.4)$$

$$\delta(t - t_0) \xleftrightarrow{\ \text{LT}\ } e^{-s\,t_0}.$$

$$b.\ X_2(s) = \int_0^{\infty} u(t)e^{-st}dt = \frac{1}{s}e^{-st}\Big|_{t=0^-}^{\infty}$$
$$\qquad (5.6.5)$$

$$= \frac{1}{s} \ \rightarrow u(t) \xleftrightarrow{\ \text{LT}\ } \frac{1}{s},$$

$$c.\ X_3(s) = \int_0^{\infty} e^{-at}e^{-st}dt = \int_0^{\infty} e^{-(s+a)t}dt$$

$$= \frac{1}{s+a}, e^{-at}u(t) \xleftrightarrow{\ \text{LT}\ } \frac{1}{s+a}. \qquad (5.6.6) \ \blacksquare$$

Example 5.6.2 Find the unilateral Laplace transforms of the following functions:

$$a. x_1(t) = 2\sqrt{\frac{t}{\pi}}, \ b. x_2(t) = \sinh(t).$$

Solution:

$$a. X_1(s) = \frac{2}{\sqrt{\pi}} \int_0^{\infty} t^{1/2}e^{-st}dt.$$

Using the change of variable $\alpha = t^{1/2}, d\alpha = (1/2)t^{-1/2}dt$ and the integral tables result in

$$X_1(s) = \frac{4}{\sqrt{\pi}} \int_0^{\infty} \alpha^2 e^{-s\,\alpha^2}d\alpha = \frac{1}{s^{3/2}}.$$

$$b.\ X_2(s) = L[\sinh(t)] = \frac{1}{2} \int_0^{\infty} \left[e^{-(s-a)t} - e^{-(s+a)t} \right] dt$$

$$= \frac{a}{s^2 - a^2}. \qquad (5.6.7)$$

The two transform pairs in this example are given by

$$2\sqrt{\frac{t}{\pi}} \xleftrightarrow{\ \text{LT}\ } \frac{1}{s^{3/2}}, \quad \sinh(at) \xleftrightarrow{\ \text{LT}\ } \frac{a^2}{s^2 - a^2}. \qquad (5.6.8) \ \blacksquare$$

As in the case of F-transforms we can derive the properties of the L-transforms. Most of the proofs are similar in these cases. We will not go through the details but will point out some of the important facets of the properties. In cases where they are different we will go through the proofs. Following is a table of some of the one-sided Laplace transforms theorems. Note that the regions of convergence are not identified.

5.6.1 Properties of the One-Sided Laplace Transform

Unilateral Laplace transforms properties are given in Table 5.6.1.

5.6.2 Comments on the Properties (or Theorems) of Laplace Transforms

The proof of the *linearity* property of the Laplace transforms is straightforward as the integral of a

Table 5.6.1 One-sided Laplace transform properties

Superposition (linearity):

$$x(t) = \sum_{i=1}^{N} a_i x_i(t) \overset{\text{LT}}{\longleftrightarrow} \sum_{i=1}^{N} a_i X_i(s). \tag{5.6.9}$$

Time delay:

$$x(t - \tau)u(t - \tau) \overset{\text{LT}}{\longleftrightarrow} e^{-s\tau} X(s); \ \tau > 0. \tag{5.6.10}$$

Complex frequency shift (times-exponential):

$$e^{-at} x(t) \overset{\text{LT}}{\longleftrightarrow} X(s + a). \tag{5.6.11}$$

Time scaling:

$$x(at) \overset{\text{LT}}{\longleftrightarrow} \frac{1}{|a|} X(s/a); \ a \neq 0, \ a \text{ is a constant}. \tag{5.6.12}$$

$$x(at)e^{\pm s_0 t} \overset{\text{LT}}{\longleftrightarrow} \frac{1}{|a|} X(s \mp s_0). \tag{5.6.13}$$

Convolution in time:

$$x_i(t) * x_j(t) \overset{\text{LT}}{\longleftrightarrow} X_i(s) X_j(s). \tag{5.6.14}$$

Multiplication in time

$$x_1(t) x_2(t) \overset{\text{LT}}{\longleftrightarrow} \frac{1}{2\pi j} [X_1(s) * X_2(s)]. \tag{5.6.15}$$

Times-t:

$$(t)^n x(t) \overset{\text{LT}}{\longleftrightarrow} (-1)^n \frac{d^n X(s)}{ds^n}. \tag{5.6.16a}$$

Times-$(1/t)$:

$$\frac{x(t)}{t} \overset{\text{LT}}{\longleftrightarrow} \int_s^\infty X(\alpha) d\alpha \tag{5.6.16b}$$

Derivative:

$$\frac{d^n x(t)}{dt^n} \overset{\text{LT}}{\longleftrightarrow} s^n X(s) - s^{n-1} x(0^-) - s^{n-2} x^{(1)}(0^-) - \dots x^{n-1}(0^-);$$

$$x^{(i)}(0^-) = \frac{d^i x(t)}{dt^i} \big|_{t=0^-}. \tag{5.6.17}$$

Integration:

$$\int_{-\infty}^{t} x(\alpha) d\alpha \overset{\text{LT}}{\longleftrightarrow} \frac{1}{s} \int_{-\infty}^{0^-} x(\alpha) d\alpha + \frac{X(s)}{s}. \tag{5.6.18}$$

Initial Value:

$$x(0^+) = \lim_{s \to \infty} [sX(s)]; \ (X(s) \text{ is proper}). \tag{5.6.19}$$

Final value:

$$\lim_{t \to \infty} x(t) = \lim_{s \to \infty} [sX(s)]; \ (\text{Poles of } X(s) \text{ lie in left half s } - \text{ plane}) \tag{5.6.20}$$

Switched periodic ($x_T(t)$ is periodic with period T):

$$x(t) = x_T(t)u(t) \overset{\text{LT}}{\longleftrightarrow} \frac{X(s)}{1 - e^{-sT}}. \tag{5.6.21}$$

Differentiation with respect to a second independent variable:

$$\frac{\partial x(t, r)}{\partial r} \overset{\text{LT}}{\longleftrightarrow} \frac{\partial X(s, r)}{\partial r}; \ (r \text{ is independent of } s \text{ and } t). \tag{5.6.22}$$

Integration with respect to a second independent variable:

$$\int_{r_0}^{r} X(s, r) dr \overset{\text{LT}}{\longleftrightarrow} \int_{r_0}^{r} x(t, \beta) d\beta; \ (r \text{ is independent of } s \text{ and } t). \tag{5.6.23}$$

sum is equal to the sum of the integrals. The *time delay* property can be shown by using a change of variable in the transform integral. The *complex shift* property can be shown by the following:

$$\int_0^\infty [e^{s_0 t} x(t)] e^{-st} dt = \int_0^\infty x(t) e^{-(s+a)t} dt = X(s+a).$$

The *time-scaling* and *frequency-shifting* properties follow by combining the time-scaling and the complex frequency-shifting properties.

Convolution property: The convolution of two functions was defined in Chapter 2. Assuming the functions start at $t = 0$, we can write

$$x_1(t) * x_2(t) = \int_0^\infty x_1(\alpha) x_2(t - \alpha) d\alpha$$

$$= \int_0^\infty x_2(\beta) x_1(t - \beta) d\beta. \qquad (5.6.24)$$

The transform of this function is given by

$$L[x_1(t) * x_2(t)] = \int_0^\infty \left[\int_0^\infty x_1(\alpha) x_2(t - \alpha) d\alpha \right] e^{-st} dt$$

$$= \int_0^\infty x_1(\alpha) \left[\int_0^\infty x_2(t - \alpha) e^{-st} dt \right] d\alpha.$$

Using the change of variable $\xi = t - \alpha$ and simplifying the integral, we have

$$L[x_1(t) * x_2(t)] = \int_0^\infty x_1(\alpha) \left[\int_{-\alpha}^\infty x_2(\xi) e^{-s(\xi + \alpha)} d\xi \right] d\alpha.$$

We are considering only positive time functions, so $x_2(\xi) = 0$ for $\xi < 0$, which allows us to change the lower limit on the second integral in the above equation and

$$L[x_1(t) * x_2(t)] = \left[\int_0^\infty x_1(\alpha) e^{-s\alpha} d\alpha \right]$$

$$\left[\int_0^\infty x_2(\xi) e^{-s\xi} d\xi \right] = X_1(s) X_2(s). \qquad (5.6.25)$$

This proves the convolution theorem.

Complex frequency shift: This is also referred to as *times-exponential* property and

$$L\{e^{-at} x(t)\} = \int_0^\infty e^{-at} x(t) e^{-st} dt$$

$$= \int_0^\infty x(t) e^{-(s+a)t} dt = X(s+a). \quad (6.5.26)$$

Times-t property: This follows from the following equation:

$$\frac{dX(s)}{ds} = \int_0^\infty x(t) \frac{de^{-st}}{ds} dt = \int_0^\infty [-tx(t)] e^{-st} dt. \quad (5.6.27)$$

We can generalize this result by repeated derivatives of the transform.

Example 5.6.3 In Example 5.6.2 we have derived the Laplace transform of the hyperbolic sine function. Use the times-t property to show that the following is true:

$$y(t) = t \sinh(at) \overset{LT}{\longleftrightarrow} \frac{2as}{(s^2 - a^2)^2} = Y(s). \quad (5.6.28)$$

Solution: Taking the derivative of the transform function in (5.6.8), we have

$$\frac{dX(s)}{ds} = \int_0^\infty [(-t) \sinh(at)] e^{-st} dt = \frac{d}{ds} \frac{2as}{(s^2 - a^2)}$$

$$= -\frac{2as}{(s^2 - a^2)^2}.$$

The result in (5.6.28) follows by identifying the term in the integrand and its transform on the right in the above equation. ∎

Times-$1/t$ property or complex integration property: This is

$$\int_{s}^{\infty} X(\beta)d\beta = L\left\{\frac{x(t)}{t}\right\} \text{ provided}$$
(5.6.29)

$$\lim_{t \to 0}[x(t)/t] \text{ exists.}$$

This can be shown by

$$\int_{s}^{a} X(\beta)d\beta = \int_{s}^{a}\left[\int_{0}^{\infty} e^{-\beta t}x(t)dt\right]d\beta$$

$$= \int_{0}^{\infty} x(t)\left[\int_{s}^{a} e^{-\beta t}d\beta\right]dt$$

$$= \int_{0}^{\infty} \frac{x(t)}{t}[e^{-st} - e^{-at}]dt$$

$$= \int_{0}^{\infty} \frac{x(t)}{t}e^{-st}dt - \int_{0}^{\infty} \frac{x(t)}{t}[e^{-at}]dt.$$

Assuming $x(t)/t$ has a limit as $t \to 0$ and for $a \to \infty$, the second integral on the right in the above equation goes to zero resulting in the times-($1/t$) property

Derivative property: Assuming the constraint $\lim_{t \to \infty}[x(t)e^{-st}] = 0$, we can write

$$L\left\{\frac{dx(t)}{dt}\right\} = \int_{0^-}^{\infty} \frac{dx(t)}{dt}e^{-st}dt$$

$$= x(t)e^{-st}\Big|_{t=0^-}^{t=\infty} + s\int_{0^-}^{\infty} x(t)e^{-st}dt$$

$$= -x(0^-) + s\int_{0^-}^{\infty} x(t)e^{-st}dt$$

$$= sX(s) - x(0^-).$$
(5.6.30)

The transform would *not* exist if the constraint is not satisfied. We can generalize this property to the nth derivative by integrating n times by parts. The convolution and the derivative properties are the most used properties in linear systems theory.

Integration property: This property can be seen using the integration by parts.

$$L\left\{\int_{-\infty}^{t} x(\alpha)d\alpha\right\} = \int_{0^-}^{\infty}\left[\int_{-\infty}^{t} x(\alpha)d\alpha\right]e^{-st}dt.$$

Assuming $u = \int_{-\infty}^{t} x(\alpha)d\alpha, du = x(t)dt,$

$$dv = e^{-st}dt, v = -\frac{1}{s}e^{-st}, \text{ we have}$$

$$L\left\{\int_{-\infty}^{t} x(\alpha)d\alpha\right\} = \left[-\frac{e^{-st}}{s}\int_{-\infty}^{t} x(\alpha)d\alpha\right]$$

$$\Big|_{0}^{\infty} + \frac{1}{s}\int_{0}^{\infty} x(t)e^{-st}dt, \quad x^{(-1)}(0^-) = \int_{-\infty}^{0} x(\alpha)d\alpha,$$

$$\Rightarrow L\left\{\int_{-\infty}^{t} x(\alpha)d\alpha\right\} = \frac{1}{s}\int_{0^-}^{\infty} x(t)e^{-st}dt + \frac{1}{s}\int_{-\infty}^{0^-} x(\alpha)d\alpha$$

$$= \frac{1}{s}X(s) + \frac{1}{s}\left[x^{(-1)}(0^-)\right].$$
(5.6.31)

Initial value theorem: This theorem states that $x(t)$ and its derivative $x'(t)$ are *both* Laplace transformable, then

$$x(0^+) = \lim_{s \to \infty} sX(s).$$
(5.6.32)

The proof of the initial value theorem can be seen by starting with $L\{x'(t)\} = sX(s) - x(0^-)$ and

$$\lim_{s \to \infty}\left[L\left\{\frac{dx(t)}{dt}\right\}\right] = \lim_{s \to \infty}\int_{0^-}^{\infty} x'(t)e^{-st}dt$$

$$= \lim_{s \to \infty}[sX(s) - x(0^-)].$$

Second,

$$L\{x'(t)\} = \int_{0^-}^{\infty} x'(t)e^{-st}dt = \int_{0^-}^{0^+} x'(t)e^{-st}dt$$

$$+ \int_{0^+}^{\infty} x'(t)e^{-st}dt = x(0^+) - x(0^-)$$

$$+ \int_{0^+}^{\infty} x'(t)e^{-st}dt. \qquad (5.6.33)$$

$$\Rightarrow sX(s) = x(0^+) + \int_{0^+}^{\infty} \frac{dx}{dt}e^{-st}dt.$$

Taking the limit results in

$$\lim_{s\to\infty}[sX(s)] = x(0^+) + \int_{0^+}^{\infty} \frac{dx}{dt}\lim_{s\to\infty}[e^{-st}]dt = x(0^+).$$

Notes: The initial value theorem should be applied only if $X(s)$ is *strictly proper*. That is, when $M < N$ in the function, then

$$X(s) = \frac{n_0 s^M + n_1 s^{M-1} + \dots + n_M}{s^N + d_1 s^{N-1} + \dots + d_N} = \frac{N(s)}{D(s)}. \qquad (5.6.34)$$

If $M \geq N$, then $\lim_{s\to\infty} sX(s)$ does not exist. However, $X(s)$ can be written in the following form:

$$X(s) = \left[\sum_{i=0}^{M-N} a_i s^i\right] + \frac{B(s)}{D(s)}. \qquad (5.6.35)$$

$[B(s)/D(s)]$ is now a strictly proper rational function. The inverse transform of the first part in (5.6.35) produces only impulses and their derivatives, which occur only at $t = 0$. The final value of the function depends only on the second part. That is,

$$x(0^+) = \lim_{s\to\infty}\left[\frac{sB(s)}{D(s)}\right]. \qquad (5.6.36) \blacksquare$$

Example 5.6.4 Use the transform of the unit step function $u(t)$ to find its initial value.

Solution: The unit step function has a jump discontinuity at $t = 0$ and

$$u(0^+) = \lim_{s\to\infty}[sL\{u(t)\}] = \lim_{s\to\infty}\left\{s\frac{1}{s}\right\} = 1 \qquad \blacksquare$$

Final value theorem: The proof of this is very similar to the initial value theorem and

$$L\left\{\frac{dx}{dt}\right\} = sX(s) - x(0^-),$$

$$\text{and } \lim_{s\to0}\int_0^{\infty} \frac{dx(t)}{dt}e^{-st}dt = x(\infty) - x(0^-).$$

From this it follows:

$$x(\infty) - x(0^-) = \lim_{s\to0}\{sX(s) - x(0^-)\}$$

$$\to \lim_{t\to\infty} x(t) = \lim_{s\to0} sX(s). \qquad (5.6.37)$$

The final value theorem is used effectively to compute the steady-state values of the output responses and is valid only when the limit on the right in (5.6.37) exists.

Notes: The final value theorem applies only if the poles of $sX(s)$ are in the *left half-plane*. The only pole permitted on the imaginary axis is a simple pole at $s = 0$. $x(\infty)$ is indeterminate if there are conjugate pole pairs on the imaginary axis. This implies $x(t)$ has sinusoids, and their final values are indeterminate. $\blacksquare$

Switched periodic functions and their Laplace transforms: Consider the periodic function $x_T(t) = x_T(t + T)$. Defining the function for one period, $x(t) = x_T(t + nT), 0 < t < T$ and zero everywhere else, we can write the *one-sided Laplace* transform of the function:

$$L\{x_T(t)u(t)\} = L\left\{\sum_{n=0}^{\infty} x(t-nT)\right\}$$

$$= \sum_{n=0}^{\infty} L\{x(t-nT)\} = X(s)\sum_{n=0}^{\infty} e^{-nsT}$$

$$= \frac{X(s)}{1-e^{-sT}}\left(\text{note:} \sum_{n=0}^{\infty} e^{-nsT} = \frac{1}{1-e^{-sT}}\right).$$

$$\qquad (5.6.38)$$

Example 5.6.5 Find the one-sided Laplace transform of the square wave given by

$$x(t) = \begin{cases} 1, & 0 < t < T/2 \\ -1, & T/2 < t < T \end{cases}, x_T(t) = x(t + nT).$$

$$(5.6.39a)$$

Solution: Using the above results we have

$$X(s) = \frac{1}{(1 - e^{-sT})} \int_0^T x(t) e^{-st} dt$$

$$= \frac{1}{(1 - e^{-sT})} \left[\int_0^{T/2} e^{-st} dt - \int_{T/2}^T e^{-st} dt \right] \frac{1}{(1 - e^{-sT})}$$

$$= \frac{1}{(1 - e^{-sT})} \left[\frac{1}{s}(-e^{-st})\Big|_0^{T/2} + \frac{1}{s}(e^{-st})\Big|_{T/2}^T \right]$$

$$= \frac{1}{(1 - e^{-sT})} [1 - e^{-sT/2} - e^{-sT/2} + e^{-sT}],$$

$$= \frac{1}{s} \frac{[1 - e^{-sT/2}]^2}{[1 - e^{-sT}]} = \frac{1}{s} \frac{[1 - e^{-sT/2}]}{[1 + e^{-sT/2}]}. \quad (5.6.39b) \ \blacksquare$$

Example 5.6.6 Find the Laplace transform of the square wave defined by

$$x(t) = u(t) - 2u\left(t - \frac{T}{2}\right) + 2u(t - T) - \dots \quad (5.6.40a)$$

Solution: Using the transform of the step function and the delay property, we have

$$L[x(t)] = \frac{1}{s}(1 - 2e^{-sT/2} + 2e^{-sT} - \dots)$$

This can be reduced by using the closed-form expression for the sum and

$$X(s) = \frac{1}{s} \left[\frac{2}{1 + e^{-sT/2}} - 1 \right] = \frac{1}{s} \frac{1 - e^{-sT/2}}{1 + e^{-sT/2}}.$$

$$(5.6.40b) \ \blacksquare$$

Example 5.6.7 Consider the two functions $x(t)$ and $y(t)$ given below. Find their Laplace transforms and sketch $y(t)$ assuming $T = 5$.

$$x(t) = \Pi[t - .5],$$

$$y(t) = \begin{cases} \sum_{n=0}^{\infty} x(t - nT), & t \geq 0 \\ 0, & \text{otherwise} \end{cases} \quad (5.6.41a)$$

Solution: *a*. First,

$$x(t) = u(t) - u(t-1) \xleftrightarrow{LT} \frac{1}{s} - \frac{e^{-s}}{s} = \frac{1}{s}[1 - e^{-s}].$$

$$(5.6.41b)$$

The function $y(t)$ is on for a second and off for 4 s for $t \geq 0$. The Laplace transform of this function is

$$Y(s) = L\left\{ \sum_0^\infty x(t - nT) \right\} = X(s) \sum_{n=0}^\infty e^{-nsT}$$

$$= X(s) \frac{1}{(1 - e^{-sT})} = \frac{(1 - e^{-s})}{s(1 - e^{-sT})}. \quad (5.6.41c)$$

First part of the result in (5.6.41c) is the transform of the single pulse given by $X(s)$ and the second part $[1/(1 - e^{-sT})]$ takes care of the switching part. $\blacksquare$

Example 5.6.8 Find the Laplace transform of the convolution $y(t) = t * e^{-at}, a > 0$.

Solution: Using the convolution theorem and the times-t property, we have

$$Y(s) = L\{tu(t)\} L\{e^{-at}\} = (1/s^2)(1/(s + a)).$$

$$(5.6.42) \ \blacksquare$$

Notes: In computing the one-sided Laplace transforms of arbitrary time function $x(t)$, we are really computing the transform of $x(t)u(t)$.

Example 5.6.9 Find the one-sided Laplace transforms of the following functions:

a. $x_1(t) = e^{\pm j\omega_0 t}$,

b. $x_2(t) = \cos(\omega_0 t)$,

c. $x_3(t) = \sin(\omega_0 t)$

d. $x_4(t) = e^{-at} \cos(\omega_0 t)$,

e. $x_5(t) = e^{-at} \sin(\omega_0 t)$,

f. $x_6(t) = te^{-at} u(t)$

g. $x_7(t) = te^{-at} \cos(bt)$,

h. $x_8(t) = te^{-at} \sin(bt), a > 0$.

Solution:

a. $X_1(s) = \int_0^\infty e^{\pm j\omega_0 t)} e^{-st} dt = \int_0^\infty e^{-(s \mp j\omega_0)t} dt$

$$= \frac{1}{(s \mp j\omega_0)}. \quad (5.6.43a)$$

b. $x_2(t) = \dfrac{1}{2}[e^{j\omega_0 t} + e^{-j\omega_0 t}]$

$= \cos(\omega_0 t) \overset{LT}{\longleftrightarrow} \dfrac{1}{2}\left[\dfrac{1}{s - j\omega_0} + \dfrac{1}{s + j\omega_0}\right]$

$= \dfrac{s}{s^2 + \omega_0^2},$ (5.6.43b)

c. $x_3(t) = \dfrac{1}{2j}[e^{j\omega_c t} - e^{-j\omega_c t}]$

$= \sin(\omega_0 t) \overset{LT}{\longleftrightarrow} \dfrac{1}{2j}\left[\dfrac{1}{s - j\omega_0} - \dfrac{1}{s + j\omega_0}\right]$

$= \dfrac{\omega_0}{s^2 + \omega_0^2},$ (5.6.43c)

d. and *e.* We will find the transforms in these cases by using the results in parts *b* and *c* and use times-exponential, i.e., the complex shift property:

$e^{-at}\cos(\omega_0 t) \overset{LT}{\longleftrightarrow} \dfrac{(s + a)}{(s + a)^2 + \omega_0^2},$ (5.6.44)

$e^{-at}\sin(\omega_0 t) \overset{LT}{\longleftrightarrow} \dfrac{\omega}{(s + a)^2 + \omega^2},$ (5.6.45)

f. Considering $e^{-at}u(t) \overset{LT}{\longleftrightarrow} 1/(s + a)$ and the times-*t* property, we have

$L\{te^{-at}\} = -\dfrac{d(1/(s + a))}{ds} = \dfrac{1}{(s + a)^2},$ (5.6.46)

g and *h.* As in the last part, we can make use of the times-*t* property for both these parts. The results are given below and the details are left as homework problems:

$te^{-at}\cos(bt) \overset{LT}{\longleftrightarrow} \dfrac{(s + a)^2 - b^2}{\{(s + a)^2 + b^2\}^2}, a > 0,$ (5.6.47)

$te^{-at}\sin(bt) \overset{LT}{\longleftrightarrow} \dfrac{2b(s + a)}{\{(s + a)^2 + b^2\}^2}, a > 0.$ (5.6.48)

The squared terms in the denominators follow from the times-*t* property. ∎

Example 5.6.10 Show that the following Laplace transform relationship is true:

$x(t) = 1/\sqrt{\pi t}\, u(t) \overset{LT}{\longleftrightarrow} \dfrac{1}{\sqrt{s}}.$

$\left(\text{use the identity} \displaystyle\int_0^\infty e^{-at^2 dt} = \dfrac{1}{2}\sqrt{\dfrac{\pi}{a}}\right)$

(5.6.49)

Solution: Considering $t = \alpha^2/\pi$ in the transform integral and simplifying, we have

$X(s) = \displaystyle\int_0^\infty x(t)e^{-st}dt = \int_0^\infty \dfrac{1}{\sqrt{\pi t}}e^{-st}dt$

$= \dfrac{2}{\pi}\displaystyle\int_0^\infty e^{-s\alpha^2/\pi}d\alpha = \dfrac{1}{\sqrt{s}}.$ ∎

Example 5.6.11 What can you say about the transform $L\{e^{t^2}u(t)\}$?

Solution: It does not exist since $\lim\limits_{t\to\infty} e^{(t-a)t} \to \infty.$ ∎

Example 5.6.12 Using the result $\displaystyle\int_s^\infty X(\alpha)d\alpha = L\{x(t)/t\}$, show that

$x(t) = \dfrac{\sin(\omega_0 t)}{\omega_0 t}u(t) \overset{LT}{\longleftrightarrow} \dfrac{1}{\omega_0}\tan^{-1}\left[\dfrac{\omega_0}{s}\right] = X(s).$ (5.6.50)

Solution: Using the times-$(1/t)$ property, integral tables and (5.6.44), we have

$L\{x(t)\} = \displaystyle\int_s^\infty \dfrac{1}{\alpha^2 + \omega_0^2}d\alpha = \dfrac{1}{\omega_0}\tan^{-1}(\alpha/\omega_0)\Big|_s^\infty$

$= \dfrac{1}{\omega_0}\left[\dfrac{\pi}{2} - \tan^{-1}(s/\omega_0)\right] = \dfrac{1}{\omega_0}\tan^{-1}\left(\dfrac{\omega_0}{s}\right).$

This can also be seen by using the Laplace transform property of integration with respect to a second independent variable. See Problem 5.6.5 at the end of the chapter. ∎

In Chapter 6 *linear systems* will be studied. They can be described by *linear constant coefficient differential equations*. Finding solutions of these is one of the important steps in the analysis of these systems. One way to obtain the solutions of the differential equations is by using Laplace transforms. The differential equation consists of an input $x(t)$ and the output $y(t)$ and their derivatives. The input function is assumed to be known and is Laplace transformable. The transform of the output and its inverse, i.e., the output time function is desired.

Example 5.6.13 Consider the linear constant coefficient differential equation

$\dfrac{d^2 y(t)}{dt^2} + 3\dfrac{dy(t)}{dt} + y(t) = x(t) + \dfrac{dx(t)}{dt}.$ (5.6.51)

Using the linearity property, the derivative property, and the time-integral property of the Laplace transforms, derive the expression for the Laplace transform of the above equation assuming arbitrary initial conditions.

Solution: Using the linearity property and derivative properties of the Laplace transform of the equation in (5.6.51), we have by using the initial conditions (see (5.6.17)).

$$[s^2 Y(s) - sy(0^-) - y'(0^-)] + 3[sY(s) - y(0^-)]$$
$$+ 2Y(s) = X(s) + sX(s) - x(0^+). \qquad (5.6.52)$$

We will come back to this example shortly. Instead of differentials we can have integrals in computing the Laplace transforms. These appear when dealing with circuits.   ∎

Example 5.6.14 Using the linearity property, the derivative property, and the time-integral property of the Laplace transforms, derive the expression for the Laplace transform of the following equation assuming that $y(0^-) = 1$ and $y^{(-1)}(0^-) = 2$:

$$\frac{dy}{dt} + 3y(t) + 2\int_{-\infty}^{t} y(\alpha)d\alpha = x(t) = u(t). \qquad (5.6.53)$$

Solution: Now consider the time-integration property, see (5.6.31).

$$L\left\{\int_{-\infty}^{t} y(\alpha)d\alpha\right\} = \frac{1}{s}\int_{0^-}^{\infty} y(t)e^{-st}dt + \frac{1}{s}\left[\int_{-\infty}^{0^-} y(\alpha)d\alpha\right]$$
$$= \frac{Y(s)}{s} + \frac{1}{s}\left[y^{(-1)}(0^-)\right] = \frac{Y(s)}{s} + \frac{2}{s}$$
$$(5.6.54)$$

Using the linearity, derivative, and the time-integral properties of the Laplace transforms, we can now write the Laplace transform of the equation in (5.6.53).

$$sY(s) - 1 + 3Y(s) + \frac{Y(s)}{s} + \frac{2}{s}$$
$$= \frac{1}{s} \rightarrow (s^2 + 3s + 1)Y(s) = (s - 1),$$

$$\Rightarrow Y(s) = \frac{(s-1)}{s^2 + 3s + 1}. \qquad (5.6.55)$$

The Laplace transform of an *integral–differential equation* is an algebraic equation. The Laplace transform of a function is *unique*. The output is $y(t) = L^{-1}\{Y(s)\}$. Finding such a function requires the inverse Laplace transform, which will be discussed shortly.   ∎

5.7 Rational Transform Functions and Inverse Laplace Transforms

The inverse Laplace transform of $X(s)$ is

$$x(t) = \frac{1}{2\pi j}\int_{\sigma - j\infty}^{\sigma + j\infty} X(s)e^{st}dt. \qquad (5.7.1)$$

The integration needs to be performed in the s-plane along a line $\text{Re}\{s\} = \sigma$, where σ is a constant factor chosen to ensure the convergence of the integral

$$\int_{-\infty}^{\infty} |x(t)|e^{-\sigma t}dt < \infty. \qquad (5.7.2)$$

Equation (5.7.1) is valid for both one-sided and two-sided transforms. In the case of the two-sided Laplace transform, the ROC must be specified before the inverse transform can be determined uniquely. The integration process involves complex integration and it requires knowledge of complex variables. This is beyond the scope here.

Only one-sided Laplace transforms are considered below and the lower limit in (5.7.2) is 0 in this case. The transform is assumed to be $L[x(t)] = X(s)$ and $x(t) = L^{-1}[X(s)]$ is unique, i.e., there is a one-to-one correspondence between the direct (forward) and the inverse transforms, see McGillem and Cooper (1991). In some pathological cases, this is not true, see McGillem and Cooper (1991). We can determine the time function $x(t)$ corresponding to its transform $X(s)$ by first writing it in terms of a sum of simple functions of s using partial fraction expansion. Then determine the inverse transform of the given function using the simple functions and tables of Laplace transform functions.

Table 5.6.2 One-sided Laplace tranform pairs

Function $x(t) = \dfrac{1}{2\pi j}\displaystyle\int_{\sigma-j\infty}^{\sigma+j\infty} X(s)e^{st}ds$	Laplace Transform $X(s) = \displaystyle\int_{0^-}^{\infty} x(t)e^{-st}dt$
$\delta(t)$	1
$u(t)$	$1/s$
$t^n u(t);\ n > 0$ is an integer.	$\dfrac{n!}{s^{n+1}}$
$e^{-\alpha t}u(t);\ \alpha > 0.$	$\dfrac{1}{s+\alpha}$
$t^n e^{-\alpha t}u(t);\ n > 0$ is an integer and $\alpha > 0.$	$\dfrac{n!}{(s+\alpha)^{n+1}}$
$\cos(\alpha t)u(t)$	$\dfrac{s}{s^2+\alpha^2}$
$\sin(\alpha t)u(t)$	$\dfrac{\alpha}{s^2+\alpha^2}$
$e^{-\alpha t}\cos(\beta t)u(t);\ \alpha > 0.$	$\dfrac{s+\alpha}{(s+\alpha)^2+\beta^2}$
$e^{-\alpha t}\sin(\beta t)u(t);\ \alpha > 0.$	$\dfrac{\beta}{(s+\alpha)^2+\beta^2}$
$te^{-\alpha t}\cos(\beta t)u(t);\ \alpha > 0.$	$\dfrac{(s+\alpha)^2-\beta^2}{\left[(s+\alpha)^2+\beta^2\right]^2}$
$te^{-\alpha t}\sin(\beta t)u(t);\ \alpha > 0.$	$\dfrac{2\beta(s+\alpha)}{[(s+\alpha)^2+\beta^2]}$
$\dfrac{1}{a^2}(1-\cos(at))$	$\dfrac{1}{s(s^2+a^2)}$
$\dfrac{1}{a}\sinh(at)$	$\dfrac{1}{s^2-a^2}$
$\dfrac{1}{(b-a)}\left[e^{-bt}-e^{-at}\right]$	$\dfrac{1}{(s+a)(s+b)}$
$\dfrac{1}{\sqrt{\pi t}}$	$\dfrac{1}{\sqrt{s}}$
$\dfrac{\sin(\omega_0 t)}{\omega_0 t}u(t)$	$\dfrac{1}{\omega_0}\tan^{-1}\left[\dfrac{\omega_0}{s}\right]$
$u(t)-2u(t-\tfrac{T}{2})+2u(t-T)-...$ (Switched square wave)	$\dfrac{1}{s}\cdot\dfrac{1-e^{sT/2}}{1+e^{-sT/2}}$

5.7.1 Rational Functions, Poles, and Zeros

A rational function $H(s)$ can be written in the form

$$H(s) = \frac{n_0 s^M + n_1 s^{M-1} + ... + n_M}{s^N + d_1 s^{N-1} + ... + d_N}$$
$$= \frac{K(s-z_1)(s-z_2)...(s-z_M)}{(s-p_1)(s-p_2)...(s-p_N)} = \frac{Y(s)}{X(s)}. \quad (5.7.3)$$

Note that in expressing the function $H(s)$, we assumed that the denominator polynomial $D(s)$ had its initial coefficient equal to 1. The degrees of the numerator and the denominator polynomials in (5.7.3) are, respectively, equal to M and N. The roots of the polynomial $Y(s)$ are denoted by $z_i, i = 1, 2, ..., M$ and are called the zeros. The roots of denominator polynomial are denoted by $p_k, k = 1, 2, ..., N$ and are referred to as poles of the function. We will assume that the coefficients of the polynomials are real, and therefore the roots of polynomials are real or complex with their conjugate pairs. Some of the roots may be repeated. In writing the expression in (5.7.3), it is customary to cancel out any poles and zeros that are common between the numerator and the denominator.

The roots identified by z_i and p_k are finite zeros and poles. If we include the values of infinity, we need to consider poles and zeros in that region as well. In the case of our function $X(s)$, it has $(M - N)$ poles at infinity if $M > N$ and $(N - M)$ zeros at infinity if $N > M$. This points out the fact that if we *include the poles and zeros at infinity*, then the number of poles is equal to the number of zeros. Pole–zero plots give some insight to the transform functions. Finite zeros are identified by circles 0s and finite poles by Xs or crosses. Poles and zeros at infinity are usually not identified in the pole–zero plots.

Example 5.7.1 Consider the function $X(s)$ given below and sketch the pole–zero plots associated with this function. Identify the poles (or zeros) that are located at infinity.

$$X(s) = \frac{K(s - 1)(s^2 + 1)}{s(s + 3)(s^2 + s + 1)}.$$

Solution: The other zeros are located at $s = 1$, $\pm j1$, and at infinity. The poles are located at $s = 0, -3, -(1/2) \pm j(\sqrt{3}/2)$. The pole–zero plot is given in Fig 5.7.1. Zero at infinity is not shown. ∎

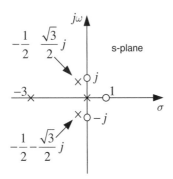

Fig. 5.7.1 Pole–zero plot, Example 5.7.1

Rational transform functions to differential equations: If we *ignore* the *initial conditions*, then

$$\frac{d^{\,k}x(t)}{dt^k} \xrightarrow{\text{LT}} s^k X(s).$$

This result can be used to determine the differential equation corresponding to the equation in (5.7.3). Cross multiplying terms in (5.7.3) results in

$$\left[s^N + d_1 s^{N-1} + \cdots + d_N\right] Y(s)$$
$$= \left[n_0 s^M + n_1 s^{M-1} + \cdots + n_M\right] X(s). \quad (5.7.4)$$

Now replace $s^k Y(s)$ by $\frac{d y^k(t)}{dt^k}$ and $s^k X(s)$ by $\frac{d^k x(t)}{dt^k}$ in (5.8.4) resulting in

$$\frac{d^N y}{dt^N} + d_1 \frac{d^{N-1} y}{dt^{N-1}} + \ldots + d_N y$$
$$= n_0 \frac{d^M x}{dt^M} + n_1 \frac{d^{M-1} x}{dt^{M-1}} + \ldots + n_M x. \quad (5.7.5)$$

Note the correspondence between (5.7.4) and (5.7.5) and one can derive these one from the other by inspection.

5.7.2 Return to the Initial and Final Value Theorems and Their Use

The initial value theorem predicts the initial value $x(0^+)$ from its strictly proper part of the transform as

$$x(0^+) = \lim_{s \to \infty} s X(s). \quad (5.7.6)$$

If $X(s)$ is strictly not proper, i.e., if the degree of the numerator is greater than or equal to the degree of the denominator, then the initial value can be determined by first using long division and then using the *strictly proper part of the function*. Transforms can be used to find the initial value of the successive derivatives of a function whose transform is known. For example,

$$\lim_{t \to 0^+} x'(t) = \lim_{s \to \infty} s L\{x'(t)\} = \lim_{s \to \infty} s[s X(s) - x(0^+)]$$
$$= \lim_{s \to \infty} [s^2 X(s) - s x(0^+)] \quad (5.7.7a)$$

$$\lim_{t \to 0^+} x''(t) = \lim_{s \to \infty} s L\{x''(t)\}$$
$$= \lim_{s \to \infty} [s^3 X(s) - s^2 x(0^+) - s x(0^+)] \quad (5.7.7b)$$

provided the limits exist. In systems and control theory one of the interesting functions we deal with is the error function. The initial values of this function and its derivatives provide important measures in analysis and design.

Example 5.7.2 Find the values of $y(t)$ and $y'(t)$ at $t = 0^+$ given the function and verify the results by using the inverse transform.

$$Y(s) = \frac{1}{s^2 + s + 1} = \frac{1}{(s + (1/2))^2 + (\sqrt{3}/2)}$$

$$\overset{LT}{\longleftrightarrow} \frac{1}{(\sqrt{3}/2)} e^{-(1/2)t} \sin((\sqrt{3}/2)t)u(t). \quad (5.7.8)$$

Solution: By (5.7.6) and (5.8.7a), we have

$$y(0^+) = \lim_{s \to \infty} \frac{s}{s^2 + s + 1} = 0, y'(0^+)$$

$$= \lim_{s \to \infty} \left[\frac{s^2}{s^2 + s + 1} - 0 \right] = 1. \quad (5.7.9)$$

Verification:

$$y(t)|_{t=0^+} = \frac{1}{(\sqrt{3}/2)} e^{-(1/2)t} \sin((\sqrt{3}/2)t)u(t)|_{t=0^+} = 0$$

$$y'(t)|_{t=0^+} = \frac{2}{\sqrt{3}} \left\{ [e^{-(1/2)t}u(t)] \left(\frac{\sqrt{3}}{2}t \right) \cos\left(\frac{\sqrt{3}}{2}t \right)|_{t=0^+} \right.$$

$$\left. + \left[\sin\left(\frac{\sqrt{3}}{2}t \right) \frac{d(e^{-(1/2)t}u(t))}{dt} \right]|_{t=0^+} \right\} = 1$$

∎

Notes: In the above example the initial conditions for the given function and its derivative were calculated. If the initial conditions are desired for several derivatives, a better way of doing it is by long division. First assume $Y(s)$ has a zero at infinity. If it does not, (5.7.6) cannot be used as the limit does not exist. Assuming it has zero at infinity, by long division,

$$Y(s) = \frac{a_1}{s} + \frac{a_2}{s^2} + \frac{a_3}{s^3} + ... \quad (5.7.10)$$

From (5.8.4) and (5.8.5a), we have

$$y(0^+) = \lim_{s \to \infty} \left[a_1 + \frac{a_2}{s} + \frac{a_3}{s^2} + ... \right] = a_1, \quad (5.7.11a)$$

$$y'(0^+) = \lim_{s \to \infty} \left[a_1 s + a_2 + \frac{a_3}{s} + ... - a_1 s \right] = a_2, \quad (5.7.11b)$$

$$y''(0^+) = \lim_{s \to \infty} \left[a_1 s^2 + a_2 s + a_3 + ... - a_1 s^2 - a_2 s \right] \quad (5.7.11c)$$

$$= a_3.$$

In the above example, the function $Y(s)$ can be written by long division in the form

$$Y(s) = \frac{1}{s^2 + s + 1} = \frac{1}{s^2} - \frac{1}{s^3} + ...$$

Comparing the terms in this equation with (5.7.9), $a_1 = 0, a_2 = 1, a_3 = -1$. For additional discussion on this topic, see Close (1966). ∎

Example 5.7.3 Use the final value theorem to find $x(\infty)$.

$$X(s) = \frac{(s + a_1)}{s(s + b_1)(s^2 + a_2 s + b_2)}. \quad (5.7.12a)$$

Solution:

$$x(\infty) = \lim_{t \to \infty} x(t) = \lim_{s \to 0} sX(s) = a_1/b_1 b_2.$$

$$(5.7.12b) \quad ∎$$

Example 5.7.4 Find the initial values of the functions from their transforms given below:

a. $X_1(s) = \dfrac{K(s^2 + a_1)}{(s + b_1)(s^2 + b_2 s + b_3)}$,

$$(5.7.13)$$

b. $X_2(s) = \dfrac{(2s + 1)}{(s + 1)}$.

c. In Part b, verify the result using the inverse transform and then evaluate it at $t = \infty$.

Solution: a. Noting that the $X_1(s)$ is a strictly proper function, we have

$$x(0^+) = \lim_{t \to 0} x(t) = \lim_{s \to \infty} sX(s) = K. \quad (5.7.14)$$

b. Noting that $X_2(s)$ is not a strictly proper function, we first write

$$X_2(s) = 2 - 1/(s + 1) = 2 + Y(s),$$

$$(5.7.15)$$

$(Y(s) -$ a strictly proper function).

The strictly proper part gives the initial value of $x(t)$ and is given by

$$x(0^+) = \lim_{s \to \infty} s Y(s) = -1. \qquad (5.7.16)$$

c. $x(t)|_{t=0^+} = x(0^+) = L^{-1}\{Y(s)\}|_{t=0^+}$
$$= -e^{-t}u(t)|_{t=0^+} = -1.$$
$$\qquad (5.7.17) \blacksquare$$

Example 5.7.5 This example illustrates the final value theorem of the one-sided transform using generalized functions studied in Chapter 1. Consider the Laplace transform pair and find the final value of the function $x(t) = u(t)\cos(\omega_0 t)$ using (5.6.44).

Solution: In an ordinary sense, $\cos(\omega_0 t)$ has *no defined final value*. However, as a *generalized function* or as a *distribution function* (see Papoulis (1962)),

$$\lim_{t \to \infty} \cos(\omega_0 t) = 0 \to \lim_{t \to \infty} [\cos(\omega_0 t)u(t)]$$
$$\qquad\qquad (5.7.18) \blacksquare$$
$$= \lim_{s \to 0} \frac{s^2}{s^2 + \omega_0^2} = 0.$$

Finding the solution of a linear constant coefficient differential equation is an important application of Laplace transforms. The steps in this process are: 1. find the transform of the differential equation, 2. solve for the unknown transform function, and 3. find the inverse transform of this function.

5.8 Solutions of Constant Coefficient Differential Equations Using Laplace Transforms

Consider the constant coefficient differential equation given in (5.7.5) and is

$$\frac{d^N y}{dt^N} + d_1 \frac{d^{N-1} y}{dt^{N-1}} + \cdots + d_N y = n_0 \frac{d^M x}{dt^M} + n_1 \frac{d^{M-1} x}{dt^{M-1}}$$
$$+ \cdots + n_M x.$$
$$\qquad\qquad (5.8.1)$$

For future discussion, $x(t)$ and $y(t)$ are the input and the output of a system, and d's and n's are some constants. Note also the coefficient of $y^{(N)}(t)$

on the left side of the equation is taken as 1. Since $x(t)$ is assumed to be a known input, the right side of the equation in (5.8.1) is the *forcing function* $F(t)$.

$$\frac{d^n y}{dt^n} + d_1 \frac{d^{n-1} y}{dt^{n-1}} + \ldots + d_N y = F(t). \qquad (5.8.2)$$

If $F(t)$ is identically zero, this equation is reduced to the *homogeneous* equation

$$\frac{d^n y}{dt^n} + d_1 \frac{d^{n-1} y}{dt^{n-1}} + \ldots + d_N y = 0. \qquad (5.8.3)$$

The solution of this equation is the *zero input response* and also called the *natural response* or the *complementary solution identified by* $y_n(t)$ in the following. The general solution of the *non-homogeneous differential equation* in (5.8.1) has two parts

$$y(t) = y_n(t) + y_f(t). \qquad (5.8.4)$$

The term $y_f(t)$ satisfies (5.8.1) and is the *forced response*. $y_n(t)$ is the solution of the homogeneous equation in (5.8.3). This is the response when there is no input, i.e., the zero input response. Taking the Laplace transform of a constant coefficient differential equation introduces initial conditions into the picture, see (5.6.17).

Example 5.8.1 Solve the following differential equation:

$$\frac{d^2 y(t)}{dt^2} + 3\frac{dy(t)}{dt} + y(t) = x(t) + \frac{dx(t)}{dt}. \qquad (5.8.5)$$

Solution: Taking the Laplace transform of the above equation, we have

$$[s^2 Y(s) - sy(0^-) - y'(0^-)] + 3[sY(s) - y(0^-)]$$
$$+ 2Y(s) = X(s) + sX(s) - x(0^+). \qquad (5.8.6)$$

Equation (5.8.6) can be written in two parts on the right, one is due to the initial conditions on $y(t)$ and the other one is based on the input function.

$$s^2 Y(s) + 3sY(s) + 2Y(s)$$
$$= [sy(0^-) + y'(0^-)] + [X(s) + sX(s) - x(0^+)]. \qquad (5.8.7)$$

The next step is to solve for $Y(s) = Y_n(s) + X(s)$ using the equations

$$s^2 Y_n(s) + 3sY_n(s) + 2Y_n(s) = [sy(0^-) + y'(0^-)]$$
$$(5.8.8)$$

$$s^2 Y_f(s) + 3sY_f(s) + 2Y_f(s) = [X(s) + sX(s) - x(0^+)]$$
$$(5.8.9)$$

$$\Rightarrow Y_n(s) = \frac{1}{s^2 + 3s + 2}[sy(0^-) + y'(0^-)] \quad (5.8.10a)$$

$$Y_f(s) = \frac{1}{s^2 + 3s + 2}[X(s) + sX(s) - x(0^+)],$$
$$(5.8.10b)$$

$$Y(s) = Y_n(s) + Y_f(s). \quad (5.8.10c)$$

If all the initial conditions are zero, then from above, we have

$$Y(s) = \frac{s+1}{s^2 + 3s + 2}X(s) = H(s)X(s). \quad (5.8.10d) \blacksquare$$

Notes: In most applications the initial conditions on the input signal are assumed to be zero. That is $x^{(k)}(0^-) = 0$. The polynomial

$$s^n + d_1 s^{n-1} + \dots + d_N = 0 \quad (5.8.11)$$

is called the characteristic polynomial of the system. If all the initial conditions are zero and taking the Laplace transform on both sides of the equation in (5.8.1) gives the equation in (5.7.3) and is given below in a slightly different form:

$$Y(s) = H(s)X(s)$$
$$= \frac{n_0 s^M + n_1 s^{M-1} + \dots + n_M}{s^N + d_1 s^{N-1} + \dots + d_N}X(s), \quad (5.8.12)$$

$$H(s) = \frac{n_0 s^M + n_1 s^{M-1} + \dots + n_M}{s^N + d_1 s^{N-1} + \dots + d_N}. \quad (5.8.13)$$

The transform $H(s)$ is called the *transfer function* relating the input to the output in the transform domain. Clearly if the input function is an impulse function, then the output transform and its inverse are, respectively, given by

$$Y(s) = H(s)X(s) = H(s)L\{\delta(t)\}$$
$$= H(s) \xleftrightarrow{LT} h(t) = y(t). \quad (5.8.14)$$

$h(t)$ is called the *impulse response*, i.e., the response to the impulse input. The last step in the solution of the differential equation is finding the inverse Laplace transform of the transform function $Y(s)$:

$$y(t) = L^{-1}[Y_n(s)] + L^{-1}[Y_f(s)] = y_n(t) + y_f(t).$$
$$(5.8.15)$$

$y_n(t)$ depends on the characteristic polynomial and the initial conditions on $y(t)$. If all the initial conditions are zero, then the output is the zero-state response.

$$y(t) = L^{-1}[Y_f(s)], \, Y_f(s) = L[y_n(t)].$$

5.8.1 Inverse Laplace Transforms

The inverse transform is defined by

$$x(t) = L^{-1}[X(s)] = \frac{1}{2\pi j}\int_{\sigma-j\infty}^{\sigma+j\infty} X(s)e^{st}dt. \quad (5.8.16)$$

Evaluation of (5.9.8) involves complex integration and is accomplished by contour integration. It requires understanding of complex variable theory, especially Cauchy integral theorem, which involves the residues of the complex function $X(s)e^{st}$. For a discussion on this topic, see Chirlian (1973). This approach is beyond the scope of our study. In our discussion we will determine the inverse transforms of rational functions by simplifying the transform into a sum of simpler terms by using partial fraction expansion. We will then use transform tables to compute the corresponding time functions.

5.8.2 Partial Fraction Expansions

For our discussion, we will assume that the transform functions are real rational functions and have the form given in (5.7.3) corresponding to N poles. Furthermore, the poles are assumed to be known in factored form. If there are common factors between the numerator and the denominator polynomials, they are deleted before we start the expansion. In addition we will assume that if $N \geq M$, then by long division we can express $Y(s)$ as a sum of a polynomial and a *proper rational function*:

$$X(s) = A(s) + X_p(s),$$

$$A(s) = \sum_{i=0}^{N-M} \alpha_i s^i, X_p(s) - \text{a proper function.} \quad (5.8.17)$$

The inverse transform of $L^{-1}\{Y(s)\} = L^{-1}\{A(s)\} + L^{-1}\{Y_p(s)\}$. If the numerator of $Y(s)$ has exponentials of the form $e^{-s\tau}$, they indicate time delays. First find the inverse transform without this factor and then use the property $L[y(t-\tau)] = Y(s)e^{-s\tau}$. Partial fraction expansion is simpler if the poles are simple. This is considered first.

Case of distinct poles: If $X_p(s)$ has only simple poles (i.e., $p_i \neq p_j, j \neq i$) and is a proper function. Then the function can be expressed in terms of a partial fraction expansion

$$Y_p(s) = \frac{N(s)}{(s+p_1)(s+p_2)...(s+p_N)}$$

$$= \frac{A_1}{(s+p_1)} + \frac{A_2}{(s+p_2)} + \cdots$$

$$+ \frac{A_k}{(s+p_k)} + \cdots + \frac{A_N}{(s+p_N)}.$$

The constants $A'_k s$ are called the residues. These can be evaluated by multiplying the last equation by $(s+p_k)$ and evaluating both sides of the equation at the poles $s = -p_k$.

$$(s+p_k)Y_p(s)\big|_{s=-p_k} = \sum_{i=1 i \neq k}^{N} \frac{A_i(s+p_k)}{(s+p_i)}\big|_{s=-p_k} + A_k = A_k. \quad (5.8.18)$$

Since we assumed the function is a real rational function, the residue A_k is real if the corresponding pole $s = -p_k$ is real. In the case of a complex pole, the corresponding residue A_k will be complex. In addition, since the function is a function with real coefficients, the complex roots of a real polynomial appear as complex conjugates. The corresponding residues will also be complex conjugates. That is, we need to compute only one residue for each complex pole. That is, if the residue corresponding to the pole $s = -\alpha_k - j\beta_k$ is given by $r_k = A_k + jB_k$, then

the residue corresponding to the pole $s = -\alpha_k + j\beta_k$ is given by $r_k^* = A_k - jB_k$.

Case of multiple poles: Consider the function with a simple pole and a pole at $s = -\rho$ of order k.

$$X_p(s) = \frac{N(s)}{(s+p_1)(s+\rho)^k}. \quad (5.8.19)$$

Partial fraction expansion of such a function can be written as

$$X_p(s); = \frac{K_1}{(s+p_1)} + \frac{A_0}{(s+\rho)^k} + \frac{A_1}{(s+\rho)^{k-1}} + \cdots$$

$$+ \frac{A_{k-1}}{(s+\rho)}. \quad (5.8.20)$$

The subscripts l on the coefficients A_l are in ascending order, the corresponding powers $(k-l)$ on $(s+\rho)$ are in descending order and they add up to k. The coefficient K_1, corresponding to the simple pole can be determined as before by

$$K_1 = (s+p_1)X_p(s)\big|_{s=-p_1}. \quad (5.8.21)$$

The coefficients $A_i, i = 0, 1, 2$ can be computed by the following procedure:

$$A_0 = (s+\rho)^k X_p(s)\big|_{s=-\rho},$$

$$A_1 = \frac{d[(s+\rho)^k X(s)]}{ds}\big|s = -\rho,$$

$$A_2 = \frac{1}{2!}\frac{d^2[(S+\rho)^k X_p(s)]}{ds^2}\big|_{s=-\rho}.$$

In general,

$$A_m = \frac{1}{m!}\frac{d^m[(s+\rho)X_p(s)]}{ds^m}, m = 0, 1, 2, \ldots, k-1. \quad (5.8.22)$$

Example 5.8.2 Using the above formulas find the partial fraction expansion of

$$Y_p(s) = \frac{(s+2)}{s(s+1)^2}.$$

Solution: Partial fraction expansion of this function is

$$Y_p(s) = \frac{K_1}{s} + \frac{A_0}{(s+1)^2} + \frac{A_1}{(s+1)}, \qquad (5.8.23)$$

$$\Rightarrow K_1 = sX(s)|_{s=0} = \frac{(s+2)}{(s+1)}\Big|_{s=0} = 2,$$

$$A_0 = (s+1)^2 X(s) = \frac{(s+2)}{s}\Big|_{s=-1} = -1,$$

$$A_1 = \frac{1}{1!}\frac{d(s+1)^2 X(s)}{ds}\Big|_{s=-1} = \frac{d[(s+2)/s]}{ds}\Big|_{s=-1}$$

$$= \left[\frac{s-(s+2)}{s^2}\right]\Big|_{s=-1} = -2 \Rightarrow Y_p(s)$$

$$= \frac{2}{s} + \frac{-1}{(s+1)^2} + \frac{-2}{(s+1)}. \qquad (5.8.24) \ \blacksquare$$

Notes: The above procedure is tedious. A simpler way to achieve the partial fraction expansion is by repeated application of the simple pole case. Consider $Y_p(s)$ and write

$$X_p(s) = \frac{1}{(s+1)}\left[\frac{(s+2)}{s(s+1)}\right]. \qquad (5.8.25a)$$

The term inside the bracket can be expanded by the partial fraction expansion and

$$\frac{(s+2)}{s(s+1)} = \frac{A_1}{s} + \frac{A_2}{(s+1)}, A_1 = 2, A_2 = -1.$$

Using this expansion in (5.8.22) and using the partial fraction expansion again, we have

$$X_p(s) = \frac{2}{s(s+1)} + \frac{-1}{(s+1)^2}$$

$$= \frac{2}{s} + \frac{(-2)}{(s+1)} + \frac{(-1)}{(s+1)^2}. \qquad (5.9.25b) \ \blacksquare$$

MATLAB uses the "residue," a routine that determines the partial fraction expansion of a function $Y_1(s)$ that can have multiple real or complex poles.

$$Y_1(s) = \frac{n_0 s^{M_1} + n_1 s^{M_1-1} + \dots + n_{M_1}}{s^N + d_1 s^{N-1} + \dots + d_N}$$

$$= Y_0(s) + Y_p(s), \ Y_0(s) \qquad (5.8.26a)$$

$$= \sum_{n=0}^{N-M_1} k_n s^n, N \geq M_1,$$

$$Y_p(s) = \frac{N(s)}{D(s)} = \frac{n_0 s^{M_1} + n_1 s^{M_1-1} + \dots + n_{M_1}}{s^N + d_1 s^{N-1} + \dots + d_N}, M_1 < N,$$

$$(5.8.26b)$$

$$Y_p(s) = \left[\frac{c_{11}}{(s-p_1)} + \frac{c_{12}}{(s-p_1)^2} + \dots + \frac{c_{1,m_1}}{(s-p_1)^{m1}}\right]$$

$$+ \left[\frac{c_{2,1}}{(s-p_2)} + \frac{c_{2,2}}{(s-p_2)^2} + \dots + \frac{c_{2,m_2}}{(s-p_2)^{m_2}}\right]$$

$$+ \left[\frac{c_{r,1}}{(s-p_r)} + \frac{c_{r2}}{(s-p_r)^2} + \dots + \frac{c_{1,m_r}}{(s-p_r)^{m_r}}\right]$$

$$(5.8.26c)$$

$Y_1(s)$ can have multiple real or complex poles. The routine first determines $Y_0(s)$. It then computes the partial fraction expansion coefficients and the poles corresponding to the strictly proper function $X_p(s)$ given in (5.4.11). The MATLAB statement for computing the partial fraction expansion of (5.8.26a) is

$$B = [n_0, n_1, \dots, n_{M_1}]; A = [1, d_1, d_2, \dots, d_N];$$
$$[r, p, k] = \text{residue}(B, A). \qquad (5.8.27)$$

There are three output vectors: $-r, p$, and k. The vector r contains the coefficients $c_{i,1}, c_{i,2}, \dots, c_{i,m_i}$, $i = 1, 2, \dots, r$; the vector p contains the values of poles $p_i, i = 1, 2, \dots, r$; and the vector k contains the entries $k_i, i = 0, 1, 2, \dots, (M_1 - N)$.

The inverse transforms of each of the terms can be found from tables. We generally do not come across multiple complex poles. For a transform table, see McCollum and Brown (1965). A short list of transforms and their inverses are listed below. (Table 5.8.1). $\blacksquare$

Example 5.8.3 Consider the differential equation with $d_1 = 1, d_0 = 1, b_1 = 1, b_0 = 1$.

Table 5.8.1 Typical rational replace transforms and their inverses

$$X(s) \xleftrightarrow{\text{LT}} x(t),$$

$$\frac{A}{(s+a)} \xleftrightarrow{\text{LT}} Ae^{-at}u(t), \qquad\qquad\qquad\qquad\qquad\qquad (5.8.28)$$

$$\frac{A}{(s+a)^n} \xleftrightarrow{\text{LT}} \frac{A}{(n-1)!}t^{n-1}e^{-at}u(t), \qquad\qquad\qquad\qquad (5.8.29)$$

$$\frac{As+B}{(s+\alpha)^2 + \beta^2} \xleftrightarrow{\text{LT}} e^{-\alpha t}\left[A\cos(\beta t) + \frac{B - \alpha A}{\beta}\sin(\beta t)\right]u(t), \qquad (5.8.30)$$

$$\left[\frac{A+jB}{s+\alpha+j\beta} + \frac{A-jB}{s+\alpha-j\beta}\right] \xleftrightarrow{\text{LT}} 2e^{\alpha t}[A\cos(\beta t) + B\sin(\beta t)]u(t), \qquad (5.8.31a)$$

$$\left[\frac{Me^{j\theta}}{s+\alpha+j\beta} + \frac{Me^{-j\theta}}{s+\alpha-j\beta}\right] \xleftrightarrow{\text{LT}} 2Me^{-\alpha t}\cos(\beta t - \theta)u(t),\ M = \sqrt{A^2 + B^2}, \theta = \tan^{-1}(B/A). \qquad (5.8.31b)$$

$$\frac{d^2y(t)}{dt^2} + d_1\frac{dy(t)}{dt} + d_0y(t) = b_1\frac{dx(t)}{dt} \qquad (5.8.32)$$
$$+ b_0x(t), x(t) = u(t),$$

$$y(0^-) = 1 \text{ and } dy(t)/dt|_{t=0^-} = 1. \qquad (5.8.33)$$

$$\Rightarrow Y_f(s) = \frac{(s+1)}{s(s^2 + s + 1)} = \frac{1}{s} - \frac{s}{s^2 + s + 1},$$

$$Y_n(s) = \left[\frac{(s+2)}{(s^2 + s + 1)}\right],$$

$$Y(s) = \frac{1}{s} + \frac{1}{s^2 + s + 1}. \qquad (5.8.35)$$

a. Using Laplace transforms determine $Y(s)$ and identify the corresponding transforms of the natural and forced responses.

b. Determine the natural and forced responses.

c. Use the initial and the final value theorems of the Laplace transforms to verify the initial and final values determined in part *b*.

Solution: *a.* Taking the transform on both sides of the equation in (5.8.32) results in

$$s^2 Y(s) - sy(0^-) - \frac{dy(t)}{dt}\bigg|_{t=0^-} + sY(s) - \frac{dy(t)}{dt}\bigg|_{t=0^-}$$
$$+ Y(s) = (s+1)X(s),$$

$$(s^2 + s + 1)Y(s) - [sy(0^-) + y'(0^-)]$$
$$- y(0^-) = (s+1)X(s),$$

$$Y(s) = \left[\frac{(s+1)}{(s^2 + s + 1)}\right]X(s)$$
$$+ \left[\frac{sy(0^-) + y'(0^-) + d_1y(0^-)}{(s^2 + s + 1)}\right] \qquad (5.8.34)$$
$$= Y_f(s) + Y_n(s),$$

In dealing with real polynomials, the complex roots appear as a pair of complex-conjugate roots. If there is only a pair of complex poles and the other poles are real, then it is simpler to compute the partial fraction expansion for the real poles first and then subtract that portion from the given function, which results in the partial fraction expansion for the pair of complex poles. This is what was done to compute the partial fraction expansion for $Y_f(s)$. $Y_n(s)$ has the appropriate form to find the inverse transform from a table of Laplace transforms.

b. The respective time responses are given by

$$y_f(t) = u(t) - e^{-(1/2)t}[\cos(\sqrt{(3/4)}t)]u(t), \qquad (5.8.36a)$$

$$y_n(t) = e^{-(1/2)t}\left[\cos(\sqrt{(3/4)}) + \frac{1.5}{\sqrt{3/4}}\sin(\sqrt{(3/4)})\right]u(t), \qquad (5.8.36b)$$

$$y(t) = y_f(t) + y_n(t). \qquad (5.8.36c)$$

c. From the initial and the value theorems, it follows that

$$y_f(0^+) = \lim_{s \to \infty} [sY_f(s)] = 0, y_n(0^+)$$

$$= \lim_{s \to \infty} [sY_n(s)] = 1, y(0^+) \quad (5.8.37a)$$

$$= \lim_{s \to \infty} [sY(s)] = 1,$$

$$\lim_{t \to \infty} y_f(t) = \lim_{s \to 0} [sY_f(s)] = 0, \lim_{t \to \infty} y_n(t)$$

$$= \lim_{s \to 0} [sY_n(s)] = 1, \lim_{t \to \infty} y(t) \quad (5.8.37b) \blacksquare$$

$$= \lim_{s \to 0} [sY(s)] = 1.$$

The transform functions are assumed to be rational and strictly proper. If they are not, then the inverse transforms will have impulses or the derivatives of the impulses.

Example 5.8.4 Find

$$L^{-1}\{Y(s)\} = L^{-1}\left(\frac{s^2 + s + 1}{(s + 1)}\right). \quad (5.8.38a)$$

Solution: The function is improper. First divide the numerator by the denominator, i.e.,

$$Y(s) = s + [1/(s + 1)], \quad (5.8.38b)$$

$$\Rightarrow y(t) = \frac{d\delta(t)}{dt} + e^{-t}u(t). \quad (5.8.38c) \blacksquare$$

It is not uncommon to have $e^{-s\tau}$ terms in the transform functions. See Examples 5.6.5 and 5.6.6. If e^{-sT} terms are in the denominator, they indicate the existence of a switching function. First, separate the switching part of the transform from the pulse transform. Find the inverse transforms of the pulse transform and the switching function. Then combine these results to find the inverse transform of the function. As an example, let $L\{x(t)\} = X(s)$. Then

$$Y(s) = X(s)\frac{1}{(1 - e^{-sT})}$$

$$= X(s)\sum_{n=0}^{\infty} e^{-nsT} \xrightarrow{LT} \sum_{n=0}^{\infty} x(t - nT) = y(t).$$

Example 5.8.5 Find the inverse Laplace transform of this function

$$Y(s) = \frac{(3s + 4)e^{-s} + (2s + 3)}{(s + 1)(s + 2)}. \quad (5.8.39)$$

Solution: First,

$$Y(s) = Y_1(s)e^{-s} + Y_2(s),$$

$$Y_1(s) = \frac{3s + 4}{(s + 1)(s + 2)}, y_1(t - 1) \xrightarrow{LT} Y_1(s)e^{-s},$$

$$Y_2(s) = \frac{2s + 3}{(s + 1)(s + 2)}. \quad (5.8.40)$$

Expanding $Y_i(s), i = 1, 2$ by partial fraction expansions and then finding inverses results in

$$Y_1(s) = \frac{1}{(s + 1)} + \frac{2}{(s + 2)}, Y_2(s) = \frac{1}{(s + 1)} + \frac{1}{(s + 2)},$$

$$y_1(t) = (e^{-t} + 2e^{-2t})u(t), y_2(t) = (e^{-t} + e^{-2t})u(t).$$

Noting that $y_1(t - 1) \xrightarrow{LT} Y_1(s)e^{-s}$, the time function is

$$y(t) = y_1(t - 1) + y_2(t) = (e^{-(t-1)} + 2e^{-2(t-1)})$$

$$\times u(t - 1) + (e^{-t} + e^{-2t})u(t). \quad (5.8.41) \blacksquare$$

5.9 Relationship Between Laplace Transforms and Other Transforms

We have developed Fourier transforms starting with Fourier series. Fourier series is a sum of weighted harmonics and the transform describes a signal as the integral of weighted harmonics. The Fourier transform has some drawbacks. The Fourier transform cannot handle exponentially growing signals. Furthermore it cannot handle initial conditions. The Laplace transform tries to overcome these problems by introducing a convergence factor $e^{-\sigma t}$. That is $x(t)$ is replaced by $[x(t)e^{-\sigma t}]$ in the Fourier transform integral, thus allowing for existence of Laplace transforms of functions that do not have Fourier transforms. Neither can handle functions such as $e^{\alpha t^2}$. Fourier transforms and Laplace transforms are related if the regions of convergence include the imaginary axis.

$$e^{-at}u(t) \xleftrightarrow{\text{FT}} \frac{1}{(1+j\omega)}, a>0,$$

$$x(t) = e^{-at}u(t) \xleftrightarrow{\text{LT}} \frac{1}{(s+a)}, \text{ROC} - \sigma > -a,$$

$$(5.9.1)$$

$$e^{-a|t|} \xleftrightarrow{\text{FT}} \frac{2a}{a^2 + \omega^2}, a>0,$$

$$e^{-a|t|} \xleftrightarrow{\text{LT}_{\text{II}}} \frac{2a}{(s^2 - a^2)}, a>0, \text{ROC}: -a<\sigma<a.$$

$$(5.9.2)$$

In the above cases, the one-sided Laplace transform exists if the region of convergence includes the imaginary axis, i.e., the $j\omega$-axis. The Fourier transform can be obtained by simply replacing the Laplace transform variable s by $j\omega$. Note that the Laplace transform functions in these examples do *not* have any poles on the *imaginary axis*. Consider the following two interesting cases: one is a unit step function and other one is a one-sided cosine function. The Fourier and the Laplace transforms of these two functions are given below:

$$u(t) \xleftrightarrow{\text{FT}} \pi\delta(\omega) + \frac{1}{j\omega}, \quad u(t) \xleftrightarrow{\text{LT}} \frac{1}{s} \quad (5.9.3)$$

$$u(t)\cos(\omega_c t) \xleftrightarrow{\text{FT}} \frac{j\omega}{(\omega_c^2 - \omega^2)} + \frac{\pi}{2}[\delta(\omega - \omega_c)$$
$$+ \delta(\omega + \omega_c)] \quad (5.9.4a)$$

$$u(t)\cos(\omega_c t) \xleftrightarrow{\text{LT}} \frac{s}{s^2 + \omega_c^2}. \quad (5.9.4b)$$

If the Laplace transforms of the functions have *simple poles on the imaginary axis*, then the Fourier transforms of these exist in the limiting sense and they have impulses.

5.9.1 Laplace Transforms and Fourier Transforms

We now consider finding the Fourier transforms of functions from their *one-sided Laplace transforms*.

The Laplace transforms of such functions can be expressed by

$$X(s) = X_0(s) + \sum_n \frac{r_n}{s - j\omega_n}. \quad (5.9.5)$$

In (5.9.5) $X_0(s)$ represents the portion of $X(s)$ with *no poles on the imaginary axis* and the second term represents only the terms in $X(s)$ with *simple poles on the imaginary axis*, each of these poles occurs at a frequency ω_n. The corresponding F- transform is

$$X(j\omega) = X(s)\Big|_{s=j\omega} + \pi \sum_n r_n \delta(\omega - \omega_n). \quad (5.9.6)$$

The above equivalence is *valid if the poles on the imaginary axis are of first order*.

Relationship between the Fourier transforms and Laplace transforms: Either one of these transforms is not a generalization of the other. The F-transform of a causal function $x(t)$ can be obtained from the one-sided Laplace transform of this function with s replaced by $j\omega$ with the following constraints:

1. The signal is causal, i.e., $x(t) = 0, t<0$
2. The signal is absolutely integrable. This implies that the region of convergence includes the imaginary axis.
3. Fourier transforms can also be obtained from Laplace transforms for causal signals that are *not* absolutely integrable, such as the unit step function by including impulse functions in the Fourier transform. Laplace transforms of the causal signals can be obtained from the non-impulsive part of the F-transform.
4. The operational properties of the Fourier and Laplace transforms are very similar in most cases; in some cases, they are quite different. For example, superposition, shifting, scaling, convolution, and products of functions have similar relationships. The derivative property needs to be modified from the Fourier to Laplace transform to include the initial conditions. The integrands of the Fourier transform and its inverse essentially have the same structure and thus allow for the symmetry or duality properties. This makes it very nice in the digital

Table 5.9.1 One sided Laplace transforms and Fourier transforms

Sequence	Laplace Transform	Fourier Transform
$\delta(t)$	1	1
$e^{-at}u(t);\ a > 0.$	$\dfrac{1}{(s+a)}$	$\dfrac{1}{(j\omega + a)}$
$te^{-at}u(t);\ a > 0.$	$\dfrac{1}{(s+a)^2}$	$\dfrac{1}{(j\omega + a)^2}$
$e^{-at}\cos(bt)u(t);\ a > 0.$	$\dfrac{s+a}{(s+a)^2 + b^2}$	$\dfrac{j\omega + a}{(j\omega + a)^2 + b^2}$
$e^{-at}\sin(bt)u(t);\ a > 0.$	$\dfrac{b}{(s+a)^2 + b^2}$	$\dfrac{b}{(j\omega + a)^2 + b^2}$
$u(t)$	$\dfrac{1}{s}$	$\dfrac{1}{j\omega} + \pi\delta(\omega)$
$\cos(bt)u(t)$	$\dfrac{s}{s^2 + \omega^2}$	$\dfrac{j\omega}{b^2 - \omega^2} + \pi[\delta(\omega + b) + \delta(\omega - b)]$
$\sin(bt)u(t)$	$\dfrac{b}{s^2 + \omega^2}$	$\dfrac{b}{b^2 - \omega^2} + j\pi[\delta(\omega + b) - \delta(\omega - b)]$

computation of transforms and their inverses. That is, one algorithm can be used to compute both the forward and inverse Fourier transforms with few modifications. There is no symmetry property of Laplace transforms.

5. Noting that $X(s)$ is a function of $s = \sigma + j\omega$, a complex quantity, it can only be plotted as a surface plot. On the other hand, the Fourier transform is a function of $j\omega$ and therefore it is the cross section of the surface plot along the $j\omega$-axis.

6. Noting Item 5 above, the circuits and systems literature use the Laplace transform to compute the frequency characteristics of the function by simply substituting $s = j\omega$ in the Laplace transform. The complex function is generally written in terms of the magnitude and phase frequency characteristics of the signal.

5.9.2 Hartley Transforms and Laplace Transforms

Hartley transform is the symmetrical form of the Fourier transform. It can be derived from the one-sided Laplace transforms. A few examples are given below for the Laplace transform functions $X(s)$ with poles in the left half-plane only and later with poles in the left half-plane and with poles on the imaginary axis.

Example 5.9.1 Find the Hartley transform of $x(t)$ using its Laplace transform.

$$x(t) = e^{-at}u(t) \xrightarrow{LT} \frac{1}{(s+a)} \to \frac{1}{j\omega + a}$$

$$= \text{Re}\left[\frac{1}{j\omega + a}\right] + j\,\text{Im}\left[\frac{1}{j\omega + a}\right].$$

Solution: The Hartley transform of $x(t)$ can be obtained from

$$x(t) \xleftrightarrow{Hart} \text{Re}\left[\frac{1}{j\omega + a}\right] - \text{Im}\left[\frac{1}{j\omega + a}\right] = \left[\frac{a + \omega}{\omega^2 + a^2}\right].$$

(5.9.7) ∎

Example 5.9.2 Find the Hartley transform of the function $u(t)$ from its Laplace transform.

$$x(t) = u(t) \xleftrightarrow{LT} X(s) \Rightarrow X(j\omega) = \pi\delta(\omega) - (j/\omega).$$

(5.9.8)

Solution: The Hartley transform of the function is

$$X_H(\omega) = \pi\delta[\omega] + (1/\omega). \qquad (5.9.9) \quad\blacksquare$$

See the chapter by Olejniczak in Poularikis, ed. (1996) for an extensive discussion on the relationship between the Laplace and Hartley transforms.

5.10 Hilbert Transform

Another transform that is closely related to the Fourier transform is the Hilbert transform. It is used in the theoretical descriptions and implementations of analog and digital Hilbert transformers. A device called the Hilbert transformer is basic and has important applications in single sideband modulation of signals and in digital signal processing. Hilbert transforms can be introduced with Euler's formula $e^{j\omega t} = \cos(\omega t) + j\sin(\omega t)$. We will see shortly that the Hilbert transform of $\cos(\omega t)$ is $\sin(\omega t)$. Hilbert transforms became an important area with *analytic signals* that are complex valued with one-sided spectrum. These have the form $x_a(t) = x(t) + j\hat{x}(t)$, where $\hat{x}(t)$ is the Hilbert transform of $x(t)$. Analytic signals are considered in Section 5.10.3. Also, the real and imaginary parts of transfer functions of systems are tied together by Hilbert transforms.

5.10.1 Basic Definitions

There are two ways of introducing the Hilbert transforms. One is by using an integral and the other by using the Fourier transform of the function. It is simpler to view it starting with the transform and derive the integral that defines the Hilbert transform. To start with assume $x(t)$ is the input and $y(t)$ is the output and $F[x(t)] = X(j\omega)$ and $F[y(t)] = Y(j\omega)$. The output transform $Y(j\omega)$ is assumed to be related to the input transform by

$$Y(j\omega) = H(j\omega)X(j\omega). \quad (5.10.1)$$

The function $H(j\omega)$ is called the Hilbert transformer and is defined by

$$H(j\omega) = -j\,\mathrm{sgn}(\omega) = \begin{cases} e^{-j\pi/2}, & \omega > 0 \\ e^{j\pi/2}, & \omega < 0 \end{cases}, \quad (5.10.2)$$

$$h(t) \xleftrightarrow{\text{FT}} H(j\omega).$$

Noting that multiplication in the frequency domain corresponds to the time-domain convolution, we have

$$y(t) = h(t) * x(t), h(t) = F^{-1}[H(j\omega)]. \quad (5.10.3)$$

From Chapter 4, $F[\mathrm{sgn}(t)] = 2/(j\omega)$. See (4.4.36). Using the symmetry or the duality property of the Fourier transforms, it follows that

$$h(t) = \frac{1}{\pi t}. \quad (5.10.4)$$

Using the convolution integral and (5.10.4) results in

$$y(t) = h(t) * x(t) = \int_{-\infty}^{\infty} x(\alpha)h(t - \alpha)d\alpha \overset{\Delta}{=} \hat{x}(t). \quad (5.10.5)$$

$\hat{x}(t)$ is the *Hilbert transform* of the function $x(t)$. Note the hat in $\hat{x}(t)$. Hilbert transform is a convolution operation and is a *function of time*. This is symbolically represented by

$$x(t) \xleftrightarrow{\text{HT}} \hat{x}(t) = H[x(t)] = y(t) \xleftrightarrow{\text{HT}} Y(j\omega)$$
$$= H(j\omega)X(j\omega), H(j\omega) = -j\,\mathrm{sgn}(\omega). \quad (5.10.6)$$

Hilbert transforms can be computed directly by the convolution in (5.10.5) or by using the transforms in (5.10.6). If $\hat{x}(t)$ is known, how do we compute $x(t)$ from $\hat{x}(t)$, if $\hat{x}(t)$ is *not* identically zero? It turns out that the Hilbert transform of $x(t)$ is equal to $[-x(t)]$. This can be seen from

$$\hat{\hat{x}}(t) = [x(t) * h(t) * h(t)] \xleftrightarrow{\text{FT}} H(j\omega)H(j\omega)X(j\omega)$$
$$= (-j\,\mathrm{sgn}(\omega))^2 X(j\omega) = -X(j\omega). \quad (5.10.7)$$

It implies that *if* $\hat{x}(t) \neq 0$ then we have the *inversion formula*

$$F[\hat{\hat{x}}(t)] = -X(j\omega) \rightarrow \hat{\hat{x}}(t)$$
$$= F^{-1}[-X(j\omega)] = -x(t). \quad (5.10.8)$$

Example 5.10.1 Find their Hilbert transforms of the following functions:

a. $x_1(t) = e^{-j\omega_0 t}, a_0 > 0,$ *b.* $x(t) = \cos(\omega_0 t),$

c. $y(t) = \sin(\omega_0 t), d. x_4(t) = A,$ a constant

Solution: *a.* The Fourier transforms of these functions are given by

$$F[e^{\pm j\omega_0 t}] = 2\pi\ \delta(\omega \mp \omega_0). \qquad (5.10.9)$$

Figure 5.10.1a,b gives the Fourier transforms of the functions in (5.10.9). Figure 5.10.1c gives $H(j\omega) = -j\ \text{sgn}(\omega)$. It follows that

$$[H(j\omega)(2\pi\delta(\omega + \omega_0))]$$
$$= -j\ 2\pi\text{sgn}(\omega)\delta(\omega + \omega_0) = j2\pi\delta(\omega + \omega_0). \qquad (5.10.10)$$

From (5.10.10),

$$\hat{x}(t) = F^{-1}[j2\pi\delta(\omega + \omega_0)]$$
$$= je^{-j\omega_0 t} = e^{-j(\omega_0 t - (\pi/2))}. \qquad (5.10.11)$$

Changing the sign in the exponent is a minor matter and we have

$$[e^{\pm j\omega_0 t}] \overset{\text{HT}}{\longleftrightarrow} \left[e^{\pm j(\omega_0 t - (\pi/2))}\right] \qquad (5.10.12)$$

b. The Hilbert transform of the cosine function can be obtained by using Euler's formula

$$\cos(\omega_0 t) = \left[\frac{1}{2}e^{j\omega_0 t} + \frac{1}{2}e^{-j\omega_0 t}\right]$$
$$\overset{\text{HT}}{\longleftrightarrow} \left[\frac{1}{2}e^{j(\omega_0 t - (\pi/2))} + \frac{1}{2}e^{-j(\omega_0 t - (\pi/2))}\right]$$
$$= \cos(\omega_0 t - (\pi/2)).$$

This can be simplified and the corresponding Hilbert transform pair is

$$\cos(\omega_0 t) \overset{\text{HT}}{\longleftrightarrow} \sin(\omega_0 t). \qquad (5.10.13)$$

Hilbert transform operation is an integral operation and the Hilbert transform of a sum is equal to the sum of the Hilbert transforms.

c. We can repeat the above process and show that

$$\sin(\omega_0 t) \overset{\text{HT}}{\longleftrightarrow} \sin(\omega_0 t - (\pi/2)) = -\cos(\omega_0 t). \qquad (5.10.14)$$

Note also

$$x(t) \overset{\text{HT}}{\longleftrightarrow} \hat{x}(t), \hat{x}(t) \overset{\text{HT}}{\longleftrightarrow} -x(t), \hat{x}(t) \neq 0. \qquad (5.10.15)$$

From the above we note that the Hilbert transform of a sine or a cosine function can be obtained by adding a phase shift of $-(\pi/2)$.

d. In the case of a constant, the transform is an impulse function. Note that $\text{sgn}(\omega)|_{\omega=0} = 0$. That is, $H(j\omega)|_{\omega=0} = 0$ and $Y(j\omega)|_{\omega=0} = 0$ in (5.10.1). Since the Fourier transform of a constant is an impulse function, it follows that the Hilbert transform of a constant is zero.　∎

We can generalize the above results and state that any periodic function with zero average value can be written in terms of Fourier cosine and sine series and therefore the Hilbert transform of such a periodic function $x_T(t)$ is

$$x_T(t) \overset{\text{HT}}{\longleftrightarrow} x_T\left(t - \frac{\pi}{2}\right). \qquad (5.10.16)$$

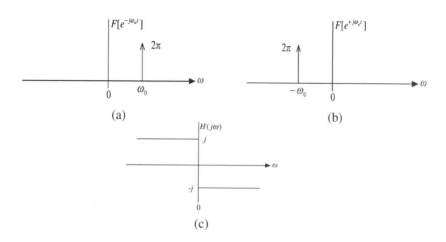

Fig. 5.10.1 (a)$F[e^{-j\omega_0 t}]$,
(b) $F[e^{j\omega_0 t}]$,
(c) $H(j\omega) = -j\text{sgn}(\omega)$

In the next example we will make use of the integral to compute the Hilbert transform.

Example 5.10.2 Find the Hilbert transform of the following functions:

$$a.\ x(t) = \Pi[t/\tau],\quad b.\ x_1(t) = x(t - \tau/2),\quad c.\ y(t) = A.$$
$$(5.10.17)$$

Solution: a. Using the integral expression, the Hilbert transform is given by

$$\hat{x}(t) = \int_{-\infty}^{\infty} \frac{x(\alpha)}{\pi(t-\alpha)}d\alpha = \frac{1}{\pi}\int_{-\tau/2}^{\tau/2}\frac{d\alpha}{(t-\alpha)}$$

$$= -\frac{1}{\pi}[\ln(t-\tau/2) - \ln(t+\tau/2)] = \frac{1}{\pi}\ln\left|\frac{t+\tau/2}{t-\tau/2}\right|$$
$$(5.10.18)$$

The pulse and its Hilbert transform are shown in Fig. 5.10.2. The pulse we started with is time limited, whereas the time-width of the Hilbert transform pulse is infinite.

b. Hilbert transform of the delayed pulse can be obtained by changing the variable of integration in (5.10.18) by $\beta = t - \tau/2$ and following the above procedure results in

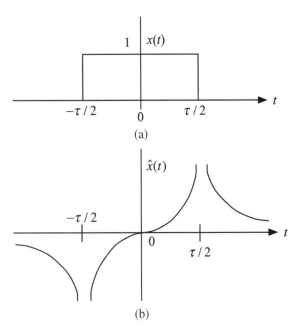

Fig. 5.10.2 (a) Pulse function and (b) Hilbert transform of the pulse

$$\Pi\left[\frac{t-(\tau/2)}{\tau}\right] \xrightarrow{\text{HT}} \frac{1}{\pi}\ln\left|\frac{t}{t-\tau}\right|. \qquad (5.10.19)$$

c. Hilbert transform of a constant is zero. This can be seen from (5.6.18) at the limit $\tau \to \infty$, as $\ln(1) = 0$, which verifies our earlier result. ∎

Example 5.10.3 Show that a. the energy (or the power) in an energy signal (or a power signal) $x(t)$ and its Hilbert transform $\hat{x}(t)$ are equal and b. the signal and its Hilbert transform are orthogonal.

Solution: The results are shown for energy signals and the results for power signals are left as exercises. a. The energies in the two functions are given by

$$E_{\hat{x}} = \frac{1}{2\pi}\int_{-\infty}^{\infty}|F[\hat{x}(t)]|^2 d\omega$$

$$= \frac{1}{2\pi}\int_{-\infty}^{\infty}|-j\text{sgn}(\omega)|^2|X(j\omega)|^2 d\omega$$

$$= \frac{1}{2\pi}\int_{-\infty}^{\infty}|X(j\omega)|^2 d\omega = E_x. \qquad (5.10.20)$$

b. The two functions are orthogonal by the generalized Parseval's theorem.

$$\int_{-\infty}^{\infty}x(t)\hat{x}(t)dt = \frac{1}{2\pi}\int_{-\infty}^{\infty}X(j\omega)[F[\hat{x}(t)]]^* dt$$

$$= \frac{1}{2\pi}\int_{-\infty}^{\infty}j\text{sgn}(\omega)|X(j\omega)|^2 d\omega.$$
$$(5.10.21) ∎$$

5.10.2 Hilbert Transform of Signals with Non-overlapping Spectra

In Chapter 10 single-sided modulation schemes will be studied, where Hilbert transforms play an important role. Consider the signals $x(t)$ and $g(t)$ with their spectra defined by

$$x(t) \xrightarrow{\text{FT}} X(j\omega),\quad |X(j\omega)| = 0, |\omega| > W$$
$$(5.10.22)$$
$$g(t) \xrightarrow{\text{FT}} G(j\omega),\quad G(j\omega) = 0, |\omega| < W.$$

That is, $x(t)$ is a low-pass signal and $g(t)$ is a high-pass signal. Such is the case in the single sideband modulation. We state that

$$H[x(t)g(t)] = x(t)H[g(t)] = x(t)\hat{g}(t). \quad (5.10.23)$$

This can be proven by using the following steps:

1. Write the time function $x(t)g(t)$ using the transform convolution integral in the form and then using the Hilbert transform, we have

$$x(t)g(t) = \frac{1}{2\pi}\int_{-\infty}^{\infty} X(\omega - \alpha)G(\alpha)d\alpha, \quad (5.10.24)$$

$$H[x(t)g(t)] \overset{FT}{\longleftrightarrow} -j\mathrm{sgn}(\omega)$$
$$\left[\frac{1}{2\pi}\int_{-\infty}^{\infty} X(\omega - \alpha)G(\alpha)d\alpha\right]. \quad (5.10.25)$$

2. Write the time-domain product $x(t)\hat{g}(t)$ in terms of the Fourier convolution integral.
3. Noting the non-overlapping spectra of the two functions, we can relate the two results and then show (5.10.23). The details are left as an exercise.

Example 5.10.4 Let $m(t)$ be a low-pass signal with $M(\omega) = 0, |\omega| > W$ and $\omega_c > W$, then

$$H[m(t)\cos(\omega_c t)] = m(t)\sin(\omega_c t),$$
$$H[m(t)\sin(\omega_c t)] = -m(t)\cos(\omega_c t). \quad (5.10.26)$$

Solution: These follow from (5.10.23) and the Hilbert transforms of the sine and cosine functions. We will use results in studying single sideband modulations in Chapter 10. ∎

5.10.3 Analytic Signals

The Hilbert transform is used to define an analytic signal of the real signal $x(t)$ by

$$x_a(t) = \frac{1}{2}[x(t) + j\hat{x}(t)]. \quad (5.10.27)$$

Some authors do not use the constant $(1/2)$ in the definition of the analytic signal. The real significance of the analytic signal is its spectrum and is

$$F[x_a(t)] = \frac{1}{2}F[x(t) + j\hat{x}(t)] = \frac{1}{2}[1 + \mathrm{sgn}(\omega)]X(j\omega)$$
$$= \begin{cases} X(j\omega), & \omega > 0 \\ 0, & \omega < 0 \end{cases}. \quad (5.10.28)$$

The spectrum of the analytic signal $x_a(t)$ is the *positive portion* of the spectrum of the real signal $x(t)$. This property will be useful in the development of single sideband modulation scheme in Chapter 10. Some authors use the symbol $x_+(t)(= x_a(t))$ for analytic signals. A real signal $x(t)$ can be written in terms of analytic signals

$$x(t) = [x_a(t) + x_a^*(t)]/2 \overset{HT}{\longleftrightarrow} [x_a(t) - x_a^*(t)]/2. \quad (5.10.29)$$

Use of this gives

$$\cos(\omega_0 t) = [e^{j\omega_0 t} + e^{-j\omega_0 t}]/2,$$
$$\sin(\omega_0 t) = [e^{j\omega_0 t} - e^{-j\omega_0 t}]/2j. \quad (5.10.30)$$

Narrowband noise signals: Although statistical description of noise is beyond our scope here, we will study signals with their spectra centered at a frequency f_c with a bandwidth $B \ll f_c$. For example, the output of an amplitude modulated signal is $x_c(t) = m(t)\cos(\omega_c t)$ with $m(t)$ being a low-pass signal with its bandwidth much and much smaller than the carrier frequency f_c. Such signals are *narrowband (NB)* signals. These are expressed in terms of the *envelope* $R(t)$, a slowly varying function and the *phase* $\phi(t)$ written in the form

$$n(t) = R(t)\cos(\omega_c t + \phi(t)), R(t) \geq 0. \quad (5.10.31)$$

In most cases, $R(t)$ and $\phi(t)$ are not transformable.

Example 5.10.5 Find the envelope and the complex envelope of the NB signal in terms of two NB signals $n_c(t)$ and $n_s(t)$ given by

$$n(t) = n_c(t)\cos(\omega_0 t) - n_s(t)\sin(\omega_0 t). \quad (5.10.32)$$

Table 5.10.1 Hilbert transform pairs

Sinusoids:

$$\sin(\omega_0 t) \xrightarrow{\text{HT}} -\cos(\omega_0 t); \quad \cos(\omega_0 t) \xrightarrow{\text{HT}} \sin(\omega_0 t)$$

Exponential:

$$e^{j\omega t} \xrightarrow{\text{HT}} -j\,\text{sgn}(\omega)e^{j\omega t}$$

Rectangular pulse:

$$\Pi\left[\frac{t}{\tau}\right] \xrightarrow{\text{HT}} \frac{1}{\pi}\ln\left|\frac{t + \tau/\tau 22}{t - \tau/\tau 22}\right|$$

Impulse:

$$\delta(t) \xrightarrow{\text{HT}} 1/\pi t$$

Solution: First,

$$n(t) = R(t)\cos(\omega_0 t + \phi(t)),\, R(t) = \sqrt{n_c^2(t) + n_s^2(t)},$$
$$\phi(t) = \tan^{-1}[n_s(t)/n_c(t)]. \qquad (5.10.33)$$

From this representation, the analytic signal can be obtained and is

$$n_a(t) = n(t) + j\hat{n}(t) = n_c(t)\cos(\omega_0 t) - n_s(t)\sin(\omega_0 t)$$
$$+ jn_c(t)\sin(\omega_0 t) + jn_s(t)\cos(\omega_0 t)$$
$$= n_c(t)[\cos(\omega_0 t) + j\sin(\omega_0 t)] + jn_s(t)[\cos(\omega_0 t)$$
$$+ j\sin(\omega_0 t)] = [n_c(t) + jn_s(t)]e^{j\omega_0 t}.$$

The *envelope and the complex envelopes* are, respectively, given by

$$|n_a(t)| = \sqrt{n_c^2(t) + n_s^2(t)},$$
$$\tilde{n}(t) = n_c(t) + jn_s(t). \qquad (5.10.34) \quad \blacksquare$$

Ziemer and Tranter (2002) give interesting applications of the Hilbert transforms. Refer table 5.10.1 for a short table of Hilbert transforms, see Hahn Poularikis, ed. (1996).

5.11 Summary

In this chapter Fourier transform integral is used to discuss and derive some of the related transforms. These include cosine, sine, Hartley, Laplace, and Hilbert transforms. Bulk of this chapter deals with the one-sided Laplace transforms. Specific topics are:

- Fourier cosine and sine and Hartley transforms
- Laplace transforms and their inverses; regions of convergence
- Basic properties of Laplace transforms; initial and final value theorems
- Partial fraction expansions
- Solutions of constant coefficient differential equations using Laplace transforms
- Relationship between Laplace and Fourier transforms
- Hilbert transforms and their inverses
- Various tables listing some simple time functions and their transforms

Problems

5.2.1 Derive the following properties using $X_c(\omega) = \text{FCT}[x(t)]$:

a. $X_c(t) \xrightarrow{\text{FCT}} (\pi/2)x(\omega),$

b. $x(at)\cos(bt) \xrightarrow{\text{FCT}} \frac{1}{2a}\left[X_c\left(\frac{\omega + b}{a}\right) + X_c\left(\frac{\omega + b}{a}\right)\right],$

   $a > 0, b > 0,$

c. $t^2 x(t) \xrightarrow{\text{FCT}} -\frac{d^2 X_c(\omega)}{d\omega^2},$

d. $x''(t) \xrightarrow{\text{FCT}} -\omega^2 X_c(\omega) - x'(0^+).$

   (Assume $x(t)$ and $x'(t)$ vanish as $t \to \infty$.).

5.2.2 Show

a. $x(t) = \dfrac{1}{a^2 + t^2} \overset{\text{FCT}}{\longleftrightarrow} \dfrac{\pi}{2a} e^{-a\omega}$,

b. $\dfrac{1}{a^2 - t^2} \overset{\text{FCT}}{\longleftrightarrow} \dfrac{\pi}{2a} \sin(a\omega)$,

c. $\dfrac{\beta}{(t-a)^2 + \beta^2} + \dfrac{\beta}{(t+a)^2 + \beta^2} \overset{\text{FCT}}{\longleftrightarrow} \pi \sin(a\omega) e^{-\beta\omega}$,

d. $\dfrac{d^2[e^{-at}u(t)]}{dt^2} \overset{\text{FCT}}{\longleftrightarrow} \dfrac{a^3}{a^2 + \omega^2}$.

5.2.3 Derive the following associated with sine transforms using $X_s(\omega) = \text{FST}[x(t)]$:

a. $x(at)\cos(\beta t) \overset{\text{FST}}{\longleftrightarrow} \dfrac{1}{2a}\left[X_s\left(\dfrac{\omega+\beta}{a}\right) + X_s\left(\dfrac{\omega-\beta}{a}\right)\right]$,

b. $x(at) \overset{\text{FST}}{\longleftrightarrow} \dfrac{1}{a}X_s(\omega/a), a > 0$

5.2.4 Show the following are valid:

a. $\dfrac{1}{\sqrt{t}} \overset{\text{FST}}{\longleftrightarrow} \sqrt{\dfrac{\pi}{2\omega}}$,

b. $\dfrac{\beta}{\beta^2 + (t-a)^2} - \dfrac{\beta}{\beta^2 + (t-a)^2} \overset{\text{FST}}{\longleftrightarrow} \pi \cos(a\omega) e^{-\beta\omega}$, $\beta > 0$.

5.3.1 The energy spectral density of a signal $x(t) \overset{\text{FT}}{\longleftrightarrow} X(j\omega)$ can be expressed by

$$(1/2\pi)|X(j\omega)|^2 = (1/2\pi)\left\{[\text{Re}(X(j\omega))]^2\right.$$

$$\left. + [\text{Im}(X(j\omega))]^2\right\}.$$

Derive this in terms of the Hartley transform $X_H(\omega)$. Also derive the expression for the phase angle of the spectrum in terms of the Hartley transform.

5.3.2 Derive the Hartley transforms of the following functions using Fourier transforms:

a. $x_1(t) = x(at), a \neq 0$,

b. $x_2(t) = x(t)\cos(\omega_0 t)$,

c. $x_3(t) = x(t)\cos(\omega_0 t)$.

5.3.3 Derive an expression for the Hartley transform of the convolution $y(t) = x(t) * h(t)$.

5.3.4 Show the following transforms are true.

a. $e^{-\alpha t}\sin(\omega_0 t)u(t) \overset{\text{Hart}}{\longleftrightarrow} \dfrac{\omega_0(\alpha^2 + \omega_0^2 - \omega^2) + 2(\alpha\omega)}{(\alpha^2 + \omega_0^2 - \omega^2) + 2(\alpha\omega)^2}$,

b. $e^{-\alpha t}\cos(\omega_0 t)u(t) \overset{\text{Hart}}{\longleftrightarrow}$

$$\dfrac{(\alpha - \omega)(\alpha^2 + \omega_0^2 - \omega^2) + 2\omega(\alpha + \omega)}{(\alpha^2 + \omega_0^2 - \omega^2) + 2(\alpha\omega)^2},$$

c. $\displaystyle\sum_{n=-\infty}^{\infty} \delta(t - nT) \overset{\text{Hart}}{\longleftrightarrow} \dfrac{\pi}{T}\sum_{k=-\infty}^{\infty} \delta(\omega - (k/T))$,

d. $e^{j\omega_0 t} \overset{\text{Hart}}{\longleftrightarrow} \pi\delta(\omega - \omega_0)$.

5.4.1 Find the two-sided Laplace transforms and their ROCs of the following functions:

a. $x_1(t) = \Pi[t]$, b. $x_2(t) = te^{-a|t|}, a > 0$.

The following problems are concerned with *one-sided Laplace transforms*.

5.4.2 Find the Laplace transforms of the following functions:

a. $x_1(t) = 2\sqrt{t/\pi}$. b. $x_2(t) = \Lambda[t]$, c. $x_3(t) = \cosh(\beta t)$.

5.4.3 Find the Laplace transform of the function $x(t) = tu(t)$ by the following methods: a. Use the times-t property and the transform of $u(t)$ to show that $L[t\,u(t)] = 1/s^2$.

b. Use the result in part a to show by *induction* that $t^n u(t) \overset{\text{LT}}{\longleftrightarrow} n!/(s^{n+1})$.

The proof by induction uses the following procedure. The result is first shown to be true for the case of $n = 1$. Then verify that if the result is true for n then it is also true for $n + 1$.

5.6.1 Find the L-transform of $x(t) = |\sin(t)|, t \geq 0$.

5.6.2 a. Find $X_1(s) = L\{x(t)\}$ by using the transform of the unit step function.

$$x_1(t) = \dfrac{1}{T}\Pi\left[\dfrac{t - T/2}{T}\right] \overset{\text{LT}}{\longleftrightarrow} X_1(s).$$

b. Take the limit of the transform as $T \to 0$ and identify the corresponding transform pair.

5.6.3 Show the following transform pair is true by a. using the integral of the transform and by b. using the second derivative of the Laplace transform to show the above result.

$$x(t) = \Lambda[t-1]$$

$$= \begin{cases} t, 0 < t < 1 \\ (2-t), 1 < t < 2 \\ 0, \text{otherwise} \end{cases} \xrightarrow{\text{LT}} \frac{1}{s^2}(1 - 2e^{-s} + e^{-2s}).$$

5.6.4 a. Use the differentiation with respect to the second variable property of the Laplace transforms to show that $L[te^{-at}] = [1/(s+a)^2]$.

5.6.5 Determine the transform of the Laplace transform of the following function $x(t)$:

$$x(t) = \frac{\sin(\omega_0 t)}{t}, \quad \int_0^{\omega_0} \cos(\omega t) d\omega = \frac{\sin(\omega_0 t)}{t}.$$

5.7.1 Verify the following the transform pairs by a. evaluating the transform directly and by b. using the partial fraction expansion and then identify term by term from tables.

$$\frac{1}{a^2}(1 - \cos(at)) \xrightarrow{\text{LT}} \frac{1}{s(s^2 + a^2)}.$$

5.7.2 Verify the following by using the Laplace transforms properties:

a. $x_1(t) = (1/a)e^{bt} \sinh(at) \xrightarrow{\text{LT}} 1/[(s-b)^2 - a^2]$

$$= X_1(s),$$

b. $x_2(t) = (1/2a)t\sin(at) \xrightarrow{\text{LT}} s/[(s^2 + a^2)^2] = X_2(s).$

c. $x_3(t) = t/T, 0 < t < T, x_3(t + nT)$

$$= x(t), n \geq 0 \xrightarrow{\text{LT}} \frac{1}{as^2} - \frac{e^{-as}}{s(1 - e^{-as})}.$$

5.8.1 Find the solution of

$$\frac{dy}{dt} + 3y = t\cos(t); y(0^-) = 1.$$

5.8.2 Using Laplace transforms to find the solution of the differential equation

$$\frac{d^2x(t)}{dt^2} - x(t) = \sinh(t), \quad x(0) = 0 \text{ and } x'(0) = 0.$$

5.9.1 Find

$$x_i(t) = L^{-1}[X_i(s)] \text{ with}$$

a. $X_1(s) = \dfrac{1 - e^{-s}}{s(s+1)}$, b. $X_2(s) = \dfrac{1}{s^2(s^2+4)}$,

c. $X_3(s) = \dfrac{1 - e^{-s}}{s(1 - e^{-2s})}.$

5.9.2 Show the residues A_i can be determined by

$$X(s) = \frac{N(s)}{D(s)} = \frac{N(s)}{(s+s_1)(s+s_2)\ldots(s+s_N)}$$

$$= \sum_{i=1}^N \frac{A_i}{(s+s_i)}, \quad A_i = \left[\frac{N(s)}{dD(s)/ds}\right]\Big|_{s=-s_i}.$$

a. Use this result to find $L^{-1}\{X(s)\} = L^{-1}\{1/[s(s+3)]\}$.

b. Now consider $Y(s) = (1/s)X(s)$. Use Part a. to generalize this.

5.9.3 Assuming the regions of convergence are a. $\sigma > 1$, b. $\sigma < -2$, find

$$L^{-1}\{X_{II}(s)\} = L^{-1}\left\{\frac{2s+1}{(s^2 + s - 2)}\right\}.$$

5.9.4 Assuming the following Laplace transforms, find the corresponding Fourier transforms:

a. $X_{II}(s) = \dfrac{1}{(s+a)^2} - \dfrac{1}{(s-a)^2}, |\sigma| < a$, b. $X(s)$

$$= \frac{s + \omega_0}{s^2 + \omega_0^2}.$$

5.10.1 Show

a. $\dfrac{1}{1+t^2} \xrightarrow{\text{HT}} \dfrac{t}{(1+t^2)}$,

b. $\delta(t) \xrightarrow{\text{HT}} \dfrac{1}{\pi t}$,

c. $X_s[0] + \displaystyle\sum_{k=1}^K X_s[k]\cos(k\omega_0 t)$

$$+ \theta[k]) \xrightarrow{\text{HT}} \sum_{k=1}^K X_s[k]\sin(k\omega_0 t + \theta[k]).$$

Chapter 6
Systems and Circuits

6.1 Introduction

In this chapter we will consider systems in general, and in particular linear systems. Most systems are inherently nonlinear and time varying. A human being is a good example. He can run fast for a while and then speed comes down. If you plot speed versus time, the plot is not going to be a straight line, i.e., the function speed versus time is not linear. Humans are nonlinear and also time-varying systems. For example, if you want to ask your dad for a new car, you do not ask him when he is not happy. Moods change with time. These considerations are important in, for example, speaker identification. Human beings are not only nonlinear but also time-varying complicated systems. Nonlinear time-varying systems are very hard to deal with. Even though many of the systems may have nonlinear behavior characteristics, they can be approximated to be linear systems and they allow for transform analysis. In addition we are interested in systems that operate in the same manner every time we use them. That is, the systems must be independent of time. Linear time-invariant system analysis and design is the basis of present day system analysis and design. Transfer functions associated with these systems are discussed. In addition the frequency analysis makes it very attractive for the design of systems. Majority of the discussion in this chapter is on linear time-invariant systems. These allow for transfer function analysis. The study of the amplitude and phase frequency responses of linear time-invariant systems is one of the important topics. When a signal through some media, it is modified by the media. In the frequency domain we can say that some frequencies are amplified and some are attenuated. In addition different frequencies are delayed differently. Our goal is to filter frequencies with appropriate attenuations of the input frequencies with a constant delay at all frequencies in the frequency band of interest. The delay response is related to the phase response of a system. If the delay is not constant, then delay compensation may be required.

One of the topics we will be interested is filter circuits. Toward this goal ideal low-pass, high-pass, band-pass, and band-elimination filter functions are introduced in this chapter. In addition to these simple examples of a differentiator, integrator, and a delay circuit are illustrated. A brief introduction to nonlinear systems is included later. The topic of linear systems is one of the topics every undergraduate student in electrical engineering program goes through. See the books Haykin and Van Veen (1999), Lathi (1998), Oppenheim et al. (1997), Nillsson and Riedel (1996), Poularikas and Seely (1991), Carlson (2000) and others.

6.2 Linear Systems, an Introduction

Our study starts with a system that has an input and an output. It is symbolically represented by a block diagram shown in Fig. 6.2.1. The T inside the box is some transformation that converts the input signal $x(t)$ into the output signal $y(t)$ and

$$y(t) = T[x(t)] (T \text{ maps } x(t) \text{ into } y(t)). \quad (6.2.1)$$

R.K.R. Yarlagadda, *Analog and Digital Signals and Systems*, DOI 10.1007/978-1-4419-0034-0_6,
© Springer Science+Business Media, LLC 2010

$x(t)$ 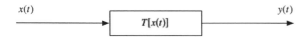 $y(t)$

Fig. 6.2.1 Block diagram of a system

Some use $L[x(t)]$ to represent a linear system. The system described by (6.2.1) is called a *linear system* if the transformation given by $T[x(t)]$ satisfies the following conditions:

Principle of additivity: If $T[x_1(t)] = y_1(t)$ and $T[x_2(t)] = y_2(t)$, then the *superposition* property states that

$$T[x_1(t) + x_2(t)] = T[x_1(t)] + T[x_2(t)] = y_1(t) + y_2(t)$$
$$(6.2.2)$$

Principle of proportionality: If $T[x(t)] = y(t)$, then for any constant α

$$T[\alpha x(t)] = \alpha T[x(t)] = \alpha y(t). \qquad (6.2.3)$$

This property is also referred to as the *homogeneity* property. We can combine the two properties into one and state that a *linear system* satisfies the following property

$$T[\alpha_1 x(t) + \alpha_2 x_2(t)] = \alpha_1 y_1(t) + \alpha_2 y_2(t) \qquad (6.2.4)$$

for any pair of constants α_1 and α_2. Otherwise, the system is called a *nonlinear system*.

Example 6.2.1 Show that a system described by $y(t) = x^2(t)$ is a nonlinear system.

Solution: For two inputs $x_1(t)$ and $x_2(t)$, the corresponding outputs are, respectively, given by $y_1(t) = x_1^2(t)$ and $y_2(t) = x_2^2(t)$. If the input is $x_1(t) + x_2(t)$, then the output is $(x_1(t) + x_2(t))^2 \neq x_1^2(t) + x_2^2(t)$ and therefore the system is nonlinear. We can make a general statement that if the output of a system is a power of the input, and the power is not equal to one, then the system is nonlinear. Other examples of nonlinear systems include

$$y_1(t) = \log(x(t)), y_2(t) = |x(t)|. \qquad \blacksquare$$

Example 6.2.2 Two of the many modulation schemes that we will be interested in are given

below. What can you say about the linearity of theses schemes?

$$a.\ y_1(t) = m(t)\cos(\omega_c t),$$
$$b.\ y_2(t) = (A + m(t))\cos(\omega_c t), A \neq 0. \qquad (6.2.5)$$

Solution: *a.* For the inputs $m_1(t)$ and $m_2(t)$, the outputs are, respectively, given by $m_1(t)\cos(\omega_c t)$ and $m_2(t)\cos(\omega_c t)$. For the input $\alpha_1 m_1(t) + \alpha_2 m_2(t)$, the output is the sum of the two individual outputs $[\alpha_1 m_1(t) + \alpha_2 m_2(t)]\cos(\omega_c t)$. Therefore the system is linear.

b. For the inputs $x_1(t)$ and $x_2(t)$, the outputs are, respectively, given by $(A + m_1(t))\cos(\omega_c t)$ and $(A + m_2(t))\cos(\omega_c t)$. For the input $(\alpha_1 m_1(t) + \alpha_2 m_2(t))$, the output is given by $[A + \alpha_1 m_1(t) + \alpha_2 m_2(t)]\cos(\omega_c t)$ which is not equal to

$$(A + m_1(t))\cos(\omega_c t) + (A + m_2(t))\cos(\omega_c t).$$

Therefore the system is nonlinear. Note that the Fourier transform is a linear operation as

$$F[\alpha_1 x_1(t) + \alpha_2 x_2(t)] = \alpha_1 F[x_1(t)] + \alpha_2 F[x_2(t)].$$
$$(6.2.6) \qquad \blacksquare$$

The systems described by the following equations are linear systems and the reader is encouraged to go through the proofs:

$$a.\ y(t) = kx(t) \qquad \text{(Amplifier)} \qquad (6.2.7a)$$

$$b.\ y(t) = k\frac{dx(t)}{dt} \qquad \text{(Differentiator)} \qquad (6.2.7b)$$

$$c.\ y(t) = \int x(\beta)d\beta \qquad \text{(Integrator)} \qquad (6.2.7c)$$

$$d.\ y(t) = x(t - \tau), \tau \geq 0 \quad \text{(Delay device).} \quad (6.2.7d)$$

6.3 Ideal Two-Terminal Circuit Components and Kirchhoff's Laws

In this section we will consider two-terminal passive and active components and laws that pertain to the interconnection of elements. These are two powerful laws, referred to as Kirchhoff's voltage and current laws first formulated by Kirchhoff (pronounced as kear-koff) in 1847. One gives equations in terms of voltages across components and the other gives equations in terms of currents flowing through the components. Component equations and the Kirchhoff's laws provide us with circuit analysis tools.

6.3.1 Two-Terminal Component Equations

Simple circuits include ideal sources, voltage, and current sources and three types of components resistors, inductors, and capacitors. The symbols for the sources are shown in Fig. 6.3.1. An ideal voltage source is a two-terminal component whose voltage across the two terminals is a constant or a function of time regardless of what the current through the component is. Examples of voltage sources are wall outlets, where we assume that the voltage is $v_s(t) = V_m \cos(\omega_m t)$, and batteries, where the voltage across is a constant. The first source we refer to as an alternating current (AC) source and the second one is a constant voltage source (DC). The positive sign on top of the ideal voltage source indicates the higher potential whenever the source voltage is positive. Most generators are voltage sources. The ideal current source is a two-terminal component whose current is a constant or a function of time, regardless of what the voltage across it is. Transistors and many other electronic devices act more like a current source rather than a voltage source. It is important to notice the voltage signs and the direction of the currents through the sources. This convention shows that the sources provide power.

The three basic passive components are the resistor, inductor, and the capacitor. The *Lumped parameter models* are shown in Fig. 6.3.2. These are passive elements, i.e.,

The three basic passive components are the resistor, inductor, and the capacitor. The *lumped parameter models* are shown in Fig. 6.3.2. These are passive elements, i.e., they do not produce any power. Therefore the notation for the three components is that the current flows from the positive terminal to the negative terminal. The resistance is measured in Ohms, the inductor in Henries, and the capacitor in Farads. The voltage across a resistor is related to the current by the Ohm's law and is given by

$$v_R(t) = R i_R(t). \tag{6.3.1}$$

The voltage across an inductor is given by

$$v_L(t) = L \frac{d i_L(t)}{dt}. \tag{6.3.2}$$

Since the voltage across an inductor is the derivative of current, it is zero for a constant current. The inductor stores energy in a magnetic field produced by current through a coil of wire. Inductor is an energy storage device and the instantaneous stored energy, measured in Joules, is

$$w_L = \frac{1}{2} L i_L^2(t). \tag{6.3.3}$$

The current in the inductor can be computed from (6.3.2) and is

$$i_L(t) = i_L(t_0) + \frac{1}{L} \int_{t_0}^{t} v_L(\beta) d\beta. \tag{6.3.4}$$

The term $i_L(t_0)$ corresponds to the initial conditions on the current through the inductor at time $t = t_0$. The current through a capacitor is given by

$$i_C(t) = C \frac{d v_C(t)}{dt} \tag{6.3.5}$$

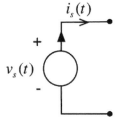

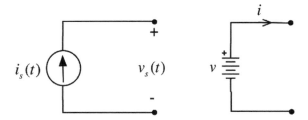

Fig. 6.3.1 Voltage, current, and a constant voltage source

Fig. 6.3.2 Resistor, inductor, and the capacitor

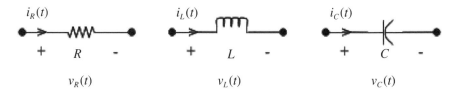

The gap in the capacitor symbol reflects that when the voltage across the capacitor is constant, then the capacitor acts like an open circuit. Capacitor is an energy storage device and the instantaneous energy, measured in Joules, is

$$w_C = \frac{1}{2}Cv_C^2. \tag{6.3.6}$$

The current in the capacitor is

$$v_C(t) = v_C(t_0) + \frac{1}{C}\int_{t_0}^{t} i_C(t)dt. \tag{6.3.7}$$

The voltage $v_c(t_0)$ corresponds to the initial voltage across the capacitor at time $t = t_0$. Initial conditions are necessary when we consider transient analysis. In the design of systems we generally do not consider the initial conditions, as the systems are supposed to work for any initial conditions. We assume that the initial conditions on the capacitors and the inductors are assumed to be zero with t_0 is equal to $-\infty$. These allow us to use both the Laplace and the Fourier transforms. This is especially true in network synthesis, as the design specifications are given in terms the sinusoidal steady state. Once we have the designs, we can always test the systems with initial conditions and any possible changes in the responses of systems associated with the initial conditions. The component equations for the three two-terminal components discussed above can be expressed in terms of Laplace transforms. To avoid any confusion from the inductor values we will make use of the symbol $L\{.\}$ for the Laplace transform. These are

$$V_R(s) = L\{v_R(t)\} = RL\{i_R(t)\} = RI_R(s) \tag{6.3.8}$$

$$V_L(s) = L\{v_L(t)\} = L\left\{L\frac{di_L(t)}{dt}\right\}$$

$$= sLI_L(s) - Li_L(0^-) \text{ or } I_L(s) \tag{6.3.9}$$

$$= \frac{i_L(0^-)}{s} + \frac{1}{sL}V_L(s)$$

$$V_c(s) = L\{v_C(t)\} = L\left\{\frac{dv_c(t)}{dt}\right\}$$

$$= \frac{v_C(0^-)}{s} + \frac{1}{sC}I_C(s) \text{ or } I_C(s) \tag{6.3.10}$$

$$= sCV_C(s) - Cv_C(0^-).$$

Assuming the *initial conditions are zero*, we have the component equations in terms of the Laplace transformed variable s for the three components

$$L\{v_R(t)\} = V_R(S), L\{i_R(t)\} = I_R(s),$$
$$L\{v_L(t)\} = V_L(s), F\{i_L(t)\} = I_L(s)$$
$$L\{v_c(t)\} = V_c(s), L\{i_c(t)\} = I_c(s)$$
$$V_R(s) = RI_R(s), V_L(s) = Lsi_L(s),$$
$$V_c(s) = (1/Cs)I_C(s) \tag{6.3.11}$$

The component equations in terms of the Fourier transform variable $j\omega$ are

$$V_R(j\omega) = F[v_R(t)], V_L(j\omega) = F[v_L(t)], V_c(j\omega) = F[v_c(t)] \tag{6.3.12a}$$

$$I_R(j\omega) = F[i_R(t)], i_L(j\omega) = F[i_L(t)], i_c(j\omega)s = F[i_c(t)] \tag{6.3.12b}$$

$$V_R(j\omega) = RI_R(j\omega), V_L(j\omega) = j\omega LI_L(j\omega)$$
$$V_c(j\omega) = (1/j\omega C)I_c(j\omega). \tag{6.3.12c}$$

In the case of zero initial conditions either Laplace or Fourier transform variables can be used. One can be obtained from the other. Also, the voltage to the current transform ratio is called as an *impedance* of the component under consideration and its inverse as the *admittance* of that component. The impedances of the three components are, respectively, given by R, Ls (or $j\omega L$) and $1/Cs$ (or $1/j\omega C$).

6.3.2 Kirchhoff's Laws

Circuit analysis is based on the *Kirchhoff's current* and *voltage* laws and the *component equations*. The Kirchhoff's current law (KCL) states that the sum of the currents going into a junction, a *node*, is equal to the sum of the currents going out of that junction. In other words, the algebraic sum of the currents going into a node is equal to zero. The dual to the current law is the voltage law and is stated for a loop in a circuit. A loop is any path that goes from one node to another node and returns to the starting node. The Kirchhoff's voltage law (KVL) states that the sum of the voltage drops around any loop is equal to the sum of the voltage rises. Or, the algebraic sum of voltages around a loop is equal to zero. Examples are given in Fig. 6.3.3.

Example 6.3.1 Consider the simple *RC* circuit shown in Fig. 6.3.4a. Derive the differential equation relating the input and the out put of the circuit.

Solution: Using the KVL, we have the input voltage $x(t)$ is equal to the sum of the voltages across the resistor and the capacitor.

$$x(t) = v_R(t) + v_C(t). \qquad (6.3.13)$$

Assuming that the current flowing through the output node is zero, i.e., the *circuit is not loaded*, we have

$$i_R = i_C, \text{ and } i_R = \frac{x(t) - v_C(t)}{R} = \frac{x(t) - y(t)}{R}. \qquad (6.3.14)$$

Using the component equation for the capacitor and the relation in (6.3.14) and relating input and output results in

$$i_C = C\frac{dv_C}{dt} = C\frac{dy}{dt}, \quad C\frac{dy}{dt} = \frac{x(t) - y(t)}{R}. \qquad (6.3.15a)$$

$$RC\frac{dy}{dt} + y(t) = x(t). \qquad (6.3.15b)$$

It is a linear combination of the output and the derivative of the output related to the input. The system described by this differential equation is a linear system. In terms of the Laplace and the Fourier transforms at each step and simplifying, the output transforms can be expressed as follows:

$$x(t) \overset{LT}{\longleftrightarrow} X(s), y(t) \overset{LT}{\longleftrightarrow} Y(s), x(t) \overset{FT}{\longleftrightarrow} X(j\omega) \text{ and}$$

$$y(t) \overset{FT}{\longleftrightarrow} Y(j\omega) \qquad (6.3.16)$$

$$RC(s)Y(s) + Y(s) = X(s) \text{ or}$$
$$(RCj\omega + 1)Y(j\omega) = X(j\omega). \qquad (6.3.17)$$

$$Y(s) = \frac{(1/Cs)}{R + (1/Cs)}$$

$$X(s) = \frac{(1/RC)}{s + (1/RC)}X(s),$$

$$Y(j\omega) = \frac{1}{1 + j\omega RC}X(j\omega). \qquad (6.3.18) \quad \blacksquare$$

A simple integrator and a differentiator: The circuit in Fig. 6.3.4a can be used as an *integrator* in the low-frequency range. The integral form of the equation in (6.3.15b) is

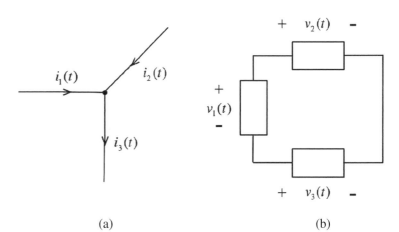

Fig. 6.3.3 Illustration of Kirchhoff's current and voltage laws
(a) $i_1(t) + i_2(t) - i_3(t) = 0$,
(b) $v_1(t) - v_2(t) - v_3(t) = 0$

(a)   (b)

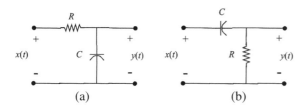

Fig. 6.3.4 *RC* circuits

$$RCy(t) + \int_{-\infty}^{t} y(\alpha)d\alpha = \int_{-\infty}^{t} x(\alpha)d\alpha. \quad (6.3.19)$$

If the time constant RC is large enough that the integral on the left side of the equation in (6.3.19) is dominated by $RCy(t)$, then (6.3.19) can be approximated by an integral, a *smoothing operation,* and is

$$RCy(t) \approx \int_{-\infty}^{t} x(\alpha)d\alpha \text{ or } y(t) = \frac{1}{RC}\int_{-\infty}^{t} x(\alpha)d\alpha.$$

$$(6.3.20)$$

Example 6.3.2 Relate the input and the output transforms of the circuit in Fig. 6.3.4b.

Solution: Using the Kirchhoff's voltage law, we have

$$x(t) = v_c(t) + y(t) \rightarrow \frac{dx(t)}{dt}$$
$$= \frac{dv_c(t)}{dt} + \frac{dy(t)}{dt} = \frac{i_c(t)}{c} + \frac{dy(t)}{dt}$$
$$\frac{dx(t)}{dt} = \frac{dy(t)}{dt} + \frac{1}{RC}y(t). \quad (6.3.21)$$

If the time constant (RC) is small enough that the second term dominates the first term on the right in the last equation, we can approximate (6.3.21) by

$$y(t) \approx RC\frac{dx}{dt}. \quad (6.3.22)$$

The circuit approximates the derivative operation or it acts like a *differentiator.* Taking the transform of the equation in (6.3.21) and solving for $Y(s)$, we have

$$Y(s) = s/[(s + (1/RC))]X(s) = H(s)X(s),$$
$$H(s) = s/[(s + (1/RC))]. \quad (6.3.23a)$$

The above two examples provide simple circuits for low-pass and high-pass filters. The amplitude response $|H(j\omega)|$ can be approximated for small frequencies and

$$|H(j\omega)|^2 = \frac{|\omega|^2}{(1/RC) + \omega^2} \approx RC|\omega^2|, |\omega| \ll \sqrt{\frac{1}{RC}}$$
$$\rightarrow H(j\omega) \approx j\omega(RC). \quad (6.3.23b) \quad \blacksquare$$

The networks containing ideal resistors (R's), inductors (L's), and capacitors (C's) result in constant coefficient differential equations.

6.4 Time-Invariant and Time-Varying Systems

For any system of use, we like the system to respond every time the same way when we switch the system on. That is, if we switch today or tomorrow, the system should respond exactly the same. Such a system is called a *time-invariant* system.

Time-invariant system: If the response of a system is $y(t)$ to the input $x(t)$, i.e., $y(t) = T[x(t)]$, then the system is called a time invariant or a fixed system if $T[x(t - t_0)] = y(t - t_0)$. Otherwise, it is a time-varying system.

Linear time-invariant system: A system is linear time invariant *(LTI)* if it is linear and time invariant.

Example 6.4.1 The systems described by constant coefficient differential equations are linear time-invariant systems. RLC networks are linear time-invariant systems. Circuits containing diodes, transistors, and other electronic components are nonlinear. $\quad \blacksquare$

Example 6.4.2 Consider the model of a carbon microphone shown in Fig. 6.4.1. The resistance R

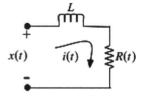

Fig. 6.4.1 A time-varying system

is a function of the pressure generated by sound waves on the carbon granules of the microphone, which is a function of time. The circuit has only one loop and using the Kirchhoff's voltage law (KVL) and the component equations, we can write

$$x(t) = v_R + v_L = R(t)i(t) + L\frac{di(t)}{dt}. \quad (6.4.1)$$

The resistance is a function of time and the resulting differential equation has coefficients that vary with time and the system is a time-varying system. ∎

Earlier we have indicated that a human being is a nonlinear time-varying system. The speech signal is a time-varying signal, albeit, a slowly time-varying signal. To analyze a slowly time-varying signal, we segment the speech signal by using windows and find the needed information for each segment. Obviously the spectral characteristics of different phonemes are different. In the earlier chapters we defined causal signals that are zero for $t < 0$. We can similarly define causal systems.

Causal systems: Causal systems do not respond until the input is applied. That is, they do not anticipate the input. For a causal system, if the input

$x(t) = 0$ for all $t \le T$, then the output
$y(t) = 0$ for all $t \le T$ $\qquad$ (6.4.2)

Memory and memoryless systems: A system is called *memoryless* if the output of the system at a particular time depends only on the input at that time. The resistor is memoryless since $v(t) = Ri(t)$ and the voltage and the current pertaining to this component are related at each value of t. The capacitor voltage is related to the current by an integral and the inductor voltage is related to the voltage by an integral. Capacitors and inductors have memory and initial conditions can be assigned on these. The relations are

$$v_C(t) = \frac{1}{C}\int_{-\infty}^{t} i_C(\alpha)d\alpha, \; i_L(t) = \frac{1}{L}\int_{-\infty}^{t} v_L(\alpha)d\alpha.$$

$$(6.4.3)$$

Invertibility: A system is said to be invertible if the input of the system can be recovered from the output of the system. Consider that we have the response of the system given by

$$y(t) = T[x(t)]. \quad (6.4.4)$$

The system is invertible if there is a transformation T^{-1} such that

$$T^{-1}[y(t)] = T^{-1}[T[x(t)]] = x(t). \quad (6.4.5)$$

That is $(T^{-1}T) = I$, the identity operator. Simple examples of non-invertible systems include $y_i(t) = x^2(t), y_2(t) = u(x(t))$ and many others. In each of these cases we cannot determine the function $x(t)$ from $y_i(t)$.

Example 6.4.3 Give an expression for the derivative of the current in an inductor.

Solution: The current in an inductor is

$$i_L(t) = \frac{1}{L}\int_{-\infty}^{t} v_L(t)dt \Rightarrow v_L(t) = L\frac{di_L}{dt}. \quad (6.4.6)$$

Derivative and the integral are inverse operations. ∎

Now consider one of the important concepts in system analysis, i.e., the signal and the system interaction.

6.5 Impulse Response

Consider the block diagram of a *LTI* system in Fig. 6.5.1 with input $x(t)$ and the corresponding output is $y(t)$. We like to find a relationship between the input and the output and the system characteristics. If $x(t) = \delta(t)$, an impulse, then the output, the response of the impulse, is the *impulse response* of the LTI system identified by

Fig. 6.5.1 A linear time-invariant system

$$y(t) = h(t) = T[\delta(t)]. \qquad (6.5.1)$$

Therefore the response for the input $\alpha_i \delta(t - t_i)$ is $\alpha_i h(t - t_i)$, i.e., $T[\alpha_i \delta(t - t_i)] = \alpha_i h(t - t_i)$ where α_i's are some constants. If the input is a linear combination of impulses, then the response will be a linear combination of the corresponding impulse responses. That is,

$$T[\sum_{i=N_1}^{N_2} \alpha_i \delta(t - t_i)] = \sum_{i=N_1}^{N_2} \alpha_i h(t - t_i). \qquad (6.5.2)$$

We can tie this relationship to an arbitrary input using the approximation of an impulse and relate the output to the input in terms of a sum of delayed impulse responses. Impulse functions were represented in the limit by (see Section 1.4.)

$$\delta(t - t_i) = \lim_{\Delta t \to 0} \frac{1}{\Delta t} \Pi \left[\frac{t - t_i}{\Delta t} \right]. \qquad (6.5.3)$$

Consider an arbitrary signal $x(t)$ shown in Fig. 6.5.2a. There is no specific significance for the shape of this function. Now divide the time into intervals of Δt seconds apart as shown in Fig. 6.5.2b. The strip centered at $t = n\Delta t$ with a width of Δt can be approximated by $x(n\Delta t)\Pi[(t - n\Delta t)/\Delta t]$.

If Δt is negligibly small, the pulse can be assumed to be a rectangular pulse and the above approximation is good. The function $x(t)$ can now be approximated by

$$x(t) \cong \sum_{n=N_1}^{N_2} [x(n\Delta t)\Delta t] \frac{1}{\Delta t} \Pi \left[\frac{t - n\Delta t}{\Delta t} \right]. \qquad (6.5.4)$$

Note the multiplication and division by Δt in (6.5.4). The term $x(n\Delta t)\Delta t$ approximates the area of the pulse centered at $t = n\Delta t$. Now

$$\lim_{\Delta t \to 0} x(n\Delta t)\left[(1/\Delta t)\Pi \left[\frac{t - n\Delta t}{\Delta t} \right] \right] = x(t_n)\delta(t - t_n).$$
$$(6.5.5)$$

The time instant $t_n = n\Delta t$ is at some point on the time axis. As $\Delta t \to 0$, $n\Delta t$ approaches a continuous variable β, the sum becomes an integral and Δt becomes a differential and

$$x(t) = \int_{-\infty}^{\infty} x(\beta)\delta(t - \beta)d\beta. \qquad (6.5.6)$$

This is valid provided $x(t)$ is continuous for all t. It is the convolution of the two functions $\delta(t)$ and $x(t)$. See the equation in (2.2.2a) in Chapter 2. That is,

$$x(t) = x(t) * \delta(t). \qquad (6.5.7)$$

In a similar manner, the output expression can be derived. Since the system is a time-invariant system, the input $x(n\Delta t)\delta(t - n\Delta t)$ produces an output $x(n\Delta t)h(t - n\Delta t)$. Combining all the responses, we can pictorially identify

$$\sum_{n=N_1}^{N_2} x(n\Delta t)[\delta(t - n\Delta t)\Delta t] \xrightarrow[\text{the output}]{\text{Produces}} y(t)$$

$$\cong \sum_{n=N_1}^{N_2} x(n\Delta t)h(t - n\Delta t)\Delta t. \qquad (6.5.8)$$

In the limit, i.e., when $\Delta t \to 0$, $n\Delta t$ becomes a continuous variable β. The time interval Δt becomes a differential $d\beta$ and the summation becomes an integral. Noting the limits on the sum are arbitrary, the sum can be taken as over all positive and negative integers and the integral correspondingly goes from $-\infty$ to $+\infty$.

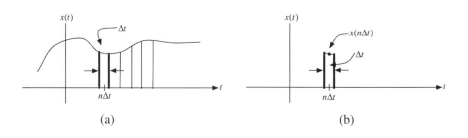

Fig. 6.5.2 (a) $x(t)$(b) Pulse centered at $t = n\Delta t$

(a)   (b)

$$y(t) = \int_{-\infty}^{\infty} x(\beta)h(t-\beta)d\beta. \qquad (6.5.9)$$

This integral is a superposition or a convolution integral of two functions, input and the impulse response of the linear time-invariant (LTI) system. The response of the LTI system to *any* input $x(t)$ is $y(t)$. Symbolically, it can be written in the form

$$y(t) = x(t) * h(t) = \int_{-\infty}^{\infty} x(\beta)h(t-\beta)d\beta$$

$$= \int_{-\infty}^{\infty} h(\alpha)x(t-\alpha)d\alpha = h(t) * x(t).$$

$$(6.5.10)$$

$$\Rightarrow Y(s) = H(s)X(s). \qquad (6.5.11)$$

The function $H(s)$ is the transform of the impulse response, $H(s) = L[h(t)]$ and is called the transfer function of the LTI system in Laplace transform domain or s-domain. It is symbolically represented by the block diagram in Fig. 6.5.3. In the Fourier domain, (6.5.11) is expressed by

$$Y(j\omega) = H(j\omega)X(j\omega), H(j\omega) = F[h(t)]. \qquad (6.5.12)$$

The input, the output, and their Fourier (and Laplace) transforms are identified on the block diagram along with impulse response $h(t)$ and its transform $H(j\omega)$ or $H(s)$.

Notes: Input–output energy spectral density relations of a linear system: The output energy spectral density is

$$|Y(j\omega)|^2 = |H(j\omega)|^2 |X(j\omega)|^2. \qquad (6.5.13)$$

Correspondingly, the output autocorrelation (AC) can be expressed in terms of the input AC as shown below.

$$\phi_x(\tau) = x(\tau) * x(-\tau) \xleftrightarrow{\text{FT}} |X(j\omega)|^2, \phi_h(\tau)$$
$$(6.5.14a)$$
$$= h(\tau) * h(-\tau) \xleftrightarrow{\text{FT}} |H(j\omega)|^2$$
$$\phi_y(\tau) = y(\tau) * y(-\tau) \xleftrightarrow{\text{FT}} = |Y(j\omega)|^2, \phi_y(\tau)$$
$$= h(\tau) * h(-\tau) * \phi_x(\tau). \qquad (6.5.14b)$$

In a similar manner, the output power spectral density of a periodic (or a random signal) can be expressed in terms of the input power spectral density by

$$R_x(\tau) \xleftrightarrow{\text{FT}} S_x(\omega), R_y(\tau) \xleftrightarrow{\text{FT}} S_y(\omega),$$
$$S_y(\omega) = |H(j\omega)|^2 S_x(\omega). \qquad (6.5.15)$$

These operations are basic to the study of linear systems, as they provide a simple way of expressing how the energy (or power) of an input signal is distributed at the output.

From Examples (6.3.1) and (6.3.2), the respective transfer functions and the corresponding impulse responses are given by

$$H(s) = \frac{(1/RC)}{s + (1/RC)} \xleftarrow{\text{LT}} \frac{1}{RC}e^{-t/RC}u(t) = h(t)$$
$$(6.5.16a)$$

$$H(j\omega) = \frac{1/RC}{(j\omega + (1/RC))} \xleftrightarrow{\text{FT}} \frac{1}{RC}e^{-(1/RC)t}u(t)$$
$$(6.5.16b)$$

$$H(s) = \frac{s}{(s + 1/RC)} \xleftarrow{\text{LT}} \delta(t) - e^{-t/RC}u(t) = h(t)$$
$$(6.5.17a)$$

$$H(j\omega) = \frac{j\omega}{(j\omega + 1/RC)} \xleftrightarrow{\text{FT}} \delta(t) - e^{-t/RC}u(t) = h(t).$$
$$(6.5.17b) \quad \blacksquare$$

Notes: A single input–single output of a linear time-invariant system is described by its transfer function $H(s) = Y(s)/X(s)$ in the Laplace domain

Fig. 6.5.3 Inputs, outputs and transfer functions

$x(t)$		$y(t) = h(t) * x(t)$
$X(j\omega)$	$h(t)$	$Y(j\omega) = H(j\omega)X(j\omega)$
$X(s)$	$H(j\omega), H(s)$	$Y(s) = H(s)X(s)$

or in the Fourier domain by $H(j\omega)$. Its impulse response is $h(t) = L^{-1}[H(s)]$ or $h(t) = F^{-1}[H(j\omega)]$. $|H(j\omega)|$ and $\angle H(j\omega)$ are the amplitude and the phase responses of the linear system.   ∎

A **causal continuous-time LTI system is memoryless** if and only if $h(t) = c\delta(t)$, where c is a constant. Noting that the output of a *causal* continuous-time LTI system is described in terms of the input $x(t)$ and its impulse response $h(t)$, it is expressed by

$$y(t) = \int_{-\infty}^{\infty} h(\alpha)x(t-\alpha)d\alpha$$

$$= \int_{-\infty}^{\infty} c\delta(\alpha)x(t-\alpha)d\alpha = cx(t).$$

6.5.1 Eigenfunctions

The transfer function of a linear time-invariant system can be expressed in terms of its impulse response $h(t)$ with the input of the form $x(t) = e^{st}$. Then the system output is

$$y(t) = T[x(t)] = h(t) * x(t) = \int_{-\infty}^{\infty} h(\alpha)x(t-\alpha)d\alpha$$

$$= \int_{-\infty}^{\infty} h(\alpha)e^{s(t-\alpha)}d\alpha = e^{st}H(s).$$

$$(6.5.18)$$

An equation satisfying

$$T\{e^{st}\} = H(s)e^{st}. \qquad (6.5.19)$$

is called an *eigenfunction* (or a *characteristic function*) and $H(s)$ is the *eigenvalue* (or the *characteristic value*). That is, e^{st} is the eigenfunction and the eigenvalue is defined as the system function. In terms of Fourier transforms,

$$y(t) = H(j\omega)e^{j\omega t}. \qquad (6.5.20)$$

It is the response of a linear time-invariant system with a transfer function $H(j\omega)$ to an input $e^{j\omega t}$ and the relationship holds for each ω. The responses for

each frequency, corresponding to a linear time-invariant (LTI) system, are tied by (6.5.20). The response of a LTI system with a transfer function $H(j\omega)$ to a unit input $H(0)$.

6.5.2 Bounded-Input/Bounded-Output (BIBO) Stability

BIBO stability of a *LTI* system is tied to its impulse response $h(t)$. Consider the output of the LTI system described by the convolution integral to a bounded input $x(t)$ with $|x(t)| \leq M$. It can be shown that $y(t)$ is bounded provided the impulse response $h(t)$ is absolutely integrable. That is,

$$y(t) = \int_{-\infty}^{\infty} x(t-\beta)h(\beta)d\beta \Rightarrow |y(t)|$$

$$\leq \int_{-\infty}^{\infty} |x(t-\beta)||h(\beta)|d\beta \leq M \int_{-\infty}^{\infty} |h(\beta)|d\beta.$$

$$(6.5.21)$$

$$\int_{-\infty}^{\infty} |h(\beta)|d\beta = K < \infty \Rightarrow |y(t)| \leq MK = N, \quad (6.5.22.)$$

If (6.5.22) is satisfied, then the output is bounded and the system is *BIBO stable*.

Example 6.5.1 Determine if the system described by its impulse response $h(t) = e^{-at}u(t), a > 0$ is BIBO stable.

Solution: The system is BIBO stable since

$$\int_{-\infty}^{\infty} |h(\beta)|d\beta = \int_{0}^{\infty} e^{-2a\beta}d\beta = -\frac{1}{2a}e^{-2a\beta}\Big|_{0}^{\infty} = \frac{1}{2a} < \infty.$$

                                                                                                     ∎

BIBO stability requires that the transfer function is strictly proper. That is, the degree of the numerator polynomial of the transfer function is less than the degree of the denominator polynomial. Otherwise, the impulse response will contain the derivatives of the impulse function, which are not absolutely integrable. The ideal differentiator is

$$y(t) = dx(t)dt \xleftrightarrow{\text{LT}} sX(s), H(s) = s, h(t) = \delta'(t).$$

The derivative of an impulse function is not absolutely integrable. The transfer function is not strictly proper. Similarly the ideal integrator has a transfer function $H(s) = 1/s$ has a simple pole on the $j\omega$ axis at the origin and is marginally stable.

Stability analysis is an important topic in all areas of systems engineering, especially in control systems. The literature is extensive in this area and the discussion here is limited to simple ideas. A linear time-invariant system is stable if every root of its characteristic equation, i.e., the poles of the transfer function $H(s)$ have negative real parts. The natural and forced responses of these systems can be described by seeing the properties of the inverse Laplace transforms of response functions with poles at various locations on the s plane. As mentioned earlier the responses of systems have two parts, one is a natural response that is due to the system itself and the other one is the response of the system when there is input. If a characteristic root is a simple real and negative, then the response corresponding to this pole is exponentially decaying. If a root is multiple and real, then the response is a polynomial in t multiplied by an exponentially decaying response. If we have a simple complex conjugate poles on the imaginary axis, then the corresponding response is oscillatory. A system with simple finite poles on the imaginary axis is called *wide sense stable* or *marginally stable*. The ideal integrator has a transfer function $H(s) = 1/s$ has a simple pole on the $j\omega$ axis at the origin. It is marginally stable. If we have a pair of complex conjugate poles on the left half plane, the corresponding response is exponentially decaying oscillatory response. If the poles are multiple complex conjugate, then the time response has the form of a polynomial multiplied by an oscillatory decaying response.

The systems are stable if its transfer function has all poles on the left half of the s plane. The behavior of the impulse response depends on the poles closest to the imaginary axis. For most systems these are simple complex poles, referred to as *dominant poles*. Complicated systems with transfer function given by $H(s)$ generally have many poles and are approximated by a reduced-order system $H_R(s)$ by keeping only the poles near the imaginary axis, referred to as a *model reduction*.

Example 6.5.2 Illustrate the model reduction of the system with the transfer function

$$H(s) = 100\left[\frac{1}{(s+1)} - \frac{1}{(s+5)}\right] \xleftrightarrow{\text{LT}} 100e^{-t}u(t)$$
$$- 100e^{-5t}u(t) = h(t)$$

Solution: Noting that the pole at $s = -5$ is farther away than the pole at $s = -1$, the transfer function $H(s)$ can be approximated by a reduced-order function

$$H_{R,1}(s) = \frac{100}{(s+1)} \xleftrightarrow{\text{LT}} 100e^{-t}u(t) = h_{R,1}(t).$$

Another way is ignore the poles away from the imaginary axis. Then

$$H(s) = \frac{400}{5(s+1)(.2s+1)} \Rightarrow H_{R,2}(s)$$
$$= \frac{80}{(s+1)} \xleftrightarrow{\text{LT}} 80e^{-t}u(t) = h_{R,2}(t). \quad \blacksquare$$

6.5.3 Routh–Hurwitz Criterion (R–H criterion)

The *R–H criterion* Kuo (1987) provides a test if the roots of a polynomial given below are on the left half plane without actually factoring the polynomial.

$$D(s) = d_n s^n + d_{n-1}s^{n-1} + \cdots + d_1 s + d_0 \quad (6.5.23)$$

Without loosing any generality the coefficient of s^n is assumed to be 1. Furthermore, if $d_0 = 0$, i.e., the polynomial has a root at $s = 0$, the polynomial can be divided by s and test the resulting polynomial for the stability. The *Routh array* starts by arranging two rows consisting of the coefficients of the polynomial in the following form:

Note that $(n - k)^{th}$ row starts with the coefficient (s^{n-k}). If n is even (odd), then d_0 is the last entry in row n $(n-1)$. The next step is construct row $(n-2)$ by using rows n and $n-1$.

Row n : (s^n) : d_n d_{n-2} d_{n-4} ...

Row $n-1$: (s^{n-1}) : d_{n-1} d_{n-3} d_{n-5} ...

$$(6.5.24)$$

Row $n-2$: (s^{n-2}) : $\dfrac{d_{n-1}d_{n-2} - d_n d_{n-3}}{d_{n-1}}$

$$\dfrac{d_{n-1}d_{n-4} - d_n d_{n-5}}{d_{n-1}} \quad \dots \qquad (6.5.25)$$

The entries in row $n-2$ can be written in terms of determinants.

$$\text{Row} \quad n-2 : -\frac{1}{d_{n-1}}\begin{vmatrix} d_n & d_{n-2} \\ d_{n-1} & d_{n-3} \end{vmatrix}$$

$$-\frac{1}{d_{n-1}}\begin{vmatrix} d_n & d_{n-4} \\ d_{n-1} & d_{n-5} \end{vmatrix} \quad \dots \quad (6.5.26)$$

Row $n-3$ is computed in a similar manner using rows $n-1$ and $n-2$. The procedure is continued until we reach row 0. The R–H criterion states that all the roots of the polynomial $D(s)$ lie on the left half of the s-plane if all the entries in the left most column of the Routh array are nonzero and have the same sign. The number of sign changes in the leftmost column is equal to the number of roots of $D(s)$ in the right half s-plane.

Example 6.5.3 Determine the number of roots that are on the right half s-plane of $D(s)$.

$$D(s) = s^4 + 2s^3 + 3s^2 + 4s + 5. \qquad (6.5.27)$$

Solution: Routh array is given below.

Row 4 :(s^4) :   1   3   5

Row 3 :(s^3) :   2   4   0

Row 2 :(s^2) : $-\dfrac{\begin{vmatrix} 1 & 3 \\ 2 & 4 \end{vmatrix}}{2} = -1$ $-\dfrac{\begin{vmatrix} 1 & 0 \\ 2 & 5 \end{vmatrix}}{2} = -\dfrac{5}{2}$ 0

Row 1 :(s^1) : $-\dfrac{\begin{vmatrix} 2 & 4 \\ -1 & -(5/2) \end{vmatrix}}{-1} = -1$ 0   0

Row 0 :(s^0) : $-\dfrac{\begin{vmatrix} -1 & -5/2 \\ -1 & 0 \end{vmatrix}}{-1} = 5/2$ 0   0

Entries in the first column can be written by $[1, 2, -1, -1, 5/2]$. There are two sign changes indicating that there are two roots on the right half s-plane. Using MATLAB, the roots of the polynomial can be computed by using the following:

$d = [1\ 2\ 3\ 4\ 5]$: coefficents of the polynomial

$r = roots(d)$: Gives the roots : $\begin{bmatrix} .2878 \pm j1.4161 \\ -1.288 \pm j.8579 \end{bmatrix}$

$d = poly(r)$: Gives the coefficeients of the polynomial

Routh array gives the information about the number of roots on the right half of the s-plane and not the actual roots. In the Routh array, there are divisions. If these division terms are zero, then a different technique is needed to overcome this. ∎

Special cases:

1. Routh array has a zero in the first column of a row.
2. Routh array has an entire row of zeros.

1. If the first entry in the row $(n-i), i \neq 0$ or 1 is zero, to compute the entries in the row $(n-i+1)$, a problem of division by zero arises. To alleviate this problem, ε is assigned to 0. ε is allowed to approach zero either $\varepsilon \to 0^+$ or 0^-.

Example 6.5.4 Consider the polynomial $D(s) = s^4 + s^3 + 2s^2 + 2s + 1$. Use the Routh array to determine the number of roots on the right half of the s plane.

Solution: Routh array is given by

Row 4 : (s^4) :   1   2   1
Row 3 : (s^3) :   1   2   0
Row 2 : (s^2) : $0(\varepsilon)$ 1   0
Row 1 : (s^1) : $-\frac{1-2\varepsilon}{\varepsilon}$ 0   0
Row 0 : (s^0) :   1

In the next step the entries in the first column are written by

First column: $[1,1,\varepsilon,-(1-2\varepsilon)/\varepsilon,1]$
$\rightarrow \{\varepsilon=0^+ \Rightarrow [+,+,+,-,+]\}, \varepsilon=0^- \Rightarrow [+,+,-,+,+]\}$

Using either of the two cases, there are two sign changes. There are two roots on the right half of the s-plane. Using MATLAB, the roots of $D(s)$ are $0.1247 \pm j1.3066$, $-0.6217 \pm j0.4406$. The polynomial has two roots in the right half of the s plane.∎

2. Next consider the case that an entire row in the Routh array consists of zeros. To illustrate this consider the following possibilities:

   a. Roots are located on the imaginary axis
   b. Roots with symmetry about origin
   c. Roots with quadrant symmetry

Each of these implies the following type of factors in the polynomial $D(s)$:

a. $(s \pm j\beta) \rightarrow (s^2 + \beta^2)$
b. $(s \pm \alpha) \rightarrow (s^2 - \alpha^2)$ (6.5.28a)

c. $(s \pm \alpha \pm j\beta) \rightarrow s^4 + [2(\alpha^2 + \beta^2) - 4\alpha^2]s^2 + (\alpha^2 + \beta^2)^2$ (6.5.28b)

These roots produce even polynomials resulting in a row of zeros in the Routh array. The row before the row of zeros in the array gives the even polynomial identified here as $D_2(s)$ and is called the *auxiliary equation*. That is $D(s) = D_1(s)D_2(s)$. To complete the Routh array, take the derivative of the auxiliary equation and replace the row of zeros by the row obtained from the coefficients of the derivative of the auxiliary equation.

Example   6.5.5   Consider   the   polynomial $D(s) = s^4 + s^3 - s^2 + s - 2$. Show that the system described by this characteristic polynomial is unstable using the Routh array.

Solution: Routh array is given by

Row 4 : (s^4) :   1 -1 -2
Row 3 : (s^3) :   1   1   0
Row 2 : (s^2) : $-(1 - (-1)) = -2$ -2 0
Row 1 : (s^1) :   0   0

Noting that the row 1 has all zeros, the auxiliary equation can be written from row 2 and $D_2(s) = -(s^2 + 1) = 0$. Since the (–) sign is irrelevant for the roots, the sign can be ignored and written as $D(s) = D_1(s)(-D_2(s))$ and $D_1(s) = s^2 + s - 2$. The auxiliary polynomial $D_2(s)$ has a pair of imaginary roots at $s = \pm j1$. In this simple example, the polynomial $D_1(s)$ can be factored and its roots are located at $s = 1, -2$ indicating that $D(s)$ has one root on the right half of the s-plane. If the number of roots $D_1(s)$ is higher than 2, then the Routh array can be continued in the following manner.

Row 4 : (s^4) :   1 -1 -2
Row 3 : (s^3) :   1   1   0
Row 2 : (s^2) : -2 -2 0   (Auxiliary polynomial, $D_2(s) = -(2s^2 + 2)$)
Row 1 : (s^1) : -4 0   0   ($D_2'(s)$)
Row 0 : (s^0) : -2

Entries of the first column in the above Routh array are $[1, 1, -2, -4, -2]$ indicating there is one root inside the right half of the s-plane. As mentioned before, Routh array does not provide the roots of the polynomial. It merely identifies the number of roots in the right half s-plane. Routh array is frequently used in feedback control systems to determine the condition of stability of a control system. Note that if *all* the coefficients of the characteristic polynomial $D(s)$ do not have the same sign, the polynomial has some roots on the right half s-plane and the corresponding system is unstable.

Example 6.5.6 Using the Routh array determine the range of values for K for which all the roots of the polynomial $D(s) = s^3 + 3s^2 + 3s + K$ are located inside the left half plane.

Solution: The Routh array is

$$
\begin{array}{lll}
\text{Row 3} : (s^3) : & 1 & 3 \\
\text{Row 2} : (s^2) : & 3 & K \\
\text{Row 1} : (s^1) : & -\dfrac{(K-9)}{3} & 0 \\
\text{Row 0} : (s^0) : & K
\end{array}
$$

To have all the roots of the polynomial on the left half plane, the coefficients in the first column in the Routh array must have the same signs. This implies that $(9 - K) > 0$ and $K > 0$. All the roots are on the right half plane if $0 < K < 9$, which gives the range of values of K to keep the system stable. When $K = 9$, row 1 has all zeros. Correspondingly, the auxiliary polynomial is $D_2(s) = (3s^2 + 9)$, indicating polynomial has a pair of roots on the imaginary axis. In the case of $K = 0$, there is a root at $s = 0$.

In the case of a root at $s = 0$, it is evident from the polynomial that $D(s) = sD_1(s)$ and the Routh array can be determined starting with $D_1(s)$. ∎

Notes: A polynomial $D(s)$ with all its roots on the left half s-plane is called a *strictly Hurwitz* polynomial. If it has all its roots on the left half s-plane and in addition, it has simple poles on the imaginary axis, then it is called a *pseudo-Hurwitz* polynomial. ∎

6.5.4 Eigenfunctions in the Fourier Domain

In terms of the Fourier domain, we have from (6.5.20) that

$$
y(t) = T\{e^{j\omega t}\} = H(j\omega)e^{j\omega t} = \{|H(j\omega)|e^{j\phi(\omega)}\}e^{j\omega t}
$$
$$
= |H(j\omega)|e^{j[\omega t + \phi(\omega)]}. \tag{6.5.30}
$$

For a particular value of $\omega = \omega_0$, (6.5.30) reduces to

$$
T\{e^{jk\omega_0 t}\} = H(jk\omega_0)e^{jk\omega_0 t} = |H(jk\omega_0)|e^{j[k\omega_0 t + \phi(k\omega_0)]} \tag{6.5.31}
$$

Since the system under consideration is a LTI system, the response to several frequencies can be determined by (6.4.31). The system with the frequency response $H(j\omega)$ acts like a gate to allow certain frequencies fully or partially through or attenuated or eliminated.

Example 6.5.7 Consider a LTI system with a transfer function $H(j\omega)$. Use (6.5.30) to find the responses to the real periodic inputs given by

$$
a. \ x_T(t) = \sum_{k=-\infty}^{\infty} X_s[k]e^{jk\omega_0 t},
$$

$$
b. \ x_T(t) = X_s[0] + \sum_{k=1}^{\infty} d[k]\cos(k\omega_0 t + \theta[k]). \tag{6.5.32}
$$

Solution: *a.* Using (6.5.30), the output of the linear time-invariant (LTI) system is

$$
y_T(t) = H(j\omega)\sum_{k=-\infty}^{\infty} X_s[k]e^{jk\omega_0 t}
$$
$$
= \sum_{k=-\infty}^{\infty} H(jk\omega_0)X_s[k]e^{jk\omega_0 t} = \sum_{k=-\infty}^{\infty} Y_s[k]e^{jk\omega_0 t}. \tag{6.5.33}
$$

If the input to a LTI system is periodic, then the output is also periodic with the same period and the F-series coefficients of the output and the input are related by

$$
Y_s[k] = H(jk\omega_0)X_s[k] = |H(jk\omega_0)||X_s([k])|e^{j(\phi(k\omega_0) + \theta[k])} \tag{6.5.34a}
$$

$$
|Y_s[k]| = |X_s[k]||H(jk\omega_0)|, \ \angle Y_s[k] = \angle X_s[k] + \phi(k\omega_0). \tag{6.5.34b}
$$

b. $y_T(t) = H(0)X_s[0]$

$$+ \sum_{k=1}^{\infty} (|H(k\omega_0)||d[k]|) \cos(k\omega_0 t + \phi(k\omega_0)$$

$$+ \theta(k\omega_0)). \qquad (6.5.35) \quad \blacksquare$$

Notes: The response given in (6.5.33) is the *steady-state response* of the linear system to a periodic input. A linear time-invariant system does *not* produce any new frequencies. The output amplitudes and the phases of the harmonics are different from the amplitudes and the phases of the input signal harmonics and are determined by (6.5.34b), for a real periodic input. $\qquad \blacksquare$

Example 6.5.8 Use the Fourier transforms to derive the output given in (6.5.33).

Solution:

$$x_T(t) = \sum_{k=-\infty}^{\infty} X_s[k]e^{jk\omega_0 t} \xrightarrow{\text{FT}} X(j\omega)$$

$$= \sum_{k=-\infty}^{\infty} \pi X_s[k]\delta(\omega - k\omega_0)$$

$$Y(j\omega) = H(j\omega) \sum_{k=-\infty}^{\infty} \pi X_s[k]\delta(\omega - k\omega_0)$$

$$= \sum_{k=-\infty}^{\infty} \pi\{H(jk\omega_0)X_s[k]\}\delta(\omega - k\omega_0).$$

Taking the inverse transform of the transform of $Y(j\omega)$ in (6.4.35), we have

$$y_T(t) = \sum_{k=-\infty}^{\infty} \{H(jk\omega_0)X_s[k]\}e^{jk\omega_0 t}. \qquad (6.5.36) \quad \blacksquare$$

Example 6.5.9 Find the output $y_T(t)$ of the *RC* circuit in Fig. 6.5.4b corresponding to the periodic pulse signal shown in Fig. 6.5.4a with a period equal to $T = 2\pi$.

Solution: The Fourier series of the input waveform is given by

$$x_T(t) = \frac{4}{\pi}\left[\frac{\cos(t)}{1} - \frac{\cos(3t)}{3} + \frac{\cos(5t)}{5} - \cdots\right],$$

$$T = 2\pi, \quad \omega_0 = 1. \qquad (6.5.37)$$

The transfer function, the amplitude, and phase responses are given by

$$H(j\omega) = \frac{1}{(1 + j\omega RC)}, |H(j\omega)| = \frac{1}{\sqrt{1 + (\omega RC)^2}},$$

$$\angle - \tan(\omega RC). \qquad (6.5.38)$$

The kth harmonic term and the steady-state output response are, respectively, given by

$$|H(jk\omega_0)| = \frac{1}{\sqrt{1 + (k\omega_0 RC)^2}}, \angle H(jk\omega_0)$$

$$= -\tan^{-1}(k\omega_0 RC). \qquad (6.5.39)$$

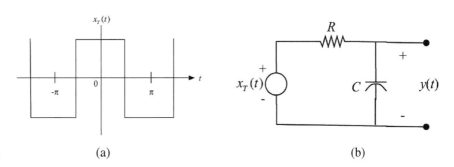

Fig. 6.5.4 (a) Periodic pulse waveform and (b) RC circuit

(a)

(b)

$$y_T(t) = \frac{4}{\pi}\left[\frac{\cos(t - \tan^{-1}(1))}{1\sqrt{2}} - \frac{\cos(3t - \tan^{-1}(3))}{3\sqrt{10}} + \frac{\cos(5t - \tan^{-1}(5))}{5\sqrt{26}} - \ldots\right]. \tag{6.5.40}$$

Note the kth harmonic input produces the kth harmonic output illustrated below.

$$\frac{4}{\pi}\cos(k\omega_0 t)\big|_{\omega_0=1} \to \frac{4}{\pi\sqrt{1 + k^2\omega_0^2(R^2C^2)}}\cos(k\omega_0 t - \tan^{-1}(k\omega_0 RC))\big|_{\omega_0=1}. \tag{6.5.41}$$

Attenuation is proportional to $(1/k)$ and the phase shift is $-\tan^{-1}(k\omega_0 RC)$ in the kth harmonic term. The RC circuit is a low-pass filter. The low frequencies have smaller attenuations and higher frequencies are significantly attenuated. ∎

6.6 Step Response

The step response of a continuous time LTI system is the response to a step input $x(t) = u(t)$. The step response $s(t)$, is related to the impulse response $h(t)$ (see (6.5.10)) and

$$s(t) = h(t) * u(t) = \int_{-\infty}^{\infty} h(\beta)u(t-\beta)d\beta = \int_{-\infty}^{t} h(\beta)d\beta. \tag{6.6.1}$$

In the above equation the variable of integration is β not t and $u(t - \beta) = 0$ for $t < \beta$. The step response can be obtained by integrating the impulse response and the impulse response can be obtained by differentiating the step response. That is,

$$h(t) = \frac{ds(t)}{dt}. \tag{6.6.2}$$

Example 6.6.1 Determine the step response of the RC circuit in Example 6.3.1 from the impulse response and vice versa.

Solution: From (6.5.16a), the impulse response is

$$h(t) = \frac{1}{RC}e^{-t/RC}u(t). \tag{6.6.3a}$$

The step response is

$$s(t) = \int_{-\infty}^{t} h(\beta)d\beta$$

$$= \int_{-\infty}^{t} \frac{1}{RC}e^{-(\beta/(RC))}u(\beta)d\beta = (1 - e^{-(t/RC)})u(t). \tag{6.6.3b}$$

The impulse response from the step response by

$$h(t) = \frac{ds(t)}{dt} = \frac{d(1 - e^{-t/RC})u(t)}{dt}$$

$$= \delta(t)(1 - e^{-t/RC}) + u(t)\left[\frac{1}{RC}e^{-t/RC}\right]$$

$$= (1/RC)e^{-t/RC}u(t).$$

Note $(1 - e^{-t/RC})$ is continuous at $t = 0$ and $\delta(t)(1 - e^{-t/RC})$ is zero, see (1.4.5). The impulse and the step responses are sketched in Fig. 6.6.1. The *rise time* of the RC circuit is the time required for a unit step response to go from 10 to 90% of its final value. It is given by

$$t_r = t_2 - t_1, s(t_1) = (1 - e^{-t_1/RC}) = .1,$$
$$s(t_2) = 1 - e^{-t_2/RC}$$

$$0.9 = e^{-t_1/RC}, \text{ and } 0.1 = e^{-t_2/RC}, t_r = (t_2 - t_1)$$
$$= RC\ln(9) = 2.197RC. \tag{6.6.4}$$

Fig. 6.6.1 (a) Impulse response and (b) step response

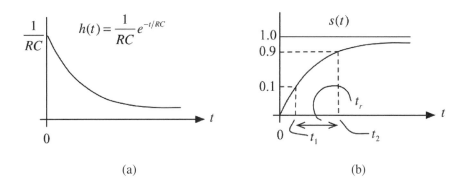

(a)   (b)

Rise time is a measure of how fast the system responds to an input. It is related to the bandwidth of the circuit and we will discuss this shortly. ∎

Rise time and the 3 dB bandwidth: The output transform and the transfer function of the RC circuit given Fig. 6.3.4a are

$$Y(j\omega) = \left[\frac{1}{1 + j\omega RC}\right] X(j\omega) = H(j\omega)X(j\omega),$$

$$H(j\omega) = \left[\frac{1}{1 + j\omega RC}\right]. \qquad (6.6.5)$$

The amplitude and the phase responses of the transfer function are

$$20og|H(j\omega)| = 20\log\frac{1}{\sqrt{1 + (\omega RC)^2}}\, dB,$$

$$\angle H(j\omega) = -\tan^{-1}(\omega RC). \qquad (6.6.6)$$

The responses are shown in Fig. 6.6.2 for positive frequencies. Note the amplitude response is even and the phase response is odd. The amplitude at $\omega = 0$ is 1 and in dB, it is 0 dB. At $\omega = 1/RC$, the magnitude is equal to $1/\sqrt{2}$ and in dB this is -3 dB. The *3 dB frequency* (or *the half-power*) is

$$\omega_{3dB} = 1/RC, \text{ or } f_{3dB} = 1/2\pi RC \text{ Hertz} \quad (6.6.7)$$

The amplitude response $|H(j\omega)|$ decreases smoothly for higher frequencies and goes to zero at infinity. The rise time is related to the 3 dB bandwidth and is

$$t_r = 2.197/(2\pi f_{3dB}) = .35/f_{3dB}. \qquad (6.6.8)$$

In summary, the RC circuit is a simple low-pass filter passing frequencies between 0 and $f_{3\,dB}$ with small attenuations and all the higher frequencies are attenuated significantly. The phase response is zero at $\omega = 0$. At the 3 dB frequency, it is equal to $(-\pi/4)$ and at the infinite frequency the phase response reaches $(-\pi/2)$ rad or $-90°$. ∎

Ideal integrator: The transfer function of the ideal integrator is $H(s) = 1/s$. The amplitude and the phase responses are, respectively, given by

$$H(j\omega) = (1/j\omega) = (-j/\omega), |H(j\omega)| = 1/|\omega|,$$
$$\angle H(j\omega) = -\pi/2, \omega > 0. \qquad (6.6.9)$$

This function represents an ideal integrator by noting that if the input is a sinusoid, say $\cos(\omega t)$, the output of the integrator and its transform are

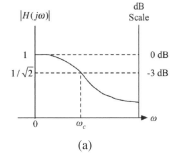

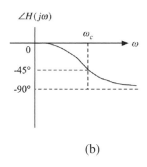

Fig. 6.6.2 (a) Amplitude response and (b) phase response

(a)   (b)

$$y(t) = \int_{-\infty}^{t} x(t)dt = \frac{1}{\omega}\sin(\omega t) = \frac{1}{\omega}\cos\left(\omega t - \frac{\pi}{2}\right)$$

$$\tag{6.6.10}$$

$$Y(j\omega) = H(j\omega)X(j\omega), H(j\omega) = |H(j\omega)|e^{j\angle H(j\omega)},$$
$$|H(j\omega)| = 1/|\omega|, \angle H(j\omega) = -j\,\text{sgn}(\omega). \tag{6.6.11}$$

The amplitude response is inversely proportional to $|\omega|$; the phase response for $\omega > 0$ is $(-\pi/2)$, a constant. Since the amplitude gain of an ideal integrator is $(1/|\omega|)$, it suppresses the higher frequency components and enhances the low-frequency components. The noise signals contain mostly high-frequency components, and the integrator reduces the size of the high-frequency components.

Ideal differentiator: The transfer function of an ideal differentiator is

$$H(s) = s, H(j\omega) = j\omega. \tag{6.6.12}$$

The amplitude and phase responses are given by

$$|H(j\omega)| = |\omega|\text{and } \angle H(j\omega) = \frac{\pi}{2}, \omega > 0. \tag{6.6.13}$$

Consider that a sinusoidal function $x(t) = \cos(\omega t)$ is passed through a differentiator, then

$$y(t) = \frac{dx(t)}{dt} = \frac{d\cos(\omega t)}{dt} = -\omega\sin(\omega t)$$
$$= \omega\cos\left(\omega t + \frac{\pi}{2}\right). \tag{6.6.14}$$

Note the amplitude response increases linearly with frequency ω and phase response is constant and is equal to $(\pi/2)$ rad for positive frequencies. From the amplitude response expression, we see that the high frequencies are enhanced. Most corrupted signals contain noise components that are of high frequencies. Using a derivative function enhances the noise signal much more than a low-frequency signal. Derivative function is used to sharpen a signal. For example, to sharpen an image at the edges, we use a derivative function. Note the discontinuity in the phase response at $\omega = 0$.

Example 6.6.2 Show the circuit shown in Fig. 6.6.3 can be used as a differentiator.

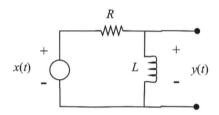

Fig. 6.6.3 *RL* circuit

Solution: The output transform is

$$Y(j\omega) = \frac{j\omega L}{R + j\omega L}X(j\omega), |H(j\omega)| = \frac{|\omega|L}{\sqrt{R^2 + (\omega L)^2}},$$

$$\angle H(\omega) = \frac{\pi}{2} - \tan^{-1}\left(\frac{\omega L}{R}\right). \tag{6.6.15}$$

For small frequencies, i.e., $|\omega| = (R/L)$, the output transform can be approximated. Noting the F-transform derivative theorem, it follows that the circuit acts like a differentiator. That is,

$$Y(j\omega) \approx j\omega[Lx(j\omega)] \xrightarrow{\text{FT}} L\frac{dx(t)}{dt} \approx y(t). \tag{6.6.16} \blacksquare$$

Example 6.6.3 Find the response of the *RC* circuit in Fig. 6.3.4a to the input pulse

$$x(t) = A\Pi\left[\frac{t - \frac{T}{2}}{T}\right]. \tag{6.6.17}$$

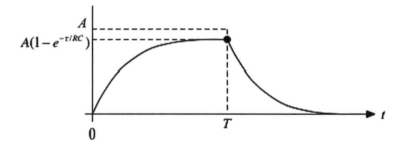

Fig. 6.6.4 *RC* circuit
response to a pulse input

Solution: The transfer function $H(j\omega)$, its impulse response $h(t)$, output frequency response and the response $y(t)$ using the convolution integral, we have

$$Y(j\omega) = H(j\omega)X(j\omega) \xleftrightarrow{\text{FT}} h(t) * x(t) = y(t), \quad (6.6.18)$$

$$H(j\omega) = \frac{1}{(1+j\omega RC)} \xleftrightarrow{\text{FT}} \frac{1}{RC}e^{-t/RC}u(t) = h(t)$$
$$(6.6.19)$$

$$y(t) = \int_{-\infty}^{\infty} h(t-\beta)x(\beta)d\beta, \, h(t-\beta)$$

$$= \begin{cases} \frac{1}{RC}e^{-(t-\beta)/RC}, & \beta < t \\ 0, & \beta > t \end{cases}. \quad (6.6.20)$$

$$y(t) = \begin{cases} 0, & t<0 \\ \int_0^t \frac{A}{RC}e^{-(t-\beta)/RC}d\beta, & 0<t<T \\ \int_0^T \frac{A}{RC}e^{-(t-\beta)/RC}d\beta, & t>T \end{cases}$$

$$= \begin{cases} 0, & t<0 \\ A(1-e^{-t/RC}), & 0<t<T. \\ A(1-e^{-T/RC})e^{-(t-T)/RC}, & t>T \end{cases}$$
$$(6.6.21)$$

Note that the input pulse is the same as in Example 2.2.4, except the pulse is of width T instead of $2T$ and the pulse started at $t=0$ rather than at $t=-T$. The function in (6.6.21) is sketched in Fig. 6.6.4. The response can be visualized by the following argument. For $t<0$, the input is zero and the output is zero as well. At $t=0$ we have a step input and the capacitor voltage cannot charge instantaneously and the voltage across the capacitor starts at 0 and increases exponentially with a time constant RC. At $t=T$, the input becomes zero and for $t>T$ the charge across the capacitor discharges through the resistor and the capacitor voltage decreases exponentially from the peak value of $A(1-e^{-T/RC})$ to zero as $t \to \infty$. Another way to derive (6.6.2) is that the input pulse function is $A\Pi[(t-(T/2))/T] = Au(t) - Au(t-T)$.

Assuming the unit step input response is $h_u(t)$, the delayed step input response is $h_u(t-T)$. The response to the pulse input is $A(h_u(t) - h_u(t-T))$. ∎

Simple frequency analysis of the *RC* circuit in the last example: The Fourier transforms of the input, the transfer function and the output transform are

$$X(j\omega) = F\left\{A\Pi\left[\frac{t-\frac{T}{2}}{T}\right]\right\} = A\tau\frac{\sin(\omega T/2)}{(\omega T/2)}e^{-j\omega\frac{T}{2}},$$

$$H(j\omega) = \frac{1}{1+j\omega RC}. \quad (6.6.22)$$

$$Y(j\omega) = \frac{A\tau\,\text{sinc}(\omega T/2)e^{-j\omega T/2}}{(1+j\omega RC)}. \quad (6.6.23)$$

We will sketch the amplitude of the output transform by considering two special cases:

a. Pulse width T is very large compared to the time constant $\tau = RC$ (i.e., $T \gg RC$)
b. Time constant is very small compared to the time constant (i.e., $T \ll RC$). Now

$$|Y(j\omega)| = |H(j\omega)||X(j\omega)|. \quad (6.6.24)$$

For the two special cases, the functions $|H(j\omega)|$ and $|X(j\omega)|$ are sketched in Fig. 6.6.5a,b. In case a, the 3 dB bandwidth is assumed to be much larger than the main lobe width of the response. That is, $(1/(2\pi)f_{3dB}) = (1/RC) \ll 1/T$ or $T \gg RC$ and the function $|H(j\omega)|$ is essentially flat in the range $|\omega| < 1/T$. In this frequency band the amplitude of the output transform is approximately equal to the magnitude of the input transform and we can approximate and $|Y(\omega)| \approx k|X(\omega)|, k$ – a constant. The output pulse will be a good approximation of the input pulse. In case b, $(1/(2\pi f_{3dB})) \gg 1/T$ or $T \ll RC$. From Fig. 6.6.5 we see that $|X(j\omega)|$ is essentially flat in the 3 dB frequency range. That is, $|Y(j\omega)| \approx |H(j\omega)|$ in this range. This indicates that the amplitude of the output transform looks more like the magnitude of the system transform in this case. We are interested in the input signal transform, not the transform of the system. When a signal is passed through a system, the bandwidth of the system must be much larger than the

Fig. 6.6.5 Frequency
analysis of an RC
circuit with a pulse
input(a) $T = RC$,
(b) $T = RC$

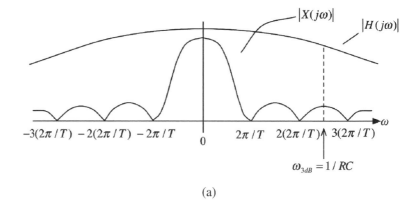

(a)

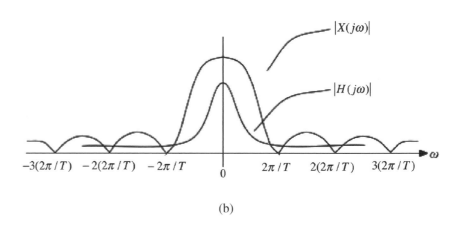

(b)

bandwidth of the input signal in order the output response to have some resemblance of the input.

Let us now consider some simple ideas about the response of a sequence of pulses. The detection of the existence of the pulse at the output can be improved by increasing the amplitude of the input pulse or increasing the width of the pulse or both. Increasing the amplitude increases the power requirements on the input. Increasing the pulse width implies that the number of pulses that can be transmitted per unit time has to be reduced. In addition the pulses are not band limited. Since the RC circuit is a LTI system, it is conceivable that if the input consists of a set of pulses, the output response will be a sum of the individual responses of the pulses with appropriate delays in the pulse responses. The sum may become unbounded.

Next we will consider the process of removing the effects of the RC circuit. This type of a situation appears in measuring the voltage across a

component, seeing a picture through a lens and many others. In these measurements, the signal is affected by the measuring device or the system we visualize with. If the bandwidth of these devices is much, much larger than the signal bandwidths, then the effect of the measuring devices is minimal. Removing the effects of the transmission system from the received signal is an important problem and this process is called the deconvolution and is discussed next.

Deconvolution: Let the output response of a LTI system with the transfer function $H(j\omega)$, and the corresponding impulse response $h(t)$, is given by

$$y(t) = h(t) * x(t) \overset{FT}{\longleftrightarrow} H(j\omega)X(j\omega) = Y(j\omega).$$
$$(6.6.25)$$

To recover the signal, $x(t)$ from $y(t)$, consider Fig. 6.6.6. The first block represents a system and

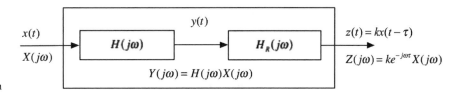

Fig. 6.6.6 Deconvolution

the second block identified by $H_R(\omega)$ represents a system to recover the original signal. The output transfrom, $Z(j\omega)$ is related to the input transform, assuming no loading effects, is

$$Z(j\omega) = H_R(j\omega)H(j\omega)X(j\omega). \quad (6.6.26)$$

To recover the signal, it is desired to have $|H(j\omega)||H_R(j\omega)| \approx k$, a gain constant in the frequency range of interest. Since there is an inherent delay in every system, say τ seconds, this delay can be incorporated and $z(t) = kx(t - \tau)$. This implies that

$$H(j\omega)H_R(j\omega) = k\,e^{-j\omega\tau} \text{ or } H_R(j\omega) = k\,e^{-j\omega\tau}/H(j\omega). \quad (6.6.27)$$

A circuit that gives the transfer function $H_R(j\omega)$ in (6.6.27) may *not* always be possible. For example, if $H(j\omega) = 0$ at some frequency $\omega = \omega_i$, the function $H_R(j\omega)$ goes to infinity at this frequency. Therefore, $H_R(j\omega)$ can only be approximated. In terms of the time domain, in the ideal case, the inverse transform of $Z(j\omega)$ is given by

$$\begin{aligned} z(t) &= F^{-1}[Z(j\omega)] = F^{-1}[H(j\omega)H_R(j\omega)X(j\omega)] \\ &= h_R(t) * h(t) * x(t) \end{aligned} \quad (6.6.28)$$

There is perfect deconvolution if $h(t) * h_R(t) = \delta(t)$ and $z(t) = \delta(t) * x(t) = x(t)$.

6.7 Distortionless Transmission

A system is called *distortionless* if the output is the same as the input except the signal may be attenuated by the same amount for *all* frequencies along with a delay of t_0 seconds. A distortionless system has the output

$$\begin{aligned} y(t) &= H_0 x(t - t_0) \\ &\quad (H_0 \text{ and } t_0 > 0 \text{ are some constants.}). \end{aligned} \quad (6.7.1)$$

For simplicity, assume $H_0 > 0$. We essentially tried to obtain a distortionless signal in using the deconvolution process in Fig. 6.6.6. Taking the transform of $y(t)$, the output transform and the transfer functions are as follows:

$$Y(j\omega) = H_0 e^{-j\omega t_0} X(j\omega) \rightarrow H(j\omega) = H_0 e^{-j\omega t_0}. \quad (6.7.2)$$

The amplitude and the phase responses of the distortionless system are

$$|H(j\omega)| = |H_0|, \angle H(j\omega) = -\omega t_0. \quad (6.7.3)$$

These functions are shown in Fig. 6.7.1, where H_0 is assumed to be positive. This implies that all frequencies are attenuated (or amplified) by the same amount. It is referred to as an all-pass system. The phase response is linear. The delay associated with an *ideal delay line,* a LTI system, can be seen by considering a sinusoidal input $x(t) = \cos(\omega t)$. The output is $H_0 \cos(\omega(t - t_0))$. The amplitude response is the same for all frequencies. In addition, the output is $H_0 \cos(\omega(t - t_0)) = H_0 \cos(\omega t - \omega t_0)$. That is, the time shift is t_0 and the phase shift is ωt_0 and the *phase* is *linearly proportional* to the frequency ω with a slope of $(-t_0)$

6.7.1 Group Delay and Phase Delay

The phase response in (6.7.3) and the delay are respectively given by

$$\theta(\omega) = \angle H(j\omega) = -\omega t_0 \quad (6.7.4)$$

Fig. 6.7.1 Amplitude and phase responses of a distortionless system

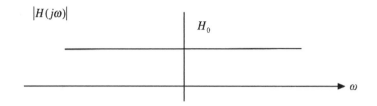

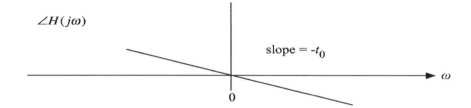

$$t_0 = -\frac{d\angle H(j\omega)}{d\omega} = -\frac{d\theta(\omega)}{d\omega} = \frac{d(\omega t_0)}{d\omega}. \quad (6.7.5)$$

Group delay: The group delay deals with a group of frequencies (usually referred as delay for simplicity). The result in (6.7.5) can be used to find the delay associated with a transfer function.

$$H(j\omega) = |H(j\omega)|e^{j\theta(\omega)}. \quad (6.7.6)$$

The group delay deals with a group of frequencies (usually referred as delay for simplicity). It is a non-linear function of frequency and is defined by

$$T_g(\omega) = -\frac{d\theta(\omega)}{d\omega}. \quad (6.7.7)$$

Example 6.7.1 Find the group delay associated with the transfer function

$$H(j\omega) = \frac{1}{(c - a\omega^2) + b(j\omega)}, a, b, c > 0.$$

$$\left(\text{Use the identity } \frac{d\tan^{-1}(x)}{dy} = \frac{1}{1 + x^2}\frac{dx}{dy}.\right).$$

$$(6.7.8)$$

Solution: The amplitude, the phase response, and the group delay are given by

$$|H(j\omega)| = \frac{1}{\sqrt{(c - a\omega^2) + (b\omega)^2}},$$

$$\theta(\omega) = \angle H(j\omega) = -\tan^{-1}\left[\frac{b\omega}{c - a\omega^2}\right].$$

$$(6.7.9)$$

$$T_g(\omega) = -\frac{d\theta(\omega)}{d\omega} = \frac{d\tan^{-1}\left(b\omega/(c - a\omega^2)\right)}{d\omega}. \quad (6.7.10)$$

$$= \frac{1}{(b\omega/(c - a\omega^2))^2 + 1}\frac{(c - a\omega^2)b - b\omega(-2a\omega)}{(c - a\omega^2)^2}$$

$$= \frac{b(c + a\omega^2)}{(b\omega^2) + (c - a\omega^2)^2}. \quad (6.7.11)$$

Figure 6.7.2 illustrates the delay function in (6.7.11) and is not constant for all ω.   ∎

In a later section, filters will be designed that have transfer functions with nonlinear phase. When signals passed through such filters, different frequencies are delayed differently. This is not critical for speech signals, as the human ear compensates for *small delays*. It is a problem in transmitting data and compensation is necessary so that the *filter delay equalizer* combination has approximately a constant delay in the desired frequency range. A second-order delay equalizer has the form

$$H_{ci}(j\omega) = \frac{(b_i - \omega^2) - ja_i\omega}{(b_i - a_i\omega^2) + ja_i\omega}, a_i, b_i > 0. \quad (6.7.12)$$

The amplitude and phase responses are

$$|H_{ci}(j\omega)| = \frac{\sqrt{(b_i - \omega^2)^2 + (a_i\omega)^2}}{\sqrt{(b_i - \omega^2)^2 + (a_i\omega)^2}} = 1,$$

$$\angle H_{ci}(j\omega) = -2\tan^{-1}\frac{a_i\omega}{b_i - \omega^2}. \quad (6.7.13)$$

Fig. 6.7.2 Group delay characteristics in Example 6.7.1

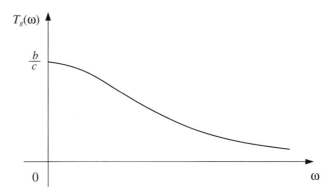

Noting that the phase angles of a product of transfer functions add, the amplitude and phase response of a cascade of n second-order sections result in

$$H_c(j\omega) = \prod_{i=1}^{n} H_{ci}(j\omega), |H_c(j\omega)| = \prod_{i=1}^{n} |H_{ci}(j\omega)| = 1$$

$$\angle H_c(\omega) = \sum_{i=1}^{N} \angle H_{ci}(\omega). \quad (6.7.14)$$

The parameters $a_i's$ and $b_i's$ can be adjusted so that the amplitude response of the filter cascaded by the delay equalizer has the same magnitude as the filter function $H_F(j\omega)$ and the corresponding phase angle has approximately linear phase (i.e., approximately constant delay) in the *desired frequency range*. That is,

$$H(j\omega) = H_F(j\omega)H_c(j\omega), |H(j\omega)| \approx |H_F(j\omega)|$$

$$\angle H(j\omega) = \angle H_F(j\omega) + \angle H_c(j\omega), \quad (6.7.15a)$$

$$-\frac{d}{d\omega}\angle H(j\omega) \approx \text{constant}. \quad (6.7.15b)$$

Phase delay: Consider the input $x_T(t) = A\cos(\omega_0 t)$ to a LTI system with a transfer function

$$H(j\omega) = |H(j\omega)|e^{j\angle H(j\omega)}, \angle H(j\omega) = \theta(\omega). \quad (6.7.16)$$

If the input to a LTI system is a sinusoid then the output is also a sinusoid at the same frequency, although the output may have a different amplitude and phase. See Example 6.5.6. Let the output of the system to the input $x_T(t)$ is

$$y_T(t) = A|H(j\omega_0)|\cos(\omega_0 t + \theta_0)$$
$$= A|H(j\omega_0)|\cos[\omega_0(t + (\theta_0/\omega_0))]. \quad (6.7.17)$$

To have a distortionless transmission, the output must have the form

$$y_T(t) = B\cos[\omega_0(t - t_0)]. \quad (6.7.18)$$

Comparing this with (6.7.17), the time delay between the input and the output at the frequency $f_0 = \omega_0/2\pi$ is $t_0 = -\theta_0/\omega_0$. For a single frequency, phase delay is appropriate. The *phase delay $T_p(\omega)$* in terms of the system phase response $\theta(\omega) = \arg(H(\omega))$ is defined by

$$T_p(\omega) = -\theta(\omega)/\omega. \quad (6.7.19)$$

Notes: For rational transfer functions, $\theta(\omega)$ is a transcendental function, whereas the group delay is a *rational function* of ω^2 making it easier for filter design. ∎

Example 6.7.2 Find the phase delay and the group delay of the transfer function

$$H(j\omega) = -[(1 - j\omega)/(1 + j\omega)]. \quad (6.7.20a)$$

Solution: The phase, the group delay, and the phase delay responses are

$$\theta(\omega) = \pi - 2\tan^{-1}(\omega) \rightarrow T_g(\omega) = -\frac{d\theta(\omega)}{d\omega}$$

$$= \frac{2}{1 + \omega^2}, T_p(\omega) = -\frac{\theta(\omega)}{\omega}. \quad (6.7.20b) \quad ∎$$

Earlier, we have seen that the delay associated with a system is a function of its phase response. Transfer functions of *stable* systems can have the *same amplitude* response with *different phase and delay* responses. There are three important systems to consider. These are *minimum phase, mixed phase,* and *maximum phase systems.*

Consider the transfer function of a system

$$H(s) = K \frac{\prod_k (s + z_k)}{\prod_m (s + p_m)}. \qquad (6.7.21)$$

For stability reasons all the poles located at $s = -p_k$ are located on the *left half of the s-plane.* If the zeros of the transfer function $s = -z_k$ are on the *negative half of the s-plane,* then the system is a *minimum phase* system. If some zeros are on the *right half s-plane* and some are on the *left half s-plane*, the system is a mixed phase and if all the zeros are located on the *right half of the s-plane*, then the system is a *maximum phase* system.

6.8 System Bandwidth Measures

In Sections 4.2.2, bandwidth measures of a signal were briefly studied. The concentration was on the time–bandwidth product and illustrated examples, wherein the bandwidth is inversely proportional to time width of the signal. Similar ideas can be used using the duration of the impulse response of a system and the system bandwidth. In Section 6.3, a simple, but a practical measure, the half-power or the 3 dB bandwidth was considered. This width is the range of frequencies over which the magnitude of the function exceeds $(1/\sqrt{2})$ of its maximum. Half-power bandwidth comes from the fact that the square of the magnitude is power and $20\log(1/\sqrt{2}) = -3\text{dB}$. There are different measures that are used and some of these are considered.

6.8.1 Bandwidth Measures Using the Impulse Response h(t) and Its Transform H(jω)

The time and the frequency durations of the impulse response $h(t)$ are defined by

$$T_0 = \frac{1}{h(0)} \int_{-\infty}^{\infty} h(t)dt,$$

$$B_0 = \frac{1}{H(0)} \int_{-\infty}^{\infty} H(j\omega)df, \omega = 2\pi f. \qquad (6.8.1)$$

From the central ordinates theorems of the F-transforms, the time width times the bandwidth is equal to 1. This definition for the bandwidth makes use of the spectrum on both sides. That is,

$$T_0 B_0 = 1. \qquad (6.8.2)$$

For the one side case, which is what we mostly use, divide the frequency width by 2.

Example 6.8.1 Using the above measures show that $T_0 = 1$ and $B_0 = 1$ for the following:

a. $x_1(t) = \Pi(t) \overset{FT}{\longleftrightarrow} \text{sinc}\left(\frac{\omega}{2}\right) = X_1(j\omega),$

b. $x_2(t) = e^{-at}u(t) \overset{FT}{\longleftrightarrow} \frac{1}{a + j\omega} = X_2(\omega),$

c. $x_3(t) = e^{-at^2} \overset{FT}{\longleftrightarrow} \sqrt{\frac{\pi}{a}} e^{-\omega^2/4a} \qquad (6.8.3)$

Solution: *a.* In Chapter 4, the areas of the rectangular and the sinc functions were considered and the results are $T_0 = 1, B_0 = 1$.
b. Noting that $h(0) = 1/2$ ($h(t)$ is discontinuous at $t = 0$) and $H(0) = 1/a$, it follows that

$$2 \int_{-\infty}^{\infty} e^{-at}u(t)dt = \frac{2}{a} \text{ and } \frac{a}{2\pi} \int_{-\infty}^{\infty} \frac{1}{a + j\omega} d\omega$$

$$= \frac{a^2}{2\pi} \int_{-\infty}^{\infty} \frac{1}{a^2 + \omega^2} d\omega$$

$$= \frac{a^2}{\pi} \int_{0}^{\infty} \frac{1}{a^2 + \omega^2} d\omega = \frac{a}{2},$$

$$T_0 = (2/a), B_0 = (a/2), T_0 B_0 = 1. \qquad (6.8.4)$$

c. By making use of integral tables, the time width and the bandwidth of the Gaussian pulses both come out to be one and the product is one. ∎

Time functions can take both positive and negative values, some authors use the magnitudes or the squares of the time functions $h(t)$ in defining the time width in (6.8.2a). Others use moments to define the time and bandwidths. In the following, the bandwidth measures that are simple and practical will be considered.

6.8.2 Half-Power or 3 dB Bandwidth

In Section 6.6, an RC circuit was considered. On the amplitude spectrum, the 3 dB frequency was identified (see Fig. (6.6.2a)). The half-power or the 3 dB bandwidth is a practical measure and is widely used in systems and circuit theory, especially in filter designs. In identifying the 3 dB bandwidths, only *positive frequencies* are considered.

Example 6.8.2 Show the 3 dB bandwidth of the Gaussian function is W (Carlson (1975)).

$$H(j\omega) = e^{-(\ln(2/2)(\omega/W))} \quad (\text{note}, H(j0) = 1).$$

Solution: The half-power frequency is equal to W since

$$|H(j\omega_{3dB})|^2 = \frac{1}{2} = (e^{-2(\ln(2)/2)(\omega_{3dB}/W)^2})|_{\omega_{3dB}=W}$$

$$\Rightarrow \frac{1}{2} = e^{-\ln(2)}. \qquad (6.8.5) \quad \blacksquare$$

Although the 3 dB bandwidth is the most common one, we could obviously define 6 dB bandwidth or any other value for the bandwidth measure. In summary the 3 dB bandwidth computes the width by considering the peak value of the spectrum and a value (or values) of the $(1/\sqrt{2})$ below the maximum

value of the spectrum. This measure is simple and it does *not* take into consideration any ripples in $|H(j\omega)|$ function between the two 3 dB frequencies. A more generalized measure that takes into consideration the ripples by making use of integrals in computing the bandwidths. These methods are used in random signal analysis, as the spectrum of noisy signals have many peaks. For a good discussion on this topic, see Peebles (2001). These measures have been developed using signals rather than systems. To make it uniform in our discussion we will use $X(j\omega)$ rather than $H(j\omega)$. When we consider examples of transfer functions, $H(j\omega)$ will be used.

6.8.3 Equivalent Bandwidth or Noise Bandwidth

The equivalent noise bandwidth is obtained by equating the areas contained in the signal energy spectrum with a pulse spectrum of bandwidth $W_{eq} = W_N$ rad/s. Figure 6.8.1 illustrates a signal spectrum and noise equivalent is computed as follows:

$$\int_{-\infty}^{\infty} |X(j\omega)|^2 d\omega = |X(j\omega)|^2_{max} 2W_{eq}$$

$$\Rightarrow W_{eq} = \frac{\int_{0}^{\infty} |X(j\omega)|^2 d\omega}{|X(j\omega)|^2_{max}}, B_{eq} = \frac{W_{eq}}{2\pi}. \qquad (6.8.6)$$

If the system bandwidth of the transfer function $H(j\omega)$ is of interest, replace $X(j\omega)$ by $H(j\omega)$ in (6.8.6).

Example 6.8.3 Determine the noise equivalent bandwidth of the filter transfer function

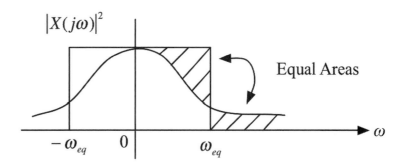

Fig. 6.8.1 Noise equivalent bandwidth

$$|H(j\omega)|^2 = \frac{1}{1 + (\omega RC)^2} = [1/(1 + (\omega/W)^2)],$$

$$\omega = 2\pi f, W = 1/RC.$$

Solution: Using (6.8.6) and $|H(j\omega)|_{\max} = 1$, we have

$$W_{eq} = \frac{1}{2\pi} \int_0^\infty \frac{W^2}{(W^2 + \omega^2)} d\omega = W\tan^{-1}(\omega/W)\big|_0^\infty$$

$$= W\pi/2. \tag{6.8.7}$$

$$\Rightarrow B_{eq} = [1/4RC]\,\text{Hz}. \tag{6.8.8}$$

The 3 dB frequency of the RC circuit, $f_{3dB} = 1/2\pi RC$ is related to the equivalent noise bandwidth and $B_{eq} = 1.57 B_{3dB}$. The equivalent noise bandwidth works as well for signals that have spectrum in the middle, such as the band-pass spectrum. In such cases, using the center frequency ω_0 for the peak in the amplitude response of the band-pass filter, the equivalent bandwidth is

$$W_{eq} = \frac{\int_0^\infty |H(j\omega)|^2 d\omega}{|H(j\omega_0)|^2}, \quad B_{eq} = \frac{W_{eq}}{2\pi}\,\text{Hz}. \tag{6.8.9} \quad \blacksquare$$

6.8.4 Root Mean-Squared (RMS) Bandwidth

The RMS bandwidth comes from the statistical measures, where the variance is a measure of the spread of a density function. Consider the low-pass energy spectral density shown in Fig. 6.8.1. The area under this function is the energy

$$E = \frac{1}{2\pi} \int_{-\infty}^\infty |X(j\omega)|^2 d\omega. \tag{6.8.10}$$

Now define the normalized energy spectral density function by

$$|X_{no}(j\omega)|^2 = |X(j\omega)|^2/E. \tag{6.8.11}$$

It is real, even, and positive and the area under the function is 1. It has the same properties as a probability density function (PDF). In the PDF case, we define the variance as a measure of the spread of the density function. In this case the spread is measured

by the bandwidth. We can define the root mean square (RMS) bandwidth as

$$W_{RMS}^2 = \int_{-\infty}^\infty \omega^2 |X_{no}(j\omega)|^2 d\omega = \frac{\int_{-\infty}^\infty \omega^2 |X(j\omega)|^2 d\omega}{\int_{-\infty}^\infty |X(j\omega)|^2 d\omega}. \tag{6.8.12}$$

Example 6.8.4 Compute the RMS bandwidth (see Peebles (2001).) and compare it with the 3 dB frequency which is given by

$$|X(j\omega)|^2 = \frac{10}{[1 + (\omega/10)^2]^2}. \tag{6.8.13}$$

Solution: Using integral tables, we have

$$\int_{-\infty}^\infty \frac{10}{[(1 + (\omega/10)^2]^2} d\omega = 10^5 \int_{-\infty}^\infty \frac{d\omega}{[100 + \omega^2]^2} = 50\pi. \tag{6.8.14a}$$

$$\int_{-\infty}^\infty \omega^2 |X(j\omega)|^2 d\omega = 10^5 \int_{-\infty}^\infty \frac{\omega^2}{[100 + \omega^2]^2} d\omega = 5000\pi. \tag{6.8.14b}$$

$$\omega_{RMS}^2 = \frac{5000\pi}{50\pi} = 100, \omega_{RMS} = 10\,rad/s,$$

$$F_{RMS} = 1.5915\,Hz.$$

$$|X(j\omega_{3dB})|^2 = \tfrac{1}{2}|H(0)|^2 = \frac{1}{2}10 = 5 = \frac{10}{[1 + (\omega_{3dB}/10)^2]^2}.$$

$$\Rightarrow \omega_{3dB} = 6.436 \text{ or } f_{3dB} = 1.243\text{Hz}.$$

In this special case, the 6 dB bandwidth comes out to be the same as the RMS bandwidth. The concepts of RMS bandwidth can be easily extended to band-pass spectra. Assuming that most of the spectra is around $\pm\omega_0$, the RMS bandwidth is given by

$$W_{RMS}^2 = \frac{4\int_0^\infty (\omega - \omega_0)^2 |X(j\omega)|^2 d\omega}{\int_0^\infty |X(j\omega)|^2 d\omega} \tag{6.8.15}$$

RMS bandwidth is more general than the 3 dB bandwidth, as it can handle general spectra with several peaks and valleys in the passband of the energy spectral density. $\quad \blacksquare$

6.9 Nonlinear Systems

In this section we will consider simple nonlinear systems and illustrate the difficulties in the spectral analysis of the responses. A nonlinear system is described by a time domain relationship between the input and the output. This can be expressed in the form of a graphical representation or in terms of a general output function $y(t) = g(x(t))$, where the output function $g(.)$ is a complicated closed form expression or in terms of a power series of $x(t)$. A system is nonlinear if it has components that have nonlinear characteristics, such as a diode. In many cases, nonlinear systems are approximated by linear systems, as they are easier to handle, see Ziemer and Tranter (2002).

The system described by a polynomial function of the input $x(t)$, such as

$$y(t) = \sum_{i=0}^{n} a_i x^i(t). \qquad (6.9.1)$$

is linear if all a_i's are zero except a_1. If any of the other a_i's are nonzero, then the system is nonlinear. An example is a device that has saturation nonlinearity. It has a voltage to current characteristic that is linear within a range and outside that range, the voltage saturates, see Fig. 6.9.1a. A device that may have this type of a characteristic is a resistor. The Ohm's law says that $v = Ri$ is valid in a certain range of currents and voltages. Outside this range, the resistor is a nonlinear component. A hard limiter is an important example. The output voltage is 1 if the input voltage is positive and -1 if the input voltage is negative (see Fig. 6.9.1b.).

Example 6.9.1 Find the output $y(t)$ defined below and sketch the one-sided line spectra of the input $x(t) = B_1 \cos(\omega_1 t) + B_2 \cos(\omega_2 t)$, $B_1, B_2 > 0$ with $\omega_i = 2\pi f_i$, $i = 1, 2$ and $f_2 > f_1$ and $y(t)$, see Ziemer and Tranter (2002).

$$y(t) = a_0 + a_1 x(t) + a_2 x^2(t), a_i \neq 0, x(t)$$
$$= B_1 \cos(\omega_1 t) + B_2 \cos(\omega_2 t), \ B_1, B_2 > 0. \ (6.9.2)$$

Solution: The output is

$$y(t) = a_0 + a_1 B_1 \cos(2\pi f_1 t) + a_1 B_2 \cos(2\pi f_2 t)$$
$$+ a_2 B_1^2 \cos^2(\omega_1 t) + a_2 B_2^2 \cos^2(\omega_2 t)$$
$$+ a_2(2B_1 B_2) \cos(\omega_1 t) \cos(\omega_2 t) \qquad (6.9.3)$$
$$= [a_0 + \frac{1}{2} a_2 B_1^2 + \frac{1}{2} a_2 B_2^2] \qquad \text{DC offset term}$$
$$+ [a_1 B_1 \cos(\omega_1 t) + a_1 B_2 \cos(\omega_2 t)] \qquad \text{Linear terms}$$
$$+ \frac{1}{2} a_2 [B_1^2 \cos(2\omega_1 t) + B_2^2 \cos(2\omega_2)t] \quad \text{Harmonic terms}$$
$$+ a_2 B_1 B_2 [\cos(\omega_1 + \omega_2)t + \cos(\omega_2 - \omega_1)t] \quad \text{Inter}$$
$$\text{modulation terms.}$$

$$(6.9.4)$$

Figure 6.9.2 gives the input and the output one-sided amplitude line spectra. System nonlinearity created a DC term, linear terms (frequencies f_1 and f_2), harmonic distortion terms (frequencies, $2f_1$ and $2f_2$), and inter modulation terms (sums and differences of the input frequencies, $(f_2 - f_1)$ and $(f_1 + f_2)$). Note that the output of a linear time-invariant system has the same frequencies as the input with possible changes in the amplitudes and phases. *Nonlinear system* generates new frequencies. ∎

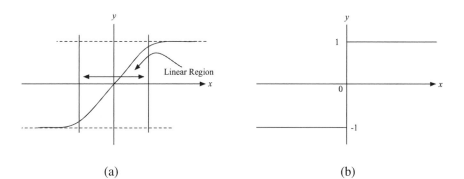

Fig. 6.9.1 Examples of nonlinear input–output characteristics

(a)

(b)

Fig. 6.9.2 Example 6.9.2,
(a) Input line spectra and
(b) Output line spectra

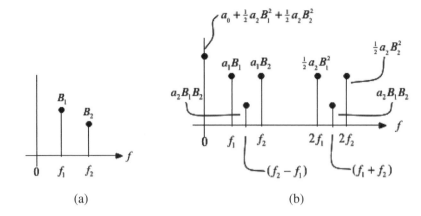

(a)   (b)

6.9.1 Distortion Measures

Most systems have inherent nonlinear components. It may be desirable to operate them in the linear region, if possible. Amplification is a good example, where the nonlinearities may be small and the distortions will be small enough that they can be tolerated. The next question is how does one measure the distortions due to a nonlinear system? A simple way is compare a nonlinear system time response to a linear system time response. To achieve these measures, start with a single input, a sinusoid, say $x(t) = \cos(\omega_0 t)$ and measure the distortion due to the nonlinearities in the system described $x(t)$ by

$$y(t) = \sum_i a_i x^i(t), \text{ where } a_i' \text{ s some constants.}$$
$$(6.9.5)$$

The powers of the cosine functions can be expressed in terms of sine and cosine terms, where the frequencies will be multiples of the input frequency, i.e., we will have harmonics. The output can be written in terms of trigonometric Fourier series

$$y(t) = Y[0] + \sum_{k=1}^{\infty} A[k]\cos(k\omega_0 t) + \sum_{k=1}^{\infty} B[k]\sin(k\omega_0 t).$$
$$(6.9.6)$$

The constants $Y[0]$, $A[k]'s$, and $B[k]'s$ are functions of the constants $a_i's$ and the powers of the input sinusoid. The kth distortion term is measured by the ratio

$$D[k] = \sqrt{\frac{A^2[k] + B^2[k]}{A^2[1] + B^2[1]}}.$$
$$(6.9.7)$$

Obviously the frequency of interest $f_0 = \omega_0/2\pi$ should be in the passband of the signal. Manufacturers of stereo systems provide literature that gives these numbers in terms of dBs for their systems. Clearly, if the distortion terms $D[k]'s, k \neq 1$ are negligible, then the nonlinear system comes close to a linear system.

6.9.2 Output Fourier Transform of a Nonlinear System

In the following example, a system with a polynomial nonlinearity is considered and illustrates the effect of the nonlinearities in terms of the input and the output frequencies.

Example 6.9.2 Let the input $x(t)$ and the output $y(t)$ in terms of the input are as given below. Noting $F[x^2(t)] = (1/2\pi)[X(j\omega) * X(j\omega)]$, sketch the output spectrum assuming

$$x(t) \overset{FT}{\longleftrightarrow} X(j\omega) = \prod\left[\frac{\omega}{W}\right],$$

$$y(t) = a_0 + a_1 x(t) + a_2 x^2(t) \overset{FT}{\longleftrightarrow} Y(j\omega). \quad (6.9.8)$$

$$Y(j\omega) = a_0 2\pi\delta(\omega) + a_1 X(j\omega) + (a_2/2\pi)[X(j\omega) * X(j\omega)]. \quad (6.9.9)$$

Solution: Convolution of two identical rectangular pulses is a triangular pulse (see Example 2.3.1.) and

$$Y(j\omega) = a_0 2\pi\delta(\omega) + a_1 \prod\left[\frac{\omega}{W}\right] + (a_2/2\pi)\Lambda\left[\frac{\omega}{W}\right].$$
$$(6.9.10)$$

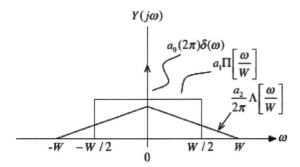

Fig. 6.9.3 Output transform of a nonlinear system

The three parts are explicitly shown in Fig. 6.9.3. Note that the width of the triangular pulse is $2W$. Therefore, the bandwidth of the output signal is two times that of the input signal. If a system has a second-order nonlinearity, then the frequency width of the output signal will be doubled that of the input signal. A system with a nth order nonlinearity, then the bandwidth of the output signal will increase from B Hz to (nB) Hz. Most signals are not band limited and the transmission systems have a limited bandwidth and filtering is necessary. ∎

Frequency analysis of a nonlinear system is difficult and may not even be possible. If it can be approximated by a linear system, then frequency domain analysis provides useful information. Time domain analysis is simpler for nonlinear systems.

6.9.3 Linearization of Nonlinear System Functions

Nonlinear systems are hard to deal with in general terms. A function $g(x)$ can be approximated about a point $x = x_0$ using Taylor series expansion

$$g(x) = g(x_0) + \left[\frac{dg}{dx}\right]\Big|_{x=x_0}(x - x_0)$$
$$+ \left[\frac{d^2 g}{dx^2}\right]\left[\frac{(x - x_0)^2}{2!}\right] + \ldots$$

$$\delta g(x) = g(x) - g(x_0) \approx \left[\frac{dg}{dx}\right]_{x=x_0}(x - x_0) = m|_{x=x_0}\delta x.$$

$$(6.9.12)$$

Note that the approximation is valid for small excursions of x from x_0 and we can neglect higher-order terms. It is a linear relationship between *small changes* in both the *input* and *output* related by the *slope of the function m*at $x = x_0$, see Nise (1992).

6.10 Ideal Filters

In this section we will consider the basics of low-pass, high-pass, band-pass, band-elimination filters, and the ideal delay line filters. The filters are specified based on a transfer function $H(j\omega)$ in terms of its amplitude, phase, or delay responses, $|H(j\omega)|$, phase $\angle H(j\omega)$ or $-[dH(j\omega)/d\omega]$. Finding $H(j\omega)$ from the specifications is the first step. The next step involves the synthesis. We will consider here the ideal filter functions that describe their functions, and *simple circuits* that can be used as filters.

Low-pass filters allow low frequencies to pass through with small attenuation and attenuate or eliminate high frequencies; high-pass filters eliminate or attenuate low frequencies and allow high frequencies go through with possibly small attenuations; band-pass filters allow a band of frequencies to go though with small attenuation and attenuate or eliminate frequencies that are outside this band; band-elimination or band-reject filters let the low and high frequencies pass through and attenuate or eliminate a band of frequencies somewhere in the middle. Delay line filters are primarily used in cascade with filters so that the cascaded filter delay line combination has an approximate linear phase characteristics. Filters are used in every communication system. If the frequencies of the two signals are disjoint, then we can remove the undesired signal by using a band-pass filter that allows the desired signal to go through with a small attenuation and attenuate or eliminate the undesired signal. Tuning to a particular radio station involves eliminating, i.e., filtering out all the other signals from the other stations all available at the front end of the radio or TV receivers. The DC component can be removed by using a high-pass filter or a simple bias removal component, a capacitor.

Example 6.10.1 Illustrate the *Bell System TOUCH-TONE telephone dialing scheme.*

Solution: Bell systems TOUCH-TONE telephone dialing scheme uses some of the filters. The discussion follows that of Daryanani (1976). The filters are used in the detection of signals generated by a push button telephone. As we dial a telephone number, i.e., by pushing a button on the telephone, a unique set of two-tone signals are generated and transmitted to the telephone central office, where the signals are processed to identify the number that is transmitted. The buttons and the tone assignments of a TOUCH-TONE telephone are shown in Fig. 6.10.1. It has 12 buttons. These correspond to 10 decimal digits, a star button, and a pound button. The letters are also identified on the buttons. For example, on the button identified by 2 has the letters ABC indicating that the number 2 represents A, B, and C as well as 2. Operator button (0) is used for zero. The star (*) button and the pound (#) button are used for other special purposes, such as responding to queries from an answering machine. There are four other buttons that are not shown and are used for special purposes. The signaling code provides 16 distinct signals that use 4 low and high frequencies given by

Low : (697 Hz, 770 Hz, 852 Hz, 941 Hz),

High : (1209 Hz, 1336 Hz, 1477 Hz, 1633 Hz)

Pressing one of the buttons generates a pair of unique frequencies, one lower and the other higher. The fourth high-band frequency, 1633 Hz, is for special services. The block diagram shown in Fig. 6.10.2 illustrates the detection scheme in the telephone office. The received two tones are amplified first and the two tones are then separated into two groups by the low-pass and the high-pass filters. The low-pass filters pass the low frequencies with very low attenuation and block the high frequencies. Similarly the high-pass filters pass the high frequencies with very little attenuation and block the low frequencies. The separated tones are then converted to square waves of fixed amplitudes by using limiters. Signals are then passed through eight band-pass filters. Each of these passes only one tone and rejects the others. The band-pass filter characteristics are such that there is very little attenuation for the particular frequency and a significant attenuation to block the other frequencies. For a detailed discussion on the amplitude characteristics of low-pass, band-pass, and high-pass filters, see Daryanani (1976). The outputs of the band-pass filters are fed into detectors. The detectors are energized when their input voltage exceeds a set threshold value and the outputs of the detector provides the required dc switching level to connect the caller to the party being called. ∎

Filters can be implemented either in terms of analog or digital domain. Next we will consider each of the filter types in a more formal fashion and discuss the generation of simple transfer functions that allow for the analysis of these filters.

6.10.1 Low-Pass, High-Pass, Band-Pass, and Band-Elimination Filters

The words low-pass means that when the signal $x(t)$ is passed through a low-pass filter, only low

697 Hz →	1	ABC 2	DEF 3	
770 Hz →	GHI 4	JKL 5	MNO 6	
852 Hz →	PRS 7	TUV 8	WXY 9	
941 Hz →	*	Oper 0	#	
	↑ 1209 Hz	↑ 1336 Hz	↑ 1477 Hz	↑ 1633 Hz

Low-band frequencies

High-band frequencies

Fig. 6.10.1 Tone assignments for TOUCH-TONE dialing

Fig. 6.10.2 Block diagram of detection scheme in the telephone office

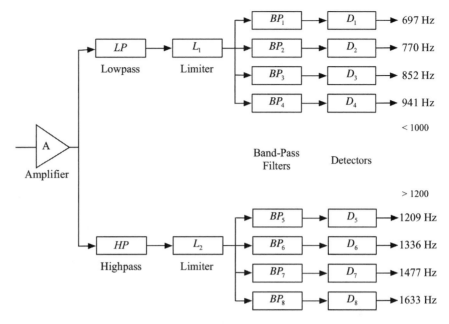

frequencies, say 0 to $f_c = \omega_c/2\pi$, are passed and block all frequencies above the *cutoff frequency f_c*. The amplitude of the *ideal low-pass filter* transfer function is (H_0 is assumed to be positive)

$$|H_{LP}(j\omega)| = H_0\Pi\left[\frac{\omega}{2\omega_c}\right] = \begin{cases} H_0, & |\omega| \leq \omega_c \\ 0, & |\omega| > \omega_c \end{cases}, \omega_c = 2\pi f_c.$$

$$(6.10.1)$$

Every transmission system takes time, i.e., the signal will be delayed. It is ideal to have this delay to be a constant, say t_0 s for *all frequencies,*which may not be possible. In terms of frequency domain, the output transform of such a system is

$$Y(j\omega) = H_{Lp}(j\omega)X(j\omega), \ H_{LP}(j\omega) = H_0\Pi\left[\frac{\omega}{2\omega_c}\right]e^{-j\omega t_0},$$

$$|H_{Lp}(j\omega)| = H_0\Pi(\omega/2\omega_c), \text{ and } \angle H_{Lp}(j\omega) = -\omega t_0.$$

$$(6.10.2)$$

The amplitude and the phase response plots are shown with respect to the frequency variable ω in Fig. 6.10.3. On the magnitude plot the band of frequencies from 0 to f_c as the passband and the band of frequencies from f_c to ∞ as the stopband are shown. Since the amplitude spectrum of a real signal is even and the phase spectrum is odd, the discussion can be limited to only positive frequencies. The phase response is assumed to be linear, i.e., slope is constant. The group delay is

$$t_0 = -d\angle H_{Lp}(j\omega)/d\omega. \qquad (6.10.3)$$

Can we design a real circuit that has the transfer function $H_{Lp}(j\omega)$? For a physically realizable system, the impulse response $h(t) = 0$ for $t < 0$, i.e., the system is causal. For a realizable system, the output cannot exist before the input is applied. The impulse response of the ideal low-pass filter can be

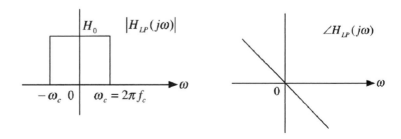

Fig. 6.10.3 Amplitude and phase responses of an ideal low-pass filter

determined from the results in Chapter 4 (see (4.3.28).). It is repeated below.

$$\frac{\sin(a(t_0 - t_0))}{\pi(t - t_0)} \xleftrightarrow{\text{FT}} \prod\left[\frac{\omega}{2a}\right]e^{-j\omega t_0} \qquad (6.10.4)$$

$$h_{Lp}(t) = F^{-1}[H_{LP}(j\omega)] = F^{-1}\left[H_0 \prod\left[\frac{\omega}{2\omega_c}\right]e^{-j\omega t_0}\right]$$

$$= H_0(2f_c)\frac{\sin(\omega_c(t - t_0))}{\omega_c(t - t_0)}. \qquad (6.10.5)$$

In Fig. 6.10.4, the input, an impulse function $\delta(t)$, applied at $t = 0$, the block diagram representing the ideal low-pass filter and the impulse response are identified. The impulse response, a sinc function, peaks at $t = t_0$, giving a value of $(2f_cH_0)$ at this time. From the figure we can see that the response is nonzero for $t < 0$. That is, there is output before the input is applied. The *ideal low-pass filter* is *not causal* and is *physically unrealizable*. We can also see this from the Payley–Wiener criterion stated below.

For a causal system, the impulse response $h(t) = 0$ for $t < 0$, as it does not respond before the input is applied. The causality condition can be stated in terms of the transfer function $H(j\omega)$. It is called the *Paley–Wiener criterion* Papoulis (1962) and is given in terms of the inequality

$$\int_{-\infty}^{\infty} \frac{|\ln|H(j\omega)||}{(1 + \omega^2)}d\omega < \infty. \qquad (6.10.6)$$

If $|H(j\omega)| = 0$ over a finite frequency band, then the above integral becomes infinite. $|H(j\omega)|$ can be zero at isolated frequencies and still satisfy the criterion. The criterion describes the physical reliability conditions and is not of practical value.

That is, if $|H(j\omega)| = 0$ over any band of frequencies, the Payley–Wiener criterion states that the system is physically unrealizable. Ideal low-pass filter violates the condition. We can make a general statement that if the amplitude spectrum is a brick wall type function, the corresponding transfer function is physically unrealizable. Since the ideal low-pass filter function is physically unrealizable, the next best thing is find a function that approximates the ideal filter characteristics. First, consider the simple RC circuit in Fig. 6.3.4a. The transfer function of this circuit is

$$H_{Lp}(j\omega) = 1/(1 + j\omega RC).$$

The frequency amplitude characteristic is shown in Fig. 6.6.2a with the cutoff frequency $f_c = f_{3dB} = (1/(2\pi RC))$. The input and the output transforms are related by $Y(j\omega) = H_{Lp}(j\omega)X(j\omega)$. At a particular frequency f_i,

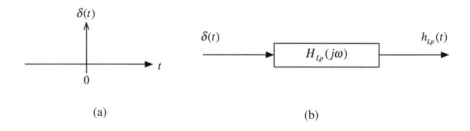

(a)

(b)

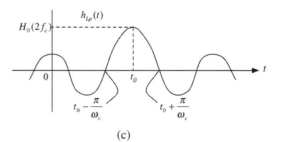

Fig. 6.10.4 (a) Impulse input, (b) block diagram of a low-pass filter, and (c) impulse response

(c)

$$|Y(j\omega_i)| = |H_{Lp}(j\omega_i)||X(j\omega_i)|, \omega_i = 2\pi f_i. \quad (6.10.7)$$

The output spectral amplitudes at frequencies $f = f_i$ are attenuated from the input magnitude spectral amplitudes by the factor of $|H_{Lp}(j\omega_i)|$. For frequencies between $0 \leq f \leq f_c = (1/2\pi RC)$, the output amplitude spectrum is within 3 dB of the input amplitude spectrum, whereas for $f > f_c$, the output amplitude spectrum is significantly reduced or attenuated. The simple RC low-pass filter allows the low frequencies 0 to f_{3dB} go through with a small attenuation and the frequencies from f_{3dB} to ∞ are attenuated significantly. The circuit has low-pass filter characteristics. To raise the amplitude characteristics in the passband and, at the same time, lower the amplitude characteristics in the stopband, the following amplitude response function would work:

$$|H_{Lp}(j\omega)| = \frac{1}{\sqrt{1 + (\omega/\omega_c)^{2n}}}. \quad (6.10.8)$$

Since ω/ω_c is less than 1 in the passband, by taking the power of this by $(2n)$, we are decreasing the value of the denominator in the passband, thus increasing the amplitude in the passband. On the other hand, in the stopband, i.e., the band above the cutoff frequency $\omega > \omega_c$, the denominator in (6.10.8) increases as ω increases above the cutoff frequency, and the value of the function reduces in this range. Figure 6.10.5 gives two sketches for n, say n_1 and $n_2, n_2 > n_1$. In the limit, i.e., when $n \to \infty$, the filter characteristics approach the ideal

characteristics. Equation (6.10.8) can be generalized to control the passband attenuation by choosing

$$|H_{Lp}(j\omega)| = \frac{1}{\sqrt{1 + \varepsilon^2(\omega/\omega_c)^{2n}}}. \quad (6.10.9)$$

There are two parameters in (6.10.9), ε and n. ε controls how far the amplitude characteristics will go down to from a value of 1 at $\omega = 0$ to when $\omega = \omega_c$. The value of n controls how fast the attenuation of the amplitude characteristics in the stop-band region.

The amplitude characteristic goes from 1 to $1/\sqrt{(1 + \varepsilon^2)}$ corresponding to the frequencies 0 and f_c, respectively. By using the power series, for small ε, we can write

$$1 - [1/\sqrt{1 + \varepsilon^2}] \approx \varepsilon^2/2. \quad (6.10.10)$$

The function in (6.10.9) is the *Butterworth filter function*. It has interesting properties. At $\omega = 0$, $(2n - 1)$ derivatives of the function $1/[1 + \varepsilon^2(\omega/\omega_c)^{2n}]$ are zero, identified as a *maximally flat* amplitude response. For $|\omega/\omega_c| \gg 1$, the high-frequency roll-off of an nth order Butterworth function is $20n$ dB/decade. The proofs of these are left as exercises. As $n \to \infty$, the filter response has the ideal low-pass characteristics. In the low-pass filter specifications, three bands, namely passband $(0 \to \omega_c)$, transition band $(\omega_c \to \omega_r)$, and stopband $(\omega_r \to \infty)$ are identified. The transition band is not

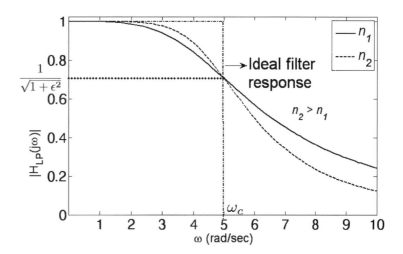

Fig. 6.10.5 Butterworth amplitude filter response $n_1 = 2, n_2 = 3, \varepsilon = 1, \omega_c = 5$

shown in Fig. 6.10.6, as it depends on required attenuations at the edges of the transition band.

Returning to the simple *RC* low-pass filter, the element values of one of the two components *R* or *C* can be determined from the given cutoff frequency, see (6.6.7). Normally the capacitor value is selected, as the number of available capacitor values is much smaller than that of the available resistor values. As a final step, the time response of these filters, for the two simple first-order low-pass *RC* and *RL* filters is shown in Figs. 6.3.4a and 6.6.3. The response is determined by the time

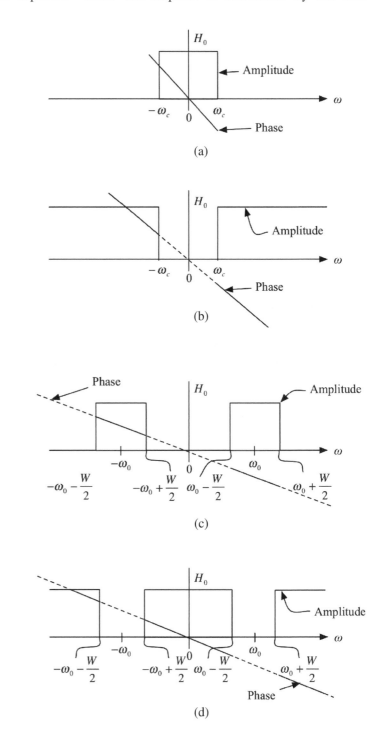

Fig. 6.10.6 Amplitude and phase plots of ideal filters: (a) low-pass, (b) high-pass, (c) band-pass, and (d) band elimination

constants, $\tau = RC$ for the RC circuit and $\tau = L/R$ for the RL circuit. The output of an LTI system with input $x(t)$ and the impulse response $h(t)$ is given assuming it zero for $t < 0$ by the convolution integral

$$y(t) = \int_0^t h(\alpha)x(t-\alpha)d\alpha. \qquad (6.10.11)$$

The impulse response $h(t)$ is the *circuit weighting function*. It gives the amount of *memory* the circuit has. For example, if $h(t) = \delta(t)$, it gives zero weight to the past values of the input function as

$$y(t) = \int_0^t h(\alpha)x(t-\alpha)d\alpha = \int_0^t \delta(\alpha)x(t-\alpha)d\alpha = x(t). \qquad (6.10.12)$$

If $h(t) = u(t)$, then the circuit has a perfect memory giving equal weights and

$$y(t) = \int_0^t h(\alpha)x(t-\alpha)d\alpha = \int_0^t u(\alpha)x(t-\alpha)d\alpha$$

$$= \int_0^t x(t-\alpha)d\alpha. \qquad (6.10.13)$$

Ideal filter frequency responses of the low-pass (Lp), high-pass (Hp), band-pass (Bp), and the band-elimination (Be) filters are given below. The amplitude and the phase response plots of these are given for *positive frequencies* in Fig. 6.10.6a,b,cd, respectively. Note the phase responses of these ideal filter functions are shown as linear.

$$H_{Lp}(j\omega) = H_0 \Pi\left[\frac{\omega}{2\omega_c}\right] e^{-j\omega\, t_0} \qquad (6.10.14a)$$

$$H_{Hp}(j\omega) = H_0\left[1 - \Pi\left[\frac{\omega}{2\omega_c}\right]\right] e^{-j\omega\, t_0} \qquad (6.10.14b)$$

$$H_{Bp}(j\omega) = H_0\left[\Pi\left[\frac{\omega - \omega_0}{W}\right] + \Pi\left[\frac{\omega - \omega_0}{W}\right]\right] e^{-j\omega\, t_0} \qquad (6.10.14c)$$

$$H_{Be}(j\omega) = H_0\left[1 - \left[\Pi\left[\frac{\omega - \omega_0}{W}\right] + \Pi\left[\frac{\omega - \omega_0}{W}\right]\right]\right] e^{-j\omega\, t_0}. \qquad (6.10.14d)$$

The Lp and Hp filter responses have one passband and one stopband. The Bp filter response has one passband and two stopbands. The Be filter response has two passbands and one stopband.

6.11 Real and Imaginary Parts of the Fourier Transform of a Causal Function

The real and the imaginary parts of the Fourier transform of a *causal* function $x(t)$ are shown to be related below. Let

$$x(t) \overset{FT}{\longleftrightarrow} X(j\omega) = \text{Re}(X(j\omega)) + j\,\text{Im}(X(j\omega)).$$

By noting $\text{Re}[X(j\omega)]$ is even and $\text{Im}[X(j\omega)]$ is odd and integral of an odd function over a symmetric interval is zero, we have

$$x(t) = \frac{1}{2\pi}\int_{-\infty}^{\infty} X(j\omega)e^{j\omega t}d\omega$$

$$= \frac{1}{2\pi}\int_{-\infty}^{\infty} (\text{Re}[X(j\omega)] + j\,\text{Im}[X(j\omega)])$$

$$[\cos(\omega t) + j\sin(\omega t)]d\omega$$

$$= \frac{1}{\pi}\int_0^{\infty} \text{Re}[X(j\omega)]\cos(\omega t)d\omega$$

$$- \frac{1}{\pi}\int_0^{\infty} \text{Im}[X(j\omega)]\sin(\omega t)d\omega. \qquad (6.11.1)$$

Noting that $x(t)$ is causal, i.e., $x(-t) = 0, t > 0$, results in

$$x(-t) = \frac{1}{\pi}\int_0^{\infty} \text{Re}[X(j\omega)]\cos(\omega t)d\omega$$

$$- \frac{1}{\pi}\int_0^{\infty} \text{Im}[X(j\omega)]\sin(-\omega t)d\omega = 0.$$

Since $\cos(\omega t)$ and $\sin(\omega t)$ are defined everywhere, $\text{Re}[X(j\omega)]$ and $\text{Im}[X(j\omega)]$ are the real and imaginary parts of the transform of the causal function. That is,

$$\frac{1}{\pi}\int_0^{\infty} \text{Re}[X(j\omega)]\cos(\omega t)d\omega$$

$$= -\frac{1}{\pi}\int_0^{\infty} \text{Im}[X(j\omega)]\sin(\omega t)d\omega, t > 0.$$

This implies that a causal signal $x(t)$ can be expressed in terms of either $\text{Re}[X(j\omega)]$ or $\text{Im}[X(j\omega)]$ and

$$x(t) = \frac{2}{\pi} \int_0^\infty \text{Re}[X(j\omega)] \cos(\omega t) d\omega,$$

$$x(t) = -\frac{2}{\pi} \int_0^\infty \text{Im}[X(j\omega)] \sin(\omega t) d\omega \quad (6.11.2)$$

These are true as long as there are no impulses at $t = 0$ and they imply that $\text{Re}[X(j\omega)]$ and $\text{Im}[X(j\omega)]$ cannot be specified independently. Using the transform and solving for real and the imaginary parts of the transform, we have

$$\text{Re}[X(j\omega)] = -\frac{2}{\pi} \int_0^\infty \int_0^\infty \text{Im}[X(jv)] \sin(vt) \cos(\omega t) dv dt$$

$$(6.11.3a)$$

$$\text{Im}[X(j\omega)] = \frac{2}{\pi} \int_0^\infty \int_0^\infty \text{Re}[X(jv)] \cos(vt) \sin(\omega t) dv dt.$$

$$(6.11.3b)$$

That is, for a causal signal, the real and imaginary parts of the transform can be expressed in terms of the other. The results in (6.11.3a and b) are difficult to use. A more elegant way of expressing these relations is by using Hilbert transforms discussed in Chapter 5.

6.11.1 Relationship Between Real and Imaginary Parts of the Fourier Transform of a Causal Function Using Hilbert Transform

Consider impulse response of a realizable function and its transform

$$h(t) = h_e(t) + h_0(t) \xleftrightarrow{\text{FT}} H(j\omega)$$
$$= \text{Re}[H(j\omega)] + j\text{Im}[H(j\omega)] \quad (6.11.4)$$

$$h_e(t) = [h(t) + h(-t)]/2, h_o(t) = [h(t) - h(-t)]/2.$$
$$(6.11.5)$$

Noting that $h(-t) = 0, t > 0$, the following interesting relations result:

$$h_e(t) = \left\{ \begin{array}{l} h_0(t), t > 0 \\ -h_0(t), t < 0 \end{array} \right\} = h_0(t)\text{sgn}(t) \text{ or}$$
$$h_0(t) = h_e(t)\text{sgn}(t). \quad (6.11.6)$$

Using $F[\text{sgn}(t)] = 2/j\omega$ and using the Fourier time multiplication theorem,

$$h_0(t)\text{sgn}(t) \xleftrightarrow{\text{FT}} \text{Re}[H(j\omega)], h_e(t)\text{sgn}(t)j\text{Im}[H(j\omega)].$$
$$(6.11.7)$$

The real and the imaginary parts of the function can be related by using the frequency convolution theorem studied in Chapter 4. The frequency convolution theorem corresponding to the two functions is given in (6.11.8) using $F[x_i(t)] = X_i(j\omega)$, $i = 1, 2$.

$$x_1(t)x_2(t) \xleftrightarrow{\text{FT}} \frac{1}{2\pi}[X_1(j\omega) * X_2(j\omega)]$$
$$= \frac{1}{2\pi} \int_{-\infty}^\infty X_1(j\alpha)X_2(j(\omega - \alpha)) d\alpha. \quad (6.11.8)$$

Using these in (6.11.7) the following results:

$$\text{Re}[H(j\omega)] = \frac{1}{2\pi}\left\{ j\,\text{Im}[H(j\omega)] * \frac{2}{j\omega} \right\},$$

$$j\,\text{Im}[H(j\omega)] = \frac{1}{2\pi}\left\{ \text{Re}[H(j\omega)] * \frac{2}{j\omega} \right\}. \quad (6.11.9)$$

$$\text{Im}[H(j\omega)] = -\frac{1}{\pi} \int_{-\infty}^\infty \frac{\text{Im}[H(j\omega)]}{(\omega - \alpha)} d\alpha,$$

$$\text{Re}[H(j\omega)] = \frac{1}{\pi} \int_{-\infty}^\infty \frac{\text{Im}(H(j\omega))}{(\omega - \alpha)} d\alpha. \quad (6.11.10)$$

Note that the causal signal does not contain an impulse at $t = 0$ is assumed. If it does, then it adds a constant to its transform. Let $h(t) = K\delta(t) + h_1(t)$, where $h_1(t)$ does not have an impulse at $t = 0$. The impulse at $t = 0$ appears in the transform and

$$K = \lim_{\omega \to \infty} H(j\omega) = \text{Re}[H(j\infty)]. \quad (6.11.11)$$

If there is an impulse at $t = 0$, the real and the imaginary parts of the transform of the causal signal are related by the following relations in terms of Hilbert transforms:

$$\text{Im}[H(j\omega)] = -\frac{1}{\pi} \int_{-\infty}^{\infty} \frac{\text{Im}[H(j\omega)]}{(\omega - \alpha)} d\alpha,$$

$$\text{Re}[H(j\omega)] = \text{Re}[H(j\infty)] + \frac{1}{\pi} \int_{-\infty}^{\infty} \frac{\text{Im}(H(j\omega))}{(\omega - \alpha)} d\alpha.$$

$$(6.11.12)$$

Notes: From these two equations we note that the real and the imaginary parts of a realizable transfer function $H(j\omega)$ are tied together by the Hilbert transform. This implies that $H(j\omega)$ can be found from its real part alone, referred to as the *real-part sufficiency*. A physically realizable transfer function can also be found from its magnitude spectrum alone, which is referred to as a *minimum phase transfer function* and was briefly mentioned in Section 6.7. A linear system with a transfer function $H(s)$ has *no zeros or poles on the right halfs − plane* is called a *minimum phase system*. The relations between amplitude and phase responses of causal functions are referred to as *Bode relations*. Detailed study of these is beyond the scope here. For a discussion on this topic and other relations, see Bode (1945). Finding an impedance function $Z(s)$ from $\text{Re}[Z(j\omega)]$ has been investigated by many authors, see Weinberg (1962). Here, finding the minimum phase transfer function $H(s)$ from the given amplitude spectrum $|H(j\omega)|$ is of interest. The following gives a simpler procedure compared to the above results. ■

6.11.2 Amplitude Spectrum $|H(j\omega)|$ to a Minimum Phase Function $H(s)$

Given $(|H(j\omega)|)^2$ find the minimum phase function $H(s) = KN(s)/D(s)$ that is stable. Starting with $(|H(j\omega)|)^2$, we have

$$|H(j\omega)|^2 = H(j\omega)H^*(j\omega)$$

$$= H(j\omega)H(-j\omega) = K^2 N(\omega^2)/D(\omega^2)$$

$$|H(j\omega)|^2 = H(s)H(-s)|_{s=j\omega} = K^2 \frac{N(\omega^2)}{D(\omega^2)} \quad \text{or}$$

$$H(s)H(-s) = K^2 \frac{N(\omega^2)}{D(\omega^2)}|_{\omega^2 = -s^2}. \quad (6.11.13)$$

Poles and zeros of the function $H(s)H(-s)$ have *quadrantal symmetry and mirror symmetry* about the $j\omega$ axis giving a choice in selecting the poles and zeros of $H(s)$.

1. Choose only the left half plane roots of $D(\omega^2)|_{\omega^2 = -s^2}$. This gives $D(s)$.
2. To have a minimum phase system, choose only the left half plane roots of $N(s)N(-s)$. Obviously there are other choices and those do not result in minimum phase functions.
3. Select a value $K > 0$ to match the value of the amplitude function $|H(j\omega)|$ at a desirable frequency.

Example 6.11.1 Find the minimum phase stable transfer function $H(s)$ given the amplitude-squared spectrum below.

$$|H(j\omega)|^2 = \frac{9(\omega^2 + 4)}{(\omega^4 + 10\omega^2 + 9)}. \quad (6.11.14)$$

Solution: Substituting $\omega^2 = -s^2$, and using the above procedure, results in

$$|H(j\omega)|^2|_{-\omega^2 = s^2} = \frac{9(4 - s^2)}{(9 - 10s^2 + s^4)} = H(s)H(-s)$$

$$= K^2 \frac{(s+2)}{(s+1)(s+3)} \frac{(-s+2)}{(-s+1)(-s+3)}$$

$$\Rightarrow H(s) = K\frac{(s+2)}{(s+1)(s+3)}, K = \sqrt{|H(j\omega)|^2}|_{\omega=0} = 3.$$

■

6.12 More on Filters: Source and Load Impedances

In this section simple passive analog filters are considered. In the next chapter, the design of various types of filters starting from the specifications to the synthesis using passive and active elements will be considered. In the simple examples considered so far we assumed only one resistor and one inductor (or capacitor) in the filter circuit. The filter problem is illustrated in Fig. 6.12.1a, where we have three boxes, one represents a source, second one represents a filter, and the third one represents the load. Using the Thevenin's equivalent circuit, we can replace the boxes represented by the source by the source plus the source impedance and the box representing the load by the load impedance. This is shown in Fig. 6.12.1b. The source and load impedances are generally assumed to be resistive in the frequency range of interest. This is a standard assumption in most filter design problems as we are operating in a small range of filter frequencies. In stead of Thevenin's equivalent circuit we could use the Norton's equivalent circuit shown in Fig. 6.12.1c. That is,

replace the series circuit consisting of source and source resistance in Fig. 6.12.1b by a current source in parallel with the source impedance.

This procedure allows the designer to separate the work associated with the filters from any designs associated with the left of the filter, i.e., the source and to the right of the filter, i.e., the load. We might also add that the source box and the load box may include several parts and the filter designer does not have to worry about those parts. In a later chapter when we consider two-port circuit analysis we will come back to this. For now let us consider a simple example illustrating the effect of the load. In the following we will derive the transfer functions in the Laplace transform domain. The frequency responses can be derived by replacing $s = j\omega$ in the transfer functions.

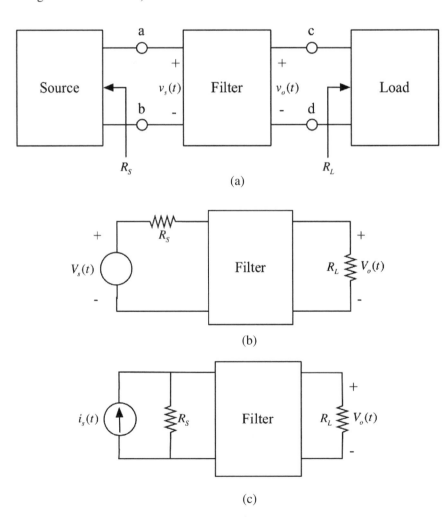

Fig. 6.12.1 (a) Filter with source and load resistors, (b) filter using Thevenin's source equivalent circuit, and (c) filter using source Norton's equivalent circuit

6.12.1 Simple Low-Pass Filters

Example 6.12.1 Consider the RC circuit shown in Fig. 6.12.2 with the source and the load resistors. Derive the transfer function and sketch the amplitude characteristic function for the two cases. a. $R_L = \infty$ and b. $R_L = R_s$.

Solution: The transfer functions are given by

$$\left(\frac{Y(s)}{X(s)}\right)_a = H_{LP}(s) = \frac{Z_2}{Z_1 + Z_2},$$

$$Z_2 = \frac{R_L/Cs}{R_L + (1/Cs)} = \frac{R_L}{1 + R_L Cs}, \quad Z_1 = R_s \quad (6.12.1a)$$

$$\left(\frac{Y(s)}{X(s)}\right)_b = H_{LP}(s) = \frac{R_L/R_s R_L C}{s + [(R_s + R_L)/R_s R_L C]}$$

$$= \frac{K}{s + \omega_c}, K = 1/R_s C, \omega_c = \frac{R_s + R_L}{R_s R_L C}.$$
$$(6.12.1b)$$

In case $a.$, the load resistance is infinite, i.e., the circuit is *not loaded*. In case $b.$ $R_s = R_L$. For the two cases the corresponding transfer functions are

$$a.\ H_{LP, R_L \to \infty}(s) = \frac{(1/R_s C)}{(s + (1/R_s C))},$$

$$b.\ H_{LP, R_s = R_L}(s) = \frac{(1/R_s C)}{s + (2/R_s C)}. \quad (6.12.2)$$

In both cases, the gain constant is the same. However, the cutoff frequency is increased in the case of a load resistance. Note that the peak value of the amplitude response function in case $b.$ is $(1/2)$. So, the 3 dB frequency corresponds to the value of the magnitude of the function equal to $(1/2)(1/\sqrt{2})$. At $\omega = 0$ the filter circuit is transparent; at $\omega = \infty$

there is no signal transmission; in between these frequencies, the output signal amplitude attenuation is determined by the equation $|Y(j\omega)| = |H(j\omega)||X(j\omega)|$. At the 3 dB frequency, ω_{3dB}

$$|Y(j\omega)|_{\omega=\omega_{3dB}} = (1/\sqrt{2})|X(j\omega)|_{\omega=\omega_{3dB}}. \qquad \blacksquare$$

Notes: For simplicity, generic functions $x(t)$ for the input and the output voltage $y(t)$ are used. Usually, $v_i(t)$ (or $v_s(t)$) for the input and $v_0(t)$ for the output are common. $\qquad\blacksquare$

6.12.2 Simple High-Pass Filters

In the ideal low- and high-pass filter cases shown in Fig. 6.10.7a, and b, it can be see that

$$|H_{Lp}(\omega)|_{\omega=0} = |H_{HP}(j\omega)|_{\omega=\infty} = 1 \text{ and}$$
$$|H_{Hp}(j\omega)|_{\omega=0} = |H_{Lp}(j\omega)|_{\omega=\infty} = 0. \quad (6.12.3)$$

In addition, the amplitudes of these functions transition at the frequency $\omega = \omega_c$ and the change in the amplitudes are as follows:

$$1(\text{low-pass}) \to 0\ (\text{high-pass}) \text{ or}$$
$$0\ (\text{high-pass}) \to 1\ (\text{low-pass}).$$

A logical conclusion is that $\omega \to (1/\omega)$ (i.e., $s \to (1/s)$) provides a transformation that gives a way to find a high-pass filter function from a low-pass filter function. Noting that the impedance of an inductor is $(j\omega L)$ and the impedance of a capacitor is $(1/j\omega C)$, a high-pass filter can be obtained from a

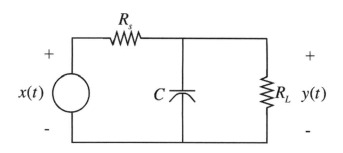

Fig. 6.12.2 Example 6.12.1

low-pass filter by replacing a capacitor by an inductor and an inductor by a capacitor. Since the impedance of the resistor R is independent of frequency, no change is necessary in the case of resistors. Using the RC and the RL low-pass circuits studied earlier in the low-pass case, we have two simple high-pass filters shown in Fig. 6.12.3a,b, one is a RL and the other one is a RC circuit. The filter is a low-pass if the inductor is in the series arm and the capacitor in the shunt arm. Similarly the filter acts as a high-pass filter if the capacitor is in the series arm and the inductor is in the shunt arm. The transfer functions corresponding to the two circuits in Fig. 6.12.3 are

$$H_{Hp_a}(s) = \frac{s}{s + (1/RC)}, H_{Hp_b}(s) = \frac{s}{s + (R/L)}$$

$$H_{HP_a}(j\omega) = \frac{j\omega}{j\omega + (1/RC)}, H_{HP_b}(j\omega) = \frac{j\omega}{j\omega + (R/L)}.$$

$$(6.12.4)$$

The amplitude and the phase response characteristics of these are as follows:

$$\left| H_{Hp_a}(j\omega) \right| = \frac{|\omega|}{\sqrt{\omega^2 + (1/RC)^2}},$$

$$\angle H_{Hp_a}(\omega) = 90^0 - \tan^{-1}(\omega RC) \qquad (6.12.5a)$$

$$\left| H_{Hp_b}(j\omega) \right| = \frac{|\omega|}{\sqrt{\omega^2 + (R/L)^2}},$$

$$\angle H_{Hp_b}(\omega) = 90^0 - \tan^{-1}(\omega L/R). \qquad (6.12.5b)$$

The amplitude and phase responses are shown in Fig. 6.12.4 for $\omega > 0$. The maximum value of the amplitude response is 1 or 0 dB. The 3 dB frequencies can be computed by equating the amplitude response function to $(1/\sqrt{2})$ and solving for ω. That is,

$$\frac{1}{\sqrt{2}} = \frac{|\omega_{ca}|}{\sqrt{\omega_{ca}^2 + (1/RC)^2}}, \frac{1}{\sqrt{2}} = \frac{|\omega_{cb}|}{\sqrt{\omega_{cb}^2 + (R/L)^2}}.$$

$$(6.12.6)$$

$$\Rightarrow \omega_{ca} = (1/RC), \omega_{cb} = (R/L). \qquad (6.12.7)$$

As in the low-pass case, given the cutoff frequency ω_c, one of the reactive component (inductor or capacitor) values can be solved by selecting the resistor value. The high-pass filter is transparent from the input to the output at infinite frequency and no signal transmission at zero frequency. Note the low and high-frequency behavior of the low-pass and the high-pass filter functions $H_{Lp}(s)$ (or $H_{Hp}(s)$) at $s = 0$ and $s = \infty$.

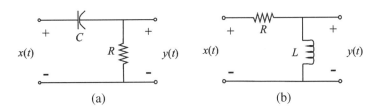

Fig. 6.12.3 Simple high passive filters (a) RC circuit, (b) RL circuit

(a)   (b)

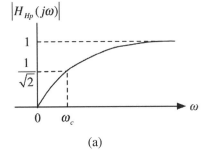
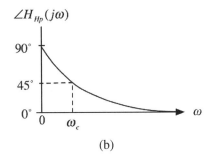

Fig. 6.12.4 Simple high-pass filter responses: (a) amplitude and (b) phase

(a)   (b)

6.12.3 Simple Band-Pass Filters

These filters pass a band of frequencies called the passband and attenuate or eliminate the frequencies outside the passband, called the stopband. The simplest band-pass filter has a second-order transfer function. The ideal band-pass filters have two cutoff (or 3 dB) frequencies ω_{low} and ω_{high}. These frequencies are defined as the frequencies for which the magnitude of the transfer function is equal to $\max(1/\sqrt{2})|H_{Bp}(j\omega)|$. In addition to these, a new frequency referred as the *center or the resonant frequency* ω_0, is of interest. It is defined as the frequency at which the transfer function of the circuit $H_{Bp}(j\omega)$ is purely real. The center frequency is *not* in the middle of the passband. It is the *geometric center* of the pass-band edges. It is related to the 3 dB frequencies by

$$\omega_0 = \sqrt{\omega_{low}\omega_{high}}. \qquad (6.12.8)$$

The second parameter of interest is the *3 dB bandwidth* given by

$$\beta = \omega_{high} - \omega_{low}. \qquad (6.12.9)$$

The third parameter, the *quality factor* is the ratio of the center frequency to the 3 dB bandwidth. It is given by

$$Q = \frac{\omega_0}{(\omega_{high} - \omega_{low})}. \qquad (6.12.10)$$

A *second-order function* that has the *band-pass characteristics* is

$$H_{Bp}(s) = \frac{H_0 a s}{s^2 + as + b} = \frac{H_0(\omega_0/Q)s}{s^2 + (\omega_0/Q)s + \omega_0^2}. \qquad (6.12.11)$$

For simplicity the gain constant is assumed to be $H_0 = 1$ in the following. The transfer function has a *zero* at the *origin* $(s = 0)$ and at *infinity* $(s = \infty)$ indicating that the function goes to zero at $\omega = 0$ and at $\omega = \infty$. For $Q > 1/2$, it has a pair of complex poles given by

$$s_1, s_2 = -\frac{\omega_0}{2Q} \pm j\omega_0\sqrt{1 - \frac{1}{4Q^2}}. \qquad (6.12.12)$$

The corresponding transfer function, the amplitude, and phase responses are given by

$$\begin{aligned} H_{Bp}(j\omega) &= \frac{(\omega_0/Q)\omega}{(\omega_0^2 - \omega^2) + j(\omega_0/Q)\omega} \\ &= \frac{1}{\left[1 + jQ\left(\frac{\omega}{\omega_0} - \frac{\omega_0}{\omega}\right)\right]\left[1 + jQ\left(\frac{\omega}{\omega_0} - \frac{\omega_0}{\omega}\right)\right]} \end{aligned}$$
$$(6.12.13)$$

$$|H_{Bp}(j\omega)| = 1/\sqrt{1 + Q^2\left(\frac{\omega}{\omega_0} - \frac{\omega_0}{\omega}\right)^2},$$
$$\angle H_{Bp}(j\omega) = -\tan^{-1}\left[Q\left(\frac{\omega}{\omega_0} - \frac{\omega_0}{\omega}\right)\right]. \qquad (6.12.14)$$

The amplitude and the phase responses are sketched in Fig. 6.12.5 for positive values of ω. From the amplitude response, the peak of the amplitude appears at the center frequency ω_0. The peak magnitude is 1 at $\omega = \omega_0$. The phase angle starts at $(\pi/2)$, crosses the frequency axis at $\omega = \omega_0$ and it asymptotically reaches $(-\pi/2)$ as $\omega \to \infty$. Higher the value of Q is, the more *peaked* the amplitude response is and steeper the phase response is around $\omega = \omega_0$. The 3 dB or half-power bandwidth can be determined by assuming $\omega_{low} < \omega_0$ and $\omega_{high} > \omega_0$. These frequencies can be computed from

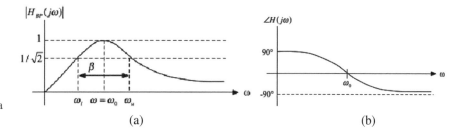

Fig. 6.12.5 (a) Amplitude and (b) phase responses of a band-pass filter

$$\left|H_{Bp}(j\omega)\right|^2\big|_{\omega=\omega_l,\omega_u}=\frac{1}{2}\rightarrow Q^2\left(\frac{\omega}{\omega_0}-\frac{\omega_0}{\omega}\right)^2=1\Rightarrow$$

$$(\omega^2-\omega_0^2)=\pm\frac{\omega_0\omega}{Q},\omega^2\mp\frac{1}{Q^2}\omega\omega_0-\omega_0^2=0.$$

There are four roots of this equation, two for positive and two for negative frequencies.

$$\omega_u,-\omega_l=\frac{\omega_0}{2Q}\pm\frac{1}{2}\sqrt{\left(\frac{\omega_0}{Q}\right)^2+4\omega_0^2},$$

$$-\omega_u,\omega_l=-\frac{\omega_0}{2Q}\mp\frac{1}{2}\sqrt{\left(\frac{\omega_0}{Q}\right)^2+4\omega_0^2}.$$

Assuming that $Q>(1/2)$, the *positive roots* are given by

$$\omega_l,\omega_u=\omega_0\sqrt{1+\frac{1}{4Q^2}}\mp\frac{\omega_0}{2Q}. \qquad (6.12.15)$$

The 3 dB bandwidth (B or β) and ω_0 are respectively given by

$$\beta=(\omega_{high}-\omega_{low})=\omega_0/Q,$$
$$\omega_0^2=\omega_l\omega_u\rightarrow\omega_0=\sqrt{\omega_l\omega_u}. \qquad (6.12.16)$$

Clearly from the first equation in (6.12.16) the bandwidth is inversely proportional to the value of Q. That is, the bandwidth decreases as Q increases and vice versa. The filter is assumed to be *narrowband* if ω_0 is very large compared to the bandwidth of the filter, i.e., $\omega_0\gg B$. As a rough measure, we assume the filter is a narrowband filter if

$$Q=\omega_0/\beta\geq10. \qquad (6.12.17a)$$

For the narrowband case the edges of the passband and ω_0 are given below.

$$\omega_l,\omega_u=\omega_0\mp(\omega_0/2Q). \qquad (6.12.17b)$$

In this case, ω_0 is in the *middle* of the 3 dB frequencies and is the *center frequency*.

Example 6.12.2 Consider the circuit shown in Fig. 6.12.6. Find the transfer function and derive

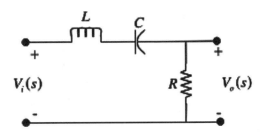

Fig. 6.12.6 A simple band-pass filter

the expressions for the center frequency, bandwidth, and the quality factor.

Solution: The transfer functions are

$$H_{Bp}(s)=\frac{V_0(s)}{V_i(s)}=\frac{(R/L)s}{s^2+(R/L)s+(1/LC)},$$

$$H_{Bp}(j\omega)=\frac{(R/L)j\omega}{[(1/LC)-\omega^2]+j\omega(R/L)]}. \qquad (6.12.18)$$

The corresponding amplitude and phase responses are given for *positive* ω by

$$\left|H_{Bp}(j\omega)\right|=\frac{(R/L)\omega}{\sqrt{((1/LC)-\omega^2)^2+((R/L)\omega)^2}},$$

$$\angle H_{Bp}(j\omega)=90^o-\tan^{-1}\frac{((R/L)\omega)}{[(1/LC)-\omega^2)]}. \qquad (6.12.19)$$

From (6.12.19), by using (6.12.16) results in

$$\omega_0=\sqrt{1/LC}, \beta=\text{Bandwidth}=R/L \text{ and}$$

$$Q=\omega_0/B=\sqrt{L/CR^2}. \qquad (6.12.20)$$

There two equations had three unknowns. In the design of a second-order band-pass filter, use the following steps. From the given second-order function, find ω_0 and Q. Select one of the element values, say C, and then use the other two equations to solve for the element values R and L. These are $L=1/\omega_0^2C, R=\beta L$. Note that the gain constant is assumed to be $H_0=1$. If the gain is higher than 1, then circuit needs amplification.

So far we have not considered of having both source and load impedances in the band-pass case. Also, inductors are never ideal and can be modeled by a resistor in series with an ideal inductor shown in Fig. 6.12.7 resulting in the impedance of the non-ideal inductor as R_i+sL. The new transfer function and the amplitude response are

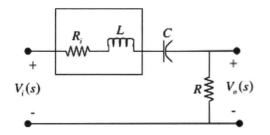

Fig. 6.12.7 A simple band-pass filter with a nonideal inductor

$$H_{Bp}(s) = \frac{(R/L)s}{s^2 + [(R + R_i)/L]s + (1/LC)} \quad (6.12.21)$$

$$\left|H_{Bp}(j\omega)\right| = \frac{(R/L)|\omega|}{\sqrt{((1/LC) - \omega^2)^2 + \omega((R + R_i)L)}}. \quad (6.12.22)$$

The center frequency is the same as before. The maximum value of the amplitude response is now $R/(R_i + R)$. Also the new bandwidth is $(R + R_i)/L$. The nonideal inductor reduces the peak value and increases the bandwidth of the filter response. Correspondingly, the Q value is reduced. Nonideal inductors make the amplitude response less peaked with an increase in the bandwidth, i.e., the amplitude response becomes lower and broader.

A band-pass filter can be viewed as a cascaded low-pass and a high-filter combination with the cutoff frequency of the low-pass filter greater than the cutoff frequency of the high-pass filter, $\omega_{c,LP} > \omega_{c,HP}$. An obvious question is, can we obtain a band-pass filter function from a low-pass filter function? In the two simple low-pass circuits considered earlier, the RL circuit, with the replacement of the inductor by a series LC circuit results in the circuit studied above. By replacing the capacitor in the RC low-pass circuit by a parallel LC circuit results in a band-pass circuit shown in Fig. 6.12.7. The element values in the band-pass case need to be determined using the center

frequency and the required bandwidths. The figure illustrates only the concepts here. In Chapter 7 we will come back to the frequency transformations that involve changing cutoff frequencies of filters, converting various filter specifications into *a proto-type* low-pass filter specification, finding the appropriate filter function, the appropriate frequency transformations, synthesizing the filter function and finally scaling the circuit to fit the given specifications. Figure 6.12.8a gives a low-pass circuit. A band-pass circuit can be obtained by replacing the capacitor by an LC circuit as shown in Fig. 6.12.8b.

Inductors tend to be bulky and nonideal. That is, they have a resistive part R_w, thus reducing the quality factor of such coils. Impedance of the inductor needs to be replaced from $j\omega L$ to $R_w + j\omega L$. The quality factor of the coil is given by $Q = (\omega L / R_w)$, a *function of frequency*. Designing inductors with high Q values is a difficult process. In addition, since the field of operation associated with coils is the magnetic field, there is coupling between different inductors in a circuit. This can be reduced by either shielding one inductor from another and/or by placing in a manner shown in Fig. 6.12.9 requiring more space on the circuit board.

6.12.4 Simple Band-Elimination or Band-Reject or Notch Filters

The amplitude response of a band-reject filter has the shape of a notch and is used to remove a band of frequencies somewhere in the middle of the frequency band and pass the low and high frequencies outside this band. A second-order notch filter has a transfer function of the form

$$H_{Be}(s) = \frac{(s^2 + 2bs + \omega_0^2)}{s^2 + (\omega_0/Q)s + \omega_0^2}, \quad b << \frac{\omega_0}{2Q}$$

$$(6.12.23)$$

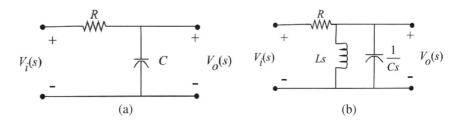

Fig. 6.12.8 (a) Low-pass and (b) band-pass filter circuits

(a)　　　　　　　　(b)

Fig. 6.12.9 Placement of two inductors to reduce magnetic coupling

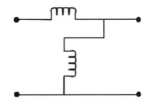

$$\underset{s\to 0}{Lim}\, H_{Be}(s) = 1 \text{ and } \underset{s\to\infty}{Lim}\, H_{Be}(s) = 1. \quad (6.12.24)$$

The second-order band-pass and the notch filter functions have the same denominator, see (6.12.11) and (6.11.23). Notch filter passes low and high frequencies of the input signal without much attenuation and attenuates (or eliminates) a band of frequencies in the middle. This can be seen by first computing the zeros of the transfer function.

$$z_1, z_2 = -b \pm j\sqrt{\omega_0^2 - b^2} \Rightarrow$$

$$z_1, z_2 \approx -b \pm j\omega_0 \text{ (In the case of } \omega_0 \gg b).$$

$$\quad (6.12.25)$$

A special and an interesting case is when $b = 0$ and for this case

$$|H_{Be}(j\omega)| \cong \frac{|\omega_0^2 - \omega^2|}{\sqrt{(\omega_0^2 - \omega^2)^2 + [(\omega_0/Q)\omega]^2}},$$

$$\angle H_{Be}(j\omega) = -\tan^{-1}\left[\frac{(\omega_0/Q)\omega}{(\omega_0^2 - \omega^2)}\right]. \quad (6.11.26)$$

Note there is no output at the *notch frequency* ω_0 as $|H_{Be}(j\omega_0)| = 0$. The amplitude and phase responses of the notch filter are shown in Fig. 6.12.10 for $\omega > 0$. We can see that

$$|H_{Be}(j\omega)|_{\omega=0} = 1 \text{ and } \lim_{\omega\to\infty}|H_{Be}(j\omega)| = 1.$$

The 3 dB frequencies are obtained by equating $|H(\omega_{3dB})| = 1/\sqrt{2}$. The two frequencies are

$$\omega_l, \omega_u = \omega_0 \pm (\omega_0/2Q), \beta = \omega_0/Q. \quad (6.12.27)$$

If $b \neq 0$, then

$$|H_{Be}(j\omega)|_{\omega=\omega_0} = Qb/\omega_0. \quad (6.12.28)$$

The attenuation will be significant at the notch frequency as ω_0 is usually large. Also, note the phase reversal at $\omega = \omega_0$ in the phase response.

Notch filters are used wherever a narrowband of frequencies needs to be eliminated from a received signal. In any electronic device, 60 Hz undesired hum, is ever present and a notch filter can be used to remove this. There are many applications in the telephone industry. In a long-distance call, a single frequency is transmitted from the caller to the telephone office until the end of the dialing of the number. After the party answers, the tone signal ceases and billing of the call begins and it continues as long as the signal tone is absent until the call is complete. A different application is toll-free long-distance calls that are not billed. For these, the signal tone is transmitted to the telephone office for the entire period of the call. Since the signal is within the voice frequency band, it must be removed from the voice signal before being transmitted from the telephone office to the listener. A simple second-order notch filter could be used for such an application.

Notch filters are used wherever a narrowband of frequencies need to be eliminated from a received signal. In any electronic device, 60 Hz, an undesired hum, is ever present and a notch filter can be used to remove this. There are many applications in the telephone industry.

Fig. 6.12.10 (a) $|H_{Be}(j\omega)|$ and (b) $\angle H_{Be}(j\omega)$

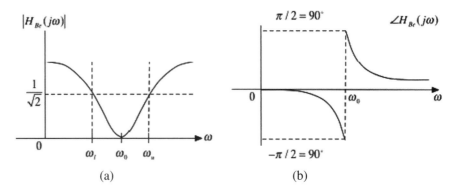

(a)

(b)

Example 6.12.3 Band-elimination filters can be derived from low-pass filters by replacing an inductor (capacitor) by a parallel LC (series LC) circuit. Consider the circuit in Fig. 6.12.11 with a nonideal inductor with the equivalent impedance of the coil equal to $(R_w + j\omega Ls)$. Assuming the circuit is not loaded, the output current is zero. Derive the transfer function and show it corresponds to a band-elimination filter.

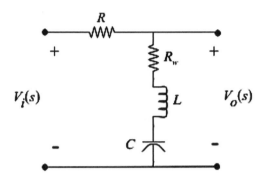

Fig. 6.12.11 A band-elimination filter with a nonideal inductor

Solution: The transfer function, its amplitude and phase responses are

$$H_{Be}(s) = \frac{R_w + Ls + (1/Cs)}{(R + R_w) + Ls + (1/Cs)}$$

$$= \frac{(1 + R_w s + LCs^2)}{(1 + LCs^2) + (R + R_w)Cs} \quad (6.12.29)$$

$$H_{Be}(j\omega) = \frac{(R_w + j\omega L + (1/(j\omega C)))}{[(R + R_w) + j\omega L + (1/(j\omega C))]}$$

$$= \frac{(1 - \omega^2 LC) + j\omega R_w C}{(1 - \omega^2 LC) + j\omega(R + R_w)C} \quad (6.12.30)$$

$$|H_{Be}(j\omega)| = \frac{\sqrt{(1 - LC\omega^2)^2 + (R_w C\omega)^2}}{\sqrt{(1 - \omega^2 LC)^2 + (\omega(R + R_w)C)^2}}$$

$$(6.12.31a)$$

$$\angle H_{Be}(j\omega) = \arctan[R_w C\omega/(1 - \omega^2 LC)]$$
$$- \arctan[\omega(R + R_w)C/(1 - \omega^2 LC)].$$

$$(6.12.31b)$$

The transfer function has the band-elimination characteristics as

$$\lim_{\omega \to 0}|H_{Be}(j\omega)| = 1, \quad \lim_{\omega \to \infty}|H_{Be}(j\omega)| = 1, \text{ and}$$

$$|H_{Be}(j\omega)|_{\omega = 1/\sqrt{LC}} = R_w/(R + R_w). \quad (6.12.31c)$$

If the winding resistance R_w is small, then the peak value of the amplitude response is close to 1. Using (6.12.27) and (6.12.28), we have $2\beta = R_w/LC$, $\omega_0 = 1/\sqrt{LC}$ and $(\omega_0/Q) = (R + R_w)C$. The notch bandwidth is $B = [\omega_0/Q] = [(R + R_w)C]$. In the ideal case, a second-order notch filter can be obtained from a second-order band-pass filter by replacing the parallel (series) LC circuit by a series (parallel) LC circuit. ∎

Passive filter designs use resistors, inductors (and transformers), and capacitors. Four ladder forms of low-pass, high-pass, band-pass, and band-elimination filter circuits with source and load resistances are shown in Fig. 6.12.12. Source with source resistance is identified by a circle and the load resistance by a square. Between the source and the load, a lossless circuit is inserted, i.e., lossless coupling between the source and load.

Figure 6.12.12a: Low-pass filter: When $\omega = 0$, inductors will be short and the capacitors will be open and the source is directly connected to the source and the output is $v_0 = [R_L/(R_i + R_L)]v_i$. When $\omega = \infty$, inductors will be open and the capacitors will be short and the load is disconnected from the source and the output is zero.

Figure 6.12.12b: High-pass filter: When $\omega = 0$, inductors will be open and the capacitors will be short and the load is disconnected from the source and the output is zero. For $\omega = \infty$, inductors will be short and the capacitors will be open and the source is directly connected to the load and $v_0 = [R_L/(R_i + R_L)]v_i$.

Figure 6.12.12c: Band-pass filter: At $\omega = 0$, inductors will be short and the capacitors will be open and there is no output. At $\omega = \infty$, there is no output either. At the center frequency ω_0, if $(L_{si}C_{si} = 1/\omega_0^2)$, the series arm is short since

$$L_{sei}s + (1/C_{sei}s)|_{s^2 = -(1/L_{sei}C_{sei})} =$$
$$(L_{sei}C_{si}s^2 + 1)/C_{sei}s|_{s^2 = -(1/L_{sei}C_{sei})} = 0.$$

In a similar manner we can show that at the frequency ω_0, the shunt arm is open since

$$[L_{shi}s/(L_{shi}C_{shi}s^2 + 1)]|_{s^2 = -(1/L_{shi}C_{shi})} = \infty.$$

Figure 6.12.12d: In the band-elimination case, we can show that $v_0 = [R_L/(R_i + R_L)]v_i$ at $\omega = 0$ and $\omega = \infty$. At $\omega = \omega_0$, the output is zero.

The four filters have the desired transfer characteristics values at $\omega = 0, \infty$ (and, in addition $\omega = \omega_0$, in

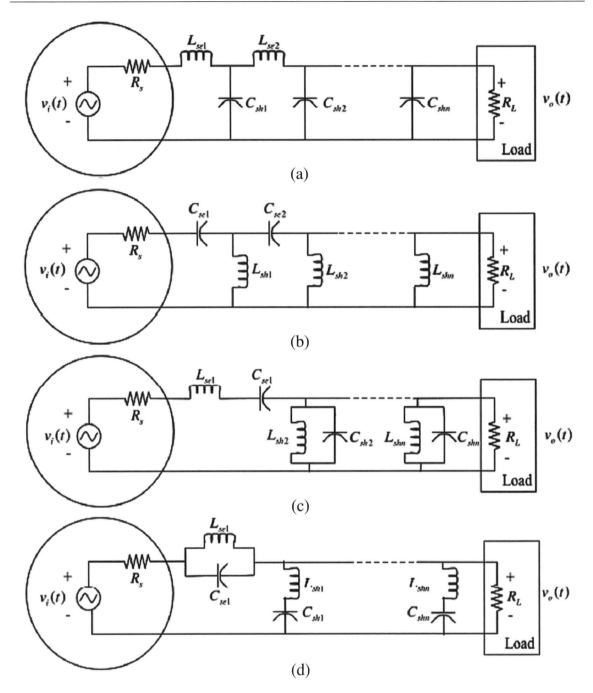

Fig. 6.12.12 Passive filters

the case of band-pass and band-elimination filters. The exact characteristics at other frequencies cannot be determined since the actual filter element values are not known. In the passive filter design, and in general in communication theory, power transfer between the source and the load is important.

6.12.5 Maximum Power Transfer

If there is no filter in Fig. 6.12.12, i.e., the source resistance is connected directly to the load, the maximum power is available at the output provided

$$R_s = R_L. \qquad (6.12.32a)$$

This is the *maximum power transfer theorem and* can be proven by first expressing the power delivered to the load without the filter. That is,

$$p_L = R_L i^2, i = \frac{v_s}{(R_s + R_L)} = \frac{R_L}{(R_s + R_L)^2} v_s^2.$$

$$(6.12.32b)$$

Taking the derivative of p_L with respect to R_L, then solving for R_L results in (6.12.32a).

If the filter is inserted between the source and the load, then there will be less power available at the load, which will vary with frequency. This is defined as *insertion loss*. The design using these concepts is beyond the scope here, see Weinberg (1962).

6.12.6 A Simple Delay Line Circuit

In Section 6.4 we considered the properties of a delay function. In this part of the section we will consider a simple circuit that has a *constant amplitude response* and a phase response that can be adjusted. Consider the circuit shown in Fig. 6.12.13. Assuming the output current is equal to zero, the output voltage can be expressed by

$$V_0(s) = V_c(s) - V_x(s) = \frac{(1/Cs)}{(R + (1/Cs))} V_{in}(s) - \frac{1}{2} V_{in}(s)$$

$$= \frac{1 - RCS}{2(RCs + 1)} V_{in}(s).$$

$$(6.12.33)$$

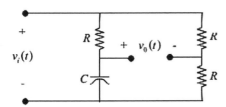

Fig. 6.12.13 A simple delay line circuit

The transfer function in the s domain and in the frequency domain, the corresponding magnitude and phase responses and the group delay function are respectively given by

$$H(s) = \frac{1}{2} \frac{(a - s)}{(a + s)}, H(j\omega) = \frac{1}{2} \frac{(a - j\omega)}{(a + j\omega)}, a = \frac{1}{RC}$$

$$(6.12.34a)$$

$$|H(j\omega)| = 1/2, \angle H(j\omega) = -2 \tan^{-1}(\omega/a),$$

$$T_g(\omega) = -\frac{d\angle H(j\omega)}{d\omega} = \frac{a}{a^2 + \omega^2}. \qquad (6.12.34b)$$

The transfer function, the amplitude, phase, and the group delay responses are

$$H(s) = \frac{1}{2} \frac{(a - s)}{(a + s)}, H(j\omega) = \frac{1}{2} \frac{(a - j\omega)}{(a + j\omega)}, a = \frac{1}{RC}$$

$$|H(j\omega)| = 1/2, \angle H(j\omega) = -2 \tan^{-1}(\omega/a), \quad (6.12.35a)$$

$$T_g(\omega) = -\frac{d\angle H(j\omega)}{d\omega} = \frac{a}{a^2 + \omega^2}. \qquad (6.12.35b)$$

Since the amplitude response is (1/2), a constant for all frequencies, this function is referred to as an *all-pass* function. All-pass filters are used in cascade with the filters to provide the overall phase of the filter delay line combination and have an approximate linear phase characteristics. Additional phase due to the all-pass circuit adds to the filter delay.

6.13 Summary

In this chapter we have started with basics of systems analysis and circuits. The circuits considered are simple. Specific topics that were covered in this chapter are given below.

- Linear systems and their properties
- Two-terminal components: resistors, inductors, capacitors, voltage, and current sources
- Classification of systems based on linearity, time-invariance, and other concepts
- Impulse response of a linear system and the output in terms of the convolution integral
- Transfer functions along with examples of simple circuits
- System stability concepts and Routh's stability test
- Distortionless systems and distortion measures for nonlinear systems
- Group delay and phase delay responses
- System bandwidth measures similar to signal bandwidth
- Relations between real and imaginary parts of a Fourier transform of a causal function
- Derivation of the minimum phase transfer function from a given magnitude function
- Ideal low-pass, high-pass, band-pass, band-elimination filters along with delay lines

Problems

6.2.1 Consider the systems described by the following input–output relations. In each case, determine whether the system satisfies the following: 1. memoryless, 2. causal, 3. stable, 4. linear, and 5. time invariant.

$$a.\ y(t) = x(1-t), b.\ y(t) = x(t/2),$$
$$c.\ y(t) = \sin(x(t)).$$

6.3.1 Show that the systems $a.\ y(t) = x^2(t), b.\ y(t) = \text{sgn}(t)$ are not invertible.

6.3.2 Determine whether the system $y(t) = T[x(t)] = tx(t)$ is 1. memoryless, 2. causal, 3. stable, 4. linear, and 5. time invariant.

6.3.3 Consider the amplitude modulated function (discussed in Chapter 10) $y(t) = A_c[A + m(t)] \cos(\omega_c t)$ with cases $a.\ A = 0, b.\ A \neq 0$. Determine whether the system described by this equation is 1. memoryless, 2. causal, 3. linear, 4. time invariant, and 5. BIBO stable.

6.3.4 Repeat Problem 6.3.3 assuming the output of the system is $y(t) = x(t - \tau)$.

6.3.5 Determine whether the system described by the following is $a.$ linear or nonlinear, $b.$ time invariant or time varying

$$y(t) = \sum_{k=-\infty}^{\infty} x(t)\delta(t - kt_s) = \sum_{k=-\infty}^{\infty} x(kt_s)\delta(t - kt_s).$$

6.3.6 Consider the system described by $y(t) = x(\beta t)$. Determine for what values of β the system is $a.$ causal, $b.$ linear, and $c.$ time invariant.

6.4.1 In Fig. 6.4.1 we have considered a simple RL time-varying circuit. Consider the RC time-varying circuit shown in Fig. P6.4.1. Assume that the time constant is large enough to justify that the circuit can be used as an integrator in an approximate sense. Identify the approximation used in considering this circuit acts like an integrator.

6.4.2 Apply the periodic pulse waveform shown in Fig. 6.5.4a to a simple RL circuit obtained by

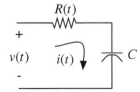

Fig. P6.4.1 An RC time-varying circuit

replacing the capacitor by an inductor in Fig. 6.5.4b. Give the corresponding steady-state response.

6.4.3 Determine the impulse responses of the ideal low-pass, high-pass, band-pass, and band-elimination filters defined in (6.10.14a, b, c, and d) and the ideal delay line function in (6.7.1).

6.4.4 Determine the system responses for the inputs $a.\ x_1(t) = u(t), b.\ x_2(t) = \Pi[t - .5]$ assuming the system impulse response is

$$h(t) = te^{-t}u(t).$$

6.4.5 Determine the stability of the integrator and a differentiator given below. Find their impulse responses and then use BIBO stability condition to see their stability.

$$y(t) = \int_{-\infty}^{t} x(\alpha)d\alpha, \qquad y(t) = \frac{dx(t)}{dt}.$$

6.4.6 Using very simple functions with the properties as identified to show the responses become unbounded. Consider the transfer functions $H(s)$. $a.\ H(s)$ has a pole on the right half s-pane, $b.H(s)$ has multiple poles on the imaginary axis, $c.\ H(s)$ has poles on the imaginary axis and the input function has a pole at this location. Explain why the responses become unbounded with the aid of the inverse Laplace transforms.

6.4.7 Consider the transfer function given by

$$T(s) = \frac{1}{1 + H(s)}, H(s) = \frac{K}{(1 + .1s)(a + s)}.$$

$a.$ Assuming $a = 1$, use the Routh array to determine the range of K for which the system is stable. $b.$ Repeat the problem in Part $a.$ assuming $K = 1$ and determine the range of K for which the system is stable. $c.$ Using the Routh array to determine the range of K for which the system has only poles to the left of $s = -1$ in the s-plane.

6.4.8 Use the Routh array and factor the polynomial $D(s) = s^4 + s^3 + 2s^2 + s + 1$.

6.4.9 Use the Routh array to find the number of right half s-plane roots of the polynomial $D(s) = s^4 + s^2 + s + 1$.

6.5.1 Give an RC circuit that approximates a differentiator and sketch the circuit's amplitude and phase responses.

6.5.2 The transfer function of a linear time-invariant system is

$$H(s) = \frac{Y(s)}{X(s)} = \frac{(s+1)}{(s^2 + s + 1)}.$$

Assuming $x(t) = \cos(\omega_0 t)$, find the steady-state response $y(t)$ of the system.

6.5.3 Find the response of the RL circuit shown in Fig. 6.6.3 for input pulses

$$a.\, x_a(t) = \Pi\left[\frac{t - (T/2)}{T}\right], T = 2\pi,$$

$$b.\, x_b(t) = \sin(t)x_a(t).$$

6.5.4 Consider a circuit that has the response $y(t) = x(t) - x(t - T)$ with the input $x(t)$. Give the system impulse response and the expressions $H(j\omega)$ and $|H(j\omega)|$.

6.5.5 Consider the following two differential equations with $x(t) = e^{j\omega t}$:

$$a.\, LC\frac{d^2y}{dt} + \frac{L}{R}\frac{dy}{dt} + y(t) = x(t)$$

$$b.\, LC\frac{d^2y}{dt^2} + y(t) = x(t).$$

Derive the transfer functions $H(j\omega) = Y(j\omega)/X(j\omega)$ in each case assuming $x(t) = e^{j\omega t}$.

6.5.6 Determine in each case below if the system described by its impulse response is stable or realizable, or both. Explain your results.

$$a.h_a(t) = \delta(t), \quad b.h_b(t) = e^{-t}u(t),$$
$$c.h_c(t) = \delta(t) - e^{-t}u(t).$$

6.5.7 Consider the impulse response and the corresponding transfer function $h_{Lp}(t) = e^{-t}u(t)$, $L[h_{Lp}(t)] = H_{Lp}(s) = 1/(s+1)$. What can you say about the deconvolution filter $H_R(s)$ and $h_R(t) = L^{-1}\{H_R(s)\}$? Is this function realizable?

6.6.1 Consider the following functions with $D(s) = (s+1)(s+4)(s+5)$. Classify these as minimum or mixed or maxi phase systems. Sketch their amplitude and phase responses.

$$a.\, H_1(s) = (s-3)(s+2)/D(s)$$
$$b.\, H_2(s) = (s+3)(s+2)/D(s)$$
$$c.\, H_3(s) = (s-3)(s-2)/D(s).$$

6.6.2 Sketch the amplitude and phase responses of

$$a.\, H_1(s) = 1/(s^2 + \sqrt{2}s + 1),$$
$$b.\, H_2(s) = 1/(s^2 + 3s + 3).$$

6.6.3 What can you say about the group delays associated with an all-pass functions at $\omega = 0$ and $\omega = \infty$ in Problem 6.6.2? Can you draw any general conclusions?

6.6.4 The second-order Butterworth function is given in Problem 6.6.2a.
a. Find its impulse response and the corresponding step response. b. Find the 10–90% rise time. c. Find the expression for the group delay of this function.

6.6.5 Assume the following node equations of a 3 nodes plus a reference node is given by

$$[(V_A - V_1)Y_1 + V_A Y_3 + (V_A - V_0)Y_2 = 0,$$
$$(V_0 - V_A)Y_2 + (V_0 - V_1)Y_4 = 0.$$

Give a circuit that has these node equations. Derive the transfer function V_0/V_1 assuming $Y_1 = 1/sL_1$, $Y_2 = 1/sL_2, Y_3 = sC_2, Y_3 = sC_3$.

6.6.6 Find the impulse and step responses of the following transfer functions with $a > 0$.

$$a.\, H_d(s) = \frac{1 - as}{1 + as}, a > 0,$$

$$b.\, H_{LP}(s) = \frac{a}{s^2 + as + a},$$

$$c.\, H_{BP}(s) = \frac{s}{s^2 + s + 1}$$

$$d.\, H_{Be}(s) = \frac{s^2 + a^2}{s^2 + bs + a^2},$$

$$e.\, H_{HP}(s) = \frac{s^2}{s^2 + s + 1}.$$

Give the amplitude and phase responses of these filters.

6.7.1 Consider the impulse response of a system $h(t) = e^{-t}u(t)$. Derive the expressions for its group and the phase delays. Sketch these functions on the same plot.

6.8.1 Show that the noise bandwidth of a band-pass function with center frequency ω_0 is

$$W_N = \int_0^\infty \left[|H(j\omega)|^2/|H(j\omega_0)|^2\right] d\omega.$$

6.8.2 Determine the RMS and the equivalent bandwidth of $H(j\omega)H^*(j\omega) = \Lambda[\omega/W]$.

6.9.1 Sketch the input and the output line spectra of a nonlinear circuit described by its input–output relationship $y(t) = x^2(t)$ assuming $x(t) = 2\cos(2\pi(60)t) + \sin(2\pi(60)t)$.

6.9.2 Assuming $y(t) = x(t) + x^2(t)$, sketch $Y(j\omega)$ by assuming

$$X(j\omega) = \Pi[(\omega + \omega_0)/W] + \Pi[(\omega - \omega_0)/W].$$

6.9.3 Assume the input is $x(t) = \cos(\omega_0 t)$ to a nonlinear system described by $y(t) = x(t) + .1x^2(t)$ determine the second-order distortion term.

6.9.4 Consider the systems described by the following input–output relations. In each case determine whether the system is *a.* memoryless, *b.* causal, *c.* stable, *d.* linear or non-linear, and *e.* time-invariant system: *a.* $y(t) = x(1 - t)$, *b.* $y(t) = x(t/2)$, *c.* $y(t) = \sin(x(t))$.

6.10.1*a.* Show the high-frequency slope of the nth order low-pass Butterworth function

$$|H(j\omega)| = \sqrt{1/(1 + \varepsilon^2 \omega^{2n})} \text{ is } 20n \, dB/decade.$$

b. Show that the first $(2n - 1)$ derivatives of $|H(j\omega)|^2$ are zero at $\omega = 0$. Use long division and then compare that to a power series and identify the

corresponding derivatives. *c.* Show (6.10.10). *d.* Assuming $n = 2$, determine the corresponding second-order Butterworth transfer function $H(s)$ and find its impulse response.

6.10.2 Sketch normalized function

$$|H(j\omega/\omega_c)| = \sqrt{1/(1 + \varepsilon^2(\omega/\omega_c)^{2n})}$$

for $n = 3$ and 4.

6.11.1 Determine the minimum phase transfer function corresponding to the functions

$$a. \ |H(j\omega)|^2 = [1/(1 + \omega^2)],$$

$$b. \ |H(j\omega)|^2 = [1/(1 + \omega^4)].$$

6.12.1 Consider the circuits shown in Fig. P6.12.1. For each of these cases determine the corresponding transfer function. In the case of band-pass or band-elimination filters, determine the center frequencies. Give the 3 dB cutoff frequencies in each case. Give the expression for the quality factor of the circuits wherever appropriate. In addition, identify the type of the filter in each case. Simplify the expressions by assuming $R_w = 0$. Sketch the amplitude responses for each of the cases and identify the important values.

6.12.2 Prove the maximum power transfer theorem. Sketch the power delivered to the resistor R_L. Assume the source resistance is R_s.

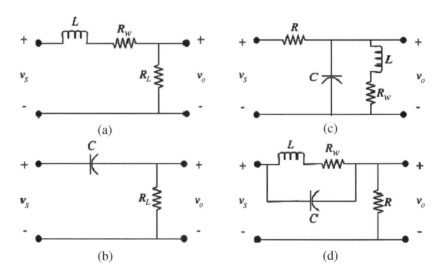

Fig. P6.12.1 Circuits to determine the transfer functions

Chapter 7
Approximations and Filter Circuits

7.1 Introduction

In the first part of this chapter we will consider a graphical representation of the transfer function in terms of its frequency response $H(j\omega) = |H(j\omega)|e^{\angle H(j\omega)}$. *Bode diagrams or plots* consist of two separate plots, the amplitude $|H(j\omega)|$ and the phase angle $\angle H(j\omega)$, with respect to the frequency ω on a logarithmic scale. These plots are named after Bode, in recognition of his pioneering work Bode (1945). Bode's basic work was based upon approximate representation of amplitude and phase response plots of a communication system. Wide range of frequencies of interest in a communication system dictated the use of the logarithmic frequency scale. Bode plots use the asymptotic behavior of the amplitude and the phase responses of simple functions by straight-line segments and are then approximated by smooth plots with ease and accuracy. Bode plots can be created by using computer software, such as MATLAB. The topic is mature and can be found in most circuits, systems, and control books. For example, see Melsa and Schultz (1969), Lathi (1998), Close (1966), Nilsson and Riedel (1966), and many others.

Filter approximations will be considered in the second part of this chapter. In Section 6.10, Butterworth approximation of an ideal low-pass filter amplitude response was introduced and the amplitude squared Butterworth function is

$$|H_{Bu}(j\omega)|^2 = \frac{1}{1 + \varepsilon^2(\omega/\omega_c)^{2n}}. \qquad (7.1.1)$$

The value of this function at $\omega = 0$ is 1, at $\omega = \infty$, the function goes to zero, and, in between these

frequencies, the function decays. The low-pass filter passes frequencies between 0 and ω_c with small attenuation and blocks or attenuates the frequencies above ω_c, the cut-off frequency. In Section 6.11, we have considered deriving the transfer function $H(s)$ from $|H(j\omega)|^2$. In the next stage we are interested in coming up with a circuit that has the given transfer function. The circuit may consist of passive elements, such as resistors, inductors, capacitors, and transformers. Early filter designs were done exclusively with passive networks. Mathematics associated with passive network synthesis is elegant. There is very little leeway in the designs. See, for example, Weinberg (1962), and others for the passive filter limtations. Another problem of passive network synthesis is the use of inductors and transformers, as these are not ideal components in reality.

Last part of the chapter deals with active filter synthesis using operational amplifiers, resistors, and capacitors. Active filter synthesis avoids the use of coils. Mathematical sophistication in active filter synthesis is much lower than passive filter synthesis. Active filter synthesis is based upon coming up with circuits with different topologies consisting of operational amplifiers, resistors, and capacitors. The circuit is then analyzed in terms of the $R's$ and $C's$, assuming the operational amplifier is *ideal*. Comparing the derived and the given transfer function and equating the corresponding coefficients of s in the two transfer functions result in a set of equations with more unknowns than equations. As a result, we have infinite number of solutions for the component values. This gives a good deal of leeway for a circuit designer to optimize the circuits. One of the optimization criteria is

R.K.R. Yarlagadda, *Analog and Digital Signals and Systems*, DOI 10.1007/978-1-4419-0034-0_7, © Springer Science+Business Media, LLC 2010

minimization of the *sensitivity* of a network with respect to changes in the component values.

Introduction to sensitivity: In introducing sensitivity, Bode was concerned about the changes in the transfer function resulting from large changes in the element values in the transmission systems that included vacuum tubes. Even though we are in the era of integrated circuits, we are still interested in the effect of changes in the component values on the transfer function. This effect may be in the form of a shift in a pole frequency ω_p or change in the quality factor Q or any other system parameter with respect to a component value. Pole sensitivity is defined as the per-unit change in the pole frequency, $\Delta\omega_p/\omega_p$, caused by a per-unit change in the desired component value $\Delta x/x$. The sensitivity of the parameter ω_p with respect a component value x is defined by

$$S_x^{\omega_p} = \lim_{\Delta x \to 0} \left\{ \left[\Delta\omega_p/\omega_p \right] / \left[\Delta x/x \right] \right\} = \frac{x}{\omega_p} \frac{\partial\omega_p}{\partial x}$$

$$= \frac{\partial \ln(\omega_p)}{\partial \ln(x_i)}. \tag{7.1.2}$$

The parameter can be any that is important to the circuit's function. We will assume the function of interest is Y_i as a function of x. Formulas for sensitivities of simple functions can be seen by inspection. These are given in Table 7.1.1, where $Y_i = Y_i(x)$ and x is a variable and c is a constant. From the sensitivity equation (7.1.2), we have

$$-S_x^{1/Y} = -\frac{\partial(\ln(1/Y))}{\partial(\ln x)} = -\frac{\partial(-\ln(Y))}{\partial(\ln(x))} = S_x^Y. \tag{7.1.3}$$

Example 7.1.1 The transfer function (TF), $H(s) = V_2(s)/V_1(s)$ of the circuit in Fig. 7.1.1 is as follows:

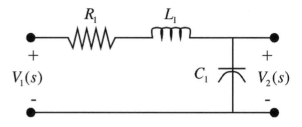

Fig. 7.1.1 Example 7.1.1

$$H(s) = \frac{V_2(s)}{V_1(s)} = \frac{1/L_1 C_1}{s^2 + (R_1/L_1)s + (1/L_1 C_1)}$$

$$= \frac{\omega_0^2}{s^2 + (\omega_0/Q)s + \omega_0^2},$$

$$\omega_0 = \frac{1}{\sqrt{L_1 C_1}}, \quad Q = \frac{\omega_0 L_1}{R_1} = \frac{1}{R_1}\sqrt{L_1/C_1}. \tag{7.1.4}$$

Derive the sensitivities of the functions $\omega_0 = L_1^{-1/2} C_1^{-1/2}$ and Q with respect to the element values R_1, L_1, and C_1.

Solution: We note that

$$S_{L_1}^{\omega_0} = S_{L_1}^{L_1^{-1/2} C_1^{-1/2}} = \frac{(L_1 C_1)^{-1/2}}{1/\sqrt{L_1 C_1}} \frac{\partial(L_1^{-1/2})}{\partial L_1}$$

$$= -\frac{(L_1^{3/2} L_1^{-3/2})}{2} = -\frac{1}{2}. \tag{7.1.5}$$

Similarly,

$$S_{C_1}^{\omega_0} = (-1/2), S_{R_1}^{\omega_0} = 0,$$

$$S_{L_1}^Q = -(1/2), S_{R_1}^Q = -1. \tag{7.1.6}$$

Note that $S_{R_1}^{\omega_0} = 0$ since ω_0 is not a function of R. From (7.1.6), we have a 1% change in any one of the three component values in the RLC circuit in the example resulting in either (1/2)% or 0% change in ω_0 and (1/2)% or 1% change in Q. The sign of the values indicates whether the change is increasing or decreasing. ∎

Transfer function of a circuit is a function of a set of parameters that are functions of the circuit components. For example, ω_0 is a function of L_1 and C_1 in Example 7.1.1. The change in ω_0, $\Delta\omega_0$ can be approximated by using Taylor's series written in terms of the variables x_i in the form

Table 7.1.1 Formulae for computing sensitivities

$S_x^{Y_1 Y_2} = S_x^{Y_1} + S_x^{Y_2}$	$S_x^{Y_1/Y_2} = S_x^{Y_1} - S_x^{Y_2}$
$S_{x^n}^Y = (1/n)S_x^Y$	$S_x^{Y^n} = nS_x^Y$
$S_x^{Y_1 + Y_2} = \frac{Y_1 S_x^{Y_1} + Y_2 S_x^{Y_2}}{Y_1 + Y_2}$	$S_x^{c\,Y(x)} = S_x^{Y(x)}, S_x^c = 0$

$$\Delta\omega_0 = \frac{\partial\omega_0}{\partial x_1} + \frac{\partial\omega_0}{\partial x_2} + \dots + \frac{\partial\omega_0}{\partial x_n}$$
$$+ \text{ second and higher order terms.} \quad (7.1.7)$$

See Tomovic and Vukobratovic (1972). If the change in the element values is assumed to be small, then the second-order and higher-order terms can be ignored.

$$\Delta\omega_0 \cong \sum_{i=1}^{n} \frac{\partial\omega_0}{\partial x_i}\partial x_i = \sum_{i=1}^{n} \left[\frac{\partial\omega_0}{\partial x_i}\frac{x_i}{\omega_0}\right]\frac{\partial x_i}{x_i}\omega_0. \quad (7.1.8)$$

$$\frac{\Delta\omega_0}{\omega_0} = \sum_{i=1}^{n} S_{x_i}^{\omega_0}\left[\frac{\Delta x_i}{x_i}\right], \quad S_{x_i}^{\omega_0} = \left[\frac{\partial\omega_0}{\partial x_i}\frac{x_i}{\omega_0}\right]. \quad (7.1.9)$$

In Example 7.1.1 there are two important parameters, one is ω_0 and the other one is quality factor Q. Per-unit changes in Q in terms of its sensitivities can be expressed with respect to the parameters as well. In turn, the sensitivities of gain and phase of a transfer function in terms of the frequency ω_0 and Q can be determined.

Block diagrams: In system control, system stability is one of the most important properties to be dealt with and closed-loop feedback control is basic to many systems. A simple feedback loop is shown in Fig. 7.1.2, where we have the Laplace transforms of the input signal with $L[r(t)] = R(s)$, error or actuating signal $L[e(t)] = E(s)$, control signal $L[m(t)] = M(s)$, controlled output $L[c(t)] = C(s)$, and primary feedback signal $L[b(t)] = B(s)$. The blocks identified by the transforms $G_1(s), G_2(s)$, and $H(s)$ represent control elements, plant or process, and feedback elements, respectively. The transfer function of the feedback system can be computed by writing the appropriate equations and solving for the output in terms of the input. These are as follows:

$$e(t) = r(t) \mp b(t) \overset{\text{LT}}{\longleftrightarrow} R(s) \mp B(s) = E(s)$$

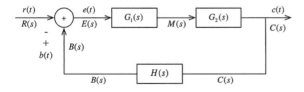

Fig. 7.1.2 A simple feedback system

$$M(s) = G_1(s)E(s) = G_1(s)(R(s) \mp B(s)),$$
$$C(s) = G_2(s)M(s), \quad B(s) = H(s)C(s)$$

$$C(s) = G_2(s)G_1(s)[R(s) \mp B(s)]$$
$$= G_2(s)G_1(s)[R(s) \mp H(s)C(s)]$$

$$C(s)[1 + G_2(s)G_1(s)H(s)] = G_1(s)G_2(s)R(s)$$

Transfer function:

$$T(s) = \frac{C(s)}{R(s)} = \frac{G_1(s)G_2(s)}{1 + G_1(s)G_2(s)H(s)}. \quad (7.1.10)$$

The product $G(s) = G_1(s)G_2(s)$ is the *direct transfer function*, $H(s)$ is the *feedback transfer function*, the product $G(s)H(s)$ is the *loop transfer function* or the *open-loop transfer function*, and $T(s)$ is the *closed-loop transfer function*. There are several books (for example, DiStefano et al. (1990).) that cover the block diagram algebra that gives simplifications in deriving the transfer functions. A useful transfer function that can be written in a special form is given by DiStfano et al. and the sensitivity of this function with respect to K is as follows:

$$T(s) = \frac{A_1(s) + KA_2(s)}{A_3(s) + KA_4(s)}$$
$$(K \text{ is independent of } A_i(s), i=1,2,3,4. \quad (7.1.11)$$

$$\Rightarrow S_K^T = \frac{K[A_2(s)A_3(s) - A_1(s)A_4(s)]}{[A_3(s) + KA_4(s)][(A_1(s) + KA_2(s)]}. \quad (7.1.12)$$

The transfer function can be expressed as a ratio of two polynomials in the form

$$T(s) = \frac{N(s)}{D(s)}. \quad (7.1.13)$$

Example 7.1.2 Let *a.* $T_1(s) = G(s)H(s) = [K/(s^2 + s + 1)]$, *b.* $T_2(s) = [K/(s^2 + s + 1 + K)]$.

Determine the sensitivities of these functions to the parameter K.

Solution: *a.* Using (7.1.11), we have $A_1(s) = 0$, $A_2(s) = 1$, $A_3(s) = s^2 + s + 1$, $A_4(s) = 0$. Using these in (7.1.12), it follows that $S_K^{T_1(s)} = 1$ for all K.

b. Again, using (7.1.11) and (7.1.12), we have

$$A_1(s) = 0, A_2(s) = 1,$$
$$A_3(s) = s^2 + s + 1, A_4(s) = 1,$$

$$S_K^{T_2(s)} = \frac{K(s^2 + s + 1)}{(s^2 + s + 1 + K)K}$$
$$= \frac{1}{1 + K/(s^2 + s + 1)}. \quad (7.1.14)$$

Sensitivity of the open-loop function $S_K^{T_1(s)}$ is 1 for all values of K and the sensitivity of the closed-loop function $S_K^{T_2(s)}$ is a function of K and s. Using $s = j\omega$, we can observe that for small values of ω and $K = 1$, $S_K^{T_2(s)} \cong .5$. The feedback system is less sensitive than the open-loop system with respect to K. In Section 7.12 amplitude and phase sensitivities will be considered. ∎

One of the main topics in this chapter is active filter synthesis. The first step in the active filter design is the analysis of a circuit with the appropriate topology. In Chapter 6 we considered computing the transfer function of a given circuit with two-terminal components using the Kirchhoff's current and voltage laws and the component equations. Active filter circuits include multiple terminal components, including operational amplifiers (or op amps). These active devices are represented by controlled or dependent sources in the analysis. Kirchhoff's laws, two-terminal component equations, and the controlled source representations of active devices provide a way to analyze circuits. A brief discussion on two-port representations of circuits is included by making use of the *indefinite admittance matrix* (Mitra, 1969). Other topics include scaling, frequency normalization, and adjustment of gain constants of the filter.

7.2 Bode Plots

In this section we will study the basic concepts associated with a pictorial representation of a rational function, say $H(s)$ or $H(j\omega)$,

$$H(s) = [N(s)/D(s)], H(s)|_{s=j\omega} = |H(j\omega)|e^{j\theta(\omega)}. \quad (7.2.1a)$$

Consider

$$H(s) = Ks^d \frac{\prod\limits_{m=1}^{M} (1 + (1/z_m)s)}{\prod\limits_{n=1}^{N} (1 + (1/p_n)s)} \quad (7.2.1b)$$

$$H(j\omega) = H(s)|_{s=j\omega}$$
$$= K(j\omega)^d \frac{(1+j\omega/z_1)(1+j\omega/z_2)...(1+j\omega/z_M)}{(1+j\omega/p_1)(1+j\omega/p_2)...(1+j\omega/p_N)}$$
$$= |H(j\omega)|e^{j\theta(\omega)}. \quad (7.2.1c)$$

Since the transfer is a ratio of real polynomials, the complex poles (or zeros) exist as complex-conjugate pairs and are *usually* simple. Multiple poles and zeros are possible and d is usually negative. In most cases, only a reasonable estimate of the system behavior and that to only at a very few frequencies is desired. The amplitude and phase responses at ω_i are given by

$$A(\omega_i) = |H(j\omega_i)| = |K||\omega_i|^d \frac{\prod\limits_{m=1}^{M} Z_m}{\prod\limits_{n=1}^{N} P_n},$$

$$Z_m = |1 + j\omega_i/z_m|, \quad P_n = |1 + j\omega_i/p_n| \quad (7.2.1d)$$

$$\theta(\omega_i) = \angle K + d(90°) + \sum_{m=1}^{M} \tan^{-1}(\omega_i/z_m)$$

$$- \sum_{n=1}^{N} \tan^{-1}(\omega_i/p_n). \quad (7.2.1e)$$

Although this approach is simple to see, finding these values and sketching them is time consuming. An alternate one is to obtain approximate sketches for the amplitude $|H(j\omega)|$ and the phase response $\theta(\omega)$ using Bode plots using the following factors:

1. Constant term, K.
2. Poles or zeros at the origin, $s^{\pm k}$.
3. Real poles or zeros, $(\tau s + 1)^{\pm k}$
4. Complex-conjugate poles or zeros, $(\tau^2 s^2 + 2\xi\tau s + 1)^{\pm k}$, where ξ is the damping ratio and $0 < \xi < 1$.

We need to study only simple poles and zeros, as the extensions to the multiple pole cases are simple. Note that log denotes base 10 and (ln) denotes base e.

$$20\log|B(j\omega)|^N = N\log|B(j\omega)| \text{ and } \angle[B(j\omega)]^N$$
$$= N\angle[B(j\omega)]. \tag{7.2.2a}$$

$$\log(\alpha + j\beta) = \log|\alpha + j\beta| + j\arg(\alpha + j\beta)$$
$$= \log\left(\sqrt{\alpha^2 + \beta^2}\right)$$
$$+ j\tan^{-1}(\beta/\alpha). \tag{7.2.2b}$$

Note that the term τ is used instead of the explicit poles or zeros and have combined the complex poles and their conjugates. Substituting $s = j\omega$ in the transfer function $H(s)$, we have $H(j\omega) = |H(j\omega)|$ $\angle H(j\omega)$ or $|H(j\omega)|_{dB} = 20\log|H(j\omega)|$ and $\phi(\omega) = \angle H(j\omega)$. We are *only* interested in sketching for *positive frequencies*. In addition, we will consider only poles that are on the negative half *s*plane, including the imaginary axis and the zeros can be anywhere. In most applications, the poles and zeros are simple, with the exception that may include multiples at the origin. Since the log of a product is the sum of the corresponding logs of the terms, we can sketch the magnitude function by using the simple functions. The phase responses of the terms in the transfer function can be added to obtain the total phase response. The dB magnitude versus $\log(\omega)$ plot, i.e., logarithmic magnitude frequency response plot is called the *Bode amplitude plot*, and the phase angle versus $\log(\omega)$ is called *Bode phase phase plot* (or *Bode diagrams*). Logarithmic scale for the ω-axis makes the sketches simple and allows sketches over a *wider range of frequencies* than the *linear scale*. Let the logarithmic frequency

variable be defined by $u = \log(\omega)$ or $\omega = 10^u$. The frequencies ω_1 and ω_2 are separated by an *octave* if $\omega_2 = 2\omega_1$ and by a *decade* if $\omega_2 = 10\omega_1$. Note

$$u_2 - u_1 = \log(\omega_2) - \log(\omega_1) = \log(\omega_2/\omega_1). \tag{7.2.3}$$

$$\text{Octaves} = \log_2(\omega_2/\omega_1) = [\log_{10}(\omega_2/\omega_1)/\log_{10}(2)],$$
$$\text{Decades} = \log_{10}(\omega_2/\omega_1). \tag{7.2.4}$$

The amplitude and phase plots of $H(j\omega)$ are considered using the four possible factors of a transfer function.

Constant K: The logarithm of a constant is a constant with respect to ω. The plot of $20\log(|K|)$ versus $\log(\omega)$ is a horizontal line. The phase angle is either $0°$ or $-180°$ depending upon whether the K is positive or negative.

The factor $(j\omega)^N$:

$$20\log|j\omega|^N = 20N\log|\omega|, \angle(j\omega)^N$$
$$= N(\pi/2). \tag{7.2.5}$$

Noting that $\log_{10}(2) = .3013$, if $\omega_1 = \alpha$ and $\omega_2 = 2\alpha$, the amplitude in (7.2.5) has increased by $6N$dB/octave or $20N$dB/decade. The function $20N\log|j\omega|$ plots as a straight line on the Bode plot and has a slope equal to $6N$ dB/octave or $20N$ dB/decade. The slope of the line is positive (negative) depending on whether N is positive (negative). The magnitude and phase plots are shown in Fig. 7.2.1a,b for $(1/j\omega)$. It is simple to obtain the plots for multiple poles.

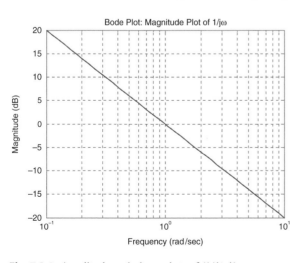

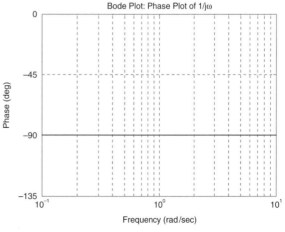

Fig. 7.2.1 Amplitude and phase plots of $(1/(j\omega))$

The factor $[1/(j\omega\tau + 1)]$: Noting that $A_1(\omega) = -20\log|(1 + j\omega\tau)| = -10\log[1 + (\omega\tau)^2]$, we have for small and for large values of ω, the amplitudes can be approximated by

$$(\omega\tau) \ll 1, \ A(\omega) \approx -20\log(1) = 0\,\mathrm{dB},$$
$$(\omega\tau) \gg 1, \ A(\omega) \approx -20\log(|\omega\tau|)\,\mathrm{dB} . \qquad (7.2.6)$$

These are the asymptotes to the true curve corresponding to the very small and very large frequencies. The first asymptote is a horizontal line and the second asymptote is a straight line with a slope of –6 dB/octave or –20 dB/decade. The two asymptotes intersect at the *corner frequency or the break frequency* $\omega = 1/\tau$. The actual value of the magnitude function at this frequency is equal to $-20\log(\sqrt{2})\mathrm{dB} \approx -3\mathrm{dB}$. It is simple to draw the asymptotic and the actual curves using the following guidelines:

1. The constant τ is the break point (or the corner frequency) in the asymptotic plot.
2. From the break point, draw the two asymptotes, one with a zero slope toward the ω small and the other one with a –6 dB/octave slope extending toward $\omega \to \infty$.
3. At the break point, the true response is displaced by –3 dB. In addition, an octave below and above the break point, the true curve is separated by –1 dB. A sketch of the amplitude response using the table and the above guidelines is shown in Fig. 7.2.2a. Note the frequency is

plotted using the log scale. The *phase angle* of the term $[1/(1 + j\omega\tau)]$ is equal to $\phi_1(\omega) = -\tan^{-1}(\omega\tau)$ radians or $-[57.3\tan^{-1}(\omega\tau)]$ degrees. It can be approximated using the power series expansion (Spiegel, 1966):

$$\tan^{-1}(\omega\tau) = (\omega\tau) - (1/3)(\omega\tau)^3$$
$$+ (1/5)(\omega\tau)^5 - ..., |\omega\tau| > 1, \qquad (7.2.7a)$$

$$\tan^{-1}(\omega\tau) = \pm(\pi/2) - [1/(\omega\tau) - (1/3)(1/\omega\tau)^3$$
$$+ (1/5)(1/\omega\tau)^5 - ...], + \text{ if } \omega\tau \ge 1,$$
$$- \text{ if } \omega\tau \le -1. \qquad (7.2.7b)$$

$$\tan^{-1}(\omega\tau) = \pi/4, \quad \omega\tau = 1. \qquad (7.2.7c)$$

Figure 7.2.2 gives the Bode amplitude and phase plots. The phase angle plot approaches $0°$ as $(\omega\tau) \to 0$ and $-90°$ as $(\omega\tau) \to \infty$. Noting (7.2.7c), we can see that the phase angle is $-45°$ at the break frequency $\omega = 1/\tau$. These two asymptotes can be connected by drawing a line from the $0°$ asymptote starting at one decade below the break frequency $(.1/\tau)$ with $0°$ phase and draw a line with a slope of $45°$/decade passing through $-45°$ at the break frequency and continuing to $-90°$ one decade above the break frequency $(10/\tau)$. For the zeros, the amplitude and phase response sketches can be similarly drawn since only the signs need to be altered.

Quadratic factors: The Bode plots corresponding to a pair of complex poles are usually given in terms of the damping factor $\xi \le 1$ by

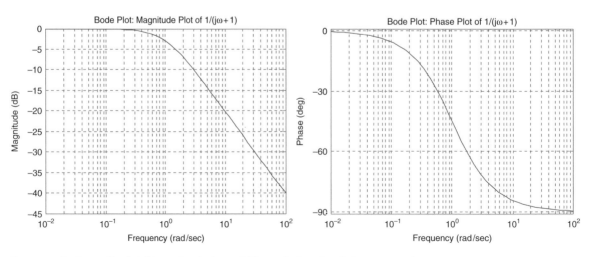

Fig. 7.2.2 Bode amplitude $1/|1 + j\omega|$ and phase $\angle 1/(1 + j\omega)$ plots, break frequency = 1

$$H_2(s) = \frac{1}{[1 + 2\xi\tau s + \tau^2 s^2]}$$

$$= \frac{1}{[1 + (Q_p/\omega_p)s + (1/\omega_p^2)s^2]}, 0 \le \xi \le 1. \quad (7.2.8)$$

$$|H_2(j\omega)|^2 = 1/[(1 - \omega^2\tau^2)^2 + 4(\xi\tau\omega)^2],$$
$$A_2(\omega) = 20\log(|H_2(j\omega)|) \quad (7.2.9a)$$

$$\phi_2(\omega) = -\tan^{-1}[2\xi\omega\tau/(1 - \omega^2\tau^2)] \quad (7.2.9b)$$

It is common in the filter designs to use the quality factor Q_p and ω_p in (7.2.8). The peak of the magnitude squared function can be found by taking the derivative of the denominator in (7.2.9a) with respect to ω and equating to zero. That is,

$$\frac{d[1/|H_2(j\omega)|^2]}{d\omega} = 2(1 - \omega^2\tau^2)(-2\omega\tau^2) + 2(4\xi^2\tau^2\omega)$$

$$= 0 \to \omega = \frac{\sqrt{1 - 2\xi^2}}{\tau}. \quad (7.2.10)$$

$$1 - 2\xi^2 > 0 \text{ or } \xi < (1/\sqrt{2}) = .707 \ (\omega \text{ is real}). \ (7.2.11)$$

Asymptotic approximations – second-order case:
$H_2(j\omega) = 1/[1 + j2\xi\tau\omega - \tau^2\omega^2]$

1. For low and high frequencies we can write:

$$(\omega\tau) \ll 1, \ 20\log|H_2(j\omega)| = 0 \, \text{dB}(\omega\tau)$$
$$\gg 1, \ 20\log|H_2(j\omega)| \approx$$
$$- 40\log(\omega\tau) \, \text{dB}. \quad (7.2.12)$$

The high frequency asymptote is a straight line with a slope of –12 dB/octave (–40 dB/decade).

2. The break frequency, i.e., the intersection of the low-frequency and the high-frequency asymptotes is located at the frequency $\omega = (1/\tau)$, which can seen from the fact that $40\log(\omega\tau) = 0$, $\omega = 1/\tau$ for all ξ. At the break frequency, we have

$$20\log|H_2(j\omega)|\big|_{\omega=1/\tau} = -20\log(2\xi) \, \text{dB}. \quad (7.2.13)$$

As an example at $\xi = .2$, we have the value –7.958 dB and for $\xi = 1/2$, the above equation reduces to 0 dB. A few values are given below for (7.2.13).

ξ	0	.05	.1	.2	.3	.4	.5	.6	.707	1
$-20\log(2\xi)$dB	∞	20	14	8	4.5	2	0	–1.5	–3	–6

Now consider the phase asymptotic plots of the second-order function from the $\phi_2(\omega)$ given in (7.2.9b). At the break frequency,

$$\phi_2(\omega)\big|_{\omega=(1/\tau)} = -\tan^{-1}(\infty)$$
$$= -90° \text{ for all } \xi. \quad (7.2.14)$$

The phase starts at $0°$ at low frequencies. At $\omega = (.1/\tau)$, it starts to decrease at a rate of $-90°/$decade. At $\omega = (1/\tau)$, it is $-90°$. It continues to decrease and reaches $-180°$ as $\omega \to \infty$. The amplitude and phase responses are plotted for a few values of ξ in Fig. 7.2.3. The phase response will

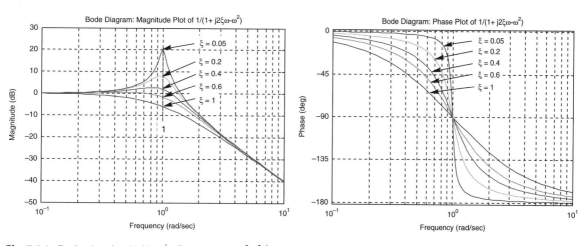

Fig. 7.2.3 Bode plots $|H_2(j\omega)| = |1/[1 + j2\xi\tau\omega - \tau^2\omega^2]|$, $\angle H_2(j\omega)$, $0 \le \xi \le 1$

have a discontinuity going from 0 to $-180°$ at the break point corresponding to $\xi = 0$.

The above discussion is given in terms of poles. For the zeros of a transfer function, multiply the dB and the phase angle values by -1. Bode plots of the transfer function can be constructed by summing the log magnitudes and phase angle contributions of each pole and zero (or pairs of complex and their conjugates of poles and zeros). DiStefano III et al. (1990), Nise (1992), and others give systematic procedures for sketching the Bode plots with examples to illustrate the construction process. Thaler and Brown (1960) give a tabulation of typical control system transfer functions, with associated polar plots, Bode plots, and root locus plots.

There are a few ways of computing the amplitude versus frequency on the Bode plot. Construct each factor separately and at selected values of ω add the amplitudes. Then, sketch the amplitude function through these points. It is simpler to use the asymptotes. For this purpose we need to have the transfer function in factored form. Next arrange the transfer function with increased values of poles and zeros. Many of the control system transfer functions have poles at the origin with a multiplicity of l, $l \geq 0$. Such a system is called a *type l system*. The system amplitude plot has a slope at low frequencies of $-10\,l$ dB/decade or ($-6l$ dB/octave). This slope is maintained until the first corner frequency is reached. At the first corner frequency the slope is changed by ∓ 10 dB/decade for a first-order pole or zero. If it is a second-order function, then the slope is changed by ∓ 20 dB/decade. This procedure is used to sketch the asymptotic plot. We can construct the composite asymptote provided the exact location of the lowest frequency segment can be located.

For $l = 0$, the lowest-frequency asymptote with a constant gain K is $20 \log(K)$ dB. For $l = 1$, locate the point $\omega = K$ on the 0 dB axis. The lowest frequency asymptote passes through this point with a slope of -10 dB/decade. For $l = 2$, locate the point $\omega = \sqrt{K}$ on the 0 dB axis. This frequency asymptote passes through this point with a slope of -20 dB/decade. It is very rare to have more than a double pole at the origin. Noting that for $\omega T \ll 1$, all the factors of the form $(j\omega T + 1)$ reduce to 1 in this region. Then, we have

$$20 \log|H(j\omega)| \cong 20 \log(K) - 20 \log\left|(j\omega^l)\right|. \quad (7.2.15a)$$

For example, for $l = 2$, the above equation represents a straight line with a slope of -20 dB/decade and the corresponding intercept is determined by

$$20 \log(K) - 20 \log\left|(j\omega^2)\right| = 0 \rightarrow \omega = \sqrt{K}. \quad (7.2.15b)$$

Example 7.2.1 Sketch the Bode plots for the following transfer function:

$$H(s) = \frac{10(1+s)}{s^2(1+(s/4)+(s/4)^2)},$$

$$H(j\omega) = \frac{10(1+j\omega)}{(j\omega)^2(1+(1/4)j\omega - ((1/4)).25\omega)^2)}. \quad (7.2.16)$$

Solution: The corresponding amplitude in terms of dB and the phase responses are

$$20 \log|H(j\omega)| = 20 \log(10) + 20 \log|(1+j\omega)| - 40 \log|j\omega|$$
$$- 20 \log\left|1/[(1+j(\omega/4) - (\omega/4)^2]\right|, \quad (7.2.17)$$

$$\angle H(j\omega) = \angle(1+j\omega) + \angle(1/j\omega)^2$$
$$+ \angle[1/(1+j\omega/4 - (\omega/4)^2)]. \quad (7.2.18)$$

The asymptotic amplitude plot is obtained by adding the asymptotic plots of each of the terms. The first term on the right in (7.2.17) is equal to 20 dB for all values of ω. The third term corresponds to a double pole at the origin and the asymptote goes through the corner frequency $\omega = 1$ with a slope of -40 dB/decade. The high-frequency asymptote of the second term in (7.2.17) starts at the corner frequency $\omega = 1$ and has a slope of 20 dB/decade. The fourth term corresponds to a pair of complex poles. Noting $\tau = 1/4$ and $2\xi\tau = .25$ ($\xi = .5$, damping factor). The high-frequency asymptote of the complex pair of poles starts at the corner frequency $\omega = 4$ with a slope of -40 dB/decade. All of these are sketched in Fig. 7.2.4 using MATLAB software.

Before we obtain the composite amplitude asymptotic Bode plot of the transfer function, we need to locate the lowest frequency asymptote. Using (7.2.15b), the corresponding asymptote is a line through the point $\sqrt{K} = \sqrt{10}$ at a slope of -40 dB/decade on the 0 dB axis. We are now ready to sketch the asymptotic Bode magnitude plot of the

Fig. 7.2.4 (a) Bode amplitude and (b) Bode phase plots of individual factors

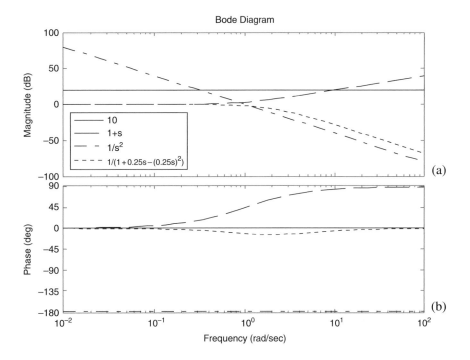

transfer function by adding the asymptotic plots of each of the four terms. Start at a low frequency say $\omega = .1$, follow the low-frequency asymptote that passes through $\omega = \sqrt{10}$ on the 0 dB axis to the first corner frequency $\omega = 1$ corresponding to the $(j\omega + 1)$ factor. Since this is a numerator factor, the next asymptote changes from –40 dB/decade to –20 dB/decade and continues to the next corner frequency located at $\omega = 4$. The next asymptote takes into consideration of the second-order factor in the denominator. The last high-frequency asymptote starts at $\omega = 4$ with a change in slope to –60 dB/decade. The individual

amplitude and phase plots of the transfer function are shown in Fig. 7.2.4a and b. We can obtain the actual amplitude plot by correcting for the errors in the asymptotic plots and the true functions. An easy way to do is find the amplitudes at the corner frequencies and then sketch the function through the computed values. Bode plots are only sketches. If accurate plots are desired, software packages, such as MATLAB, need to be used to obtain the desired results. Phase approximations can be used for the phase angle asymptotic plots. To get a sketch, it is easier to make use of arctangent function to obtain the

MATLAB code for Example 7.2.1

```
%Plot individual terms (Figure 7.2.4 Example 7.2.1)
sys1=tf (10,1 ); sys2=tf ([ 1 1],1); sys3=tf(1,[ 1 0 0] );
sys4=tf(1,[ -0.25 0.25 1] );
bode(sys1,' k' ,sys2,' k--' ,sys3,' k-.' ,sys4,' k:' ,{ 0.01,100} )
legend(' 10' ,' 1+s' ,' 1/s^2' ,' 1/(1+0.25s-(0.25s)^2)' )

%Plot the whole Bode plot (Figure 7.2.5, Example 7.2.1)
num=10*[ 1 1];
den=[ -(1/4) (1/4) 1 0 0];
w=logspace(-2,2,100);
bode(num,den,w)
grid
```

Fig. 7.2.5 Bode amplitude and phase plots of the composite function

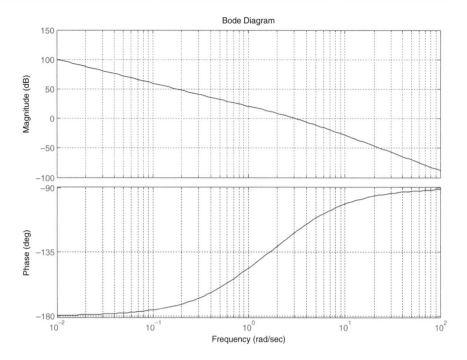

phase angles at some important frequencies, such as at corner frequencies, and sketch the function using these values. The amplitude and phase plots for the composite are given in Fig. 7.2.5a,b using the MATLAB code given below.

7.2.1 Gain and Phase Margins

We would like to consider two important topics that are used in the stability analysis of feedback control systems. Our discussion will be brief. The characteristic polynomial of a feedback control system is $D(s) = 1 + H_0(s) = 0$ with $H_0(s) = G_1(s)G_2(s)H(s)$ (see 7.1.10). Practicing engineers use the *gain margin* (G_M) and the *phase margin* (Φ_M), see, for example, Melsa and Schultz (1969). Graphical analysis is more appealing to engineers than analytical analysis. Gain and phase margins are measures of relative stability of the feedback control system. These are defined at the phase and margin crossover frequencies ω_π and ω_c.

$$G_M = \frac{1}{|H_0(j\omega_\pi)|} \text{ with } \angle H(j\omega_\pi) = -180^0,$$

$\omega_\pi =$ phase crossover frequency,

$$(7.2.19)$$

$$\Phi_{PM} = \left[180^0 + \angle H_0(j\omega_c)\right] \text{ with } H(j\omega_c) = 1,$$

$\omega_c =$ gain crossover frequency. $\quad (7.2.20)$

They are measures on how closely $H_0(j\omega)$ approaches a magnitude of unity and a phase of –

MATLAB code for Example 7.2.2

```
n=5 ;d=[1 3 4 2];
w=logspace(-2,2,100),
[mag, phase]=bode
(n,d,w);margin(mag, phase,w)
```

$180°$ quantifying the relative stability of the system. They can be read using the Bode plots. Negative phase margin implies instability. Most engineers use the criteria that a phase margin of $30°$ and a gain margin of 6 dB are safe margins. Analytical computation of gain and phase margins may not be possible since it requires factoring polynomials.

Example 7.2.2 Using the following methods obtain the gain and phase margins:

a. Analytical methods and *b.* MATLAB for the following function:

$$H_0(s) = \frac{5}{s^3 + 3s^2 + 4s + 2} \Rightarrow H(j\omega)$$

$$= \frac{5}{(2 - 3\omega^2) + j\omega(4 - \omega^2)}. \quad (7.2.21)$$

Solution: *a.* Equate the imaginary part to zero, i.e., $\text{Im}(H_0(j\omega)) = 0$. Solving for ω_π and then evaluating the real part at this frequency, we have

$$\text{Im}(H_0(j\omega)) = 0, \omega_\pi = 2 \rightarrow \text{Re}(H_0(j\omega_\pi))$$
$$= -1/2. \qquad (7.2.22)$$

We can increase the gain by 2 before the real part becomes –1. The gain margin in dB is

$$G_M = 20 \log(2) \cong 6 \, \text{dB}. \qquad (7.2.23)$$

We could also solve for ω_π by noting that the characteristic polynomial

$$D(s) = 1 + \alpha H_0(s) \Rightarrow 0 \rightarrow s^3 + 3s^2 + 4s + 12$$
$$= (s^2 + 4)(s + 3) = 0$$

has imaginary roots given by $s = \pm j2$. See the discussion on Routh table Chapter 6. For phase margin, we need to equate $|H_0(j\omega)| = 0$ and solve for $\omega = \omega_c$. This requires software, such as MATLAB. These result in

$$25 = (2 - 3\omega^2)^2 + \omega^2(4 - \omega^2)^2$$
$$\rightarrow \omega^6 + \omega^4 + 4\omega^2 - 21 = 0.$$

We can use MATLAB command roots ([1,0,1,0,4,0,–21]) and obtain the real positive root of the polynomial given by $\omega_c = 1.4315$ resulting in

$$\angle H_0(j\omega)|_{\omega=\omega_c=1.4315}$$

$$= \angle\left[5/(2 - 3\omega^2) + j\omega(4 - \omega^2)\right]_{\omega=\omega_c=1.4315} \approx -146^0.$$

The phase margin, the difference between this angle and –180°, is

$$\Phi_M \approx 180^\circ - 146^\circ = 34^\circ. \qquad (7.2.24)$$

b. Analytical computation may not be possible and computational tools, such as MATAB, are good to use. For this example the code is given below. MATLAB Bode plots are given in Fig. 7.2.6. The gain and the phase margins are shown. ∎

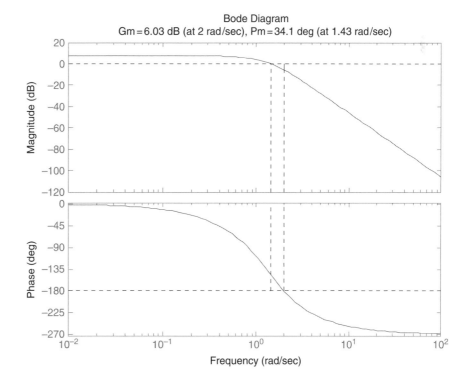

Fig. 7.2.6 Illustration of gain and phase margins, Example 7.2.2

7.3 Classical Analog Filter Functions

7.3.1 Amplitude-Based Design

Noting $|H(j\omega)|$ is even, start with

$$|H(j\omega)|^2 = H(j\omega)H(-j\omega)$$

$$= \frac{\sum\limits_{l=0}^{M} a_l \omega^{2l}}{1 + \sum\limits_{l=1}^{N} b_l \omega^{2l}}, N \geq M. \qquad (7.3.1)$$

The goal is to determine the coefficients a_l and b_l so that (7.3.1) satisfies a given set of specifications on the amplitude response. Letting $\omega \to s/j$, we have the product $H(s)H(-s)$. The *minimum phase transfer function* is obtained by assigning the left half-plane poles and zeros of $H(s)H(-s)$ to $H(s)$.

In this section we will consider Butterworth, Chebyshev I, and Chebyshev II filter functions.

The *ideal low-pass filter* function was defined in Chapter 6 and is

$$H_{Lp}(j\omega) = \Pi[\omega/2\omega_c]e^{-j\omega t_0},$$

$$\omega_c = 2\pi f_c, \text{ cut - off frequency.} \qquad (7.3.2)$$

The ideal low-pass filter response in (7.3.2) is not physically realizable, see Section 6.10.1. The next best thing is approximate an ideal filter response. We have seen that a simple RC circuit approximates a low-pass filter. There are four types of filters identified by

1. $H_{Lp}(j\omega)$, low-pass function
2. $H_{Hp}(j\omega)$, High-pass function
3. $H_{Bp}(j\omega)$, Band-pass function
4. $H_{Be}(j\omega)$, Band-elimination function

Low-pass filter specifications:

$$\left.\begin{cases} \text{Pass band} : 0 \leq |\omega| \leq \omega_c, \frac{1}{1+\varepsilon^2} \leq |H_{Lp}(j\omega)|^2 \leq 1 \text{ or } -X\text{dB} \leq 20\log|H_{Lp}(j\omega)| \leq 0\text{dB} \\ \text{Stop band} : \quad |\omega| \geq \omega_r, \quad |H_{Lp}(j\omega)|^2 \leq (1/A^2), \text{ or } 20\log|H_{Lp}(j\omega)| \leq -Y\text{dB}, \omega_c < \omega_r \end{cases}\right\} \qquad (7.3.3a)$$

High-pass filter specifications:

$$\left.\begin{cases} \text{Pass band} : \omega_c \leq |\omega| \leq \infty, \quad \frac{1}{1+\varepsilon^2} \leq |H_{Hp}(j\omega)|^2 \leq 1 \text{ or } -X\,\text{dB} \leq 20\log|H_{Hp}(j\omega)| \leq 0\,\text{dB} \\ \text{Stop band} : \quad 0 \leq |\omega| \leq \omega_r, \quad |H_{Hp}(j\omega)|^2 \leq (1/A^2), \text{ or } 20\log|H_{Hp}(j\omega)| \leq -Y\,\text{dB}, \omega_r < \omega_c. \end{cases}\right\} \qquad (7.3.3b)$$

Band-pass filter specifications:

$$\left.\begin{cases} \text{Pass band} 0 : \omega_l \leq |\omega| \leq \omega_h, \quad \frac{1}{1+\varepsilon^2} \leq |H_{Bp}(j\omega)|^2 \leq 1 \text{ or } -X\,\text{dB} \leq 20\log|H_{Bp}(j\omega)| \leq 0\,\text{dB} \\ \text{Stop bands} : 0 \leq |\omega| \leq \omega_1 \text{ and } \omega_2 \leq |\omega| \leq \infty, \quad |H_{Bp}(j\omega)|^2 \leq (1/A^2), \\ \qquad \text{or } 20\log|H_{Bp}(j\omega)| \leq -Y\,\text{dB}, \omega_1 < \omega_l, \omega_h < \omega_2. \end{cases}\right\} \qquad (7.3.3c)$$

Band elimination filter specifications:

$$\left.\begin{cases} \text{Pass bands} : \ 0 \leq |\omega| \leq \omega_l, \omega_h \leq \omega \leq \infty, \quad (1/(1+\varepsilon^2)) \leq |H_{Be}(j\omega)|^2 \leq 1 \\ \qquad \text{or } -X\,\text{dB} \leq 20\log|H_{Be}(j\omega)| \leq 0\,\text{dB} \\ \text{Stopband} : \quad \omega_1 \leq |\omega| \leq \omega_2, \quad 0 \leq |H_{Be}(j\omega)|^2 \leq (1/(1+\varepsilon^2)), \\ \qquad \text{or } -Y\,\text{dB} \geq 20\log|H_{Be}(j\omega)|, \omega_l \leq \omega_1, \omega_h \geq \omega_2. \end{cases}\right\} \qquad (7.3.3d)$$

Fig. 7.3.1 Analog filter amplitude specifications: (a) low pass, (b) high pass, (c) band pass, and (d) band elimination

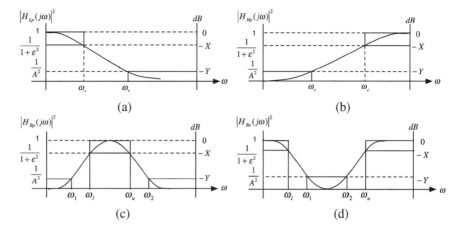

These are illustrated in Fig. 7.3.1a,b,c,d in terms of the squares of amplitudes and in dB scale on all the figures. For simplicity the responses are assumed to be smooth.

7.3.2 Butterworth Approximations

A simple RC circuit was considered in Chapter 6 with a transfer function (see 6.5.16b)

$$H(j\omega) = \frac{1}{j\omega RC + 1}, \quad |H(j\omega)| = H(s)|_{s=j\omega}$$

$$= \frac{1}{\sqrt{(1 + (\omega/\omega_c)^2}}, \quad \omega_c = 1/RC. \quad (7.3.4)$$

It has a single parameter ω_c controlling the amplitude response. Assuming the input and the output transforms as $X(j\omega)$ and $Y(j\omega)$, we have $Y(j\omega) = H(j\omega)X(j\omega)$. This implies $|Y(j\omega)| = |H(j\omega)||X(j\omega)|$. The filter acts like a gate in the sense that low frequencies are passed with very little attenuation and the high frequencies are attenuated significantly. The shape of the amplitude response function $|H(j\omega)|$ controls what frequencies are allowed through and what frequencies are attenuated or eliminated. To be more specific, in the low-pass filter design, we assume we have three bands defined by

$$\text{Pass band}: 0 \leq \omega \leq \omega_c,$$
$$\text{transition band}: \omega_c < \omega < \omega_r,$$
$$\text{stop band}: \omega_r \leq \omega \leq \infty. \quad (7.3.5)$$

The frequencies $0 \leq \omega \leq \omega_c$ in the input will be allowed to pass through the low-pass filter without much attenuation and therefore we call this band of frequencies as the pass band. The band of frequencies $\omega_r \leq \omega \leq \infty$ in the input signal will be attenuated by the low-pass filter significantly and this band is called the stop band. In between these two bands the filter amplitude response will have to be tapered or gradual and the input frequencies are attenuated gradually. A popular function that can act as a low-pass filter that satisfies the above criterion is a Butterworth function defined by

$$|H_{Bu}(j\omega)| = \frac{1}{\sqrt{1 + \varepsilon^2(\omega/\omega_c)^{2n}}}. \quad (7.3.6)$$

The subscript Bu on H is usually not shown and seen from the context. The ε term controls how far the filter amplitude characteristic will go down to when $\omega = \omega_c$ from 1 at $\omega = 0$ and the value of n controls how fast the magnitude characteristic attenuates in the stop-band region. We have shown that the ripple amplitude in the pass band having the value for *small* ε is $1 - (1/\sqrt{(1 + \varepsilon^2)}) \approx \varepsilon^2/2$ (see 6.10.10). The pass-band and stop-band specifications for the low-pass filter are given in (7.3.3a) and the specifications are identified in Fig. 7.3.1a. It is common to specify these two in terms of the dB scale, as identified in this figure. Using the edges of the frequency bands ω_c and ω_r, we can write

$$|H_{Bu}(j\omega)|^2|_{\omega=\omega_c} = 10\log(1/(1+\varepsilon^2))$$

$$= -X\,\mathrm{dB} \Rightarrow \varepsilon$$

$$= \sqrt{10^{(x/10)}-1}. \qquad (7.3.7)$$

$$|H_{Bu}(j\omega)|^2|_{\omega=\omega_r} = 10\log(1/A^2)$$

$$= -Y\,\mathrm{dB}.10\log\frac{1}{1+\varepsilon^2(\omega_r/\omega_c)^{2n}}$$

$$= -Y\,\mathrm{dB} \Rightarrow (n)_{\mathrm{integer}}$$

$$\geq \frac{\log[(10^{.1Y}-1)/(10^{.1X}-1)]}{2\log(\omega_r/\omega_c)}. \quad (7.3.8)$$

$$(n)_{\mathrm{integer}} \geq \frac{\log\sqrt{(A^2-1)}-\log(\varepsilon)}{\log(\frac{\omega_r}{\omega_c})}$$

$$\cong \frac{\ln(A/\varepsilon)}{(\omega_r/\omega_c)-1}. \qquad (7.3.9)$$

Now compute the transfer function $H_{Bu}(s)$ from (7.3.6). That is,

$$|H_{Bu}(j\omega)|^2 = H_{Bu}(s)H_{Bu}(-s)|_{\omega^2=-s^2}. \quad (7.3.10)$$

Since the amplitude response function has one in the numerator, we need to compute only the poles of the transfer function. They can be found by solving

$$1+\varepsilon^2(\frac{\omega}{\omega_c})^{2n}|_{\omega^2=-s^2} = 1+\varepsilon^2\left[\frac{(-s^2)^n}{\omega_c^{2n}}\right] = 0 \quad (7.3.11)$$

$$\Rightarrow s_{2v+1} = \left(\frac{\omega_c}{\varepsilon^{1/n}}\right)e^{j(2v+1+n)\frac{\pi}{2n}},$$

$$v = 0,1,2,\ldots,2n-1. \qquad (7.3.12)$$

The left half-plane poles of $H_{Bu}(s)$ in the exponential and trignometric forms are

$$s_{2v+1} = \left(\omega_c/\varepsilon^{1/n}\right)e^{j(2v+1+n)\frac{\pi}{2n}},$$

$$v = 0,1,\ldots,n-1. \qquad (7.3.13a)$$

$$s_{2v+1} = \left(\omega_c/\varepsilon^{1/n}\right)[-\sin(\frac{2v+1}{2n}\pi)$$

$$+j\cos(\frac{2v+1}{2n}\pi)], v = 0,1,\ldots,n-1. \quad (7.3.13b)$$

Since $|s_{2v+1}| = (\omega_c/\varepsilon^{1/n})$, poles of the Butterworth function are equally spaced on a circle of radius $(\omega_c/(\varepsilon^{1/n}))$ on the $s$plane. Selecting the poles on the left half of the s plane, the transfer function is

$$H_{Bu}(s) = \frac{(-s_1)(-s_3)\ldots(-s_{2n-1})}{(s-s_1)(s-s_3)\ldots(s-s_{2n-1})}.$$

(Note $H_{Bu}(0) = 1$). $\qquad (7.3.14)$

Example 7.3.1 Compute the values of ε and n and derive Butterworth transfer function for the specifications given in Fig. 7.3.2 with $-X = -2\,\mathrm{dB}$ and $-Y = -15\mathrm{dB}$.

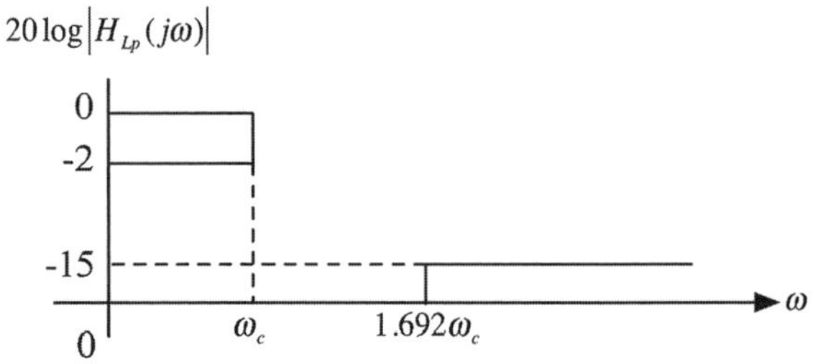

$20\log|H_{Lp}(j\omega)|$

Fig. 7.3.2 Specifications in Example 7.3.1

Solution: From (7.3.7) and in (7.3.8), $\varepsilon = .7648$ and $n \geq 3.76$. Since n has to be an integer, it follows that $n = 4$. From (7.3.13b), the left-half s-plane poles and the transfer function are

$$s_{1,7} = (\omega_c/0.93516)(-0.38268 \pm j0.92388),$$

$$s_{3,5} = (\omega_c/0.93516)(-0.92388 \pm j0.38268)$$

$$H_{Bu}(s) = \frac{(-s_1)(-s_3)(-s_5)(-s_7)}{(s-s_1)(s-s_3)(s-s_5)(s-s_7)}. \quad (7.3.15)\blacksquare$$

Maximally flat amplitude property of the Butterworth function: Consider

$$|H_{Bu}(j\omega)|^2 = \frac{1}{1+\varepsilon^2(\omega/\omega_c)^{2n}}.$$

$$\left(\text{Even function of } (\omega/\omega_c)^2 = \lambda.\right). \quad (7.3.16)$$

That is,

$$|H_{Bu}(j\lambda)|^2 = |H_{Bu}(j\omega)|^2\Big|_{(\omega/\omega_c)^2=\lambda} = \frac{1}{1+\varepsilon^2\lambda^n}$$

Expanding this function in power series in the neighborhood of $\lambda = 0$, we have

$$\frac{1}{1+\varepsilon^2\lambda^n} = 1 + (0)\lambda + \ldots + (0)\lambda^{n-1} + (-\varepsilon^2)\lambda^n + \ldots$$

$$(7.3.17)$$

A simple way to see this is divide the numerator (1) by the denominator $(1 + \varepsilon^2\lambda^n)$. The coefficients for the terms λ^k, $k = 1, 2, \ldots, n-1$ are identically zero. That is, $(n-1)$ derivatives of the Butterworth function are equal to zero. This is called the *maximally flat property*. Butterworth approximation starts with 1 at $\omega = 0$ and monotonically goes to zero as $\omega \to \infty$. It has the maximally flat response in both pass and stop bands. In the pass band the magnitude of the function goes from 1 to $(1/\sqrt{(1+\varepsilon^2)})$. Zero error at $\omega = 0$ and maximum error at the cutoff frequency ω_c is approximately equal to $\varepsilon^2/2$. It may be of interest to distribute the error throughout the pass band and the Chebyshev approximation achieves that.

7.3.3 Chebyshev (Tschebyscheff) Approximations

The nth $(n > 0)$ order Chebyshev polynomial is defined by the transcendental function

$$C_n(\alpha) = \cos(n\cos^{-1}\alpha)$$
$$= \begin{cases} \cos(n\cos^{-1}(\alpha)), & |\alpha| \le 1 \\ \cosh(n\cosh^{-1}(\alpha)), & |\alpha| > 1 \end{cases}. \quad (7.3.18)$$

It can be expressed as a polynomial. To show this, let

$$n\phi = \cos^{-1}(\alpha) \text{ and } C_n(\alpha) = \cos(n\phi).$$
$$(C_0(\alpha) = 1 \text{ and } C_1(\alpha) = \alpha). \quad (7.3.19a)$$

Using the trigonometric identities, we can write

$$c_{n+1}(\alpha) = \cos[(n \pm 1)\phi]$$
$$= \cos(n\phi)\cos(\phi) \mp \sin(n\phi)\sin(\phi).$$

$$C_{n+1}(\alpha) = 2\alpha C_n(\alpha) - C_{n-1}(\alpha),$$
$$C_0(\alpha) = 1, C_1(\alpha) = \alpha. \quad (7.3.19b)$$
$$C_n^2(\alpha) = .5[C_{2n}(\alpha) + 1].$$

Using (7.3.19b) the Chebyshev polynomials can be derived. First few of these are

$$C_0(\alpha) = 1, \qquad C_1(\alpha) = \alpha$$
$$C_2(\alpha) = 2\alpha^2 - 1, \qquad C_3(\omega) = 4\alpha^3 - 3\alpha$$
$$C_4(\alpha) = 8\alpha^4 - 8\alpha^2 + 1, \; C_5(\alpha) = 16\alpha^5 - 20\alpha^3 + 5\alpha$$

$$(7.3.19c)$$

These are sketched for $n = 1, 2, 3, 4$ in Fig. 7.3.3 in the range $0 \le \alpha$. Since the Chebyshev polynomials $C_n(\alpha)$ have even (odd) powers of α only for *n*even

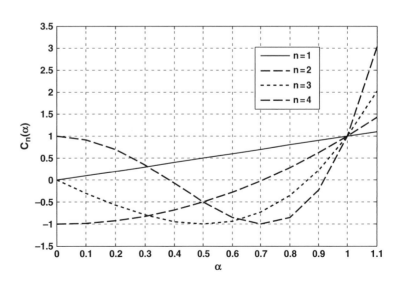

Fig. 7.3.3 Chebyshev polynomials, $c_n(\alpha), n = 1, 2, 3, 4$

(odd), sketching the polynomials for negative α is straightforward. For large values of α,

$$C_n(\alpha) \approx 2^{n-1}\alpha^n, \quad \alpha \gg 1. \qquad (7.3.19\text{d})$$

Properties of the Chebyshev polynomials (see Scheid (1968)):

1. Since $\cosh(\alpha)$ is never zero for real ω, it follows that $C_n(\alpha) = 0$ only for $|\alpha| \le 1$. The roots of $C_n(\alpha) = 0$, $\alpha_k, k = 1, 2, .., n$ are real and $|\alpha_k| < 1$.
2. Since $C_n(\alpha) = \cos(n\cos^{-1}(\alpha))$, $|\alpha| \le 1$, it follows that $|C_n(\alpha)| \le 1$ for $|\alpha| \le 1$.
3. For $|\alpha| > 1$, $C_n(\alpha)$ increases monotonically consistent with the degree n.
4. $C_n(\alpha)$ is an odd (even) polynomial if n odd (even).
5. The polynomial $C_n(\alpha)$ oscillates with *an equiripple* character varying between a maximum of $+1$ and a minimum of -1 for $|\omega| \le 1$.

6. $$\left[\begin{array}{l} n-\text{even}: C_n(0) = (-1)^{n/2}, C_n(-1) = 1 \\ n-\text{odd}: C_n(0) = 0, C_n(\pm 1) = \pm 1 \, (\text{respectively}) \end{array} \right].$$

$$C_n(\alpha) = 0, \ \alpha = \cos((2k+1)\pi/2n),$$
$$k = 0, 1, 2, ..., n-1, \ -1 \le \alpha \le 1. \qquad (7.3.20\text{a})$$

$$C_n(\alpha) = (-1)^k, \ \alpha = \cos(k\pi/n),$$
$$k = 0, 1, 2, ..., n, \ -1 \le \alpha \le 1. \qquad (7.3.20\text{b})$$

7. Slope $$\frac{dC_n(\alpha)}{d\alpha}\Big|_{\alpha=1} = n^2. \qquad (7.3.20\text{c})$$

Equations (7.3.20a) and (7.3.20b) follow from

$$0 = \cos(n\cos^{-1}(\alpha)) \Rightarrow n\cos^{-1}(\alpha)$$
$$= k\pi/2, \ k\text{-odd}, \qquad (7.3.21\text{a})$$

$$\pm 1 = \cos(n\cos^{-1}(\alpha)) \Rightarrow \cos^{-1}(\alpha) = k\pi/n. \quad (7.3.21\text{b})$$

The Chebyshev polynomial has n roots and they are located in the range $-1 \le \alpha \le 1$. Outside this range, $C_n(\alpha)$ is monotonically increasing (or decreasing in the case of negative α) function. Since $C_n(\alpha)$ is either an even or an odd function, it follows that $C_n^2(\omega)$ is an even function. Chebyshev polynomial gives the *best* approximation in the sense that it minimizes the maximum magnitude of the error for a given value of n.

Chebyshev 1 approximation: Noting the characteristics of the Chebyshev polynomials in the range $-1 \le \alpha \le 1$, $\varepsilon^2 C_n^2(\alpha)$ varies between 0 and ε^2 in the interval $|\alpha| \le 1$ and increases rapidly for $|\alpha| > 1$ consistent with n. With these properties in mind, a low-pass amplitude response function can be defined, so that the response swings between 1 and $1/\sqrt{(1+\varepsilon^2)}$, in the pass band and monotonically decreasing property in the stop band. Such a function is the Chebyshev 1 function given with a subscript $(c1)$ with the argument (ω/ω_c) is given below, see, for example, in Figs. 7.3.4 and 7.3.5.

$$|H_{c1}(j\omega)|^2 = \frac{1}{1 + \varepsilon^2 C_n^2(\omega/\omega_c)}. \qquad (7.3.22\text{a})$$

For $|\omega/\omega_c| \le 1$, $|H_{c1}(j\omega)|$ oscillates between 1 and $1/(1+\varepsilon^2)$ with equal ripple character.

$$|H_{c1}(0)|^2 = \begin{cases} (1/(1+\varepsilon^2)), & n-\text{even}, \\ 1, & n-\text{odd}, \end{cases} \qquad (7.3.22\text{b})$$

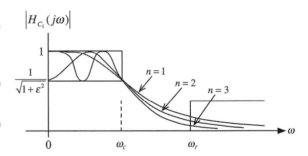

Fig.7.3.4 $|H_{c1}(j\omega)|, n = 1, 2, 3$

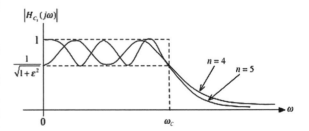

Fig. 7.3.5 $|H_{c1}(j\omega)|, n = 4, 5$

$|H_{c1}(j\omega_c)|^2 = (1/(1+\varepsilon^2)), n - \text{even and } n - \text{odd}.$

$$(7.3.22c)$$

The number of *peaks* ($|H_{c1}(j\omega)| = 1$) plus the number of *valleys* ($|H_{c1}(j\omega_c)|$) in the positive frequency range of the pass band is equal to n. This is referred to as the *equal-ripple* property. Fig. 7.3.5 illustrates this for $n = 4, 5$. For $|\omega| > |\omega_c|$, $|H_{c1}(j\omega)|$ decreases rapidly consistent with the value of n. For ε small, the width of the ripple in the pass band can be approximated and is $\varepsilon^2/2$ (see 6.10.10). From the filter specifications, ε gives the permissible range of amplitudes of the Chebyshev 1 response in the pass band and the stop-band attenuation constant A gives a measure of acceptable attenuation in the stop band. The range of frequencies between ω_c and ω_r is the transition band. Chebyshev 1 transfer function can be computed from

$$H_{c1}(s)H_{c1}(-s) = H_{c1}(j\omega)H_{c1}(-j\omega)|_{\omega=s/j}$$

$$= \frac{1}{1+\varepsilon^2 c_n^2\left(\frac{\omega}{\omega_c}\right)}\Big|_{\omega=s/j}. \qquad (7.3.23)$$

Solving for the roots of the equation $C_n(\omega/\omega_c)|_{\omega=s/j} = \pm j/\varepsilon$ and selecting the left-half-plane roots results in the following poles of the transfer function (see Weinberg, 1962):

$$s_i = \omega_c\left[-\sinh\left(\frac{1}{n}\sinh^{-1}\left(\frac{1}{\varepsilon}\right)\right)\sin\left(\frac{2i+1}{2n}\pi\right)\right.$$

$$\left. +j\cosh\left(\frac{1}{n}\sinh^{-1}\left(\frac{1}{\varepsilon}\right)\right)\cos\left(\frac{2i+1}{2n}\pi\right)\right],$$

$$i = 0, 1, 2, ..., n-1. \qquad (7.3.24)$$

Chebyshev-1 low-pass transfer functions:

$$H_{c1}(s) = \begin{cases} \dfrac{1}{\sqrt{1+\varepsilon^2}}\dfrac{(-s_1)(-s_2)\ldots(-s_n)}{(s-s_1)(s-s_2)\ldots(s-s_n)}, & n-\text{even} \\[3mm] \dfrac{(-s_1)(-s_2)\ldots(-s_n)}{(s-s_1)(s-s_2)\ldots(s-s_n)}, & n-\text{odd} \end{cases},$$

$$H_{c1}(s)|_{s=0} = \begin{cases} 1/\sqrt{1+\varepsilon^2}, & n-\text{even} \\ 0, & n-\text{odd}. \end{cases} \qquad (7.3.25)$$

Design parameters: ε controls the ripple width in the pass band and n controls the attenuation in the stop band. That is,

$$\frac{1}{\sqrt{1+\varepsilon^2 C_n^2\left(\frac{\omega_r}{\omega_c}\right)}}\Big|_{\omega_r=\omega_c} = \frac{1}{\sqrt{1+\varepsilon^2}},$$

$$\frac{1}{A} = \frac{1}{\sqrt{1+\varepsilon^2 C_n^2\left(\frac{\omega_r}{\omega_c}\right)}}, \quad \omega_r > \omega_c. \qquad (7.3.26)$$

Noting $C_n(\alpha) = \cosh(n\cosh^{-1}(\alpha))$, $|\alpha| > 1$, the integer n must satisfy

$$n \geq \frac{\cosh^{-1}((\sqrt{A^2-1})/\varepsilon)}{\cosh^{-1}(\omega_r/\omega_c)}$$

$$\approx \frac{\ln(2A/\varepsilon)}{[(2/\omega_c)(\omega_r - \omega_c)]^{1/2}}. \qquad (7.3.27)$$

The approximation follows from $\cosh^{-1}(x) = \ln(x+\sqrt{x^2-1}) \approx \ln(2x), x \geq 1$. If the constraints are given in terms of dB, then

$$n \geq \frac{\cosh^{-1}[(10^{.1Y}-1)/(10^{.1X}-1)]}{\cosh^{-1}(\omega_r/\omega_c)}. \qquad (7.3.28)$$

Example 7.3.2 Find the Chebyshev 1 transfer function that has $X = 2$ dB ripple in the pass band and a minimum attenuation in the stop band of $Y = 15$ dB.

Solution: Noting $C_n^2(1) = 1$,

$$10\log\left[\frac{1}{1+\varepsilon^2 C_n^2\left(\frac{\omega}{\omega_c}\right)}\right]\Big|_{\omega=\omega_c} = -2 \text{ dB}$$

$$\Rightarrow \varepsilon = \sqrt{(10^{-2}-1)}$$

$$= 0.7648. \qquad (7.3.29)$$

This is the same as in Example 7.3.1. The value of n is determined from

$$10\log\left[\frac{1}{1+\varepsilon^2 C_n^2\left(\frac{\omega}{\omega_c}\right)}\right]\Big|_{\omega=\omega_r=1.69196\omega_c} = -15 \text{ dB}$$

$$\Rightarrow C_n(1.69196) = \sqrt{(10^{1.5}-1)/(10^{.2}-1)}$$

$$= 7.2358. \qquad (7.3.30)$$

$$C_n(1.69196) = \cosh(n\cosh^{-1}(1.69196)) = 7.2358,$$

$$n \geq \frac{\cosh^{-1}(7.2358)}{\cosh^{-1}(1.69196)} = 2.387$$

It follows that $n = 3$ since n must be an integer. The maximum attenuation in the stop band and the transfer function can be determined. These are

$$10\log\left[\frac{1}{1 + \varepsilon^2 C_3(1.69196)}\right] = -20.81\text{dB} \qquad (7.3.31)$$

$$H_{c1}(s) = \frac{(-s_0)(-s_1)(-s_2)}{(s - s_0)(s - s_1)(s - s_2)},$$
$$s_{0,2} = \omega_c(-0.184445 \pm j0.92078),$$
$$s_1 = -\omega_c(0.36891). \qquad (7.3.32)$$

Figure 7.3.6 shows the specifications and the derived amplitude response. ■

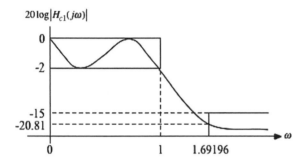

Fig. 7.3.6 Example 7.3.2

Chebyshev 1 response has equal ripple property and is monotonic in the stop band. It has a steep transition compared to the Butterworth approximation. Also, the phase response of the Butterworth filter is better with good delay properties. The idea is to find a maximally flat pass-band response to improve the delay performance and retain the steep transition like the Chebyshev 1. Such a case is Chebyshev 2 approximation and is discussed below.

Chebyshev 2 or inverse Chebyshev approximation: The Chebyshev 2 approximation function can be derived from Chebyshev 1 function by first considering the normalized Chebyshev 1 function with the cut-off frequency of one and a

transformation that takes the zero frequency to ∞ and vice versa, which is referred to as low-pass to high-pass transformation. Consider the Chebyshev 1 function with $\omega_c = 1$ in the form

$$|H_{c1}(jv)|^2 = \frac{1}{1 + \varepsilon^2 C_n^2(v)}. \qquad (7.3.33a)$$

The low-pass to high-pass transformation $v \rightarrow (1/v)$ translates the ripples in the pass band to the stop band in the region $v > 1$ and monotonic response in the region of $v \leq 1$.

$$|H_{c1}(j/v)|^2 = \frac{1}{1 + \varepsilon^2 C_n^2(1/v)}. \qquad (7.3.33b)$$

The transformation translates the ripples in the pass band to the stop band in the region $v > 1$ and gives monotonic response in the region of $v \leq 1$. The low-pass function is

$$|H_{C2}(jv)|^2 = 1 - |H_{C1}(j/v)|^2 = 1 - \frac{1}{1 + \varepsilon^2 C_n^2(1/v)}$$
$$= \frac{\varepsilon^2 C_n^2(1/v)}{1 + \varepsilon^2 C_n^2(1/v)} \qquad (7.3.34)$$

For $n = 2$ and 3,

$$C_2(\alpha) = 2\alpha^2 - 1 \text{ and } C_3(\alpha) = 4\alpha^3 - 3\alpha.$$

With $\alpha = 1/v$, we have

$$C_2(1/v) = 2(1/v)^2 - 1 = (2 - v^2)/v^2,$$
$$C_3(1/v) = 4(1/v)^3 - 3(1/v) = (4 - 3v^2)/v^3.$$

Therefore,

$$|H_{c2,2}(jv)|^2 = \frac{\left[\varepsilon\left(\dfrac{2 - v^2}{v^2}\right)\right]^2}{1 + \left[\varepsilon\left(\dfrac{2 - v^2}{v^2}\right)\right]^2}$$
$$= \frac{[\varepsilon^2(4 - 4v^2 + v^4)]}{[\varepsilon^2(4 - 4v^2 + v^4)] + v^4} \qquad (7.3.35a)$$

$$|H_{c2,3}(jv)|^2 = \frac{[\varepsilon^2(16 - 24v^2 + 9v^4)]}{[\varepsilon^2(16 - 24v^2 + 9v^4)] + v^6}. \qquad (7.3.35b)$$

The responses have the maximally flat property at $v = 0$. They are monotonic in the range $0 \leq v \leq 1$

Fig. 7.3.7 Chebyshev 2 low-pass amplitude response (a) $n = 2$, (b) $n = 3$

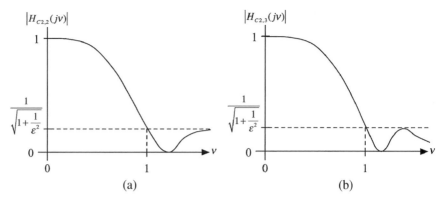

and have ripples in the range $1 < v < \infty$. The two functions are sketched in Fig. 7.3.7a,b. ε can be computed from

$$|H_{c2}(jv)|^2_{v=1} = \frac{\varepsilon^2}{1 + \varepsilon^2} = \frac{1}{1 + (1/\varepsilon^2)}.$$

The term $(1/\varepsilon^2)$ is related to the ripple width in the stop band similar to the term ε^2 used in determining the ripple width in the pass band in the Chebyshev 1. Note $|H_{c2}(0)| = 1$ for all values of n, $C_n^2(0) = 0$ for n–odd and $C_n^2(0) = 1$ for n–even. Therefore,

$$\lim_{v \to \infty} |H_{c2}(jv)| = \lim_{v \to \infty} \frac{\varepsilon^2 C_n^2(1/v)}{1 + \varepsilon^2 C_n^2(1/v)}$$

$$= \begin{cases} 0, & n - \text{odd} \\ [\varepsilon^2/(1 + \varepsilon^2)], & n - \text{even} \end{cases}. \quad (7.3.36)$$

The Chebyshev 2 normalized transfer function can be computed from

$$H_{c2}(\hat{s})H_{c2}(-\hat{s}) = |H_{c2}(jv)|^2 \big|_{v=s/j}. \quad (7.3.37)$$

For the derivation of the Chebyshev 2 function, see Weinberg (1962). A summary is given below. Chebyshev 2 function has a maximally flat response in the pass band as in the Butterworth approximation. Note $H_{c2}(j\omega)\big|_{\omega=0} = 1$. The left half-plane poles s_v and the zeros z_m of the Chebyshev 2 function normalized to the frequency 1 are determined using the constant ε_I obtained from the edges of

the stop-band and the pass-band frequencies ω_r and ω_c in terms of ε computed from the pass-band edge specification.

1. $n \geq \dfrac{\cosh^{-1}[(10^{.1Y} - 1)/(10^{.1X} - 1)]}{\cosh^{-1}(\omega_r/\omega_c)}$

   (Same as in Chebyshev 1.) $\qquad (7.3.38a)$

2. $\varepsilon_I = [1/\varepsilon C_n(\omega_r/\omega_c)]$

3. $\hat{s}_v = -\sinh\left(\dfrac{1}{n}\sinh^{-1}\dfrac{1}{\varepsilon_I}\right)\sin\dfrac{2v+1}{2n}\pi$

   $\qquad + j\cosh\left(\dfrac{1}{n}\sinh^{-1}\dfrac{1}{\varepsilon_I}\right)\cos\dfrac{2v+1}{2n}\pi,$

   $v = 0, 1, ..., n - 1 \qquad (7.3.38c)$

4. Poles: $s_v = 1/\hat{s}_v \qquad (7.3.38d)$

5. Zeros: $\hat{z}_m = j\sec((2m+1)\pi/2n),$

   $\begin{cases} m = 0, 1, ..., (n-1)/2, & n - \text{odd} \\ m = 0, 1, ..., (n/2) - 1, & n - \text{even} \end{cases}.$

   $\qquad (7.3.38e)$

The function C_{n2} is obtained by substituting $\hat{s} = s/\omega_r$ in the normalized function.

Example 7.3.3 Find the Chebyshev 2 transfer function that has attenuation of $X = 2$ dB at the edge of the pass band and the minimum attenuation in the stop band of $Y = 15$ dB. Note the pass-band and stop-band specifications are the same as in Example 7.3.2.

Solution: From Example 7.3.2, $\varepsilon = .76478$ and $n = 3$. From (7.3.38e), the zeros of the transfer function are $\hat{z}_1, \hat{z}_1^* = \pm j(1.1547)$. The other zero is at infinity. The constant ε_I is

$$\varepsilon_I = [1/\varepsilon C_3(\omega_r/\omega_c)]$$
$$= (1/.76478)(1/14.2998) \cong .09144.$$

Using (7.3.38d), the poles are $\hat{s}_{0,2} = (-.6103 \pm j1.3665)$, $\hat{s}_1 = -(1.2206)$. The normalized transfer function is

$$H_{c2}(\hat{s}) = K\frac{(\underline{s} - z_1)(\underline{s} - z_1^*)}{(\underline{s} - s_0)(\underline{s} - s_1)(\underline{s} - s_2)},$$
$$K = \frac{(-s_0)(-s_1)(-s_2)}{(-z_1)(-z_2)} \text{ (note } H_{c2}(0) = 1.).$$

The denormalized transfer function is obtained from

$$H_{c2}(s) = H_{c2}(\underline{s})\big|_{\underline{s} = s/\omega_r}.$$

The amplitude response function is sketched in Fig. 7.3.8. ∎

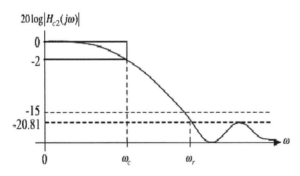

$20\log|H_{c2}(j\omega)|$

Fig. 7.3.8 Amplitude response of the Chebyshev 2 transfer function in Example 7.3.3

Elliptic filter approximations: Elliptic filter functions have equal ripple in both bands. Elliptic functions are beyond the scope here (see Storer, 1957). For a given set of filter amplitude response specifications, the order of the filter for the Butterworth (n_{Bu}), Chebyshev 1 and 2 ($n_{C1 \text{ and }} n_{C2}$), and elliptic filters (n_E) satisfy (see Storer, (1957).):

$$n_{Bu} \geq n_{C1} = n_{C2} \geq n_E. \qquad (7.3.39)$$

7.4 Phase-Based Design

A system is distortionless if its output is the same as the input except it is attenuated by the same amount for all frequencies with a constant delay (see Section 6.11). The transfer function of a linear time-invariant (LTI) system is given by

$$H(j\omega) = |H(j\omega)|e^{j\theta(\omega)}. \qquad (7.4.1)$$

If $\theta(\omega)$ is linear, then

$$\theta(\omega) = -\omega\tau. \qquad (7.4.2)$$

Since linear phase analog filters are not realizable, they are approximated. The group delay and the phase delays were defined by (6.7.7) and (6.7.19).

$$T_g(\omega) = -\frac{d\theta(\omega)}{d\omega}, \quad T_p(\omega) = -\frac{\theta(\omega)}{\omega}. \qquad (7.4.3)$$

Linear phase implies that the group delay in (7.4.3) is a constant. Since we are more interested in the phase angle, we can write (7.4.1) in the form below and solve for $\theta(\omega)$.

$$\ln[H(j\omega)] = \ln|H(j\omega)| + j\theta(\omega)$$
$$= (1/2)\ln|H(j\omega)|^2 + j\theta(\omega)$$
$$= (1/2)\ln[H(j\omega)H(-j\omega)] + j\theta(\omega). \qquad (7.4.4)$$

$$\theta(\omega) = (1/j)\ln[H(j\omega)] - (1/2j)\ln[H(j\omega)]$$
$$\quad - (1/2j)\ln[H(-j\omega)]$$
$$= (1/2j)\ln[H(j\omega)] - (1/2j)\ln[H(-j\omega)]$$
$$= (-j/2)\ln\left[\frac{H(j\omega)}{H(-j\omega)}\right]. \qquad (7.4.5a)$$

The *generalized phase function* is defined by

$$\theta(s) = -.5\ln[H(s)/H(-s)].$$

From (7.4.3), and using the chain rule given below, the group delay is given by

$$T_g(\omega) = -\frac{d\theta(\omega)}{d\omega}$$

$$= -\frac{1}{2j}\frac{d(\ln[H(j\omega)])}{d\omega} + \frac{1}{2j}\frac{d(\ln[H(-j\omega)])}{d\omega}$$

$$= -\left\{\left[\frac{1}{2}\frac{d\ln[H(j\omega)]}{d(j\omega)}\right] + \left[\frac{1}{2}\frac{d\ln[H(-j\omega)]}{d(-j\omega)}\right]\right\}.$$

$$(7.4.5b)$$

$$\left(\text{Chain rule}: \frac{d\ln[H(j\omega)]}{d\omega} = \frac{d\ln[H(j\omega)]}{d(j\omega)} \cdot \frac{d(j\omega)}{d\omega} = j\frac{d\ln[H(j\omega)]}{d(j\omega)}\right).$$

Noting the complex-conjugate terms inside the brackets {.} in (7.4.5b), the group delay is

$$T_g(\omega) = -\text{Re}\left\{\frac{d\ln[H(j\omega)]}{d(j\omega)}\right\}$$

$$= -\text{Ev}\left\{\frac{d\ln[H(s)]}{ds}\right\}\Big|_{s=j\omega}. \quad (7.4.6)$$

Notes: The symbol $\text{Ev}\{X(s)\}$ is the even part of $X(s)$ and is

$$\text{Ev}\{X(s)\} = \frac{1}{2}[X(s) + X(-s)]$$

$$= \frac{1}{2}\int_{-\infty}^{\infty} x(t)e^{-st}dt + \frac{1}{2}\int_{-\infty}^{\infty} x(t)e^{st}dt,$$

$$\text{Ev}\{X(s)\}\Big|_{s=j\omega} = [X(j\omega) + X^*(j\omega)]/2$$

$$= \text{Re}\{X(j\omega)\}.$$

$$(7.4.7) \quad \blacksquare$$

We should note that the group delay function is an even function. Assuming that the transfer function $H(s) = P(s)/Q(s)$ is a ratio of two polynomials, we have

$$\frac{d\ln[H(s)]}{ds} = \frac{d\ln[P(s)]}{ds} - \frac{d\ln[Q(s)]}{ds} = \frac{P'(s)}{P(s)} - \frac{Q'(s)}{Q(s)}.$$

The *generalized phase* and the *group delay* can be defined in terms of the variable s by

$$\theta(s) = -\frac{1}{2}\ln\left[\frac{H(s)}{H(-s)}\right],$$

$$\theta(\omega) = -j\theta(s)\Big|_{s=j\omega} \rightarrow T_g(s) = -\frac{d\theta(s)}{ds},$$

$$T_g(\omega) = T_g(s)\Big|_{s=j\omega}. \quad (7.4.8)$$

Using the property that $\ln[H(s)/H(-s)] = \ln[H(s)] - \ln[H(-s)]$ the group delay can be

expressed in terms of the transform variable s given below, which is useful in computing the delay associated with a transfer function and prime (′) denotes differentiation and

$$T_g(s) = -\frac{1}{2}\left[\frac{P'(s)}{P(s)} + \frac{P'(-s)}{P(-s)} - \frac{Q'(s)}{Q(s)} - \frac{Q'(-s)}{Q(-s)}\right],$$

$$(T_g(s) = T_g(-s)). \quad (7.4.9)$$

This results in the group delay that is real and an even function of ω.

Example 7.4.1 Compute the generalized phase and the group delay functions for the transfer function $H(s) = 1/[as^2 + bs + c]$.

Solution: From (7.4.8), the generalized phase and group delays are as follows:

$$\text{Phase}: \theta(s) = -(1/2)\ln[H(s)/H(-s)],$$

$$\theta(\omega) = -j\theta(s)\Big|_{s=j\omega}$$

Group delay:

$$P(s) = 0; \ P'(s) = 0; \ P'(-s) = 0;$$

$$Q(s) = as^2 + bs + c; \ Q(-s) = as^2 - bs + c$$

$$Q'(s) = 2as + b, \ Q'(-s) = -\frac{d}{ds}Q(-s)$$

$$= -\frac{d}{ds}[as^2 - bs + c]$$

$$= -[2as - b] = -2as + b$$

$$\Rightarrow T_g(s) = -\frac{1}{2}\left[0 + 0 - \frac{2as + b}{as^2 + bs + c} - \frac{-2as + b}{as^2 - bs + c}\right]$$

$$= \frac{bc - abs^2}{(as^2 + c)^2 - b^2s^2}. \quad (7.4.10a)$$

$$T_g(\omega) = T_g(s)\Big|_{s=j\omega} = \frac{bc + ab\omega^2}{(c - a\omega^2)^2 + b^2\omega^2}$$

(function is real and even). $\quad (7.4.10b)\blacksquare$

7.4.1 Maximally Flat Delay Approximation

If we assume in (7.4.1) that $|H(j\omega)| = 1$, then

$$H(j\omega) = e^{-j\omega t_0}, |H(j\omega)| = 1, \theta(\omega) = \angle H(j\omega) = -\omega t_0,$$

$$\tau(\omega) = -\frac{d\angle H(\omega)}{d\omega} = t_0. \quad (7.4.11)$$

It has flat amplitude, linear phase, and a constant group delay with respect to ω. The system described by (7.4.11) is distortionless. Using the *analytic continuation*(see Balbanian et al., 1969), the transfer function can be written in the Laplace transform domain by replacing $j\omega$ by s. Let us define a normalized transfer function using

$$H(\underline{s}) = H(s)\big|_{\underline{s}=st_0} = e^{-st_0}\big|_{\underline{s}=st_0} = e^{-\underline{s}}. \quad (7.4.12)$$

$$= \frac{1}{\cosh(\underline{s}) + \sinh(\underline{s})}\bigg|_{\underline{s}=st_0}$$

$$= \frac{1}{[(\cosh(\underline{s})/\sinh(\underline{s})) + 1]\sinh(\underline{s})}. \quad (7.4.13)$$

Storch (1954) approximates $e^{-\underline{s}}$ by an nth order rational function of the form in (7.4.13) below using the power series approximation of the hyperbolic sine and cosine functions.

$$H(\underline{s}) \approx H_n(\underline{s}) = \frac{b_0}{B_n(\underline{s})} = \frac{b_0}{b_n\underline{s}^n + b_{n-1}\underline{s}^{n-1} + \ldots + b_0},$$

$$b_n = 1, H_n(0) = 1. \quad (7.4.14)$$

$B_n(\underline{s})$ are the *Bessel polynomials* and they can be derived using (see Spiegel, 1968.)

$$B_n(\underline{s}) = (2n - 1)B_{n-1}(\underline{s}) + s^2 B_{n-2}(\underline{s}),$$

$$B_0(\underline{s}) = 1, B_1(\underline{s}) = 1 + \underline{s}. \quad (7.4.15)$$

For $n = 0, 1, 2, 3, 4$, these are

$$B_0(\underline{s}) = 1, \quad B_1(\underline{s}) = 1 + \underline{s},$$

$$B_2(\underline{s}) = 3 + 3\underline{s} + \underline{s}^2, \quad B_3(\underline{s}) = 15 + 15\underline{s} + 6\underline{s}^2 + \underline{s}^3,$$

$$B_4(s) = 105 + 105\underline{s} + 45\underline{s}^2 + 10\underline{s}^3 + \underline{s}^4 \quad (7.4.16)$$

The roots of the polynomials can only be determined numerically. The transfer function $H_n(j\omega)$

has maximally flat delay characteristics. The transfer function, the phase, and the group delay responses are given for the Bessel transfer function (in terms of frequency $\underline{\omega}$) by

$$H_n(j\underline{\omega}) = |H_n(\underline{\omega})|e^{j\theta_n(\underline{\omega})}, \theta_n(\underline{\omega}) = \angle H_n(\underline{\omega}),$$

$$\tau_n(\underline{\omega}) = -d\theta_n(\underline{\omega})/d\underline{\omega}. \quad (7.4.17)$$

Example 7.4.2 Show the maximally flat property of the group delays of $H_n(\underline{s})$, $n = 1, 2$.

$$H_1(\underline{s}) = \frac{1}{\underline{s} + 1}, H_2(\underline{s}) = \frac{3}{\underline{s}^2 + 3\underline{s} + 3}. \quad (7.4.18)$$

Solution: The phase and the delay responses are given by

$$\theta_1(\underline{\omega}) = -\tan^{-1}(\underline{\omega}),$$

$$\tau_1(\underline{\omega}) = -\frac{d\theta_1}{d\underline{\omega}} = -\frac{d(-\tan^{-1}(\underline{\omega}))}{d\underline{\omega}} = \frac{1}{1 + \underline{\omega}^2}, \quad (7.4.19a)$$

$$\theta_2(\underline{\omega}) = -\tan^{-1}\left(\frac{3\underline{\omega}}{3 - \underline{\omega}^2}\right),$$

$$\tau_2(\underline{\omega}) = -\frac{d\theta_2}{d\underline{\omega}} = \frac{(9 + 3\underline{\omega}^2)}{(9 + 3\underline{\omega}^2) + \underline{\omega}^4} \quad (7.4.19b)$$

Expressing these in terms of Maclaurin power series in the neighborhood of $\underline{\omega} = 0$, we can show that the first $(2n - 1)$ derivatives of the group delay function vanish at the zero frequency and the maximally flat property follows. This is valid for all n. ∎

7.4.2 Group Delay of Bessel Functions

Baher (1990) gives a relationship between an all pole rational function and its group delay and is summarized below in terms of a Bessel transfer function $H_n(\underline{s})$. First, we can write the transfer function (see (7.4.13)) in the form

$$H_n(j\underline{\omega}) = \frac{b_0}{(b_0 - b_2\underline{\omega}^2 + b_4\underline{\omega}^4 - \ldots) + j(b_1\underline{\omega} - b_3\underline{\omega}^3 + b_5\underline{\omega}^5 - \ldots)} = \frac{b_0}{E_n(\underline{\omega}) + jO_n(\underline{\omega})}. \quad (7.4.20)$$

The amplitude, phase, and the corresponding group delay responses are

$$|H_n(j\underline{\omega})|^2 = \frac{b_0^2}{E_n^2(\omega) + O_n^2(\omega)},$$

$$\theta_n(\underline{\omega}) = -\tan^{-1}\left[\frac{O_n(\omega)}{E_n(\omega)}\right]. \qquad (7.4.21a)$$

$$\tau_n(\underline{\omega}) = -\frac{d\theta_n(\omega)}{d\omega}$$

$$= -\left[\frac{E_n(\omega)\frac{dO_n(\omega)}{d\omega} - O_n(\omega)\frac{dE_n(\omega)}{d\omega}}{E_n^2(\omega) + O_n^2(\omega)}\right].$$

$$\Rightarrow \tau_n(\underline{\omega}) = 1 - \omega^{2n}|H_n(j\omega)|^2(1/b_0^2). \qquad (7.4.21b)$$

Example 7.4.3 Verify the results in (7.4.19a and b) using (7.4.21b).

Solution: These can be shown by

$$n = 1 : 1 - \omega^2|H_1(j\omega)|^2(1/b_0^2)$$

$$= 1 - \frac{\omega^2}{1 + \omega^2} = \frac{1}{1 + \omega^2} = \tau_1(\underline{\omega}) \qquad (7.4.22a)$$

$$n = 2 : 1 - \omega^4|H_2(j\omega)|^2(1/b_0^2)$$

$$= 1 - \frac{\omega^4}{(9 + 3\omega^2) + \omega^4} = \frac{(9 + 3\omega^2)}{(9 + 3\omega^2) + \omega^4}$$

$$= \tau_2(\underline{\omega}). \qquad (7.4.22b) \quad \blacksquare$$

Notes: Note $H_n(0) = 1$ and the group delay has the maximally flat response with $\tau_n(0) = 1$. The design involves finding the n for a set of specifications including maximum attenuation in the pass band in dB and a constant delay within a prescribed tolerance in the pass band. The group delay can be approximated by using the first two terms in the series and the approximation is good for $n > 3$ (see Temes and Mitra, 1973). Assuming the frequency is normalized by t_0, that is $\underline{\omega} = \omega t_0$, we have

$$\tau_n(\underline{\omega}) = \left[1 - \left(\frac{\omega^n}{b_0}\right)^2 + -\ldots\right]$$

$$\cong \left[1 - \left(\frac{(2^n n!)^2}{(2n)!}\right)(\omega)^{2n}\right]. \qquad (7.4.23a)$$

The amplitude response of a Bessel filter function is Gaussian. The attenuation for a filter of order $n > 3$, the attenuation and the 3 dB frequency can be approximated by

$$-20\log|H_n(j\underline{\omega})| \cong 4.3429\omega^2/(2n - 1)) \qquad (7.4.23b)$$

See Problem 7.4.5 for its use of this. $\blacksquare$

Example 7.4.4 Determine a. the 3 dB frequency and b. the frequency at which the group delay deviates by 1% for a second-order Bessel function.

$$H_2(\underline{s}) = \frac{3}{\underline{s}^2 + 3\underline{s} + 3},$$

$$|H_2(j\underline{\omega})|^2 = \frac{9}{[(3 - \underline{\omega}^2)^2 + 9\underline{\omega}^2]}. \qquad (7.4.24a)$$

Solution: a. It follows that

$$|H_2(j\underline{\omega}_{3dB})|^2 = \frac{1}{2} = \frac{9}{[(3 - \underline{\omega}_{3dB}^2)^2 + 9\underline{\omega}_{3dB}^2}$$

$$\Rightarrow \underline{\omega}_{3dB} = 1.36. \qquad (7.4.24b)$$

b. The frequency at which the group delay deviates is computed using (7.4.22b)

$$(\tau_2(\omega))_{.99} = 1 - \left[\frac{\omega_{.99}^4}{9 + 3\underline{\omega}_{.99}^2 + \underline{\omega}_{.99}^4}\right]$$

$$= .99 \Rightarrow \underline{\omega}_{.99} = .56. \qquad (7.4.24c) \quad \blacksquare$$

In this example, the 3 dB frequency and the frequency at which certain percent deviation in $\tau_n(\underline{\omega})$ from 1 can be analytically computed. For an arbitrary n, these can be computed either by (7.4.23) or by tables (see Weinberg, 1962.). Table 7.4.1 gives the

Table 7.4.1 Normalized frequencies, $\underline{\omega} = \omega t_0$. Time delay and a loss table giving the normalized frequency $\underline{\omega}$ at which the zero frequency delay and loss values deviate by specified amounts for Bessel filter functions

n	1	2	3	4	5	6	7	8	9	10	11
$\underline{\omega}_{3dB}$	1	1.36	1.75	2.13	2.42	2.70	2.95	3.17	3.39	3.58	3.77
$\underline{\omega}_{1\%deviation}$	0.1	0.56	1.21	1.93	2.71	3.52	4.36	5.22	6.08	6.96	7.85
$\underline{\omega}_{10\%deviation}$	0.34	1.09	1.94	2.84	3.76	4.69	5.64	6.59	7.55	8.52	9.48

Fig. 7.4.2 Example 7.4.5: (a) amplitude and (b) group delay response specifications

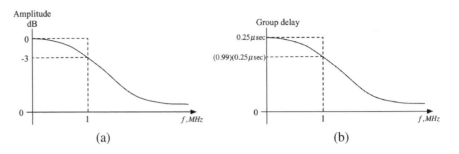

(a)   (b)

normalized 3 dB frequency and the frequencies at which the $\tau(\underline{\omega})$ deviates 1 and 10% from 1. Compare the results in (7.4.24b and c) to the table.

Example 7.4.5 Find n for the Bessel filter specifications in Fig. 7.4.2a (for the amplitude) and Fig. 7.4.2b (for the delay) with

1. A delay of $\tau_0 = .25$ μs up to 1 MHz within 1% deviation
2. A loss of less than 3 dB up to 1 MHz.

Solution: From the specifications, the pass-band edge of the normalized frequency is

$$\underline{\omega}_{3dB} = \omega_{3dB} t_0 \big|_{\omega = 2\pi(10^6)} = 2\pi(10^6)(.25(10^{-6})) \cong 1.57.$$
$$\text{(7.4.25)}$$

From the first condition, using Table 7.4.1, we have $n \geq 4$. To satisfy the second condition, again using Table 7.4.1, n must be at least equal to 3, as $1.36 < 1.57 < 1.75$. Selecting the higher value $n = 4$, the normalized transfer function and the transfer function with the desired delay are, respectively, given by (see 7.4.16). See Fig. 7.4.2

$$H(\underline{s}) = \frac{105}{[\underline{s}^4 + 10\underline{s}^3 + 45\underline{s}^2 + 105\underline{s} + 105]},$$

$$H(s) = H(\underline{s})|\underline{s} = s/t_0, \ t_0 = .25(10^{-6}). \quad \text{(7.4.26)} \quad \blacksquare$$

7.5 Frequency Transformations

Since the low-pass filter's cut-off frequency is a design parameter, it is simpler to use a *normalized* low-pass (LP) filter with a cut-off frequency of 1. The amplitude response of an nth order Butterworth LP filter, the normalized filter function s and the s-domain function are

$$\left|H_{Lp}(j\omega)\right| = \frac{1}{\sqrt{1 + \varepsilon^2(\omega/\omega_c)}},$$

$$\left|H_{Lpn}(j\underline{\omega})\right| = \frac{1}{\sqrt{1 + \varepsilon^2(\underline{\omega})^{2n}}}, \underline{\omega} = \frac{\omega}{\omega_c}. \quad \text{(7.5.1)}$$

$$H_{Lp}(s) = H_{Lpn}(\underline{s})\big|_{\underline{s} = s/\omega_c}. \quad \text{(7.5.2)}$$

Normalized functions are used as a first step in designing the filters. If the given cut-off frequency is used, the transfer function coefficients will be large. Normalization results in smaller coefficents. Filter transformations are shown in Fig. 7.5.1a,b,c,d.

The normalized Lp (Lpn) to Lp filter transformation is *frequency scaling*, as ω is scaled by ω_c. Figure 7.5.1a illustrates the frequency transfor-mation from the normalized Lp to Lp and vice versa. Figure 7.5.1b,c,d illustrates the transformations from the normalized Lp to high pass (Hp), band pass (Bp), and band elimination *(Be)*, and vice versa.

7.5.1 Normalized Low-Pass to High-Pass Transformation

The ideal Hp function can be obtained by frequency inversion from the Lp function.

$$\underline{s} = (\omega_c/s) \Rightarrow \underline{s} = j\underline{\omega}$$
$$= \omega_c/j\omega \text{ or } \underline{\omega} = -\omega_c/\omega \text{ and } H_{Hp}(\underline{s})$$
$$= H_{Lpn}(\omega_c/s). \quad \text{(7.5.3)}$$

This transformation is a non-linear transformation. Figure 7.5.1b gives the frequency transformations from normalized low-pass specifications to high-pass specifications and vice versa. If we have a pole in the normalized low-pass function at $\underline{s} = -p_1$, then

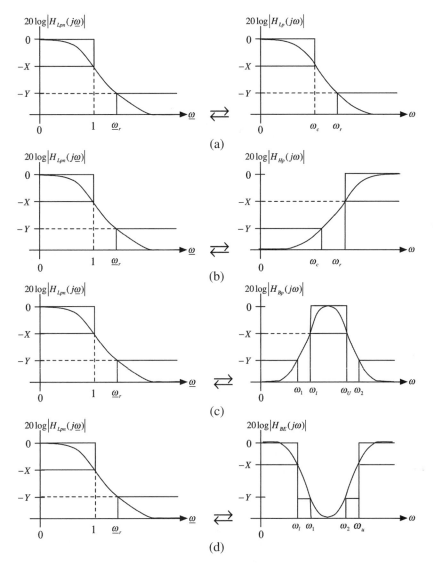

Fig. 7.5.1 Illustration of analog-to-analog frequency transformations: (a) normalized low pass ⟷ low pass, $\omega_r = \underline{\omega}_r \omega_c, \underline{\omega}_r = \omega_r/\omega_c$, (b) normalized low pass ⟷ high pass, $\underline{\omega}_r = \omega_c/\omega_r, \omega_r = \omega_c/\underline{\omega}_r$, (c) normalized low pass ⟷ band pass, $\omega_r = \min\{\omega_{r_1}, \omega_{r_2}\}$, $\omega_{r_1} = \frac{\omega_u \omega_l - \omega_1^2}{(\omega_u - \omega_l)\omega_1}$, $\omega_{r_2} = \frac{\omega_2^2 - \omega_u \omega_l}{(\omega_u - \omega_l)\omega_2}$, (d) normalized low pass ⟷ band elimination, $\omega_r = \min\{\omega_{r_1}, \omega_{r_2}\}, \omega_{r_1} = \frac{(\omega_u - \omega_l)\omega_1}{\omega_u \omega_l - \omega_1^2}, \omega_{r_2} = \frac{(\omega_u - \omega_l)\omega_2}{\omega_u \omega_l - \omega_2^2}$. (7.5.2)

$$[1/(\underline{s} + p_1)]_{\underline{s} = \omega_c/s} = (s/p_1)/(s + (\omega_c/p_1)). \quad (7.5.4)$$

The normalized low-pass transformation to high-pass transformation converts the pole from $s = -p_1$ into a zero at the origin and a pole at $s = -\omega_c/p_1$.

Example 7.5.1 Find the transfer function of the high-pass filter function with the cut-off frequency

ω_c using the normalized second-order Butterworth filter function given below.

$$H_{Lpn}(\underline{s}) = \frac{1}{[(\underline{s}^2 + \sqrt{2}\underline{s} + 1)]}. \quad (7.5.5a)$$

Solution: The high-pass filter function is given below. See Fig. 7.5.1b for a sketch of the amplitude response of a high-pass function.

$$H_{Hp}(s) = H_{Lpn}(\underline{s})\big|_{\underline{s}=\omega_c/s} = \frac{s^2}{s^2 + \sqrt{2}\omega_c s + \omega_c^2},$$

$$|H_{Hp}(j\omega)| = \frac{|\omega|^2}{\sqrt{(\omega_c^2 - \omega^2)^2 + 2\omega_c^2\omega^2}}. \qquad (7.5.5b) \quad \blacksquare$$

7.5.2 Normalized Low-Pass to Band-Pass Transformation

Band-pass filter eliminates or significantly attenuates both low and high frequencies. We can interpret the *Bp* filter operation as passing a signal through a *Hp* filter with a cut-off frequency equal to ω_l and then pass the resulting signal through a *Lp* filter with a cut-off frequency equal to ω_h, $\omega_l < \omega_h$. The transformation is

$$\underline{s} = (s^2 + \omega_0^2)/Bs, \quad B = (\omega_h - \omega_l),$$
$$\omega_0^2 = \omega_l\omega_h, \quad \underline{\omega} = (\omega^2 - \omega_0^2)/B\omega. \qquad (7.5.6a)$$

The frequencies ω_l and ω_h are the low-edge and high-edge frequencies of the band-pass filter. ω_0 is the *geometric mean* of the two frequencies ω_l and ω_h. It is interesting to point out that if frequency is plotted on a logarithmic scale, ω_0 falls midway between ω_l and ω_h. The BP function can be obtained from the normalized *Lp* function by

$$H_{Bp}(s) = H_{Lpn}(\underline{s})\big|_{\underline{s}=(s^2+\omega_0^2)/Bs} = H_{Lpn}\left(\frac{s^2 + \omega_0^2}{Bs}\right). \qquad (7.5.6b)$$

The transformation in (7.5.6a) is a *non-linear transformation* mapping the frequencies

$$\underline{\omega} = 0 \to \omega = \pm\omega_0, \quad \underline{\omega} = \pm\infty \to \omega = \pm\infty$$
$$\underline{\omega} = 1 \to \omega = -\omega_l \text{ and } \omega_h,$$
$$\underline{\omega} = -1 \to \underline{\omega} = -\omega_h \text{ and } \omega_l$$

For example, for $\underline{\omega} = 1$,

$$1 = \frac{\omega^2 - \omega_0^2}{B\omega} \Rightarrow \omega^2 - (\omega_h - \omega_l)\omega - \omega_h\omega_l = 0$$

$$\omega = \frac{\omega_h - \omega_l}{2} \pm \frac{\sqrt{(\omega_h - \omega_l)^2 + 4\omega_h\omega_l}}{2} \Rightarrow \omega = \omega_h, -\omega_l.$$

Similarly we can show for $\underline{\omega} = -1$. The normalized LP zero frequency is mapped to the BP frequencies $\omega = \pm\infty$. Each LP pole transforms into a pair of BP poles under this transformation and the order of the BP function is twice that of LP function. Figure 7.5.1c illustrates the transformations from BP to normalized LP specifications and vice versa.

Example 7.5.2 Find the BP function using the transformation in (7.5.6b) from the normalized LP function and identify the zero and the poles in the BP function.

$$H_{Lpn}(\underline{s}) = 1/(\underline{s} + p_1). \qquad (7.5.7)$$

Solution: The band-pass function is

$$H_{Bp}(s) = \frac{1}{[(s^2 + \omega_0^2)/Bs] + p_1} = \frac{Bs}{[(s^2 + \omega_0^2) + Bp_1 s]}.$$

LP simple pole $\Rightarrow$ zero at $s = 0$ and poles at $s = (1/2)\left(-Bp_1 \pm \sqrt{B^2 p_1^2 - 4\omega_0^2}\right).$ $\blacksquare$

7.5.3 Normalized Low-Pass to Band-Elimination Transformation

A band-elimination (*Be*) or a band-reject filter eliminates or attenuates a band of frequencies somewhere in the middle of the frequency band and passes outside this band with a small attenuation. The transformation from the normalized LP function to the BE function is

$$\underline{s} = Bs/(s^2 + \omega_0^2), \quad B = (\omega_u - \omega_l), \quad \omega_0^2 = \omega_l\omega_u,$$
$$\underline{\omega} = B\omega/(\omega_0^2 - \omega^2). \qquad (7.5.8)$$

The frequencies ω_l and ω_h are the pass-band edge frequencies of the band-elimination filter. ω_0 is the geometric mean of the two frequencies ω_l and ω_h. The corresponding band-stop function can be obtained from the normalized low-pass function by

$$H_{Be}(s) = H_{Lpn}(\underline{s})\big|_{\underline{s}=Bs/(s^2+\omega_0^2)} = H_{Lpn}\left(Bs/(s^2 + \omega_0^2)\right).$$

The transformation in (7.5.8) maps the frequencies in the following manner.

Transformations summary: $\underline{\omega} = B\omega/(\omega^2 - \omega_0^2)$, $\omega_0^2 = \omega_u\omega_l$, $B = \omega_u - \omega_l$

$\underline{\omega} = 0 \Rightarrow \omega = 0$, $\underline{\omega} = \pm\infty \Rightarrow \omega = \pm\infty$

$\underline{\omega} = 1 \Rightarrow \omega = -\omega_l$ and ω_u,

$\underline{\omega} = -1 \Rightarrow \underline{\omega} = -\omega_u$ and ω_l.

Some authors refer to B as the *notch width* of the notch filter. As in the band-pass case, ω_0 is the geometric mean of the two frequencies ω_l and ω_u. Each low-pass pole transforms into a pair of BE filter poles and the order of the BE function is twice that of the LP function. Figure 7.5.1d gives the transformations from BE filter specifications to normalized LP filter specifications and vice versa.

Example 7.5.3 Consider the normalized Lp function given in (7.5.7). Find the corresponding Be function using the transformation given in (7.5.8).

Solution: The transformation creates a pair of complex-conjugate zeros on the imaginary axis and a complex pole pair on the left half of the s plane. The function is

$$H_{Be}(s) = \frac{1}{[(Bs/(s^2 + \omega_0^2)) + p_1]}$$

$$= \frac{(s^2 + \omega_0^2)}{(s^2 + \omega_0^2)p_1 + Bs}. \qquad \blacksquare$$

In Chapter 6, we considered simple first-order Lp and Hp filters and second-order Bp and Be filters using resistors, inductors, and capacitors (see Fig. 6.12.12). The design of passive filters is primarily based on reactive components, i.e., inductors (may include transformers) and capacitors. The transformations given earlier can be used to derive the Lp, Hp, Bp, and Be filter circuits from a normalized Lp filter circuit. This is discussed next. The cut-off frequency of the normalized Lp filter is $\omega_c = 1$ rad/s.

These transformations in table 7.5.1 allow for finding an Lp or a Hp, or a Bp or a Be function from a normalized Lp function of an RLC filter. It also gives the changes in the corresponding component values. The inductor–capacitor series and parallel pairs of components resonate at the frequency $\omega_0 = 1/\sqrt{(LC)}$ in the Bp and Be cases.

Table 7.5.1 Frequency transformations

Transformation	Variable transformation	Passive element transformation
Normalized low-pass → Low-pass filter with a cut-off frequency of ω_c.	$\underline{s} = (s/\omega_c);$ $L\underline{s} = (L/\omega_c)s;$ $(1/C\underline{s}) = (\omega_c/Cs).$	$L \rightarrow L/\omega_c$; $C \rightarrow C/\omega_c$
Normalized low-pass → High-pass filter with a cut-off frequency of ω_0.	$\underline{s} = (\omega_c/s);$ $L\underline{s} = (L\omega_c/s);$ $C\underline{s} = C(\omega_c/s).$	$L \rightarrow 1/L\omega_c$; $C \rightarrow C\omega_c$
Normalized low-pass → Band-pass with center frequency ω_0 and bandwidth B.	$\underline{s} = \dfrac{s^2 + \omega_0^2}{Bs};$ $L\underline{s} = \dfrac{L}{B} \cdot \dfrac{s^2 + \omega_0^2}{s};$ $\dfrac{1}{C\underline{s}} = \dfrac{B}{C} \cdot \dfrac{s}{s^2 + \omega_0^2}.$	$L \rightarrow L/B$, $B/L\omega_0^2$ (series); $C \rightarrow C/B$, $B/C\omega_0^2$ (parallel)
Normalized low-pass → Band elimination with center frequency ω_0 and notch width B.	$\underline{s} = \dfrac{Bs}{s^2 + \omega_0^2};$ $L\underline{s} = \dfrac{s}{(s^2 + \omega_0^2)} \cdot LB;$ $\dfrac{1}{C\underline{s}} = \dfrac{(s^2 + \omega_0^2)}{s} \cdot \dfrac{1}{CB}.$	$L \rightarrow LB/\omega_0^2$, $1/LB$ (parallel); $C \rightarrow CB/\omega_0^2$, $1/CB$ (series)

Fig. 7.5.2 Example 7.5.4

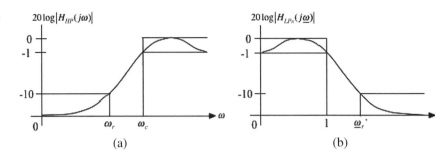

(a)   (b)

7.5.4 Conversions of Specifications from Low-Pass, High-Pass, Band-Pass, and Band Elimination Filters to Normalized Low-Pass Filters

Given the Lp, Hp, Bp, and Be filter specifications in Fig. 7.5.1, how do we transform these into a normalized Lp specification? Notation wise, we use the following:

ω **for an arbitrary filter function** and $\underline{\omega}$ **for the normalized** Lp **filter function**.

Low pass to a normalized low-pass specification: The transformation from a low-pass filter specification to a normalized low-pass specification is rather simple as the amplitudes remain the same at the critical frequencies, i.e., at the cut-off frequency and the edge of the stop-band frequency. The frequencies are of course normalized in the normalized low-pass filter specifications. These can be summarized by

$$\left|H_{Lp}(\omega_c)\right| = \left|H_{Lpn}(1)\right|,\ \left|H_{Lp}(\omega_r)\right| = \left|H_{Lpn}\left(\frac{\omega_r}{\omega_c}\right)\right|,$$

$\underline{\omega}_c = 1$ and $\underline{\omega}_r = \omega_r/\omega_c > 1$ and $\omega_r = \underline{\omega}_r\omega_c$. (7.5.9)

Figure 7.5.1a gives the transformations from the Lp to the normalized Lp and vice versa.

High pass to the normalized Lp **specifications:** The Hp filter specifications can be transformed to the normalized Lp filter specifications by

$\underline{\omega}_c = 1$ and $\underline{\omega}_r = (\omega_c/\omega_r)$ or $\omega_r = \omega_c/\underline{\omega}_r$. (7.5.10)

Figure 7.5.1b gives the transformation from the Hp filter specifications to the normalized low-pass filter specifications and vice versa. Note $\omega_c > \omega_r$ (high-pass filter).

Example 7.5.4 Use the Chebyshev 1 approximation assuming for the specifications given in Fig. 7.5.2a with $\omega_c = 2\pi(100\,k)$ and $\omega_r = 2\pi(50\,k)$.

Solution: First, using the high-pass to normalized low-pass frequency transformation,

$$\underline{\omega} = [2\pi(\omega_c)/\omega] = [2\pi(100\,k)/\omega],$$

$$\underline{\omega}_c = \omega\big|_{\omega=2\pi(100k)} = 2\pi(100k) = 1,\ \underline{\omega}'_r\big|_{\omega=2\pi(50k)} = 2.$$

(7.5.11a)

The normalized Lp filter specifications are shown in Fig. 7.5.2b. The normalized Chebyshev 1 function is

$$\left|H_{Lpn}(j\underline{\omega})\right|^2 = \frac{1}{1 + \varepsilon^2 C_n^2(\underline{\omega})}$$ (7.5.11b)

At $\underline{\omega} = 1$, we have $-1 = 10\log(1/10^{-1}) = 10\log[1/(1 + \varepsilon^2 C_n^2(1))]$. Noting that $C_n^2(1) = 1$, we have $\varepsilon \cong .509$. At $\underline{\omega} = 2$,

$$-10 = 10\log(1/10) = 10\log\left[\frac{1}{1 + \varepsilon^2 C_n^2(2)}\right]$$

$$\rightarrow 10 = 1 + .259 C_n^2(2)$$

$$\rightarrow C_n(2) \cong 5.9$$

$$n \geq \frac{\cosh^{-1}(5.9)}{\cosh^{-1}(2)} = 1.87 \Rightarrow n = 2\,(\text{see } (7.3.29a \text{ and } b)).$$

(7.5.11c)

Using (7.3.19d),

$$C_n(2) \cong (2 + \sqrt{4 - 1})^n/2 = 5.9 \Rightarrow n \geq 1.88 \Rightarrow n = 2.$$

(7.5.11d)

Noting that $C_2(\underline{\omega}) = 2\underline{\omega}^2 - 1$ and $C_2^2(\underline{\omega})\big|_{\underline{\omega}=0} = 1$, the low-pass normalized amplitude squared response function and the left half-plane poles are, respectively, given by

$$|H_{Lpn}(\underline{\omega})|^2 = \frac{1}{1 + \varepsilon^2(2\underline{\omega}^2 - 1)^2}|H_{Lpn}(\underline{\omega})|^2_{\underline{\omega}=0}$$

$$= \frac{1/(1 + \varepsilon^2)}{1 + \varepsilon^2(2\underline{\omega}^2 - 1)^2};$$

$$p_v = -\sinh\phi_2\sin\frac{(2v+1)\pi}{4} + j\cosh\phi_2\cos\frac{(2v+1)\pi}{4},$$

$$v = 0, 1; \ \phi_2 = (1/2)\sinh^{-1}(1/\varepsilon).$$

The Chebyshev 1 normalized second-order transfer function is

$$H_{Lpn}(\underline{s}) = \frac{1}{\sqrt{1 + \varepsilon^2}}\frac{(p_0p_1)}{(\underline{s} - p_0)(\underline{s} - p_1)}. \quad (7.5.11e)$$

The scaling factor at $\underline{s} = 0$ is not 1, as we are dealing with an even Chebyshev function. That is $H_{Lpn}(\underline{s})|_{\underline{s}=0} = (1/(\sqrt{1 + \varepsilon^2}))$. The desired high-pass filter transfer function is

$$H_{Hp}(s) = H_{Lpn}(\underline{s})|_{\underline{s}=\omega_c/s}. \quad \blacksquare$$

Band-pass to normalized low-pass specification: BP filter specifications can be transformed to normalized LP filter specifications by

$$\underline{s} = \frac{s^2 + \omega_0^2}{Bs}, \ j\underline{\omega} = -j\frac{\omega_0^2 - \omega^2}{B\omega} = j\frac{\omega^2 - \omega_0^2}{B\omega},$$

$$B = \omega_u - \omega_l, \omega_0^2 = \omega_l\omega_u. \quad (7.5.12a)$$

The normalized low-pass critical frequencies are as follows:

$$\omega = \omega_l \Rightarrow \frac{\omega_l^2 - \omega_u\omega_l}{(\omega_u - \omega_l)\omega_l} = -1 = -\underline{\omega}_c,$$

$$\omega = j\omega_u \Rightarrow \frac{\omega_u^2 - \omega_u\omega_l}{(\omega_u - \omega_l)\omega_u} = 1 = \underline{\omega}_c. \quad (7.5.12b)$$

$$\omega = \omega_1 \Rightarrow \frac{\omega_1^2 - \omega_u\omega_l}{(\omega_u - \omega_l)\omega_1} = -\underline{\omega}_{r1} > 0,$$

$$\omega = \omega_2 \Rightarrow \frac{\omega_2^2 - \omega_u\omega_l}{(\omega_u - \omega_l)\omega_2} = \underline{\omega}_{r2} > 0. \quad (7.5.12c)$$

$$\underline{\omega}_c = 1 \text{ and } \underline{\omega}_r = \min\{\omega_{r1}, \omega_{r2}\}. \quad (7.5.12d)$$

Note the use of the function, $\min\{\omega_{r1}, \omega_{r2}\}$, as the two sides of the band-pass spectrum result in two cut-off frequencies (one positive and the other one is negative) in the normalized low-pass specification. Noting that the magnitude spectrum is even, we need to use the tighter specification. Figure 7.5.1c illustrates the transformation from the band-pass filter specifications to the normalized low-pass filter specifications and vice versa.

Example 7.5.5 Use the Butterworth approximation to find the transfer function assuming for the specifications of a band-pass filter shown in Fig. 7.5.3a. The pass-band and the stop-band edge frequencies are

$$\omega_1 = 2\pi(200\,k), \ \omega_l = 2\pi(300\,k), \ \omega_u = 2\pi(400\,k),$$

$$\omega_2 = 2\pi(800\,k). \quad (7.5.13a)$$

Solution: Using the transformations in (7.5.12b), the normalized LP frequency is $\underline{\omega}_c = 1$. With (7.5.12c), we have

$$\underline{\omega}|_{\omega=\omega_1} = \frac{(2\pi)^2(200\,k)^2 - (300\,k)(400\,k)}{(2\pi)^2(400\,k) - (300\,k))(200\,k)}$$

$$= -4 = -\omega_{r1}. \quad (7.5.13b)$$

$$\underline{\omega}|_{\omega=\omega_2} = \frac{(2\pi)^2(800\,k)^2 - (300\,k)(400\,k)}{(2\pi)^2\ 400\,k - (300\,k)(800\,k)}$$

$$= 6.5 = \omega_{r2}. \quad (7.5.13c)$$

Note the $(2\pi)^2$ cancel out. From (7.5.10) $\min\{4, 6.5\} = 4$. The tighter specifications correspond to $\underline{\omega}_r = 4$ and we can ignore the other one. The normalized low-pass filter specifications are shown in Fig. 7.5.3b. The normalized Butterworth low-pass function is given by

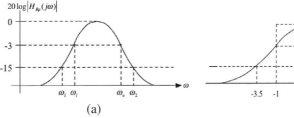

Fig. 7.5.3 Example 7.5.5

$$|H_{Lpn}(j\underline{\omega})|^2 = \frac{1}{1 + \varepsilon^2 \underline{\omega}^{2n}}. \qquad (7.5.13d)$$

These result in

$$\underline{\omega}_c = 1, \ 20\log(|H_{Lpn}(j1)|) = -3\,dB \Rightarrow \varepsilon \cong 1,$$
$$20\log|H_{Lpn}(\underline{\omega})|\big|_{\underline{\omega}=4} = 20\log(1/(1+4^{2n}))$$
$$= -15\text{dB} \rightarrow 1 + 4^{2n}$$
$$= 10^{1.5} \Rightarrow n = 2$$

$$(n)_{\text{integer}} \geq \frac{\log(10^{15/10}-1)}{2\log(4/1)} = 1.23 \rightarrow n = 2$$
$$(\text{Using } (7.3.10a))$$

$$|H_{Lpn}(\underline{\omega})|^2 = \frac{1}{1+\underline{\omega}} \Rightarrow H_{Lpn}(-\underline{s})\frac{1}{1+\underline{s}^4}.$$
$$\Rightarrow H_{Lpn}(\underline{s}) = \frac{1}{\underline{s}^2 + \sqrt{2}\,\underline{s} + 1} \qquad (7.5.13e)$$

Applying the low-pass to band-pass transformation, the band-pass transfer function is

$$H_{Bp}(s) = H_{Lpn}(\underline{s})\Big|_{\underline{s}=\frac{s^2+\omega_l\omega_u}{s(\omega_u-\omega_l)}}. \qquad (7.5.13f) \quad \blacksquare$$

Substituting the values for ω_i's and simplifying the expression gives the final result.

Band-elimination filter to normalized low-pass filter specifications: The band elimination filter specifications can be transformed to normalized low-pass filter specifications similar to the band-pass case. These are summarized below.

$$\underline{s} = \frac{Bs}{s^2 + \omega_0^2}, \ j\underline{\omega} = j\frac{B\omega}{\omega_0^2 - \omega^2},$$
$$B = \omega_u - \omega_l, \omega_0^2 = \omega_l\omega_u. \qquad (7.5.14a)$$

$$\omega = \omega_l \Rightarrow \frac{(\omega_u - \omega_l)\omega_l}{\omega_u\omega_l - \omega_l^2} = 1 = \underline{\omega}_c, \quad \omega = \omega_u$$
$$\Rightarrow \frac{(\omega_u - \omega_l)\omega_u}{\omega_u\omega_l - \omega_u^2} = -1 = -\underline{\omega}_c. \qquad (7.5.14b)$$

$$\omega = \omega_1 \Rightarrow \frac{(\omega_u - \omega_l)\omega_1}{\omega_u\omega_l - \omega_1^2} = \underline{\omega}_{r1},$$
$$\omega = \omega_2 \Rightarrow \frac{(\omega_u - \omega_l)\omega_2}{\omega_u\omega_l - \omega_2^2} = -\underline{\omega}_{r2}. \qquad (7.5.14c)$$

$$\underline{\omega}_c = 1, \ \text{and} \ \underline{\omega}_r = \min\{\omega_{r1}, \omega_{r2}\} \qquad (7.5.14d)$$

Tighter constaraint results in (7.5.14d). Figure 7.5.1d illustrates the transformation from the BE filter specifications to the normalized low-pass filter specifications and vice versa.

Example 7.5.6 Use the Butterworth approximation to find an expression for the BE filter specifications in Fig. 7.5.5a with

$$\omega_l = 5 \text{ kHz}, \omega_u = 2\pi(40) \text{ kHz},$$
$$\omega_1 = (2\pi)10 \text{ kHz}, \omega_2 = (2\pi)20 \text{ kHz}. \qquad (7.5.15a)$$

Solution: Using the specifications, the normalized filter attenuation specifications are

$$\underline{\omega}\big|_{\omega=\omega_1} = \frac{(\omega_u - \omega_l)\omega_1}{(\omega_u\omega_l - \omega_1^2)} = \frac{(2\pi)^2(40-5)(10)}{(2\pi)^2(200-100)}$$
$$= 3.5 = \omega_{r1} \qquad (7.5.15b)$$
$$\underline{\omega}\big|_{\omega=\omega_2} = \frac{(\omega_u - \omega_l)\omega_2}{(\omega_u\omega_l - \omega_2^2)} = \frac{(40-5)(20)}{(200-400)}$$
$$= -3.5 = -\omega_{r2}. \qquad (7.5.15c)$$

We should note that the transformation of the stopband edge frequencies gave $\omega_{r1} = |-\omega_{r2}| = 3.5$. This results since $\omega_0^2 = \omega_1\omega_2 = \omega_l\omega_u$ in our example. The corresponding normalized low-pass filter specifications are shown in Fig. 7.5.4b. The order of (n) for the Butterworth low-pass filter with $\varepsilon = 1$ results in

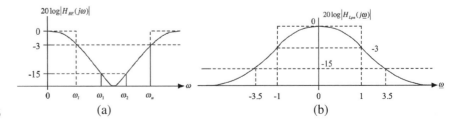

Fig. 7.5.4 Example 7.5.6

$$(n)_{\text{int}} \geq \left[\log(10^{1.5} - 1)/2\log(3.5)\right] \cong 1.366 \Rightarrow n = 2.$$

The normalized low-pass and the corresponding band-elimination filter functions are

$$H_{Lpn}(\underline{s}) = \frac{1}{\underline{s}^2 + \sqrt{2}\underline{s} + 1},$$

$$H_{Be}(s) = H_{Lpn}(\underline{s}) \Big|_{\underline{s} = \frac{(\omega_u - \omega_l)s}{s^2 + \omega_l\omega_u}}. \quad (7.5.16) \quad \blacksquare$$

7.6 Multi-terminal Components

In Chapter 6 we have studied two-terminal components including ideal voltage and current sources, passive components, resistors, inductors, and capacitors. Passive filters require inductors and they are bulky and not pure. There is coupling between adjacent inductors. One alternative to passive filters is active filters with components that have more than two terminals. The basic component in these is the operational amplifier (op amp), a multi-terminal component. Other useful multi-terminal components include transistors, operational amplifiers, controlled, or dependent sources. We are interested in the input–output characteristics of these multi-terminal components.

7.6.1 Two-Port Parameters

The block diagram of a four terminal or a two-port network is shown in Fig. 7.6.1a. Figure 7.61b,c gives two networks, one has only a series element and the other has only a shunt element. There are two voltage variables and two current variables in time identified as $v_1(t), v_2(t), i_1(t)$ and $i_2(t)$ or their transforms $V_1(s), V_2(s), I_1(s)$ and $I_2(s)$. Variable s may not be shown explicitly.

$$v_1(t) \xleftrightarrow{\text{LT}} V_1(s), \ v_2(t) \xleftrightarrow{\text{LT}} V_2(s),$$

$$i_1(t) \xleftrightarrow{\text{LT}} I_1(s), \ \text{and} \ i_2(t) \xleftrightarrow{\text{LT}} I_2(s).$$

Two-port models:

$$\begin{bmatrix} V_1(s) \\ V_2(s) \end{bmatrix} = \begin{bmatrix} Z_{11}(s) & Z_{12}(s) \\ Z_{21}(s) & Z_{22}(s) \end{bmatrix} \begin{bmatrix} I_1(s) \\ I_2(s) \end{bmatrix}, Z_{ij}(s) - \left\{ \begin{array}{l} Z \text{ or impedance or} \\ \text{open circuit parameters} \end{array} \right\}, \quad (7.6.1)$$

$$\begin{bmatrix} I_1(s) \\ I_2(s) \end{bmatrix} = \begin{bmatrix} Y_{11}(s) & Y_{12}(s) \\ Y_{21}(s) & Y_{22}(s) \end{bmatrix} \begin{bmatrix} V_1(s) \\ V_2(s) \end{bmatrix}, Y_{ij}(s) - \left\{ \begin{array}{l} Y \text{ or admittance} \\ \text{or short circuit parameters} \end{array} \right\}, \quad (7.6.2)$$

$$\begin{bmatrix} V_1(s) \\ I_2(s) \end{bmatrix} = \begin{bmatrix} H_{11}(s) & H_{12}(s) \\ H_{21}(s) & H_{22}(s) \end{bmatrix} \begin{bmatrix} I_1(s) \\ V_2(s) \end{bmatrix}, H_{ij}(s) - \text{Hybrid or } H \text{ parameters}, \quad (7.6.3a)$$

$$\begin{bmatrix} I_1(s) \\ V_2(s) \end{bmatrix} = \begin{bmatrix} G_{11}(s) & G_{12}(s) \\ G_{21}(s) & G_{22}(s) \end{bmatrix} \begin{bmatrix} V_1(s) \\ I_2(s) \end{bmatrix}, G_{ij}(s) - \text{Hybrid or } G \text{ parameters}, \quad (7.6.3b)$$

$$\begin{bmatrix} V_1(s) \\ I_1(s) \end{bmatrix} = \begin{bmatrix} A & B \\ C & D \end{bmatrix} \begin{bmatrix} V_2(s) \\ -I_2(s) \end{bmatrix}, ABCD \text{ or transmission or cascade or } F \text{ parameters}. \quad (7.6.4)$$

The term, open (short) circuit parameters, is used in the sense that a parameter is determined by assuming the current (voltage) variable to be zero. For example,

$$Z_{21}(s) = [V_2(s)/I_1(s)]\big|_{I_2(s)=0} \text{ and}$$

$$Y_{21}(s) = [I_2(s)/V_1(s)]\big|_{V_2(s)=0}. \quad (7.6.5)$$

We can obtain one set of parameters from the others *if* the second set of parameters exists. The two-port Z and Y parameters are related by

$$Y_{11}(s) = \frac{Z_{22}(s)}{\Delta Z}, \ Y_{12}(s) = -\frac{Z_{12}(s)}{\Delta Z},$$

$$Y_{21}(s) = -\frac{Z_{21}(s)}{\Delta Z}, \ Y_{22}(s) = \frac{Z_{11}(s)}{\Delta Z} \quad (7.6.6)$$

$$\Delta Z = Z_{11}(s)Z_{22}(s) - Z_{12}(s)Z_{21}(s) \neq 0.$$

Fig. 7.6.1 Two ports: (a) general, (b) with only a series element, and (c) with only a shunt element

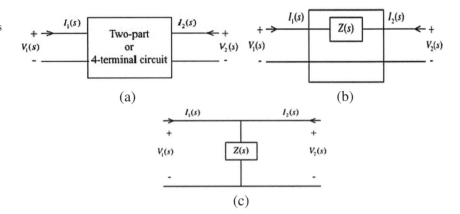

(a)   (b)

(c)

The Y parameters *exist only if* $|\Delta Z| \neq 0$. See Appendix A for a review on matrix algebra. The $ABCD$ parameters are related to the Z parameters of a two port if $Z_{21} \neq 0$ by (proof is left as an exercise).

$$\begin{bmatrix} V_1 \\ I_1 \end{bmatrix} = \begin{bmatrix} A & B \\ C & D \end{bmatrix} \begin{bmatrix} V_2(s) \\ -I_2(s) \end{bmatrix}$$

$$= \begin{bmatrix} Z_{11}/Z_{21} & \Delta Z/Z_{21} \\ 1/Z_{21} & Z_{22}/Z_{21}(s) \end{bmatrix} \begin{bmatrix} V_2 \\ -I_2 \end{bmatrix}. \quad (7.6.7a)$$

The $ABCD$ parameters are related to the Y parameters of a two port by (proof is left as an exercise.)

$$\begin{bmatrix} V_1(s) \\ I_1(s) \end{bmatrix} = \begin{bmatrix} A & B \\ C & D \end{bmatrix} \begin{bmatrix} V_2(s) \\ -I_2(s) \end{bmatrix}$$

$$= \begin{bmatrix} -Y_{22}(s)/Y_{21}(s) & -1/Y_{21}(s) \\ (Y_{12}Y_{21} - Y_{11}Y_{22})/Y_{21} & Y_{22}(s)/Y_{21}(s) \end{bmatrix}$$

$$\begin{bmatrix} V_2(s) \\ -I_2(s) \end{bmatrix}. \quad (7.6.7b)$$

Notes: A *reciprocal two-port* system is *symmetric* if its ports can be interchanged without changing the values of the terminal currents and voltages. The impedance and admittance matrices, i.e., the coefficient matrices in (7.6.1) and (7.6.2) of a reciprocal network are symmetric matrices. In a similar manner we can define a two-terminal reciprocal component. Resistors, inductors, and capacitors are reciprocal, whereas electronic circuits, such as a diode, are not. RLC networks

are *reciprocal*, linear, and time-invariant systems. They satisfy the following relations for a two-port network:

$$Z_{12} = Z_{21}; \quad Y_{12} = Y_{21}; \quad H_{12} = -H_{21},$$
$$G_{12} = -G_{21}; \quad AD - BC = 1. \quad (7.6.8) \quad \blacksquare$$

Example 7.6.1 *a*. Show the circuit in Fig. 7.6.1b cannot be described by Z parameters.
b. Derive its $ABCD$ parameters.

Solution: *a*. Noting that $I_1(s) = -I_2(s)$ and $I_1(s) = (V_1(s) - V_2(s))/Z(s)$, we have $Y_{11}(s) = (1/Z(s))$, $Y_{12}(s) = -1/Z(s)$, $Y_{21}(s) = -1/z(s)$, $Y_{22}(s) = 1/Z(s)$ and

$$\begin{bmatrix} I_1(s) \\ I_2(s) \end{bmatrix} = \begin{bmatrix} (1/Z(s)) & -1/Z(s) \\ -1/Z(s) & 1/Z(s) \end{bmatrix} \begin{bmatrix} V_1(s) \\ V_2(s) \end{bmatrix},$$

$$\Delta = \begin{vmatrix} 1/Z(s) & -1/Z(s) \\ -1/Z(s) & 1/Z(s) \end{vmatrix} = 0.$$

We cannot solve for voltages and the circuit cannot be described by the Z parameters.
b. The $ABCD$ parameters can be derived in terms of the Y parameter by using (7.6.7b) and

$$\begin{bmatrix} V_1(s) \\ I_1(s) \end{bmatrix} = \begin{bmatrix} 1 & Z(s) \\ 0 & 1 \end{bmatrix} \begin{bmatrix} V_2(s) \\ -I_2(s) \end{bmatrix}. \quad (7.6.9a) \quad \blacksquare$$

Example 7.6.2 *a*. Show that the circuit in Fig. 7.6.1c cannot be described Y parameters.
b. Derive its $ABCD$ parameters.

Solution: Using the Kirchhoff's voltage law, we have $V_1(s) = V_2(s)$ and using the Kirchhoff's current law, we have $V_1(s) = Z(s)[I_1(s) + I_2(s)]$. As before we can see the following and short-circuit parameters do not exist since $\Delta = 0$:

$$\begin{bmatrix} V_1(s) \\ V_2(s) \end{bmatrix} = \begin{bmatrix} Z(s) & Z(s) \\ Z(s) & Z(s) \end{bmatrix} \begin{bmatrix} I_1(s) \\ I_2(s) \end{bmatrix},$$

$$\Delta = \begin{vmatrix} Z(s) & Z(s) \\ Z(s) & Z(s) \end{vmatrix} = 0.$$

Using (7.6.7a), we have

$$\begin{bmatrix} V_1(s) \\ I_1(s) \end{bmatrix} = \begin{bmatrix} 1 & 0 \\ Y(s) & 1 \end{bmatrix} \begin{bmatrix} V_2(s) \\ -I_2(s) \end{bmatrix}. \quad (7.6.9\text{b}) \quad \blacksquare$$

Ideal transformer: Its symbol is shown in Fig. 7.6.2. Magnetic coupling provides interaction between the primary and the secondary coils. There is *no* direct electrical connection from the primary to the secondary. The behavior of an ideal transformer with time-varying excitation depends only on the number of turns in the two coils. Let the primary and the secondary coils have N_1 and N_2 turns, respectively, and let $a = N_2/N_1$. The voltages and the currents are related by the equations $v_2(t) = av_1(t), i_2(t) = -i_1(t)/a$. The relative voltage potentials are indicated by the *dot convention* in Fig. 7.6.2. The equations for an ideal transformer can be written in a matrix form in the time or transform domain by

$$\begin{bmatrix} v_1(t) \\ i_2(t) \end{bmatrix} = \begin{bmatrix} 0 & a \\ -a & 0 \end{bmatrix} \begin{bmatrix} i_1(t) \\ v_2(t) \end{bmatrix} \Rightarrow \begin{bmatrix} V_1(s) \\ I_2(s) \end{bmatrix}$$

$$= \begin{bmatrix} 0 & a \\ -a & 0 \end{bmatrix} \begin{bmatrix} I_1(s) \\ V_2(s) \end{bmatrix}. \quad (7.6.10)$$

If $a > 0$ then both dotted terminals have the same reference voltage polarity in Fig. 7.6.2. Ideal transformer cannot be described by either Z or Y parameters. Note the ideal transformer is a *lossless network* since $v_1(t)i_1(t) + v_2(t)i_2(t) = 0$.

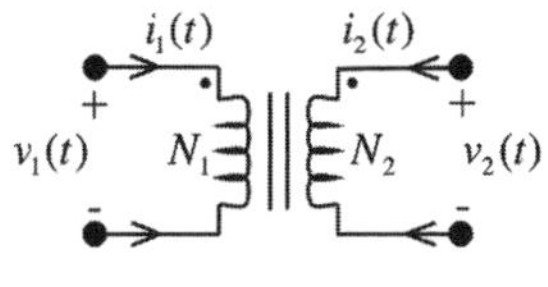

Fig. 7.6.2 Ideal transformer

Example 7.6.3 The *cascade* or the *ABCD* parameters of a two-port network is useful in the analysis of cascaded systems. Derive the matrix expression relating the input–output voltages and currents using Fig. 7.6.3. The voltages and the appropriate currents are given in terms of the *ABCD* parameters (variable s is not shown) as

$$\begin{bmatrix} V_1(s) \\ I_1(s) \end{bmatrix} = \begin{bmatrix} A_1 & B_1 \\ C_1 & D_1 \end{bmatrix} \begin{bmatrix} V_2(s) \\ -I_2(s) \end{bmatrix}, \quad \begin{bmatrix} V_3(s) \\ I_3(s) \end{bmatrix}$$

$$= \begin{bmatrix} A_2 & B_2 \\ C_2 & D_2 \end{bmatrix} \begin{bmatrix} V_4(s) \\ -I_4(s) \end{bmatrix}. \quad (7.6.11)$$

Solution: With $V_2(s) = V_3(s)$ and $I_2(s) = -I_3(s)$, we have

$$\begin{bmatrix} V_1(s) \\ I_1(s) \end{bmatrix} = \begin{bmatrix} A_1 & B_1 \\ C_1 & D_1 \end{bmatrix} \begin{bmatrix} A_2 & B_2 \\ C_2 & D_2 \end{bmatrix} \begin{bmatrix} V_4(s) \\ -I_4(s) \end{bmatrix}$$

$$= \begin{bmatrix} A_1A_2 + B_1C_2 & A_1B_2 + B_1D_2 \\ C_1A_2 + D_1C_2 & C_1B_2 + D_1D_2 \end{bmatrix}$$

$$\begin{bmatrix} V_4(s) \\ -I_4(s) \end{bmatrix} \quad (7.6.12) \quad \blacksquare$$

Example 7.6.4 Consider the circuit in Fig. 7.1.1 with $R_1 = 1\,\Omega$ and the load resistance is assumed to be *infinite* at the output. Derive the transfer function and the element values L_1 and C_1 using the *ABCD* parameters assuming the transfer function given below.

$$\frac{V_2}{V_1}\Big|_{I_2(s)=0} = H_{Bu}(s) = \frac{1}{s^2 + \sqrt{2}s + 1}. \quad (7.6.13\text{a})$$

Solution: Using the the impedance of the series arm is $Z_1(s) = R_1 + L_1 s$ and the admittance of the shunt arm is $Y_2(s) = C_1 s$, we can use the *ABCD* parameters of these from (7.6.9a and b) in (7.6.12) resulting in

$$\begin{bmatrix} V_1(s) \\ I_1(s) \end{bmatrix} = \begin{bmatrix} 1 & Z_1(s) \\ 0 & 1 \end{bmatrix} \begin{bmatrix} 1 & 0 \\ Y_2(s) & 1 \end{bmatrix} \begin{bmatrix} V_2(s) \\ -I_2(s) \end{bmatrix}.$$

$$= \begin{bmatrix} 1 & (R_1 + L_1 s) \\ 0 & 1 \end{bmatrix} \begin{bmatrix} 1 & 0 \\ C_1 s & 1 \end{bmatrix} \begin{bmatrix} V_2(s) \\ -I_2(s) \end{bmatrix}$$

$$= \begin{bmatrix} 1 + (R_1 + L_1 s)C_1 s & (R_1 + L_1 s) \\ C_1 s & 1 \end{bmatrix} \begin{bmatrix} V_2(s) \\ -I_2(s) \end{bmatrix}.$$

$$\Rightarrow H(s) = V_2(s)/V_1(s)|_{I_2=0}$$
$$= 1/[1 + R_1C_1s + L_1C_1s^2]. \quad (7.6.13b)$$

Equating the coefficients in (7.6.13a and b) with $R_1 = 1$, we have $C_1 = \sqrt{2}$ and $L_1 = 1/\sqrt{2}$.

Use of *ABCD* parameters is a simple way to analyze *ladder networks* with series and shunt arms, as in the last example. In these cases, the input–output equations can be written using a product of matrices. In the example, we have assumed a particular circuit, found its transfer function, equated to the given transfer function, and solved for the element values. It illustrated the method *coefficient matching*.

Loading effects on transfer functions: Consider the cascaded system shown in Fig. 7.6.3. If $I_2(s) = 0$, we say the System 2 does not load System 1. System 2 is not effected by any load if $I_4(s) = 0$. These are illustrated by the following example.

the other systems in the cascaded systems. Use of op amps, discussed in the next section, makes this possible and loading effects can be *neglected*. ∎

Doubly terminated two-port networks: Following parameters are of interest for the network shown in Fig. 7.6.4, see Nilsson and Riedel (1996).

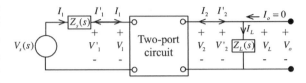

Fig. 7.6.4 Double terminated two-port network

1. Input impedance $Z_i = V_1/I_1$
2. Output impedance $Z_L = V_L/I_L$
3. Current gain I_2/I_1
4. Voltage gains V_2/V_1 and V_2/V_g.

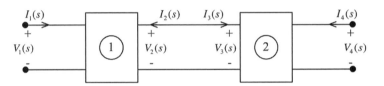

Fig. 7.6.3 Example 7.6.3, Systems 1 and 2 in cascade

Example 7.6.5 Find the transfer function $V_4(s)/V_1(s)$ of the cascaded system in Fig. 7.6.3 for the two cases *a.* $I_4(s) = 0$, $I_2(s) \neq 0$. *b.* $I_2 = 0$, $I_4 = 0$.

Solution: *a.* From (7.6.12),

$$\frac{V_4(s)}{V_1(s)}\Big|_{I_4(s)=0} = \frac{1}{A_1A_2 + B_1C_2}.$$

b. Using (7.6.11) and (7.6.12), with $I_2(s) = I_4(s) = 0$, we have

$$\frac{V_2(s)}{V_1(s)} = \frac{1}{A_1} \text{ and}$$
$$\frac{V_4(s)}{V_3(s)} = \frac{1}{A_2} \Rightarrow \frac{V_4(s)}{V_1(s)}\Big|_{I_2(s)=0,I_4(s)=0} = \left[\frac{1}{A_1}\right]\left[\frac{1}{A_2}\right].$$

Comparing the two cases, in Case *a*, part of the transfer includes the product B_1C_2, which is not there in Case *b*. That is, the transfer function of the cascaded system is equal to the product of the two transfer functions. This allows for ease in the design of individual systems without considering

Now consider the derivation of some of the other models by using the equations in (7.6.1) relating the indicated variables in Fig. 7.6.4.

$$V_1(s) = Z_{11}(s)I_1(s) + Z_{12}(s)I_2(s) \text{ and}$$
$$V_2(s) = Z_{21}(s)I_1(s) + Z_{22}(s)I_2(s). \quad (7.6.14a)$$
$$V_1(s) = V_i(s) - Z_i(s)I_1(s). \quad (7.6.14b)$$
$$V_L(s) = V_2(s) = -Z_L(s)I_L(s). \quad (7.6.14c)$$

In the following, the *L*-transform variable *s* will be dropped for simplicity. In terms of the *ABCD* parameters of the two-port network, we have

$$\begin{bmatrix} V_s \\ I_s \end{bmatrix} = \begin{bmatrix} 1 & Z_s \\ 0 & 1 \end{bmatrix}\begin{bmatrix} V'_1 \\ -I'_1 \end{bmatrix},$$
$$\begin{bmatrix} V_1 \\ I_1 \end{bmatrix} = \begin{bmatrix} A & B \\ C & D \end{bmatrix}\begin{bmatrix} V_2 \\ -I_2 \end{bmatrix},$$
$$\begin{bmatrix} V_3 \\ I_3 \end{bmatrix} = \begin{bmatrix} 1 & 0 \\ 1/Z_L & 1 \end{bmatrix}\begin{bmatrix} V_0 \\ -I_0 \end{bmatrix}.$$

Noting that $V'_1 = V_1, I'_1 = -I_1, V_2 = V'_2, I'_2 = -I_L$, and $V'_2 = V_0$, we have

$$\begin{bmatrix} V_s \\ I_s \end{bmatrix} = \begin{bmatrix} 1 & Z_s \\ 0 & 1 \end{bmatrix} \begin{bmatrix} A & B \\ C & D \end{bmatrix} \begin{bmatrix} 1 & 0 \\ 1/Z_L & 1 \end{bmatrix} \begin{bmatrix} V_0 \\ -I_0 \end{bmatrix}$$

$$= \begin{bmatrix} 1 & Z_s \\ 0 & 1 \end{bmatrix} \begin{bmatrix} A+(B/Z_L) & B \\ C+(D/Z_L) & D \end{bmatrix} \begin{bmatrix} V_0 \\ -I_0 \end{bmatrix}$$

$$= \begin{bmatrix} A+Z_sC+(B/Z_L)+(DZ_s/Z_L) & B+DZ_s \\ C+(D/Z_L) & D \end{bmatrix} \begin{bmatrix} V_0 \\ -I_0 \end{bmatrix}.$$

$$\Longrightarrow \frac{V_0}{V_i}\Big|_{I_0=0} = \frac{1}{A+Z_sC+(B/Z_L)+(DZ_s/Z_L)}.$$

Substituting the values for the $ABCD$ parameters in terms of the open circuit parameters given in (7.6.7), we have the following result, which can be derived by solving directly

$$\frac{V_0}{V_s} = \frac{Z_{21}Z_L}{Z_{11}Z_L + Z_sZ_L + Z_{11}Z_{22} - Z_{12}Z_{21} + Z_{22}Z_s}$$

$$= \frac{Z_{21}Z_L}{(Z_{11}+Z_s)(Z_{22}+Z_L) - Z_{12}Z_{21}}. \quad (7.6.15a)$$

In addition to this we have other important relations. Some of these are given below:

$$\frac{V_1}{I_1} = Z_{11} - \frac{Z_{12}Z_{21}}{Z_{22}+Z_L}, \quad (7.6.15b)$$

$$\frac{I_L}{V_s} = \frac{Z_{21}}{(Z_{11}+Z_s)(Z_{22}+Z_L) - Z_{12}Z_{21}}, \quad (7.6.15c)$$

$$V_2|_{I_2=0} = \frac{Z_{21}}{Z_{11}+Z_s} V_i, \quad (7.6.15d)$$

$$\frac{V_2}{I_2}\Big|_{V_i=0} = Z_{22} - \frac{Z_{12}Z_{21}}{Z_{11}+Z_s}, \quad (7.6.15e)$$

$$\frac{I_2}{I_1} = \frac{-Z_{21}}{Z_{22}+Z_L}, \quad (7.6.15f)$$

$$\frac{V_2}{V_1} = \frac{Z_{21}Z_L}{Z_LZ_{11} + Z_{11}Z_{22} - Z_{12}Z_{21}}. \quad (7.6.15\,g)$$

Other parameters of interest include *scattering parameters* (see Seshu and Balabanian, 1959) originated in the study of transmission lines. These cannot be obtained by rearranging the voltages and currents in describing the two-port circuits. However, they are as good as other parameters and are indirectly specified using the relationships between external voltages and currents in terms of incident and reflected voltages at each of the ports. *Image parameters* grew out of the study of wave propagation.

7.6.2 Circuit Analysis Involving Multi-terminal Components and Networks

In circuit analysis we write a set of equations with respect to a reference point or a node. An N node network has $(N-1)$ *independent* equations and we write a set of equations with respect to a reference point or a node. If we desire to come up with $(N-1)$ equations with a difference reference node, node equations make it very simple.

Example 7.6.6 Figure 7.6.5 shows a three-terminal network and its description of its Y parameters in terms of $Y_{ij}^{(3)}(s)$ in (7.6.16) indicating node 3 is the reference node.

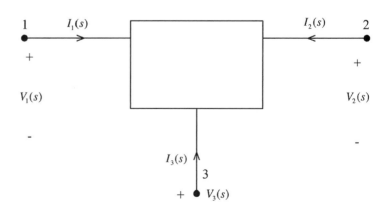

Fig. 7.6.5 A three-terminal node network

$$\begin{bmatrix} I_1(s) \\ I_2(s) \end{bmatrix} = \begin{bmatrix} Y_{11}^{(3)}(s) & Y_{12}^{(3)}(s) \\ Y_{21}^{(3)}(s) & Y_{22}^{(3)}(s) \end{bmatrix} \begin{bmatrix} V_1(s) \\ V_2(s) \end{bmatrix}. \qquad (7.6.16)$$

Derive the two-port short-circuit parameters assuming node 2 is grounded.

Solution: Assuming the network is a *super node*, we have $I_1(s) + I_2(s) + I_3(s) = 0$ and

$$\begin{bmatrix} I_1(s) \\ I_2(s) \\ I_3(s) \end{bmatrix} = \begin{bmatrix} Y_{11}^{(3)}(s) & Y_{12}^{(3)}(s) & -[Y_{11}^{(3)}(s) + Y_{12}^{(3)}(s)] \\ Y_{21}^{(3)}(s) & Y_{22}^{(3)}(s) & -[Y_{21}^{(3)}(s) + Y_{22}^{(3)}(s)] \\ -[Y_{11}^{(3)}(s) + Y_{21}^{(3)}(s)] & -[Y_{12}^{(3)}(s) + Y_{22}^{(3)}(s)] & [Y_{11}^{(3)} + Y_{12}^{(3)} + Y_{21}^{(3)} + Y_{22}^{(3)}] \end{bmatrix} \begin{bmatrix} V_1(s) \\ V_2(s) \\ V_3(s) \end{bmatrix}. \qquad (7.6.17)$$

The coefficient matrix in (7.6.17) is a *singular indefinite admittance matrix* of the three-terminal network by noting that the three rows (or the three columns) add to a row of zeros (or columns) and therefore the determinant of the coefficient matrix is zero.

Equation (7.6.17) reduces to (7.6.16) by deleting the third row and considering the node 3 as the reference node, i.e., $V_3(s) = 0$. If we change the reference node from 3 to 2, we can obtain the short-circuit parameter equations with reference to node 2 by

$$\begin{bmatrix} I_1(s) \\ I_3(s) \end{bmatrix} = \begin{bmatrix} Y_{11}^{(3)}(s) & -[Y_{11}^{(3)}(s) + Y_{12}^{(3)}(s)] \\ -[Y_{11}^{(3)}(s) + Y_{21}^{(3)}(s)] & [Y_{11}^{(3)}(s) + Y_{12}^{(3)}(s) + Y_{21}^{(3)}(s) + Y_{22}^{(3)}(s)] \end{bmatrix} \begin{bmatrix} V_1(s) \\ V_3(s) \end{bmatrix}$$

$$= \begin{bmatrix} Y_{11}^{(2)}(s) & Y_{12}^{(2)}(s) \\ Y_{21}^{(2)}(s) & Y_{22}^{(2)}(s) \end{bmatrix} \begin{bmatrix} V_1(s) \\ V_3(s) \end{bmatrix}. \qquad (7.6.18) \quad \blacksquare$$

Problem (7.6.3) makes use of this procedure.

7.6.3 Controlled Sources

Controlled source is a unidirectional, non-autonomous active two port having a pair of input terminals and a pair of output terminals, one controlled by the other. The unidirectional property of these sources provides the important property that the controlling terminal-pair variables are insensitive (or independent) of the controlled terminal-pair variables. These idealized two ports are usually referred to as *controlled sources* or *transducers* illustrated in Fig. 7.6.6a,b,c,d. They are referred to as transducers or *dependent sources*. There are two variables, voltage and current, at the input and at the output. We need to consider four cases of controlled-source models. Note circles are used for both dependent and independent sources.

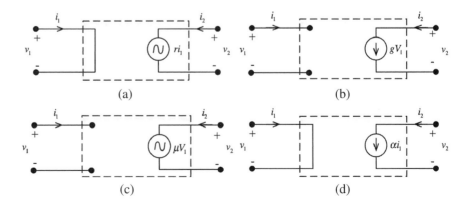

Fig. 7.6.6 Controlled sources (a) CVT, (b) VCT, (c) VVT, and (d) CCT

Current-to-voltage transducer (CVT): Output voltage is proportional to the input current.

Voltage-to-current transducer (VCT): Output current is proportional to the input voltage.

Voltage-to-voltage transducer (VVT): Output voltage is proportional to the input voltage.

Current-to-current transducer (CCT): Output current is proportional to the input current.

$$\begin{bmatrix} v_1 \\ v_2 \end{bmatrix} = \begin{bmatrix} 0 & 0 \\ r & 0 \end{bmatrix} \begin{bmatrix} i_1 \\ i_2 \end{bmatrix} (\text{CVT}) \qquad (7.6.19\text{a})$$

$$\begin{bmatrix} i_1 \\ i_2 \end{bmatrix} = \begin{bmatrix} 0 & 0 \\ g & 0 \end{bmatrix} \begin{bmatrix} v_1 \\ v_2 \end{bmatrix} (\text{VCT}) \qquad (7.6.19\text{b})$$

$$\begin{bmatrix} i_1 \\ v_2 \end{bmatrix} = \begin{bmatrix} 0 & 0 \\ \mu & 0 \end{bmatrix} \begin{bmatrix} v_1 \\ i_2 \end{bmatrix} (\text{VVT}) \qquad (7.6.19\text{c})$$

$$\begin{bmatrix} v_1 \\ i_2 \end{bmatrix} = \begin{bmatrix} 0 & 0 \\ \alpha & 0 \end{bmatrix} \begin{bmatrix} i_1 \\ v_2 \end{bmatrix} (\text{CCT}) \qquad (7.6.19\text{d})$$

7.7 Active Filter Circuits

7.7.1 Operational Amplifiers, an Introduction

Operational amplifiers (or **op amps** or **op-amps**) appeared in the market during the 1940s. With integrated circuits, they are the most used components in circuit design. For a review on op amps, see Mitra (1969), Van Valkenburg (1982), Daryanani (1976), and others. Since op amps are available in packaged forms, the analysis of circuits involving operational amplifiers is presented in terms of op amp idealized form. Figure 7.7.1 shows its standard symbol and the equivalent circuit symbolic models. Op amps provide high-gain amplification of the difference between two input voltages v^+ and v^-. The requirements for an ideal operational amplifier are

$$v_0 = A(v^+ - v^-) = -Av_i,$$
$$v_0 \longrightarrow 0 \text{ when } v_i = v^- - v^+ \longrightarrow 0. \qquad (7.7.1)$$

The two inputs v^- and v^+ are at the inverting and non-inverting terminals, respectively, and v_0 is the resulting output signal. When both terminals are used, the op amp is called a *differential input op*

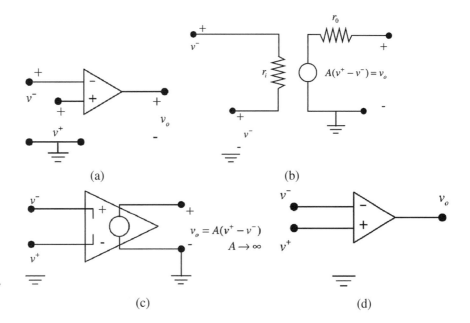

Fig. 7.7.1 (a) Op amp symbol, (b) op amp model, (c) ideal op amp, and (d) simplified symbol

amp. The ideal op amp acts like a voltage-to-voltage transducer (VVT) or simply a controlled voltage source described by the relation in (7.6.24c). To be general, the gain of the op amp is a function of frequency. For our filter applications, frequencies are below 10 kHz. In this range, the gains are over 10,000. The input impedance of an op amp is around 500 k Ω and the output impedance is around 300 Ω. For our analysis we will assume that the op amp is an ideal component and is assumed to have an infinite gain and small output impedance. For a finite output signal $v_0(t)$, we must have $v^+ \approx v^-$. This property is referred to as the *virtual ground*. The input impedance is assumed to be infinite and therefore the input currents to the ideal op amp are zero. Also, the output impedance of the ideal op amp is zero. These allow for the active filter synthesis of a transfer function using cascaded sections. To be realistic, we know that all voltages and currents must be finite. In addition, the op amp is used as part of a feedback loop and not used in an open loop. The VVT model of the op amp is shown in Fig. 7.7.1c. Figure 7.7.1d shows the simplified symbol of the op amp.

7.7.2 Inverting Operational Amplifier Circuits

Figure 7.7.2 illustrates a single-loop feedback circuit, where the non-inverting input terminal of the op amp is grounded. Noting $I_1(s) = -I_2(s)$, we have the following:

$$I_1(s) = \frac{V_i(s)}{Z_1(s)} \simeq -\frac{V_0(s)}{Z_2(s)} = -I_2(s), \qquad (7.7.2)$$

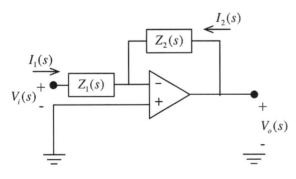

Fig. 7.7.2 Single-loop feeback cicuit with an op amp

$$\frac{V_0(s)}{V_i(s)} = H(s) = -\frac{Z_2(s)}{Z_1(s)}$$

(inverting circuit transfer function)   (7.7.3)

Example 7.7.1 Find the output voltage in terms of the input voltage for each of the cases in Fig. 7.7.3a,b,c,d,e by assuming the op amps as ideal devices.

Solution: *a.* The current into the ideal op amp is zero implies $I_3(s) = -(I_1(s) + I_2(s))$ and

$$V_0(s) = Z_F(s) I_3(s) = -\left[\frac{V_1(s)}{Z_1(s)} + \frac{V_2(s)}{Z_2(s)}\right] Z_F(s)$$

$$= -\left[\frac{Z_F(s)}{Z_1(s)}\right] V_1(s)$$

$$- \left[\frac{Z_F(s)}{Z_2(s)}\right] V_2(s). \qquad (7.7.4)$$

If $Z_i(s) = R, i = 1, 2$ and $Z_F(s) = R$,   then $V_0(s) = -(V_1(s) + V_2(s))$.

b. The voltage across the input terminals of the op amp is approximately zero and follows that $V_0(s) \approx V_i(s)$, a special case of a non-inverting amplifier or a *voltage follower* and can be used as a buffer between a high-resistance source and a low-resistance load.

c. The node equation at point a in Fig. 7.7.3c is $(V_i(s)/R_a) + (V_0(s)/R_b) = 0$ and the gain is an inverting gain and is an *inverting amplifier*.

d. The node equation at point *a* in Fig. 7.7.3d

$$\left[\frac{1}{R_1} + \frac{1}{R_2}\right] V_1(s) - \frac{1}{R_2} V_2(s) = 0 \Longrightarrow V_2(s)$$

$$= [1 + (R_2/R_1)] V_1(s) = K V_1(s),$$

$$K = 1 + (R_2/R_1) \geq 1. \qquad (7.7.5)$$

Unity gain, i.e., $K = 1$ if $R_1 \to \infty$ or $R_2 = 0$. It is a non-inverting amplifier. It is symbolically represented in Fig. 7.7.3d and $V_0(s) = K V_i(s)$.

e. The circuit in Fig. 7.7.3e gives the difference of two voltages $V_0(s) = V_1(s) - V_2(s)$. Verification is left as an exercise.   ∎

Inverting first-order low-pass and high-pass RC filter circuits:

Example 7.7.2 Consider the two RC circuits shown in Fig. 7.7.4a,b. Derive the transfer functions and

Fig. 7.7.3 Example 7.7.1

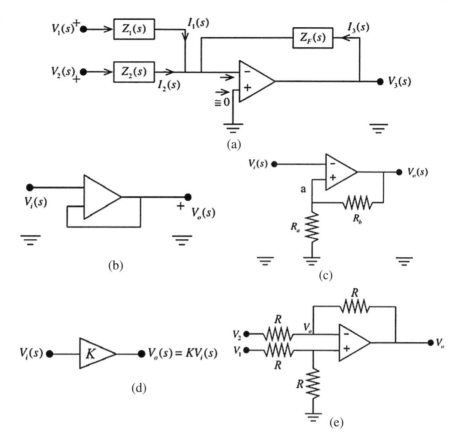

(a)

(b)

(c)

(d)

(e)

identify the type of filter by considering these at $s = 0$ and ∞.

Solution: With (7.7.3), the following results:

$$H_a(s) = \frac{V_0(s)}{V_i(s)} = -\frac{(1/R_1C)}{s + (1/R_2C)}$$

$$= -\frac{R_2}{R_1} \frac{(1/R_2C)}{s + (1/R_2C)}$$

$$= K_a \frac{(1/R_2C)}{s + (1/R_2C)} \text{(low pass)} \qquad (7.7.6)$$

$$H_b(s) = \frac{V_0(s)}{V_i(s)} = -\frac{R_2}{R_1 + (1/Cs)}$$

$$= -\frac{R_2}{R_1} \frac{s}{s + (1/R_1C)}$$

$$= K_b \frac{s}{s + (1/R_1C)} \text{(high pass)}. \qquad (7.7.7)$$

The low-frequency gain in the LP case is $K_a = -(R_2/R_1)$ and the high-frequency gain in the HP case is $K_b = -R_2/R_1$. The value of R_1 in both cases should include the input resistance of the source.

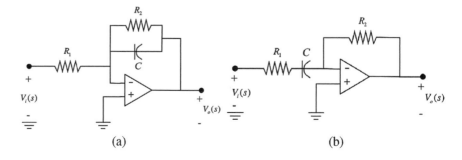

Fig. 7.7.4 (a) First-order low-pass filter and (b) first-order high-pass filter

(a)

(b)

The inverting amplifier draws input current in these circuits and a voltage follower can be inserted as a buffer between the source and the filter. ∎

7.7.3 Non-inverting Operational Amplifier Circuits

The circuit shown in Fig. 7.7.5a is a generalization of the circuit used in Example 7.7.1 (see Part c.) with R_i replaced by $Z_i(s)$, $i = 1, 2$. The transfer function is

$$H(s) = V_0(s)/V_i(s) = 1 + (Z_2(s)/Z_1(s)). \quad (7.7.8)$$

It is a non-inverting op amp filter that synthesizes a transfer function with a real zero and a pole. The transfer functions $K/(s + p_1)$, K/s and Ks are not realizable by this circuit.

Example 7.7.3 Find an active RC circuit that has the transfer function

$$H(s) = (s + a)/(s + b)$$
$$= 1 + [(a - b)/(s + b)], a > b. \quad (7.7.9)$$

Solution: The circuit in Fig. 7.7.5b has this transfer function with

$Z_1(s) = 1$ and

$$Z_2(s) = \frac{(a - b)}{(s + b)} = \frac{1}{[s/(a - b)] + [b/(a - b)]}. \quad ∎$$

Example 7.7.4 Derive the transfer functions of the circuits in Fig. 7.7.6ab.

Solution: The input currents into the op amp are zero and from Fig. 7.7.6a, we have

$$V_1(s) = [R_A/(R_A + R_B)]V_0(s) \text{ or}$$
$$V_0(s) = [1 + R_B/R_A]V_1(s). \quad (7.7.10)$$

$$\frac{V_i(s) - V_1(s)}{R} = CsV_1(s) \Rightarrow H_a(s)$$
$$= \frac{V_0(s)}{V_i(s)} = \frac{K/CR}{(s + (1/CR))}, K = 1 + \frac{R_B}{R_A}. \quad (7.7.11)$$

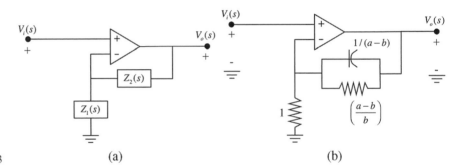

Fig. 7.7.5 Example 7.7.3   (a)   (b)

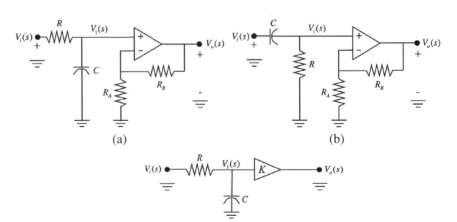

Fig. 7.7.6 Non-inverting op amp circuits: (a) low-pass circuit, (b) high-pass circuit, and (c) simplified low-pass circuit

Figure 7.7.6c gives the simplified version of the circuit in Fig. 7.7.6b, where the resistors R_A and R_B are not explicit. The circuit is a low-pass filter. The circuit in Fig. 7.7.6c is a high-pass filter. The transfer function is

$$H_b(s) = \frac{V_0(s)}{V_i(s)} = \frac{Ks}{(s + (1/CR))},$$
$$K = 1 + \frac{R_B}{R_A}. \qquad (7.7.12) \quad \blacksquare$$

First-order all-pass functions: The circuit transfer functions in Fig. 7.7.7a,b are

$$H_a(s) = \frac{V_0(s)}{V_i(s)} = \frac{s - (1/RC)}{s + (1/RC)},$$
$$H_b(s) = \frac{V_0(s)}{V_i(s)} = -\frac{1}{2}\frac{s - (1/RC)}{s + (1/RC)}. \qquad (7.7.13)$$

A nice way to find the transfer function is by noting that the output voltage V_0 is the sum of two voltages V_{01} and V_{02}. The voltage V_{01} is determined using the circuit in Fig. 7.7.7c. Similarly, the voltage V_{02} is determined by using the circuit in Fig. 7.7.7d. Using Fig. 7.7.7c and d, we have

$$V_{01} = -\frac{R}{R}V_i, \; V_{02} = \frac{R}{R + (1/Cs)}\left[1 + \frac{R}{R}\right]$$
$$V_i = \frac{2R}{R + (1/Cs)}V_i \qquad (7.7.14a)$$

$$V_0 = V_{01} + V_{02} = -V_i + \frac{2R}{R + (1/Cs)}V_i$$
$$= \frac{(s - 1/RC)}{(s + 1/RC)}V_i. \qquad (7.7.14b)$$

Equation (7.7.13) now follows:

Notes: Table 7.7.1 gives a set of guidelines for selecting passive component values. See Van Valkenburg (1982) for additional information. The input and feedback resistances connected to an op

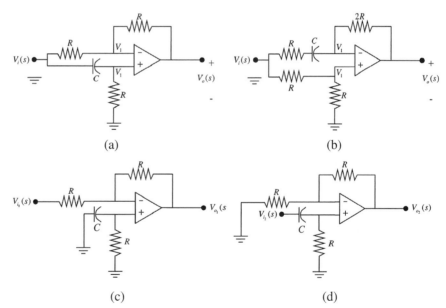

Fig. 7.7.7 (a) and (b) First-order all-pass circuits. (c) and (d) Illustration of the principle of superposition by using these circuits

Table 7.7.1 Guidelines for passive components

Capacitors			Inductors			Resistors	
	Largest	Smallest		Largest	Smallest	Preferred	$1-100\,\text{k}\Omega$
Readily realizable	$1\,\mu\text{F}$	$5\,\text{pF}$	Readily realizable	$1\,\text{mH}$	$1\,\mu\text{H}$	Lower limit	$.1-1\,\text{k}\Omega$
Practical	$10\,\mu\text{F}$	$.2\,\text{pF}$	Practical	$10\,\text{mH}$	$.1\,\mu\text{H}$	Upper limit	$100-500\,\text{k}\Omega$

amp should be within the range of 1 kΩ to 100 kΩ. The cut-off frequency or the center frequency f_c of an active RC filter is determined by capacitors and resistors. Capacitors are more expensive than resistors and are selected first. A *rule of thumb* is to select the capacitor values in the range of (see Carlson, 2000)

$$10^{-6}/f_c < C < 10^{-4}/f_c. \qquad (7.7.15)$$

7.7.4 Simple Second-Order Low-Pass and All-Pass Circuits

Example 7.7.5 Derive the transfer function of the circuit in Fig. 7.7.8a. It is one class of popular low-pass active filter circuits given by Sallen and Key (1955).

Solution: Considering the voltage division at node 3 and the Kirchhoff's current law,

Node 1:

$$V_1(s) = V_3(s), \ V_1(s) = \frac{R_A}{R_A + R_B} V_0(s) \longrightarrow \frac{V_0(s)}{V_1(s)} = 1 + \frac{R_B}{R_A}, \ V_3(s) = \frac{V_0(s)}{K} \qquad (7.7.16a)$$

Node 2:

$$\frac{1}{R_2}\left[V_2(s) - \frac{V_0(s)}{K}\right] + C_1 s[V_2(s) - V_0(s)]$$
$$+ \frac{1}{R_1}[V_2(s) - V_i(s)]$$
$$= 0 \qquad (7.7.16b)$$

Node 3:

$$\frac{1}{R_2}\left[\frac{V_0(s)}{K} - V_2(s)\right] + \frac{V_0(s)}{K} C_2 s = 0. \qquad (7.7.16c)$$

Eliminating $V_2(s)$ results in

$$H(s) = \frac{V_0(s)}{V_i(s)} = \frac{K(1/R_1 R_2 C_1 C_2)}{s^2 + [(1/R_1 C_1) + (1/R_2 C_1) + (1/R_2 C_2) - (K/R_2 C_2)]s + (1/R_1 R_2 C_1 C_2)}, K \geq 1. \quad (7.7.17)$$

$$= K\omega_0^2/[s^2 + (\omega_0/Q)s + \omega_0^2],$$
$$\omega_0^2 = 1/R_1 R_2 C_1 C_2. \qquad (7.7.18)$$

The design strategy is *coefficient matching* by equating the coefficients in (7.7.17) with (7.7.18). That is, determine K and the four element values from the given parameters ω_0 and Q. As an example, consider $R_1 = R_2 = 1$ and $C_1 = C_2 = 1$ and substitute these values in (7.7.17) resulting in

$$H(s) = \frac{K}{[s^2 + (3 - K)s + 1]}. \qquad (7.7.19)$$

The designer has the choice of selecting $K = 1 + (R_B/R_A)$ depending only on the ratio of the resistors R_B and R_A. From (7.7.18), it follows that

$$Q = 1/(3 - K). \qquad (7.7.20)$$

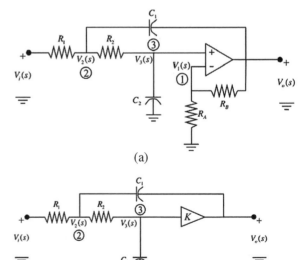

Fig. 7.7.8 (a) Sallen–Key low-pass filter and (b) simplified circuit

The gain constant K is adjusted to satisfy the Q value and the design requirements of the circuit. Van Valkenburg (1982) gives several strategies in selecting the values of the parameters and its effect on the pole locations. A special case is when $K = 1$.   ∎

Notes: The resistors R_A and R_B are not shown in the circuit in Fig. 7.7.8b. This simplifies the analysis, as $V_0(s) = KV_3(s)$. There are three constants $R_B, R_A,$ and K, one of them depends on the other

two constants. A solution is $R_B = R_A(K - 1)$. The gain constant is positive and it is a circuit with a *positive feedback topology*. A second-order low-pass circuit that has a negative gain constant, i.e., with a *negative feedback topology* is given in Problem 7.7.10.   ∎

The transfer function of the second-order delay circuit shown in Fig. 7.7.9 (see Delyiannis, 1969) is given by (analysis is left as an exercise)

$$\frac{V_0(s)}{V_i(s)} = H(s) = K\frac{s^2 + [(2/R_2C) + (1 - (1/K))/R_1C]s + (1/R_1R_2C^2)}{s^2 + (2/R_2C)s + (1/R_1R_2C^2)}, K = \frac{R_4}{R_3 + R_4}. \qquad (7.7.21)$$

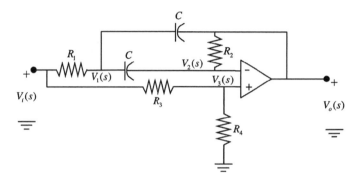

Fig. 7.7.9 A second-order delay line circuit

The *condition* for *the all-pass* and the corresponding second-order all-pass function are

$$\frac{2}{R_2C} + \frac{(1 - (1/K))}{R_1C} = -\frac{2}{R_2C},$$

$$H(s) = K\frac{s^2 - \omega_0/Q + \omega_0^2}{s^2 + \omega_0/Q + \omega_0^2}. \qquad (7.7.22)$$

See Van Valkenburg (1982) for selecting the component values, see also Problem 7.7.6.

7.8 Gain Constant Adjustment

The filter circuit provides a gain, which may be different than the required. The gain adjustment (attenuation and gain enhancement) can be achieved by first using *attenuation* circuit at the *input end* and later, *enhancement* circuit at the *output end,* see Daryanani (1976) and Van Valkenburg (1982).

Consider the block diagram shown in Fig. 7.8.1a corresponding to a given transfer function $H(s)$. The input to the active circuit is through the impedance Z_1. We would like to keep the impedance Z_1 the same as before and change the input voltage to the active RC circuit attenuated by α. That is, our goal is to modify the input end of the circuit so that the transfer function is given by $\alpha H(s), \alpha < 1$. This is achieved by the circuit in Fig. 7.8.1b. The input and the voltage divider can be replaced by a Thevenin's voltage source $V_{Th}(s)$ and the resultant Thevenin's impedance Z_{Th}. These are

$$V_{Th}(s) = \frac{Z_{12}}{Z_{11} + Z_{12}} V_i(s) = \alpha V_i(s),$$

$$\alpha = \frac{Z_{12}}{Z_{11} + Z_{12}} < 1. \qquad (7.8.1a)$$

$$Z_{Th} = Z_{11}Z_{12}/(Z_{11} + Z_{12}) = Z_1. \qquad (7.8.1b)$$

$$\Rightarrow Z_{11} = Z_1/\alpha \text{ and } Z_{12} = [\alpha/(1 - \alpha)]$$
$$Z_{11} = Z_1/(1 - \alpha). \qquad (7.8.1c)$$

Fig. 7.8.1 (a) Active RC
circuit with impedance Z_1,
(b) circuit with input
attenuation, and (c) circuit
with Thevenin's input and
Thevenin's impedance

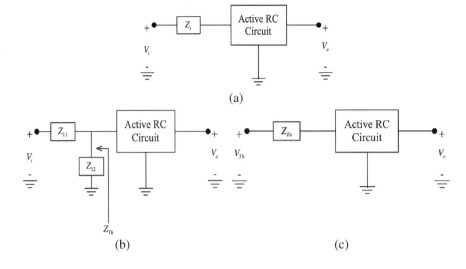

The equations are valid for impedances and, of course as a special case, for resistances. Figure 7.8.1c gives the block diagram with the Thevenin's impedance explicitly shown. The transform variable s is omitted in the above equations. In the case of *resistive gain enhancement*, the inverse of the earlier approach is used at the *output end*. Our goal is to modify the given circuit with the transfer function $H(s)$ to $\beta H(s)$, $\beta > 1$.

Figure 7.8.2a,b gives the Sallen–Key low-pass and the gain attenuation circuits. Figure 7.8.2c gives the resistive gain enhancement circuit. The derivation of the transfer function of the circuit in (7.8.2b) is left as an exercise. In this circuit, a portion of the output voltage V_0, kV_2 is fed back through the capacitor C_1 with $k = r_{12}/(r_{11} + r_{12})$. The transfer function is

$$H(s) = \frac{K(1/R_1 R_2 C_1 C_2)}{s^2 + [1/R_1 C_1 + 1/R_2 C_1 + 1/R_2 C_2 - Kk/R_2 C_2]s + 1/R_1 R_2 C_1 C_2}. \tag{7.8.2a}$$

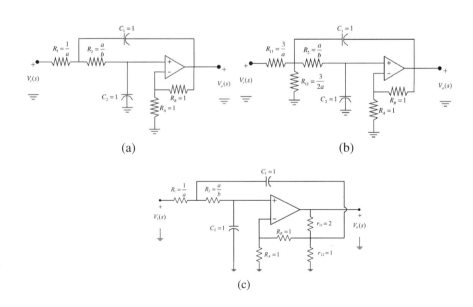

Fig. 7.8.2 Example 7.8.1,
Gain attenuation and
enhancement

As an example, let $C_1 = C_2 = C$ and $R_1 = R_2 = R$. Using (7.7.18), the Q of the circuit is

$$Q = 1/(3 - kK). \qquad (7.8.2b)$$

To keep the same Q, K can be made larger provided k is made smaller at the same time. This is called the *resistive gain adjustment*. Note that the enhancer circuit consists of two resistors r_{11} and r_{12} at the output end. Lower case symbols are used to separate from the attenuation case. If the current through the capacitor C_1 is small enough, then the corresponding gain enhancement factor is

$$\beta = [r_{11} + r_{12}]/r_{12}. \qquad (7.8.2c)$$

Daryanani (1976) discusses the limitations on the gain enhancement, which depends on the maximum current the op amp can deliver.

Example 7.8.1 *a*. Use the circuit in Example 7.7.5 to synthesize the transfer function

$$H_1(s) = 2b/[s^2 + as + b]. \qquad (7.8.3)$$

b. Use the attenuation method to synthesize the transfer function $H_1(s)/3$.

c. Use the gain enhancement method to synthesize the transfer function $3H_1(s)$.

Solution: *a*. The element values corresponding to the transfer function in (7.8.3) can be obtained by comparing it to the equation in (7.7.17) with $K = 2$. A solution is

$$C_1 = C_2 = 1F, R_A = R_B = 1\Omega, R_1 = (1/a)\Omega,$$
$$R_2 = (a/b)\Omega. \qquad (7.8.4)$$

The corresponding circuit is shown in Fig. 7.8.2a. Computations are left as an exercise. *b*. We may be tempted to recompute the element values for this case. This will *not* be possible since the new gain constant $K = 2/3 < 1$. The attenuation here is $\alpha = 1/3$. Using the above procedure, i.e., replace the source and the resistor R_1 in Fig. 7.8.2a by the circuit shown in Fig. 7.8.2b. Using (7.8.1c) and $\alpha = 1/3$, we have $R_{11} = R_1/\alpha = (3/a)$ and the shunt resistance is $R_{12} = R_{11}/(1 - \alpha) = 3R_1/2 = 3/2\alpha$. We now have two new resistors and the other values remain the same.

c. In this case, we have $\beta = 3 > 1$ and we are interested in gain enhancement. The transfer function is given below. The new circuit is shown in Fig. 7.8.2c.

$$H_2(s) = 3H_1(s) = \frac{6b}{[s^2 + as + b]}. \qquad (7.8.5)$$

The resistors r_{11} and r_{12} at the *output satisfy* $\beta \cong (r_{11} + r_{12})/r_{12} = 3 > 1$. In this case we could have gone back and recompute and come up with a new set of element values with $K = 6 > 1$. Instead, the example asks for a modification of the original circuit in Part *a*. to illustrate the gain enhancement. Select $r_{11} = 2, r_{12} = 1$. ∎

7.9 Scaling

In Chapter 6 we have used circuits with element values with small numbers, such as $1\Omega, 1H$, and $1F$. Even though these are unrealistic and impractical, they simplify the computations. Once the designs are made with small values, the designer can transform these into realistic values using *scaling*. It is used in everyday life. Maps are drawn using scaling, such as 1 in. represents 1 mile. Frequency scaling that may correspond to 1 rad represents ω_a radians. In Chapter 4 frequency compression and expansion were considered. For example, if $v(t)$ is replaced by $v(10t)$, then v is compressed in time by a factor of 10. Similarly, if $v(t)$ is replaced by $v(t/10)$, then v is expanded in time by 10. In a similar manner, if $V(j\omega)$ is replaced by $V(j10\omega)$ (or $V(j\omega/10)$), then V has been compressed (or expanded) by a factor of 10 in frequency.

7.9.1 Amplitude (or Magnitude) Scaling, RLC Circuits

From our earlier discussion on circuits the impedance of a two-terminal network is a ratio of input voltage transform $V(j\omega)$ to the input current transform $I(j\omega)$ defined by

$$Z(j\omega) = V(j\omega)/I(j\omega). \qquad (7.9.1)$$

If the magnitude of the impedance $|Z(j\omega)|$ is replaced by $k_m Z(j\omega)$, then we say that the impedance is scaled up or down depending upon $k_m > 1$ or $k_m < 1$. Consider the three passive components with the element values $R_{old}\,\Omega$, $L_{old}\,$H, and $C_{old}\,$F. The impedances of these components and the corresponding magnitudes of these impedances are given by

$$Z_R = R_{old}, Z_L = j\omega L_{old}, \text{ and } Z_c = 1/j\omega C_{old}$$
$$\Rightarrow Z_R = R_{old}, |Z_L| = |\omega| L_{old}, |Z_C| = 1/|\omega| C_{old}.$$
$$(7.9.2)$$

Since the component values are positive and multiplying them by k_m results in

$$k_m Z_R = k_m R_{old}, \quad k_m |Z_L| = \omega(k_m L_{old}),$$
$$k_m |Z_C| = 1/[\omega(C_{old}/k_m)]. \quad (7.9.3a)$$

The new elements with subscripts "new" corresponding to the scaled values are

$$R_{new} = k_m R_{old}, \quad L_{new} = k_m L_{old}, \quad \text{and}$$
$$C_{new} = C_{old}/k_m. \quad (7.9.3b)$$

If the elements are changed according to these, the input impedance of a circuit is scaled in magnitude by k_m. Using the subscripts old and new, the impedances are related by

$$Z_{new} = k_m Z_{old}. \quad (7.9.3c)$$

7.9.2 Frequency Scaling, RLC Circuits

We would like to scale the frequency without altering the magnitude scaling discussed above. Since the resistor value is independent of the frequency, it is unchanged by any frequency scaling. The magnitudes of the impedance of the inductor and the capacitor are preserved by scaling the frequency and, at the same time, scaling the element values. That is,

$$|Z_L| = \omega L_{old} = (k_f\omega)(L_{old}/k_f) = (k_f\omega)L_{new}, \quad (7.9.4a)$$

$$|Z_c| = \frac{1}{\omega C_{old}} = \frac{1}{(k_f\omega)((1/k_f)C_{old})]}$$
$$= \frac{1}{(k_f\omega)(C/k_f)}. \quad (7.9.4b)$$

These two equations point out that if we increase the frequency by k_f, then we must reduce the inductor and the capacitor values by the factor k_f to keep the two impedances invariant. Correspondingly the new element values after frequency scaling are given by

$$L_{new} = (1/k_f)L_{old}, \quad C_{new} = (1/k_f)C_{old}. \quad (7.9.5)$$

Magnitude and frequency scaling, RLC circuits: Simultaneous magnitude and frequency scaling results in the following *scaling equations*:

$$R_{new} = k_m R_{old}, \quad L_{new} = (k_m/k_f)L_{old},$$
$$C_{new} = (1/k_m k_f)C_{old}. \quad (7.9.6)$$

7.9.3 Amplitude and Frequency Scaling in Active Filters

The active RC filter design is based on starting with a circuit, finding its transfer function, and matching the circuit transfer function coefficients with that of desired transfer function coefficients. This provides more unknowns than equations allowing for multiple solutions. In our examples, the element values are chosen so that the element values come out to be easy to handle, but not practical. Amplitude scaling is used so that the designed values are practical and is illustrated by the following example.

Example 7.9.1 The following transfer function needs to be synthesized by the circuit used in Example 7.7.5 with some practical element values:

$$H(s) = \frac{Kb}{s^2 + as + b} = \frac{10,000}{s^2 + 100s + 5,000}, \quad K = 2. \quad (7.9.7a)$$

a. Give the appropriate equations in terms of the element values in the active RC network and the parameters given in (7.9.7a).

b. Assuming $a = 100$ and $b = 5000$, determine a set of element values that can be used in synthesizing the filter.

Solution: The transfer function is

$$H(s) = \frac{K(1/R_1R_2C_1C_2)}{s^2 + [(1/R_1C_1) + (1/R_2C_1) + (1/R_2C_2) - (K/R_2C_2)]s + (1/R_1R_2C_1C_2)}. \tag{7.9.7b}$$

Comparing the coefficients in (7.9.7a) and (7.9.7b), we have

$$H_0 = K(b),$$

$$a = \frac{1}{R_2C_2}(1 - K) + \frac{1}{R_2C_1} + \frac{1}{R_1C_1},$$

$$b = \frac{1}{R_1R_2C_1C_2}, \quad K = 1 + \frac{R_b}{R_a}. \tag{7.9.7c}$$

Noting that $K = 2$, it follows that $R_A = R_B$. There are an infinite number of possible solutions. Selecting the capacitor values equal to 1, we have a simple solution given by

$$R_B = R_A = 1\,\Omega, C_1 = C_2 = 1F, R_1 = 1/a, R_2 = a/b.$$

$$\Rightarrow R_1 = 0.01\,\Omega, R_2 = 100/5000 = 0.02\,\Omega.$$

Clearly, the element values are not practical. One *property* of *active* RC filters is that the resistors $R_i's$ and the capacitors $C_j's$ with *number subscripts* always *appear* as R_iC_j products. For example, see (7.9.7b). The resistors with *letter subscripts* appear as a *ratio of resistances* in the transfer function. The components used in the gain attenuation and enhancement appear as ratios. These allow for *amplitude scaling* in the active filter design. If we increase all the $R_i's$ by a factor α and, at the same time decrease all the capacitor values by this factor, then the products will remain the same and the transfer function is unchanged. As an example, consider $\alpha = 10^6$, i.e., $R_i \Rightarrow R_i(10^6)$ and $C_i \Rightarrow C_i(10^{-6})$ then the corresponding practical values are

$$C_1 = C_2 = 1\mu F, R_1 = (.01)10^6 = 10\,k\Omega,$$

$$R_2 = .02(10^6) = 20\,k\Omega, \quad R_B = R_A = 10\,k\Omega. \tag{7.9.8}$$

The two resistors R_A and R_B appear as a ratio and can be scaled independently with respect to the other elements and one choice is given in (7.9.8). ∎

Frequency scaling is used to shift the frequency response from one part of the frequency axis to another part allowing for the filter designs at a normalized frequency. For a given set of specifications, look up tables can be generated for a particular type of filter with a cut-off frequency equal to 1 rad/s. In the low-pass filter design we have seen that if a filter that satisfies all the design criteria corresponding to a cut-off frequency equal to 1 rad/s, then the transfer function corresponding to a cut-off frequency of ω_c rad/s can be obtained by replacing s by s/ω_c in the normalized function. Suppose the specifications require that the transfer function given above is to be shifted up along the frequency axis by a factor of 10, then the corresponding transfer function is

$$H(s/\omega_c) = H(s/10)7 = 10,000/(s/10)^2$$
$$+ 100(s/10) + 5,000. \tag{7.9.9}$$

From (7.9.7c),

$$H(s/10) = \frac{K(1/R_1R_2C_1C_2)}{(s/10)^2 + [(1/R_1C_1) + (1/R_2C_1) + (1/R_2C_2) - (K/R_2C_2)](s/10) + (1/R_1R_2C_1C_2)}$$

$$= \frac{K(10)(10)/R_1R_2C_1C_2}{s^2 + [(10/R_1C_1) + (10/R_2C_1) + (10/R_2C_2) - (10\,K/R_2C_2)]s + (1/R_1R_2C_1C_2)(10)(10)}.$$

Frequency scaling is accomplished *by decreasing all the resistors by a factor of 10* or *by decreasing all the capacitors by 10, not both.* By decreasing all the capacitor values by a factor of 10, the corresponding element values can be computed from (7.9.8), we have the new element values for the frequency-scaled network:

$$C_1 = C_2 = .1\,\mu\,F, \quad R_1 = 1\,k\Omega,$$
$$R_2 = 2\,k\Omega, \quad R_B = R_A = 10\,k\Omega \tag{7.9.10}$$

Note that R_A and R_B are not altered. We can see now that amplitude and frequency scaling can be used to synthesize active filter circuits that give practical element values in the circuit.

Notes: Amplitude scaling effects the impedances and admittances. Voltage transfer functions are not affected by amplitude scaling. The transfer function of the frequency-scaled network is obtained by replacing s by s/k_f in the unscaled network. ∎

7.9.4 Delay Scaling

In Chapter 4 we have seen that time and frequency goes hand in hand. Consider

$$x(t) = \sin(\omega_0 t) = \sin(t/(1/\omega_0)) = \sin(t/t_0),$$
$$t_0 = 1/\omega_0. \qquad (7.9.11)$$

The inverse relationship between time and frequency given in the last part of the above equation, $\omega_0 = 1/t_0$ indicates that time compression implies frequency expansion and vice versa. We further have seen that the time–bandwidth product of a system is a constant. Time scaling in a system can be interpreted as speeding up or slowing down the response of a system. How do we incorporate time scaling in the description of the system? The transfer function and the output of a system with a pure delay of t_0 s are given by

$$H(j\omega) = e^{-j\omega t_0} = e^{-j(\omega/\omega_0)(1)}, \ \omega_0 = 1/t_0. \quad (7.9.12a)$$

$$y(t) = x(t - t_0) = \sin(\omega(t - t_0)) = \sin[(\omega/t_0)(t - 1)]. \qquad (7.9.12b)$$

That is, scaling the frequency by t_0 corresponds to the delay of 1 s for the frequency-scaled system. We can make use of element-scaling equations given in (7.9.6a, b, and c) in the delay scaling and write

$$R_{\text{new}} = k_m R_{\text{old}}, \quad L_{\text{new}} = (k_m k_D) L_{\text{old}},$$
$$C_{\text{new}} = (k_D/k_m) C_{\text{old}}, \quad k_D = (1/t_0) \qquad (7.9.13)$$

Note the similarity of (7.9.6) to these equations.

Notes: Although filter designs are based on the frequency domain, they are also evaluated on their transient responses. They are important in controls systems. Optimal filter functions generally have complex poles on the left half of the s-plane close to the imaginary axis. Systems are evaluated on the basis of a pair of complex poles of its transfer function near the imaginary axis and the corresponding step response. The step response of such a sytem can be expressed in the transform domain of the form

$$C(s) = \frac{1}{s} \frac{\omega_0^2}{s^2 + 2\xi\omega_0 s + \omega_0^2}. \qquad (7.9.14)$$

The main interest in the transient response is for $0 < \xi < 1$. The corresponding time response is

$$c(t) = 1 - \frac{e^{-\xi\omega_0 t}}{\sqrt{(1 - \xi^2}} \sin\left(\omega_d t + \tan^{-1}\frac{\sqrt{1 - \xi^2}}{\xi}\right) u(t),$$
$$\omega_d = \omega_0 \sqrt{1 - \xi^2}.$$

The time response starts at 0 and steadily rises to the peak value of $\max(c(t)) > 1$ and then it oscillates around its final value 1. Several specifications are used in evaluating the transient response (see Ogata, 2004). Some of the related ones to our study are introduced below.

The delay time t_d is the time the response takes to reach half the value the very first time. The rise time t_r is the time required for the response to rise from 10 to 90%. See Section 6.6 for a simple example using an RC circuit. Peak time is the time required for the response to reach the first peak of the overshoot. The maximum overshoot $M_p = \max(c(t))$ is the maximum peak value. Section 6.6 illustrates the time response of a simple RC circuit and the rise time is proportional to the time constant RC (see (6.6.4)). In the second-order case, the rise time is inversely proportional to ω_d.

Filters with poles near the imaginary axis produce more ringing with a shorter rise time than the systems with poles away from the imaginary axis. For example, Chebyshev filter responses have a larger overshoots and a shorter rise time compared to Butterworth filters. These have a smaller rise time compared to Bessel filter responses. Bessel filter responses have minimal overshoots and are popular for one of the reasons. ∎

Example 7.9.2 The Sallen–Key circuit shown Fig. 7.7.8 is used in the design of delay line filters.

Fig. 7.9.1 Example 7.9.2

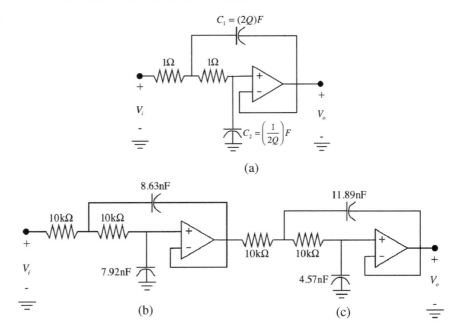

(a)

(b) (c)

a. Find its transfer function using (7.7.18) by assuming $K = 1$ and $\omega_0^2 = 1$. Use the scaling method to obtain a set of realistic component values.

b. Determine the cascaded circuit using Example 7.4.4 with practical component values resulting in a delay circuit with a delay of $D = .25\,\mu$ s.

Solution: *a.* With $R_1 = 1, R_2 = 1, C_1 = 2Q, C_2 = 1/2Q$ and $K = 1$, (7.7.17) reduces to

$$H(s) = 1/[s^2 + (1/Q)s + 1]. \qquad (7.9.15a)$$

This function can be realized using the circuit shown in Fig. 7.9.1a. Note the gain is equal to 1. The analysis is left as an exercise.

b. Example 7.4.4 considered the transfer functions using Bessel polynomial of order 4. The transfer function is given by (see (7.4.18) for the Bessel polynomial)

$$H(s) = \frac{9.14013}{(s^2 + 5.792425s + 9.14013)} \frac{11.4878}{(s^2 + 4.207585s + 11.4878)}. \qquad (7.9.15b)$$

Using the standard expressions in terms of ω_{0i} and Q_i, we have

$$H(s) = H_1(s)H_2(s), H_i(s) = \frac{\omega_{0i}^2}{s^2 + (\omega_{0i}/Q_i)s + \omega_{0i}^2}$$

$$= \frac{1}{(s/\omega_{0i})^2 + (1/Q_i)(s/\omega_{0i}) + 1}. \quad \omega_{01} = 3.023,$$

$$Q_1 = .522, \omega_{02} = 3.389, Q_2 = .806. \qquad (7.9.15c)$$

For the delay filter, $K = 1$. There are two parameters for each section, ω_{0i} and Q_i, $i = 1, 2$ and element values can be solved from these. There are an infinite number of solutions. A solution is

Stage 1: $R_1 = R_2 = 1, C_1 = 1.044, C_2 = .958$,
$$k_{f1} = 3.023, k_{m1} = 10^4$$
Stage 2: $R_1 = R_2 = 1, C_1 = 1.612, C_2 = .620$,
$$k_{f2} = 3.389, k_{m2} = 10^4$$

Frequency scaling can be used for each section by replacing each capacitor C_j by C_j/k_{fi}, where k_{fi} is the frequency scale factor. To have practical element values, magnitude scaling can be used by a factor k_{mi}. That is, replace each resistor R_j by $R_j k_{mi}$. Then scale the element values to meet the delay specification of $D = .25(10^{-6})$ s. The new capacitor and the resistor values are $C_{new} =$

$[D/k_m k_f] C_{old}$ and $R_{new} = k_m R_{old}$. Final element values are given by the following:

Stage 1: $R_1 = R_2 = 10^4\,\Omega, C_1 = (1.044)\dfrac{.25(10^{-6})}{(3.023)(10^4)}$

$$= 8.63\text{nF}, C_2 = (.958)\dfrac{.25(10^{-6})}{3.023(10^4)} = 7.92\text{nF}$$

Stage 2: $R_1 = R_2 = 10^4\,\Omega,$

$$C_1 = (1.612)\dfrac{.25(10^{-6})}{(3.389)(10^4)} = 11.89\text{nF},$$

$$C_2 = (.620)\dfrac{.25(10^{-6})}{(3.389)(10^4)} = 4.57\text{nF}$$

The circuit is shown in Fig. 7.9.1b. The two cascaded sections are arranged arbitrarily. For a brief discussion on the arrangement, see Section 7.11. ∎

Frequency transformations were used in Section 7.5 to transform a transfer function from a normalized low pass to other types.

7.10 RC–CR Transformations: Low-Pass to High-Pass Circuits

A normalized low-pass filter with a cut-off frequency $\underline{\omega} = 1$ can be transformed into a normalized high-pass filter with a cut-off frequency $\omega = 1$ by using the relationship:

$$\omega = 1/\underline{\omega}. \qquad (7.10.1)$$

We can generalize this and write it in the s-domain assuming the cut-off frequency is equal to ω_c rather than 1 by modifying the active RC low-pass filter by replacing

$$R_i(\text{resistor}) \xrightarrow{\text{replaced by}} C_i(\text{capacitor}) = (1/R_i)$$
$$C_j(\text{capacitor}) \xrightarrow{\text{replaced by}} R_j(\text{resistor}) = 1/C_j$$
$$(7.10.2)$$

These replacements are made with respect to the circuit in Fig. (7.7.8a) and *not* with the transfer function, such as the one in (7.7.17). The gain constants K, R_A and R_B are *not* affected by this transformation.

Example 7.10.1 Use the RC– CR transformation to determine the second-order high-pass filter transfer function from the Sallen–Key low-pass filter circuit given in Example 7.7.4.

Solution: First using the correspondence in (7.10.2), the high-pass circuit derived from the low-pass circuit in Fig. 7.7.8 is shown in Fig. 7.10.1a. The resistors R_A and R_B need not be shown explicitly and write the transfer function in terms of $K = 1 + (R_B/R_A)$ (see Fig. 7.10.1b). The transfer function is

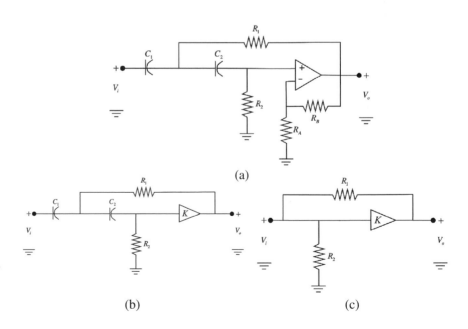

(a)

Fig. 7.10.1 (a) Sallen–Key high-pass circuit, (b) simplified version, and (c) circuit at $\omega = \infty$

(b)   (c)

$$H(s) = \frac{V_0(s)}{V_i(s)} = \frac{Ks^2}{s^2 + [(1/R_2C_1) + (1/R_2C_2) + (1-K)(1/R_1C_1)]s + (1/R_1R_2C_1C_2)}, K = 1 + \frac{R_B}{R_A}. \quad (7.10.3)$$

It is good to verify the transfer function at $s = 0$ and ∞. At dc, the capacitors are open and no signal will go through and the output is zero. At $\omega = \infty$, the capacitors act as shorts and the corresponding circuit is shown in Fig. 7.10.1c. A circuit that realizes a negative gain constant uses a negative feedback topology (see Problem 7.7.5). ∎

Example 7.10.2 Consider the normalized second-order low-pass Butterworth function

$$H_{Lp}(s) = \frac{1}{s^2 + \sqrt{2}s + 1}$$

$$= \frac{K\omega_0^2}{s^2 + (\omega_0/Q)s + \omega_0^2}. \quad (7.10.4a)$$

a. Determine a Sallen and Key low-pass circuit corresponding to this function.

b. Use the $RC \rightarrow CR$ transformation to determine the corresponding high-pass circuit.

Solution: a. Noting $K = 1, Q = 1/\sqrt{2}$ and $\omega_0 = 1$ and selecting $R_1 = R_2 = 1$, the low-pass function is

$$H_{Lp}(s) = \frac{1}{s^2 + (1/Q)s + 1} = \frac{1/C_1C_2}{s^2 + (2/C_1)s + 1/C_1C_2}. \quad (7.10.4b)$$

$$\Rightarrow C_1 = 2Q = \sqrt{(2)}, C_2 = 1/2Q = 1/\sqrt{(2)}.$$

The corresponding low-pass circuit is shown in Fig. 7.10.2a.

b. Using the transformations in (7.10.2), the element values for the high-pass circuit are

$$R_1 = 1/2Q = 1/\sqrt{2}, R_2 = 2Q = \sqrt{2},$$
$$C_1 = 1, \text{ and } C_2 = 1.$$

We now have the high-pass transfer function given below and the corresponding high-pass Sallen and Key circuit is shown in Fig. 7.10.2b.

$$H_{Hp}(s) = \frac{s^2}{(s^2 + (1/Q)s + 1)}. \quad (7.10.5) \quad ∎$$

One of the main drawbacks of passive filters is the use of inductors. Tellegen (1948) devised a frequency-independent electronic two port called a *gyrator* that can be used to replace an inductor by a gyrator and a capacitor (see Problem 7.10.4). The symbolic representation of a gyrator is shown in Fig. 7.10.3 and is characterized by (7.10.6).

$$V_1 = -ri_2, \quad V_2 = ri_1 \quad (7.10.6)$$

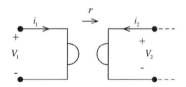

Fig. 7.10.3 Gyrator

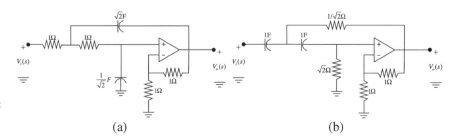

Fig. 7.10.2 Example 7.10.2:
(a) low-pass circuit and
(b) high-pass circuit

7.11 Band-Pass, Band-Elimination and Biquad Filters

For the design of second-order band-pass filter circuits, see Delyiannis (1968). Earlier we have considered a normalized low-pass to high-pass, to band-pass, and to band-elimination filter transformations on the network function. For passive filters, network transformations exist to achieve these (see Fig. 7.5.3). There is no such network transformation that exists to modify an active low-pass filter to a band-pass filter (Mitra, 1969). We can consider a circuit with a topology that has the band-pass characteristic. Problem 7.7.6 uses an alternate way to design a band-pass filter by cascading a low-pass filter and a high-pass filter with proper isolation.

Example 7.11.1 *a*. Derive the transfer function (TF) of the circuit in Fig. 7.11.1.

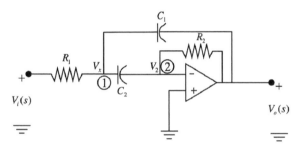

Fig. 7.11.1 Example 7.11.1

b. Assume $C_1 = C_2 = C$ and $R_1 = 1$. Reduce this to the TF below by selecting R_2 and C.

$$H(s) = \frac{-(1/Q)s}{[s^2 + (1/Q)s + 1]}. \quad (7.11.1a)$$

Solution: Kirchhoff's current laws at nodes 1 and 2 result in

$$(V_2 - V_x)C_2s + (V_2 - V_0)/R_2 = 0,$$
$$V_2 = 0 \rightarrow V_0 = -R_2C_2sV_x, \quad (7.11.1b)$$

$$(1/R_1)(V_x - V_i) + C_2sV_x + (V_x - V_0)C_1s = 0. \quad (7.11.1c)$$

Note that the equation $V_2 = -R_2C_2sV_x$ simply corresponds to the inverting op amp circuit studied earlier and could have been written by inspection. Substituting V_x from (7.11.1b) into (7.11.1c) and solving for the transfer function results in

$$H(s) = \frac{V_0}{V_i} = \frac{(-1/R_1C_1)s}{s^2 + (1/R_2C_1 + 1/R_2C_2)s + 1/R_1R_2C_1C_2}. \quad (7.11.1d)$$

Comparing (7.11.1d) with (7.11.1a) and equating the coefficients, we have

$$\omega_0 = 1/C\sqrt{R_1R_2}, \quad Q = (1/2)\sqrt{R_2/R_1},$$
$$\text{bandwidth} = \omega_0/Q = 2/R_2C. \quad (7.11.1e)$$

$$\Rightarrow C_1 = C_2 = C, \text{ and } R_1 = 1,$$
$$R_2 = 4Q^2, \quad C = 1/2Q. \qquad \blacksquare$$

Example 7.11.2 Derive the transfer function and the component values for the *twin-T notch* circuit in Fig. 7.11.2 (see Nilsson and Riedel (1966)).

Solution: Using the Kirchhoff's current law the the three node equations are as follows:

Node 1: $[V_1(s) - V_i(s)]Cs + [V_1(s) - V_0(s)]Cs$
$$+ (2/R)[V_1(s) - KV_0(s)] = 0 \quad (7.11.2a)$$

Node 2: $(1/R)[V_2(s) - V_i(s)] + 2Cs[V_2(s) - KV_0(s)]$
$$+ (1/R)[V_2(s) - V_0(s)] = 0 \quad (7.11.2b)$$

Node 3: $Cs[V_0(s) - V_1(s)] + (1/R)[V_0(s) - V_2(s)] = 0$
$$(7.11.2c)$$

$$\begin{bmatrix} RCsV_i(s) \\ V_i(s) \\ 0 \end{bmatrix} = \begin{bmatrix} 2(RCs+1) & 0 & -(RCs+2K) \\ 0 & 2(RCs+1) & -(2KRCs+1) \\ -RCs & -1 & (RCs+1) \end{bmatrix} \begin{bmatrix} V_1(s) \\ V_2[s] \\ V_0(s) \end{bmatrix} \quad (7.11.3)$$

Fig. 7.11.2 Twin-T notch circuit

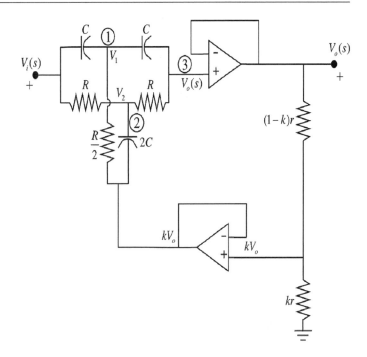

Using Cramer's rule (see Section A.3.2) we have

$$\Delta = \begin{vmatrix} 2(RCs+1) & 0 & -(RCs+2K) \\ 0 & 2(RCs+1) & -(2KRCs+1) \\ -RCs & -1 & (RCs+1) \end{vmatrix},$$

$$\Delta_3 = V_i(s) \begin{vmatrix} 2(RCs+1) & 0 & RCs \\ 0 & 2(RCs+1) & 1 \\ -RCs & -1 & 0 \end{vmatrix}$$

$$\frac{V_0(s)}{V_i(s)} = \frac{\Delta_3}{\Delta} = \frac{s^2 + (1/RC)^2}{s^2 + [4(1-K)/RC]s + (1/RC)^2}$$

$$= \frac{s^2 + \omega_0^2}{s^2 + Bs + \omega_0^2}, B = \frac{\omega_0}{Q}. \qquad (7.11.4)$$

$$\Rightarrow \omega_0^2 = 1/(R^2C^2), \quad B = 4(1-K)/RC. \quad (7.11.5)$$

There are two equations with three unknowns $(R, C, \text{ and } K)$. One of the parameters, usually the capacitors, can be selected, as these are fewer commercially available components compared to resistors. Then, solve for R and K in terms of C, B, and Q (see (7.11.6)) as follows:

$R = (1/\omega_0 C)$ and $K = [1-(B/4\omega_0)] = [1-(1/4Q)]$.
$B = \omega_0/Q$. \qquad (7.11.6) ∎

Example 7.11.3 Find the *Bainter* (1975) circuit transfer function in Fig. 7.11.3a.

Solution: The circuit has two feedback paths and can be separated as shown in Fig. 7.11.3b,c, see Van Valkenburg (1982). Now use the *principle of superposition*.

$$V_2 = -(1/R_4C_1s)V_0' - (1/R_3C_1s)V_1. \quad (7.11.7a)$$

$$(V_0' - V_i')C_2s + (V_0/R_6) + [(V_0 - V_2)/R_5)] = 0. \qquad (7.11.7b)$$

In addition, we have from the first op amp $V_1 = -(R_2/R_1)V_i$. Using the circuit, we can see that $V_0' = V_0$ and $V_i' = V_i$ and

$$V_2 = -(1/R_4C_1s) + (1/R_3C_1s)(R_2/R_1)V_i, \qquad (7.11.8a)$$

$$V_2 = [R_5C_2s + (R_5/R_6) + 1]V_0 - R_5C_2sV_i. \qquad (7.11.8b)$$

Fig. 7.11.3 (a) Bainter (1975) circuit and (b) and (c) feedback paths of the circuit

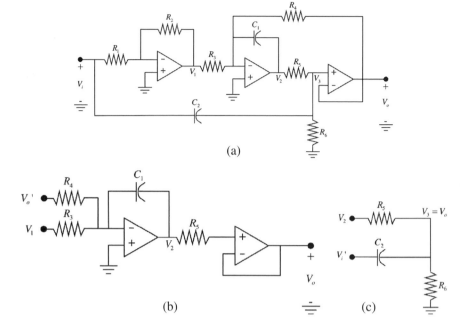

Upon simplifying, the transfer function is

$$\frac{V_0}{V_i} = \frac{s^2 + [R_2/R_1R_3R_5C_1C_2]}{s^2 + [(R_5 + R_6)/R_5R_6C_2]s + 1/R_4R_5C_1C_2}$$

$$= \frac{(s^2 + \omega_z^2)}{s^2 + (\omega_0/Q)s + \omega_0^2}. \tag{7.11.9}$$

Note that $H(0) = \omega_z^2/\omega_0^2$, $H(j\omega_z) = 0$ and $H(j\infty) = 1$. ∎

Example 7.11.4 Synthesize the function by an active filter circuit.

$$\frac{V_0}{V_i} = H(s) = \frac{-H}{s^2 + (1/Q)s + 1} \Rightarrow \left[\frac{(s^2 + (1/Q)s + 1)}{s(s + (1/Q))} \right] V_0$$

$$= -\frac{H}{s(s + (1/Q))} V_i. \tag{7.11.10a}$$

$$V_0 = \left[\frac{-1}{s + (1/Q)} V_0 + \frac{-H}{s + (1/Q)} V_i \right] \left(-\frac{1}{s} \right)(-1). \tag{7.11.10b}$$

Solution: With (7.11.10b), we can synthesize the transfer function by using the following steps and the three corresponding circuits shown in Fig. 7.11.4:

1. Inverting circuit of gain (–1)
2. Inverting amplifier with a transfer function $-(1/s)$
3. Circuit that can generate a single real pole, a lossy integrator, and a circuit that produces a sum of voltages. ∎

Transfer functions of the three circuits in Fig. 7.11.4 with a proper feedback results in a second-order Lp function. Figure 7.11.5 gives such a circuit.

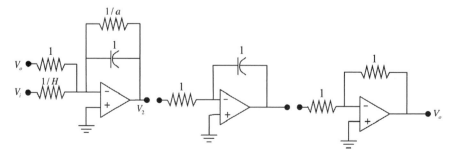

Fig. 7.11.4 Individual parts of the Tow-Thomas biquad circuit

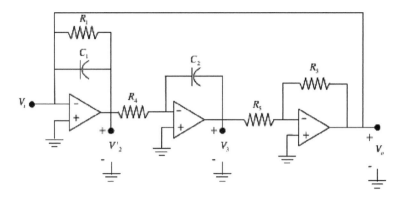

Fig. 7.11.5 Tow–Thomas biquad circuit

This circuit's transfer function can be determined by noting the output voltage $V_0 = V_1$ and using the three separate transfer functions in (7.11.4) with arbitrary element values (see Problem 7.11.2.).

$$\frac{V_0}{V_i} = -\frac{1/R_3 R_4 C_1 C_2}{s^2 + (1/R_1 C_1)s + 1/R_2 R_4 C_1 C_2}. \quad (7.11.11a)$$

If the output V_2 is taken, then the corresponding transfer function has the BP filter form

$$\frac{V_2}{V_i} = \frac{-(1/R_3 C_1)s}{s^2 + (1/R_1 C_1)s + 1/R_2 R_4 C_1 C_2}. \quad (7.11.11b)$$

Tapping at different locations provides different transfer functions. See the rich history on analog computers (Scott, 1960). This circuit is referred to as a *Tow–Thomas biquad* or simply a biquad circuit, see Tow (1968) and Thomas (1971).

Synthesis of the general *biquadratic* function of the form below is useful.

$$H(s) = \frac{ms^2 + cs + d}{ns^2 + as + b}, m = 1 \text{ or } 0 \text{ and}$$

$$n = 1 \text{ or zero.} \quad (7.11.12)$$

See Daryanani (1976) and Deliyannis et al. (1999). The same topology can be used in generating all the filter types. The transfer function of the Tow–Thomas three op amp biquad (see Fig. 7.11.6) is given by (see Tow (1968) and Thomas (1971).). Deliyannis et al. (1999) provide a summary of the results and are given below.

$$H(s) = \frac{V_0(s)}{V_i(s)} = \frac{(C_1/C)s^2 + (1/RC)[(R/R_1) - (r/R_3)]s + (1/C^2 RR_2)}{s^2 + (1/CR_4)s + (1/C^2 R^2)}. \quad (7.11.13)$$

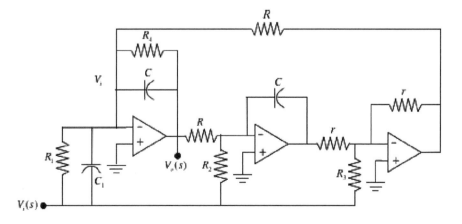

Fig. 7.11.6 The Tow–Thomas three op amp biquad

This *biquad* can be used to *synthesize different second-order filter functions* by an appropriate choice of the component values. That is,

Low pass : $C_1 = 0, R_1 = R_3 = \infty$ (7.11.14a)

High pass : $C_1 = C, R_1 = R_2 = R_3 = \infty$ (7.11.14b)

Band pass : $C_1 = 0, \ R_1 = R_2 = \infty$ (positive sign)
(7.11.14c)

Band pass : $C_1 = 0, R_2 = R_3 = \infty$ (negative sign)
(7.11.14d)

Band elimination : $C_1 = C, R_1 = R_3 = \infty$ (7.11.14e)

All pass : $C_1 = C, R_1 = \infty, r = R_3/Q$ (7.11.14f)

Cascade realization: In terms of biquadratic functions, we can write

$$V_0(s) = H(s)V_i(s), \ H(s) = \prod_{i=1}^{M} H_i(s),$$

$$H_i(s) = K_i \frac{m_i s^2 + c_i s + d_i}{n_i s^2 + a_i s + b_i}. \quad (7.11.15)$$

Each of the quadratic functions can be synthesized and cascaded as shown in Fig. 7.11.7. One of the terms in (7.11.15) may be a first-order function. The sections being connected in cascade are *isolated* in the sense that each successive circuit does not load the previous circuit. Since the ideal op amp is assumed to have infinite input impedance and zero output impedance, active filter realizations lend themselves for good implementations in terms of cascaded networks. If we use the cascade method to derive the filter structure in terms of second structures, the following steps are used in deciding the sequence of the second-order structures:

1. Pole–zero pairing
2. Cascade sequencing, and
3. Gain distribution.

1. How to assign poles and zeros for each $H_i(s)$ among many possibilities is of interest. The idea is

$V_i(s) \longrightarrow \boxed{H_1(s)} \longrightarrow \boxed{H_2(s)} \vdash \cdots \rightarrow \boxed{H_N(s)} \longmapsto V_o(s)$

Fig. 7.11.7 $H(s) = \prod_{i=1}^{N} H_i(s)$

select a pole–zero pair such that $\max |H_i(j\omega)| - \min |H_i(j\omega)|$ is as small as possible in the filter passband. Lueder (1970) presented a nice algorithm on this topic.

2. Having a maximum dynamic range is good. This is achieved by selecting the neighboring biquads having the frequencies of maxima as far apart as possible. For example, if there are three biquads that have a low-pass, band-pass, and high-pass characteristics, the order should be low-pass first, band-pass next, and the high-pass last. Halfin (1970) presented an efficient algorithm in this case.

3. Distribute the overall gain of the filter among the stages, so that the product of all the gains is equal to the overall gain. Sedra and Bracket (1978) use the idea of distributing the overall filter gain so that the maximum voltage at the output of each stage is the same. The goal of any active filter designer is to maximize the dynamic range, minimize the signal-to-noise ratio due to coefficient quantization, and transmission sensitivity minimization. For a survey, see Deliyannis et al. (1999).

7.12 Sensitivities

The transfer function of a filter is a function of its element values. They are available only within $\pm$ few percent of the desired. The components change due to aging, environment, etc. These may result in a different filter performance than it is designed for. In an extreme case, the poles of the filter transfer function may be shifted to the right half-plane, resulting in instability. Sensitivity measures can be used to reduce these effects. Section 7.1 illustrated two simple examples. Design of the active filters using the sensitivity measures is beyond our scope. Discussion is limited to the analysis here.

Active filter synthesis uses the following steps:

1. Find a transfer function that satisfies a given set of specifications.
2. Select a circuit with a topology that has the desired transfer function.
3. Select the element values that minimizes some sensitivity measure(s), or, simply minimize the

transmission sensitivity. Designing active filters using component sensitivity is a major area in active filter synthesis, as there are infinite number of possible element values for a given transfer function, see Daryanani (1976) and Van Valkenburg (1982).

Filter designers use computer aids to evaluate the gain changes to the statistical variations in the component element values. Consider the biquadratic function and the corresponding gain function

$$H(s) = K \frac{s^2 + (\omega_z/Q_z)s + \omega_z^2}{s^2 + (\omega_p/Q_p)s + \omega_p^2}. \qquad (7.12.1)$$

$$\begin{aligned} G(\omega) &= 20\log|H(j\omega)| \\ &= 10\log\left[(\omega_z^2 - \omega^2)^2 + (\omega_z\omega/Q_z)^2\right] \\ &\quad - 10\log\left[(\omega_p^2 - \omega^2)^2 + (\omega_p\omega/Q_p)^2\right] \\ &\quad + 20\log(K). \qquad (7.12.2) \end{aligned}$$

Sensitivity functions with respect to amplitude and phase: Since the transfer function is a complex function $H(j\omega) = |H(j\omega)|e^{j\theta(\omega)}$, sensitivities can be derived with respect to the amplitude and phase angle. Using sensitivity of $H(s)$ and $H(j\omega)$ with respect to x, we have

$$S_x^{H(s)} = \frac{x}{H(s)} \frac{\partial H(s)}{\partial x}. \qquad (7.12.3)$$

$$S_x^{H(j\omega)} = \frac{x}{H(j\omega)} \frac{\partial H(j\omega)}{\partial x} = \frac{x}{|H(j\omega)|e^{j\theta(\omega)}} \frac{\partial |H(j\omega)|e^{j\theta(\omega)}}{\partial x}. \qquad (17.12.4a)$$

$$S_x^{H(j\omega)} = \frac{xe^{j\theta(\omega)}}{|H(j\omega)|e^{j\theta(\omega)}} \frac{\partial |H(j\omega)|}{\partial x} + \frac{x|H(j\omega)|}{|H(j\omega)|e^{j\theta(\omega)}} \frac{\partial e^{j\theta(\omega)}}{\partial x}.$$

$$= \frac{x}{|H(j\omega)|} \frac{\partial |H(j\omega)|}{\partial x} + jx \frac{\partial e^{j\theta(\omega)}}{\partial x}. \qquad (7.12.4b)$$

The sensitivities of the transfer function amplitude and the phase are

$$S_x^{|H(j\omega)|} = \mathrm{Re}\left[S_x^{H(j\omega)}\right], \quad S_x^{\theta(\omega)} = [1/\theta(\omega)]\mathrm{Im}[S_x^{H(j\omega)}]. \qquad (7.12.5)$$

These are functions of ω and can be evaluated at a particular frequency $\omega = \omega_i$.

Example 7.12.1 Use (7.12.5) to determine the amplitude and phase sensitivities of the following function with respect to Q and ω_0:

$$H_{LP}(s) = \frac{\omega_0^2}{s^2 + (\omega_0/Q)s + \omega_0^2}. \qquad (7.12.6a)$$

Solution:

$$\begin{aligned} S_Q^{H_{LP}(s)} &= \frac{Q}{H_{Lp}(s)} \frac{\partial H_{Lp}(s)}{\partial Q} = -\frac{(\omega_0/Q)s}{s^2 + (\omega_0/Q)s + \omega_0^2} \\ &\Rightarrow S_Q^{H(j\omega_0)} = \frac{-(\omega_0^2/Q)j}{\omega_0^2 - \omega_0^2 + j(\omega_0^2/Q)}. \qquad (7.12.6b) \end{aligned}$$

$$\Rightarrow S_Q^{|H_{Lp}(j\omega_0)|} = \mathrm{Re}S_Q^{H_{Lp}(j\omega_0)} = -1. \qquad (7.12.7a)$$

The sensitivity function with respect to ω_0 is given by

$$S_{\omega_0}^{H_{Lp}(s)} = \frac{\omega_0}{H_{Lp}(s)} \frac{\partial H_{Lp}(s)}{\partial \omega_0} = \frac{\omega_0}{H_{Lp}(s)} \frac{(s^2 + (\omega_0/Q)s + \omega_0^2)(2\omega_0) - \omega_0^2[(1/Q)s + 2\omega_0]}{(s^2 + (\omega_0/Q)s + \omega_0^2)^2}$$

$$S_{\omega_0}^{H_{Lp}(j\omega_0)} = \frac{-2\omega_0^2 + j\omega_0^2/Q}{-\omega_0^2 + j(\omega_0^2/Q) + \omega_0^2} = 1 + j2Q. \qquad (7.12.7b) \quad \blacksquare$$

Sensitivities can be used to measure the deviations in pole frequency, gain, etc.

Multi-element deviations: In Section 7.1, for small component deviation in a component x_j, identified

by Δx_j, the deviation in a filter parameter ω_0 was given by (see (7.1.9).)

$$\Delta\omega_0 = \sum_{i=1}^{m} S_{x_i}^{\omega_0} \frac{\Delta x_i}{x_i} \omega_0. \qquad (7.12.8a)$$

The variations can be positive or negative.

Worst-case analysis: In a system or a circuit, each component affects in its own way on the transfer characteristics. The specifications are generally specified by the design engineer that any appreciable departure from the nominal value of the component is unusable. Let the component values vary or change within the range $x_i - \Delta x_i \leq x_i \leq x_i + \Delta x_i$ and $\Delta x_i / x_i$ is the percentage change. If there are n elements, then n sensitivity functions need to be considered with each parameter, such as a pole, zero. As an example, if a pole $s = -p_1$ is under consideration, then the percentage change in $\Delta p_1 / |p_1|$ with respect to n elements identified below as e_i is

$$\frac{\Delta p_1}{|p_1|} = \sum_{i=1}^{n} S_{e_i}^{p_1} \left(\frac{\Delta e_i}{e_i} \right). \qquad (7.12.8b)$$

If all components can change within $\pm 1\%$ of their normal value, in the worst case,

$$\frac{\Delta p_1}{|p_1|} \Big|_{max} = \max \sum_{i=1}^{n} S_{e_i}^{p_1}(\pm .01) = \pm .01 \sum_{i=1}^{n} |S_{e_i}^{p_1}|.$$
$$(7.12.8c)$$

This results in the maximum deviation possible.

Gain sensitivity functions: The gain sensitivity is defined as the change in gain in dB due to a per-unit change in the parameter x_j by keeping all the other parameters fixed. The gain sensitivity expression and the corresponding gain variation are as follows:

$$S_x^{G(\omega)} = \frac{\partial G(\omega)}{\partial x / x} = \frac{\partial (20 \log |H(j\omega)|)}{\partial x / x} = 8.686 S_x^{|H(j\omega)|}$$
$$(7.12.9a)$$

$$\Delta G = \sum_j S_{x_j}^{G(\omega)} [\Delta x_j / x_j]. \qquad (7.12.9b)$$

The deviated gains are then compared to the nominal gain at all the frequencies. The procedure is repeated for each element. This gives information on gain changes due to *small changes* in component values. A second approach is a statistical approach, unlike the above method is not restricted to small changes in the component values and is referred to as a *Monte Carlo technique* (see Semmelman et al.,

1971). The input to the computer program consists of a topological description of the circuit along with the nominal element values, the atmospheric conditions, such as temperature and humidity, and others including the component aging aspects and manufacturing statistical tolerances. The tolerances are described in terms of distributions, such as uniform, Gaussian, and the means (μ_i), and standard deviations (σ_i) of the components. General computer circuit analysis programs are used to analyze the circuit and are statistically evaluated for its suitability and performance. Statistical techniques are generally computer intensive. These are beyond our scope here.

Pole (zero) displacements due to parameter variations: The poles of the transfer functions of highly optimal filters are close to the imaginary axis. Any incremental change in one of the network parameters causes a change in the pole locations and such a change can make a system even unstable. Dahlquist and Bjorck (1974) give an interesting example illustrating the effect of a slight change in the coefficient of a polynomial and the corresponding change in the roots. The polynomial is

$$D(\lambda) = (\lambda - 1)(\lambda - 2)...(\lambda - 20)$$
$$= \lambda^{20} - 210\lambda^{19} + ... + 20!. \qquad (7.12.10)$$

Suppose the coefficient -210 is changed to $-(210 + 2^{-23})$ in the above polynomial and the remaining coefficients are unchanged, the roots of the new polynomial change greatly and, as an example, a pair of the roots of this new polynomial are given by (correct to nine decimal places) $16.730737466 \pm j2.812624894$. The moral here is factor a higher order polynomial first using a general purpose computer and then use second-order systems in implementing filters. Quantifying the displacement of poles (or zeros) in a transfer function due to the incremental variations of one or more network parameters is of interest. Mitra (1969) and Budak (1974) give nice presentations on this topic. The above analysis is based on the assumption that the change in the parameter values is *small*. Truxal (1955) considers sensitivity functions for larger variations of a parameter.

7.13 Summary

In this chapter we have discussed some of the classical approximation methods that are useful in circuits, systems, and communication theory. These are Bode plots and the classical Butterworth and Chebyshev filter functions. Simple designs that are based on amplitude and phase responses are discussed. Active filters make use of op amps that are multi-terminal components. Two-port models of circuits are introduced. Active filter synthesis is introduced by first finding the transfer function of a circuit with a specific topology and then solving for component values along with a brief discussion on sensitivity measures. Amplitude and frequency scaling is introduced. Specific topics are

- sketches of the amplitude and phase responses from transfer functions using Bode plots;
- analog filter approximations based on Butterworth, Chebyshev, and Bessel functions;
- concepts associated with amplitude- and phase-based designs;
- frequency transformations relating the transfer functions of low-pass, high-pass, band-pass, and band-elimination filter functions to prototype low-pass filter functions;
- two-port open circuit, short circuit, hybrid, and *ABCD* parameters;
- indefinite admittance matrices;
- active filter synthesis including a brief review of sensitivity analysis.

Problems

7.1.1 Show the formulas given in Table 7.1.1 and the results in (7.1.6).

7.1.2 *a.* Show the equality in (7.1.14). An inverting amplifier circuit has a transfer function given by

$H(s) = -KR_2/[(R_1 + R_2) + KR_1]$. Use the results from Part *a.* to determine the sensitivity functions: *b.* S_K^H and *c.* $S_{R_1}^H$.

7.2.1 This problem deals with sketching the Bode plots for three simple RC networks that have important applications in control and communication systems. Let the transfer function in each case be expressed by $H(j\omega) = |H(j\omega)|\angle H(j\omega)$. *a.* Consider the phase lead RC network shown in Fig. P7.2.1a. Derive the expression for the transfer function. Sketch the Bode amplitude and phase plots of this function. For this network, the phase angle satisfies the constraint $0^0 \le \angle H(j\omega) \le 90^0$. *b.* Consider the phase lag RC network $(\angle H(j\omega) < 0)$ shown in Fig. P7.2.1b. Derive the expression for the transfer function. Sketch the amplitude and phase plots of this function. For this network, the phase angle satisfies the constraint $-90^0 \le \angle H(j\omega) \le 0^0$. *c.* Consider the phase lag-lead RC network shown in Fig. P7.2.1c. Derive the expression for the transfer function. Sketch the amplitude and phase plots of this function. For this network, the phase angle satisfies the constraint $-90^0 \le \angle H(j\omega) \le 90^0$.

7.2.2 Sketch the Bode amplitude and phase plots of the transfer function.

$$H(s) = (1 + s\tau_a)/[s(1 + s\tau_1)], \tau_a > \tau_1,$$
$$H(j\omega) = |H(j\omega)|e^{j\theta(\omega)}.$$

7.3.1 Sketch the amplitude response of the Butterworth amplitude frequency response function $|H_{Bu}(j\omega)| = 1/\sqrt{[1 + \varepsilon^2(\omega/\omega_c)^{2n}]}$ and then use the Bode plot approximations. Butterworth functions provide slopes of integer multiples of 10 dB/decade; they are useful in approximating an arbitrary magnitude function, see Weinberg (1962).

7.3.2 Show the following results for the Chebyshev polynomials:

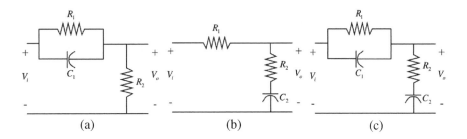

Fig. P.7.2.1 Simple RC circuits

(a)   (b)   (c)

a. n − even: $C_n(0) = (-1)^{n/2}$, $C_n(-1) = 1$, n-odd: $C_n(0) = 0, C_n(-1) = -1$.

b. $C_n(\alpha) = 0$, $\alpha = \cos((2k+1)\pi/2n)$, $k = 0, 1, 2, ..., n-1, |\omega| \leq 1$.

c. $C_n(\alpha) = 1$, $\alpha = \cos(i\pi/n), i = 0, 1, 2, ..., n$

7.3.3 Show the slope of the n th-order Chebyshev low-pass approximation at the pass band edge is n times that of the Butterworth approximation assuming that they both satisfy the same pass-band requirements. Do the following steps in your solution:

a. By induction, show that $[dC_{n+1}(\alpha)/d\alpha]|_{\alpha=1} = (n+1)^2$.

b. Find the slopes of the two normalized functions at $\omega = 1$.

$$1/|H_{Bu}(j\omega)| = \sqrt{1 + \varepsilon^2(\omega)^{2n}} \text{ and}$$

$$1/|H_{c1}(j\omega)| = \sqrt{1 + \varepsilon^2 C_n^2(\omega)}.$$

7.3.4 A low-pass filter is specified by the specifications ω_c, $\omega_r = 2\omega_c$, $X = 3\,\mathrm{dB}$, $Y = 15\,\mathrm{dB}$.

  a. Determine the transfer function assuming that the amplitude response has a maximally flat response.

  b. Determine the corresponding group delay function and sketch the amplitude and the group delay frequency responses.

7.3.5 Repeat Problem 7.3.4 using the Chebyshev 1 and Chebyshev 2 approximations.

7.3.6 a. Starting from the expression $10\log[1/(1+\varepsilon^2(\omega_r/\omega_c)^{2n}] = -Y\mathrm{dB}$, derive the equation given in (7.3.8).

  b. Starting with the equality $1/[1 + \varepsilon^2(\omega_r/\omega_c)^{2n}] = 1/A^2$, derive the approximation given in (7.3.9). Use the power series expansion of $\ln(1+\alpha)$ and $\ln(\omega_r/\omega_c) \cong (\omega_r - \omega_c)/\omega_c$.

  c. Derive the approximate expression in (7.3.27).

  d. Show that the value of n is the same for the Chebyshev 1 and Chebyshev 2 approximations.

7.4.1 Consider the second-order all-pass function expressed in terms of ω_0 and Q by

$$H(s) = \frac{s^2 - (\omega_0/Q) + \omega_0^2}{s^2 + (\omega_0/Q) + \omega_0^2}, H_2(j\omega) = H_2(s)|_{s=j\omega},$$

$$H(j\omega) = |H(j\omega)|\angle H(j\omega).$$

a. Give the function $\angle H(j\omega)$ and express it in terms of $\omega = \omega/\omega_0$.

b. Derive the group delay $D_2(\omega)$.

c. Find the maximum value of this function by differentiating the function $\omega_0 D_2(\omega)$ with respect to ω, equating it to zero and then using this value in obtaining the maximum value of the group delay. Our interest is for $Q \geq 1$. Give an approximate value of this function for $Q = 1$. For a good discussion on this topic, see Blinchikoff and Zverev (1976), Budak (1974), and Van Valkenburg (1982).

7.4.2 Show that the delay function $\tau(\omega)$ associated with $H(j\omega)$ given below satisfies the shown integral value given below.

$$H(j\omega) = \frac{[(b - \omega^2) - ja\omega]}{[(b - \omega^2) - ja\omega]}, \quad \int_0^\infty \tau(\omega)d\omega = 2\pi.$$

7.4.3 The magnitude response function of a filter is given by $|H(j\omega)| = (1 - \omega^2)/\sqrt{(1 + \omega^6)}$. Noting that the function $1/(1 + \omega^6)$ relates to the Butterworth function, determine the corresponding minimum phase function $H(s)$.

7.4.4 Consider the function $H(s) = [(s - a)/(s + a)], a > 0$. Sketch $\angle H(j\omega)$.

7.4.5 a. Find the value of n of the Bessel function approximation of group delay of 2 μs with a maximum error of 20% in the 0–.4 MHz frequency band.

  b. Determine the approximate loss using (7.4.23b) at the frequency .4 MHz

7.5.1 Give the pole–zero plots corresponding to the normalized low-pass functions given below and the corresponding high-pass functions using the transformation $s \rightarrow \omega_c/s$.

  a. $H_1(s) = 1/(s + p_1)$, $p_1 > 0$,

  b. $H_2(s) = 1/(s^2 + as + b)$, $a, b > 0$.

7.5.2 a. Use the normalized low-pass to band-pass transformation to convert the low-pass function in the last problem. Find the locations of the band-pass filter poles.

  b. Find the approximate locations of the band-pass poles for the narrow-band case, i.e., $\omega_0 \gg B$ and sketch the pole–zero plot.

7.5.3 *a.* Use the normalized low-pass to band-elimination transformation for the function in Problem 7.5.1. Find the locations of the band-elimination filter poles.

b. Find the approximate locations of the band-elimination filter poles for the narrowband case, i.e., $\omega_0 \gg B$ and sketch the pole–zero plot.

7.5.4 Find the lowest order Butterworth approximation function $H_{Be}(s)$ that satisfies the high-pass filter requirements with a cut-off frequency of $\omega_c = 2000\,\text{rad/s}$ and the stop-band frequency is $\omega_r = (1/2)\omega_c$. The attenuation at the edge of the pass band is 3 dB and the minimum attenuation at the edge of the stop band is 15 dB. Sketch $|H_{Be}(j\omega)|$.

7.5.5 Band-pass filters are used to recover a single frequency from a multiple number of frequencies (McGillem and Cooper, 1991). The squared amplitude Butterworth band-pass function with the center frequency ω_0, $\varepsilon = 1$, and B the bandwidth is

$$\left|H_{Bp}(j\omega)\right|^2 = \frac{1}{\left(1 + [(\omega - \omega_0)/\pi B]^{2n}\right)}.$$

Filter the waveform $x(t) = \cos(2\pi(100)t) + \cos(2\pi(500)t)$ and recover the first term.

a. Assume $B = 200\,\text{Hz}$ and solve for the integer value of n so that the second term in $x(t)$ is attenuated by 20 dB.

b. Assuming $n = 2$, give the 3 dB B to attenuate the second term by 20 dB.

7.5.6 Find the Chebyshev 1 approximation satisfying the band-pass filter specifications.

$$X = 1\text{dB}, Y = 20\text{dB}, \omega_l = 2\pi(300),$$
$$\omega_h = 2\pi(3000), \omega_2 = 2\pi(9000).$$

a. Noting that one side of the stop band of the band-pass filter constraints are not given, give the value of ω_1 that keeps the symmetry of the stop band requirements.

b. Sketch the normalized low-pass amplitude response function $\left|H_{Lp_n}(j\omega)\right|$ as well as the band-pass amplitude response function $\left|H_{Bp}(j\omega)\right|$.

7.5.7 Consider the second-order narrowband notch filter to suppress $f_0 = 60$ Hz hum.

$$H_{Be}(s) = \frac{s^2 + 2\beta s + \omega_0^2}{s^2 + (\omega_0/Q)s + \omega_0^2}, \quad 0 < \beta \ll \frac{\omega_0}{2Q}.$$

a. Give the amplitude of the function at $\omega = \omega_0$ for $\beta \neq 0$.

b. Show that the complete rejection of the input signal at the output at ω_0 provided $\beta = 0$. Compute the positive 3 dB frequencies. Sketch the amplitude response function $|H_{Be}(j\omega)|$ using the following steps. Express the poles of the function in complex form. That is, $s = -\omega_0 e^{j\theta} = -\omega_0 \cos(\theta) \pm j\omega_0 \sin(\theta)$ with $0 < \theta < \pi/2$. Relate θ to the pole Q used in the transfer function. Sketch the amplitude response function for $\theta = 45^0$ and 60^0.

c. Make a qualitative statement about the amplitude response in terms of θ and Q.

7.5.8 An electronic audio circuit suffers from a 2.32 kHz signal. Find a band elimination filter transfer function that can be used for this purpose. The pass bands are assumed to be in the ranges 0–1800 Hz and 3000–∞ Hz. The attenuation should be less than 3 dB or less in the pass band. The stop band is assumed to be in the range 2100–2400 Hz and the maximum attenuation in this range is 10 dB. Find the transfer function of a filter satisfying these requirements and sketch the amplitude response $|H_{Be}(j\omega)|$ in dB.

7.6.1 *a.* Give the controlled-source model of the ideal transformer.

b. Show that it is a lossless two port by showing $v_1 i_1 + v_2 i_2 = 0$.

c. Find the input impedance of the ideal transformer terminated by a resistor of R Ohms.

7.6.2 Figure P7.6.2 gives symbols for a common-emitter and a common-base transistors. Determine the hybrid parameters of the transistor in the common-base orientation h_{ijb} in terms of h_{ije} given

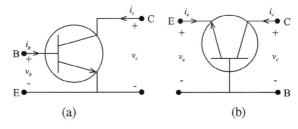

(a)   (b)

Fig. P7.6.2 (a) Common-emitter and (b) common-base transistors below by using the indefinite admittance matrices. Mitra (1969) gives a procedure illustrating the derivation of the hybrid parameters of a common base transistor from the hybrid parameters of a common emitter transistor.

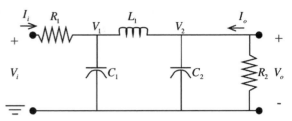

Fig. P7.6.6 A Ladder network

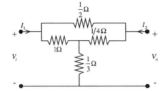

Fig. P7.6.5 A bridge circuit

1. Find the short-circuit admittance parameters for the common emitter from the hybrid parameters.
2. Find the 3×3 indefinite admittance matrix, see Section 7.6.2.
3. Delete the row and column corresponding to the emitter. These results can be extended for an arbitrary N node circuit.

$$\begin{bmatrix} v_B \\ i_C \end{bmatrix} = \begin{bmatrix} h_{11E} & h_{12E} \\ h_{21E} & h_{22E} \end{bmatrix} \begin{bmatrix} i_B \\ v_C \end{bmatrix},$$

$$\begin{bmatrix} v_E \\ i_C \end{bmatrix} = \begin{bmatrix} h_{11B} & h_{12B} \\ h_{21B} & h_{22B} \end{bmatrix} \begin{bmatrix} i_E \\ v_C \end{bmatrix}.$$

7.6.3 Assuming $Z_{21}(s) = Z_{12}(s)$ in (7.6.1) derive *a.* hybrid parameters, *b.* ABCD parameters and show that $H_{12}(s) = -H_{21}(s)$ and $AD - BC = 1$.

7.6.4 Consider the doubly terminated network shown in Fig. 7.6.7. Derive the equations given in (7.6.15b)–(7.6.15 g).

7.6.5 *a.* Determine the short-circuit parameters for the bridge circuit in Fig. P7.6.5.

b. Determine the open circuit parameters of the circuit from the results in Part *a.*

c. Find the ABCD parameters of this circuit and verify the equality $AD - BC = 1$.

7.6.6 Determine the transfer function of the network in Fig. P7.6.6, which is given using *a.* ABCD parameters, *b.* n ode, and *c.* loop equations, see Problem 7.9.2 for the answer.

7.6.7 Find the input impedance of the infinite ladder network in Fig. P7.6.7 by equating the input impedances identified in the figure. That is $Z_i = Z_1$. Using this, show that the *golden ratio is defined by* $Z_i/R = [(1 + \sqrt{5})/2]$.

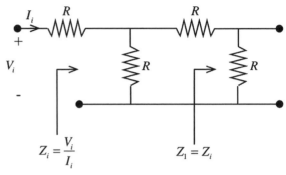

$$Z_i = \frac{V_i}{I_i} \qquad\qquad Z_1 = Z_i$$

Fig. P7.6.7 Ladder network of resistors

7.7.1 Verify the equations in *a.* (7.7.6) and *b.* (7.7.7). *c.* Show the transfer function of the circuit given in Fig. P7.7.1 is $H(s) = -(C_1/C_2)[s + (1/R_1C_1)]/(s + (1/R_2C_2))$.

7.7.2 Consider the circuit given in Fig. P7.7.2.

a. Write the differential equation using $v_0(0^-) = v_0$ when the op amp is ideal.

b. Show that the circuit can be used as an integrator by expressing the ouput in terms of an integral of the input.

7.7.3 Derive the transfer functions of the circuits in Fig. P7.7.3a,b,c.

d. Show that the transfer function in (7.7.12) corresponds to the circuit in Fig. (7.7.6b).

e. Derive the transfer function of the circuit shown in Fig. P7.7.3d.

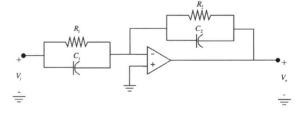

Fig. P7.7.1 A simple active RC filter with two capacitors

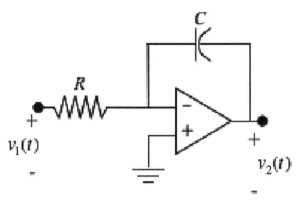

Fig. P7.7.2 A simple active RC circuit

$$\frac{V_0(s)}{V_i(s)} = H(s) = \frac{-Kb}{s^2 + as + b}, K = \frac{R_2}{R_1},$$

$$b = \frac{1}{R_2 R_3 C_1 C_2}, a = \frac{1}{C_2}\left[\frac{1}{R_1} + \frac{1}{R_2} + \frac{1}{R_3}\right].$$

7.7.6 *a.* Show the transfer function in (7.7.21) given in Example 7.7.6 is valid.

b. Noting that there are infinite number of solutions, Van Valkenberg (1982) suggested several solutions. One of these is as follows. Assume $\omega_0 = 1$, $R_1 = 1$, $R_2 = 4Q^2$, and $R_3 = 1$. Give the values for R_4 and C.

7.7.7 Consider the circuit shown in Fig. P7.7.5.

a. Show that the transfer function of the circuit in P7.7.5 is a product of a low-pass function and a high-pass function of the form $H(s) = -K[\omega_l/(s + \omega_l)][s/(s + \omega_u)]$.

b. Give the gain constant K in terms of R_1 and R_2. Do the same in terms of C_1, C_2, ω_l and ω_u. This can be used as a *nonresonant* inverting *wideband band-pass* filter if $\omega_l \ll \omega_u$, see Carlson (2000). We are faced with a 60 Hz hum and a

7.7.4 *a.* Use Kirchhoff's current law to show that the transfer function $H_a(s)$ in (7.7.13) corresponds to the circuit given in Fig. 7.7.7a.

b. Use the principle of superposition to show that the transfer function $H_b(s)$ in (7.7.13) of the circuit is shown in Fig. 7.7.7b.

7.7.5 *a.* Show that $V_0(s)/V_i(s)$ in Fig. P7.7.4 (see Nilsson and Riedel (1976).) is

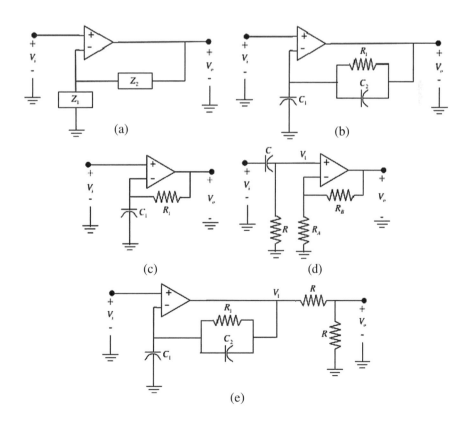

Fig. P7.7.3 Simple active RC filters

Fig. P7.7.4 A simple active
RC filter

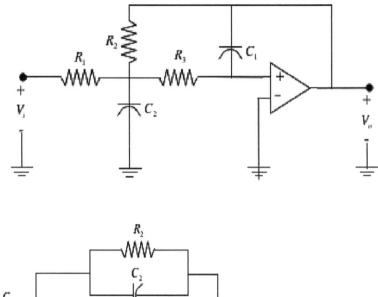

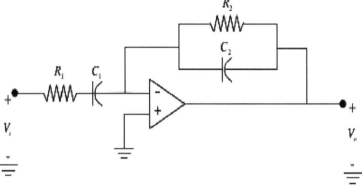

Fig. P7.7.5 A simple active RC filter

15 kHz whistle in an audio system. Use this circuit in designing a band-pass filter for the above specifications. Use a lower and upper cut-off frequencies $f_l = 200$ Hz and $f_u = 4$ kHz.

c. Use the thumb rule given in (7.7.15) to find the capacitor values.

d. Show that $|H(j\omega_l)| \cong |H(j\omega_u)| \cong K/\sqrt{2}$ and $|H(j\omega_0)| \cong K$.

7.8.1 Determine transfer function of the active RC circuits shown in Fig. P7.8.1a,b. In Fig. P7.8.1b, the blocks identified by Z_is are impedances.

7.8.2 a. Show the transfer function in 7.8.2a is true. b. Reduce it with $R_1 = R_2 = R$ and $C_1 = C_2 = C$. Express the pole Q in terms of R, C, k, and K and comment that K can be made larger provided k is made smaller for a given Q.

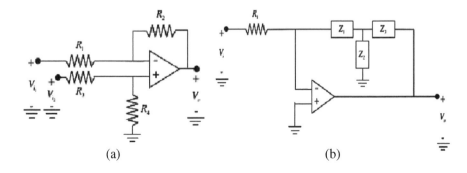

Fig. P7.8.1 Simple active
RC filter

(a)   (b)

7.8.3 Consider the Sallen–Key low-pass circuit corresponding to the transfer function

$$H(s) = \frac{K\omega_0^2}{[s^2 + (\omega_0/Q) + \omega_0^2]}, \quad K = 1, \ \omega_0 = 1.$$

a. Assuming $R_1 = R_2 = 1$, solve for the capacitor values in terms of the Q value and give the corresponding circuit with some of the elements values given in terms of Q.

b. Use the resistive voltage divider at the input such that the Thevenin's input resistance remains to be 1 and, at the same time, the voltage gain is reduced by 2.

c. Use the gain enhancement and Part a so the voltage gain is increased by 2.

d. Redesign the Sallen–Key circuit corresponding to the parts c and d if possible. Identify the constraints under which the resistive gain enhancement is valid in Part c.

7.9.1 a. Determine the center frequency and the bandwidth of the TF (see Fig. 6.12.6).

$$H(s) = \frac{V_0}{V_i} = \frac{Rs}{LCs^2 + RCs + R}$$
$$= \frac{K(\omega_0/Q)s}{s^2 + (\omega_0/Q)s + \omega_0^2},$$
$$R = 1, L = 1, C = 1.$$

b. Use magnitude and frequency scaling that gives a circuit with the same Q, center frequency of 2 kHz, and a 1 µF capacitor.

7.9.2 Figure P7.6.6 gives an RLC circuit with the element values as $R_1 = R_2 = 1, C_1 = 1.255$, $L_1 = .5528$ and $C_2 = .1922$. The transfer function of that circuit is

$$V_0/V_i = R_2/[L_1C_1C_2R_1R_2s^3 + (R_1L_1C_1 + L_1C_2R_2)s^2 + (R_1R_2C_1 + R_1R_2C_2 + L_1)s + (R_1 + R_2)].$$

a. Show that the transfer function results in an approximate Bessel function.

b. Denormalize the circuit so that the delay line filter has the dc delay equal to 2 µs with the terminating resistors equal to 1 $k\Omega$.

7.10.1 a. Find the transfer function of the circuits in Fig. P7.10.1a and b.

b. Find a set of component values for the active HP filter by assuming $C_1 = C_2 = 1$ for

$$H(s) = \frac{s^2}{(s^2 + (\sqrt{2})s + 1)}.$$

c. Considering only the dc and infinite frequencies, (see Fig. P7.10.1c) comment on the circuit.

d. Use the amplitude and frequency scaling discussed in Section 7.9 and derive the scaled version of the transfer function given in (7.10.3). That is $H'(s) = H(s/k_f)$.

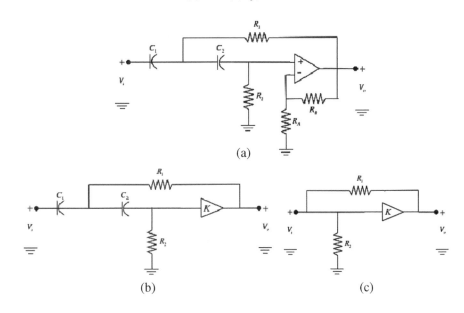

Fig. P7.10.1 Active RC circuit

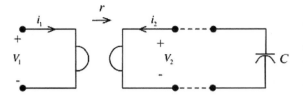

Fig. P7.10.3 Synthesis of an inductor using a gyrator and a capacitor

7.10.2 Using the RC to CR transformation, convert the LP in Fig. P7.7.5 to a HP.

7.10.3 Show that *a.* the input impedance of the circuit in Fig. P7.10.3 is $Z(s) = (r^2 C)s$.

b. Show that the ideal gyrator is a lossless network by showing that $v_1 i_1 + v_2 i_2 = 0$.

7.10.4 If an odd-ordered active high-pass filter is to be implemented then we need a first a simple first-order RC circuit and a buffer. Give this RC circuit along with the buffer.

7.10.5 Find an active RC high-pass filter with a 3 dB cut-off rerquency of $2\pi(200)$. At the edge of the stop band at $2\pi(60)$, the attenuation must be at least 20 dB.

7.11.1 Show the transfer function of the circuit in Fig. P7.11.1 is

$$H(s) = \frac{(KG_1/C_1)s}{s^2 + [\{(1/C_1)(G_1 + G_2 + G_3) - KG_2\} + (G_3/C_2)]s + [(1/C_1 C_2)(G_3(G_1 + G_2))]}, \quad G_i = 1/R_i.$$

7.11.2 Show the transfer functions of the circuit in Fig. 7.11.5 are as follows:

$$\frac{V_0(s)}{V_i(s)} = -\frac{(1/R_3 R_4 C_1 C_2)}{s^2 + (1/R_1 C_1)s + (1/R_2 R_4 C_1 C_2)},$$
$$\frac{V_2(s)}{V_i(s)} = \frac{(1/R_3 C_1)}{s^2 + (1/R_1 C_1)s + (1/R_2 R_4 C_1 C_2)}$$

7.11.3 In Problem 7.3.1 the high-frequency asymptotic slopes of Butterworth filter functions were considered. Nilsson and Reidel (1966) pose a problem that illustrates the comparison between an n th-order Butterworth function and a product of n first-order Butterworth functions. Compare the slopes of the two functions at the corner frequencies given below.

$$H_{Bu}(s) = [1/(1 + s^n)] \text{ and } H_{Bu_n}(s) = [1/(1 + s)^n].$$

$$x = 20 \log|H_{Bu}(j\omega)|(\omega = \omega),$$
$$y = 20 \log|H_{Bu_n}(j\omega)|^n, y = 20 \log|H_{Bu_n}(j\omega)|^n$$

$$[1/(1 + \omega_x^{2n})] = 1/2, \ [1/(1 + \omega_c^2)^n] = 1/2. \text{ Show}$$

a. $\omega_x = 1$ and $\omega_c = \sqrt{(2^{1/n} - 1)}.$

b. $\dfrac{dx}{d\log(\omega)} = \dfrac{-20n\omega^{2n}}{[1 + (1/\omega^{2n})]\omega^{2n}}$ dBs/decade,

c. $\dfrac{dy}{d\log(\omega)} = -\dfrac{20n\omega^2}{1 + \omega^2}$ dBs/decade.

d. Slopes at the corner frequencies: $[dx/d\log(\omega)]|_{\omega=1} = -10n$ dBs/decade

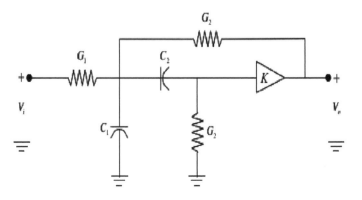

Fig. P7.11.1 A simple active RC filter

e. Illustrate the difference in these functions by the slopes for $n = 1, 2, 3$, and ∞ at these frequencies.

f. Comment on the slopes and the suitability of using the two functions for filtering.

g. In Part *a*., we have seen that the bandwidth is given by $\omega_c = \sqrt{(2^{1/n} - 1)}$. Show that for large n, the bandwidth can be approximated for large n by $\omega_c \cong .833/\sqrt{n}$. In the derivation, use the first two terms in the power series of $2^{1/n} = e^{(1/n)\ln(2)}$, see Lee (2004).

7.11.4 Design a unity gain band-reject filter for the following specifications using the twin-T circuit given in Fig. 7.11.2 with $\omega_0 = 2000$ radians/s and the quality factor $Q = 5$. Use $C = 1\mu F$. Determine the values of K and R.

7.12.1 Derive the expressions for S_Q^H and $S_{\omega_0}^H$ for the second-order band-pass function in Problem 7.9.1 with $K = 1$.

7.12.2 Consider the transfer function given in Example 7.1.1. Compute the worst-case per-unit change in ω_0 and Q assuming component values can change by $\pm 1\%$. Sensitivities of the components with respect to ω_0 and Q are given in Example 7.1.1.

Chapter 8
Discrete-Time Signals and Their Fourier Transforms

8.1 Introduction

So far in this book we have concentrated on the continuous-time signals. These included continuous periodic and aperiodic time signals and the corresponding Fourier series and transform representations. We divided the continuous signals based on their energy and power properties. In the case of periodic signals, the Fourier series coefficients are discrete. In this chapter, we will start with continuous signals and their sampled versions. The advances in computers and the ease in implementing discrete algorithms using personal computers (PCs) made this as an essential area every electrical engineer should be interested in. Most discrete-time signals come from sampling continuous signals, such as speech, seismic, sonar, images, biological, and other signals. These days, telephone along with a computer forms an integral part of most communication systems. The advances in telemetry allow us to monitor remotely located patients. The ease of processing discrete-time signals made discrete-time implementations of analog operations, such as filtering, made it very popular. The analog signals are first converted to digital signals by making use of a device referred to as an *analog-to-digital (A/D) converter*. The reverse process of reconstructing an analog signal from a digital signal is achieved by a device referred to as a *digital-to-analog (D/A) converter*. Obviously if the source is an analog device and the end user requires an analog signal, the use of a digital processor requires the A/D and the D/A converters. Although user signals are usually analog, there are many situations wherein the discrete-time signals are source signals. For example, the stock prices, the temperatures at a particular time

in a city, grades of students and many others are digital in nature. The transform study of the discrete-time signals is basic to our study.

In Chapters 3 and 4 we considered continuous periodic and aperiodic time signals and the corresponding Fourier series and transform representations. We divided the continuous and discrete-time signals based on their energy and power properties. In the case of periodic signals, the Fourier series coefficients are discrete. In this chapter we will see that when the function is discrete in time and periodic, then it is described by discrete-time Fourier series (DTFS). When the time signal is discrete and aperiodic, then the transform is continuous and periodic. Table 8.1.1 gives the relationship between time and frequency representations of periodic and aperiodic discrete and continuous signals. Most discrete-time signals come from sampling continuous signals, such as speech, seismic, sonar, images, biological, and other signals.

In the first part we will consider a signal $x(t)$ that is band limited to B Hertz. It is sampled at periodic intervals of time $t = nt_s$, where t_s is the sampling interval, resulting in the sampled values of the signal $x(nt_s)$ and t_s is tied to the bandwidth of the signal. Assuming t_s is a constant a discrete-time signal can be defined by

$$x[n] = x(nt_s). \qquad (8.1.1)$$

It is defined at discrete times by a sequence of numbers, denoted by $\{x_n\}$ or $x[n]$, where n is an integer. Transforming one set of numbers to another set may correspond to filtering or any transformation to be done by a discrete system. To achieve these goals, we need to learn discrete

R.K.R. Yarlagadda, *Analog and Digital Signals and Systems*, DOI 10.1007/978-1-4419-0034-0_8,
© Springer Science+Business Media, LLC 2010

Table 8.1.1 Fourier representations of discrete-time and continuous-time signals

Time property	Periodic	Aperiodic
Continuous (t)	Fourier series	Fourier transform
Discrete [n]	Discrete-time	Discrete-time
	Fourier series	Fourier transform

Fourier transforms of discrete data, fast computation of the transforms, learn about other transforms, especially the z-transforms (considered in the next chapter). See Ambardar (2007), Brigham (1974), Cartinhour (2000), Hsu (1995), McClellan et al. (2003), Oppenheim and Schafer (1975), Strum and Kirk (1998), Ziemer and Tranter (2002) and others.

8.2 Sampling of a Signal

An analog signal $x(t)$ is to be converted to a digital signal by selecting periodically spaced samples of the signal. Consider the example of an analog signal shown in Fig. 8.2.1a and the periodically sampled ideal version of the analog signal $x_s(t)$. The samples of the signal at discrete values $t = k\,t_s$ are $x(k\,t_s)$ where the sampling interval is t_s. The values $x(k\,t_s)$ have infinite precision. The following questions need to be answered:

1. How often must we sample, i.e., what is the value of the sampling interval t_s to get an accurate representation of the signal $x(t)$?
2. Given the set of sample values $x(k\,t_s)$, how do we reconstruct the analog signal from the sampled values?

In the first case, an analog signal is converted into a digital signal. That is *analog-to-digital conversion (A/D)*. In the second case, a digital signal is converted into an analog signal. That is *digital-to-analog conversion (D/A)*.

8.2.1 Ideal Sampling

Consider the periodic impulse sequence illustrated in Fig. 8.2.1a with period t_s

$$\delta_{t_s}(t) = \sum_{n=-\infty}^{\infty} \delta(t - nt_s). \qquad (8.2.1)$$

The idealized sampler, shown in Fig. 8.2.1b, consists of a device that multiplies two inputs $x(t)$ and $\delta_{t_s}(t)$ resulting in the product of these two signals. The sampled signal is given by

$$x_s(t) = x(t)\delta_{t_s}(t) = x(t) \sum_{n=-\infty}^{\infty} \delta(t - nt_s)$$

$$= \sum_{n=-\infty}^{\infty} x(nt_s)\delta(t - nt_s). \qquad (8.2.2)$$

$x_s(t)$ in (8.2.2) is an *ideal instantaneous sampling waveform* and $x(t)$ is assumed to be continuous at $t = nt_s$. The sampled values $x(nt_s)$ have infinite precision. These will be later converted and quantized for transmission. That will be considered in Chapter 10. The Fourier transform of the periodically sampled signal $x_s(t)$ is given by

$$X_s(j\omega) = F\left[x_s(t)\right] = F\left[x(t) * \delta_{t_s}(t)\right]$$
$$= F\left[x(t)\right] * F\left[\delta_{t_s}(t)\right]. \qquad (8.2.3)$$

Using the Fourier transform of the periodic impulse sequence (see (4.4.17).) and the convolution theorem, the transform of the sequence $x_s(t)$ is given below:

$$x(t)\overset{\text{FT}}{\longleftrightarrow}X(j\omega), \delta_{t_s}(t)\overset{\text{FT}}{\longleftrightarrow}\delta_{\omega_s}(\omega)=\frac{2\pi}{t_s}\sum_{k=-\infty}^{\infty}\delta(\omega-k\omega_s),$$

$$X_s(j\omega) = F[x_s(t)] = [X(j\omega) * \delta_{\omega_s}(\omega)]$$
$$= \frac{2\pi}{t_s}\left[\frac{1}{2\pi}\right]\int_{-\infty}^{\infty} X(j\alpha)\delta_{\omega_s}(\omega - \alpha)d\alpha,$$

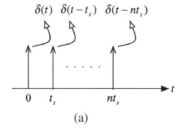

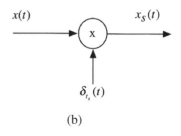

Fig. 8.2.1 Ideal sampling, (a) δ_{t_s}, (b) Idealized sampler

(a)

(b)

$$= \frac{1}{t_s} \sum_{k=-\infty}^{\infty} \int_{-\infty}^{\infty} X(j\alpha)\delta(\omega - \alpha - k\omega_s)d\omega$$

$$= \frac{1}{t_s} \sum_{k=-\infty}^{\infty} X(j(\omega - k\omega_s))$$

$$= \frac{1}{t_s} \sum_{m=-\infty}^{\infty} X(j(\omega + m\omega_s)) = X_s(j\omega). \qquad (8.2.4)$$

These follow from the sifting property of the impulse functions. Note the *spectrum of the ideally sampled signal* $X_s(j\omega)$ is *periodic* with period $\omega_s = 2\pi f_s = 2\pi/t_s$.

A special case: Consider a causal function $x(t)$, i.e., $x(t) = 0$, $t < 0$ and has a discontinuity at $t = 0$. In such a case, the ideally time sampled signal and its transform are

$$x_s(t) = (1/2)x(0^+)\delta(t) + \sum_{n=1}^{\infty} x(nt_s)\delta(t - nt_s) \xrightarrow{\text{FT}} X_s(j\omega)$$

$$= \frac{1}{2}x(0^+) + \frac{1}{t_s} \sum_{n=-\infty}^{\infty} X(j(\omega + n\omega_s)). \qquad (8.2.5)$$

Important questions: Can $x(t)$ be recovered from $x_s(t)$? Or, in other words, can the transform $X(j\omega)$ recovered from $X_s(j\omega)$? If not, why? Or, at least, can an approximate version of $x(t)$ be recovered from the sampled signal?

Consider an arbitrary signal $x(t)$ and its transform $X(j\omega)$ given in Fig. 8.2.2a. For the present analysis, the shape of the signal or the spectrum is not critical, whereas its bandwidth is. For simplicity, $X(j\omega)$ is assumed to be real and the signal is band limited to $W = 2\pi B$ rads/s. Figure 8.2.2b gives the periodic impulse sequence and its transform. Note the spectrum of the periodic impulse sequence is also a periodic impulse sequence with period equal to the sampling frequency ω_s. Figure 8.2.2c gives the ideally sampled signal $x_s(t)$ and its transform. In sketching the spectrum of the ideally sampled signal, it is assumed *implicitly* that $\omega_s \geq 2(2\pi B)$. That is,

$$\omega_s \geq 2\,W = W_N = 2\pi f_N. \qquad (8.2.6)$$

The term $f_N = 2B$ is called the *Nyquist frequency* or the *Nyquist rate*, where B is the highest frequency

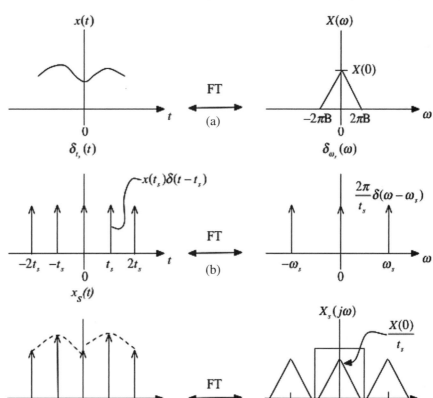

Fig. 8.2.2 (a) Signal $x(t)$ and its transform; (b) periodic impulse sequence and its transform; and (c) ideally sampled signal and its transform

Fig. 8.2.3 (a) Sampling and the reconstruction process; (b) $|H_{Lp}(j\omega)|$

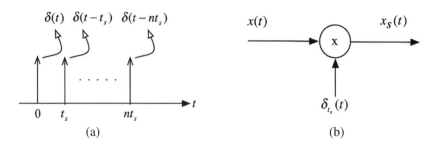

(a)   (b)

in the input signal. With the constraint in (8.2.6) and from Fig. 8.2.2c, the adjacent spectra, i.e., $X(j(\omega + k\omega_s))$ and $X(j(\omega + (k-1)\omega_s))$ do not overlap. The spectrum of the original signal can be extracted from the sampled spectrum by passing the ideally sampled signal through an ideal low-pass filter. Figure 8.2.3a illustrates the ideal sampling and reconstruction process of the original signal using an *ideal low-pass filter*. The filter transfer function and the output frequency response are as follows:

$$H_{Lp}(j\omega) = H_0 \Pi\left[\frac{\omega}{W_N}\right] e^{-j\omega t_0},$$

$$\Pi\left[\frac{\omega}{W_N}\right] = \begin{cases} 1, & |\omega| < W_N/2 \\ 0, & 0 \text{ Otherwise} \end{cases},$$

$$\begin{aligned} Y(j\omega) &= H_{LP}(j\omega)X_s(j\omega) \\ &= (H_0/t_s)X(j\omega)e^{-j\omega t_0}. \end{aligned}$$

(8.2.7a)

(8.2.7b)

The amplitude of the filter is shown in Fig. 8.2.3b, where the gain is H_0 and $W_N/2 = 2\pi B$ rad/s or B Hz is the bandwidth of the ideal LP filter. The exponential term $(e^{-j\omega t_0})$ is included in the transfer function to accommodate for the delay caused by the ideal filter. Noting the properties of the LTI systems,

$$y(t) = F^{-1}\{Y(j\omega)\} = (H_0/t_s)x(t-t_0). \quad (8.2.7c)$$

Note $y(t)$ in (8.2.7c) is the same as the analog signal $x(t)$, except it is delayed by t_0s and scaled by a gain factor of (H_0/t_s). For ideal reconstruction, the gain constant is $H_0 = t_s$. For simplicity, the delay will be assumed to be $t_0 = 0$. The sampling interval t_s tends to be small and therefore the gain constant (H_0/t_s) tends to be large. Note that the signal $x(t)$ is assumed to be band limited to B Hz and did not distinguish whether it is a low pass or a band pass signal. Sampling band pass signals will be considered in Section 8.2.6. These results are stated in the following theorem.

8.2.2 Uniform Low-Pass Sampling or the Nyquist Low-Pass Sampling Theorem

A real valued signal $x(t) \xleftrightarrow{\text{FT}} X(j\omega)$ that has *no* frequency components above B Hz, i.e., $|X(j\omega)| = 0$, $\omega = 2\pi f$, $f > B$Hz is uniquely determined by the samples taken at the uniform rate of $2B$ samples per second or greater. The signal can be recovered from the sampled signal if the time between samples, i.e., the sampling interval t_s is no greater than $(1/2B)$ and it does not specify where to sample. This will be considered shortly.

Example 8.2.1 Consider the signal

$$x(t) = A\cos(2\pi f_1 t)\cos(2\pi f_2 t),$$
$$f_1 = 200 \text{ Hz}, \ f_2 = 500 \text{ Hz}.$$

(8.2.8)

Assuming the signal is ideally sampled with a sampling rate of 2000 Hz find the spectrum of the sampled signal. Sketch the spectrum of the signal and determine the range of the bandwidth of the ideal low-pass filter to recover the signal.

Solution: Using trigonometric relations result in the following:

$$\begin{aligned} x(t) = .5A[&\cos(2\pi(500-200)t) \\ &+ \cos(2\pi(500+200)t)], \end{aligned}$$

(8.2.9a)

$$\begin{aligned} X(j\omega) = .5A\pi[&\delta(\omega - 2\pi(300)) + \delta(\omega + 2\pi(300)) \\ &+ \delta(\omega - 2\pi(700)) + \delta(\omega + 2\pi(700))]. \end{aligned}$$

(8.2.9b)

The transform of the ideally sampled signal is (see (8.2.4) with $\omega_s = 2\pi(2000)$.)

$$X_s(j\omega) = \frac{1}{t_s}\sum_{k=-\infty}^{\infty} X(j(\omega + k\omega_s)),$$

$$= A\pi(1/2t_s)\{\ldots + [\delta(\omega + 2\pi(2700)) + \delta(\omega + 2\pi(2300))$$
$$+ \delta(\omega + 2\pi(1700)) + \delta(\omega + 2\pi(1300))]$$
$$+ [\delta(\omega + 2\pi(700)) + \delta(\omega + 2\pi(300)) + \delta(\omega - 2\pi(300))$$
$$+ \delta(\omega - 2\pi(700))] + [\delta(\omega - 2\pi(1300))$$
$$+ \delta(\omega - 2\pi(1700)) + \delta(\omega - 2\pi(2300))$$
$$+ \delta(\omega - 2\pi(2700))] + \cdots\}. \tag{8.2.10}$$

See Fig. 8.2.4 for the transforms of the signal and sampled version. For simplicity we assumed $A > 0$. The terms corresponding to $k = -1, 0, 1$ in the expression in (8.2.10) are explicitly shown in Fig. 8.2.4b. The highest frequency in the input signal is $B = 700$ Hz. The Nyquist frequency is therefore 1400 Hz. Note that we are (essentially) using the *entire frequency band* between DC and the highest frequency 700 Hz. Using the Nyquist sampling theorem, the minimum sampling rate is $f_s \geq f_N = 2B = 1400$ Hz. So, a sampling rate of $f_s = 2000$ Hz or 2000 samples per second is a *valid sampling rate*. From Fig. 8.2.4b, the highest positive frequency in the term corresponding to $k = 0$ in (8.2.10) is 700 Hz. The nearest frequency in the term corresponding to $k = 1$ is (-1300) Hz, and the lowest frequency in the term corresponding to $k = -1$ is 1300 Hz. To recover $X(j\omega)$, a low-pass filter is needed with a bandwidth of $B = B_0$ Hz. The constraint on the bandwidth of the filter is 700 Hz $< B_0 <$ 1300 Hz. ∎

Proof of the uniform sampling theorem: It will be proved that when the sampled signal

$$x_s(t) = \sum_{n=-\infty}^{\infty} x(nt_s)\delta(t - nt_s). \tag{8.2.11}$$

is passed through an ideal low-pass filter, the signal $x(t)$ can be recovered. Consider the block diagram shown in Fig. 8.2.3a. The transfer function of the ideal low-pass filter is assumed to be equal to $H_{Lp}(j\omega)$ (see (8.2.7a)) and its inverse transform is given by

$$H_{Lp}(j\omega) = H_0\Pi\left[\frac{\omega}{2\pi W_N}\right]e^{-j\omega t_0} \overset{FT}{\longleftrightarrow} h_{Lp}(t), \tag{8.2.12}$$

$$W_N = 2B_0,$$
$$h_{Lp}(t) = (2B_0H_0)\frac{\sin(2\pi B_0(t - t_0))}{2\pi B_0(t - t_0)}$$
$$= (2B_0H_0)\text{sinc}(2\pi B_0(t - t_0)). \tag{8.2.13}$$

See (1.2.15) for the sinc function. The output of the ideal filter (see Fig. 8.2.3a) is

$$y(t) = x_s(t) * h_{Lp}(t) \overset{FT}{\longleftrightarrow} X_s(j\omega)H_{LP}(j\omega) = Y(j\omega). \tag{8.2.14}$$

By using the leniarity of the convolution integral, we have

$$y(t) = x_s(t) * h_{Lp}(t) = \left[\sum_{n=-\infty}^{\infty} x(nt_s)\delta(t - nt_s)\right] * h_{Lp}(t),$$

$$= \sum_{n=-\infty}^{\infty} x(nt_s)[h_{Lp}(t) * \delta(t - nt_s)]$$

$$= \sum_{n=-\infty}^{\infty} x(nt_s)h_{Lp}(nt_s),$$

$$= \sum_{n=-\infty}^{\infty} x(nt_s)(2B_0H_0)\text{sinc}(2\pi B_0(t - t_0 - nt_s)). \tag{8.2.15}$$

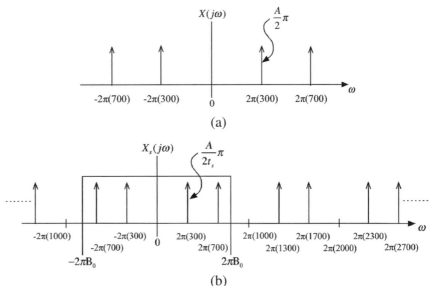

Fig. 8.2.4 (a) $X(j\omega)$;
(b) spectrum of the ideally
sampled signal

This is the *representation of a band-limited signal by its sampled values*. To illustrate this, consider the special case

$$B_0 = f_s/2, H_0 = t_s, \ t_0 = 0, \ \text{or}$$
$$2B_0H_0 = f_s \, t_s = 1, \qquad (8.2.16)$$

$$\Rightarrow y(t) = \sum_{n=-\infty}^{\infty} x(nt_s) \sin c(\pi(f_s t - n)). \quad (8.2.17)$$

The function $y(t)$, i.e., the reconstructed form of $x(t)$ from the sample values of $x(t)$, $x(nt_s)$. Equation (8.2.17) is an *interpolation formula*. At $t = nt_s$ with $f_s t_s = 1$,

$$y(t)|_{t=k\,t_s} = \sum_{n=-\infty}^{\infty} x(n\,t_s)\mathrm{sinc}(f_s t_s \pi(k-n))$$
$$= \sum_{n=-\infty}^{\infty} x(nt_s)\mathrm{sinc}(\pi(k-n)) = x(kt_s),$$
$$(8.2.18)$$

$$\text{Note}: \ \mathrm{sinc}(\pi(k-n)) = \frac{\sin(\pi(k-n))}{\pi(k-n)} = \begin{cases} 1, & k=n \\ 0, & k \ne n \end{cases}.$$

Figure 8.2.5 illustrates the interpolation. What can we say about the values of $y(t)$ at $t \ne kt_s$? For example, when $t = (2k+1)t_s/2$, i.e., at the mid-point in the interval $kt_s \le t < (k+1)t_s$, (8.2.17) allows for determination of the values of $x(t)$ in that interval. They are *interpolated values* and are given by

$$y(t)|_{t=(2k+1)(t_s/2)} = \sum_{n=-\infty}^{\infty} x(nt_s)\mathrm{sinc}[\pi(f_s((2k+1)t_s/2) - n)],$$
$$= \sum_{n=-\infty}^{\infty} x(nt_s)\mathrm{sinc}\left(\left(\frac{2k+1}{2} - n\right)\pi\right) = \sum_{n=-\infty}^{\infty} x(nt_s)\mathrm{sinc}\left[\frac{2(k-n)+1}{2}\pi\right],$$
$$= \cdots + x((k-1)t_s)\mathrm{sinc}(3\pi/2) + x(kt_s)\mathrm{sinc}(\pi/2) + x((k+1)t_s)\mathrm{sinc}(-\pi/2) + \cdots. \quad (8.2.19)$$

In (8.2.19), the terms corresponding to $n = k - 1$, $k, k+1$ are identified. Noting that $\mathrm{sinc}(-\pi\alpha) = \mathrm{sinc}(\pi\alpha)$, an even function, we can simplify (8.2.19) further. At the sample points, say at $t = nt_s$, all the terms in the summation go to zero, except the nth term and the result is $y(t)|_{t=nt_s} = x(n t_s)$.

Example 8.2.2 Let $x(nt_s)$ be the sampled values of the band-limited function $x(t)$ to $B = f_s/2$ and are given by

$$x(nt_s) = \begin{cases} 2, \ n = 1 \\ -1, \ n = 2 \\ 0, \ \text{otherwise} \end{cases}. \quad (8.2.20)$$

Evaluate the function $y(t)$ at $t = .5t_s, t_s, 1.5t_s, 2t_s$ using the interpolation formula in (8.2.17) and the sampled values of the function $x(t)$ in (8.2.20).

Solution: By using the interpolation formula and noting that $\mathrm{sinc}(\pi f_s t)$ is even, we have

$$y(t) = 2\mathrm{sinc}(\pi(f_s t - 1)) - \mathrm{sinc}(\pi(f_s t - 2)), \quad (8.2.21)$$

$$y(t_s) = 2\mathrm{sinc}(\pi(f_s t_s - 1)) - \mathrm{sinc}(\pi(f_s t_s - 2))$$
$$= 2\mathrm{sinc}(0) - \mathrm{sinc}(-\pi) = 2 = x(t_s),$$

$$y(2t_s) = 2\mathrm{sinc}((2t_s f_s - 1)\pi) - \mathrm{sinc}((f_s(2t_s) - 2)\pi)$$
$$= -1 = x(2t_s),$$

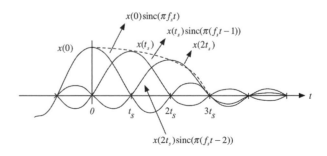

Fig. 8.2.5 Interpolation using three sinc functions

Fig. 8.2.6 Example 8.2.2

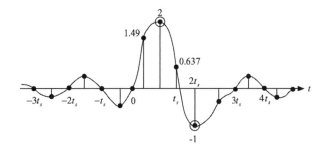

$$y(t_s/2) = 2\text{sinc}(\pi(f_s(t_s/2) - 1))$$
$$- \text{sinc}(\pi(f_s(t_s/2) - 2))$$
$$= 2\text{sinc}(\pi/2) - \text{sinc}(-3\pi/2)$$
$$= 1.2732 - (-.212207) \cong 1.4854$$

$$y(1.5t_s) = 2\text{sinc}(\pi/2) - \text{sinc}(-\pi/2) \cong .637.$$
See figure 8.2.6   ∎

Notes: The interpolated function $y(t) = 0$ for $t = kt_s$, k is an integer and $k \neq 1, 2$. The interpolating function $y(t)$ has oscillating tails that die out and is shown in Fig. 8.2.6. $x(t)$ is band limited to $B_0 = f_s/2$ and therefore it cannot be time limited as the product of the spectral width and the time duration of the function cannot be less than a certain minimum value. See the uncertainty principle in Fourier analysis in Section 4.7.3.   ∎

8.2.3 Interpolation Formula and the Generalized Fourier Series

The interpolation formula is the *generalized Fourier series expansion* (see Section 3.3) with the orthogonal basis function set consisting of sinc functions.

$$\{\text{sinc}(\pi(f_s t - n)),$$
$$n = \cdots -2, -1, 0, 1, 2, \ldots\}., -\infty < t < \infty. \quad (8.2.22)$$

In the first step, the set in (8.2.22) is shown to be an *orthogonal basis set* over the interval $-\infty < t < \infty$. That is,

$$\int_{-\infty}^{\infty} \text{sinc}(\pi(f_s t - n))\text{sinc}(\pi(f_s t - m))dt$$
$$= \begin{cases} k_n = (1/f_s), & n = m \\ 0, & n \neq m \end{cases}. \quad (8.2.23)$$

The constant $k_n = (1/f_s)$ is the energy contained in each of the sinc functions. To show this, consider the two functions and their transforms given by

$$x_1(t) = \text{sinc}(\pi f_s(t - nt_s)) \xleftrightarrow{\text{FT}} \frac{1}{f_s}\Pi\left[\frac{\omega}{2\pi f_s}\right]e^{-j\omega(nt_s)}$$
$$= X_1(j\omega), \quad (8.2.24a)$$

$$x_2(t) = \text{sinc}(\pi f_s(t - mt_s)) \xleftrightarrow{\text{FT}} \frac{1}{f_s}\Pi\left[\frac{\omega}{2\pi f_s}\right]e^{-j\omega(mt_s)}$$
$$= X_2(j\omega). \quad (8.2.24b)$$

Using generalized Parseval's theorem assuming $n \neq m$ with $\omega_s t_s = 2\pi$ and using the transforms of the sinc functions, we have

$$\int_{-\infty}^{\infty} x_1(t)x_2(t)dt = \frac{1}{2\pi}\int_{-\infty}^{\infty} X_1(j\omega)X_2^*(j\omega)d\omega$$
$$= \frac{1}{2\pi(f_s)^2}\int_{-\infty}^{\infty}\left\{\Pi\left[\frac{\omega}{2\pi f_s}\right]\right\}^2 e^{-j\omega(n-m)t_s}d\omega,$$
$$= \frac{1}{2\pi(f_s)^2}\int_{-\omega_s/2}^{\omega_s/2} e^{-j\omega(n-m)t_s}d\omega$$
$$= \frac{1}{2\pi(f_s)^2}\left[\frac{e^{-j\omega(n-m)t_s}}{-(n-m)t_s}\right]_{\omega=-\omega_s/2}^{\omega=\omega_s/2},$$
$$= \frac{1}{2\pi(f_s)^2(n-m)t_s}\left[e^{j(n-m)\pi} - e^{-j(n-m)\pi}\right] = 0.$$
$$(8.2.25)$$

For $n = m$,

$$\int_{-\infty}^{\infty} x_1(t)x_2(t)dt = \frac{1}{(2\pi f_s)^2}\int_{-\infty}^{\infty}\left\{\Pi\left[\frac{\omega}{2\pi f_s}\right]\right\}^2 e^{-j\omega(n-m)t_s}d\omega$$
$$= \frac{1}{2\pi(f_s)^2}\int_{-\omega_s/2}^{\omega_s/2} d\omega = \frac{1}{f_s}. \quad (8.2.26)$$

From (8.2.25) and (8.2.26), it follows that the set in (8.2.22) is an orthogonal basis set. Therefore, the generalized Fourier series expansion of $y(t)$ is

$$y(t) = \sum_{k=-\infty}^{\infty} Y_s[k]sinc(\pi f_s(t - kt_s)). \quad (8.2.27)$$

The generalized Fourier series coefficients can be determined from

$$Y_s[k] = f_s \int_{-\infty}^{\infty} y(t)sinc(\pi f_s(t - kt_s))dt. \quad (8.2.28)$$

Noting the transform of the sinc pulse is a rectangular pulse (see (4.3.28)) and the generalized Parseval's theorem (see (8.2.25)) and $F[y(t)] = F[x(t)] = X(j\omega)$ results in the following:

$$sinc(\pi f_s(t - nt_s))$$

$$= \frac{\sin(\pi f_s(t - nt_s))}{\pi f_s(t - nt_s)} \overset{FT}{\longleftrightarrow} \begin{cases} \frac{1}{f_s}e^{-n\omega t_s}, |\omega| < \frac{\omega_s}{2} \\ 0, \text{otherwise} \end{cases}, \quad (8.2.29)$$

$$\Rightarrow Y_s[k] = f_s \int_{-\infty}^{\infty} x(t)sinc(\pi(f_s t - n))dt$$

$$= \frac{1}{2\pi} \int_{-\omega_s/2}^{\omega_s/2} X(j\omega)e^{j\omega n t_s}d\omega,$$

$$= \frac{f_s}{2\pi(f_s)} \int_{-\infty}^{\infty} X(j\omega)e^{j\omega t}d\omega|_{t=kt_s} = x(t)|_{t=kt_s} = x(kt_s).$$

$$(8.2.30)$$

Since the transform of the function $x(t)$ is band limited to $\omega_s/2$, the limits on the transform integral can be changed from $((-\omega_s/2), (-\omega_s/2))$ to $(-\infty, \infty)$ in (8.2.30).

Example 8.2.3 Let $x(t) = \sin(2\pi(1)t)$ shown in Fig. 8.2.7. It is sampled at the Nyquist rate of $f_N = 2(1) = 2$ samples per second and sampled at $t = 0, .5, 1, 1.5, \ldots$. The sampled values are equal to zero indicating that the signal cannot be recovered from the samples. *Nyquist theorem does not identify where to sample.* ∎

The sampling rate has to be larger than the Nyquist rate. Its selection is signal dependent and cost-

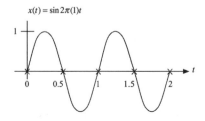

Fig. 8.2.7 $x(t) = \sin 2\pi(1)t$, Sampled two times per second

effectiveness, as the analog-to-digital (A/D) converters are expensive at both the low and the high sampling rate. Sampling a function at much higher than the Nyquist rate does not help. Recovering an analog signal from the samples requires the computation using more samples than necessary and the errors in computation nullifies any advantage used in high sampling. As a guide, the sampling rate is more than the Nyquist rate, about 2.5–10 times the highest frequency in the signal. For seismic signals, the frequencies of interest are in few hundred Hertz range. In these cases, higher sampling rates are used. For speech, the frequency range of interest is from a few Hertz to 3.5 kHz. The sampling rate is taken as 8 kHz or 10 kHz. For CDs, the frequency range of the input signal is from a low frequency of few Hertz to 20 kHz. The sampling rate is taken as 44.1 kHz and the standard sampling rate for studio quality audio is 48 kHz. The compact disc recording system samples each of the two stereo signals with a 16-bit A/D converter at 44.1 kHz (Haykin and Van Veen (2003)).

Example 8.2.4 Consider signal $x(t)$ band limited to $(2\pi B)$ rad/s. Determine the Nyquist rates for the functions: $a. y_1(t) = x(2t), b. y_2(t) = x(t)\cos(\omega_0 t)$

Solution: a. Note $y_1(t)$ is formed from $x(t)$ by compressing the time axis by a factor of 2. From the Fourier scale change theorem (see Section 4.3.4), we have the following:

$$y_1(t) = x(at) \overset{FT}{\longleftrightarrow} \frac{1}{|a|}X(j\omega/a), a \neq 0$$

$$\Rightarrow x(2t) \overset{FT}{\longleftrightarrow} \frac{1}{2}X(j\omega/2) = Y_1(j\omega).$$

Time compression by a factor of 2 results in expansion in frequency by a factor of 2. The Nyquist rate is given by $\omega_{s1} = 2\pi(2(2B))$. It is like playing an audio tape fast. b. The signal is a modulated signal with a center frequency ω_0 with a bandwidth of $2B$ Hz and

$$F[y_2(t)] = F[x(t)\cos(\omega_0 t)] = 0.5X(j(\omega - \omega_0))$$
$$+ 0.5X(j(\omega + \omega_0)).$$

The highest frequency in the modulated signal is $\omega_0 + 2\pi B = (\omega_0 + \omega_s)/2$. The Nyquist rate is $\omega_{s2} = \omega_s + 2\omega_0$. ∎

The sinc interpolation function is not the best way to approximate the function from its sample values, as it decays only at a rate of $(1/t)$. There are other better functions. The function $x(t)$ is known at $t = nt_s$. The interpolation formula can be expressed in terms of a function $h_i(t)$ that is 0 at all the sampling instants, except at $t = 0$, where it is 1. In addition, it is absolutely integrable. Interpolation formula is given by

$$y_i(t) = \sum_{n=-\infty}^{\infty} x(nt_s)h_i(t - nt_s), h_i(kt_s - nt_s)$$

$$= \begin{cases} 1, & k = n \\ 0, & k \neq n \end{cases}.$$

Since $y_i(t)|_{t=kt_s} = x(kt_s)$, i.e., the interpolation formula gives the same values at the sampling instants and, at other times, $y_i(t)$ is an approximation of $x(t)$. Most commonly interpolating functions are step, linear, sinc, and raised cosine functions. These are given below in table 8.2.1. See Ambardar (1999) for additional discussion on the interpolation functions.

Step interpolation (zero-order-hold) uses a rectangular interpolation function and $x(nt_s)$ to produce a stepwise or a staircase approximation of $x(t)$. This is simple and does not depend on the future values of the signal. It is widely used. The reconstructed signal is (more on this in Section 8.2.5.)

$$y_c(t) = x(nt_s), nt_s \leq t < x((n + 1)t_s).$$

Linear interpolation (first-order hold) uses a linear approximation and the reconstructed signal is

$$y_l(t) = x(nt_s) + \frac{x((n + 1)t_s) - x(nt_s)}{t_s}$$

$$(t - nt_s), nt_s \leq t < (n + 1)t_s.$$

It cannot be implemented online since a future value is required. Sinc interpolation was considered earlier. *Raised cosine interpolation function* (see (4.11.9a) for the function and its transform in (4.11.9b)) uses the roll-off factor β. It reduces to the sinc interpolation function when $\beta = 0$. The raised cosine function's decaying rate is proportional to $(1/t^3)$. Faster decaying results in improved reconstruction, if the samples are not at exactly at the sampling instants (i.e., *jitter*). It requires fewer past values are needed in the reconstruction. *Polynomial-based interpolation* methods are discussed in Appendix A.9.

8.2.4 Problems Associated with Sampling Below the Nyquist Rate

Consider the functions $x_1(t)$ and $x_2(t)$ in Fig. 8.2.8. They are sampled at a rate shown. Both provide the same sample values. The function $x_1(t)$ cannot be reconstructed from the sample values. From the figure, $x_1(t)$ has a higher frequency content than $x_2(t)$. By sampling the functions at the locations shown, some of

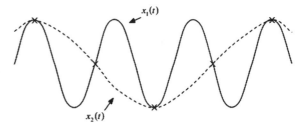

Fig. 8.2.8 Two signals sampled at the same locations

Table 8.2.1 Common interpolation functions

a. $y_c(t) = \sum_{n=-\infty}^{\infty} x(nt_s)\Pi[(t - nt_s)/t_s]$: constant or step interpolation

b. $y_l(t) = \sum_{n=-\infty}^{\infty} x(nt_s)\Lambda[(t - nt_s)/t_s]$: linear interpolation

c. $y_s(t) = \sum_{n=-\infty}^{\infty} x(nt_s)\text{sinc}[\pi(t - nt_s)/t_s]$: sinc interpolation

d. $y_{rc}(t) = \sum_{n=-\infty}^{\infty} x(nt_s)\frac{\cos(\pi\beta(t - nt_s)/t_s)}{(1 - [2\beta(t - nt_s)/t_s]^2)}\text{sinc}(\pi(t - nt_s)/t_s)$: raised cosine interpolation, β – roll-off factor, $0 \leq \beta \leq 1$.

the peaks and valleys of $x_1(t)$ are missed indicating that $x_1(t)$ needs to be sampled at a higher rate than $x_2(t)$.

A spectrum $X(j\omega)$, its ideally sampled signal spectra by assuming the sampling rates of $\omega_{s_1} \geq 2\pi(2B)$ and $\omega_{s_2} < 2\pi(2B)$ are shown in Fig. 8.2.9a,b,c. In the case of ω_{s1}, the message signal can be recovered. In the second case of ω_{s2}, the sampling rate is lower than the Nyquist rate. The resultant spectra of the ideally sampled signal will have overlaps of the adjoining spectra and the spectral components are added around half the sampling frequency and the message signal cannot be recovered by low-pass filtering the ideally sampled signal and the filtered signal will be distorted. The distortion caused by sampling below the Nyquist rate is called *aliasing*. Most signals are not band limited. Therefore, a band limiter is necessary before sampling to minimize the aliasing errors.

Example 8.2.5 Consider the amplitude spectrum of a function $x(t)$ given by

$$|X(j\omega)|^2 = \frac{2\omega_c}{(\omega)^2 + \omega_c^2}. \qquad (8.2.31)$$

The signal is sampled at the sampling frequency ω_s and an ideal low-pass filter is used to recover it from the sampled signal. Use MATLAB to quantify the effect of loosing the spectral energy outside of half the sampling frequency. For MATLAB, see Appendix B.

Solution: The energy contained in the signal is

$$E = \frac{1}{2\pi} \int_{-\infty}^{\infty} |X(j\omega)|^2 d\omega = \frac{2\omega_c}{2\pi} \int_{-\infty}^{\infty} \frac{1}{\omega^2 + \omega_c^2} d\omega$$

$$= \frac{2\omega_c}{2\pi\omega_c} \tan^{-1}(\omega/\omega_c)\big|_{-\infty}^{\infty} = \frac{1}{\pi}\pi = 1.$$

$$(8.2.32)$$

The signal is a low-frequency signal, as most of the spectral energy is concentrated around $f = 0$. It is not band limited. If it is filtered using an ideal low-pass filter with a cut-off frequency equal to half the sampling frequency $(\omega_s/2)$, then some information is lost and the loss can be measured using the spectral energy contained in the frequency range $|\omega| > \omega_s/2$ and

$$\text{Error} = \frac{1}{\pi} \int_{\omega_s/2}^{\infty} |X(j\omega)|^2 d\omega = \frac{1}{\pi} \int_{\omega_s/2}^{\infty} \frac{2\omega_c}{(\omega)^2 + \omega_c^2} d\omega$$

$$= 1 - \frac{2}{\pi} \tan^{-1}(\omega_s/2\omega_c).$$

$$(8.2.33)$$

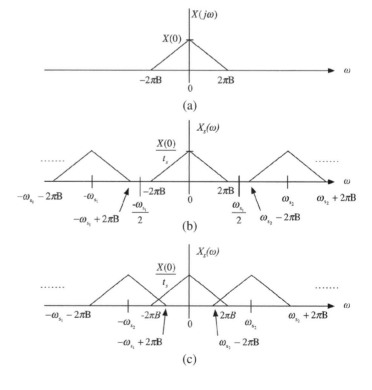

Fig. 8.2.9 (a) $X(j\omega)$, (b) $X_s(j\omega), \omega_{s_1} > 2\pi(2B)$ (sampling rate higher than the Nyquist rate), and (c) $X_s(j\omega), \omega_{s_2} < 2\pi(2B)$ (sampling rate lower than the Nyquist rate)

Fig. 8.2.10 Example 8.2.6

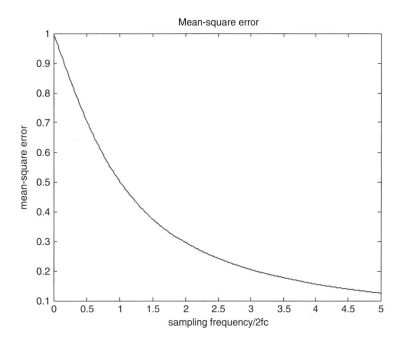

If the sampling rate goes to infinity, i.e., $\omega_s \to \infty$, then the error goes to zero as the area under the integral goes to zero. For the case of $\omega_s = 2\omega_c$ the error is 0.5 or 50%. The error slowly goes down as we increase the value of the sampling rate. A simple MATLAB routine and a sketch of the mean squared error as a function of the ratio of sampling frequency divided by $2f_c$ is given in Fig. 8.2.10. Note that $2f_c$ is *not* the Nyquist rate since the spectrum is not band limited to f_c Hz.

For the two cases $f_s/2f_c = 2$ and 4, the errors can be calculated and are 0.2952 and 0.1560, respectively. In the above example we need to have a high enough sampling rate to reduce the mean squared error. We use a pre-sampling filter that band limits the signal allowing for a decrease in sampling rate. Such a filter passes frequency components that are below the frequency $\omega_s/2$ and attenuates significantly or even suppress some of the frequency components above $\omega_s/2$. The band-limiting filter is referred to as an *anti-aliasing* filter. Even if the signal is band limited to $\omega_s/2$, an anti-aliasing filter is generally used to avoid aliasing that may result from noise that is ever present in almost all signals. The anti-aliasing filter may not be shown explicitly and is assumed to be included in the system. The bandwidth of the pre-sampling or anti-aliasing filter is signal dependent. In simple words, we state that most of the signal energy is contained within the bandwidth B Hz and the energy contained outside this band is negligible. See Section 4.7 for bandwidth measures. ∎

Most of the practical are low-pass signals that have decaying frequency response. One way to look at the aliasing error is to put a limit on the maximum aliasing error at half the sampling frequency which depends on the bandwidth of the signal. This works out nicely for signals that have decaying frequency response. The maximum error occurs at half the sampling frequency. See Spilker (1977), Ambardar (1995) and others. Another simple method is select the essential bandwidth which is taken as the frequency where the spectrum of the signal $x(t)$ given by $X(j\omega)$ reduces to say 1% of its peak value.

MATLAB Code for Fig. 8.2.10

```
x=0:.01:5;
y=1-(2/pi)*atan(x);
plot(x,y)
Title ('Mean-square error')
xlabel ('sampling frequency/2fc')
ylabel ('mean-square error')
```

Example 8.2.6 Let $x(t) = e^{-\alpha t}u(t) \overset{\text{FT}}{\longleftrightarrow} 1/(\alpha + j\omega) = X(j\omega)$, and

$$|X(j\omega)| = \frac{1}{\sqrt{\alpha^2 + \omega^2}}, X(0) = 1, \omega = 2\pi f.$$

Noting that the maximum aliasing error occurs at $\omega = \omega_s/2$, find the sampling frequency f_s using the following methods:

a. $\alpha = 1$. Maximum aliased magnitude is less than (1) 5% and (2) 1% of the peak value of the function $|X(j\omega)|$.

b. $\alpha = 2$. Use the bandwidth of $X(j\omega)$ as the frequency at which the amplitude reduces to 1% of the peak value.

Solution: a. (1). From the statement we have $|X(\omega_s/2)| \leq .05|X(0)| = .05$. Now

$$\frac{1}{\sqrt{1 + (\omega_s/2)^2}} \leq \frac{1}{20} \rightarrow [1 + (\omega_s/2)^2] < 400$$

$$\Rightarrow \omega_s = \sqrt{4(399)} \Rightarrow f_s \approx 6.36\,\text{Hz}.$$

(2). In this case, we have $|X(\omega_s/2)| \leq .01$ $|X(0)| = .01$. Correspondingly,

$$\frac{1}{\sqrt{1 + (\omega_s/2)^2}} \leq \frac{1}{100} \rightarrow [1 + (\omega_s/2)^2] > 1000$$

$$\Rightarrow \omega_s = \sqrt{(4)999} \Rightarrow f_s \geq 10.06\,\text{Hz} \quad \blacksquare$$

b. $|X(j\omega)| = 1/\sqrt{\omega^2 + 4} \Rightarrow 1\%$ of $|X(j0)| = (1/2)$ $(.01) = .005$. For $\omega \gg 2$, $|X(j\omega)| \approx 1/\omega$ and

$$|X(j\omega)| \approx 1/2\pi B = .005 \Rightarrow B = (100/\pi)\text{Hz}$$
$$\Rightarrow f_s \geq 2B = 200/\pi \approx 64\,\text{Hz}. \quad \blacksquare$$

Notes: For most signals, the actual spectrum may not be available and experimental methods may be needed to determine the bandwidth of these. $\quad \blacksquare$

Frequency sampling theorem: Since the Fourier transform and its inverse are related so closely, this theorem follows naturally the timesampling theorem. Consider a time-limited function such that $x(t) = 0, |t| > T_N$. It has a Fourier transform that can be uniquely determined from samples at frequency intervals of $n\pi/T_N$ and

$$X(j\omega) = \sum_{k=-\infty}^{\infty} X\left(\frac{jk\pi}{T_N}\right) \frac{\sin(\omega T_N - k\pi)}{(\omega T_N - k\pi)}. \quad (8.2.34)$$

For a proof and for additional discussion, see Mitra (2006). Some of the digital finite impulse response (FIR) filters are based upon frequency sampling.

Discrete-time signal bandwidth: The spectrum of an ideally sampled waveform $x_s(t)$ (see 8.2.2) is periodic with period ω_s and the measures of BW used for the continuous signals with nonperiodic spectrum cannot be used here. The ideally sampled signal is uniquely specified for frequencies in the range 0 to $f_s/2$. The bandwidth of $x_s(t)$ is the range of positive frequencies within the range $0-f_s/2$, for which the amplitude spectrum is greater than or equal to α times its maximum value, where α is a constant less than 1. The common one is $\alpha = 1/\sqrt{2}$ corresponding to the 3 dB bandwidth.

8.2.5 Flat Top Sampling

Flat top sampling uses a sample and hold device illustrated in Fig. 8.2.11 with the input is assumed to be

$$x_s(t) = \sum_{n=-\infty}^{\infty} x(nt_s)\delta(t - nt_s) \overset{\text{FT}}{\longleftrightarrow} X_s(j\omega). \quad (8.2.35)$$

The output of the summer is $x(t) - x(t - t_s)$. Let the input is $\delta(t)$. The output of the first block is $[\delta(t) - \delta(t - t_s)]$. The transfer functions of the first block and the integrator are

$$H_1(j\omega) = F[\delta(t) - \delta(t - t_s)]$$
$$= [1 - e^{-j\omega t_s}], H_2(j\omega) = 1/j\omega. \quad (8.2.36)$$

Assuming the input to the integrator is $[\delta(t) - \delta(t - t_s)]$, the output is

$$h(t) = \int_{-\infty}^{t} [\delta(\beta) - \delta(\beta - t_s)]d\beta$$
$$= \begin{cases} 1, & 0 < t < t_s \\ 0, & \text{otherwise} \end{cases} = \Pi\left[\frac{t - (t_s/2)}{t_s}\right]. \quad (8.2.37)$$

The system response to an impulse input is a rectangular pulse of width t_ss. This operation is the

Fig. 8.2.11 Zero-order-hold
(a) Use of a delay
component, a summer and
the process of integration,
(b) representation of these
using block diagrams

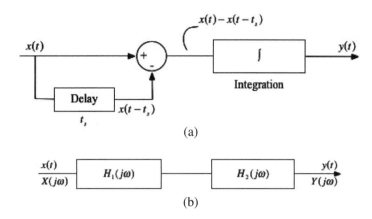

zero-order-hold (ZoH) referred to in the last section. The function $y(t)$ is approximated in the interval $nt_s \leq t < (n+1)t_s$ using the first term in the Taylor series.

$$y(t) = y(nt_s) + y'(nt_s)(t - nt_s) + \frac{1}{2!}y''$$

$$(nt_s)(t - nt_s)^2 + \cdots, nt_s \leq t < (n+1)t_s, \tag{8.2.38}$$

$$y(t) \approx y(nt_s), \ nt_s \leq t < (n+1)t_s. \tag{8.2.39}$$

It is called the zero-order-hold since the function is approximated by the constant term in the Taylor series. It can be approximated by the first two terms in the series. That is,

$$y(t) \approx y(nt_s) + y'(nt_s)(t - nt_s), nt_s \leq t < (n+1)t_s,$$

$$y'(nt_s) = \frac{dy(t)}{dt}\Big|_{t=nt_s}. \tag{8.2.40}$$

Most systems use ZoH. The transfer function of the cascaded blocks in Fig. 8.2.11 is

$$H_0(j\omega) = H_1(j\omega)H_2(j\omega) = [1 - e^{-j\omega t_s}]/j\omega,$$

$$= t_s \frac{e^{j\omega t_s/2} - e^{-j\omega t_s/2}}{2j} \frac{1}{\omega t_s/2} e^{-j\omega t_s/2}$$

$$= t_s \, \text{sinc}(\omega t_s/2)e^{-j\omega t_s/2}. \tag{8.2.41}$$

The amplitude, phase, and the delay frequency responses are

$$|H_0(j\omega)| = t_s|\text{sinc}(\omega t_s/2)|, \angle H_0(j\omega)$$

$$= \begin{cases} -(\omega t_s/2), \text{sinc}(\omega_s \tau/2) > 0 \\ -(\omega t_s/2) \pm \pi, \text{sinc}(\omega_s \tau/2) < 0 \end{cases}$$

$$\text{Group delay} = -\frac{d\angle H_0(j\omega)}{d\omega} = \frac{t_s}{2X(j\omega)},$$

$$Y(j\omega) = H_0(j\omega)X_s(j\omega). \tag{8.2.42}$$

ZoH introduces three modifications:

1. A linear phase shift corresponding to a time delay of $(t_s/2)$s.
2. The input transform is band limited to ω_m. The output transform $Y(j\omega)$ is a *distorted version* of $X(j\omega)$ affected by the curvature of the main lobe of $H_0(j\omega)$.
3. Transform of the ideally sampled signal is periodic. The envelope of the sinc function is inversely proportional to the frequency. The output transform contains distorted and attenuated versions of the images of $X(j\omega)$, centered at nonzero multiples of ω_s.

The first one follows since each sample is held constant for t_s seconds. The last two effects are caused by constant or step interpolation resulting in high frequency components. The effects of the first two items can be reduced by increasing the sampling rate. If the effects of the last two items are not acceptable, then a continuous-time *compensation filter* (an *anti-imaging filter*) in cascade with the zero-order hold is needed. The transfer function of such is given in (8.2.43). See Haykin and Van Veen (2003).

$$H_c(j\omega) \approx \begin{cases} \dfrac{\omega t_s}{2 \sin(\omega t_s/2)}, & |\omega| < \omega_m < \dfrac{\omega_s}{2} \\ 0, & |\omega| > \omega_s - \omega_m \end{cases} \tag{8.2.43}$$

Notes: The output of an ideal sampler and a "flat top" sampler are

$$x_s(t) = \sum_{n=-\infty}^{\infty} x(nt_s)\delta(t-nt_s), \qquad (8.2.44a)$$

$$x_{\text{flat top}}(t) = \sum_{n=-\infty}^{\infty} x(nt_s)\Pi\left[\frac{t-t_s/2-nt_s}{t_s}\right]. \ (8.2.44b)$$

An ideal low-pass filter is needed to recover $x(t)$ from $x_s(t)$. On the other hand, an anti-imaging filter (see (8.2.43)) is necessary to recover $x(t)$ from $x_{\text{flat top}}(t)$. In both cases, we assumed that $f_s > 2B$, $B =$ bandwidth of the signal in Hertz. The digital-to-analog (D/A) converter with a ZoH takes the sequence $x[n]$ and creates a signal $x_{\text{flat top}}(t)$.

So far, the discussion was centered on the low-pass signals. Can a sampling rate less than the twice the highest frequency of a *band-pass signal* and still recover the input signal? The answer is yes and is illustrated below. See Ziemer and Tranter (2002).

8.2.6 Uniform Band-Pass Sampling Theorem

Given a signal $x(t) \overset{\text{FT}}{\longleftrightarrow} X(j\omega)$ with

$$|X(j\omega)| = 0, \omega = 2\pi f, \begin{cases} |f| \le f_l \\ |f| \ge f_u \end{cases}.$$

$$[\text{bandwidth} = B = (f_u - f_l)\text{Hz}]. \qquad (8.2.45)$$

The signal $x(t)$ can be recovered from the sampled signal if the sampling rate is $f_s = (2f_u/m)$, where m the largest integer that is *not* exceeding (f_u/B). All higher sampling rates are *not necessarily usable* unless they exceed the Nyquist rate of $2f_u$.

Example 8.2.7 Consider the band-pass signal spectrum $X(j\omega)$ shown in Fig. 8.2.12a. with $f_l = 4$ kHz and $f_u = 5$ kHz(bandwidth is 1 kHz). Using the band pass sampling theorem, sketch the ideally sampled signal spectrum assuming a sampling rate that allows for the recovery of the original signal from the sampled signal.

Solution: Note

$$f_s = 2f_u/m, \ m = \text{Integer part of}(f_u/B) = 5,$$
$$f_s = 2(5)/5 = 2\,\text{kHz}, \qquad (8.2.46a)$$

$$X_s(j\omega) = \frac{\omega_s}{2\pi}\sum_{k=-\infty}^{\infty} X(j(\omega - k\omega_s)). \qquad (8.2.46b)$$

The band-pass signal can be recovered by noting that none of the spectra $X(j(\omega \pm k\omega_s)), k \ne 0$ over laps the spectrum of the continuous signal $x(t)$. The spectra in (8.2.46b), $X(j(\omega - k\omega_s))$ are shifted to the right by $k\omega_s$ for k positive and to the left by $k\omega_s$ for k negative. In Table 8.2.2, the frequency ranges of the terms in (8.2.46b) for $n = 0, \pm 1, \pm 2, \pm 3$ and the center frequencies of the corresponding spectra are given. For example $X(j(\omega - \omega_s))$ is centered at 2 kHz and occupies the frequency range $(-3\,kHz < f < -2\ kHz,$

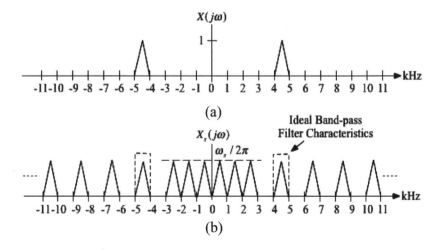

Fig. 8.2.12 (a) Band-pass spectra, (b) spectra of the ideally sampled signal

Table 8.2.2 Spectral occupancy of $X(j(\omega - n\omega_s)), \omega = 2\pi f$, $n = 0, \pm 1, \pm 2, \pm 3$

Spectra	Frequency ranges (f, kHz)	(nf_s, kHz)
$X(j\omega)$	$-5 < f < -4, 4 < f < 5$	0
$X(j(\omega - \omega_s))$	$-3 < f < -2, 6 < f < 7$	2
$X(j(\omega + \omega_s))$	$-7 < f < -6, 2 < f < 3$	-2
$X(j(\omega - 2\omega_s))$	$-1 < f < 0, 8 < f < 9$	4
$X(j(\omega + 2\omega_s))$	$-9 < f < -8, 0 < f < 1$	-4
$X(j(\omega - 3\omega_s))$	$1 < f < 2, 10 < f < 11$	6
$X_j((\omega + 3\omega_s))$	$-11 < f < -10, -2 < f < -1$	-6

$6\ kHz < f < 7\ kHz$). This can be continued with the entries given in the table and the spectra do not overlap. By using an ideal band-pass filter with a pass-band in the range $4\,kHz < f < 5\,kHz$, the original signal can be recovered. Figure 8.2.12a gives an assumed spectrum for $X(j\omega)$ and Fig. 8.2.12b gives the ideally sampled signal spectrum corresponding to the terms identified in Table 8.2.2. ∎

8.2.7 Equivalent continuous-time and discrete-time systems

Figure 8.2.13 gives flow diagrams illustrating the parts of the discrete and the corresponding continuous-time processes in terms of transfer functions. The output transforms in the two systems can be related to the input transforms and are

$$Y(j\omega) = (1/t_s)H_0(j\omega)H_c(j\omega)H(e^{j\omega t_s})H_a(j\omega)X(j\omega)$$
$$= G(j\omega)X(j\omega), \qquad (8.2.47a)$$

$$G(j\omega) = (1/t_s)H_0(j\omega)H_c(j\omega)H(e^{j\omega t_s})H_a(j\omega). \qquad (8.2.47b)$$

Note the transfer function $H(e^{j\omega t_s})$ is used to represent the discrete-data system. If the anti-aliasing and anti-imaging filters are designed to compensate for the effects of sampling and reconstruction, then

$$(1/t_s)H_0(j\omega)H_c(j\omega)H_a(j\omega) \approx 1. \qquad (8.2.47c)$$

8.3 Basic Discrete-Time (DT) Signals

Discrete-time (DT) signals are expressed as simple time sequences, such as $x[n]$, where n is an integer. It exists only at integer values of n.

Unit step sequence: The unit step sequence $u[n]$ and its delayed (or shifted) versions are defined by

$$u[n] = \begin{cases} 1, & n \geq 0 \\ 0, & n < 0 \end{cases}, \quad u[n-k] = \begin{cases} 1, & n \geq k \\ 0, & n < k \end{cases}. \qquad (8.3.1)$$

The discrete-time unit step sequence has the defined value $u[n]|_{n=0} = 1$. The unit step function $u(t)$ is not defined at $t = 0$. The unit step sequence and the delayed step sequences are illustrated in Fig. 8.3.1a,b. If the integer constant k in (8.3.1) is negative (positive), then the function is an advanced (delayed) unit step sequence.

Unit sample sequences: The unit sample sequence $\delta[n]$ and its delayed version are defined by

$$\delta[n] = \begin{cases} 1, & n = 0 \\ 0, & n \neq 0 \end{cases}, \quad \delta[n-k] = \begin{cases} 1, & n = k \\ 0, & n \neq k \end{cases}. \qquad (8.3.2)$$

(a)

(b)

Fig. 8.2.13 Flow diagrams of a discrete-time processing of continuous signals: (a) discrete-time system with filters on both ends and (b) equivalent continuous-time system.

Fig. 8.3.1 (a) $u[n]$, (b) $u[n-k]$

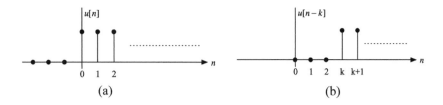

Fig. 8.3.2 (a) $\delta[n]$, (b) $\delta[n-k]$

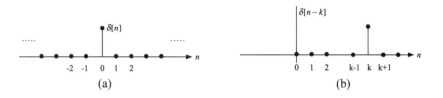

These are illustrated in Fig. 8.3.2. Note that $\delta[0] = 1$ and $\delta[n-k]$ is 1 at $n = k$. The unit sample sequence is easier to visualize than the impulse function in the continuous domain. With these definitions, a sequence of numbers can be expressed as a discrete-time sequence using the unit sample sequence. For example, if a discrete-time signal $x[n]$ is multiplied by the unit discrete-time impulse (or unit sample sequence) $\delta[n-k]$ then the result is a discrete-time impulse with its value equal to $x[k]$. That is,

$$x[n]\delta[n-k] = x[k]\delta[n-k]. \qquad (8.3.3)$$

The discrete impulse *shifts* out the value of the function $x[n]$ at $n = k$ and

$$\sum_{n=-\infty}^{\infty} x[k]\delta[n-k] = x[n]. \qquad (8.3.4)$$

In the continuous-time case we deal with integrals and in the discrete-time case we deal with summations. The time scaled discrete-time impulse function is defined for integer values of α and

$$\delta[\alpha n] = \delta[n]. \qquad (8.3.5a)$$

In the case of time scaled continuous-time impulse function was defined (see (1.4.35))

$$\delta(\alpha t) = (1/|\alpha|)\delta(t). \qquad (8.3.5b)$$

Note the scale factor α is not there in (8.3.5a). As examples, $\delta(2t) = (1/2)\delta(t)$ and $\delta[2n] = \delta[n]$.

Arbitrary sequence: Let the discrete-time sequence is given by (8.3.4). For simplicity we write the sequence in the form

$$\{x[n]\} = \{\ldots, x[-1], x[0], x[1], x[2], \ldots\} \quad (8.3.6)$$

This explicitly identifies the location of each value on the time axis. On the other hand, if the sequence is given in terms of a set of numbers, we do not know the time locations and their values. To alleviate this problem, an arrow is included on the value below the number corresponding to the value of the function at $n = 0$. The following illustrates this notation:

$$\{x[n]\} = \{\ldots, x[-2], x[-1], x[0], x[1], x[2], \ldots\}$$
$$= \{\ldots, -3, 0, 0, 1, 2, -3, \ldots\}. \qquad (8.3.7)$$

In this case, $x[0] = 2$ and the values to the left of this corresponds to the function with negative values of n and the values to the right corresponds to the positive values of n. We can use (8.3.4) to represent (8.3.7) by

$$x[n] = \cdots - 3\delta[n+4] + \delta[n+1]$$
$$+ 2\delta[0] - 3\delta[n-1] + \cdots. \qquad (8.3.8a)$$

This provides a nice way to look at discrete sequences that will lead to difference equations. The DT unit step function can be written in terms of DT impulses as

$$u[n] = \delta[0] + \delta[1] + \delta[2] + \cdots$$

$$= \sum_{k=-\infty}^{\infty} u[k]\delta[n-k] = \sum_{k=0}^{\infty} \delta[n-k], \quad (8.3.8b)$$

$$\Rightarrow \delta[n] = u[n] - u[n-1]. \quad (8.3.8c)$$

Example 8.3.1 Sketch the following functions:

(a) $x_1[n] = \delta[n-1] + 2\delta[n-2] + 3\delta[n-3]$,

(b) $x_2[n] = \alpha^n u[n]$, $0 < \alpha < 1$.

Solution: Sketches are shown in Fig. 8.3.3. ∎

Before considering the special cases, we should note that *folding or reversal* of a sequence $x[n]$ is defined by $x[-n]$. The following illustrates the time reversal sequences:

$$u[-n] = \begin{cases} 1, & n = 0, -1, -2, -3, \ldots \\ 0, & n = 1, 2, 3, \ldots \end{cases},$$

$$u[1-n] = \begin{cases} 1, & n \le 1 \\ 0, & n > 1 \end{cases}.$$

Even and odd sequences: Even sequence $x_e[n]$ and an odd sequence $x_0[n]$ satisfy the following relations:

$$x_e[-n] = x_e[n] \text{ and } x_0[-n] = -x[n]. \quad (8.3.9a)$$

Noting $x_0[0] = 0$, we have

$$\sum_{n=-K}^{K} x_e[n] = x_e[0] + \sum_{n=0}^{K} x_e[n] \text{ and}$$

$$\sum_{n=-K}^{K} x_0[n] = 0. \quad (8.3.9b)$$

An arbitrary discrete-time (DT) sequence can be expressed by

$$x[n] = x_e[n] + x_0[n] \Rightarrow x_e[n] = \{x[n] + x[-n]\}/2,$$
$$x_0[n] = \{x[n] - x[-n]\}/2. \quad (8.3.9c)$$

The unit step function can be written in terms of its even and odd parts.

$$u[n] = u_e[n] + u_0[n], u_e[n] = (1/2) + (1/2)\delta[n],$$
$$u_0[n] = (1/2)u[n] - (1/2)u[-n]. \quad (8.3.9d)$$

Discrete-time sinc sequence: It is an even sequence and exists at integer values of n.

$$\text{sinc}(n\pi/N) = \frac{\sin(n\pi/N)}{(n\pi/N)},$$
$$\text{sinc}(0) = 1, \text{sinc}(n\pi/N) = 0,$$
$$n = kN, k = \pm 1, \pm 2, \ldots. \quad (8.3.10)$$

8.3.1 Operations on a Discrete Signal

The operations on discrete-time signals are very simple and are similar to the analog case, with few exceptions. The addition of discrete-time sequences $x[n]$ and $y[n]$ is defined by the pointwise sum $x[n] + y[n]$. Similarly, the product of the two functions $y[n]$ and $x[n]$, i.e., the pairwise product is given by $y[n]x[n]$. The time shift, i.e. delay or advance of a function by k is expressed by $x[n-k]$ and $x[n+k]$, respectively. The amplitude scaling by a constant c of a function $x[n]$ is expressed by $cx[n]$. The amplitude shift of a sequence by a sequence of constants A is given by the sequence $A + x[n]$. There are two important operations that are very useful. These are *decimation and interpolation*.

The *decimation* of a sequence $x[n]$ by c is given by $x[cn]$, where c is an integer. If $c = 2$, the decimation operation deletes alternate samples resulting in compression. Decimation by a factor N, i.e., $c = N$ in $x[cn]$ is sampling the given function at

Fig. 8.3.3
Example 8.3.1 (a) $x_1[n]$, (b) $x_2[n]$

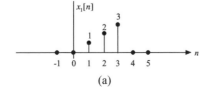

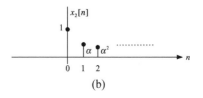

(a)   (b)

intervals of N samples resulting in $x[n] \rightarrow x[Nn]$. The operation retains every N th sample corresponding to the indices Nn and discarding others. For example

$$\{x[n]\} = \{1, -1, 1, -1, 1, -1, \ldots\} \rightarrow \{x[2n]\}$$
$$\uparrow$$
$$= \{1, 1, 1, \ldots\}. \qquad (8.3.11)$$
$$\uparrow$$

The inverse of decimation is the *interpolation*. The interpolation of a sequence $x[n]$ by c is given by $x[n/c]$, where c is an integer. In the case of $x[n/2]$, the given sequence is stretched by introducing an

interpolated value between each pair of sample values. Interpolation operation slows the signal and the new sequence is twice the length of the original sequence. How do we get the interpolated values? In Section 8.2 of this chapter we considered the interpolation of a sampled function by various methods, including the use of the Fourier transform of the signal. Here we only have a discrete set of numbers. There are several options. These include zero interpolation (a zero between samples), step interpolation (a constant equal to the previous sample), linear interpolation (average of the adjacent samples), and many others. These simple methods are illustrated below.

$$\{x[n]\} = \{\ldots, x[-1], x[0], x[1], x[2], \ldots\} \xrightarrow[n \to 2n]{\text{zero interpolation}} \{\ldots, x[-1], 0, x[0], 0, x[1], 0, x[2], \ldots\}, \qquad (8.3.12)$$

$$\{x[n]\} = \{\ldots, x[-1], x[0], x[1], x[2], \ldots\} \xrightarrow[n \to 2n]{\text{step interpolation}} \{\ldots, x[-1], x[-1], x[0], x[0], x[1], x[1], x[2], x[2], \ldots\},$$
$$(8.3.13)$$

$$\{x[n]\} = \{\ldots, x[-1], x[0], x[1], x[2], \ldots\} \xrightarrow[n \to 2n]{\text{Linear interpolation}}$$
$$\{\ldots, x[-1], .5(x[-1] + x[0]), x[0], .5(x[0] + x[1]), x[1], .5(x[1] + x[2]), x[2], \ldots\}. (8.3.14)$$

It may seem that decimation and interpolation operations are inverse operations. This is *not* true. Interpolation of the decimated sequence in (8.3.11) will not recover the original data. If we have the interpolated sequence, the original sequence can be recovered.

We can define a variable l, a function of n. For example, the sequence defined by $x[l] = x[-2n + 3]$ gives a new sequence that is obtained by shifting the sequence to the left by 3, folding the resultant sequence and then decimating the sequence by 2. This is summarized by the following:

$$\{x[n]\} \xrightarrow{\text{shift to the left}} \{x[n+3]\} \xrightarrow{\text{Fold}} x[-n+3]$$
$$\xrightarrow[n \to 2n]{\text{Decimate}} x[-2n+3].$$

Example 8.3.2 Determine the sequence $x[-2n+3]$ assuming $x[0] = 1$, $x[1] = 2$, $x[2] = 0$, $x[3] = 3$, $x[4] = 4$ and $x[n] = 0$ for other n.

Solution: First, let

$$\{n\} = \{0, 1, 2, 3\} \rightarrow \{-2n + 3\}$$
$$= \{3, 1, -1, -3, -5\},$$
$$\{x[n]\} = \{1, 2, 0, 3, 4\} \rightarrow \{x[-2n + 3]\}$$
$$\uparrow$$
$$= \{x[3], x[1], x[-1], x[-3], x[-5]\}$$
$$\rightarrow \{3, 2, 0, 0, 0\}.$$
$$\uparrow$$

Using the shift, fold, and decimation operations, the following results:

$$\{x[n]\} = \{1, 2, 0, 3, 4\} \xrightarrow[\text{the left by 3}]{\text{Shift } x[n] \text{ to}} \{1, 2, 0, 3, 4\}$$
$$\uparrow \qquad\qquad\qquad\qquad \uparrow$$
$$\xrightarrow{\text{fold}} \{4, 3, 0, 2, 1\} \xrightarrow{\text{decimate by 2}} \{3, 2\}$$
$$\uparrow$$
$$= \{x[-2n + 3]\}. \qquad (8.3.15) \quad \blacksquare$$

8.3.2 Discrete-Time Convolution and Correlation

Table 8.3.1 Properties of discrete convolution

Definition:

$$y[n] = x[n]^*h[n] = \sum_{k=-\infty}^{\infty} x[k]h[n-k] = \sum_{k=-\infty}^{\infty} h[k]x[n-k] \qquad (8.3\text{-}16)$$

Commutative property:

$$x[n]^*h[n] = x[n]^*h[n] \qquad (8.3\text{-}17\text{a})$$

Distributive property:

$$x_1[n]^*(x_2[n] + x_3[n]) = x_1[n]^*x_2[n] + x_1[n]^*x_3[n] \qquad (8.3\text{-}17\text{b})$$

Associative property:

$$x_1[n]^*(x_2[n]^*x_3[n]) = (x_1[n]^*x_2[n])^*x_3[n] \qquad (8.3\text{-}17\text{c})$$

Shifting property:

$$\text{If } y[n] = x_1[n]^*x_2[n] \text{ then } x_1[n-k]^*x_2[n-m] = y[n-(k+m)]. \qquad (8.3\text{-}17\text{d})$$

Convolution with a discrete impulse:

$$x[n]^*\delta[n] = x[n] \qquad (8.3\text{-}17\text{e}6)$$

Width property:

Given $x[n] \neq 0$ for $n = 0, 1, 2, ..., N-1$ and $h[n] \neq 0$ for $n = 0, 1, 2, ..., M-1$ (8.3-18)
If we define $y[n] = h[n]^*x[n]$, then the length of the sequence $y[n]$ is $N+M-1$.
   $y[n] = 0$ for $n < 0$ and $n \geq N + M - 1$.

The discrete time convolution of two sequences $x[n]$ and $h[n]$ is defined and its properties are listed in Table 8.3.1

Most of these properties are simple to prove. The width property will be proved shortly.

Example 8.3.3 Determine the convolution $y[n] = h[n] * x[n]$ of the sequences $x[n] = \alpha^n u[n]$, $h[n] = \beta^n u[n]$, $|\alpha| < 1$, $|\beta| < 1$, $|\alpha/\beta| < 1$ for the following cases: a. $\alpha \neq \beta$ and $|\alpha/\beta| < 1$, b. $\alpha = \beta$ using

$$\sum_{k=0}^{n} \lambda^k = \frac{1 - \lambda^{n+1}}{1 - \lambda}, |\lambda| < 1. \qquad (8.3\text{-}19\text{a})$$

Solution: a. The convolution sum is

$$y[n] = \sum_{m=-\infty}^{\infty} x[m]h[n-m] = \sum_{m=-\infty}^{\infty} \alpha^m u[m]\beta^{n-m}u[n-m]$$

$$= \beta^n \sum_{m=0}^{n}\left[\frac{\alpha}{\beta}\right]^m = \beta^n \frac{1-(\alpha/\beta)^{n+1}}{1-(\alpha/\beta)}, = \left[\frac{\alpha^{n+1}-\beta^{n+1}}{(\alpha-\beta)}\right],$$

$$\alpha \neq \beta, n \geq 0; \; y[n] = 0, n < 0., \qquad (8.3\text{-}19\text{b})$$

b. In this case,

$$y[n] = \alpha^n u[n] * \alpha^n u[n] = \sum_{m=-\infty}^{\infty} \alpha^m u[m]\alpha^{n-m}u[n-m]$$

$$= \alpha^n \sum_{m=0}^{n} 1 = (n+1)\alpha^n u[n]. \qquad (8.3\text{-}19\text{c}) \quad \blacksquare$$

Correlation of two sequences: The *cross correlation* of two sequences $x[n]$ and $h[n]$ is defined by

$$r_{xh}[k] = x[k]**h[k] = \sum_{k=-\infty}^{\infty} x[n]h[n+k]$$

$$= x[k] * h[-k], \qquad (8.3\text{-}20\text{a})$$

$$r_{hx}[k] = h[k]**x[k] = \sum_{n=-\infty}^{\infty} h[n]x[n+k]$$

$$= h[k] * x[-k]. \qquad (8.3\text{-}20\text{b})$$

Note (**) is used to define the correlation. The *autocorrelation (AC)* of $x[n]$ is even and

$$r_x[k] = x[k] **x[k] = \sum_{n=-\infty}^{\infty} x[n]x[n+k]$$

$$= x[k] * x[-k] = r_x[-k]. \tag{8.3.20c}$$

The cross-correlation functions can be expressed in terms of the convolution and

$$r_{xh}[k] = x[k] * h[-k] = r_{hx}[-k]. \tag{8.3.20d}$$

Example 8.3.4 Find the autocorrelation of the sequence $x[n] = a^n u[n]$, $|a| < 1$.

Solution: Need to find the values of $r_x[k]$ for ≤ 0, as AC function is even.

$$r_x[k] = \sum_{n=-\infty}^{\infty} a^n u[n] a^{n+k} u[n+k] = a^k \sum_{n=0}^{\infty} a^{2n}$$

$$= \frac{a^k}{1-a^2} u[k] \rightarrow r_x[k] = \frac{1}{(1-a^2)} a^{|k|}. \tag{8.3.21}$$ ∎

8.3.3 Finite duration, right-sided, left-sided, two-sided, and causal sequences

Finite duration or time-limited sequence:

$x[n] \neq 0$ for $n_1 \leq n \leq n_2$ and zero otherwise (8.3.22a)

Right-sided sequence : $x[n] = 0$ for

   $n < n_1, n_1$ is finite   (8.3.22b)

Left-sided sequence : $x[n] = 0$ for $n > n_2$,

   n_2 is finite   (8.3.22c)

Two-sided sequence : $x[n]$ can be of

   infinite extent   (8.3.22d)

Causal sequence : $x[n] = 0, n < 0$ (8.3.22e)

Discrete-time exponentials: One-sided and the two-sided exponentials are defined by

$$x_1[n] = a^n u[n], \quad x_2[n] = x_1[-n] - 1,$$

$$x_3[n] = a^{|n|}. \tag{8.3.23}$$

In (8.3.23), $x_1[n]$ is a right-sided exponential sequence, $x_2[n]$ is a left-sided sequence, and $x_3[n] = a^{|n|}$ is a two-sided sequence. If a is real in

(8.3.23), then the sequences are exponentially decaying if $a < 1$ and exponentially increasing if $a > 1$. In addition, if $a = 1$, $x_1[n]$ reduces to the unit step sequence and $x_3[n]$ reduces to the value of 1 for all integer values of n. If a is complex, then we can write the two-sided complex exponential sequence by

$$x[n] = a^n = (re^{j\theta})^n = r^n e^{jn\theta}$$

$$= r^n[\cos(n\theta) + j\sin(n\theta)]. \tag{8.3.24}$$

The real and the imaginary parts are exponentially decaying if $|a| < 1$ and exponentially growing sinusoidal sequences if $|a| > 1$.

8.3.4 Discrete-Time Energy and Power Signals

The *energy* in a *sequence* $x[n]$ is

$$E = \sum_{k=-\infty}^{\infty} |x[k]|^2. (x[k] \text{ can be real or complex}).$$

$$\tag{8.3.25}$$

If the sequence $x[n]$ is obtained from an analog signal $x(t)$ through sampling at t_s second interval, i.e., $x[n] = x(nt_s)$ then the approximate energy in $x(t)$ is $t_s(E)$. A DT signal is an *energy signal* if and only if $0 < E < \infty$. The *normalized average power* of a DT signal is

$$P = \lim_{N \to \infty} \frac{1}{2N+1} \sum_{k=-N}^{N} |x[k]|^2. \tag{8.3.26}$$

The signal $x[n]$ is a *power signal* if and only if $0 < P < \infty$. If a sequence does not satisfy the properties of energy or a power signal, it is neither an energy nor a power sequence.

Example 8.3.5 Determine if the following signals are energy or power signals or neither.

$$a. \, x_1[n] = u[n], \; b. \, x_2[n] = .5^n u[n], \; c. \, x_3[n] = e^{jn}.$$

Solution: *a.* It is a power signal since its energy infinite, whereas its power is

$$P_1 = \lim_{N \to \infty} \frac{1}{2N+1} \sum_{k=-N}^{N} |u[k]|^2$$

$$= \lim_{N \to \infty} \frac{1}{2N+1} N = \frac{1}{2} < \infty. \tag{8.3.27a}$$

b. $x_2[n]$ is an energy signal since

$$E_2 = \sum_{n=-\infty}^{\infty} .5^n u[n] = \sum_{n=0}^{\infty} .5^n = \frac{1}{1-.25}$$

$$= \frac{4}{3} < \infty \Rightarrow P_2 = 0. \qquad (8.3.27b)$$

c. $x_3[n]$ is a power signal since

$$P_3 = \frac{1}{2N+1} \sum_{n=-N}^{N} |x_3[n]|^2 = \frac{1}{2N+1} \sum_{n=-N}^{N} |e^{jn}|^2$$

$$= 1 < \infty \Rightarrow E = \infty. \qquad (8.3.27c) \quad \blacksquare$$

Periodic sequences: A DT signal $x[n]$ is *periodic* if there is a positive integer N so that

$$x[n+N] = x[n] \equiv x_N[n] \text{ for all } n. \qquad (8.3.28)$$

The *fundamental period* of $x[n]$ in (8.3.28) is the smallest positive integer N. N_0 is usually used for the fundamental period. If there is no ambiguity, it is common to use N for this. The subscript on $x_N[n]$ gives the periodicity of the DT sequence. In addition to the even and odd DT symmetric sequences, other symmetries, such as half-wave symmetry can be defined. A symmetric periodic DT sequence is half-wave symmetric if

$$x_N[n] = -x_N[n \pm \tfrac{1}{2}N]. \qquad (8.3.29)$$

The *average value* and the *average power* in $x_N[n]$ with period N are

$$x_{\text{ave}} = \frac{1}{N} \sum_{k=0}^{N-1} x_N[k], \quad P = \frac{1}{N} \sum_{k=0}^{N-1} |x_N[k]|^2. \quad (8.3.30)$$

Here after, for simplicity, we *will not show the subscript N* on periodic $x[n]$.

For *non-periodic signals*, the limiting forms of the above are

$$x_{\text{ave}} = \lim_{M \to \infty} \frac{1}{2M+1} \sum_{k=-M}^{M} x[k],$$

$$P_{\text{ave}} = \lim_{M \to \infty} \frac{1}{2M+1} \sum_{k=-M}^{M} |x[k]|^2. \qquad (8.3.31)$$

All periodic and random signals are power signals.

Example 8.3.6 Show the complex exponential sequence is periodic with period N_0.

$$x[n] = e^{j(2\pi/N_0)n} = e^{j\Omega_0 n}, \Omega_0 = 2\pi/N_0. \qquad (8.3.32)$$

Solution: It is a periodic sequence since

$$x[n+N_0] = e^{j(2\pi/N_0)(n+N_0)}$$

$$= e^{j(2\pi/N_0)n} e^{j(2\pi)} = x[n]. \qquad (8.3.33)$$

There is an *important difference* between the *continuous exponential function* $e^{j\omega_0 t}$ and the *discrete exponential sequence* $e^{j\Omega_0 n}$. In the continuous case, the functions defined by $e^{j\omega_0 t}$ are *distinct* for *distinct values* of ω_0; whereas, the sequences $e^{j\Omega_0 n}$ that differ in *frequency* by a multiple of 2π are identical since

$$e^{j(\Omega_0 + 2\pi k)n} = e^{j(\Omega_0)n}. \qquad (8.3.34) \quad \blacksquare$$

Example 8.3.7 Find the constraints on Ω_0 so that $x[n] = e^{j\Omega_0 n}$ is periodic. That is,

Solution: To be periodic,

$$x[n+N] = e^{j\Omega_0(n+N)} = e^{j\Omega_0 n} e^{j\Omega_0 N}.$$

It is equal to $x[n]$ *only if* $e^{j\Omega_0 N} = 1$ or $\Omega_0 N = m(2\pi)$. That is,

$$\Omega_0/2\pi = m/M \text{ (a rational number).} \qquad (8.3.35)$$

The fundamental period is the smallest integer

$$N_0 = \frac{N}{m} = \frac{2\pi}{\Omega_0}. \qquad (8.3.36) \quad \blacksquare$$

Using Euler's formula, we have the *sinusoidal sequence* in terms of the exponential sequence

$$x[n] = A\cos(\Omega_0 n + \theta) = (A/2)e^{j(\Omega_0 n + \theta)}$$

$$+ (A/2)e^{-j(\Omega_0 n + \theta)}.$$

The cosine function is periodic if $(\Omega_0/2\pi)$ is a rational number. This can be generalized and state that if $x_1[n]$ and $x_2[n]$ are periodic with periods N_{01} and N_{02}, respectively, then the sum of these two sequences is periodic with period N_0 if

$$x[n + N_0] = x_1[n + N_0] + x_2[n + N_0]$$

$$= x_1[n + N_{01}] + x_2[n + N_{02}].$$

This is true only when

$$mN_{01} = kN_{02} = N_0. \qquad (8.3.37)$$

If so, then the fundamental period of $x[n]$ is the *least common multiple (LCM)* of N_{01} and N_{02}, which is divisible by both N_{01} and N_{02}.

Example 8.3.8 Consider the following sequences and determine if they are periodic. If a sequence is periodic, give its fundamental period.

a. $x_1[n] = e^{j(\pi/6)n}$,

b. $x_2[n] = e^{j(1/6)n}$

c. $x_3[n] = \cos(\pi/6)n + \cos(\pi/5)n$,

d. $x[n] = x_1[n] + x_2[n]$.

Solution: a. $x_1[n] = e^{j(\pi/6)n} = e^{j\Omega_0 n} \Rightarrow \Omega_0/2\pi = 1/12$. Since $(\Omega_0/2\pi) = (\pi/6)(1/2\pi) = (1/12)$ is a rational number, it follows that $x_1[n]$ is a periodic sequence and the fundamental period is $N_0 = (2\pi/\Omega_0) = (12)$.

b. $x_2[n] = e^{j(1/6)n} = e^{j\Omega_0 n}, \Omega_0 = (1/6)$ and $\Omega_0/2\pi = 1/12\pi$. It follows that $(\Omega_0/2\pi)$ is not a rational number and therefore $x_2[n]$ is not a periodic sequence.

c. We note that $\cos(\pi/6)n$ is periodic with period $N_{01} = 12$ and $\cos(\pi/5)n$ is periodic with period $N_{02} = 10$. It follows that $x[n]$ is periodic with period N_0 and is equal to the least common multiple of 10 and 12, which is $N_0 = 60$.

d. $x_4[n]$ is not periodic since one of the functions in the sum $x_2[n]$ is not periodic. ∎

Notes: A continuous-time sinusoid $\cos(\omega_0 t + \theta)$ is always periodic, regardless of the value of ω_0 and has a unique waveform for each value of ω_0. A discrete-time sinusoid $\cos(k\Omega_0 + \theta)$ is periodic only if $(\Omega_0/2\pi)$ is a rational number and does not have a unique waveform for each value Ω_0. Note $\cos(\Omega_0 k) = \cos((\Omega_0 + 2\pi)k) = \cos((\Omega_0 + 4\pi)k) = \cdots$.

8.4 Discrete-Time Fourier Series

In Chapter 3, a set of continuous orthogonal basis functions $\{e^{jk\omega_0 t}, k = 0, \pm 1, \pm 2, \ldots\}$ were used to find the complex F-series. Similarly, a set of *discrete orthogonal basis sequences* $\varphi_k[n]$ over an integer interval $[0, N-1]$ can be defined. If any two sequences $\varphi_m[n]$ and $\varphi_k[n]$ in the basis set satisfy

$$\sum_{n=0}^{N-1} \varphi_m[n]\varphi_k^*[n] = \begin{cases} 0, & m \neq k \\ \alpha_k \neq 0, & m = k \end{cases}. \qquad (8.4.1)$$

then the set is an orthogonal discrete-time basis set. Consider

$$\varphi_k[n] = e^{jk(2\pi/N)n}, \quad k, n = 0, 1, 2, \ldots, N-1. \quad (8.4.2)$$

These sequences are functions of two variables n and k and are periodic with period N in *both* the variables n and k. That is,

$$\varphi_{k+N}[n] = e^{j(k+N)(2\pi/N)n} = e^{jk(2\pi/N)n} = \varphi_k[n]$$

$$\varphi_k[n+N] = e^{jk(2\pi/N)(n+N)} = e^{jk(2\pi/N)n} = \varphi_k[n]. \quad (8.4.3)$$

Using the summation formula in (C.6.1b) in Appendix C, the set in (8.4.2) can be shown to be is an orthogonal set, i.e.

$$\sum_{n=0}^{N-1} \varphi_m[n]\varphi_k^*[n] = \sum_{n=0}^{N-1} e^{jm(2\pi/N)n} e^{-jk(2\pi/N)n}$$

$$= \sum_{n=0}^{N-1} e^{-j(2\pi/N)(k-m)}, \qquad (8.4.4)$$

$$= \frac{1 - e^{-j(m-k)(2\pi/N)N}}{1 - e^{-j(m-k)(2\pi/N)}} = \begin{cases} 0, & m \neq k \\ N, & m = k \end{cases}. \qquad (8.4.5)$$

When $m = k$, the summation reduces to N. Let $x_s[n]$ be an N point periodic sequence, i.e., $x_s[n] = x_s[n + N]$ for all n. Using the orthogonality condition in (8.4.5), the discrete F-series coefficients can be derived by expressing $x_s[n]$ in the following form in terms of the *discrete F-series coefficients* $X_{ds}[k]$ and solving for them.

$$x_s[n] = \frac{1}{N}\sum_{m=0}^{N-1} X_{ds}[m]e^{jm(2\pi/N)n}. \qquad (8.4.6a)$$

Multiplying both sides of (8.4.6a) by $e^{jk(2\pi/N)n}$ and summing over $n = 0$ to $N - 1$ and using (8.4.5) results in the discrete Fourier series (DFS) coefficient $X_{ds}[k]$.

$$\sum_{n=0}^{N-1} x_s[n]e^{-jk(2\pi/N)n} = \frac{1}{N}\sum_{n=0}^{N-1}\sum_{m=0}^{N-1} X_{ds}[m]e^{-j(2\pi/N)n(k-m)},$$

$$= \sum_{m=0}^{N-1} X_{ds}[m]\left[\sum_{n=0}^{N-1} e^{-j(2\pi/N)(k-m)}\right]$$

$$= NX_{ds}[k], \Rightarrow X_{ds}[k]$$

$$= \frac{1}{N}\sum_{n=0}^{N-1} x_s[n]e^{-jk(2\pi/N)n}. \quad (8.4.6b)$$

The DFS coefficients, the corresponding inverse discrete F-series (IDFS) values and the symbolic relations are as follows:

$$X_{ds}[k] = \frac{1}{N}\sum_{n=0}^{N-1} x_s[n]e^{-jk(2\pi/N)n},$$

$$x_s[n] = \sum_{k=0}^{N-1} X_{ds}[k]e^{jk(2\pi/N)n}, \quad (8.4.7)$$

$$x_s[n] \overset{\text{DFS}}{\longleftrightarrow} X_{ds}[k]. \quad (8.4.8)$$

Note

$$X_{ds}[k+N] = \frac{1}{N}\sum_{n=0}^{N-1} x_s[n]e^{-j(k+N)(2\pi/N)n}$$

$$= \frac{1}{N}\sum_{n=0}^{N-1} x_s[n]e^{-jk(2\pi/N)n} = X_{ds}[k]. \quad (8.4.9)$$

The DFS coefficients $X_{ds}[k]$ and the data sequence $x_s[n]$ are both periodic with period N.

Example 8.4.1 Let $x_s[n]$ is a real periodic sequence with period N and the DFS coefficients are $X_{ds}[k]$. Show that $X_{ds}[k] = X_{ds}^*[k]$ and $X_{ds}[N/2]$ are *real* when N is *even*.

Solution: Consider

$$X_{ds}[-k] = \frac{1}{N}\sum_{n=0}^{N-1} x_s[n]e^{j(2\pi/N)nk}$$

$$= \left[\frac{1}{N}\sum_{n=0}^{N-1} x_s[n]e^{-j(2\pi/N)nk}\right]^* = X_{ds}^*[k].$$

$$(8.4.10a)$$

Since the DFS coefficients are periodic, it can be shown that

$$X_{ds}[k] = X_{ds}^*[N-k]. \quad (8.4.10b)$$

For N even,

$$X_{ds}[N/2] = \frac{1}{N}\sum_{n=0}^{N-1} x_s[n]e^{-j(2\pi/N)n(N/2)},$$

$$= \frac{1}{N}\sum_{n=0}^{N-1} x_s[n]e^{-j(n\pi)} = \frac{1}{N}\sum_{n=0}^{N-1} (-1)^n x_s[n].$$

$$(8.4.11)$$

That is, the DFS coefficient $X_{ds}[N/2]$ of a real periodic sequence is real. ∎

Example 8.4.2 Determine the DFS coefficients for the $N(=4)$ point periodic sequence $x_s[n]$ given below in (8.4.12) and verify the above results.

$$\{x_s[n]\} = \{\ldots, x_s[0], x_s[1], x_s[2], x_s[3], x_s[0], x[1],$$
$$x[2], x_s[3], \ldots\},$$
$$x_s[0] = 1, x_s[1] = 2, x_s[2] = 3, x_s[3] = 4. \quad (8.4.12)$$

Solution: Noting $\Omega_0 = 2\pi/4$ and $e^{j(2\pi/4)} = e^{j\pi/2} = j$, the DFS coefficients are as follows:

$$X_{ds}[0] = (1/4)(1+2+3+4) = 2.5,$$

$$X_{ds}[1] = \frac{1}{4}\sum_{n=0}^{3} x_s[n]e^{-j(1)(2\pi/4)n} = \frac{1}{4}\sum_{n=0}^{3} (-j)^n x[n]$$

$$= \frac{1}{4}(1-j2-3+j4) = \frac{1}{2}(-1+j),$$

$$X_{ds}[2] = \frac{1}{4}\sum_{n=0}^{3} x_s[n](-j)^{2n}$$

$$= \frac{1}{4}(1-2+3-4) = -\frac{1}{2},$$

$$X_{ds}[3] = \frac{1}{4}\sum_{n=0}^{3} x_s[n](-j)^{3n} = \frac{1}{4}(1+j2-3-4j)$$

$$= -\frac{1}{2}(1+j).$$

Since, $N = 4$, an even integer and therefore $X_{ds}[2]$ must be real and is true here. Third, from (8.4.10b) $X_{ds}[1] = X_{ds}^*[4-1] = X_{ds}^*[3]$. Also, we can verify the result for $x_s[0]$ and

$$x_s[0] = X_{ds}[0] + X_{ds}[1] + X_{ds}[2] + X_{ds}[3]$$
$$= 2.5 - (1/2) + (1/2)(-1 + j)$$
$$- (1/2)(1 + j) = 1. \qquad \blacksquare$$

8.4.1 Periodic Convolution of Two Sequences with the Same Period

It can be shown that the *periodic convolution* of two periodic sequences $x_s[n]$ and $h_s[n]$ is periodic with the *same period N* and is

$$y_s[n] = x_s[n] \otimes h_s[n] = \frac{1}{N}\sum_{k=0}^{N-1} x_s[k]h_s[n-k]$$

$$= \frac{1}{N}\sum_{k=0}^{N-1} h_s[k]x_s[n-k] = y_s[n+N]. \qquad (8.4.13)$$

Example 8.4.3 Let $x_{1s}[n]$ and $x_{2s}[n]$ be two periodic sequences with period N and $x_s[n] = x_{1s}[n]x_{2s}[n]$. Derive the expression for $X_s[k]$ in terms of $X_{1s}[k]$ and $X_{2s}[k]$. Show that $X_s[k]$ is a periodic sequence.

Solution: First note $x_{is}[n] = x_{is}[n+N]$, $i = 1, 2$ and $x_s[n] = x_s[n+N]$. The DFS expansions of the three functions are, respectively, given by

$$x_s[n] = \sum_{k=0}^{N-1} X_{ds}[k]e^{j\Omega_0 kn}, x_{1s}[n] = \sum_{k=0}^{N-1} X_{d1s}[k]e^{j\Omega_0 kn},$$

$$x_{2s}[n] = \sum_{k=0}^{N-1} X_{d2s}[k]e^{j\Omega_0 kn}, \Omega_0 = 2\pi/N. \qquad (8.4.14)$$

The DFS coefficient

$$X_{ds}[k] = \frac{1}{N}\sum_{n=0}^{N-1} x_{1s}[n]x_{2s}[n]e^{-j\Omega_0 kn}$$

$$= \frac{1}{N}\sum_{n=0}^{N-1}\left[\sum_{m=0}^{N-1} X_{1s}[m]e^{j\Omega_0 mn}\right]x_{2s}[n]e^{-j\Omega_0 kn},$$

$$= \sum_{m=0}^{N-1} X_{d1s}[m]\left[\frac{1}{N}\sum_{n=0}^{N-1} x_{2s}[n]e^{-j(k-m)\Omega_0 n}\right]$$

$$= \sum_{m=0}^{N-1} X_{d1s}[m]X_{d2s}[k-m]. \qquad (8.4.15)$$

In simplifying the above equation, the following result is used:

$$\frac{1}{N}\sum_{n=0}^{N-1} x_{2s}[n]e^{-j(k-m)\Omega_0 n} = X_{2s}[k-m].$$

From (8.4.15), the *discrete frequency domain convolution* and its symbolic form are

$$X_{ds}[k] = \sum_{m=0}^{N-1} X_{d1s}[m]X_{d2s}[k-m]$$

$$= \sum_{l=0}^{N-1} X_{d1s}[k-l]X_{d2s}[l], \qquad (8.4.15)$$

$$X_{ds}[k] = X_{d1s}[k] \otimes X_{d2s}[k]. \qquad (8.4.16)$$

Noting that $X_{d1s}[k] = X_{d1s}[k+N]$ and $X_{d2s}[k] = X_{d2s}[k+N]$, it follows that

$$X_{ds}[k+N] = \sum_{m=0}^{N-1} X_{d1s}[m]X_{d2s}$$

$$[k+N-m] = X_{ds}[k]. \qquad (8.4.17) \quad \blacksquare$$

8.4.2 Parseval's Identity

The generalized *Parseval's identity* of sequences $x_{1s}[n]$ and $x_{1s}[n]$ is

$$\frac{1}{N}\sum_{n=0}^{N-1} x_{d1s}[n]x_{d2s}[n] = \sum_{m=0}^{N-1} X_{d1s}[m]X_{d2s}[-m]. \quad (8.4.18a)$$

This follows from

$$X_{ds}[k]|_{k=0} = \frac{1}{N}\sum_{n=0}^{N-1} x_{1s}[n]x_{2s}[n]e^{-j\Omega_0 kn}|_{k=0}$$

$$= \sum_{m=0}^{N-1} X_{d1s}[m]X_{d2s}[k-m]|_{k=0}.$$

The *Parseval's identity* of a single sequence can be obtained by using $x_{2s}[n] = x_{1s}^*[n] = x_s[n]$ in (8.4.18a) resulting in

$$\frac{1}{N}\sum_{n=0}^{N-1} |x_s[n]|^2 = \sum_{k=0}^{N-1} |X_{ds}[k]|^2. \qquad (8.4.18b)$$

Example 8.4.4 Verify the Parseval's identity using the sequence in Example 8.4.2.

Solution:

$$\frac{1}{N}\sum_{n=0}^{N-1}|x_s[n]|^2 = \frac{1}{4}(1^2 + 2^2 + 3^2 + 4^2) = \frac{15}{2},$$

$$= \sum_{k=0}^{N-1}|X_{ds}[k]|^2 = (2.5)^2 + \frac{1}{4}(2) + \frac{1}{4} + \frac{1}{4}(2) = \frac{15}{2}. \blacksquare$$

8.5 Discrete-Time Fourier Transforms

Computation of the continuous F- and the inverse F-transforms involves integrals and analytical computation is possible only in a few cases. Also, the signal may be available in the form of a waveform instead of an analytical expression or in terms of a sequence. The discrete-time Fourier transform of a sequence is derived from the discrete F-series by taking the number of sample points in the discrete F-series to infinity. Presentation is *intuitive* and follows that of Haykin and Van Veen (2003).

8.5.1 Discrete-Time Fourier Transforms (DTFTs)

Let $x[n]$ be a non-periodic sequence obtained from a single period of the periodic sequence centered at the origin.

$$x[n] = \begin{cases} x_s[n], & |n| \leq M \\ 0, & |n| > M \end{cases}. \quad (8.5.1)$$

That is, one period of the periodic sequence is extracted and then it is padded with zeros outside of the period. In (8.5.1), as M increases, the periodic replicates of $x[n]$ move further and further away from the origin. The discrete-time F-series representation of the periodic signal and the DFS coefficients are as follows:

$$x_s[n] = \sum_{k=-M}^{M} X_{ds}[k]e^{jk\Omega_0 n},$$

$$X_{ds}[k] = \frac{1}{2M+1}\sum_{n=-M}^{M} x_s[n]e^{-j\Omega_0 nk}, \quad \Omega_0 = \frac{2\pi}{2M+1}. \quad (8.5.2)$$

From (8.5.1),

$$x_s[n] = x[n], \ |n| \leq M \text{ and } x[n] = 0, n > M. \quad (8.5.3)$$

Using (8.5.3), the second equation in (8.5.2) can be expressed in terms of $x[n]$ as

$$X_{ds}[k] = \frac{1}{2M+1}\sum_{n=-\infty}^{\infty} x[n]e^{-j\Omega_0 nk},$$

$$\Omega_0 = \frac{2\pi}{2M+1}. \quad (8.5.4)$$

Now define a continuous periodic function of frequency with period equal to 2π, $X(e^{j\Omega})$, so that the scaled samples of this function are the discrete-time Fourier series.

$$X(e^{j\Omega}) = \sum_{n=-\infty}^{\infty} x[n]e^{-jn\Omega}, \ \frac{1}{2M+1}X(e^{j\Omega})|_{\Omega=k\Omega_0}$$

$$= \frac{1}{2M+1}\sum_{n=-\infty}^{\infty} x[n]e^{-jnk\Omega_0}. \quad (8.5.5)$$

For real sequences, with $\Omega_0 = 2\pi/(2M+1)$, we have

$$x_s[n] = \frac{1}{2\pi}\sum_{k=-M}^{M} X(e^{jk\Omega_0})e^{jk\Omega_0 n}\Omega_0. \quad (8.5.6)$$

As M increases, the spacing between the harmonics in the discrete Fourier series decreases (see (8.5.4)). In the limit, as $M \to \infty$, $d\Omega = \Omega_0$ and $\Omega = k\Omega_0$ is some value on the frequency axis. The summation becomes an integral and

$$x[n] = \frac{1}{2\pi}\int_{-\pi}^{\pi} X(e^{j\Omega})e^{jn\Omega}d\Omega, \quad (8.5.7)$$

$$\lim_{M\to\infty}(\pm)M\Omega_0 = \lim_{M\to\infty}\frac{2\pi(\pm)M}{2M+1} = \pm\pi. \quad (8.5.8)$$

The *discrete-time Fourier transform (DTFT)* and its inverse along with their symbolic relations are as follows:

$$X(e^{j\Omega}) = \sum_{n=-\infty}^{\infty} x[n]e^{-jn\Omega},$$

$$\frac{1}{2\pi}\int_{-\pi}^{\pi} X(e^{j\Omega})e^{jn\Omega}d\Omega = x[n], \quad (8.5.9)$$

$$x[n] \xleftrightarrow{\text{DTFT}} X(e^{j\Omega}) = |X(e^{j\Omega})|e^{j\theta(\Omega)},$$

$$F\{x[n]\} = X(e^{j\Omega}). \qquad (8.5.10)$$

The transform $X(e^{j\Omega})$ of a non-periodic sequence is the discrete-time Fourier spectrum or the spectrum of the sequence $x[n]$. Some authors use $X(\Omega)$ or $X(j\Omega)$ instead of $X(e^{j\Omega})$. $X(e^{j\Omega})$ is the preferred notation, as it explicitly shows that the spectrum is periodic and it will be used here. As in the continuous case, the DTFT is in general complex. The quantities $|X(e^{j\Omega})|$ and $\theta(\Omega) = \angle X(e^{j\Omega})$ are the amplitude (or magnitude) and phase (or angle) spectra of the sequence $x[n]$. The DTFT is valid for both real and complex sequences and has interesting properties, similar to the properties of the continuous F-transform. For a real sequence $x[n]$, its amplitude spectrum is even and the phase spectrum is odd, which follows directly from the definition. A sufficient condition for the existence of $X(e^{j\Omega})$ is the sequence $x[n]$ is absolutely summable. That is,

$$\sum_{n=-\infty}^{\infty} |x[n]| < \infty. \qquad (8.5.11)$$

Note the unit step sequence does not satisfy (8.5.11).

8.5.2 Discrete-Time Fourier Transforms of Real Signals with Symmetries

A real sequence $x[n]$ can be written in terms of its even and odd parts $x_e[n]$ and $x_0[n]$ (see (8.3.9c)). Making use of the even and odd sequence properties, the DTFT of these can be written as

$$x_e[n] \xleftrightarrow{\text{DTFT}} x_e[0] + 2\sum_{n=1}^{\infty} x_e[n]\cos(n\Omega), \qquad (8.5.12a)$$

$$x_0[n] \xleftrightarrow{\text{DTFT}} -j2\sum_{n=1}^{\infty} x_0[n]\sin(n\Omega). \qquad (8.5.12b)$$

That is, if a real sequence is even, then its DTFT is real and even and if it odd, then its DTFT is pure imaginary and odd. The DTFT of an arbitrary real sequence $x[n]$ is

$$X(e^{j\Omega}) = \sum_{n=-\infty}^{\infty} x[n]e^{-jn\Omega}$$

$$= \sum_{n=-\infty}^{\infty} x_e[n]e^{-jn\Omega} + \sum_{n=-\infty}^{\infty} x_0[n]e^{-jn\Omega}, \qquad (8.5.13)$$

$$= \left[x[0] + 2\sum_{n=1}^{\infty} x_e[n]\cos(n\Omega) \right]$$

$$+ (-1)j2\left[\sum_{n=1}^{\infty} x_0[n]\sin(n\Omega) \right] = \text{Re}(X(e^{j\Omega}))$$

$$+ j\text{Im}(X(e^{j\Omega})).$$

Example 8.5.1 Find the DTFT of the following sequences $x_i[n]$, $i = 1, 2$. Sketch the responses for Part a. and identify the important values for $a = .8$.

$a.\ x_1[n] = a^n u[n], |a| < 1$ (right-side sequence)

$b.\ x_2[n] = -a^n u[-n-1]$ (left-side sequence). $\qquad (8.5.14)$

Solution: Noting that $u[n] = 0, n < 0$ and $u[n] = 1$, $n \geq 0$, and using (8.5.9), we have

$$a.\ X_1(e^{j\Omega}) = \sum_{n=-\infty}^{\infty} x_1[n]e^{-jn\Omega} = \sum_{n=0}^{\infty}(ae^{-j\Omega})^n = \frac{1}{1 - ae^{-j\Omega}}$$

$$= \frac{1}{(1 - a\cos(\Omega)) + ja\sin(\Omega)}. \qquad (8.5.15a)$$

The magnitude and the phase responses are periodic with period 2π and

$$|X_1(e^{j\Omega})| = \frac{1}{[1 + a^2 - 2a\cos(\Omega)]^{1/2}},$$

$$\angle X_1(e^{j\Omega}) = -\tan^{-1}\left[\frac{a\sin(\Omega)}{1 - a\cos(\Omega)} \right]. \qquad (8.5.15b)$$

Figure 8.5.1 gives these plots for one period, namely for $0 \leq \Omega < 2\pi$. We could plot these for $-\pi \leq \Omega < \pi$ as they are periodic. The amplitude response is even and phase response is odd. The maximum and the minimum values of the amplitude response for $a = 0.8$ can be seen by noting $|\cos(\Omega)| \leq 1$. The maximum and minimum values are located at $\Omega = 2k\pi$ and $\Omega = (2k+1)\pi$. The values are given by $1/(1-a) = 5$ and $1/(1+a) = 0.5555$, respectively. The responses are plotted in Fig. 8.5.1 for $0 \leq \Omega < 2\pi$. The phase response is a bit more complicated, as the arctangent functions is involved. First

Fig. 8.5.1 (a) $|X_1(e^{j\Omega})|$, (b) $\angle X_1(e^{j\Omega})$

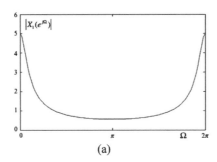

(a)

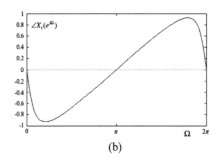

(b)

$$\angle X_1(e^{\pm jk\pi}) = 0, \ k = 0, \pm 1, \pm 2, ...$$

A simple way to find the peak phase angle is by MATLAB. For $a = .8$, the peak amplitudes and the phases are equal to 5 and 0.932 rad/s, respectively.

$$b. \ X_2(e^{j\Omega}) = -a^{-1}e^{j\Omega} - a^{-2}e^{j2\Omega} - \cdots = \sum_{n=-\infty}^{\infty} x_2[n]e^{-jn\Omega}$$

$$= -a^{-1}e^{j\Omega}[1 + a^{-1}e^{j\Omega} + a^{-2}e^{j2\Omega} + \cdots]$$

$$= \frac{-a^{-1}e^{j\Omega}}{(1 - a^{-1}e^{j\Omega})}, |a^{-1}e^{j\Omega}| < 1 \rightarrow X_2(e^{j\Omega})$$

$$= \frac{1}{1 - ae^{-j\Omega}}, |a| > 1. \tag{8.5.15c}$$

The transforms have the same forms, except the constraints on the constant $'a'$ are different. To find the time sequence from the transform, i.e. the inverse transform, we need to know whether the sequence is a right-side or a left-side sequence. ∎

The *group delay* $\tau(\Omega)$ of a sequence $x[n]$ can be defined using its transform as follows:

$$x[n] \xleftrightarrow{\text{DTFT}} X(e^{j\Omega}) = |X(e^{j\Omega})|e^{j\phi(\Omega)},$$

$$\tau(\Omega) = -d\phi(\Omega)/d\Omega. \tag{8.5.16}$$

Since $\phi(\Omega)$ is periodic with period 2π, so is the group delay $\tau(\Omega)$.

Time limited sequences: Linear phase is an important property in the digital filter design. In the following we will consider the expressions for the DTFT of a time limited real sequences $h[n] = 0, n > N, n < 0$ that have the four conditions stated below. We further assume that the samples $h[n]$ have *a.* an even symmetry about the mid point of the sequence and *b.* an odd symmetry about the mid point of the sequence. ∎

Sequences of interest (see Fig. 8.5.2):

Type 1 sequence: N–odd: Sequence with an even symmetry over its mid point.

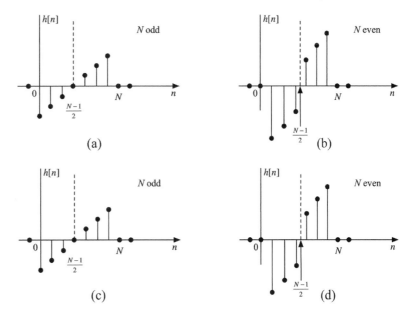

Fig. 8.5.2 Four sequences, (a) Type 1, (b) Type 2, (c) Type 3, (d) Type 4

Type 2 sequence: N–even: Sequence with an even symmetry over its mid point.

Type 3 sequence: N–odd: Sequence with an odd symmetry over its mid point.

Type 4 sequence: N–even: Sequence with an odd symmetry over its mid point.

1. Sequence has an even symmetry if

$$h[n] = h[N - 1 - n] \qquad (8.5.17)$$

2. Sequence has an odd symmetry if

$$h[n] = -h[N - 1 - n] \qquad (8.5.18)$$

We will consider Type 2 and 3 sequences and the other two are left as exercises.

Type 2 sequence: The even symmetry $h[n] = h[N - 1 - n]$ allows us to write

$$H(e^{j\Omega}) = \sum_{n=0}^{N/2-1} h[n]e^{-jn\Omega} + \sum_{n=N/2}^{N-1} h[n]e^{-jn\Omega}$$

$$= \sum_{n=0}^{N/2-1} h[n]e^{-jn\Omega} + \sum_{m=0}^{(N/2)-1} h[N-1-m]e^{-j\Omega(N-1-m)}.$$

$$= \sum_{n=0}^{(N/2)-1} h[n]\left\{ e^{-jn\Omega} + e^{jn\Omega}e^{-j(N-1)\Omega}\right\};$$

$$= \left[\sum_{n=0}^{(N/2)-1} 2h[n]\cos(((N-1)/2)-n)\Omega\right]e^{-j\Omega(N-1)/2}.$$

$$\equiv A_2(e^{j\Omega})e^{-j2\pi((N-1)/2)}. \qquad (8.5.19)$$

We have made use of the following in simplifying the above expression:

$$\left\{ e^{-jn\Omega} + e^{jn\Omega}e^{-j(N-1)\Omega}\right\}$$

$$= e^{-j(N-1)\Omega/2}\left\{ e^{j[(N-1)/2-n]\Omega} + e^{-j[(N-1)/2-n]\Omega}\right\}$$

$$= 2\cos[(N-1)/2 - n]\Omega.$$

Note $A_2(e^{j\Omega})$ is real and the phase angle is $-\{(N-1)/2\}\Omega$, which is *linear* with respect to Ω. Also, $(N-1)/2$ is *not* an integer since N is even.

Type 3 sequences: The DTFT of this sequence can be determined by noting that N is *odd* and the sequence has an *odd* symmetry over its mid point

implying the value of the sequence in the middle is zero. These result in

$$h[n] = -h[N - 1 - n] \text{ and } h[(N - 1)/2] = 0. \quad (8.5.20)$$

Splitting the sequence $h[n]$ into $\{h[0], h[1], ..., h[(N-3)/2]\}$, $h[(N-1)/2] = 0$ and $\{h[(N+1)/2)], ...h[N-1]\}$, the transform can be expressed as

$$H(e^{j\Omega}) = \sum_{n=0}^{N-1} h[n]e^{-jn\Omega} = \sum_{n=0}^{(N-3)/2} h[n]e^{-jn\Omega}$$

$$+ \sum_{n=(N+1)/2}^{N-1} h[n]e^{-j\Omega n}. \qquad (8.5.21)$$

$$\sum_{n=(N+1)/2}^{N-1} h[n]e^{-j\Omega n} = - \sum_{m=0}^{(N-3)/2} h[m]e^{-j\Omega(N-1-m)}$$

$$= - \sum_{n=0}^{(N-3)/2} h[n]e^{-j\Omega(N-1-n)}.$$

Using this result in (8.5.21), we have

$$H(e^{j\Omega}) = \sum_{n=0}^{(N-3)/2} h[n]e^{-jn\Omega} - \sum_{n=0}^{(N-3)/2} h[n]e^{-j\Omega(N-1-n)}$$

$$= 2j \sum_{n=0}^{(N-3)/2} h[n]\left\{\frac{e^{-j\Omega n} - e^{-j\Omega(N-1-n)}}{2j}\right\},$$

$$= 2j \sum_{n=0}^{(N-3)/2} h[n]e^{-j(N-1)/2}$$

$$\left[\frac{e^{j\Omega[((N-1)/2)-n]} - e^{-j\Omega[((N-1)/2)-n]}}{2j}\right],$$

$$= j\left[2 \sum_{n=0}^{(N-3)/2} \sin[(\frac{N-1}{2} - n)\Omega]\right]e^{-j\Omega(N-1)/2}$$

$$= A_3(e^{j\Omega})(je^{-j\Omega(N-1)/2}). \qquad (8.5.22)$$

The phase angle $\angle(je^{-j\Omega(N-1)/2})$ is linear. Also, $A_3(e^{j\Omega})$ is real and odd symmetric about $\Omega = 0$ and $\Omega = \pi$ and

$$A_3(e^{j\Omega}) = \sum_{n=0}^{(N-3)/2} \sin\left[\left(\frac{N-1}{2} - n\right)\Omega\right]. \quad (8.5.23)$$

Summary: Type 1 sequence:

$$H(e^{j\Omega}) = \left\{ h\left[\frac{N-1}{2}\right] + \sum_{n=0}^{(N-3)/2} 2h[n]\cos\left[\left(\frac{N-1}{2}-n\right)\Omega\right]\right\}$$

$$e^{-j\Omega(N-1)/2} = A_1(e^{j\Omega})e^{-j\Omega(N-1)/2}. \qquad (8.5.24)$$

$A_1(e^{j\Omega})$: Even symmetric about $\Omega = 0$ and $\Omega = \pi$,

$$H(e^{j0}) = h[(N-1)/2] + \sum_{n=0}^{(N-3)/2} 2\,h[n].$$

Type 2 sequence:

$$H(e^{j\Omega}) = \left[\sum_{n=0}^{N/2-1} 2h[n]\cos[n\Omega - ((N-1)/2)\Omega]\right]e^{-j\Omega(N-1)/2}$$

$$= A_2(e^{j\Omega})e^{-j\Omega(N-1)/2}, \qquad (8.5.25)$$

$$H(e^{j0}) = \sum_{n=0}^{(N/2)-1} 2\,h[n].$$

$A_2(e^{j\Omega})$: Even symmetric about $\Omega = 0$ and is odd symmetric about $\Omega = \pi \rightarrow H(e^{j\pi}) = 0$.

Type 3 sequence:

$$H(e^{j\Omega}) = j\left[2\sum_{n=0}^{(N-3)/2} h[n]\sin[((N-1)/2)-n)\Omega]\right]$$

$$e^{-j\Omega(N-1)/2} = A_3(e^{j\Omega})e^{j((\pi/2)-(N-1)\Omega/2)}, \qquad (8.5.26)$$

$A_3(e^{j\Omega})$: Odd symmetric about $\Omega = 0$ and $\Omega = \pi \rightarrow H(e^{j0}) = 0$ and $H(e^{j\pi}) = 0$.

Type 4 sequence:

$$H(e^{j\Omega}) = \left[2\sum_{n=0}^{(N/2)-1} h[n]\sin[n\Omega - ((N-1)/2)\Omega]\right]$$

$$je^{-j\Omega(N-1)/2} = A_4(e^{j\Omega})e^{j(\pi/2-(N-1)\Omega/2)}. \qquad (8.5.27)$$

$A_4(e^{j\Omega})$: Odd symmetric about $\Omega = 0$ and even symmetric about $\Omega = \pi \rightarrow H(e^{j0}) = 0$. ∎

These finite length sequences will be useful in studying *finite impulse response (FIR) filters* in Chapter 9 are considered. Specifications are given

in terms of $A_i(e^{j\Omega})$ and $h_i[n]$ can be determined given $A_i(e^{j\Omega})$. ∎

8.6 Properties of the Discrete-Time Fourier Transforms

The DTFT of a sequence and its inverse were given before and are (see (8.5.9)):

$$X(e^{j\Omega}) = \sum_{n=-\infty}^{\infty} x[n]e^{-jn\Omega}, \ x[n]$$

$$= \frac{1}{2\pi}\int_{-\pi}^{\pi} X(e^{j\Omega})e^{j\Omega n}d\Omega, \quad x[n] \overset{\text{DTFT}}{\longleftrightarrow} X(e^{j\Omega}).$$

$$(8.6.1)$$

8.6.1 Periodic Nature of the Discrete-Time Fourier Transform

The F-transform of the discrete-time signal $x[n]$ is periodic with period 2π. That is,

$$X(e^{j(\Omega+2\pi)}) = X(e^{j\Omega}). \qquad (8.6.2)$$

The continuous-time transform is defined in terms of ω in radians/second over the entire range $-\infty < \omega < \infty$. When the analog signals are sampled at a sampling frequency of f_s Hertz, the spectrum of the digitized signal is periodic with period $\omega_s = 2\pi f_s = (2\pi/t_s)$. The *normalized frequency* $\Omega = (2\pi f/f_s)$ defines the *digital frequency*.

Notes: In the continuous-time domain, periodic signals are expressed in terms of discrete F-series coefficients. In the discrete-time domain, the samples $x(nt_s)$ are located at discrete times and the DTFT is continuous and periodic with period 2π. The interest is in the digital frequency b and $|\Omega| \leq \pi$ or in the range $0 \leq \Omega < 2\pi$. ∎

Example 8.6.1 Find the DTFT of the sequence $x[n] = 1, 0 \leq n < N$. Give the expressions for the magnitude and the phase characteristics of the transform. Sketch the magnitude and phase responses for $0 < \Omega < 2\pi$ assuming $N = 21$.

Solution: The transform, its amplitude and phase responses are as follows:

$$X(e^{j\Omega}) = \sum_{n=0}^{N-1} e^{-j\Omega n} = \frac{1 - e^{-j\Omega n}}{1 - e^{-j\Omega}}; \quad \sum_{n=-\infty}^{\infty} |x[n]| = N < \infty,$$

(8.6.3)

$$= \frac{e^{-\Omega N/2}}{e^{-j\Omega/2}} \frac{(e^{j\Omega N/2} - e^{-j\Omega N/2})/2j}{(e^{j\Omega/2} - e^{-j\Omega/2})/2j}$$

$$= e^{-j\Omega(N-1)/2} \frac{\sin(\Omega N/2)}{\sin(\Omega/2)},$$

(8.6.4)

$$|X(e^{j\Omega})| = \frac{|\sin(\Omega N/2)|}{|\sin(\Omega/2)|}, \angle X(e^{j\Omega})$$

$$= -(N-1)\Omega/2 + \arg\left[\frac{\sin(\Omega N/2)}{\sin(\Omega/2)}\right]. \quad (8.6.5)$$

The amplitude $|X(e^{j\Omega})|$ is even and the phase $\angle X(e^{j\Omega})$ is odd. Both are periodic with period 2π. At $\Omega = 0$, the function is indeterminate and

$$\lim_{\Omega \to 0} \frac{\sin(\Omega N/2)}{\sin(\Omega/2)} = N. \quad (8.6.6)$$

Note $X(e^{j\Omega}) = 0$ when $\Omega = 2k\pi/N, k \neq 0$. The spacing between zero crossings is $(2\pi/N)$. The phase angle corresponding to the main lobe is $-(N-1)\Omega/2$ resulting in a value of $-(N-1)\pi/N$ at $\Omega = [2\pi/N]^-$. At $\Omega = (2\pi/N)^+$, the phase angle jumps by π rad reaching a value of π/N since $\sin(\Omega N/2)/\sin(\Omega/2)$ is positive in the main lobe and negative in the first side lobe. This process is repeated and at $\Omega = 2\pi$, the phase angle takes the value of 0 completing one period. The sequence

$h[n]$ and $20\log|H(e^{j\Omega})|$, $0 \leq \Omega \leq \pi$ are shown in Fig. 8.6.1 for $N = 21$. The side lobes of the amplitude response become smaller as Ω goes away from π on both of its sides. The peak of the first side lobe appears near the middle of the first side lobe and is approximately equal to -13.29 dB. ∎

Notes: The amplitude spectrum of a typical window is shown in Fig. 8.6.2. It is even and 2π periodic and the frequency interval of interest is $0 \leq \Omega \leq \pi$. The windows of interest have linear phase. The high frequency decay rate of the envelope of the spectrum side lobes tells how fast the spectrum *envelope* decays after the first zero crossing.

Window parameters: (See Fig. 8.6.2):

G_P = Peak gain of main lobe
$= N, G_p/N = 1 = 0$ dB
G_s = Peak side lobe gain,
$G_s/G_p \approx 0.2172 = -13.3$ dB
Ω_M = Half-width of main lobe = $2\pi/N$ (8.6.7)

$\Omega_3 = 3$ dB = half-width, $W_3/W_M = 0.44$
$\Omega_6 = 6$ dB = half-width, $W_6/W_M = 0.6$
Ω_s = Half-width of main lobe to reach $P_s, W_s/W_M = 0.81$

High-frequency attenuation = 20 dB/decade

8.6.2 Superposition or Linearity

Assuming $\text{DTFT}[x_i[n]] = X_i(e^{j\Omega})$ and a'_is are constants, the linearity property is

$$\sum_{i=1}^{M} a_i x_i[n] \overset{DTFT}{\longleftrightarrow} \sum_{i=1}^{M} a_i X_i(e^{j\Omega}). \quad (8.6.8)$$

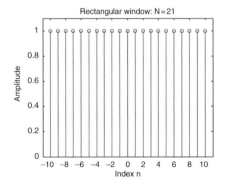

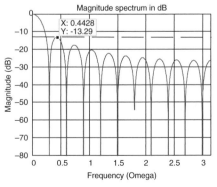

Fig. 8.6.1 Discrete rectangular window function and its amplitude spectrum

Fig. 8.6.2 Amplitude spectrum of typical windows

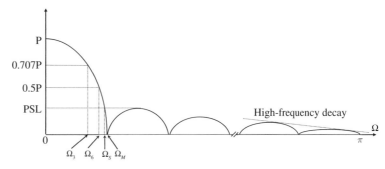

DTFT magnitude spectrum of a typical window

8.6.3 Time Shift or Delay

This property states

$$x[n-n_0] \overset{DTFT}{\longleftrightarrow} e^{-j\Omega n_0} X(e^{j\Omega}). \qquad (8.6.9)$$

This follows from

$$
\begin{aligned}
F\{x[n-n_0]\} &= \sum_{n=-\infty}^{\infty} x[n-n_0]e^{-jn\Omega}\\
&= \sum_{m=-\infty}^{\infty} x[m]e^{-j(m+n_0)\Omega}\\
&= \left[\sum_{m=-\infty}^{\infty} x[m]e^{-jm\Omega}\right]e^{-n_0\Omega}.
\end{aligned}
$$

Example 8.6.2 Show that the following relationship is true:

$$x[n]=e^{j\Omega_0 n}X(e^{j\Omega})=2\pi\delta(\Omega-\Omega_0), |\Omega_0|\le\pi. \quad (8.6.10)$$

Solution: This can be shown using the sifting property of the impulse function.

$$x[n]=\frac{1}{2\pi}\int_{-\infty}^{\infty} 2\pi\delta(\Omega-\Omega_0)e^{jn\Omega}d\Omega=e^{jn\Omega}|_{\Omega=\Omega_0}=e^{jn\Omega_0}. \quad \blacksquare$$

Example 8.6.3 Find the DTFTs of the following functions using the pair in (8.6.10) and the shift property.

$$a.\ x_1[n]=1, \quad b.\ x_2[n]=\cos(\Omega_0 n), \quad |\Omega_0|\le\pi. \quad (8.6.11)$$

Solution: *a.* Using $\Omega_0=0$ in (8.6.10), we have the result as follows:

$$x_1[n]=1 \overset{DTFT}{\longleftrightarrow} 2\pi\delta(\Omega). \qquad (8.6.12)$$

b. Using the Euler' formula and (8.6.10) results in

$$
\begin{aligned}
x_2[n] &=\cos(\Omega_0 n)=.5(e^{+j\Omega_0 n}+e^{-j\Omega_0 n}) \overset{DTFT}{\longleftrightarrow} \pi\delta(\Omega-\Omega_0)\\
&+\pi\delta(\Omega+\Omega_0)=X_2(e^{j\Omega}). \qquad (8.6.13) \quad \blacksquare
\end{aligned}
$$

8.6.4 Modulation or Frequency Shifting

The dual of time shifting is the frequency shifting and is given below

$$e^{jn\Omega_0}x[n] \overset{DTFT}{\longleftrightarrow} X(e^{j(\Omega-\Omega_0)}). \qquad (8.6.14a)$$

An extension of this is the *modulation* in time and the corresponding transform pair is

$$
\begin{aligned}
x[n]\cos(n\Omega_0) &= x[n]\left[\frac{e^{jn\Omega_0}+e^{-jn\Omega_0}}{2}\right] \overset{DTFT}{\longleftrightarrow}\\
&\frac{1}{2}\left[X(e^{j(\Omega-\Omega_0)})+X(e^{j(\Omega+\Omega_0)})\right].
\end{aligned}
$$
$$(8.6.14b)$$

8.6.5 Time Scaling

Time scaling deals with the DTFT of $x[cn]$, where "*C*" is an *integer*. For example, consider $y[n]=x[2n]$, then $y[n]$ has only the even samples of $x[n]$. This is decimation (see Section 8.3.1). To simplify the notation define the following sequence assuming *m* as an *integer*:

$$x_{(m)}[n] = \begin{cases} x[n/m] = x[k], & n = km, \ n \text{ and } k \text{ are integers} \\ 0, & n \neq km. \end{cases} \tag{8.6.15}$$

The time scaling property is

$$x_{(m)}[n] \xleftrightarrow{\text{DTFT}} X(e^{jm\Omega}). \tag{8.6.16}$$

This can be seen from

$$F[x_{(m)}[n]] = \sum_{n=-\infty}^{\infty} x_{(m)}[n]e^{-jn\Omega} = \sum_{k=-\infty}^{\infty} x[km]e^{-jk(m\Omega)}$$

$$= X(e^{jm\Omega}) \text{ (periodic with period } 2\pi/m).$$

It illustrates the inverse relationship between time and frequency. The signal spreads in time ($m > 1$) corresponds to its transform being compressed.

Time reversal: A special class of time scaling is time reversal and it results in reversal in frequency. That is,

$$F[x[-n]] = \sum_{n=-\infty}^{\infty} x[-n]e^{-jn\Omega}$$

$$= \sum_{m=-\infty}^{\infty} x[m]e^{-jm(-\Omega)} = X(e^{-j\Omega}). \tag{8.6.17}$$

Note that

$$|F[x[-n]]| = |X(e^{-j\Omega})| = |X(e^{j\Omega})| = |F[x[n]]|.$$

8.6.6 Differentiation in Frequency

This property is

$$nx[n] \xleftrightarrow{\text{DTFT}} j\frac{dX(e^{j\Omega})}{d\Omega}. \tag{8.6.18}$$

This is shown by

$$\frac{dX(e^{j\Omega})}{d\Omega} = \frac{d}{d\Omega}\left[\sum_{n=-\infty}^{\infty} x[n]e^{-jn\Omega}\right]$$

$$= \sum_{n=-\infty}^{\infty} [(-jn)x[n]]e^{-jn\Omega}.$$

Example 8.6.4 Derive the DTFT of the function $y[n] = (n+1)a^n u[n], |a| < 1$ using

$$a^n u[n] \xleftrightarrow{\text{DTFT}} \frac{1}{1 - ae^{-j\Omega}}, |a| < 1, a \neq 0. \tag{8.6.19}$$

Solution: From the differentiation property,

$$y[n] = na^n u[n] + a^n u[n] \xleftrightarrow{\text{DTFT}} j\frac{d(1/(1 - ae^{-j\Omega})}{d\Omega}$$

$$+ \frac{1}{1 - ae^{-j\Omega}} = \frac{1}{(1 - ae^{j\Omega})^2},$$

$$\Rightarrow y[n] = (n+1)a^n u[n] \xleftrightarrow{\text{DTFT}} \frac{1}{(1 - ae^{-j\Omega})^2},$$

$$|a| < 1, a \neq 0. \tag{8.6.20} \quad \blacksquare$$

Example 8.6.5 Find the DTFT of $x[n] = a^{|n|}$, $|a| < 1, a \neq 0$.

Solution: $x[n]$ can be expressed as a sum of the right-and left-side sequences in the form $x[n] = a^n u[n] + a^{-n} u[-n] - \delta[n]$. The transforms of each of these are

$$a^n u[n] \xleftrightarrow{\text{DTFT}} \frac{1}{1 - ae^{-j\Omega}}, \tag{8.6.21a}$$

$$a^{-n} u[-n] \xleftrightarrow{\text{DTFT}} \frac{1}{1 - ae^{j\Omega}}, \tag{8.6.21b}$$

$$\delta[n] \xleftrightarrow{\text{DTFT}} 1. \tag{8.6.21c}$$

Note the time reversal property of the DTFT was used to find the DTFT of $a^{-n}u[-n]$. With these and making use of the linearity property of the DTFT, we have the DTFT pair

$$x[n] = a^{|n|} \xleftrightarrow{\text{DTFT}} \frac{1}{1 - ae^{-j\Omega}}$$

$$+ \frac{1}{1 - ae^{j\Omega}} - 1, |a| < 1, a \neq 0. \tag{8.6.21d} \quad \blacksquare$$

8.6.7 Differencing

The differencing property stated below can be shown using the linearity and the time-shifting properties of the DTFT.

$$x[n] - x[n-1] \xleftrightarrow{\text{DTFT}} (1 - e^{-j\Omega})X(e^{j\Omega}). \quad (8.6.22)$$

Example 8.6.6 Find the DTFTs of the sequences

    $a.\ x_1[n] = \delta[n]$ (by the direct method),

    $b.\ x_2[n] = u[n]$

    $c.\ x_3[n] = -u[-n-1],\ d.\ x_4[n] = \text{sgn}[n].$

Solution: *a.*

$$F\{\delta[n]\} = \sum_{n=-\infty}^{\infty} \delta[n]e^{-jn\Omega} = 1. \quad (8.6.23a)$$

b. Let $U(e^{j\Omega}) = \text{DTFT}\{u[n]\}$. The unit step sequence is a limiting form of the sequence $\alpha^n u[n]$ with $\alpha \to 1$. Since $u[n]$ is *not* absolutely summable, the transform of the unit step sequence *cannot* be obtained by taking the limit of the transform as $a \to 1$ in (8.6.19). Noting that $\delta[n] = u[n] - u[n-1]$, and defining $F[u[n]] = U(e^{j\Omega})$, we have

$$F[\delta[n]] = 1 = U(e^{j\Omega}) - e^{-j\Omega}U(e^{j\Omega})$$
$$= (1 - e^{-j\Omega})U(e^{j\Omega}). \quad (8.6.23b)$$

Since $(1 - e^{-j\Omega})|_{\Omega=0} = 0$, it follows that the transform of the unit step sequence will have an impulse, in addition to the transform in (8.6.21a) with $a = 1$, resulting in

$$u[n] \xleftrightarrow{\text{DTFT}} U(e^{j\Omega}) = A\delta(\Omega) + \frac{1}{1 - e^{-j\Omega}}$$
$$= \pi\delta(\Omega) + \frac{1}{1 - e^{-j\Omega}}, |\Omega| \le \pi. \quad (8.6.24)$$

The average value of the unit step sequence is $(1/2)$ and its transform is $\pi\delta(\Omega)$. Example 4.4.10 illustrated the continuous F-transform of a unit step function.

c. The DTFT of $x_3[n]$ can be determined by

$$-u[-n-1] = u[n] - 1 \xleftrightarrow{\text{DTFT}} \pi\delta(\Omega) + \frac{1}{1 - e^{-j\Omega}}$$
$$-2\pi\delta(\Omega) = -\pi\delta(\Omega) + \frac{1}{1 - e^{-j\Omega}}. \quad (8.6.25)$$

d. $F[\text{sgn}] = F[u[n] - u[-n]] = \pi\delta(\Omega) + \dfrac{1}{1 - e^{-j\Omega}}$
$$- \pi\delta(-\Omega) - \frac{1}{1 - e^{j\Omega}} = \frac{1}{1 - e^{-j\Omega}} - \frac{1}{1 - e^{j\Omega}}. \quad (8.6.26)$$

Note the transform is given by the difference between a complex function and its conjugate illustrating the transform of an odd function and is imaginary. ∎

Inverse discrete-time Fourier transform (IDTFT): Finding the inverse transform

$$x[n] = \frac{1}{2\pi} \int_{-\pi}^{\pi} X(e^{j\Omega})e^{j\Omega n}\,d\Omega.$$

is difficult, as it involves complex integration. Simple cases are illustrated in Example 8.6.7. Alternate methods suggested below are preferable.

1. Since $X(e^{j\Omega})$ is periodic, use the F-series of this function (i.e., in the *frequency domain* Ω) and then find the corresponding Fourier series coefficients in the time domain. These methods are useful in designing filters and are discussed in Chapter 9.

2. z-Transforms, discussed in the next chapter, can be used to find the IDTFTs.

Example 8.6.7 Find the inverse DTFTs of the periodic functions with period 2π

    $a.\ X(e^{j\Omega}) = 2\pi\delta(\Omega - \Omega_0),$

    $b.\ X(e^{j\Omega}) = \begin{cases} 1, & |\Omega| \le W \\ 0, & W < |\Omega| \le \pi \end{cases}. \quad (8.6.27)$

c. Use Part *b.* to find the DTFTs of the sequences $x_1[n] = \cos(\Omega_0 n)$ and $x_2[n] = \sin(\Omega_0 n)$.

Solution: *a.* The inverse transform is

$$x[n] = \frac{1}{2\pi} \int_{-\pi}^{\pi} X(e^{j\Omega})e^{j\Omega n}\,d\Omega$$
$$= \int_{-\pi}^{\pi} \delta(\Omega - \Omega_0)e^{j\Omega n}\,d\Omega = e^{jn\Omega_0},$$

$$\Rightarrow x[n] = e^{jn\Omega_0} \xleftrightarrow{\text{DTFT}} 2\pi\delta(\Omega - \Omega_0),$$
$$- \pi < \Omega \leq \pi,$$

$$b. x[n] = \frac{1}{2\pi} \int_{-\pi}^{\pi} X(e^{j\Omega}) e^{jn\Omega} d\Omega = \frac{1}{2\pi}$$

$$\int_{-W}^{W} e^{jn\Omega} d\Omega = \frac{\sin(Wn)}{\pi n},$$

$$\Rightarrow x[n] = \frac{\sin(Wn)}{\pi n} \xleftrightarrow{\text{DTFT}} \begin{cases} 1, |\Omega| \leq W \\ 0, W < |\Omega| \leq \pi \end{cases}. \quad (8.6.29)$$

c. Using Euler's formula and the results in Part a, the DTFTs of the periodic signals are as follows with period 2π:

$$x_1[n] = \cos(\Omega_0 n) = .5(e^{j\Omega_0 n} + e^{-j\Omega_0 n}) \xleftrightarrow{\text{DTFT}}$$
$$\pi[\delta(\Omega - \Omega_0) + \delta(\Omega + \Omega_0)]$$
$$= X_1(e^{j\Omega}), -\pi < \Omega_0 \leq \pi, \quad (8.6.30)$$

$$x_2[n] = \sin(\Omega_0 n) = (1/2j)(e^{j\Omega_0 n} - e^{-j\Omega_0 n}) \xleftrightarrow{\text{DTFT}}$$
$$- j\pi[\delta(\Omega - \Omega_0) - \delta(\Omega + \Omega_0)]$$
$$= X_2(e^{j\Omega}), -\pi < \Omega_0 \leq \pi. \quad (8.6.31) \quad \blacksquare$$

8.6.8 Summation or Accumulation

The accumulation property is the discrete-time counterpart of the integration in the continuous domain. The summation property is shown later in Section (8.6.11) and is

$$\sum_{m=-\infty}^{n} x[m] \xleftrightarrow{\text{DTFT}} \pi X(e^{j0})\delta(\Omega)$$
$$+ \frac{1}{1 - e^{-j\Omega}} X(e^{j\Omega}), |\Omega| \leq \pi. \quad (8.6.32)$$

Example 8.6.8 Find the DTFT of $u[n]$ using the accumulation property.

Solution: Noting that $u[n] = \sum_{m=-\infty}^{n} \delta[m]$ (see (8.3.8b)),
$$\delta[n] \xleftrightarrow{\text{DTFT}} 1, \text{ we have}$$

$$u[n] = \sum_{m=-\infty}^{n} \delta[m] \xleftrightarrow{\text{DTFT}} \pi\delta(\Omega)$$
$$+ \frac{1}{(1 - e^{-j\Omega})} = U(e^{j\Omega}), |\Omega| \leq \pi. \quad (8.6.33) \quad \blacksquare$$

8.6.9 Convolution

Discrete-time convolution property of $x_1[n]$ and $x_2[n]$ is as follows:

$$y[n] = x_1[n] * x_2[n] = \sum_{k=-\infty}^{\infty} x_1[k]x_2[n-k]$$
$$= \sum_{k=-\infty}^{\infty} x_1[n-k]x_2[k] = x_1[n] * x_2[n], \quad (8.6.34)$$

$$y[n] = x_1[n] * x_2[n] = \sum_{k=-\infty}^{\infty} x_1[k]x_2[n-k]$$
$$\xleftrightarrow{\text{DTFT}} X_1(e^{j\Omega})X_2(e^{j\Omega}). \quad (8.6.35)$$

As in the continuous case the time-domain convolution results in the multiplication in the frequency domain and this property plays an important role in discrete-time linear systems. Using the definition of the transform, we have

$$Y(e^{j\Omega}) = F\left\{ \sum_{k=-\infty}^{n} x_1[k]x_2[n-k] \right\}$$
$$= \sum_{n=-\infty}^{\infty} \left(\sum_{k=-\infty}^{n} x_1[k]x_2[n-k] \right) e^{-j\Omega n},$$
$$= \sum_{k=-\infty}^{\infty} x_1[k] \left(\sum_{n=-\infty}^{\infty} x_2[n-k]e^{-j\Omega n} \right)$$
$$= \sum_{k=-\infty}^{\infty} x_1[k]X_2(e^{j\Omega})e^{-j\Omega k},$$
$$= X_2(e^{j\Omega}) \left[\sum_{k=-\infty}^{\infty} x_1[k]e^{-j\Omega k} \right]$$
$$= X_2(e^{j\Omega})X_1(e^{j\Omega}). \quad (8.6.36)$$

Convolution in discrete-time corresponds to multiplication in frequency. The accumulation property given in (8.6.32) can now be shown using the convolution theorem and $F[u[n]] = \pi\delta(\Omega) + [1/(1 - e^{-j\Omega})]$. That is,

$$x[n] * u[n] = \sum_{k=-\infty}^{\infty} x[k]u[n-k]$$

$$= \sum_{k=-\infty}^{n} x[k] \xleftrightarrow{\text{DTFT}} X(e^{j\Omega})U(e^{j\Omega})$$

$$= X(e^{j\Omega})\left[\pi\delta(\Omega) + \frac{1}{(1 - e^{-j\Omega})}\right]$$

$$= \pi X(e^{j0}) + \frac{1}{(1 - e^{-j\Omega})}X(e^{j\Omega}), \ |\Omega| \leq \pi.$$

$$(8.6.37)$$

Example 8.6.9 Using the DTFTs of the sequences given below (see (8.6.21a)), find the convolution $y[n] = x_1[n] * x_2[n]$ using the convolution property.

$$x_1[n] = \alpha^n u[n], x_2[n]$$
$$= \beta^n u[n], 0 < |\alpha|, |\beta| < 1, \ a. \ \alpha \neq \beta, \ b. \ \alpha = \beta.$$

$$(8.6.38)$$

Solution: *a.* The DTFTs of the two sequences are given by

$$X_1(e^{j\Omega}) = F\{x_1[n]\} = \frac{1}{1 - \alpha e^{-j\Omega}}, \ X_2(e^{j\Omega})$$

$$= F\{x_2[n]\} = \frac{1}{1 - \beta e^{-j\Omega}},$$

$$Y(e^{j\Omega}) = X_1(e^{j\Omega})X_2(e^{j\Omega})$$

$$= \frac{1}{(1 - \alpha e^{-j\Omega})(1 - \beta e^{-j\Omega})}.$$

$$(8.6.40)$$

Using the partial fraction expansion, we have

$$Y(e^{j\Omega}) = \frac{-\alpha}{(\beta - \alpha)(1 - \alpha e^{-j\Omega})} + \frac{\beta}{(\beta - \alpha)(1 - \beta e^{-j\Omega})}.$$

$$(8.6.41)$$

Finding the partial fraction expansion in terms of $p = e^{-j\Omega}$ would make it a bit easier. Using the DTFT pair in (8.6.21a), the time sequence is

$$y[n] = \frac{-\alpha}{\beta - \alpha}\alpha^n u[n] + \frac{\beta}{\beta - \alpha}\beta^n u[n]$$

$$= \frac{\alpha^{n+1} - \beta^{n+1}}{\alpha - \beta}u[n], \ \alpha \neq \beta, \ |\alpha|, |\beta| < 1.$$

$$(8.6.42)$$

b. From (8.6.20), we have $y[n] = (n + 1)\alpha^n u[n]$, $|\alpha| < 1$. ∎

8.6.10 Multiplication in Time

Dual to the convolution in time is multiplication in time and

$$y[n] = x_1[n]x_2^*[n] \xleftrightarrow{\text{DTFT}} \frac{1}{2\pi}[X_1(e^{j\Omega}) \otimes X_2(e^{-j\Omega})]$$

$$= Y(e^{j\Omega}).$$

$$(8.6.43)$$

Periodic convolution:

$$\frac{1}{2\pi}X_1(e^{j\Omega}) \otimes X_2(e^{-j\Omega})$$

$$= \frac{1}{2\pi}\int_{2\pi} X_1(e^{j\alpha})X_2(e^{-j(\Omega-\alpha)})d\alpha.$$

$$(8.6.44)$$

The transform of the product of two sequences is the *periodic convolution* of the two *transforms*. It can be seen that

$$F[y[n]] = Y(e^{j\Omega}) = \sum_{n=-\infty}^{\infty} x_1[n]x_2^*[n]e^{-jn\Omega}$$

$$= \sum_{n=-\infty}^{\infty}\left[\frac{1}{2\pi}\int_{2\pi} X_1(e^{j\alpha})e^{j\alpha n}d\alpha\right]x_2^*[n]e^{-j\Omega n}.$$

Interchanging the order of summation and integration results in

$$Y(e^{j\Omega}) = \frac{1}{2\pi} \int_{2\pi} X_1(e^{j\alpha}) \left[\sum_{n=-\infty}^{\infty} x_2^*[n] e^{-j(\Omega-\alpha)n} \right] d\alpha$$

$$= \frac{1}{2\pi} \int_{2\pi} X_1(e^{j\alpha}) \left[\sum_{n=-\infty}^{\infty} x_2[n] e^{j(\Omega-\alpha)n} \right]^* d\alpha,$$

$$= \frac{1}{2\pi} \int_{2\pi} X_1(e^{j\alpha}) X_2^*(e^{j(\Omega-\alpha)}) d\alpha$$

$$= \frac{1}{2\pi} \int_{2\pi} X_1(e^{j\alpha}) X_2(e^{-j(\Omega-\alpha)}) d\alpha.$$

8.6.11 Parseval's Identities

The discrete versions of the Parseval's identities are as follows:

$$\sum_{n=-\infty}^{\infty} x_1[n] x_2^*[n] = \frac{1}{2\pi} \int_{2\pi} X_1(e^{j\Omega}) X_2(e^{-j\Omega}) d\Omega,$$

(Generalized Parseval's identity),   (8.6.45a)

$$\sum_{n=-\infty}^{\infty} |x[n]|^2 = \frac{1}{2\pi} \int_{2\pi} |X(e^{j\Omega})|^2 d\Omega,$$

(Parseval's identity),   (8.6.45b)

These can be seen by

$$Y(e^{j\Omega})|_{\Omega=0} = \sum_{n=-\infty}^{\infty} x_1[n] x_2^*[n] e^{-jn\Omega}|_{\Omega=0}$$

$$= \frac{1}{2\pi} \int_{2\pi} X_1(e^{j\alpha}) X_2(e^{-j(\Omega-\alpha)}) d\alpha|_{\Omega=0},$$

$$= \frac{1}{2\pi} \int_{2\pi} X_1(e^{j\alpha}) X_2(e^{-j\alpha}) d\alpha$$

$$= \frac{1}{2\pi} \int_{2\pi} X_1(e^{j\Omega}) X_2(e^{-j\Omega}) d\Omega.$$

Note that if $x_2[n] = x_1[n] = x[n]$, then $X_2(e^{-j\Omega}) = X_1^*(e^{j\Omega}) = X^*(e^{j\Omega})$.

Example 8.6.10 Use the Parseval's identity and the DTFT pair in (8.6.29) to determine the energy contained in the discrete-time signal $x[n] = \sin(Wn)/\pi n$.

Solution: From the F-transform pair,

$$E = \sum_{n=-\infty}^{\infty} |x[n]|^2 = \sum_{n=-\infty}^{\infty} \frac{\sin^2(Wn)}{(\pi n)^2}$$

$$= \frac{1}{2\pi} \int_{-W}^{W} (1)^2 d\Omega = \frac{W}{\pi}.$$ (8.6.46) ∎

8.6.12 Central Ordinate Theorems

From the DTFT and the IDFT, it follows that

$$X(e^{j0}) = \sum_{n=-\infty}^{\infty} x[n], \quad x[0] = \frac{1}{2\pi} \int_{2\pi} X(e^{j\Omega}) d\Omega. \quad (8.6.47)$$

8.6.13 Simple Digital Encryption

For an introduction to data encryption, see Hershey and Yarlagadda (1986). It is a vast area and most of these techniques are based on manipulating the data in time domain. A simple spectral based encryption can be seen by using the DTFT illustrated below.

Example 8.6.11 Using $x[n] \xleftrightarrow{\text{DTFT}} X(e^{j\Omega})$, find the DTFT of $(-1)^n x[n]$.

Solution:

$$F\{(-1)^n x[n]\} = \sum_{n=-\infty}^{\infty} (-1)^n x[n] e^{-jn\Omega}$$

$$= \sum_{n=-\infty}^{\infty} x[n] e^{-jn(\Omega+\pi)} = X(e^{j(\Omega+\pi)}).$$

(8.6.50)

Multiplying the time sequence by $(-1)^n$ simply changes the sign of the data with odd indexes. Since the DTFT spectrum is periodic with period equal to 2π, this operation corresponds to the spectral inversion in the frequency band $0 \le \Omega \le \pi$. In Chapter 10, Example 10.4.1, we will consider the analog frequency band inversion.   ∎

8.7 Tables of Discrete-Time Fourier Transform (DTFT) Properties and Pairs

Table 8.7.1 Discrete-time Fourier transform (DTFT) properties

$$x_i[n] \overset{\text{DTFT}}{\longleftrightarrow} X_i(e^{j\Omega})$$

Linearity:

$$x[n] = \sum_{i=1}^{M} a_i x_i[n] \overset{\text{DTFT}}{\longleftrightarrow} \sum_{i=1}^{M} a_i X_i(e^{j\Omega}).$$

Time shift or delay:

$$x[n - n_0] \overset{\text{DTFT}}{\longleftrightarrow} e^{-j\Omega n_0} X(e^{j\Omega}); \ n_0 \text{ is an integer.}$$

Frequency shift and modulation:

$$e^{j\Omega_0 n} x[n] \overset{\text{DTFT}}{\longleftrightarrow} X(e^{j(\Omega-\Omega_0)}).$$

$$x[n] \cos(n\Omega_0) \overset{\text{DTFT}}{\longleftrightarrow} \tfrac{1}{2}\left[X(e^{j(\Omega-\Omega_0)}) + X(e^{j(\Omega+\Omega_0)})\right].$$

Conjugation:

$$x^*[n] \overset{\text{DTFT}}{\longleftrightarrow} X^*(e^{-j\Omega}).$$

Time reversal:

$$x[-n] \overset{\text{DTFT}}{\longleftrightarrow} X(e^{-j\Omega}).$$

Time scaling:

$$x_{(m)}[n] = \left\{ \begin{array}{l} x[n/m], n = km \\ 0, \ n \neq km \end{array} \right\} \overset{\text{DTFT}}{\longleftrightarrow} X(e^{jm\Omega}).$$

Times n property:

$$nx[n] \overset{\text{DTFT}}{\longleftrightarrow} j\frac{dX(e^{j\Omega})}{d\Omega}.$$

First difference:

$$x[n] - x[n-1] \overset{\text{DTFT}}{\longleftrightarrow} (1 - e^{-j\Omega})X(e^{j\Omega}).$$

Summation or accumulation:

$$\sum_{k=-\infty}^{n} x[k] \overset{\text{DTFT}}{\longleftrightarrow} \pi X(e^{j0})\delta(\Omega) + \frac{1}{1-e^{-j\Omega}} X(e^{j\Omega}); \ |\Omega|\pi.$$

Time convolution:

$$x_1[n] * x_2[n] \overset{\text{DTFT}}{\longleftrightarrow} X_1(e^{j\Omega})X_2(e^{j\Omega}).$$

Multiplication in time:

$$x_1[n]x_2^*[n] \overset{\text{DTFT}}{\longleftrightarrow} \tfrac{1}{2\pi}\left[X_1(e^{j\Omega}) \otimes X_2(e^{-j\Omega})\right], \text{periodic convolution.}$$

Even and odd parts of a real function:

$$x[n] = x_e[n] + x_o[n] \overset{\text{DTFT}}{\longleftrightarrow} Re(X(e^{j\Omega})) + jIm(X(e^{j\Omega})).$$

$$x_e[n] \overset{\text{DTFT}}{\longleftrightarrow} Re(X(e^{j\Omega})); \ x_o[n] \overset{\text{DTFT}}{\longleftrightarrow} jIm(X(e^{j\Omega})).$$

Parseval's theorem:

$$\sum_{n=-\infty}^{\infty} x_1[n]x_2^*[n] = \tfrac{1}{2\pi} \int_{2\pi} X_1(e^{j\Omega})X_2(e^{-j\Omega})d\Omega.$$

$$\sum_{n=-\infty}^{\infty} |x[n]|^2 = \tfrac{1}{2\pi} \int_{2\pi} \left|X(e^{j\Omega})\right|^2 d\Omega.$$

Central ordinate theorems:

$$X(e^{j0}) = \sum_{n=-\infty}^{\infty} x[n]; x[0] = \tfrac{1}{2\pi} \int_{2\pi} X(e^{j\Omega})d\Omega.$$

Table 8.7.2 Discrete-time Fourier transform (DTFT) pairs

Unit sample function:

$$\delta[n - n_0] \overset{\text{DTFT}}{\longleftrightarrow} e^{-j\Omega n_0}$$

Constant:

$$x[n] = A \overset{\text{DTFT}}{\longleftrightarrow} A2\pi\delta(\Omega); \ |\Omega| \leq \pi.$$

Periodic functions:

$$e^{j\Omega_0 n} \overset{\text{DTFT}}{\longleftrightarrow} 2\pi\delta(\Omega - \Omega_0); \ |\Omega|, |\Omega_0| \leq \pi.$$
$$\cos(\Omega_0 n) \overset{\text{DTFT}}{\longleftrightarrow} \pi[\delta(\Omega - \Omega_0) + \delta(\Omega + \Omega_0)]; \ |\Omega|, |\Omega_0| \leq \pi.$$
$$\sin(\Omega_0 n) \overset{\text{DTFT}}{\longleftrightarrow} -j\pi[\delta(\Omega - \Omega_0) - \delta(\Omega + \Omega_0)]; \ |\Omega|, |\Omega_0| \leq \pi.$$
$$\sum_{k=-\infty}^{\infty} \delta[n - kN] \overset{\text{DTFT}}{\longleftrightarrow} \Omega_0 \sum_{k=-\infty}^{\infty} \delta(\Omega - k\Omega_0); \ \Omega_0 = 2\pi/N.$$

Unit pulse sequences:

$$u[n] \overset{\text{DTFT}}{\longleftrightarrow} \pi\delta(\Omega) + \frac{1}{1 - e^{-j\Omega}}; \ |\Omega| \leq \pi.$$
$$-u[-n - 1] \overset{\text{DTFT}}{\longleftrightarrow} -\pi\delta(\Omega) + \frac{1}{1 - e^{-j\Omega}}; \ |\Omega| \leq \pi.$$

Exponential sequences:

$$a^n u[n] \overset{\text{DTFT}}{\longleftrightarrow} \frac{1}{1 - ae^{-j\Omega}}; \ |a| < 1.$$

$$-a^n u[-n - 1] \overset{\text{DTFT}}{\longleftrightarrow} \frac{1}{1 - ae^{-j\Omega}}; \ |a| > 1.$$

$$a^{|n|}; \ a < 1 \overset{\text{DTFT}}{\longleftrightarrow} \frac{1 - a^2}{1 - 2a\cos(\Omega) + a^2}$$

$$(n + 1)a^n u[n] \overset{\text{DTFT}}{\longleftrightarrow} \frac{1}{(1 - ae^{-j\Omega})^2}; \ |a| < 1.$$

Sinc functions:

$$x[n] = \frac{\sin(Wn)}{\pi n} \overset{\text{DTFT}}{\longleftrightarrow} \begin{cases} 1, |\Omega| \leq \pi \\ 0, W|\Omega| \leq \pi \end{cases}$$

$$x[n] = \begin{cases} 1, 0nN - 1 \\ 0, \text{otherwise} \end{cases} \overset{\text{DTFT}}{\longleftrightarrow} e^{j\Omega(N-1)/2} \frac{\sin(\Omega N/2)}{\sin(\Omega/2)}$$

8.8 Discrete-Time Fourier-transforms from Samples of the Continuous-Time Fourier-Transforms

In Section 8.2, $x(t)$ is sampled signal at a sampling rate of $f_s = (1/t_s) > 2(B)$, $B = $ Bandwith of $x(t)$. The continuous-time F-transform of $x(t)$ and its inverse are

$$X(j\omega) = \int_{-\infty}^{\infty} x(t)e^{-j\omega t}dt, x(t)$$

$$= \frac{1}{2\pi} \int_{-\infty}^{\infty} X(j\omega)e^{j\omega t}d\omega, \ x(t) \overset{\text{FT}}{\longleftrightarrow} X(j\omega).$$

$$(8.8.1)$$

The transform is then approximated using the rectangular integration formula with a sampling interval of t_s s. That is,

$$X(j\omega) \cong \sum_{n=-\infty}^{\infty} t_s[x(nt_s)e^{-j\omega nt_s}] \equiv X_{\omega_s}(j\omega), \omega_s = 2\pi f_s.$$

(8.8.2)

Note that $X_{\omega_s}(j\omega)$ is a periodic function with period $\omega_s = (2\pi/t_s)$ and is an *approximation* of $X(j\omega)$ *in the frequency range* $|\omega| \leq \omega_s/2$, as

$$X_{\omega_s}(j(\omega + (2\pi/t_s))) = t_s \sum_{n=-\infty}^{\infty} x(nt_s)e^{-j(\omega+(2\pi/t_s))nt_s}$$

$$= t_s \sum_{n=-\infty}^{\infty} x(nt_s)e^{-j\omega nt_s} = X_{\omega_s}(j\omega).$$

(8.8.3)

The transform $X(j\omega)$ is arbitrary and $X_{\omega_s}(j\omega)$ is periodic. If $x(t)$ is band limited to half the sampling rate, the signal $x(t)$ can be recovered from the sampled signal and the signal spectrum is

$$X(j\omega) = \begin{cases} X_{\omega_s}(j\omega), & |\omega| \leq \omega_s/2 \\ 0, & |\omega| > \omega_s/2 \end{cases}.$$

(8.8.4)

The discrete-time Fourier transform (DTFT) was defined earlier assuming $x[n] = x(nt_s)$ and

$$X(e^{j\Omega}) = \sum_{n=-\infty}^{\infty} x[n]e^{-jn\Omega}.$$

(8.8.5)

From this review, the following conclusions can be made:

1. The DTFT is periodic, whereas the continuous Fourier transform is not periodic.
2. The sampling interval t_s is not included in (8.8.5). Approximation to the continuous F-transform can be obtained from the DTFT of the sampled signal by multiplying it by t_s.
3. The DTFT is defined in terms of the *normalized frequency*, $\Omega = \omega t_s = \omega/f_s$ where f_s is the sampling frequency. The normalized frequency Ω is referred to as the *digital frequency* and is measured in radians/sample or in radians/cycle. Noting that the DTFT is periodic with period 2π, one

period of the DTFT gives its complete information.

4. Most of the continuous-time functions are time limited to, say $T = Nt_s$ seconds. In computing the transform, two variables need to be selected, the sampling interval t_s and the number of sample points N. Note that if $x(t)$ has discontinuities, taking its Fourier transform $X(j\omega)$ and then the inverse transform, $F^{-1}[X(j\omega)]$, gives *half-values* of the function at the discontinuities. Therefore, the sampled values at these locations are taken as half-values. For example, if $x(t) = e^{-t}u(t)$, then $x[0] = .5$.
5. The spectrum of the sampled signal in the interval $0 \leq \omega < \omega_s = (2\pi/t_s)$ is

$$X_{\omega_s}(j\omega) \cong t_s \sum_{n=0}^{N-1} x(nt_s)e^{-j(\omega t_s)n}, 0 \leq \omega < \omega_s = 2\pi/t_s.$$

(8.8.6)

6. Finally, considering Item 4 above in approximating the continuous Fourier transform using the DTFT, the number of sample points, the sampling interval, and the sample values need to be considered.

Example 8.8.1 In this example, some of the important facets associated with computing the transform $x(t) = e^{-2t}u(t)$ using DTFT are discussed. What should be the interval T before sampling the signal and the number of samples to be used? Use Example 8.2.6 Part b to find the sampling interval.

Solution: First $x(t)$ is not time limited. For discrete computations, only a finite interval of time needs to be considered, say T. Find T such that in the interval $0 \leq t < T, x(T) < < 1$. For $T = 4$ and 5, we have $x(4) = .00033$ and $x(5) = .000045$. Both are small enough, either one would be adequate and let $T = 4$. From Example 8.2.6, the sampling interval is $t_s = (1/f_s) \leq \pi/200$. The number of samples is $T/t_s \approx 255$. Discrete computations of transforms are the most efficient if the number of sample points N is a power of 2. Select $N = 256$. The next step is to identify the sample values. Noting that $x(t)$ has a jump discontinuity at $t = 0$, the first sample value is $[x(0^-) + x(0^+)]/2 = .5 = x(0)$. The sample values are

$$\{x(nt_s)\} = \{.5, e^{-2nt_s}, n = 1, \ldots, 255\}.$$

The discrete-time Fourier transform is approximated at the frequency sampled values using (8.8.6). That is,

$$X_{\omega_s}(j(k/N)\omega_s) \cong t_s \sum_{n=0}^{N-1} x_s[n] e^{-j2\pi(kn/N)},$$
$$k = 0, 1, \ldots, N - 1.$$

This is a periodic sequence with period $N = 256$. The samples in the frequency domain are spaced apart by $(1/N)f_s = (1/256)64 = .25 \, \text{Hz}$.

Notes: The DTFT is a periodic, continuous function of Ω and the sampled transform values computed from the DTFT are discrete and periodic. The spectrum of the analog signal is continuous. Increasing the product (Nt_s) implies a longer signal and the discrete transform has more frequency values. Since the sampling frequency is not changed, the effect of increasing (Nt_s) introduces more frequency values. The frequency interval between the spectral components is reduced. The sampling interval t_s $(t_s = 1/f_s)$ controls the accuracy in the approximation obtained by the DTFT compared to the actual evaluation of the continuous Fourier transform. ∎

As an example, consider that a signal of a 10-s interval that is band limited to 4 kHz. We are interested in estimating the spectrum of the segment using the above procedure with a resolution of, say 0.1 Hz in the spectral spacing. From the Nyquist sampling theorem, a sampling rate of 10 kHz would be adequate.

⇒ Number of samples $= f_s/$frequency resolution
$$= 10 \, \text{kHz}/.1 \, \text{Hz} = 100,000.$$

Transform algorithms are most efficient if the number of sample points, N is a power of 2. The next highest number that is a power of 2 is $2^{17} = 131072$. The length of the corresponding segment is equal to $T = Nt_s = N/f_s = 131,072/10,000 \approx 13.1 \, \text{s}$.

8.9 Discrete Fourier Transforms (DFTs)

In the last section, the spectrum of the sampled signal $X_{\omega_s}(\omega)$ in the interval $0 \leq \omega < \omega_s = (2\pi/t_s)$ was given in (8.8.6). If we sample this function at

intervals of $(2\pi/Nt_s)$ in one period, then we have N values $X_{\omega_s}(\omega_k)$, $\omega_k = 2\pi k/N$. That is,

$$X_{\omega_s}(j2\pi k/N) = t_s \sum_{n=0}^{N-1} x(nt_s) e^{-j(2\pi k/N)n},$$
$$k = 0, 1, 2, \ldots, N - 1. \qquad (8.9.1)$$

There are N sample values in time $x(nt_s)$ and N sample values of the spectrum in (8.9.1). These results can be applied to digital data by starting with $x[n] = x(nt_s)$, $n = 0, 1, 2, \ldots, N - 1$ and defining the *discrete Fourier transform (DFT)* by

$$X[k] \equiv \text{DFT}[x[n]] = \sum_{n=0}^{N-1} x[n] e^{-j(2\pi/N)kn},$$
$$k = 0, 1, 2, \ldots, N - 1. \qquad (8.9.2)$$

Note that multiplying $X[k]$ by t_s results in the spectral values in (8.9.1). The next question is how can the data $x[n]$ be obtained from the discrete Fourier transform coefficients $X[k]$? It turns out that these can be determined by

$$x[n] = \frac{1}{N} \sum_{k=0}^{N-1} X[k] e^{j(2\pi/N)kn},$$
$$n = 0, 1, 2, \ldots, (N - 1). \qquad (8.9.3)$$

The following shows that (8.9.3) is valid. Substituting the DFT values (see (8.9.2)) in (8.9.3) results in

$$\frac{1}{N} \sum_{k=0}^{N-1} X[k] e^{j(2\pi/N)nk}$$
$$= \frac{1}{N} \sum_{k=0}^{N-1} [\sum_{m=0}^{N-1} x[m] e^{-j(2\pi/N)mk}] e^{j(2\pi/N)kn}$$
$$= \frac{1}{N} \sum_{m=0}^{N-1} x[m] \sum_{k=0}^{N-1} [e^{j(2\pi/N)(n-m)k}]$$
$$= \frac{1}{N} \sum_{m=0}^{N-1} x[m] \left[\sum_{k=0}^{N-1} [e^{j(2\pi/N)(n-m)}]^k \right]. \qquad (8.9.4)$$

Using the summation formula for the *finite geometric series* in (C.6.1a) results in

$$\sum_{k=0}^{N-1} (e^{j(2\pi/N)(n-k)})^k = \begin{cases} 0, & n \neq m \\ N. & n = m \end{cases} . \qquad (8.9.5)$$

$$\Rightarrow \frac{1}{N}\sum_{m=0}^{N-1} x[m]\left[\sum_{k=0}^{N-1}[e^{j(2\pi/N)(n-m)}]^k\right] = x[n]. \quad (8.9.6)$$

The data $x[n]$ is now tied to the discrete Fourier transform coefficients, or DFTs identified as $X[k]s$. We can summarize the results in terms of DFTs, inverse DFTs, and the symbol for the transform pair as follows:

$$X[k] \equiv DFT[x[n]]$$

$$= \sum_{n=0}^{N-1} x[n]e^{-j(2\pi/N)kn}, k = 0, 1, 2, \ldots, N-1,$$

$$(8.9.7a)$$

$$x[n] = IDFT[X[k]] = \frac{1}{N}\sum_{k=0}^{N-1} X[k]e^{j(2\pi/N)kn},$$

$$n = 0, 1, 2, \ldots, (N-1), \quad (8.9.7b)$$

$$x[n] \overset{DFT}{\longleftrightarrow} X[k]. \quad (8.9.8)$$

The sequences $x[n], 0 \leq n \leq N-1$ and $X[k]$, $0 \leq k \leq N-1$ form a *discrete Fourier transform (DFT) pair*.

Notes: There are N equations each to determine the DFTs and the IDFTs. The exponential function $e^{\pm j(2\pi/N)}$ is periodic with period N. Indices in the DFT and IDFT variables are restricted to the principal range $0 \leq n, k \leq N-1$. The multiplicative terms in the forward and the inverse transforms $e^{\pm j(2\pi/N)kn}$ take one of the values in the set

$$\left\{1, e^{\pm j(2\pi/N)}, e^{\pm j2(2\pi/N)}, \ldots, e^{\pm j(N-1)(2\pi/N)}\right\}. \quad (8.9.9)$$

This follows from the fact that for $0 \leq k, n \leq N-1$, the product (kn) can be written as

$$kn = lN + m, 0 \leq k, n, m \leq N-1;$$
$$k, n, l, m \text{ and } N, \text{integers.} \quad (8.9.10)$$

In compact form we can use *modulo (mod) N arithmetic*. That is,

$$kn = lN + m \equiv m \bmod(N) \equiv [m]_{\bmod(N)}$$
$$= [m]_{(N)}, 0 \leq m \leq N-1. \quad (8.9.11)$$

Therefore,

$$le^{\pm j(2\pi/N)kn} = e^{\pm j(2\pi/N)[m+lN]} = e^{\pm j(2\pi/N)m}e^{\pm j(2\pi/N)(lN)}$$
$$= e^{\pm j(2\pi/N)m}, 0 \leq m \leq N-1. \quad (8.9.12)$$

Noting this, only N terms in (8.9.9) are needed to compute the DFT. The DFT and IDFT have *implied periodicity* with period N. That is,

$$X[k+N] = \sum_{n=0}^{N-1} x[n]e^{j(2\pi/N)n(k+N)}$$

$$= \sum_{n=0}^{N-1} x[n]e^{j(2\pi/N)nk}e^{j(2\pi/N)nN},$$

$$= \sum_{n=0}^{N-1} x[n]e^{j(2\pi/N)nk} = X[k]. \quad (8.9.13a)$$

$$x[n+N] = \frac{1}{N}\sum_{n=0}^{N-1} X[k]e^{j2\pi(n+N)k/N}$$

$$= \frac{1}{N}\sum_{n=0}^{N-1} X[k]e^{j2\pi(n)k/N} = x[n]. \quad (8.9.13b)$$

At a later time, time and frequency shifts will be considered, such as $x[n-n_0]$ and $X[k-k_0]$, where k_0 and n_0 are some integers. Since the integers $[n-n_0]$ and $[k-k_0]$ may fall outside of the range $[0, N-1]$, these integers need to be converted to a number in the principal range using mod N arithmetic. For example,

$$[-1]_{\bmod 16} = 15, [17]_{\bmod 16} = 1; x[[k+1]_{\bmod 16}]$$
$$= x[(k+1-16)] = x[k-15].$$

These will not be identified explicitly and implied from the context.   ∎

Notes: Modular arithmetic was introduced by *Carl Friedrich Gauss* in his work *Disquisitions Arithmetica*, see Hawking (2005). Gauss is considered to be one of the great mathematicians who ever lived. His work laid the foundation for number theory. Many of the digital coding and encryption algorithms are based on number theory. See Gilbert and Hatcher (2000), Hershey and Yarlagadda (1986), and others.   ∎

Interestingly, the *same algorithm* can be used to compute both the *forward* and the *inverse* DFT transforms, which can be seen by rewriting (8.9.7b) as

$$x[n] = \frac{1}{N} \sum_{k=0}^{N-1} [X^*[k] e^{-j(2\pi/N)nk}]^* . \qquad (8.9.14a)$$

Pictorially (8.9.8) can be described by

$$X[k] \to X^*[k] \to \text{DFT}\{X^*[k]\}$$

$$\to \frac{1}{N}(\text{DFT}\{X^*[k]\})^* = x[n]. \qquad (8.9.14b)$$

8.9.1 Matrix Representations of the DFT and the IDFT

The discrete Fourier transform (DFT) can be written in a matrix form and in compact form (see Appendix 1 for a brief review of matrices) using the equations

$$x[n] = \frac{1}{N} \sum_{n=0}^{N-1} X[k] e^{jn(2\pi/n)k} \xleftarrow{\text{DFT}} X[k]$$

$$= \sum_{k=0}^{N-1} x[n] e^{-jk(2\pi/N)n} \qquad (8.9.15)$$

These are

$$
\begin{bmatrix}
X[0] \\
X[1] \\
\cdot \\
\cdot \\
\cdot \\
X[N-1]
\end{bmatrix}
=
\begin{bmatrix}
1 & 1 & \cdot & \cdot & \cdot & 1 \\
1 & e^{-j(2\pi/N)} & \cdot & \cdot & \cdot & e^{-j(2\pi/N)(N-1)} \\
\cdot & \cdot & & & & \cdot \\
\cdot & \cdot & & & & \cdot \\
\cdot & \cdot & & & & \cdot \\
1 & e^{-j(2\pi/N)(N-1)} & \cdot & \cdot & \cdot & e^{-j(2\pi/N)(N-1)^2}
\end{bmatrix}
\begin{bmatrix}
x[0] \\
x[1] \\
\cdot \\
\cdot \\
\cdot \\
x[N-1]
\end{bmatrix}, \qquad (8.9.16a)
$$

$$\mathbf{X} = \mathbf{A}_{\text{DFT}}\mathbf{x}. \qquad (8.9.16b)$$

The vectors $\mathbf{X}$ and $\mathbf{x}$ are N-dimensional column vectors and the matrix $\mathbf{A}_{\text{DFT}}$ is a $N \times N$ complex matrix and is referred to as a *discrete Fourier transform (or DFT) matrix*. A typical entry in $\mathbf{A}_{\text{DFT}}$, say (k, n) entry, is

$$\mathbf{A}_{\text{DFT}}(k, n) = e^{-j(2\pi/N)(k-1)(n-1)} = W_N^{(k-1)(n-1)}, \quad W_N$$

$$= e^{-j(2\pi/N)}, 1 \le k \le N, 1 \le n \le N. \qquad (8.9.17)$$

The constant W_N is an Nth root of unity, as $W_N^N = 1$. From the second row or column in the $\mathbf{A}_{\text{DFT}}$ matrix we have the roots of unity. The exponent t in the entries $W_N^t = e^{-j(2\pi/N)t}$ is called the *twiddle factor or rotation factor*, see Rabiner and Gold (1975). All the entries in $\mathbf{A}_{\text{DFT}}$ can be simplified to one of the values in the following set (see 8.9.12):

$$\left\{ 1, e^{-j(2\pi/N)}, e^{-j(2\pi/N)2}, \ldots, e^{-j(2\pi/N)(N-1)} \right\}. \qquad (8.9.18)$$

The IDFT in (8.9.21b), in a matrix form and its compact from are

$$
\begin{bmatrix}
x[0] \\
x[1] \\
\cdot \\
\cdot \\
\cdot \\
x[N-1]
\end{bmatrix}
=
\frac{1}{N}
\begin{bmatrix}
1 & 1 & \cdot & \cdot & \cdot & 1 \\
1 & e^{j(2\pi/N)} & \cdot & \cdot & \cdot & e^{j(2\pi/N)(N-1)} \\
\cdot & \cdot & & & & \cdot \\
\cdot & \cdot & & & & \cdot \\
\cdot & \cdot & & & & \cdot \\
1 & e^{j(2\pi/N)(N-1)} & \cdot & \cdot & \cdot & e^{j(2\pi/N)(N-1)^2}
\end{bmatrix}
\begin{bmatrix}
X[0] \\
X[1] \\
\cdot \\
\cdot \\
\cdot \\
X[N-1]
\end{bmatrix}, \qquad (8.9.19a)
$$

$$\mathbf{x} = \frac{1}{N}\mathbf{A}_{\text{DFT}}^*\mathbf{X}. \qquad (8.9.19b)$$

A typical entry, say the (n,k) entry in $\mathbf{A}_{\text{DFT}}^*$ is

$$\mathbf{A}_{\text{DFT}}^*(n,k) = e^{j(2\pi/N)(n-1)(k-1)}$$
$$= W_N^{-(n-1)(k-1)}, \ W_N = e^{-j(2\pi/N)}. \quad (8.9.20)$$

These matrices are known as *Vandermonde matrices* (see Hohn (1958)). Note that the matrices $\mathbf{A}_{\text{DFT}}$ and $((1/N)\mathbf{A}_{\text{DFT}}^*)$ are *symmetric matrices*. That is,

$$\mathbf{A}_{\text{DFT}}(k,n) = \mathbf{A}_{\text{DFT}}(n,k) \text{ and } \mathbf{A}_{\text{DFT}}^*(k,n)$$
$$= \mathbf{A}_{\text{DFT}}^*(n,k), \qquad (8.9.21)$$

$$\Rightarrow (\mathbf{A}_{\text{DFT}})(1/\sqrt{N})(1/\sqrt{N})(\mathbf{A}_{\text{DFT}}^*)$$
$$= \mathbf{I}_N \text{ or } \mathbf{A}_{\text{DFT}}^{-1} = (1/N)\mathbf{A}_{\text{DFT}}^*. \quad (8.9.22)$$

That is, $(\sqrt{(1/N)})\mathbf{A}_{\text{DFT}}$ is a *unitary matrix*.

Example 8.9.1 Given the data $x[0] = 1, x[1] = -1,$ $x[2] = 1, x[3] = 2$, find the discrete Fourier transform (DFT) of this sequence using the DFT matrix. Using the DFT coefficients and the $\mathbf{A}_{\text{DFT}}^*$ matrix, find the corresponding data sequence.

Solution: For $N = 4$, we have $(e^{-j(2\pi/N)})^{kn} = (e^{-j(2\pi/4)})^{kn} = (e^{-j(\pi/2)})^{kn} = (-j)^{kn}$. Now, the data vector, the DFT matrix and DFT sequence (or coefficients) are

$$\mathbf{x} = \begin{bmatrix} 1 \\ -1 \\ 1 \\ 2 \end{bmatrix}, \mathbf{A}_{\text{DFT}} = \begin{bmatrix} 1 & 1 & 1 & 1 \\ 1 & -j & (-j)^2 & (-j)^3 \\ 1 & (-j)^2 & (-j)^4 & (-j)^6 \\ 1 & (-j)^3 & (-j)^6 & (-j)^9 \end{bmatrix} = \begin{bmatrix} 1 & 1 & 1 & 1 \\ 1 & -j & -1 & j \\ 1 & -1 & 1 & -1 \\ 1 & j & -1 & -j \end{bmatrix} \qquad (8.9.23)$$

$$\mathbf{X} = \begin{bmatrix} 1 & 1 & 1 & 1 \\ 1 & -j & -1 & j \\ 1 & -1 & 1 & -1 \\ 1 & j & -1 & -j \end{bmatrix} \begin{bmatrix} 1 \\ -1 \\ 1 \\ 2 \end{bmatrix} = \begin{bmatrix} 3 \\ j3 \\ 1 \\ -j3 \end{bmatrix}. \qquad (8.9.24)$$

The DFT matrix is symmetric and it contains only $N = 4$ distinct elements $1, 1, j, j^2 = -1,$ and $j^3 = -j$ corresponding to the 4-point DFT. If we can save 1 and j, we can generate the others by changing the sign of these. The data sequence can be computed from the DFT vector by

$$\mathbf{x} = \frac{1}{4}\mathbf{A}_{\text{DFT}}^*\mathbf{X} = \frac{1}{4}\begin{bmatrix} 1 & 1 & 1 & 1 \\ 1 & j & -1 & -j \\ 1 & -1 & 1 & -1 \\ 1 & -j & -1 & j \end{bmatrix} \begin{bmatrix} 3 \\ j3 \\ 1 \\ -j3 \end{bmatrix} = \begin{bmatrix} 1 \\ -1 \\ 1 \\ 2 \end{bmatrix}. \qquad (8.9.25) \ \blacksquare$$

8.9.2 Requirements for Direct Computation of the DFT

DFT is a transformation that takes a set of N complex (or real) values in time to N complex (or real) values in frequency(see (8.9.16a)). It involves the multiplication of a $N \times N$ complex matrix by a N-dimensional vector. The direct computation of DFT requires N^2 complex multiplications and $N(N-1)$ additions. *Fast Fourier transform* (FFT) algorithms, considered in the next chapter, reduces these numbers significantly. FFT algorithms are most effective when the number of data points N is a power of 2.

DFT is applicable for real and complex data and can be implemented using real multiplications. First, let $x[n] = \text{Re}[x[n]] + j\text{Im}[x[n]]$ and

$$X[k] = \sum_{n=1}^{N-1}[\text{Re}\{x[n]\}$$
$$+ j\text{Im}\{x[n]\}]\left[\text{Re}\{W_N^{kn}\} + j\text{Im}\{W_N^{kn}\}\right],$$

$$= \sum_{n=0}^{N-1}\text{Re}\{x[n]\}\text{Re}\{W_N^{kn}\} - \sum_{n=0}^{N-1}\text{Im}\{x[n]\}\text{Im}\{W_N^{kn}\}$$

$$+ j\left[\sum_{n=0}^{N-1}\text{Re}\{x[n]\}\text{Im}\{W_N^{kn}\}\right.$$

$$+ \left.\sum_{n=0}^{N-1}\text{Im}\{x[n]\}\text{Re}\{W_N^{kn}\}\right], k = 0, 1..., N-1.$$

$$= \text{Re}[X[k]] + j\text{Im}[X[k]], k = 0, 1..., N-1. \quad (8.9.26)$$

Computation of $X[k]$ requires $4N^2$ real multiplications. For fixed or floating point operations, multiplications are computationally more expensive compared to the additions or transfer of data. We will come back to this topic in Section 9.3. Interestingly,

$$x^*[n] \xrightarrow{\text{DFT}} X^*[-k]_{\text{Mod}(N)}. \qquad (8.9.27)$$

This follows from

$$\text{DFT}\{x^*[n]\} = \sum_{n=0}^{N-1} x^*[n] e^{-j2\pi nk/N} = \left[\sum_{n=0}^{N-1} x[n] e^{j2\pi nk/N}\right]^*,$$

$$= \left[\sum_{n=0}^{N-1} x[n] e^{-j2\pi n(-k)/N}\right]^*_{\text{mod}(N)} = X^*[-k]_{\text{mod}(N)}.$$

8.10 Discrete Fourier Transform Properties

DFT properties are similar to the continuous case. In proving these properties, we will assume that the discrete-time sequences are given by $x_{di}[n]$, $0 \leq n \leq N-1$, and their discrete Fourier transforms are given by $X_{di}[k], 0 \leq k \leq N-1$. That is,

$$\text{IDFT}\{X_i[k]\} = x_i[n] \xrightarrow{\text{DFT}} X_i[k]$$

$$= \text{DFT}\{x_i[n]\}, i = 1, 2, \ldots$$

$$X_i[k] = \sum_{k=0}^{N-1} x_i[n] e^{-j(2\pi/N)nk}, x_i[n]$$

$$= \frac{1}{N} \sum_{k=0}^{N-1} X_i[k] e^{j(2\pi/N)nk}. \qquad (8.10.1)$$

Notes: In the following one variable may be replaced by another variable. The variable n and k will be used to identify the *time* and the *frequency* sequences. Most of the proofs follow by using the basic definitions of the DFT (or IDFT). Both $x[n]$ and $X[k]$ are assumed to have implied periodicity with period N, i.e.

$$x[n+N] = x[n] \text{ and } X[k+N] = X[k],$$

$$x[n] = x[n]_{\text{mod}N} \equiv x[n]_N \text{ and } X[k]$$
$$= X[k]_{\text{mod}N} \equiv X[k]_N.$$

8.10.1 DFTs and IDFTs of Real Sequences

The DFT of a real sequence $x_d[n]$ has *conjugate symmetry*. That is,

$$X[-k] = X^*[k] = X[N-k]. \qquad (8.10.2)$$

These can be shown as follows:

$$X^*[k] = \left[\sum_{n=0}^{N-1} x[n] e^{-j2\pi nk/N}\right]^*$$

$$= \sum_{n=0}^{N-1} x[n] e^{-j2\pi n(-k)/N} = X[-k],$$

$$X[N-k] = \sum_{k=0}^{N-1} x[n] e^{-j(2\pi/N)n(N-k)}$$

$$= \sum_{k=0}^{N-1} x[n] e^{j(2\pi/N)nk} = X^*[k].$$

We can show that if $X[k]$ is real, then $x[-n] = x^*[n] = x[N-n]$.

Notes: The DFT of two real valued sequences $x_1[n]$ and $x_2[n]$ can be determined from the DFT of the complex sequence $x[n] = x_1[n] + jx_2[n]$ as follows:

$$x[n] = x_1[n] + jx_2[n] \xrightarrow{DFT} X[k], x[n] \xrightarrow{DFT} X[k]$$

$$x_1[n] \xrightarrow{DFT} X_1[k], x_2[n] \xrightarrow{DFT} X_2[k]$$

$$X_1[k] = \frac{1}{2}\{X[k] + X^*[N-k]\}, X_2[k]$$

$$= -\frac{1}{2}j\{X[k] - X^*[N-k]\}.$$

8.10.2 Linearity

The linearity property follows directly from the definition:

$$\sum_{i=1}^{M} \alpha_i x_i[n] \xrightarrow{DFT} \sum_{i=1}^{M} \alpha_i X_i[k]. \qquad (8.10.3)$$

Example 8.10.1 Find the $N(=8)$ point DFT of the sequence $\{x[n]\}$ and sketch the coefficients.

$$x[n] = \cos(2\pi n/N), n = 0, 1, 2, \ldots, N-1. \qquad (8.10.4)$$

Solution: Using Euler's identity and the linearity property, the DFT coefficients are

$$X[k] = \text{DFT}\{x[n]\} = \text{DFT}\{\cos(2\pi n/N))\}$$
$$= .5\text{DFT}\{e^{j(2\pi n/N)}\} + .5\,\text{DFT}\{e^{-j(2\pi n/N)}\},$$

$$= \frac{1}{2}\left[\sum_{n=0}^{N-1} e^{j(2\pi n/N)(1-k)} + \sum_{n=0}^{N-1} e^{-j(2\pi n/N)(1+k)}\right] \quad (8.10.5)$$

$$\Rightarrow X[k] = \frac{N}{2}\delta[k-1] + \frac{N}{2}\delta[k+1]$$
$$= \frac{N}{2}\delta[k-1] + \frac{N}{2}\delta[k-N+1]. \quad (8.10.6a)$$

The closed form expression for $X[k]$ is obtained by using (C.6.1b). Also, note that $\delta[k+1]$ is outside of the interval $0 \le n < N$, which can be resolved by noting the DFT coefficients have implicit periodicity with period N and write $\delta[k+1] = \delta[k-N+1]$. Thus there are two discrete-time impulses in the interval $0 \le k < N$. Here, $X_d[k]$ reduces to

$$X[k] = 4\delta[k-1] + 4\delta[k-7]. \quad (8.10.6b)$$

The transform sequence is shown in Fig. 8.10.1. Note the DFT sequence is real and has even symmetry. Furthermore, we can see that $X^*[k] = X[k] = X[N-k]$. ∎

8.10.3 Duality

The duality property is

$$x[n] \xleftrightarrow{\text{DFT}} X[k] \xrightarrow[\text{Change to}]{\frac{1}{N}} \frac{1}{N}X[n] \xleftrightarrow{\text{DFT}} x[-k].$$
$$(8.10.7)$$

To show this, start with the IDFT in terms of $X[k]$ and rewrite the function in terms of a different variable other than n and k, say l, and then replace l by $-l$. That is,

$$x[n] = \frac{1}{N}\sum_{k=1}^{N-1} X[k]e^{j(2\pi/N)nk}$$

$$\rightarrow x[l] = \frac{1}{N}\sum_{k=1}^{N-1} X[k]e^{j(2\pi/N)lk},$$

$$\Rightarrow x[-l] = \frac{1}{N}\sum_{r=0}^{N-1} X[r]e^{-j(2\pi/N)lr}.$$

Now let $l = k, r = n$, and taking the $(1/N)$ inside the summation results in the proof of the duality property as follows:

$$x[-k] = \sum_{n=0}^{N-1} (X[n]/N)e^{-j(2\pi/N)nk}, \quad (8.10.8)$$

$$\Rightarrow X[n]/N = \text{IDFT of } x[-k]$$
$$\text{or the DFT}\{X[n]/N\} = x[-k].$$

Example 8.10.2 Use the function in Example 8.10.1 to verify the duality principle.

Solution: We have

$$x[n] = \cos(2\pi n/N)n \xleftarrow{\text{DFT}} (N/2)\delta[k-1]$$
$$+ (N/2)\delta[k-(N-1)] = X(k). \quad (8.10.9)$$

Now consider $y[n] \xleftrightarrow{\text{DFT}} Y[k]$ with

$$y[n] = .5\delta[n-1]$$
$$+ .5\delta[n-(N-1)] \xleftarrow{\text{DFT}} \sum_{n=0}^{N-1} .5\{\delta[n-1]$$
$$+ \delta[n-(N-1)]\}e^{-j2\pi nk/N}$$
$$\Rightarrow Y[k] = .5[e^{-j(2\pi/N)k} + e^{j(2\pi/N)k}]$$
$$= \cos((-2\pi/N)k) = \cos(2\pi k/N)$$

Note

$$y[k] = x[-k]. \quad (8.10.10) \quad ■$$

8.10.4 Time Shift

The time shift property is

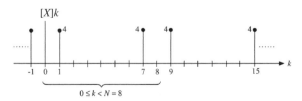

$[X]k$

$0 \le k < N = 8$

Fig. 8.10.1 Example 8.10.1

$$x[n-m]_{\text{mod}(N)} \xleftrightarrow{\text{DFT}} X[k]e^{-j(2\pi/N)km} \text{ or}$$

$$x[n-m]_{\text{mod}(N)}e^{j(2\pi/N)km} \xleftrightarrow{\text{DFT}} X[k]. \qquad (8.10.11)$$

We know that $x[n]$ and $X[k]$ can be considered as periodic sequences and $X[k]$ represents the DFT coefficients for one period of $x[n]$. By using a new variable $l = n - m$, and simplifying, we have

$$\text{DFT}\{x[n-m]\} = \sum_{n=0}^{N-1} x[n-m]e^{-j(2\pi/N)nk}$$

$$= \sum_{l=0}^{N-1} x[l]e^{-j(2\pi/N)(l+m)k}$$

$$= \left\{ \sum_{l=0}^{N-1} x[l]e^{-j(2\pi/N)lk} \right\} e^{-j(2\pi/N)mk}$$

$$= X[k]e^{-j(2\pi/N)mk}. \qquad (8.10.12)$$

8.10.5 Frequency Shift

The dual to time shift property is the frequency shift property given below and can be shown taking the IDFT of the coefficients on the right.

$$x[n]e^{j(2\pi/N)mn} \xleftrightarrow{\text{DFT}} X[k-m]_{\text{mod}(N)}. \qquad (8.10.13)$$

8.10.6 Even Sequences

If a function is even, i.e., $x[-n] = x[n] \equiv x_e[n]$, then the DFT is real and even. It can be written as

$$x_e[n] \xleftrightarrow{\text{DFT}} X_e[k] = \sum_{n=0}^{N-1} x_e[n] \cos\left(\frac{2\pi}{N}nk\right). \quad (8.10.14)$$

This can be verified using

$$x_e[-n] = x_e[N-n] = x_e[n], \qquad (8.10.15)$$

$$\Rightarrow \text{DFT}\{x_e[n]\} = \sum_{n=0}^{N-1} x_e[n]e^{-j(2\pi/N)kn}$$

$$= \sum_{n=0}^{N-1} x_e[n] \cos\left(\frac{2\pi}{N}nk\right)$$

$$-j\sum_{n=0}^{N-1} x_e[n] \sin\left(\frac{2\pi}{N}nk\right). \quad (8.10.16)$$

The second term on the right is

$$\sum_{n=0}^{N-1} x_e[n] \sin\left(\frac{2\pi}{N}nk\right) = \sum_{n=0}^{N-1} x_e[-n] \sin\left(\frac{2\pi}{N}nk\right)$$

$$= \sum_{n=0}^{N-1} x_e[N-n] \sin\left(\frac{2\pi}{N}nk\right),$$

$$= \sum_{m=0}^{N-1} x_e[m] \sin\left(\frac{2\pi}{N}(N-m)k\right),$$

$$= \sum_{m=0}^{N-1} x_e[m] \left[\sin\left(\frac{2\pi}{N}Nk\right) \cos\left(\frac{2\pi}{N}(-mk)\right) \right.$$
$$\left. + \cos\left(\frac{2\pi}{N}(Nk)\right) \sin\left(-\frac{2\pi}{N}mk\right) \right],$$

$$= \sum_{m=0}^{N-1} x_e[m] \left[\sin\left(\frac{2\pi}{N}Nk\right) \cos\left(\frac{2\pi}{N}(-mk)\right) \right.$$
$$\left. + \cos\left(\frac{2\pi}{N}(Nk)\right) \sin\left(-\frac{2\pi}{N}mk\right) \right],$$

$$= -\sum_{m=0}^{N-1} x_e[m] \sin\left(\frac{2\pi}{N}km\right)$$

$$= \sum_{n=0}^{N-1} x_e[n] \sin\left(\frac{2\pi}{N}nk\right) = 0.$$

A number is equal to its negative only when it is zero. The coefficients are real and follow from (8.10.16) and (8.10.17). Noting the periodicity of the DFT coefficients, we have

$$X_e[-k] = X_e[N-k] = \sum_{n=0}^{N-1} x_e[n] \cos\left(\frac{2\pi}{N}n(N-k)\right),$$

$$= \sum_{n=0}^{N-1} x_e[n] \left[\cos\left(\frac{2\pi}{N}kN\right) \cos\left(\frac{2\pi}{N}(kn)\right) \right.$$
$$\left. + \sin\left(\frac{2\pi}{N}kN\right) \sin\left(\frac{2\pi}{N}kn\right) \right],$$

$$= \sum_{n=0}^{N-1} x_e[n] \cos\left(\frac{2\pi}{N}nk\right) = X_e[k] \ (\Rightarrow X_e[k]$$

$$= X_e[-k], \text{ even sequence}). \qquad (8.10.18)$$

8.10.7 Odd Sequences

If a function is odd, i.e., $x[-n] = -x[n] \equiv x_0[n]$, then the DFT is real and even. The DFT coefficients of an odd function are odd and imaginary and

$$\text{DFT}[x_0[n]] = -j \sum_{k=0}^{N-1} x_0[n] \sin\left(\frac{2\pi}{N}nk\right). \quad (8.10.19)$$

The proof of this is very similar to the last property. As a final comment on this topic, we have seen that a sequence can be expressed in terms of its even and odd parts. The above two properties allow for the computation of the DFT of an arbitrary periodic sequence in terms of the sum of DFTs of even and odd sequences. That is,

$$x[n] = x_e[n] + x_0[n] \xleftrightarrow{\text{DFT}} X(e^{j\Omega})$$
$$= \text{Re}\{X(e^{j\Omega})\} + j\text{Im}\{X(e^{j\Omega})\}, \quad (8.10.20)$$

$$x_e[n] \xleftrightarrow{\text{DFT}} \text{Re}\{X(e^{j\Omega})\},$$
$$x_0[n] \xleftrightarrow{\text{DFT}} \text{Im}\{X(e^{j\Omega})\}. \quad (8.10.21)$$

It follows that if $x[n]$ is real and even, then $X(e^{j\Omega})$ is real and even and if $x[n]$ is real and odd then $X(e^{j\Omega})$ is real and imaginary.

8.10.8 Discrete-Time Convolution Theorem

In Section 8.4.1 the *periodic* (or *cyclic*) *convolution* of two functions $x[n]$ and $h[n]$ with the *same period* N was defined by

$$y[n] = x[n] \otimes h[n] = \sum_{m=0}^{N-1} x[m]h[n-m]_{\bmod(N)}$$
$$= \sum_{m=0}^{N-1} x[n-m]_{\bmod(N)}h[m]. \quad (8.10.22)$$

The time convolution theorem stated below can be proven by starting with the left side of the above equation and rearranging the terms and then simplifying it. That is,

$$x[n] \otimes h[n]_{\bmod(N)} \xleftrightarrow{\text{DFT}} X[k]H[k] \quad (8.10.23)$$

$$y[n] = \sum_{i=0}^{N-1} x[i]h[n-i] = \sum_{i=0}^{N-1} \frac{1}{N}\left[\sum_{k=0}^{N-1} X[k]e^{j(2\pi/N)ik}\right]$$
$$\left[\frac{1}{N}\sum_{m=0}^{N-1} H[m]e^{j(2\pi/N)m(n-i)}\right],$$

$$= \frac{1}{N}\sum_{k=0}^{N-1}\sum_{m=0}^{N-1} X[k]H[m]e^{j(2\pi/N)mn}$$
$$\left[\frac{1}{N}\sum_{i=0}^{N-1} e^{j(2\pi/N)ik}e^{-j(2\pi/N)im}\right], \quad (8.10.24)$$

$$= \sum_{i=0}^{N-1} x[i]h[n-i] = \frac{1}{N}\sum_{k=0}^{N-1} X[k]H[k]e^{j(2\pi/N)kn}. \quad (8.10.25)$$

Example 8.10.3 Write the periodic convolution of the following two periodic sequences with period $N = 3$. Compute these using *a*. the time sequence and *b*. the DFT.

$$\{x[n]\} = \{x[0], x[1], x[2]\} = \{1, 2, 3\}, \{h[n]\}$$
$$= \{h[0], h[1], h[2]\} = \{1, -1, 1\}, \quad (8.10.26)$$

$$y[n] = \sum_{i=0}^{N-1} x[i]h[n-i], y[n]$$
$$= x[0]h[n] + x[1]h[n-1] + x[2]h[n-2]. \quad (8.10.27)$$

Solution: *a*. Using $h[-n] = h[N-n]$, the periodic convolution values are as follows:

$$y[0] = x[0]h[0] + x[1]h[-1] + x[2]h[-2]$$
$$= x[0]h[0] + x[1]h[2] + x[2]h[1],$$

$$y[1] = x[0]h[1] + x[1]h[0] + x[2]h[2],$$
$$y[2] = x[0]h[2] + x[1]h[1] + x[2]h[0],$$

$$\text{Matrix form}: \begin{bmatrix} y[0] \\ y[1] \\ y[2] \end{bmatrix} = \begin{bmatrix} h[0] & h[2] & h[1] \\ h[1] & h[0] & h[2] \\ h[2] & h[1] & h[0] \end{bmatrix} \begin{bmatrix} x[0] \\ x[1] \\ x[2] \end{bmatrix}.$$
$$(8.10.28)$$

Note the structure of the coefficient matrix on the right in (8.10.28) has a pattern, which can be written in general terms after this example. Noting that $x[0] = 1, x[1] = 2, x[2] = 3$ and $h[0] = 1, h[1] = -1, h[2] = 1$, the convolution values are

$$\begin{bmatrix} y[0] \\ y[1] \\ y[2] \end{bmatrix} = \begin{bmatrix} 1 & 1 & -1 \\ -1 & 1 & 1 \\ 1 & -1 & 1 \end{bmatrix} \begin{bmatrix} 1 \\ 2 \\ 3 \end{bmatrix} = \begin{bmatrix} 0 \\ 4 \\ 2 \end{bmatrix}. \quad (8.10.29)$$

b. In matrix form, the transform values are

$$\begin{bmatrix} X[0] \\ X[1] \\ X[2] \end{bmatrix} = \begin{bmatrix} 1 & 1 & 1 \\ 1 & e^{-j(2\pi/3)} & e^{-j2(2\pi/3)} \\ 1 & e^{-j2(2\pi/3)} & e^{-j4(2\pi/3)} \end{bmatrix} \begin{bmatrix} x[0] \\ x[1] \\ x[2] \end{bmatrix} \cong \begin{bmatrix} 6 \\ -1.5 + j.8660 \\ -1.5 - j.8660 \end{bmatrix}, \qquad (8.10.30)$$

$$\begin{bmatrix} H[0] \\ H[1] \\ H[2] \end{bmatrix} = \begin{bmatrix} 1 & 1 & 1 \\ 1 & e^{-j(2\pi/3)} & e^{-j2(2\pi/3)} \\ 1 & e^{-j2(2\pi/3)} & e^{-j4(2\pi/3)} \end{bmatrix} \begin{bmatrix} h[0] \\ h[1] \\ h[2] \end{bmatrix} \cong \begin{bmatrix} 1 \\ 1 + j1.7321 \\ 1 - j1.7321 \end{bmatrix}. \qquad (8.10.31)$$

The product of the transform coefficients in matrix form are given by

$$\begin{bmatrix} Y[0] \\ Y[1] \\ Y[2] \end{bmatrix} = \begin{bmatrix} X[0]H[0] \\ X[1]H[1] \\ X[2]H[2] \end{bmatrix} = \begin{bmatrix} 1(6) \\ (1 + j1.7321)(-1.5 + j.8660) \\ (1 - j1.7321)(-1.5 - j.8660) \end{bmatrix} = \begin{bmatrix} 6 \\ -3 - j1.7321 \\ -3 + j1.7321 \end{bmatrix}. \qquad (8.10.32)$$

These involve complex arithmetic resulting in rounded values. The IDFT of the vector in (8.10.32) gives approximations of the results in (8.10.29). ∎

The periodic convolution can be written in general matrix and symbolic forms of two periodic sequences $x[0], x[1], ..., x[N-1]$ and $h[0], h[1], ..., h[N-1]$ as follows:

$$y[n] = x[n] \otimes h[n] = \sum_{i=0}^{N-1} x[i]h[n-i]_{\mathrm{mod}(N)}$$

$$= \sum_{i=0}^{N-1} h[i]x[n-i]_{\mathrm{mod}(N)}. \qquad (8.10.33)$$

$$\begin{bmatrix} y[0] \\ y[1] \\ y[2] \\ . \\ . \\ . \\ y[N-1] \end{bmatrix} = \begin{bmatrix} h[0] & h[N-1] & h[N-2] & . & . & . & h[1] \\ h[1] & h[0] & h[N-1] & . & . & . & h[2] \\ h[2] & h[1] & h[0] & . & . & . & h[3] \\ . & . & . & . & . & . & . \\ . & . & . & . & . & . & . \\ . & . & . & . & . & . & . \\ h[N-1] & h[N-2] & h[N-3] & . & . & . & h[0] \end{bmatrix} \begin{bmatrix} x[0] \\ x[1] \\ x[2] \\ . \\ . \\ . \\ x[N-1] \end{bmatrix}, \qquad (8.10.34a)$$

$$\mathbf{y} = \mathbf{Hx}. \qquad (8.10.34b)$$

The vectors **y** and **x** are N-dimensional column vectors and **H** is a $N \times N$ *circulant matrix* having N distinct elements with a pattern. First, $h[n]$, $n = 0, 1, 2, ..., N - 1$ is placed in column 1 in **H**. Column 2 is obtained by circularly shifting column 1 down by 1. Similarly, column 3 is obtained by circularly shifting the column 2 down by 1 and

so on. The diagonal entries are the same and the entries in each sub diagonal are the same.

8.10.9 Discrete-Frequency Convolution Theorem

The discrete-frequency convolution theorem is a dual to the time convolution theorem and is given below. The proof of this is very similar to the time convolution theorem.

$$x[n]h[n] \xleftrightarrow{\text{DFT}} \frac{1}{N}\left[X[k] \otimes H[k]\right]_{\text{mod}(N)}$$

$$= \frac{1}{N}\sum_{i=0}^{N-1} X[i]H[k-i]_{\text{mod}(N)}. \quad (8.10.35)$$

8.10.10 Discrete-Time Correlation Theorem

In Section 8.3.2 we briefly discussed the discrete cross correlation and the convolution (see (8.3.20a and b)). The discrete cross correlation of two N point sequences is

$$r_{xh}[n] = \sum_{i=0}^{N-1} x[i]h[i+n]_{\text{mod}(N)} \xleftrightarrow{\text{DFT}} \text{DFT}\{r_{xh}[n]\}.$$

$$(8.10.36)$$

Note that we are using the variable n for the cross correlation function, as we are using the variable k for the DFT function. The discrete correlation theorem is stated by

$$\sum_{i=0}^{N-1} x(i)h(n+i)_{\text{mod}(N)} \xleftrightarrow{\text{DFT}} X^*[k]H[k]. \quad (8.10.37)$$

This can be proven using the following steps:

$$\sum_{i=0}^{N-1} x[i]h[n+i]_{\text{mod}(N)} = \sum_{i=0}^{N-1}\left[\frac{1}{N}\sum_{k=0}^{N-1} X[k]e^{j(2\pi/N)ik}\right]$$
$$\left[\frac{1}{N}\sum_{m=0}^{N-1} H[m]e^{j(2\pi/N)(n+i)m}\right],$$

$$= \sum_{i=0}^{N-1}\left[\frac{1}{N}\sum_{k=0}^{N-1} X^*[k]e^{-j(2\pi/N)ik}\right]^*\left[\frac{1}{N}\sum_{m=0}^{N-1} H[m]e^{j(2\pi/N)(n+i)m}\right],$$

$$= \frac{1}{N}\sum_{k=0}^{N-1}\sum_{m=0}^{N-1} X^*[k]H[m]e^{j(2\pi/N)mn}$$
$$\left[\frac{1}{N}\sum_{i=0}^{N-1} e^{-j(2\pi/N)ik}e^{j(2\pi/N)im}\right], \quad (8.10.38a)$$

$$= \frac{1}{N}\sum_{k=0}^{N-1} X^*[k]H[k]e^{j(2\pi/N)nk}. \quad (8.10.38b)$$

Note the bracketed term in (8.10.38a) is equal to 1 if $m = k$ and zero otherwise. As in the periodic convolution, DFTs can be used to compute the cross correlation by first computing the DFTs of the two sequences and then take the IDFT of $X_d^*[k]H_d[k]$.

8.10.11 Parseval's Identity or Theorem

It states that if $x[n]$ is real, then

$$\sum_{n=0}^{N-1} x^2[n] = \frac{1}{N}\sum_{k=0}^{N-1} |X[k]|^2. \quad (8.10.39)$$

This can be shown using (8.10.36) with $x[n] = h[n]$ and is left as an exercise.

Example 8.10.4 Verify the Parseval's theorem using $h_d[n]$ in Example 8.10.3.

Solution:

$$h[0] = 1, h[1] = -1, h[2] = 1 \Rightarrow H[0] = 1,$$
$$H[1] = 1 + j(1.7321), H[2] = 1 - j1.7321,$$

$$\sum_{n=0}^{2} h^2[n] = 3, \sum_{n=0}^{2} h^2[n] = \frac{1}{3}\sum_{k=0}^{2} |H[k]|^2 = 3 \quad (8.10.40)$$

$$(1/3)\sum_{k=0}^{2} |H[k]|^2 = (1 + |1 + j(1.7321)|^2$$
$$+ |1 - j(1.7321)|^2)/3$$
$$= (1 + 2(4.0001))/3 \cong 3. \quad \blacksquare$$

8.10.12 Zero Padding

As mentioned earlier, computational complexity is significantly lower in the computation of DFT when N, the number of sample points in the data, is a power of 2, i.e., with the use of *fast Fourier transform (FFT)* algorithms discussed in the next chapter. This brings up the interesting question, what is the effect of adding zeros to the end of a sequence? Let the sequence have N_1 sample points and let N_2

zeros be added resulting in $N = N_1 + N_2$ sample points. Noting that the DFT spectrum is periodic with period 2π, the sample points are now spaced $(2\pi/(N_1 + N_2))$ instead of $(2\pi/N_1)$ apart. That is, as more zeros are added, DFT provides *closer* spaced samples of the transform of the original sequence. We should note that we do *not* have any more frequency information content than before. It gives a better display. Also, by appropriately padding a required number of zeros (N_2) so that $N = N_1 + N_2$ is a power of two, fast DFT algorithms can be used.

8.10.13 Signal Interpolation

In Chapter 1 and in an earlier part of this chapter we have made use of different interpolation functions. In this chapter we have discussed using the sinc and other functions to find interpolated values of the sampled signal. We can make use of the idea of zero padding in the frequency domain using the DFT, which is the dual of improving the spectral resolution by zero padding in the time domain discussed in the last section. Since the sampling frequency is $f_s = 1/t_s$, increasing the sampling rate reduces the sampling interval, which, in turn, increases the number of samples in the interval. Let f_{s_1} be the sampling rate used to determine N sampled values. Increasing the sampling rate from f_{s_1} to Mf_{s_1} would introduce interpolated values between samples.

Procedure: The sample sequence with N sample points with even and odd cases by

$$x[n] : x[0], x[1], x[2], x[\tfrac{1}{2}(N-1)],$$
$$\dots, x[N-1], N - \text{odd} , \qquad (8.10.41\text{a})$$

$$x[n] : x_d[0], x[1], x[2], x[\tfrac{1}{2}N],$$
$$\dots, x[N-1], N - \text{even} . \qquad (8.10.41\text{b})$$

1. Take the DFT of the given sequence. $\text{DFT}\{x[n]\} = X[k]$.
2. Insert zeros in the middle of the DFT sequence to create a MN point DFT. The cases for N even and odd are handled differently.

N-odd: Form the MN point DFT $Y[k]$ as

$$Y[k] : X[0], X[1], X[2], \dots, X[\tfrac{1}{2}(N-1)],$$
$$((MN - N) \text{ zeros}), X[\tfrac{1}{2}(N+1)], \dots, X[N-1].$$
$$(8.10.41\text{c})$$

N-even: Form the MN point DFT $Y[k]$ as

$$Y[k] : X[0], X[1], X[2], \dots, \tfrac{1}{2}X[\tfrac{1}{2}N],$$
$$((MN - N - 1) \text{ zeros}),$$
$$\tfrac{1}{2}X[\tfrac{1}{2}N], \tfrac{1}{2}X[\tfrac{1}{2}N + 1], \dots, X[N-1]. \quad (8.10.41\text{d})$$

3. Determine IDFT $[Y[k]]$ to obtain the MN point sequence $y[n]$, which may be complex. Since $x[n]$ is a real sequence, use only Re $\{y[n]\}$ and multiply by M.

Example 8.10.5 Use the above method to interpolate the two sequences given below using the factor $M = 1$ in the above procedure assuming the cases $N = 3$ and 4.

$$a.\ x[0] = 0, x[1] = 1, x[2] = 2,$$
$$b.\ x[0] = 0, x[1] = 1, x[2], x[3] = 3.$$

Solution: With the steps given above, the following results:

$a.\ N = 3$: $x[n] : 0, 1, 2;$ $X[k] : 3, -1.5 + j.866, -1.5 - j.866$

$\qquad\qquad Y[k] : 3, -1.5 + j.866, 0, 0, 0, -1.5 - j.866$
$\qquad\qquad\qquad \to x_{\text{int}}[n] : 0, 0, 1, 2, 2, 1$

$b.\ N = 4$: $x[n] : 0, 1, 2, 3;$ $X[k] : 6, -2 + j2, -2, 2 - j2$

$\qquad\qquad Y[k] : 6, -2 + j2, \tfrac{1}{2}(-2), 0, 0, 0, \tfrac{1}{2}(-2), 2 - j2$

$\qquad\qquad x_{\text{int}}[n] : 0, .0858, 1, 1.5, 2, 2.9142, 3, 1.5$

Note that $x[k] = x_{\text{int}}[2k], k = 0, 1, 2, \dots, N - 1$. The interpolated values are the values in between. Note that in the second case $x[0] = 0, x[3] = 3$, and $x[4] = 0$ indicating that the interpolated value at $x_{d,\text{int}}[7]$ will be the average value between 0 and 3, which is equal to 1.5. Similar arguments can be given for the odd case. ∎

Notes: If a band-limited signal is sampled at a rate higher than the Nyquist rate, then the interpolated sequence will be *exact* at the sampling intervals and the values between the samples will be interpolated values. In the case of periodic band-limited signals, the interpolation is exact.

Table 8.10.1 Discrete Fourier transform (DFT) properties

Linearity:

$$x[n] = \sum_{i=1}^{M} a_i x_i[n] \overset{\text{DTFT}}{\longleftrightarrow} \sum_{i=1}^{M} a_i X_i[k] = X[k]; \ a_i's \text{ are constants.}$$

Time shift or delay:

$$x[n-i]_{\text{mod}(N)} \overset{\text{DTFT}}{\longleftrightarrow} X[k]e^{-j(2\pi/N)ik}.$$

Frequency shift:

$$x[n]e^{j(2\pi/N)ni} \overset{\text{DTFT}}{\longleftrightarrow} X[k-i]_{\text{mod}(N)}.$$

Time reversal:

$$x[-n]_{\text{mod}(N)} \overset{\text{DTFT}}{\longleftrightarrow} X[-k]_{\text{mod}(N)}.$$

Alternate inversion formula:

$$x[n] = \left[\frac{1}{N} \sum_{k=0}^{N-1} X^*[k]e^{-j(2\pi/N)} \right]^*.$$

Conjugation:

$$x^*[n] \overset{\text{DTFT}}{\longleftrightarrow} X^*[-k]_{\text{mod}(N)}.$$

Duality:

$$X[n] \overset{\text{DTFT}}{\longleftrightarrow} Nx[-k]_{\text{mod}(N)}.$$

Circular convolution and correlation:

$$\sum_{i=0}^{N-1} x[n]h[n-i]_{\text{mod}(N)} = x[n] \otimes h[n]_{\text{mod}(N)} \overset{\text{DTFT}}{\longleftrightarrow} X[k]H[k].$$

$$\sum_{i=0}^{N-1} x[i]h[n+i]_{\text{mod}(N)} \overset{\text{DTFT}}{\longleftrightarrow} X^*[k]H[k].$$

$$\sum_{i=0}^{N-1} x[i]x[n+i]_{\text{mod}(N)} \overset{\text{DTFT}}{\longleftrightarrow} |X[k]|^2.$$

Multiplication:

$$x[n]h[n] \overset{\text{DTFT}}{\longleftrightarrow} \frac{1}{N}[X[k] \otimes H[k]_{\text{mod}(N)}] = \frac{1}{N} \sum_{i=0}^{N-1} x[i]H[k-i]_{\text{mod}(N)}.$$

Real sequences:

$$x[n] = x_e[n] + x_0[n] \overset{\text{DTFT}}{\longleftrightarrow} A[k] + jB[k].$$

$$x_e[n] \overset{\text{DTFT}}{\longleftrightarrow} A[k]; \ x_0[n] \overset{\text{DTFT}}{\longleftrightarrow} B_d[k].$$

Parseval's theorem:

$$\sum_{n=0}^{N-1} |x[n]|^2 = \frac{1}{N} \sum_{k=0}^{N-1} |X[k]|^2.$$

In other cases the interpolation can be poor. If the signals are not band limited, then the interpolation will be obviously poor. For $x[n]$ real, the discrete transform coefficients satisfy the conjugate symmetry property, $X[N - k] = X^*[k]$. If the procedure for insertion of the zeros discussed earlier is followed for the interpolation, the conjugate symmetry will be preserved in $Y[k]$. That is, $Y[MN - k] = Y_d^*[k]$. IDFT of $Y[k]$ will result in a real sequence, see Ambardar (1995).   ∎

8.10.14 Decimation

Decimation is an inverse operation of interpolation. It reduces the number of samples by discarding $M - 1$ samples and retaining every M th sample. Note that the corresponding new sampling rate must be above the Nyquist sampling rate to avoid aliasing. This to be of any value, the original signal is assumed to be oversampled.

8.11 Summary

This chapter started with analog signals that are sampled to obtain discrete-time signals. Fourier analysis of discrete time limited signals is discussed in terms of discrete-time and discrete Fourier transforms. The following gives a list of some of the specific topics:

- Ideal sampling of a continuous signal
- Continuous Fourier transforms of the sampled signals
- Low-pass and band-pass sampling theorems
- Basic discrete-time signals and operations, including decimation and interpolation
- Basic concepts of discrete-time convolution and correlation
- Discrete-time periodic signals and the corresponding discrete Fourier series and their properties
- Derivation of the discrete-time Fourier transform
- Properties of the discrete-time Fourier transform
- Discrete Fourier transforms and the inverse discrete Fourier transforms
- Periodic convolution and correlation and their computations directly and through DFT

- Zero-padding, interpolation, and decimation associated with discrete-time signals
- Tables of properties associated with discrete Fourier transforms.

Problems

8.2.1 Consider the function $x(t) = \cos(\omega_0 t)$. Illustrate the aliasing phenomenon by decreasing ω_s, or equivalently, increasing the sampling interval. Use a low-pass filter of bandwidth equal to $(\omega_s/2)$. In your solution use the following steps. Work out the solution using $\omega_s > 2\omega_0$ and show that the cosine function is recoverable. Now reduce the sampling frequency such that $\omega_s < 2\omega_0$. Sketch the spectrum of the ideally sampled signal and show that the signal exists in the frequency range $0 < (\omega_s - \omega_0) < \omega_s/2$.

8.2.2 Given $x(t)$ is band limited to $\omega_s/2$, determine the Nyquist rates for the functions.

$a.\ y_a(t) = \frac{dx(t)}{dt},$
$b.\ y_b(t) = x_t^2(t),$
$c.\ y_c(t) = \int\limits_{-\infty}^{\infty} x(\alpha)d\alpha,$

8.2.3 Consider the function $x(t) = \cos(\omega_0 t + \theta)$, $f_0 = \omega_0/2\pi = 200$ Hz. From the low-pass sampling theorem we know that there will not be any aliasing if $\omega_s > 2\omega_0$. Now consider that $x(t)$ is sampled at two different frequencies one below and one above the Nyquist frequency given by $a.\ f_s = 600$ Hz, $b.\ f_s = 160$ Hz. In the first case, we know that there will not be any aliasing. In the second case, the signal $x(t)$ sampled at $f_s = 160$ Hz describes a cosine function that is not the given function, but a sampled version of some other cosine function. Give the corresponding function $x_a(t) = A\cos(2\pi f_a t + \theta)$. That is, find f_a. Sketch the two functions $x(t)$ and $x_a(t)$ on the same figure and identify the points where the two functions coincide. ($x_a(t) =$ Aliased version of $x(t)$).

8.2.4 The acoustic pulse received by a receiver is represented by $x(t) = A\mathrm{sinc}^2(\omega_0 t)$. Noting the transform of this function is a triangular function, give the minimum sampling rate, the expression for the spectrum of the ideally sampled signal, and the minimum band width of the ideally low-pass filter required to reconstruct $x(t)$ from the sampled signal.

8.2.5 Find the minimum sampling rate that can be used to determine the samples that completely specify the following signals by assuming ideal sampling:

a. $x_1(t) = [\sin(2\pi(100)t)/(2\pi(100)t)]$,
b. b. $x_2(t) = \cos(2\pi(100)t + (\pi/3)) + \sin(2\pi(200)t)$

8.2.6 A signal $x(t)$ is band limited to the range $f_0 < f < 500\,\text{Hz}$. Find the minimum sampling rate for $x(t)$ without aliasing assuming a. $f_0 = 0$, b. $f_0 = 100\,\text{Hz}$

8.2.7 The signal $x(t) = A\cos(2\pi(100)t)$ is sampled at 150 Hz. Describe the corresponding signal after the sampled signal is passed through the following filters:

a. An ideal low-pass filter with a cut-off frequency of 20 Hz
b. An ideal band-pass filter with a pass band between 60 Hz and 120 Hz

8.2.8 Consider the sampled sequence $x(0) = 1$, $x(t_s) = 0, x(2t_s) = 1, x(nt_s) = 0, n \neq 0, 1, 2$. Sketch the interpolated functions using a. step, b. linear, and c. sinc interpolations.

8.2.9 Let $F[x(t)] = X(j\omega)$ with $X(j\omega) = 0, |\omega| 2\pi B$. Using the results in Section 8.2.3, show

$$\int_{-\infty}^{\infty} |x(t)|^2 dt = t_s \sum_{n=-\infty}^{\infty} x^2(nt_s), t_s = \frac{1}{2B}, f_s t_s = 1.$$

8.2.10 Use the band-pass sampling theorem to determine the possible sampling rates so that the following signal can be recovered from the sampled signal:

$$x(t) \xleftrightarrow{FT} X(j\omega) = \Pi\left[\frac{\omega + \omega_c}{2\pi(2B)}\right] + \Pi\left[\frac{\omega - \omega_c}{2\pi(2B)}\right],$$

$$B = 8\,\text{kHz}, \omega_c = 2\pi f_c = f_c = 64\,\text{kHz}.$$

Assuming the sampling rates of a. $f_{sa} = 200\,\text{kHz}$, b. $f_{sb} = 20\,\text{kHz}$, and c. $f_{sc} = 16\,\text{kHz}$, illustrate how the signal can be recovered from the sampled signals if possible.

8.3.1 Sketch the following sequences assuming $x[n] = (1-n)\{u[n] - u[n-3]\}$:

a. $y_a[n] = x[2n-1]$,
b. $y_b[n] = x[n^2 - 1]$,
c. $y_c[n] = x[1-n]$.

8.3.2 Find the even and odd parts of the functions. a. $x_a[n] = u[n], b. x_b[n] = (1/2)^n u[n]$.

8.3.3 Let $x[n] = x_e[n] + x_0[n]$. Show

$$E = \sum_{n=-\infty}^{\infty} x^2[n] = \sum_{n=-\infty}^{\infty} x_e^2[n] + \sum_{n=-\infty}^{\infty} x_0^2[n].$$

8.3.4 Derive the following identities and then simplify the results when $N \to \infty$:

a. $S = \sum_{n=0}^{N-1} \alpha^n = \dfrac{1-\alpha^N}{1-\alpha}, |\alpha| < 1,$

b. $\sum_{n=0}^{N-1} n\alpha^n = \dfrac{(N-1)\alpha^{N+1} - N\alpha^N + a}{(1-\alpha)^2},$

c. $\sum_{n=-\infty}^{\infty} e^{-\alpha|n|} = \dfrac{1 + e^{-\alpha}}{1 - e^{-\alpha}}.$

8.3.5 Find the closed form expression for $y[n] = a^n u[n] * u[n], |a| < 1$.

8.3.6 Find the cross correlation of the two sequences given by $x[n] = u[n] - u[n - n_x]$ and $h[n] = u[n] - u[n - n_h]$ for the cases: a. $n_x = n_h = 2$, b. $n_x = 2$, $n_h = 3$.

8.4.1 Determine the DTFS of the following sequences by using Euler's theorem and then by identifying the discrete Fourier series coefficients. Identify the periods.

a. $x_{sa}[n] = 1 + \sin(\pi n/2 + \theta)$,

b. $x_{sb}[n] = \cos(n\pi/20) + \sin(n\pi/40)$,

c. $x_{sc}[n] = \cos^2[n\pi/8]$

8.4.2 Find the DTFS coefficients of the N-periodic discrete-time functions

a. $x_{sa}[n] = \sum_{l=-\infty}^{\infty} \delta[n - lN]$,

b. $x_b[n] = \begin{cases} 1, 0 \le |n| \le M \\ 0, M < n < N - M \end{cases}.$

8.4.3 Determine the time-domain sequences with period $N = 7$ with the DTFS coefficients

a. $X_{sa}[k] = (1/2)$,
b. $X_{sb}[k] = \cos(2k\pi/N)$.

8.4.4 *a.* Show that $X_s[k] = X_s^*[N-k]$ for the following sequence:

$$x_s[n] = \begin{cases} 1, 0 \leq n < (N-1)/2 \\ 0, (N-1)/2 + 1 \leq n \leq N-1, \end{cases}$$
$$x_s[n] = x_s[n+N].$$

b. Given the periodic sequences

$$x_s[n] = \{0,0,1,2\}, \ h_s[n] = \{1,2,0,0\},$$
$$x_s[n] = x_s[n+4] \text{ and } h_s[n] = h_s[n+4].$$

find the DTFS of the function $y_s[n] = x_s[n]h_s[n]$. Illustrate the generalized Parseval's identity by using the DTFS of the functions $x_s[n], h_s[n]$ and $y_s[n]$.

8.4.5 Use the sequences in Problem 8.4.4b to determine $y_s[n] = x_s[n] \otimes h_s[n]$.

8.4.6 Give an example of two sinusoidal sequences that are equal. Hint: Assume $\cos(\Omega_1 \pi k + \theta) = \cos(\Omega_2 \pi k + \theta)$ with $\Omega_1 \neq \Omega_2$ and show the two functions are equal.

8.5.1 Show that

$$x^*[n] \xleftrightarrow{\text{DTFT}} X^*(e^{-j\Omega}) \text{ and } x^*[-n] \xleftrightarrow{\text{DTFT}} X^*(e^{j\Omega}).$$

8.5.2 Derive an expression for the convolution $y[n] = x[n] * x[n], \ x[n] = \alpha^n u[n]$.

8.5.3 Show that

$$\text{DFT}\{.5\delta[n] + .25\delta[n-2] + .25\delta[n+2]\} = \cos^2(\Omega).$$

8.5.4 Find the inverse transform of

$$X(e^{j\Omega}) = 1, |\Omega| \leq \Omega_c, X(e^{j\Omega}) = 0, \Omega_c < |\Omega| \leq \pi.$$

8.5.5 Consider the two-sided sequence $x[n] = a^{|n|}, |a| < 1$. Write this expression in terms of the right-side and left-side sequences. Then, derive the expression for the DTFT of this sequence using the time reversal property. Be careful about the sample point at $n = 0$.

8.5.6 Use the central ordinate theorems to evaluate the sums.

a. $\displaystyle\sum_{n=0}^{\infty} n\alpha^n, b. \sum_{n=-\infty}^{\infty} a^{|n|}, c. \sum_{n=-\infty}^{\infty} \sin\left(\frac{Wn}{(\pi n)}\right)$

8.5.7 Verify the results given in Section 8.5.2 for Type 1 and 4 sequences.

8.6.1 Prove the time reversal property in (8.6.17).

8.6.2 *a.* Determine the DTFT of the function $x[n] = (1/3)u[n]$.

b. Use the time reversal property to determine the DTFT of $(3)^n u[-n]$.

8.6.3 Find the DTFT of the function

$$y[n] = (n+1)^2 x[n].$$

8.6.4 Determine the convolutions $x_1[n] * x_2[n]$ for the following cases:

a. $x_1[n] = u[n], \ x_2[n] = u[n]$,

b. $x_1[n] = u[n], \ x_2[n] = .5^n u[n]$.

8.6.5 Determine

a. $\displaystyle\sum_{n=-\infty}^{\infty} (1/2)^{|n|}, b. \sum_{n=0}^{\infty} n(1/2)^n.$

by using the central ordinate theorems.

8.8.1 Find the DFTs of the sequences

a. $\{x[n]\} = [1, 1, -1, -1]$,

b. $\{x[n]\} = [1, -1, 1, -1]$

8.8.2 Compute the DFTs of the following N-point sequences. For Part *c.*, use Euler's formula for the cosine function in determining the DFT assuming k_0 is an integer.

a. $x[n] = \alpha^n, 0 \leq n < N$,

b. $x[n] = u[n] - u[n - n_0], 0 < n_0 < N$,

c. $x[n] = \cos(n\omega_0), \omega_0 = 2\pi k_0/N, 0 \leq n < N - 1, k_0$ is an integer.

8.8.3 Determine the 8-point DFT sequence of $x[n] = \delta[n] + 2\delta[n-3]$.

8.8.4 Consider a sequence $x[n], 0 \leq n \leq N - 1$ with $X[k] = \text{DFT}\{x[n]\}$. Find the DFTs the two sequences given below in terms of $X[k]$.

$$\{y[n]\} = \begin{cases} x[n/2], n \text{ even} \\ 0, \ n \text{ odd} \end{cases}, \{y[n]\}$$

$$= \begin{cases} x[n], n = 0, 1, 2, \ldots, N-1 \\ 0, \ n = N, N+1, \ldots, (2N-1) \end{cases}.$$

8.8.5 Compute the DFT of $x[n-3]_{\text{mod }(N)}$ directly and then using the time-shift theorem.

8.8.6 Determine $y[n] = x[n] \otimes h[n]$ directly and then using the DFT for the sequences

$$x[n] = (1/3)^n, h[n] = \sin((\pi/2)n), n = 0, 1, 2, 3.$$

8.8.7 Show that the DFT of a real sequence $x[n]$ satisfies the relation $X[N-k] = X^*[k]$. (*) denotes conjugation.

8.8.8 The DFT sequence of a real time signal is given by $\{X[k]\} = \{4, j, 0, X\}$, where X is the missing value. Use the symmetry property of DFT to determine the missing value. Find the corresponding time sequence.

8.8.9 Derive an expression DFT $[y[n]] = $ DFT$[(-1)^n x[n]]$ in terms of $X[k]$.

8.9.1 Show that for $N = 4$, $(1/N)\mathbf{A}_{\text{DFT}}\mathbf{A}_{\text{DFT}}^* = \mathbf{I}_N$ (an identity matrix)

8.9.2 Derive the matrix $(1/N)\mathbf{A}_{\text{DFT}}^2$ with $N = 4$. What can you say about this matrix?

8.9.3 Given $x(t) = \Lambda(t)$, estimate the sampling frequency and sampling interval by choosing the

bandwidth of $x(t)$ as the frequency where $|X(j\omega)|$ is 10% of its maximum.

8.10.1 Use Example 8.10.3 to compare the number of multiplications required to compute the convolution directly and by using the DFT.

8.10.2 Write the sequence $r[n]$ in matrix form

$$r[n] = \sum_{i=0}^{N-1} x[i]x[n+i]_{\text{Mod}(N)}.$$

8.10.3 Consider the two discrete N-point real sequences $x_{d1}[n]$ and $x_{d2}[n]$ and $x[n] = x_1[n] + jx_2[n], n = 0, 1, \ldots, N-1$ with $F[x_i[n]] = X_i[k]$, $i = 1, 2$ and

$$(\mathbf{x}_{d1})^T = [c\, 0123], (\mathbf{x}_{d2})^T = [c\, 2345], \mathbf{x}_d = \mathbf{x}_{d1} + j\mathbf{x}_{d2}.$$

a. First show the following in general terms and then *b.* verify this using the sequences:

$$X_1[k] = .5\{X_d[k] + X^*[N-k]\}, X_2[k] \\ = -.5j\{X[k] - X^*[N-k]\}.$$

8.10.4 Find the $N-$ point DFT of the sequences $x_1[n] = e^{j\Omega_0 n}$ for two cases:

a. $\Omega_0 = 2\pi k_0/N, b.$ $\Omega_0 \neq 2\pi k_0/N. (k_0$ is an integer).

8.10.5 Consider the sequence $x_d[n] = \{0, 1, 0, 1\}$. Compute its DFT and then use the interpolation technique discussed in Section 8.10 assuming $M = 2$ and 4.

Chapter 9
Discrete Data Systems

9.1 Introduction

In the last chapter we have discussed the concepts of discrete Fourier transforms (DFTs). In this chapter we will briefly review these and discuss its fast implementations. There are several algorithms that come under the topic-fast *Fourier transforms* (FFTs). The first FFTmethod of computing the DFT was developed by Cooley and Tukey (1965). These are innovative and useful in the signal processing area.

Continuous Fourier transforms (CFTs) in the analog and the discrete Fourier transforms (DFTs) in the discrete domains are the corner stones of signal analysis. In the continuous domain we studied the Laplace transforms, which are related to the continuous Fourier transforms. The discrete counter part of the Laplace transforms is *z*-transforms related to the discrete-time Fourier transforms (DTFTs). Table 9.1.1 summarizes the variables in the continuous-time Fourier transforms, the Laplace transforms, the discrete-time Fourier transforms, and the *z*-transforms. In this chapter we will study some of the basics associated with the *z*-transforms and its applications.

Digital filters have been popular in recent years and will continue to be in the future. In its simplest form, a digital filter is a computer program that takes a set of data and converts into another set of data. Discrete data systems may correspond to filtering or some other operation. In the analog case we have to worry about component value tolerances and the responses can change in time. The responses of analog systems cannot be duplicated, as the component values may be different from one batch to another. The responses of the filters can change if the operating conditions of the filter change. On the other hand, in the digital case, every time we process a set of data the output will be the same. Digital filters are more flexible and can be altered by simply changing the computer code. At low frequencies, analog components are bulky. We may have to deal with magnetic coupling if inductors or transforms are used as components in the analog system. Analog filters may have to be redesigned and the circuit implementations may be different if the frequencies change. On the other hand, modifying digital filters may represent a change of computer code. Digital technology is modern and powerful signal processing algorithms can be designed. Digital filters can be time shared and process several signals simultaneously. Digital integrated circuits design is much simpler compared to analog integrated circuit technology. They require lower power consumption and the digital circuitry can be fabricated in smaller packages. Digital storing is much cheaper. Searching and selecting digital information is simple and processing the data is straightforward. Digital reproduction is much more reliable and the cost of digital hardware continues to come down every year. Most source signals and the recipients are analog in nature. To replace an analog filter by a digital filter, the analog signal

Table 9.1.1 Discrete-time and continuous-time signals and their transforms

Continuous-time transform/ variable	Discrete-time transform/ variable
Continuous Fourier transform/ ω or f	Discrete time Fourier transform/Ω
Laplace transform/s	z-transform/z

R.K.R. Yarlagadda, *Analog and Digital Signals and Systems*, DOI 10.1007/978-1-4419-0034-0_9,
© Springer Science+Business Media, LLC 2010

needs to be converted to a digital signal by using an *analog-to-digital (A/D) converter*. The digital signal is then passed through a digital filter and the output of the filter needs to be converted back to analog data using a *digital-to-analog (D/A) converter*.

Digital signal processing (DSP) area has been popular during the past 30 + years. It will continue to be of interest in many areas, including seismic signal processing, speech processing, image processing, radar signal processing, and others. Telephone industry has taken the lead in the signal processing area. There are excellent texts available in the general area of signal processing. Some of these include Ambardar (2007), Strum and Kirk (1988), Mitra (1998), Oppenheim and Schafer (1975), Rabiner and Gold (1975), Cartinhour (2000), Ludeman (1986), and many others. For an excellent review on the spectral analysis, see Otnes and Enochson (1972), Marple, (1989), Press et al. (1989), and many others. For a historical survey on the spectral estimation, see Robinson (1982). MATLAB provides digital analysis and design software, see Ingle and Proakis (2007). Also, see Ramirez (1975), Smith III (2007), Smith, (2002) on FFT and its applications.

9.2 Computation of Discrete Fourier Transforms (DFTs)

Power spectrum: Most signals in practice are analog signals. Spectral analysis and estimation of these signals is basic. A simple method of power spectrum estimation of an analog signal $x(t)$ involves N values of $x(t)$ sampled every t_s s (or $f_s = 1/t_s$ samples/s) resulting in $x[n] = x(nt_s)$, $n = 0, 1, 2, \ldots, N - 1$. The DFT of the signal $x[n]$ is (see Section 8.9)

$$X[k] = \sum_{n=0}^{N-1} x[n]e^{-j(2\pi/N)nk},$$

$$k = 0, 1, 2, \ldots, N - 1. \qquad (9.2.1a)$$

The power spectrum estimate is defined at $(N/2) + 1$ frequencies by

$$P(0) = \frac{1}{N^2}|X(0)|^2,$$

$$P(\omega_s/2) = \frac{1}{N^2}|X(N/2)|^2,$$

$$\omega_s = 2\pi f_s, f_s = 1/t_s$$

$$P(\omega_k) = \frac{1}{N^2}\left[|X[k]|^2 + |X[N-k]|^2\right],$$

$$\omega_k = \frac{2\pi k}{Nt_s}, k = 1, 2, \ldots, \frac{N}{2} - 1. \qquad (9.2.1b)$$

From Chapter 4 we note that a rectangular window spectrum has a great deal of leakage into the side lobes. A tapered window $w[n]$, such as a Hamming window to be discussed later, can be used in estimating the spectrum to reduce the spectral leakage. A windowed signal $y[n] = x[n]w[n]$ is to be used in the estimation. Another popular method of spectral estimation is the *Blackman–Tukey* method, see Press et al. (1990). In its simplest form, it involves the computation of the data autocorrelation and then determining the spectrum using DFT. The spectrum of the autocorrelation is the power spectral density.

9.2.1 Symbolic Diagrams in Discrete-Time Representations

Symbolic diagrams or signal flow graphsare a network of directed branches connected at nodes is a pictorial representation of an algorithm. Figure 9.2.1 gives the flow graph symbols that are common in two different forms. Source nodes do not have any incoming braches and are used for input

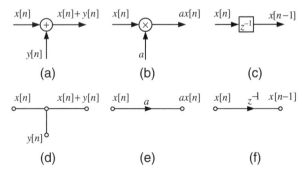

Fig. 9.2.1 Two flow graph representations: (a) and (d), summers; (b) and (e), multipliers; (c) and (f), delays

Fig. 9.2.2 Example 9.2.1

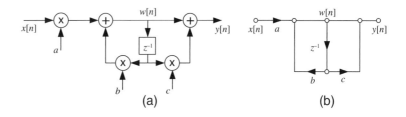

sequences. Sink node has only one entering branch and is used for the output sequence. In addition, summers and multiplier symbols are shown and are self-explanatory. The symbols z^{-1} are used to identify delay components. In Section 9.4 z-transforms will be studied. Some authors use the multiplier constant above the line and others use it below the line. If the multiplier constant a is not shown, then it is 1.

Example 9.2.1 Using the symbol representations in Fig. 9.2.1, write difference equations relating the variables $w[n]$ and $y[n]$ in terms of $x[n]$ and $w[n-1]$ in Fig. 9.2.2a,b.

Solution: The two diagrams in Fig. 9.2.2 result in the same equations and are given as follows:

$$w[n] = ax[n] + bw[n-1], \quad y[n] = w[n] + cw[n-1].$$

∎

9.2.2 Fast Fourier Transforms (FFTs)

First a brief review of the discrete Fourier transform (DFT) is given below. The discrete Fourier transform is a transformation that takes a set of N values in time to N values in frequency. First, the transform vector is given by $\mathbf{X} = \mathbf{A}_{\text{DFT}}\mathbf{x}$ (see (8.9.16a and b)), where the matrix $\mathbf{A}_{\text{DFT}}$ is a $N \times N$ matrix with its (k, n) entry being (see (8.9.17))

$$\mathbf{A}_{\text{DFT}}(k, n) = e^{-j(2\pi/N)(k-1)(n-1)}$$
$$\equiv W_N^{(k-1)(n-1)},$$
$$1 \le k, n \le N, \; W_N = e^{-j2\pi/N}. \quad (9.2.2a)$$

The vectors $\mathbf{X}$ and $\mathbf{x}$ are N-dimensional column vectors. In matrix form the DFT coefficients can be expressed in terms of $W_N^n = (e^{-j2\pi/N})^n$ by

$$
\begin{bmatrix} X[0] \\ X[1] \\ X[2] \\ \cdot \\ \cdot \\ X[N-1] \end{bmatrix} =
\begin{bmatrix}
1 & 1 & 1 & \cdot & \cdot & 1 \\
1 & W_N^1 & W_N^2 & \cdot & \cdot & W_N^{N-1} \\
1 & W_N^2 & W_N^4 & \cdot & \cdot & W_N^{2(N-1)} \\
\cdot & \cdot & \cdot & \cdot & \cdot & \cdot \\
\cdot & \cdot & \cdot & \cdot & \cdot & \cdot \\
1 & W_N^{N-1} & \cdot & \cdot & \cdot & W_N^{(N-1)^2}
\end{bmatrix}
\begin{bmatrix} x[0] \\ x[1] \\ x[2] \\ \cdot \\ \cdot \\ x[N-1] \end{bmatrix}. \quad (9.2.2b)
$$

Note W_N^n takes one of the values in the set $\{1, e^{-j(2\pi/N)}, e^{-j(2\pi/N)2}, ..., e^{-j(2\pi/N)(N-1)}\}$ for any N, see (8.9.18).

The properties in Table 9.2.1 allow for the derivation of a fast Fourier transform (FFT) algorithm. We will consider an $N(= 2^\nu)$-point *decimation-in-*

Table 9.2.1 Properties of the function $W_N = e^{-j(2\pi/N)}$

1. $W_N^{n+N} = e^{j(2\pi/N)(n+N)} = e^{j(2\pi/N)n} = W_N^n$	(9.2.3a)
2. $W_N^{n+N/2} = -e^{-j(2\pi/N)} = -W_N^n$	(9.2.3b)
3. $W_N^{kN} = e^{-j(2\pi)k} = 1, \quad k$ is an integer	(9.2.3c)
4. $W_N^{2k} = e^{-j2(2\pi/N)k} = e^{-j,(2\pi/(N/2))k} = W_{N/2}^k$	(9.2.3d)
5. $e^{-j(2\pi/N)n} \in A = \{1, e^{-j(2\pi/N)}, e^{-j(2\pi/N)2}, ..., e^{-j(2\pi/N)(N-1)}\}$	(9.2.3e)

frequency FFT algorithm. It is based on expressing one N-point DFT algorithm by two $N/2$-point DFTs, then four $(N/4)$-point DFTs, and so on. The algorithm at the end reduces to $(N/2)$-2-point transforms. The 1-point transform is trivial as $X[0] = x[0]$. The DFT of a 2-point sequence is determined by noting $W_2^1 = e^{-j2\pi/2} = -1$. The DFT values in scalar and matrix form are as follows:

$$X[0] = x[0]\,W_2^0 + x[1]\,W_2^0 = x[0] + x[1]$$
$$X[1] = x[0]\,W_2^0 + x[1]\,W_2^1 = x[0] - x[1]\,, \quad (9.2.4a)$$

$$\begin{bmatrix} X[0] \\ X[1] \end{bmatrix} = \begin{bmatrix} 1 & 1 \\ 1 & -1 \end{bmatrix} \begin{bmatrix} x[0] \\ x[1] \end{bmatrix}. \quad (9.2.4b)$$

Decimation-in-frequency FFT algorithm: Starting with the DFT of a set of data $x[n], n = 0, 1, .., N-1$ and $N = 2^v$, the DFT coefficients (see 9.2.1a.), $X[k]$, are obtained in terms of $W_N = e^{-j2\pi/N}$ as follows:

$$X[k] = \sum_{n=0}^{N-1} x[n] e^{-j(2\pi/N)nk}$$

$$= \sum_{n=0}^{N-1} x[n]\,W_N^{nk} \quad (9.2.5)$$

$$= \sum_{n=0}^{N/2-1} x[n]\,W_N^{nk} + \sum_{n=N/2}^{N-1} x[n]\,W_N^{nk}.$$

Note the variable n is used for the *time* variable and k is used for the *frequency* variable Using $m = n - (N/2)$ in the second summation in (9.2.5) and noting $W_N^{k(N/2)} = e^{-j(2\pi/N)k(N/2)} = (-1)^k$ result in

$$\sum_{n=N/2}^{N-1} x[n]\,W_N^{kn} = \sum_{m=0}^{N/2-1} x\left[\frac{N}{2}+m\right] W_N^{k(m+\frac{N}{2})}$$

$$= \sum_{m=0}^{N/2-1} x\left[\frac{N}{2}+m\right] W_N^{mk}\,W_N^{k(N/2)}, (9.2.6)$$

$$X[k] = \sum_{n=0}^{N/2-1} \left[x[n] + (-1)^k x\left[n+\frac{N}{2}\right]\right] W_N^{kn},$$
$$k = 0, 1, 2, ..., N-1. \quad (9.2.7)$$

Now separate the coefficients into $X[2k]$ and $X[2k+1]$ and use Table 9.2.1:

$$X[2k] = \sum_{n=0}^{N/2-1} \left\{x[n] + (-1)^{2k} x\left[n+\frac{N}{2}\right]\right\} W_N^{n(2k)}$$

$$= \sum_{n=0}^{N/2-1} \left\{x[n] + x\left[n+\frac{N}{2}\right]\right\} W_{N/2}^{nk}, \quad (9.2.8a)$$

$$X[2k+1] = \sum_{n=0}^{N/2-1} \left\{x[n] + (-1)^{(2k+1)} x\left[n+\frac{N}{2}\right]\right\} W_N^{n(2k+1)}$$

$$= \sum_{n=0}^{N/2-1} \left\{x[n] - x\left[n+\frac{N}{2}\right]\right\} W_{N/2}^{nk}\,] W_N^{n}, \quad (9.2.8b)$$

$$k = 0, 1, 2, \ldots, \frac{N}{2} - 1.$$

One N-point DFT is reduced to two $(N/2)$-point DFTs. One N-point DFT requires N^2 multiplications and $((N-1)$ additions), see (9.2.2b). Two $(N/2)$-point DFTs require only $2(N/2)^2$ multiplications and $2(N-1)$ additions.

Example 9.2.2 Assuming $N = 4$, show that the use of (9.2.8a and b) successively results in the DFT values. Illustrate the algorithm using the flow graph representation.

Solution: *a.* From (9.2.8a) and (9.2.8b), we have (note $W_4^0 = 1$, $W_4^1 = -j$, and $W_4^2 = -1$)

$$X[2k] = \sum_{n=0}^{1} \{x[n] + x[n+2]\} W_4^{2nk},$$

$$X[2k+1] = \sum_{n=0}^{1} \{x[n] - x[n+2]\} W_4^n W_4^{2nk}, \; k = 0, 1.$$
$$(9.2.9)$$

First, at stage 0, i.e., to start with, define $x_0[n] = x[n], n = 0, 1, 2, 3$. At stage i, identify the variables as $x_i[n]$. Algorithm has two stages corresponding to $N = 4 = 2^v$, $v = 2$. From (9.2.9), we have the following.

Direct:

$$X[0] = \{x_0[0] + x_0[2]\} + \{x_0[1] + x_0[3]\}$$
$$= \{x_1[0]\} + \{x_1[1]\} = x_2[0] = X[0]\,, \quad (9.2.10a)$$

$$X[2] = \{x_0[0] + x_0[2]\} - \{x_0[1]] + x[3]\}$$
$$= \{x_1[0]\} - \{x_1[1]\} = x_2[1] = X[2] \ , \quad (9.2.10b)$$

$$X[1] = \{x_0[0] - x_0[2]\} + \{W_4^0(x_0[1] - x[3])\}$$
$$= \{x_1[2]\} + \{x_1[3]\} = x_2[2] = X[1] \ , \qquad (9.2.10c)$$

$$X[3] = \{x_0[0] - x_0[2]\} - \{W_4^1(x_0[1] - x_0[3])\}$$
$$= \{x_1[2]\} - \{x_1[3]\} = x_2[3] = X[3] \ . \quad (9.2.10d)$$

Individual identifications at each stage from (9.2.10):
Stage 0:

$$x_0[0] = x[0], \ x_0[1] = x[1], \ x_0[2] = x[2]. \quad (9.2.11a)$$

Stage 1:

$$x_1[0] = \{x_0[0] + x_0[2]\} \ , x_1[1] = \{x_0[1] + x_0[3]\} \ ,$$
$$x_1[2] = \{x_0[0] - x_0[2]\} \ , x_1[3] = \{W_4^0(x_0[1] - x[3])\} \ . \quad (9.2.11b)$$

Stage 2:

$$x_2[0] = \{x_1[0]\} + \{x_1[1]\} \ , x_2[1] = \{x_1[0]\} - \{x_1[1]\} \ ,$$
$$x_2[2] = \{x_1[2]\} + \{x_1[3]\} \ , x_2[3] = \{x_1[2]\} - \{x_1[3]\} \ . \quad (9.2.11c)$$

End results:

$$X[0] = x_2[0], \ X[2] = x_2[1], X[1] = x_2[2], \ X[3] = x_2[3] \ . \quad (9.2.11d)$$

These equations can be used to draw the flow graph using the symbols in Fig. 9.2.3. For clarity, the multipliers are shown under the lines rather than above. Interestingly, the above equations can be seen from the flow graph in Fig. 9.2.3.

Interestingly, if the variables arein binary form, the argument k in $X[k] = X[(k_1k_0)_2]$ is related to the argument n in $x_2[(n_1n_0)_2]$ by the relation $k = (k_1 = n_0, k_0 = n_1)$. For additional information on this, see Oppenheim and Schafer (1999). ■

The above results can be extended for any $N = 2^v$ with v stages. Figure 9.2.4 gives the flow graph for $N = 8 = 2^3$. Note the multipliers are identified above and below the lines for clarity.

For a general derivation of the decimation-in-frequency algorithm and other algorithms, see Oppenheim and Schafer (1999), and others.

Notes: The decimation refers to the process of reducing the number of operations for an $N = 2^v$ point DFT, expressing the N-point DFT in terms of $2 \ (N/2) = 2^{v-1}$-point DFTs and successively expressing them in v stages with the input sequence in natural order.

Number of computations in an FFT algorithm: In Section 8.9.2 the computational aspects of discrete Fourier transforms were considered. These results are compared with FFT computational requirements. In the N-point FFTalgorithm with $N = 2^v$, we have $v = \log_2(N)$ stages. FFT computation requires $(N/2)v = (N/2)\log_2(N)$ complex multiplications and $vN = N\log_2(N)$ complex additions. Computers use real arithmetic and each complex multiplication requires four real multiplications and three real additions. The amount of effort to do multiplication is much larger than additions. We

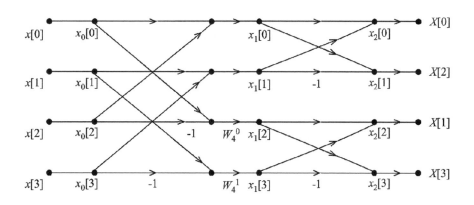

Fig. 9.2.3 Flow graph representations for $N = 4$ using the decimation-in-frequency FFT algorithm

Fig. 9.2.4 Flow graph representations for $N = 8$ using the decimation-in-frequency FFT algorithm

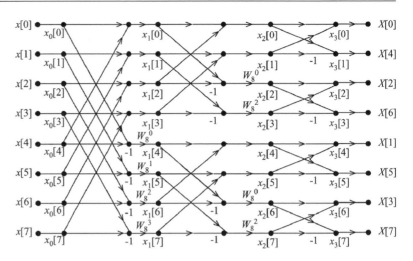

can compare the number of multiplications by the direct method versus FFT by the ratio

$$R = \frac{N^2}{.5\,N\log_2(N)} = \frac{N}{.5v} \approx \frac{N}{v}. \qquad (9.2.12)$$

For a large $N = 2^v$, the difference in the number of computations by FFT is significantly lower. Note that $.5\,N\log_2(N)$ is nearly linear, whereas N^2 is quadratic. For small N, the difference in the number of computations in computing the DFT and FFT is not that significant. As an example, consider $N = 2^{10}$, the ratio in (9.2.12) is $R \approx 204$. The DFT requires N^2 values of $W_N^{kn}, k, n = 0, 1, 2, \ldots, N-1$, whereas FFT requires at most N such values at each stage. Earlier, we have seen that $e^{-j(2\pi/N)nk} = e^{-j(2\pi/N)m}, 0 \leq m \leq N-1$. The logical way of course is to compute W_N^k once, $k = 0, 1, 2, \ldots, N-1$, store them, and use them again and again in each stage. Only *about* $(3/4)N$ of these W_N^k are distinct in the FFT algorithm, see Ambardar (2007). The FFT approach $N = 2^v$ is computationally efficient compared to the direct method only for $v > 5$ $(N > 32)$. See Wilf (1986) for an interesting discussion of algorithms and their complexity. MATLAB function for computing the DFT of a signal is the *fft* function. It can be used *for any N*. For example, to compute the DFT of a sequence x of N values, MATLAB routine is $\mathbf{X} = \text{fft}(\mathbf{x})$ to get the spectral values and the routine $\mathbf{x} = \text{ifft}(\mathbf{X})$ gives the data from the spectral values.

Just like in the continuous case, other discrete transforms related to discrete Fourier transforms can be considered, including discrete cosine, sine, Hartley, and Hilbert transforms. These are beyond the scope here. See the handbook by Poularikas (1996).

9.3 DFT (FFT) Applications

In Section 9.2, spectral analysis based on DFT was considered. Computing DFT via FFT is a *tool* to reduce the number of computations. FFT is applicable wherever DFT can be used. See, for example, Marple (1987), O'Shaughnessy (1987), Otnes and Enochson (1972), Poularikas (1996), Rabiner and Schafer (1979), Shenoi (1995), and others for FFT applications.

9.3.1 Hidden Periodicity in a Signal

Although, nothing is forever, some signals can be considered as periodic at least on a short-time basis. For example, vowel speech sounds can be considered as periodic on a short-time basis. Investors in the stock market would like to know if the price of a stock has a periodic part in the signal that is hidden. If so, the investor can sell when the stock is high and buy when it is low. For a good presentation on applying spectral analysis to various physical signals, see Marple (1987).

Example 9.3.1 Consider the sinusoid $x(t) = \cos(2\pi(100)t)$ that is sampled at twice the Nyquist rate for three full periods. Find the corresponding DFT values.

Solution: The frequency of the sinusoid is 100 Hz. The period of the signal is $T = (1/100)$ s. The Nyquist rate is 200 Hz. The corresponding sampling rate and the sampling interval are 400 Hz and $t_s = (1/400)$ s. Since three periods are used, we have four samples per period and have $N = 12$ samples to find the DFT. Note

$$\cos(2\pi f_0 t)|_{t=nt_s} = \cos(2\pi n(f_0/f_s)),$$
$$n = 0, 1, 2, \ldots, N-1 = 11 \qquad (9.3.1)$$

The sampling frequency is divided into $N = 12$ intervals and the frequency interval is $F = f_s/N = f_s/12 = 400/12 = 100/3$ referred to as the digital frequency in Section 8.6.1, where $\Omega = 2\pi(f/f_s) = 2\pi F$ was used. The DFT frequencies are

$$kF = kf_s/N = k(f_s/12),$$
$$k = 0, 1, 2, \ldots, N-1 = 11. \qquad (9.3.2)$$

Now, assume f_0 is one of these frequencies and $kF = k(f_s/N) = k(f_s/12) = f_0$ for some k. That is,

$$\frac{f_0}{f_s} = \frac{k}{N} = \frac{k}{12}, \text{ a rational number.} \qquad (9.3.3)$$

For $f_0 = 100$ Hz, $f_s = 400$ Hz, and $k = 3$, $X[3]$ gives the appropriate spectral value. This can be verified by computing DFT of the sequence

$$x[n] = \cos(2\pi(f_0/f_s)n) = \cos(2\pi(kn)/N)$$
$$= \cos(2\pi(kn)/12), n = 0, 1, 2, \ldots, 11, k = 3$$
$$\Rightarrow \{x[n]\} = \{1, 0, -1, 0, 1, 0, -1, 0, 1, 0, -1, 0\}.$$
$$(9.3.4)$$

$$\text{DFT}\{\cos(2\pi(nk)/N)\} = \tfrac{1}{2}\text{DFT}\left[e^{j2\pi nm/N} + e^{-j2\pi nm/N}\right]$$

$$= \tfrac{1}{2}\sum_{n=0}^{N-1} e^{j(2\pi/N)n(m-k)} + \tfrac{1}{2}\sum_{n=0}^{N-1} e^{-j(2\pi/N)n(m+k)} = X[k]$$

Using the summation formula for the geometric series results in the DFT values

$$X[k] = \left\{ \begin{array}{c} N/2, \ k = m \\ N/2, \ k = N - m \\ 0, \text{ otherwise} \end{array} \right\}$$
$$\Rightarrow X[k] : \{0, 0, 0, 6, 0, 0, 0, 0, 0, 6, 0, 0\}. \quad (9.3.5)$$

Noting $N = 12$ and the amplitude of the sinusoid is 1, we have $X[3] = X[9] = 6$. ∎

Notes: These results can be extended to a periodic function $x(t) = \cos(2\pi(k_0/N) + \theta)$. The DFT of $x(t)$ has only two nonzero discrete frequency values and are

$$X[k]|_{k=k_0} = (N/2)e^{j\theta} \text{ and } X[N-k]|_{k=k_0} = (N/2)e^{-j\theta}.$$

The frequency spacing $F = f_0/f_s$ needs to be a rational function so that the discrete frequency falls on the input signal frequency. FFT algorithm can be used with this in mind. Also, *leakage* results in DFT if a periodic signal is not sampled for an integer number of periods. This results in nonzero spectral components at frequencies other than the harmonic frequencies of the signal. Many times it is not easy to find the period of a signal up front. One solution to this is to use large enough number of samples. Larger is the time interval, the more closely is the spectrum sampled. That is, the spectral spacing $F = f_s/N$ is reduced, thus giving a more accurate estimate of spectrum of the given signal. Most practical signals are noise corrupted and are known only for a short time. That is, the signal is a windowed signal. As we have seen in Chapter 4, the spectrum of the windowed sinusoid is not a spike, but a sinc function indicating leakage into the side lobes. Tapered windows need to be used in any spectral analysis. If a signal contains several frequencies, then the spectrum of the windowed signal is the sum of the spectra of each sinusoid and it may not have distinct peaks as the main lobes of the sinusoids might merge. This happens if the two frequencies in the input signal are located close enough, then the frequency peaks may merge. Instead of two separate peaks, there may be only one single peak. Choosing larger

DFT lengths and good windows improves the accuracy of the spectral estimates, see Marple (1987). ∎

9.3.2 Convolution of Time-Limited Sequences

The convolution of the sequences $x[n]$ and $h[n]$ was defined by (see (8.3.16))

$$y[n] = \sum_{k=-\infty}^{\infty} x[k]h[n-k] = \sum_{m=-\infty}^{\infty} h[m]x[n-m].$$

(9.3.6)

Consider $x[n] = 0$ for $n < 0$ and $n \geq L$, $h[n] = 0$ for $n < 0$ and $n \geq M$. Equation (9.3.6) can be expressed as follows:

$$y[n] = \sum_{k=0}^{L-1} x[k]h[n-k] = \sum_{m=0}^{M-1} h[m]x[n-m]. \quad (9.3.7)$$

Sequence:

$$y[n] = 0, \; n < 0$$
$$y[0] = h[0]x[0]$$
$$y[1] = h[1]x[0] + h[0]x[1]$$
$$y[2] = h[2]x[0] + h[1]x[1] + h[0]x[2]$$
$$\cdots$$
$$y[M-1] = h[M-1]x[0] +$$
$$\cdots + h[0]x[M-1]\cdots$$
$$y[M+L-2] = h[M-1]x[L-1]$$
$$y[n] = 0, \; n \geq M+L-1.$$

(9.3.8)

Example 9.3.2 Find the sequence $y[n]$ corresponding to the convolution of the following sequences and express $y[n]$ in matrix form:

$$h[0] = 1, h[1] = -1, h[n] = 0 \text{ for } n < 0 \text{ and } n > 1,$$

(9.3.9)

$$x[0] = 1, x[1] = 1, x[n] = 0 \text{ for } n < 0 \text{ and } n > 1.$$

(9.3.10)

Solution:

$$\begin{bmatrix} y[n] = 0, & n < 0 \\ y[0] = x[0]h[0] + x[1]h[-1] = x[0]h[0] = 1 \\ y[1] = x[0]h[1] + x[1]h[0] = -1 + 1 = 0 \\ y[2] = x[0]h[2] + x[1]h[1] = x[1]h[1] = -1 \\ y[n] = 0, n \geq 3 \end{bmatrix}.$$

(9.3.11)

The equations in (9.3.11) for $y[n] \neq 0$ can be written in *two* equivalent matrix forms and

$$\mathbf{y} = \begin{bmatrix} y[0] \\ y[1] \\ y[2] \end{bmatrix} = \begin{bmatrix} h[0] & 0 \\ h[1] & h[0] \\ 0 & h[1] \end{bmatrix} \begin{bmatrix} x(0) \\ x(1) \end{bmatrix}$$

$$= \mathbf{Hx} = \begin{bmatrix} x[0] & 0 \\ x[1] & x[0] \\ 0 & x[1] \end{bmatrix} \begin{bmatrix} h[0] \\ h[1] \end{bmatrix} = \mathbf{Xh}. \quad (9.3.12)$$

The coefficient matrices $\mathbf{X}$ and $\mathbf{H}$ are (3×2) size matrices. The column vectors $\mathbf{y}$, $\mathbf{h}$, and $\mathbf{x}$ are of the dimensions 3×1, 2×1, and 2×1, respectively. The entries in the coefficient matrices $\mathbf{X}$ and $\mathbf{H}$ have a special structure. For example, the first and the second columns of the coefficient matrix $\mathbf{H}$ are, respectively, given by

$$\text{Col}(h[0], h[1], 0) = \begin{bmatrix} h[0] \\ h[1] \\ 0 \end{bmatrix},$$

$$\text{Col}(0, h[0], h[1]) = \begin{bmatrix} 0 \\ h[0] \\ h[1] \end{bmatrix}. \quad (9.3.13)$$

The second column is obtained from the first by rotating the first entry to the second, second entry to the third, and the third entry to the first. ∎

Equations in (9.3.7) can be written in matrix and symbolic forms for $y[n] \neq 0$ as follows:

$$
\begin{bmatrix} y[0] \\ y[1] \\ y[2] \\ \cdot \\ \cdot \\ \cdot \\ y[M-1] \\ y[M] \\ \cdot \\ \cdot \\ \cdot \\ y[M+L-2] \end{bmatrix} = \begin{bmatrix} h[0] & 0 & 0 & . & . & 0 \\ h[1] & h[0] & 0 & . & . & 0 \\ h[2] & h[1] & h[0] & . & . & 0 \\ & h[2] & h[1] & . & . & \\ & & . & . & . & \\ & . & . & . & . & \\ h[M-1] & h[M-2] & h[M-3] & . & . & \\ 0 & h[M-1] & h[M-2] & . & . & \\ & 0 & h[M-1] & . & . & \\ . & . & . & . & . & \\ . & . & . & . & . & h[M-2] \\ 0 & 0 & 0 & 0 & 0 & h[M-1] \end{bmatrix} \begin{bmatrix} x[0] \\ x[1] \\ \cdot \\ \cdot \\ \cdot \\ x[L-1] \end{bmatrix}, \quad \mathbf{y} = \mathbf{Hx}. \quad (9.3.14a)
$$

Note that **x** is an $L \times 1$ column matrix with the entries $x[0], x[1], \ldots, x[L-1]$. The vector **y** is a $(M + L - 1) \times 1$ column matrix and **H** is a $(M + L - 1) \times L$ matrix. The entries in the matrix **H** have the following pattern. The jth column in **H** is given by

$$
\left.\begin{cases} (j-1) \text{ zeros} \\ N \text{ coefficients written in the order} \\ h[0], h[1], \ldots, h[M-1] \\ (L-j) \text{ zeros} \end{cases}\right\}, 1 \leq j \leq L.
$$
(9.3.14b)

If one column or one row of **H** is known, the entire matrix can be constructed. Noting that the convolution is commutative, i.e., $y[n] = h[n] * x[n] = x[n] * h[n]$, equations similar to (9.3.14a and b) can be written by replacing $h[n]$ by $x[n]$ and vice versa.

Computation of convolution via DFT: In Section 8.10, it was shown that a periodic convolution can be implemented by using the DFT. This idea can be used here as well and is illustrated by a simple example. Equation in (9.3.12) can be written as

$$
\mathbf{y} = \begin{bmatrix} y[0] \\ y[1] \\ y[2] \end{bmatrix} = \begin{bmatrix} h[0] & 0 & h[1] \\ h[1] & h[0] & 0 \\ 0 & h[1] & h[0] \end{bmatrix} \begin{bmatrix} x[0] \\ x[1] \\ 0 \end{bmatrix} \quad (9.3.15)
$$

This set of equations gives the same results as the set in (9.3.12). Interestingly, (9.3.15) is the same as the one in (8.10.28), with $h[2] = 0$ corresponding to a periodic convolution. Equation (9.3.15) can be modified by adding the fourth column in the coefficient matrix and appending the data by two zeros resulting in (9.3.16).

$$
\mathbf{y}_a = \begin{bmatrix} y[0] \\ y[1] \\ y[2] \\ 0 \end{bmatrix} = \begin{bmatrix} h[0] & 0 & 0 & h[1] \\ h[1] & h[0] & 0 & 0 \\ 0 & h[1] & h[0] & 0 \\ 0 & 0 & h[1] & h[0] \end{bmatrix} \begin{bmatrix} x[0] \\ x[1] \\ 0 \\ 0 \end{bmatrix} = \mathbf{H}_a \mathbf{x}_a.
$$
(9.3.16)

The reason for using $N = 4$ for the extended sequences is that FFT can be used to find the convolution. In summary, given $h[n], n = 0, 1, \ldots, M-1$ and $x[n], n = 0, 1, 2, \ldots, L-1$, we can convolve $h[n]$ with $x[n]$ using the appended sequences as follows. Pad the sequences $h[n]$ and $x[n]$ with zeros so that they are of length $N \geq L + M - 1$ resulting in the appended sequences $h_a[n]$ and $x_a[n]$ of length N, a power of 2. Convolution of the two extended sequences results in $y_a[n] = h_a[n] * x_a[n]$. $y_a[n]$ is an appended sequence of $y[n]$ of length $2N - 1$. To determine the convolution via DFT (or FFT), do the following steps:

1. Determine $X_a[k] = \text{DFT}\{x_a[n]\}$ and $H_a[k] = \text{DFT}\{h_a[n]\}$.
2. Multiply the DFTs to form the products $Y_a[k] = H_a[k]X_a[k]$.
3. Find the inverse DFT of $Y_a[k]$. Discard the last $(N - (M + L - 1))$ data points out of $y_a[n]$ to obtain $y[n]$, $n = 0, 1, ..., N + L - 2$.

There are two methods for computing the discrete convolution, one by the convolution formula and the other by using DFTs. It may appear that the effort in computing the convolution via DFT is more computationally intensive than the direct convolution. However, using FFT, the number of computations is fewer, roughly for $N > 32$. Even though the number of computations is fewer for large N, there are difficulties with the use of DFT. The data sequence $x[n]$ may be long, say M points of data. The sequence $h[n]$ is of reasonable size, say $L \ll M$. The computation of the DFT of $x[n]$ may not be possible due to computer storage constraints involving large amount of computations resulting in a significant delay. There are two methods, overlap-add and overlap-save, which can be used for large set of data. Both section long input sequence into shorter sections. They are suitable for online implementation if the process can tolerate slight delays. The overlap-add method is discussed below; see Ambardar (2007) for the overlap-save method.

Overlap-add method: The sequence $h[n]$ of length N is assumed to start at $n = 0$. The sequence $x[n]$ is a much longer sequence of length M, also starting $n = 0$. Partition $x[n]$ into k segments each of length N (zero padding the last segment if needed). The data can be expressed in a mathematical form using a rectangular window $w_R[n]$ of length N by

$$x[n] = \sum_{i=0}^{k-1} x_i[n], \; x_i[n] = x[n]w[n - iN],$$

$$w[n] = \begin{cases} 1, & n = 0, 1, \ldots, N - 1 \\ 0, & \text{otherwise} \end{cases}. \quad (9.3.17a)$$

That is,

$$\{x[n]\} = \{\{x_0[n]\}, \{x_2[n]\}, \ldots, \{x_{k-1}[n]\}\}, x_i[n]$$
$$= \begin{cases} x[n - Ni], & i = 0, 1, \ldots, k - 1 \\ 0, & \text{elsewhere} \end{cases}.$$
$$(9.3.17b)$$

We can now write

$$y[n] = h[n] * x[n] = h[n] * \sum_{i=0}^{k-1} x_i[n]$$

$$= \sum_{i=0}^{k-1} h[n] * x_i[n] = \sum_{i=0}^{k-1} y_i[n].$$

It follows that the total convolution is the sum of the individual convolutions resulting in

$$y[n] = y_0[n] + y_1[n - N] + \ldots$$
$$+ y_{k-1}[n - (k - 1)N]. \quad (9.3.18)$$

The ith segment of the output begins at $n = iN$, as do the input segment $x_i[n]$. However, each $y_i[n]$ segment has a length equal to $(2N - 1)$ and therefore $y_i[n]$ s "overlap" each other. We can *think* of each $y_i[n]$ as having the same length of $2N - 1$ points, where each $y_i[n]$ includes zero padding before and/or after as appropriate, such that the positions of the sequences are in correct location.

Example 9.3.3 Consider the longer and shorter sequences given by $x[n] = \{1, 2, 3, 4\}$ and $h[n] = \{1, -1\}$. *a.* Use the overlap-add method to determine the convolution of the sequences. *b.* Verify the results using direct convolution.

Solution: a. Here $M = 2$ and $L = 4$. Section the sequence $x[n]$ into two sequences $x_0[n] = \{1, 2\}$ and $x_1[n] = \{3, 4\}$. Then determine $y_0[n] = x_0[n] * h[n]$ and $y_1[n] = x_1[n] * h[n]$. By the direct convolution, the sequences are as follows:

$$\begin{bmatrix} 1 & 0 \\ 2 & 1 \\ 0 & 2 \end{bmatrix} \begin{bmatrix} 1 \\ -1 \end{bmatrix} = \begin{bmatrix} 1 \\ 1 \\ -2 \end{bmatrix}, \quad \begin{bmatrix} 3 & 0 \\ 4 & 3 \\ 0 & 4 \end{bmatrix} \begin{bmatrix} 1 \\ -1 \end{bmatrix} = \begin{bmatrix} 3 \\ 1 \\ -4 \end{bmatrix}$$

$$\Rightarrow y_0[n] = \{1, 1, -2\}, y_1[n] = \{3, 1, -4\}.$$

Since the length of the sequence $y[n]$ is $(4 + 2 - 1) = 5$, the sequence $y_0[n]$ needs to be padded by two zeros at the end. Also, pad two zeros before the sequence $y_1[n]$. The result is obtained by overlapping the data and adding them at appropriate locations given below:

$$y[n] = y_0[n] + y_1[n - N]$$
$$= \{1, 1, -2, 0, 0\} + \{0, 0, 3, 1 - 4\}$$
$$= \{1, 1, 1, 1, -4\}.$$

b. The result can be verified by directly using the direct convolution given below and the result is the same in both cases. See the equations in (9.3.19a) and (9.3.19b):

$$\begin{bmatrix} y[0] \\ y[1] \\ y[0] \\ y[3] \\ y[4] \end{bmatrix} = \begin{bmatrix} x[0] & 0 \\ x[1] & x[0] \\ x[2] & x[1] \\ x[3] & x[2] \\ 0 & x[3] \end{bmatrix} \begin{bmatrix} h[0] \\ h[0] \end{bmatrix}$$

$$= \begin{bmatrix} 1 & 0 \\ 2 & 1 \\ 3 & 2 \\ 4 & 3 \\ 0 & 4 \end{bmatrix} \begin{bmatrix} 1 \\ -1 \end{bmatrix} = \begin{bmatrix} 1 \\ 1 \\ 1 \\ 1 \\ -4 \end{bmatrix}.$$ ∎

9.3.3 Correlation of Discrete Signals

Discrete cross-correlations of two sequences (see (8.3.20a and b)) were defined as follows:

$$r_{xh}[k] = x[k] * *h[k] = \sum_{n=-\infty}^{\infty} x[n]h[n + k]$$
$$= x[-k] * h[k], \qquad (9.3.19a)$$

$$r_{hx}[k] = h[k] * *x[k] = \sum_{n=-\infty}^{\infty} h[n]x[n + k]$$
$$= h[-k] * x[k]. \qquad (9.3.19b)$$

The cross-correlation of two causal sequences $x[k]$ and $h[k]$ with M and L sample points, respectively, are

$$r_{xh}[k] = x[k] * *h[k] = \sum_{n=0}^{M-1} x[n]h[n + k],$$

$$r_{hx}[k] = h[k] * *x[k] = \sum_{n=0}^{L-1} h[n]x[n + k], \qquad (9.3.19c)$$

$$r_{hx}[k] = r_{xh}[-k]. \qquad (9.3.19d)$$

The integer k represents the shift of the second sequence with respect to the first.

Example 9.3.4 Consider the data sequences given earlier in Example 9.3.2. Give the correlations of these two sequences and write them in a matrix form.

Solution: Using (9.3.20c), we have $r_{hx}[k] = 0, k \leq -2$ and $r_{hx}[k] = 0, k \geq 2$:

$$r_{hx}[-1] = h[0]x[-1] + h[1]x[0] = h[1]x[0],$$
$$r_{hx}[0] = h[0]x[0] + h[1]x[1],$$
$$r_{hx}[1] = h[0]x[1] + h[1]x[2] = h[0]x[1],$$
$$r_{hx}[2] = h[0]x[2] + h[1]x[1] = 0.$$

In matrix form $\Rightarrow$
$$\begin{bmatrix} r_{hx}[-1] \\ r_{hx}[0] \\ r_{hx}[1] \end{bmatrix} = \begin{bmatrix} h[1] & 0 \\ h[0] & h[1] \\ 0 & h[0] \end{bmatrix} \begin{bmatrix} x[0] \\ x[0] \end{bmatrix}$$

$$= \begin{bmatrix} -1 & 0 \\ 1 & -1 \\ 0 & 1 \end{bmatrix} \begin{bmatrix} 1 \\ 1 \end{bmatrix} = \begin{bmatrix} -1 \\ 0 \\ 1 \end{bmatrix}$$ ∎

The matrix equation can be generalized. The non-zero cross-correlations can be written in the following matrix and symbolic forms:

$$
\begin{bmatrix}
r_{hx}[-(M-1)] \\
r_{hx}[-(M-2)] \\
\cdot \\
\cdot \\
\cdot \\
r_{hx}[0] \\
r_{hx}[1] \\
\cdot \\
\cdot \\
\cdot \\
\cdot \\
r_{hx}[L-1]
\end{bmatrix}
=
\begin{bmatrix}
h[M-1] & 0 & . & . & . & 0 \\
h[M-2] & h[M-1] & . & . & . & 0 \\
\cdot & \cdot & \cdot & . & . & \vdots \\
\cdot & \cdot & . & . & . & h[M-1] \\
\cdot & \cdot & . & . & . & h[M-2] \\
h[0] & h[1] & . & . & . & \cdot \\
0 & h[0] & . & . & . & \cdot \\
\cdot & 0 & . & . & . & \cdot \\
\cdot & \cdot & \cdot & . & . & \cdot \\
\cdot & \cdot & \cdot & \cdot & . & \cdot \\
\cdot & \cdot & \cdot & \cdot & \cdot & \cdot \\
0 & 0 & . & . & . & h[0]
\end{bmatrix}
\begin{bmatrix}
x[0] \\
x[1] \\
\cdot \\
\cdot \\
\cdot \\
x[L-1]
\end{bmatrix}
\Rightarrow \mathbf{y} = \mathbf{H}_{\text{corr}} \mathbf{x}, \quad (9.3.20)
$$

where $\mathbf{y}$ is a column vector of dimension $(M+L-1)$, $\mathbf{x}$ is a column matrix of dimension L, and $\mathbf{H}_{\text{corr}}$ is a rectangular matrix of dimensions $(M+L-1) \times L$.

If $x(k) = h(k)$ then the cross-correlation coefficients are the autocorrelation coefficients.

Notes: In comparing (9.3.14a) and (9.3.20), for convolution, the first column of $\mathbf{H}$ has $h[n]$ in the *normal order*. For correlation, the first column of $\mathbf{H}_{\text{corr}}$ has $h[n]$ in *reverse order*. In both cases the other columns can be determined from the first column.

Computation of the cross-correlation using DFT: Given $h[n]$, $n = 0, 1, \ldots, M-1$ and $x[n]$, $n = 0, 1, \ldots, L-1$ determine the cross-correlation function

$$r_{hx}[n] = \sum_{k} h[k]x[k+n]. \quad (9.3.21)$$

Considering the equations for the convolution (see (9.3.14a)) and the cross-correlation (see (9.3.20)), we see that both have the same general form and the same computational procedure can be used for both cases. The following step-by-step procedure can be used.

1. Zero-pad both sequences to length $N \geq L + M - 1$. To use FFT, use N a power of 2.
2. Find the DFTs of $h[n]$ and $x[n]$.
3. $R_{hx}[k] = H^*[k]X[k], k = 0, 1, 2, \ldots, N-1$.
4. Find the inverse DFT of $R_{hx}[k]$.

Power spectral density: The autocorrelation (AC) sequence of $x[n]$ plays a major role in spectral estimation, as its power spectral density is $|X[k]|^2 = S_x[k]$. It is

$$
S_x[k] = \sum_{n=-(N-1)}^{N-1} r_x[n]e^{-j(2\pi/N)nk} = \sum_{n=0}^{N-1} r_x[n]e^{-j(2\pi/N)nk}
$$
$$
+ \sum_{n=0}^{N-1} r_x[n]e^{j(2\pi/N)nk} - r_x[0]. \quad (9.3.22)
$$

9.3.4 Discrete Deconvolution

We have seen in the analog domain when a signal goes through a linear time-invariant system, then the signal is modified by the impulse response of the system. The same is true in the digital domain. The convolution of two sequences that are of finite width was defined earlier and

$$y[n] = x[n] * h[n] = \sum_{k=0}^{n} h[k]x[n-k]. \quad (9.3.23)$$

There are three functions $x[n]$, $h[n]$, and $y[n]$. In finding the convolution, $x[n]$ and $h[n]$ are known and $y[n]$ is determined by (9.3.23). In the deconvolution problem, the output sequence $y[n]$ and the input data $x[n]$ are known and $h[n]$ is to be determined. There are four ways to achieve this goal. These are as follows: 1. recursion, 2. polynomial division, 3. using DFT, and 4. L_p deconvolution.

Deconvolution by recursion: From (9.3.23) $h[0] = y[0]/x[0]$. Now separate the term $h[n]$ in (9.3.23) and write in the following form and determine successively the values of $h[n]$ for $n > 0$:

$$y[n] = \sum_{k=0}^{n} h[k]x[n-k] = h[n]x[0] + \sum_{k=0}^{n-1} h[k]x[n-k].$$
(9.3.24)

Example 9.3.5 In Example 9.3.3 the convolution sequence $y[0] = 1, y[1] = 0, y[2] = -1$ was computed using the sequences $x[0] = 1, x[1] = 1$ and $h[0] = 1, h[1] = -1$. Verify the sequence $h[n]$ using the recursion method.

Solution:

$h[0] = y[0]/x[0] = 1,$
$h[1] = (1/x[0])\{y[1] - h[0]x[1]\} = 0 - 1 = -1.$
(9.3.25) ∎

This method is not practical in the presence of noise. Deconvolution using polynomial division in terms of z-transforms will be considered in Section 9.8.1. The DFT method makes use of DFTs of the sequences with $Y[k] = H[k]X[k]$. Then, $H[k] = Y[k]/X[k]$ and its inverse DFT gives $h[n]$. This procedure is similar to the one in the analog domain. It has at least two disadvantages. One of them is $X[k]$s may be zero resulting in division by zero. Also, it is sensitive to noise in the input. Fourth method is based on minimizing the L_p errors discussed in Section 3.3.

Deconvolution by L_p methods: The output is assumed to be the convolution of two sequences, say an input sequence $x[n]$, a linear discrete system response sequence given by $h[n]$, and an additive noise sequence $\varepsilon[n]$. The output is

$$y[n] = h[n] * x[n] + \varepsilon[n], n = 0, 1, 2, \ldots, N-1. \quad (9.3.26)$$

The noise signal can only be described by statistical measures. An interesting error measure is the $L_p, 1 \le p \le \infty$ measure defined by

$$|\varepsilon|^p = \sum_{n=0}^{N-1} (y[n] - (h[n] * x[n]))^p. \quad (9.3.27)$$

Minimization of this error in terms of the unknowns $h[n]$ is a difficult problem for an arbitrary p. The general solution can only be determined by iterative means, see the articles by Byrd and Payne (1979) and Yarlagadda et al. (1985). There is a simple solution when $p = 2$, which is used if the noise sequence is from a Gaussian distribution. These problems can be described under the general problem of solving a set of equations that are overdetermined and underdetermined system of equations. In Section A.6 we consider the solutions of overdetermined and underdetermined system of equations. Consider the system of equations in the symbolic matrix form

$$\mathbf{Ah} = \mathbf{y}. \quad (9.3.28a)$$

The least-squares solution to the overdetermined system in (9.3.28a) is (see (A.8.16b))

$$\mathbf{y} = \mathbf{Ah} \Rightarrow (\mathbf{A}^T\mathbf{A})\mathbf{h} = \mathbf{A}^T\mathbf{y} \Rightarrow \mathbf{h} = (\mathbf{A}^T\mathbf{A})^{-1}\mathbf{A}^T\mathbf{y}.$$
(9.3.28b)

The matrix $[(\mathbf{A}^T\mathbf{A})^{-1}\mathbf{A}]$ is a *pseudo-inverse* of the matrix $\mathbf{A}$. The MATLAB routine to compute this inverse is

$$pinv(\mathbf{A}) = (\mathbf{A}^T\mathbf{A})^{-1}\mathbf{A}^T. \quad (9.3.29)$$

The inverses of the matrix $(\mathbf{A}^T\mathbf{A})$ may not exist. In such cases, a diagonal matrix $\delta\mathbf{I}$, where δ is a small positive number, is added to the matrix $(\mathbf{A}^T\mathbf{A})$. This is called *diagonal loading*. An approximate solution of (9.3.28) is then given by

$$\mathbf{h} \cong (\mathbf{A}^T\mathbf{A} + \delta\mathbf{I})^{-1}\mathbf{Ay}. \quad (9.3.30)$$

Example 9.3.6 Solve the following set of equations using the least-squares solution:

$$\mathbf{Ah} = \begin{bmatrix} 1 & 0 \\ 2 & 1 \\ 0 & 2 \end{bmatrix} \begin{bmatrix} 1 \\ -1 \end{bmatrix} + \begin{bmatrix} \varepsilon \\ -\varepsilon \\ \varepsilon \end{bmatrix}$$

$$= \begin{bmatrix} 1 \\ 1 \\ -2 \end{bmatrix} + \begin{bmatrix} \varepsilon \\ -\varepsilon \\ \varepsilon \end{bmatrix} = \mathbf{y} + \begin{bmatrix} \varepsilon \\ -\varepsilon \\ \varepsilon \end{bmatrix}. \quad (9.3.31)$$

Solution: The pseudo-inverse of $\mathbf{A}$ and the solution vector are, respectively, given as follows:

$$(\mathbf{A}^T\mathbf{A})^{-1}\mathbf{A}^T = \frac{1}{21}\begin{bmatrix} 5 & -2 \\ -2 & 5 \end{bmatrix}\begin{bmatrix} 1 & 2 & 0 \\ 0 & 1 & 2 \end{bmatrix}$$

$$= \frac{1}{21}\begin{bmatrix} 5 & 8 & -4 \\ -2 & 1 & 10 \end{bmatrix},$$

$$\mathbf{h} = \frac{1}{21}\begin{bmatrix} 5 & 8 & -4 \\ -2 & 1 & 10 \end{bmatrix}\left\{\begin{bmatrix} 1 \\ 1 \\ -2 \end{bmatrix} + \begin{bmatrix} \varepsilon \\ -\varepsilon \\ \varepsilon \end{bmatrix}\right\}$$

$$= \frac{1}{21}\begin{bmatrix} 21 \\ -21 \end{bmatrix} + \frac{1}{21}\begin{bmatrix} -7\varepsilon \\ 7\varepsilon \end{bmatrix} = \begin{bmatrix} 1 - .3333\varepsilon \\ -1 + .3333\varepsilon \end{bmatrix}.$$

Clearly if $\varepsilon = 0$, the solution coincides with the vector $\mathbf{h}$ we started with.   ∎

Notes: Implementation of discrete algorithms generally requires multiplications, which are expensive compared to additions. The following table gives a rough comparison of how expensive the additions, multiplications, and data transfers are for fixed and floating point machines by assuming one unit of expense corresponding to an addition compared to other operations. This gives a comparison of the computational expense and not the individual machine comparison, see Stine (2003) and Swartzlander Jr. (2001).

	Multiplication	Addition	Transfer
Fixed point	10	1	0.5
Floating point	2	1	0.5

From this table one can appreciate how much FFT algorithms are cost-effective in implementing the discrete Fourier transform when the number of data points N is large.   ∎

In the following, z-transforms, the discrete-time counterpart of the L-transforms, will be presented. Theory behind z-transforms is rather sophisticated and our presentation will be simple. See Oppenheim and Schafer (1999) for a detailed discussion on this topic.

9.4 *z*-Transforms

The DTFT of the sequence $x[n]$, $X(e^{j\Omega})$ exists provided that $x[n]$ is absolutely summable (see (8.5.11), which is repeated below in (9.4.1)). This is sufficient but *not* necessary:

$$\sum_{n=-\infty}^{\infty} |x[n]| < \infty. \tag{9.4.1}$$

The DTFT pair is

$$X(e^{j\Omega}) = \sum_{n=-\infty}^{\infty} x[n]e^{-jn\Omega} \overset{\text{DTFT}}{\longleftrightarrow} \frac{1}{2\pi}\int_{-\pi}^{\pi} X(e^{j\Omega})e^{jn\Omega}d\Omega$$

$$= x[n]. \tag{9.4.2}$$

The DTFT of $x[n]e^{-n\sigma}$ and the corresponding DTFT are as follows:

$$\text{DTFT}[x[n]e^{-n\sigma}] = \sum_{n=-\infty}^{\infty} [x[n]e^{-n\sigma}]e^{-jn\Omega}$$

$$= \sum_{n=-\infty}^{\infty} x[n]e^{-j(\sigma+j\Omega)n}, \tag{9.4.3}$$

$$x[n]e^{-jn\Omega} \overset{\text{DTFT}}{\longleftrightarrow} X(e^{j(\sigma+j\Omega)}). \tag{9.4.4}$$

The convergence of the sequence $x[n]\,e^{-jn\Omega}$ can now be defined in terms of $e^{-n\sigma}$, which is similar to the convergence of L-transforms, see Section 5.4. It is *desirable* to use the notation

$$z = e^{\sigma+j\Omega} = e^{\sigma}e^{j\Omega} = re^{j\Omega} \text{ and } \ln(z)$$

$$= \sigma + j\Omega \text{ and } (1/z)dz = jd\Omega. \tag{9.4.5}$$

Using these in (9.4.1), the time sequence and the corresponding z-transform are

$$x[n] = \frac{1}{2\pi j}\oint X(z)z^{n-1}dz, \quad X(z)$$

$$= \sum_{n=-\infty}^{\infty} x[n]z^{-n}. \tag{9.4.6}$$

The z-transform of a discrete-time sequence $x[n]$ is defined in terms of a complex variable z by

$$Z\{x[n]\} = X(z) = \sum_{n=-\infty}^{\infty} x[n]z^{-n}. \tag{9.4.7a}$$

The range of values of the complex variable z for which the summation converges is called the region of convergence (ROC). The inverse z-transform and

Fig. 9.4.1 Contour of
integration on the z-plane.

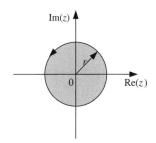

Fig. 9.4.2 Example 9.4.1:
region of convergence ($\alpha > 0$)

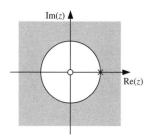

the symbolic relationship between $x[n]$ and $X(z)$ are, respectively, given by

$$x[n] = Z^{-1}[X(z)] = \left[\frac{1}{2\pi j} \oint_c X(z) z^{n-1} dz \right]$$

$$\left(\oint_c - \text{contour integral} \right), \; x[n] \overset{z}{\leftrightarrow} X(z).$$

(9.4.7b)

The contour integral is around a circle of radius r in the counterclockwise direction enclosing the origin on the z-plane, see Fig. 9.4.1. The complex domain integration requires knowledge of complex variables, which is beyond our scope.

The z-transform exists when the sum in (9.4.7a) converges. A necessary condition for convergence is absolute summability of $x[n]z^{-n}$. Let $z = re^{j\Omega}$. The absolute summability of $x[n]z^{-n}$ is

$$\sum_{n=-\infty}^{\infty} |x[n] r^{-n}| < \infty. \qquad (9.4.8)$$

The range of r for which the sum converges is the *region of convergence* (ROC).

9.4.1 Region of Convergence (ROC)

Example 9.4.1 Determine the z-transform of the right-sided sequence $x_1[n] = \alpha^n u[n]$.

Solution: The z-transform of $x_1[n]$ is

$$X_1(z) = \sum_{n=-\infty}^{\infty} x_1[n] z^{-n} = \sum_{n=-\infty}^{\infty} \alpha^n u[n] z^{-n}$$

$$= \sum_{n=0}^{\infty} (\alpha z^{-1})^n = \frac{1}{1 - \alpha z^{-1}} = \frac{z}{z - \alpha}, \; |z| > |\alpha|.$$

(9.4.9)

ROC is the range of values of z for which $|\alpha z^{-1}| < 1$ or $|z| > |\alpha|$. The transform is represented by a rational function of the complex variable z. As in the Laplace transforms, we can describe a rational function $X(z)$ in terms of its poles (the roots of the denominator) and zeros (the roots of the numerator) on the complex z-plane. There is a pole at $z = \alpha$ and a zero at $z = 0$ and these are shown in Fig. 9.4.2. The ROC is outside the circle of radius α. The boundary of the ROC is $|z| = |\alpha|$. The ROC does *not* contain any poles and is outside the circle of radius $|z| = \alpha$. ∎

Example 9.4.2 Consider the left-side sequence $x_2[n] = -\beta^n u[-n-1], \beta \neq 0$.

Solution: The z-transform is

$$X_2(z) = \sum_{n=-\infty}^{\infty} x_2[n] z^{-n} = - \sum_{n=-\infty}^{\infty} \beta^n u[-n-1] z^{-n}.$$

(9.4.10)

$u[n]$, $u[-n]$, and $u[-n-1] = u[-(n+1)]$ are sketched in Fig. 9.4.3a,b,c and

$$X_2(z) = - \sum_{n=-\infty}^{-1} (\beta/z)^n. \qquad (9.4.11)$$

Using the change of variable $m = -n$ in (9.4.12), we have

$$X_2(z) = - \sum_{m=1}^{\infty} (z/\beta)^m = 1 - \sum_{m=0}^{\infty} (z/\beta)^n$$

$$= 1 - \frac{1}{1 - (z/\beta)} = \frac{z}{z - \beta}, |z| < |\beta|. \quad (9.4.12)$$

See the pole–zero plot and the region of convergence in Fig. 9.4.4.

In the last two examples, a right-side and a left-side sequences were considered. If $\alpha = \beta$, the two

Fig. 9.4.3 (a) $u[n]$, (b) $u[-n]$, and (c) $u[-(n-1)]$

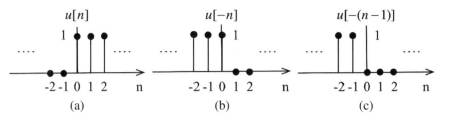

transforms are identical except the ROC is different. The z-transform of the sequence and the ROC need to be known before the sequence can be identified.

Example 9.4.3 Determine the z-transform of the two-sided sequence $y[n] = \alpha^n u[n] + \beta^n u[-n-1]$ and the ROC.

Solution: The z-transform of the sum is obtained by adding the two transforms and

$$Y(z) = \sum_{n=-\infty}^{\infty} (\alpha^n u[n] - \beta^n u[-n-1]) z^{-n}$$

$$= \sum_{n=0}^{\infty} \alpha^n z^{-n} - 1 + \sum_{m=0}^{\infty} (z/\beta)^n. \quad (9.4.13)$$

First and the second sums on the right-hand side have the ROCs $|z| > |\alpha|$ and $|z| < |\beta|$, respectively. The sum of the two functions and the corresponding ROC are given by

$$Y(z) = \frac{z}{z-\alpha} + \frac{z}{z-\beta},$$
$$\text{ROC} : \{|z| > |\alpha|\} \cap \{|z| < |\beta|\}. \quad (9.4.14)$$

The ROC in (9.4.15) exists only if there is an overlap of the regions identified by the regions $\{|z| > |\alpha|\}$ and $\{|z| < |\beta|\}$. If $|\beta| > |\alpha|$, then the transform converges in the annular region shown in Fig. 9.4.5a identified by $|\alpha| < |z| < |\beta|$. If $|\beta| < |\alpha|$, there is no region of overlap and therefore the ROC is the null set. ∎

Example 9.4.4 Give the ROC of the sequence.

$$x[n] = \left\{ \begin{array}{l} \neq 0, \ N_1 \leq n \leq N_2 \\ 0, \text{Otherwise} \end{array} \right\} \xrightarrow{z} X(z) = \sum_{n=N_1}^{N_2} x[n] z^{-n}. \quad (9.4.15)$$

Solution: For $z \neq 0$ or ∞, each term will be finite and the function $X(z)$ converges. If $N_1 < 0$ and $N_2 > 0$, the sum includes both negative and positive powers of z. As $|z| \to 0$, the terms with negative powers of z become unbounded. As $|z| \to \infty$, the terms with positive powers of z become unbounded. Therefore, the ROC of the function $X(z)$ of a finite sequence is the entire z-plane *except* for $z = 0$ and $z = \infty$. If $N_1 \geq 0$, the ROC includes $z = \infty$ and if $N_2 \leq 0$, the ROC includes $z = 0$. ∎

Example 9.4.5 Find $X(z)$ for $x[n]$ in Fig. 9.4.6 and make a pole–zero plot.

Solution: First,

$$X(z) = \sum_{n=-2}^{2} x[n] z^{-n} = 1.z^{-(-2)} + (-6).z^{-(-1)}$$
$$+ 9.z^0 + 4.z^{-1} + (-12).z^{-2}$$
$$= z^2 - 6z + 9 + 4z^{-1} - 12z^{-2}$$
$$= \frac{z^4 - 6z^3 + 9z^2 + 4z - 12}{z^2}$$
$$= \frac{(z-2)^2(z-3)(z+1)}{z^2}.$$

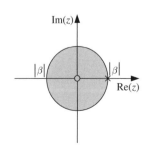

Fig. 9.4.4 Pole–zero plot and ROC of $X_2(z)$.

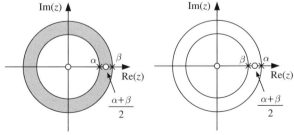

Fig. 9.4.5 (a) ROC of $Y(z)$ and (b) no region of convergence when $|\alpha| > |\beta|$

Fig. 9.4.6 (a) $x[n]$ and (b) pole–zero plot

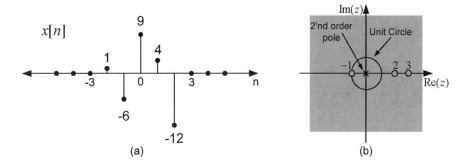

(a)

(b)

The poles of $X(z)$ are at the origin and have zeros at $z = -1, 2$, and 3. See the pole–zero plot in Fig. 9.4.6b. The ROC of this function spans the entire z-plane corresponding to the region enclosed between the poles at zero and those located at infinity. ∎

Example 9.4.6 Derive the z-transform of the sequence for the following two cases: a being arbitrary and $a = 1$:

$$x[n] = \begin{cases} a^n, & 0 \leq n \leq N-1 \\ 0, & \text{otherwise} \end{cases}. \qquad (9.4.16)$$

Solution:

$$X(z) = \sum_{n=0}^{N-1} a^n z^{-n} = \sum_{n=0}^{N-1} (az^{-1})$$
$$= \frac{1 - a^N z^{-N}}{1 - az^{-1}}, \quad \text{ROC}: |z| > 0. \qquad (9.4.17)$$

$$a = 1 \Rightarrow x[n] = \left\{ \begin{matrix} 1, \ 0 \leq n \leq N-1 \\ 0, \ \text{otherwise} \end{matrix} \right\} \overset{z}{\longleftrightarrow} \frac{1 - z^{-N}}{1 - z^{-1}}$$
$$= X(z), \quad \text{ROC}: |z| \neq 0. \qquad (9.4.18)$$

See Fig. 9.4.7 for the pole-zero plot assuming N=11 ∎

Notes on the ROC of a rational function $X(z)$: The ROC depends on the poles of the

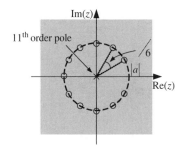

Fig. 9.4.7 Example 9.4.6: pole–zero plot ($N = 11$)

function $X(z)$ at $z = r_i e^{j\theta_i}, r_i > 0$. The maximum and minimum magnitudes of these poles are identified by $r_{\max} = \max|r_i|$ and $r_{\min} = \min|r_i|$. If the degree of the denominator of $X(z)$ is smaller than the degree of the numerator, then $X(z)$ has at least one pole at ∞.

1. The ROC does not contain any poles.
2. If the sequence is a finite sequence, then the ROC of $X(z)$ is the entire z-plane except possibly $z = 0$ or $z = \infty$.
3. If $x[n]$ is a right-side sequence, i.e., $x[n] = 0$, $n < N_1 < \infty$, and $X(z)$ converges for some values of z, the ROC is $r_{\max} < |z| \leq \infty$ with a possible exception of $z = \infty$.
4. If $x[n]$ is a left-sided sequence, i.e., $x[n] = 0$, $-\infty < N_2 < n$ and $X(z)$ converges for some values of z, then the ROC is $0 \leq |z| < r_{\min}$ with a possible exception of $z = 0$
5. If $x[n]$ is a two-sided sequence and the region of convergence of the right- and left-sided sequences are, respectively, given by $r_1 < |z|$ and $|z| < r_2$ and $X(z)$ converges for some values of z, then the ROC takes the form $r_1 < |z| < r_2$, where r_1 and r_2 are the magnitudes of the poles of $X(z)$.

Example 9.4.7 Find the z-transforms of the following sequences and their ROCs:

a. $x_1[n] = \delta[n]$,

b. $x_2[n] = u[n]$,

c. $x_3[n] = -u[-n-1]$,

d. $x_6[n] = a^{|n|}, |a| < 1$.

Solution:

$$a. \ X_1(z) = \sum_{n=-\infty}^{\infty} \delta[n] z^{-n} = z^0 = 1,$$

$$\delta[n] \overset{z}{\longleftrightarrow} 1, \text{ROC}: \text{ all } z \qquad (9.4.19)$$

b. $X_2(z) = \sum_{n=0}^{\infty} z^{-n} = \dfrac{1}{1 - z^{-1}} = \dfrac{z}{z - 1}, \text{ROC} : |z| > 1$

$$\text{(9.4.20)}$$

c. $X_3(z) = 1 - \dfrac{1}{1 - z^{-1}} = \dfrac{z}{z - 1}, \text{ROC} : |z| < 1$

$$\text{(9.4.21)}$$

d. $x_6[n] = a^{|n|} = a^n u[n] + a^n u[-n - 1]$

$$\xleftrightarrow{z} \dfrac{z}{z - a} - \dfrac{z}{z - (1/a)}$$

$$= \dfrac{(a^2 - 1)}{a} \dfrac{z}{(z - a)(z - (1/a))},$$

$$\text{ROC} : |a| < |z| < \left| \dfrac{1}{a} \right|, |a| < 1. \quad \text{(9.4.22)} \blacksquare$$

9.4.2 z-Transform and the Discrete-Time Fourier Transform (DTFT)

Note $X(z)$ is a function of the complex variable z. The point $z = re^{j\Omega}$ is located at a distance r from the origin at an angle Ω from the positive real axis. If $x[n]$ is absolutely summable, then the DTFT can be obtained from the z-transform by setting $z = re^{j\Omega}|_{r=1}$:

$$X(e^{j\Omega}) = X(z)|_{z=e^{j\Omega}} \quad \text{(9.4.23)}$$

The equation $|z| = |e^{j\Omega}| = 1$ describes a circle of unit radius ($r = 1$) centered at the origin in the z-plane. The frequency Ω in the discrete-time Fourier transform corresponds to the point on the unit circle at an angle Ω in radians with respect to the positive real axis. As the discrete-time frequency varies in the range $-\pi$ to π, in the z-plane, it corresponds to one time around the unit circle. In words, (9.4.23) states that the DTFT of a discrete-time signal $x[n]$ can be obtained from the z-transform $X(z)$ by evaluating it on the unit circle. The z-transform of the sequence is assumed to exist and the DTFT of the sequence exists provided that the *region of convergence* of $X(z)$ includes *the unit circle*. The DTFT function is represented by $X(e^{j\Omega}) = |X(e^{j\Omega})|e^{j\theta(\Omega)}$, where $|X(e^{j\Omega})|$ is called the *amplitude (or magnitude) response* and $\theta(\Omega)$ is called the *phase (or angle)*

response. The amplitude and the phase responses are *periodic* with period 2π.

Example 9.4.8 Use the z-transforms to determine the DTFT of the discrete-time function

$$x[n] = u[n] - u[n - N], N > 0 \xleftrightarrow{z} X(z)$$

$$= \sum_{n=0}^{N-1} z^{-n}, \quad \text{ROC} : |z| > 0. \quad \text{(9.4.24)}$$

Solution: Since ROC of $X(z)$ includes the unit circle, the DTFT $X(e^{j\Omega})$ exists. It is periodic with period 2π and phase response is *linear*:

$$X(e^{j\Omega}) = X(z)|_{z=e^{j\Omega}} = \sum_{n=0}^{N-1} z^{-n} = \dfrac{1 - z^{-N}}{1 - z^{-1}}|_{z=e^{j\Omega}}$$

$$= \dfrac{1 - e^{-jN\Omega}}{1 - e^{-j\Omega}} = e^{-j\Omega(N-1)/2} \dfrac{\sin(N\Omega/2)}{\sin(\Omega/2)}.$$

$$\text{(9.4.25)}$$

The amplitude and the phase responses are, respectively, given by

$$|X(e^{j\Omega})| = \left| \dfrac{\sin(N\Omega/2)}{\sin(\Omega/2)} \right|, \angle X(e^{j\Omega}) = \Omega(N-1)/2. \quad \text{(9.4.26)}$$

$\blacksquare$

9.5 Properties of the z-Transform

Let the ROC of $x_i[n]$ is R_{xi} and R' is the ROC after the appropriate operation. The ROC is stated in terms of set theory. The proofs are simple for many of these and are omitted.

9.5.1 Linearity

Let $x_i[n] \xleftrightarrow{z} X_i(z)$, then

$$x[n] = a_1 x_1[n] + a_2 x_2[n] \xleftrightarrow{z} a_1 X_1(z) + a_2 X_2(z)$$

$$= X(z), \text{ROC} : R' \supset R_{x1} \cap R_{x2},$$

$$\text{(9.5.1)}$$

where R' is the ROC of $x[n]$, which is the proper subset of the ROCs of x_1 and x_2. Note the intersection of the subsets represented by $R_{x1} \cap R_{x2}$ in (9.5.1). Expansion of the ROC takes into consideration pole cancellations with zeros in $X(z)$.

9.5.2 Time-Shifted Sequences

The z-transform of the time-shifted sequence and the corresponding ROC are

$$Z\{x[n-n_0]\} = \sum_{n=-\infty}^{\infty} x[n-n_0]z^{-n}$$

$$= \sum_{m=-\infty}^{\infty} x[m]z^{-(m+n_0)} = z^{-n_0} \sum_{m=-\infty}^{\infty} x[m]z^{-m} = z^{-n_0} X(z),$$

$$x[n-n_0] \xleftrightarrow{z} z^{-n_0} X(z), \text{ROC} : R' \supset R \cap \{0 < |z| < \infty\}.$$
(9.5.2)

Noting the multiplication factor z^{-n_0}, additional poles are introduced when $n_0 > 0$ and, at the same time, some of the poles at ∞ are deleted. In a similar manner if $n_0 < 0$ then additional zeros are introduced at $z = 0$ and some of the poles at ∞ are deleted. This implies that the points $z = 0$ and $z = \infty$ are either added or deleted from the ROC by time shifting the function. The special cases include the unit delay and advance operations:

$$x[n - 1] \xleftrightarrow{z} z^{-1}X(z), \text{ROC} :$$
$$R' = R_x \cap \{0 < |z|\},$$
(9.5.3a)

$$x[n + 1] \xleftrightarrow{z} zX(z), \text{ROC} : R'$$
$$= R \cap \{|z| < \infty\}.$$
(9.5.3b)

9.5.3 Time Reversal

If $x[n] \xleftrightarrow{z} X(z), \text{ROC} = R$, then $x[-n] \xleftrightarrow{z} X(1/z)$, ROC $:R' = 1/R$.
(9.5.4)

Reversing in time results in the transformation $z \Rightarrow (1/z)$ in the transform and, the points in the ROC R' corresponds to the inverses of the points in R. This can be shown by

$$Z\{x[-n]\} = \sum_{n=-\infty}^{\infty} x[-n]z^{-n} = \sum_{m=-\infty}^{\infty} x[m]z^m = X(1/z).$$

Example 9.5.1 Find $z[x[n]] = z[u[-n]]$ directly and then verify using the result in (9.5.4):

$$X(z) = \sum_{n=-\infty}^{\infty} x[n]z^{-n} = \sum_{n=-\infty}^{0} (z^{-1})^n = \sum_{n=0}^{\infty} z^n$$

$$= \frac{1}{1 - z}, \text{ROC} : R' = |z| < 1.$$

Solution: Noting that $u[n] \xleftrightarrow{z} (1/(1 - z^{-1}))$ and using the transformation $z \to 1/z$ results in the above equation verifying the time reversal theorem. We note that the closed-form expression for the sum is valid if $|z| < 1$. Also, the ROC of the unit step sequence is $|z| > 1$. Reversing the sequence results in the ROC from $|z| < 1$. ∎

9.5.4 Multiplication by an Exponential

If a is a complex number, then

$$Z\{a^n x[n]\} = \sum_{n=-\infty}^{\infty} a^n x[n]z^{-n} = \sum_{n=-\infty}^{\infty} x[n](z/a)^{-n}$$
$$= X(z/a),$$

$$a^n x[n] \xleftrightarrow{z} X(z/a), \text{ROC} : R' = |a|R.$$
(9.5.5)

Change in the argument of $X(z/a)$ resulted in the multiplication of ROC boundaries by $|a|$. ROC expands or contracts by the factor of $|a|$. In the special case of $a = e^{j\Omega_0 n}$:

$$e^{j\Omega_0 n} x[n] \xleftrightarrow{z} X(e^{-j\Omega_0}z), \text{ROC} :R' = R.$$
(9.5.6)

Example 9.5.2 Determine the z-transform of the real sequence $y[n] = r^n \cos(\Omega_0 n)u[n]$.

Solution: Noting $r^n u[n] \xleftrightarrow{z} z/(z - r)$, ROC : $|z| > |r|$ we can write

$$y[n] = \frac{1}{2}r^n e^{j\Omega_0 n}x[n] + \frac{1}{2}r^n e^{-j\Omega_0 n}u[n] \xleftrightarrow{z} \frac{1}{2}\frac{z}{z - re^{j\Omega_0}}$$

$$+ \frac{1}{2}\frac{z}{z - re^{-j\Omega_0}} = Y(z) ,$$

$$y[n] = r^n \cos(\Omega_0 n) u[n] \overset{z}{\longleftrightarrow} Y(z)$$

$$= \frac{z^2 - r\cos(\Omega_0)z}{z^2 - 2r\cos(\Omega_0)z + r^2}, \text{ROC}: |z| > |r|.$$

$$(9.5.7a)$$

In a similar manner,

$$r^n \sin(\Omega_0 n) u[n] \overset{z}{\longleftrightarrow} \frac{r\sin(\Omega_0)z}{z^2 - 2r\cos(\Omega_0)z + r^2},$$

$$(9.5.7b)$$

$$\text{ROC}: |z| > r.$$

■

9.5.5 Multiplication by n

The z-transform of $nx[n]$ is

$$y[n] = nx[n] \overset{z}{\longleftrightarrow} -z[dX(z)/dz] = Y(z),$$

$$(9.5.8)$$

$$\text{ROC}: R' = R.$$

This can be seen by differentiating both sides with respect to z of the following equation:

$$X(z) = \sum_{n=-\infty}^{\infty} x[n]z^{-n} \to \frac{dX(z)}{dz} = \sum_{n=-\infty}^{\infty} (-n)x[n]z^{-n-1}.$$

Multiplying both sides by $(-z)$ and identifying the appropriate time and transform terms, (9.5.8) follows. The region of convergence is the *same* for $X(z)$ and $Y(z)$.

Example 9.5.3 Determine the z-transform of the right-side sequence $x[n] = na^n u[n], a > 0$ using the multiplication by n property.

Solution: Noting that $a^n u[n] \overset{z}{\longleftrightarrow} (z/(z-a))$, ROC : $|z| > |a|$, by using (9.5.8), we have

$$na^n u[n] \overset{z}{\longleftrightarrow} -z \frac{d(z/(z-a))}{dz} = \frac{az}{(z-a)^2},$$

$$(9.5.9)$$

$$\text{ROC}: |z| > |a|.$$

■

9.5.6 Difference and Accumulation

The z-transforms of these are

$$y_1[n] = x[n] - x[n-1] \overset{z}{\longleftrightarrow} X(z)[1 - z^{-1}]$$

$$= Y_1(z), R' \supset R \cap \{|z| > 0\}, \qquad (9.5.10a)$$

$$y_2[n] = \sum_{k=-\infty}^{n} x[k] \overset{z}{\longleftrightarrow} \frac{1}{(1 - z^{-1})} X(z)$$

$$= Y_2(z), R' \supset R \cap \{|z| > 1\}. \qquad (9.5.10b)$$

9.5.7 Convolution Theorem and the z-Transform

Convolution theorem states that the convolution in the time domain corresponds to the multiplication in the z-domain and

$$y[n] = x[n] * h[n] \overset{z}{\longleftrightarrow} X(z)H(z)$$

$$= Y(z), \text{ROC}: R_y \supset (R_x \cap R_h). \qquad (9.5.11)$$

This can be shown using the expression for the convolution of two sequences $x[n]$ and $h[n]$ and then by using the transform pairs as shown below:

$$x[n] \overset{z}{\longleftrightarrow} X(z) \text{ and } h[n-k] \overset{z}{\longleftrightarrow} z^{-k}H(z),$$

$$y[n] = \sum_{k=-\infty}^{\infty} x[k]h[n-k] = \sum_{k=-\infty}^{\infty} h[k]x[n-k],$$

$$Y(z) = \sum_{n=-\infty}^{\infty} y[n]z^{-n} = \sum_{n=-\infty}^{\infty} \left[\sum_{k=-\infty}^{\infty} x[k]h[n-k] \right] z^{-n}$$

$$= \sum_{k=-\infty}^{\infty} x[k] \left[\sum_{n=-\infty}^{\infty} h[n-k]z^{-n} \right]$$

$$= \sum_{k=-\infty}^{\infty} x[k] \left[H(z)z^{-k} \right]$$

$$= H(z) \sum_{k=-\infty}^{\infty} x[k]z^{-k} = H(z)X(z)$$

The ROC of $Y(z)$ contains the intersection of the ROC of $X(z)$ and $Y(z)$. If a zero of one of the transforms cancels with a pole of the other, then the ROC of $Y(z)$ will be larger than the intersection of R_x and R_h.

Example 9.5.4 Verify the result in (9.5.10b) by using (9.5.11).

Solution: Noting $u[n - k] = 0, k > n$, $y[n]$ can be expressed by

$$y[n] = \sum_{k=-\infty}^{n} x[k] = \sum_{k=-\infty}^{\infty} x[k]u[n - k]$$

$$= x[n] * u[n] \stackrel{z}{\longleftrightarrow} X(z)z\{u[n]\}$$

$$= X(z)\frac{z}{(z - 1)}, \text{ROC} : R' \supset (R \cap \{|z| > 1\}).$$

$$\text{(9.5.12)} \blacksquare$$

Example 9.5.5 Find the z-transform of the following sequence:

$$y[n] = x[n] * h[n], x[n] = na^n u[n], h[n]$$
$$= (b)^{-n}u[-n], a = -b = -1/2. \quad \text{(9.5.13)}$$

Solution: Using the multiplication by n property, we have

$$x[n] = na^n u[n] \stackrel{z}{\longleftrightarrow} \frac{az}{(z - a)^2} = X(z),$$

$$\text{ROC} : |z| > |a|, \quad \text{(9.5.14a)}$$

$$h[n] = b^n u[n] \stackrel{z}{\longleftrightarrow} \frac{z}{(z - b)}, \text{ROC} : |z| > |b|. \quad \text{(9.5.14b)}$$

Using (9.5.14a and b), the time reversal property, and the convolution theorem, we have

$$x[n] = n(-1/2)^n u[n] \stackrel{z}{\longleftrightarrow} \frac{-(1/2)z}{(z + (1/2))^2}$$

$$= X(z), \text{ROC} : |z| > (1/2) ,$$

$$h[n] = (1/2)^{-n}u[-n] \stackrel{z}{\longleftrightarrow} \frac{1/z}{(1/z) - (1/2)} \quad \text{(9.5.14c)}$$

$$= \frac{-2}{z - 2} = H(z), \text{ROC} : |z| < 2 ,$$

$$Y(z) = H(z)X(z) = \frac{z}{(z - 2)(z + (1/2))^2},$$

$$\text{ROC} : \frac{1}{2} < |z| < 2. \quad \text{(9.5.14d)}$$

We have a pole outside and one inside the unit circle resulting in a transform that has the annular region of convergence given in (9.5.14d). See Fig. 9.5.1 for pole-zero plots and ROC. $\blacksquare$

9.5.8 Correlation Theorem and the z-Transform

In Section 8.3 the cross-correlation of two sequences $x[n]$ and $h[n]$ was defined by

$$r_{xh}[k] = x[k] * *h[k] = \sum_{n=-\infty}^{\infty} x[n]h[n + k]$$

$$= \sum_{m=-\infty}^{\infty} x[m - k]h[m]. \quad \text{(9.5.15)}$$

Correlation theorem:

$$r_{xh}[k] = \sum_{n=-\infty}^{\infty} x[n]h[n+k] \stackrel{z}{\longleftrightarrow} X(z)H(1/z)$$

$$= R_{xh}(z), \text{ROC} : R_{xh} \supset (R_x \cap R_h). \quad \text{(9.5.16)}$$

This can be seen by first noting that $r_{xh}[k] = x[k] * h[-k]$. Using the convolution and the time reversal properties, we can see the result in (9.5.16). Again there is a possibility of pole–zero cancellations and therefore $R_{xh} \supset (R_x \cap R_h)$. In the case of autocorrelation, we have $h[n] = x[n]$ and the *autocorrelation* (AC) theorem in (9.5.16) reduces to

$$r_{xx}[k] = r_x[k] = \sum_{n=-\infty}^{\infty} x[n]x[n + k] \stackrel{z}{\longleftrightarrow} X(z)X(1/z)$$

$$= R_x(z), \text{ROC} : R_{xh} \supset (R_x \cap R_h). \quad \text{(9.5.17)}$$

Fig. 9.5.1 Example 9.5.5: pole–zero plots and the ROCS: (a) $X(z)$, (b) $H(z)$, and (c) $Y(z)$

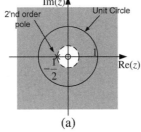

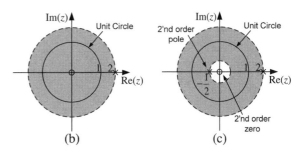

(a) (b) (c)

Example 9.5.6 Using a. the direct and b. the transform methods, find the AC of

$$x[n] = \alpha^n u[n], |\alpha| < 1. \qquad (9.5.18)$$

Solution: a. Since the AC function is even, we need to find $r_x[k], k \geq 0$ and then use $r_x[-k] = r_x[k]$. For $k \geq 0$, since $|\alpha| < 1$, we have

$$r_x[k] = \sum_{n=0}^{\infty} \alpha^n \alpha^{n+k} = \alpha^k \sum_{n=0}^{\infty} (\alpha^2)^n = \frac{1}{1-\alpha^2} \alpha^k, k \geq 0.$$

Autocorrelation is even $\Rightarrow r_x[k]$

$$= \alpha^{|k|}/(1-\alpha^2), -\infty < k < \infty. \qquad (9.5.19a)$$

b. By the autocorrelation theorem,

$$R_x(z) = X(z)X(z^{-1}) = \frac{1}{1-\alpha z^{-1}} \frac{1}{1-\alpha z}$$

$$= -\frac{1}{\alpha} \frac{z}{(z-\alpha)(z-(1/\alpha))}, \qquad (9.5.19b)$$

$$\text{ROC} : |\alpha| < |z| < \frac{1}{|\alpha|}.$$

$$\Rightarrow \alpha^{|k|} \xleftrightarrow{z} \frac{\alpha^2-1}{\alpha} \frac{z}{(z-\alpha)(z-(1/\alpha))}. \qquad (9.5.19c)$$

The region of convergence is an annular ring and the AC sequence is a two-sided sequence. Note the poles of the z-transform in (9.5.19c). ■

9.5.9 Initial Value Theorem in the Discrete Domain

If the sequence $x[n]$ is causal, i.e., $x[n] = 0, n < 0$, as $z \to \infty, z^{-n} \to 0$ for $n > 0$, we have

$$x[0] = \lim_{z \to \infty} X(z) = \lim_{z \to \infty} \sum_{n=0}^{\infty} x[n]z^{-n}$$

$$= \lim_{z \to \infty} [x[0] + x[1]z^{-1} + x[2]z^{-2} + \cdots], \qquad (9.5.20a)$$

$$x[0] = 0 \Rightarrow x[1] = \lim_{z \to \infty} zX(z). \qquad (9.5.20b)$$

9.5.10 Final Value Theorem in the Discrete Domain

The final value theorem applies only to causal sequences $x[n]$ and if all the poles of $X(z)$ lie within the unit circle, with the exception that it can have one pole at $z = 1$, then

$$\lim_{n \to \infty} x[n] = \lim_{z \to 1} (1-z^{-1})X(z) \text{ if } x[\infty] \text{exists.} \qquad (9.5.21)$$

This can be seen by

$$Z\{x[n] - x[n-1]\} = (1-z^{-1})X(z)$$

$$= \lim_{N \to \infty} \sum_{n=0}^{N} \{x[n] - x[n-1]\}z^{-n}, \qquad (9.5.22)$$

$$\lim_{z \to 1} \lim_{N \to \infty} \sum_{n=0}^{N} \{x[n] - x[n-1]\}z^{-n}$$

$$= \lim_{N \to \infty} \lim_{z \to 1} \sum_{n=0}^{N} \{x[n] - x[n-1]\}z^{-n}$$

$$\lim_{N \to \infty} [x[0] - x[-1] + x[1] - x[0] + x[2] - x[1] + \ldots]$$

$$= \lim_{N \to \infty} x[N].$$

Notes: The final value $x[\infty]$ is equal to zero if all the poles of $X(z)$ lie within the unit circle. This follows from the fact that the corresponding time function contains exponentially damped terms. It is a constant if $X(z)$ has a single pole at $z = 1$. If the poles are outside of the unit circle, then the final value theorem gives incorrect results. Also, $x[\infty]$ is indeterminate if there are complex poles on the unit circle. ■

Example 9.5.7 Find the initial and final values of $x[n]$ for the function

$$X(z) = \frac{(z-(1/3))}{(z-1)(z-(1/2))}, |z| > 1.$$

Solution: The initial and final values are, respectively, given by

$$x[0] = \lim_{z \to \infty} X(z) = \lim_{z \to \infty} \frac{(z-(1/3))}{(z-1)(z-(1/2))} = 0$$

$$\lim_{n \to \infty} x[n] = \lim_{z \to 1} (1-z^{-1})X(z)$$

$$= \lim_{z \to 1} \frac{(z-1)(z-(1/3))}{z(z-1)(z-(1/2))} = \frac{4}{3}.$$ ■

Switched periodic sequences and their z-transforms: Consider the periodic sequence $x[n]$ with the property $x[n] = x[n + N]$. Now define a causal sequence $y[n] = x[n]u[n]$. Let the z-transform of the first N-point sequence is

$$X_1(z) = \sum_{n=0}^{N-1} x[n]z^{-n}.$$

The z-transform of the function $y[n]$ is given by

$$Y(z) = X_1(z) + z^{-N}X_1(z) + z^{-2N}X_1(z) + \dots$$

$$= X_1(z)[1 + z^{-N} + z^{-2N} + \dots]$$

$$= \frac{z^N}{z^N - 1} X_1(z), \quad |z| > 1.$$

Example 9.5.8 Find the z-transform of the sequence $y[n] = x[n]u[n]$, $x[n] = \sin(n\pi/2)$.

$$y[n] = x[n]u[n] = \{0, 1, 0, -1, 0, 1, 0, -1, \dots\}.$$
$$\uparrow$$

Solution: The period of the sequence $\sin(n\pi/2)$ is 4. Therefore

$$x_1[n] = \{0, 1, 0, -1\} \overset{z}{\longleftrightarrow} z^{-1} - z^{-3} = X_1(z)$$
$$\uparrow$$

$$\Rightarrow X(z) = \frac{X_1(z)}{1 - z^{-4}} = \frac{z^{-1} - z^{-3}}{1 - z^{-4}} = \frac{z}{z^2 + 1}.$$

∎

9.6 Tables of z-Transform Properties and Pairs

Table 9.6.1 Z-transform properties

Two sided signals $x[n] \overset{z}{\longleftrightarrow} X(z), h[n] \overset{z}{\longleftrightarrow} H(z)$	
Superposition:	
$\quad ax[n] + bh[n] \overset{z}{\longleftrightarrow} aX(z) + bH(z)$	(9.6.1)
Time shift:	
$\quad x[n - n_0] \overset{z}{\longleftrightarrow} z^{-n_0} X(z)$	(9.6.2)
Scaling:	
$\quad a^n x[n] \overset{z}{\longleftrightarrow} X(z/a)$	(9.6.3)
Multiplication by $e^{jn\Omega_0}$:	
$\quad e^{jn\Omega_0} x[n] \overset{z}{\longleftrightarrow} X(e^{-j\Omega_0} z)$	(9.6.4)
Time reversal:	
$\quad x[-n] \overset{z}{\longleftrightarrow} X(1/z)$	(9.6.5)
Multiplication by n:	
$\quad nx[n] \overset{z}{\longleftrightarrow} -z\frac{dX(z)}{dz}$	(9.6.6)
Accumulation:	
$\quad \sum_{k=-\infty}^{n} x[k] \overset{z}{\longleftrightarrow} \frac{z}{z-1} X(z)$	(9.6.7)
Difference:	
$\quad x[n] - x[n-1] \overset{z}{\longleftrightarrow} (1 - z^{-1})X(z)$	(9.6.8)
Convolution:	
$\quad x[n] * h[n] \overset{z}{\longleftrightarrow} X(z)H(z)$	(9.6.9)
Cross correlation:	
$\quad x[k] * *h[k] \overset{z}{\longleftrightarrow} X(z)H(1/z)$	(9.6.10a)
Autocorrelation:	
$\quad x[n] * *x[n] \overset{z}{\longleftrightarrow} X(z)X(1/z)$	(9.6.10b)
The following properties hold for causal sequences $x[n] = 0, n < 0$.	
Initial value theorem:	
$\quad x[0] = \lim_{z \to \infty} X(z)$	(9.6.11a)
Final value theorem:	
$\quad \lim_{n \to \infty} x[n] = \lim_{z \to 1}(z - 1)X(z)$	(9.6.11b)
Switched periodic functions:	
$\quad x[n] = x[n + N], X_1(z) = \sum_{n=0}^{N-1} x[n]z^{-n}, x[n]u[n] \overset{z}{\longleftrightarrow} \frac{z^N}{z^N-1} X_1(z)$	(9.6.12)

Table 9.6.2 Z-transform pairs

Unit sample:

$$\delta[n] \overset{z}{\longleftrightarrow} 1, \quad \text{ROC : all } z. \tag{9.6.13a}$$

$$\delta[n-k] \overset{z}{\longleftrightarrow} z^{-k}; \ k > 0; \ \text{ROC : } |z| > 0. \tag{9.6.13b}$$

$$\delta[n+k] \overset{z}{\longleftrightarrow} z^{k}; \ k > 0; \ \text{ROC : } |z| < \infty. \tag{9.6.13c}$$

Unit step:

$$u[n] \overset{z}{\longleftrightarrow} \frac{z}{z-1}, \quad \text{ROC : } |z| > 1. \tag{9.6.14a}$$

$$-u[-n-1] \overset{z}{\longleftrightarrow} \frac{z}{z-1}, \quad \text{ROC : } |z| < 1. \tag{9.6.14b}$$

Exponential:

$$a^n u[n] \overset{z}{\longleftrightarrow} \frac{z}{z-a}, \quad \text{ROC : } |z| > |a|. \tag{9.6.15a}$$

$$-b^n u[-n-1] \overset{z}{\longleftrightarrow} \frac{z}{z-b}, \quad \text{ROC : } |z| < |b|. \tag{9.6.15b}$$

General type:

$$n a^n u[n] \overset{z}{\longleftrightarrow} \frac{az}{(z-a)^2}, \quad \text{ROC : } |z| > |a|. \tag{9.6.16a}$$

$$\frac{n(n-1)...(n-(k-2))a^{n-k+1}u[n]}{(k-1)!} \overset{z}{\longleftrightarrow} \frac{z}{(z-a)^k}, \quad \text{ROC :} |z| > |a|. \tag{9.6.16b}$$

$$-n a^n u[-n-1] \overset{z}{\longleftrightarrow} \frac{az}{(z-a)^2}, \quad \text{ROC : } |z| < |a|. \tag{9.6.16c}$$

$$(n+1)a^n u[n] \overset{z}{\longleftrightarrow} \frac{z^2}{(z-a)^2}, \quad \text{ROC : } |z| > a. \tag{9.6.16d}$$

Sequences involving sinusoids:

$$\cos(\Omega_0 n)u[n] \overset{z}{\longleftrightarrow} \frac{z^2 - \cos(\Omega_0)z}{z^2 - (2\cos(\Omega_0))z + 1}, \quad \text{ROC : } |z| > 1. \tag{9.6.17a}$$

$$\sin(\Omega_0 n)u[n] \overset{z}{\longleftrightarrow} \frac{\sin(\Omega_0)z}{z^2 - (2\cos(\Omega_0))z + 1}, \quad \text{ROC : } |z| > 1. \tag{9.6.17b}$$

$$r^n \cos(\Omega_0 n)u[n] \overset{z}{\longleftrightarrow} \frac{z^2 - r\cos(\Omega_0)z}{z^2 - (2r\cos(\Omega_0))z + r^2}, \quad \text{ROC : } |z| > r. \tag{9.6.17c}$$

$$r^n \sin(\Omega_0 n)u[n] \overset{z}{\longleftrightarrow} \frac{r\sin(\Omega_0)z}{z^2 - (2r\cos(\Omega_0))z + r^2}, \quad \text{ROC : } |z| > r. \tag{9.6.17d}$$

Finite sequence:

$$\left\{ \begin{array}{l} a^n, 0 \leq n \leq N-1 \\ 0, \ \text{otherwise} \end{array} \right\} \overset{z}{\longleftrightarrow} \frac{1 - a^N z^{-N}}{1 - az^{-1}}, \quad \text{ROC : } |z| > 0. \tag{9.6.18}$$

9.7 Inverse z-Transforms

In this section we will consider determining $x[n] = Z^{-1}\{X(z)\}$ by the following methods: (1) inversion formula, (2) use of z-transform tables, and (3) power series expansion.

9.7.1 Inversion Formula

The inverse z-transform is

$$x[n] = \frac{1}{2\pi j} \oint_c X(z)z^{n-1}dz. \tag{9.7.1}$$

It is a contour integral over a closed path C encircling the origin in a counterclockwise direction that lies within the region of convergence of $X(z)$ in the z-plane. The proof requires the knowledge of complex variables, which is beyond the scope here. See Churchill (1948), Poularikas (1996), and others. Let $\{a_k\}$ be the set of poles of $X(z)z^{n-1}$ inside the contour C and $\{b_k\}$ be the set of poles of $X(z)z^{n-1}$ outside the contour C in a finite region of the z-plane. Now

$$x[n] = Z^{-1}[X(z)]$$

$$= \begin{cases} \sum_k \text{Res}\{X(z)z^{n-1}, a_k), n \geq 0, & (9.7.2a) \\ \sum_k \text{Res}\{X(z)z^{n-1}, b_k\}, n < 0. & (9.7.2b) \end{cases}$$

The residue at the *multiple* and the *simple poles* at $z = p_0$ of order k are, respectively, given by

$$\text{Res}\{X(z)z^{n-1}\} = \lim_{z \to p_0} \frac{1}{(k-1)!} \frac{d^{k-1}[(z-p_0)^k X(z)z^{n-1}]}{dz^{k-1}}$$

(Multiple pole), 		(9.7.2c)

$$\text{Res}\{X(z)z^{n-1}, p_0\} = X(z)z^{n-1}(z-p_0)\big|_{z=p_0}$$

(Simple pole). 		(9.7.2d)

Example 9.7.1 Find the inverse z-transform of the following function using the residues

$$X(z) = \frac{z}{(z-.5)(z-2)},$$

$$\text{ROC}: .5 < |z| < 2. \text{ See Fig 9.7.1} \qquad (9.7.3a)$$

Solution: The function has a single pole inside and a single pole outside the unit circle. Therefore, it is a two-sided sequence. From (9.7.2c), we have

$$x[n] = \text{Res}\{X(z)z^{n-1}, .5\}$$

$$= \frac{z(z-.5)z^{n-1}}{(z-.5)(z-2)}\big|_{z=.5} = -\frac{(.5)^n}{1.5}, n \geq 0,$$

$$(9.7.3b)$$

$$x[n] = -\text{Res}\{X(z)z^{n-1}, 2\}$$

$$= -\frac{z(z-2)z^{n-1}}{(z-.5)(z-2)}\big|_{z=2} = -\frac{2^n}{1.5}, n < 0.$$

$$(9.7.3c) \blacksquare$$

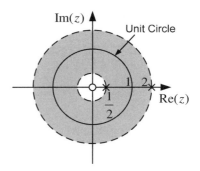

Fig. 9.7.1 Example 9.7.1: Poles, zeros, and the ROC

9.7.2 Use of Transform Tables (Partial Fraction Expansion Method)

This method is based upon expressing $X(z)$ as a sum of simple functions $X_i(z)$ by using partial fraction expansion (see Section 5.9.2.), where each one of these functions have inverse transforms that are readily available in a table. This method is limited to rational functions and will need a table of z-transform pairs that provides the appropriate transform pairs $x_i[n] \overset{z}{\longleftrightarrow} X_i(z)$. The inverse transform is given by

$$x[n] = Z^{-1}\{X(z)\} = Z^{-1}\left\{\sum_{i=1}^{M} X_i(z)\right\}$$

$$= \sum_{i=1}^{M} Z^{-1}\{X_i(z)\} = \sum_{i=1}^{M} x_i[n]. \qquad (9.7.4)$$

We considered partial fraction expansions when we studied Laplace transforms and the procedure here is the same with a slight modification. This approach provides *closed-form* solutions. From Table 9.6.2 we see that z appears in the numerator of the z-transform functions. Therefore, find $X(z)/z$, and then use the partial fraction expansion discussed in Section 5.4. The expansion of a rational function $X(z)$ can be obtained by multiplying each term in the expansion by z. The function $X(z)/z$ takes the form

$$\frac{X(z)}{z} = \frac{a_0 + a_1 z + \cdots + a_M z^M}{b_0 + b_1 z + \cdots + b_N z^N}. \qquad (9.7.5)$$

If $M > N$, then divide the numerator polynomial by the denominator polynomial and

$$[X(z)/z] = R(z) + X_0(z), R(z)$$

$$= [c_{M-N}z^{M-N} + c_{M-N-1}z^{M-N-1} + \cdots + c_1 z + c_0]$$

$$X_0(z) = \frac{d_0 + d_1 z + \cdots + d_{N-1}z^{N-1}}{b_0 + b_1 z + \cdots + b_N z^N}. \quad (9.7.6)$$

Note that the numerator polynomial in the rational function $X_0(z)$ in (9.7.6) is of degree $(N-1)$ or less. The denominator of this rational function is then factored and expanded using partial fraction expansion. The inverse z-transform of $X(z)$ can now be computed:

$$Z^{-1}[X(z)] = Z^{-1}[zR(z)]$$

$$+ Z^{-1}[z\{\text{partial fraction expansion of } X_0(z)\}]$$

$$= x_1[n] + x_2[n], \quad (9.7.7)$$

$$lZ^{-1}[c_0 z] = c_0\delta(n+1), Z^{-1}[c_1 z^2] = c_1\delta[n+2], \ldots,$$

$$Z^{-1}[c_k z^{k+1}] = c_k\delta[n+k+1], \ldots, x_1[n] = Z^{-1}[zR(z)]$$

$$= \sum_{m=N}^{M} c_{M-m}\delta[n + (M-m+1)]. \quad (9.7.8)$$

In the following we will concentrate on the second part in (9.7.7) by considering all *simple poles* and later we will consider a single multiple pole plus simple poles.

Example 9.7.2 Find the inverse z-transform of the function

$$X(z) = \frac{z[2z - (4/3)]}{[z^2 - (4/3)z + (1/3)]}, \text{ROC} : |z| > 1. \quad (9.7.9)$$

Solution: From the ROC the sequence is a right-side sequence. Now

$$\frac{X(z)}{z} = \frac{1}{z - (1/3)} + \frac{1}{z - 1}$$

$$\Rightarrow X(z) = \frac{z}{z - (1/3)} + \frac{z}{z - 1}, \text{ROC} : |z| > 1.$$

$$\frac{X(z)}{z} = \frac{1}{z - (1/3)} + \frac{1}{z-1} \Rightarrow X(z) = \frac{z}{z - (1/3)} + \frac{z}{z-1},$$

$$\text{ROC} : |z| > 1 \Rightarrow x[n] = (1/3)^n u[n] + u[n] \quad ∎$$

Multiple pole case: See Section 5.8 on the partial fraction expansion with multiple poles.

Example 9.7.3 Find $x[n] = Z^{-1}\{X(z)\}$ for the cases: $a. |z| > 1$ and $b. |z| < (1/2)$:

$$X(z) = \frac{3z^3 - (5/2)z^2}{(z - (1/2))^2 (z-1)}, \text{ROC} : |z| > 1. \quad (9.7.10)$$

Solution:

a. The sequence is a right-side sequence since the ROC is $|z| > 1$. We have a double pole at $z = (1/2)$ and a single pole at $z = 1$:

$$X_0(z) = \frac{X(z)}{z} = \frac{(3z^2 - (5/2)z)}{(z - (1/2))^2 (z-1)}$$

$$= \frac{A_{12}}{(z - (1/2))^2} + \frac{A_{11}}{(z - (1/2))} + \frac{A_3}{(z-1)}, \quad (9.7.11)$$

$$A_{12} = (z - (1/2))^2 X_0(z)\big|_{z=1/2}$$

$$= \frac{3z^2 - (5/2)z}{(z-1)}\big|_{z=1/2} = 1,$$

$$A_{11} = \frac{d}{dz}\left[\frac{3z^2 - (5/2)z}{(z-1)}\right]\big|_{z=1/2}$$

$$= \left[\frac{(z-1)(6z - (5/2)) - (3z^2 - (5/2)z)}{(z-1)^2}\right]\big|_{z=1/2} = 1,$$

$$A_3 = \frac{3z^2 - (5/2)z}{(z - (1/2))^2}\big|_{z=1} = 2,$$

$$X_0(z) = \frac{(3z^2 - (5/2)z)}{(z - (1/2))^2 (z-1)}$$

$$= \frac{1}{(z - (1/2))^2} + \frac{1}{(z - (1/2))} + \frac{2}{z-1} \quad (9.7.12)$$

$$\Rightarrow X(z) = \frac{z}{(z - (1/2))^2} + \frac{z}{z - (1/2)} + \frac{2z}{z-1}. \quad (9.7.13)$$

The last step involves determining the inverse transforms. In doing so, the ROC of the z-domain function should be kept in mind. From Table 9.6.2,

$$\frac{z}{(z-a)} \overset{z}{\longleftrightarrow} a^n u[n], \frac{z}{(z-a)^2} \overset{z}{\longleftrightarrow} na^{n-1} u[n],$$

$$\frac{z}{(z-a)^3} \overset{z}{\longleftrightarrow} \frac{1}{2!} n(n-1)a^{n-2} u[n], \dots. \quad (9.7.14)$$

The ROC is outside of the unit circle and the inverse transform is

$$x[n] = [n(1/2)^{n-1} + (1/2)^n + 2]u[n], \quad (9.7.15)$$

$$x[0] = 3, x[1] = 1 + (1/2) + 2 = 7/2, X[2]$$
$$= 1 + (1/4) + 2 = 13/4, \dots \quad (9.7.16)$$

b. Now use the transform pairs corresponding to the left-side sequences given below:

$$-u[-n-1] \overset{z}{\longleftrightarrow} \frac{z}{z-1}, |z| < 1,$$

$$-a^n u[-n-1] \overset{z}{\longleftrightarrow} \frac{z}{z-a}, \quad (9.7.17)$$

$$-na^n u[-n-1] \overset{z}{\longleftrightarrow} \frac{az}{(z-a)^2}, |z| < |a|. \quad (9.7.18)$$

$$\Rightarrow X(z) = \frac{z}{(z-(1/2))^2} + \frac{z}{z-(1/2)} + \frac{2z}{z-1}. \quad (9.7.19)$$

$$x[n] = -(2)(n(1/2)^n)u[-n-1] - (1/2)^n \quad (9.7.20a)$$
$$u[-n-1] - 2u[-n-1].$$

$$x[0] = 0, x[-1]$$
$$= 4 - 2 - 2 = 0, x[-2]$$
$$= 16 - 4 - 2 = 10, x[-3]$$
$$= 6(8) - 8 - 2 = 38, \dots, \quad (9.7.20b)$$

$$x[n] = \{\dots, 38, 10, 0, 0, \dots\}. \quad (9.7.21)$$
$$\uparrow$$

∎

Notes: It is uncommon to come across multiple poles of order more than 2. It is simple to use the repeated application of simple pole case, see Example 5.8.2. ∎

Example 9.7.4 Find the sequence $x[n] = z^{-1}\{X(z)\}$:

$$X(z) = \frac{z(z+1)}{(z+0.5)(z-2)(z-0.75)}.$$

Solution: Using the partial fraction expansion, we can write

$$X_0(z) = \frac{X(z)}{z} = \frac{(z+1)}{(z+0.5)(z-2)(z-0.75)} \frac{A_1}{(z+0.5)}$$
$$+ \frac{A_2}{(z-2)} + \frac{A_3}{(z-0.75)},$$

$$A_1 = \frac{(z+1)}{(z-2)(z-0.75)}\bigg|_{z=-0.5} = \frac{4}{25},$$

$$A_2 = \frac{(z+1)}{(z+0.5)(z-0.75)}\bigg|_{z=2} = \frac{24}{25}, \text{ and}$$

$$A_3 = \frac{(z+1)}{(z+0.5)(z-2)}\bigg|_{z=0.75} = -\frac{28}{25}$$

$$\Rightarrow X(z) = \frac{4}{25} \cdot \frac{z}{(z+0.5)} + \frac{24}{25} \cdot \frac{z}{(z-2)}$$
$$- \frac{28}{25} \cdot \frac{z}{(z-0.75)}.$$

Since the ROC is not specified, the sequence cannot be uniquely determined from the $X(z)$ alone. Therefore we will identify all possible ROCs corresponding to this function and find the sequences associated with each of them. To find the various possible ROCs, we first make a pole–zero plot as shown in Fig. 9.7.2a. Using the properties of the ROC, the different possible ROCs are shown in Fig. 9.7.2b–e.

1. ROC : $|z| < .5$: ROC extends inward to include the origin and $x[n]$ is a left-sided sequence:

$$x[n] = -\frac{4}{25} \cdot \left(-\frac{1}{2}\right)^n u[-n-1] - \frac{24}{25} \cdot (2)^n$$
$$u[-n-1] + \frac{28}{25} \cdot \left(\frac{3}{4}\right)^n u[-n-1].$$

2. ROC : $(1/2) < |z| < (3/4)$: ROC is a ring and $x[n]$ is two sided:

$$x[n] = \frac{4}{25} \cdot \left(-\frac{1}{2}\right)^n u[n] - \frac{24}{25} \cdot (2)^n$$
$$u[-n-1] + \frac{28}{25} \cdot \left(\frac{3}{4}\right)^n u[-n-1].$$

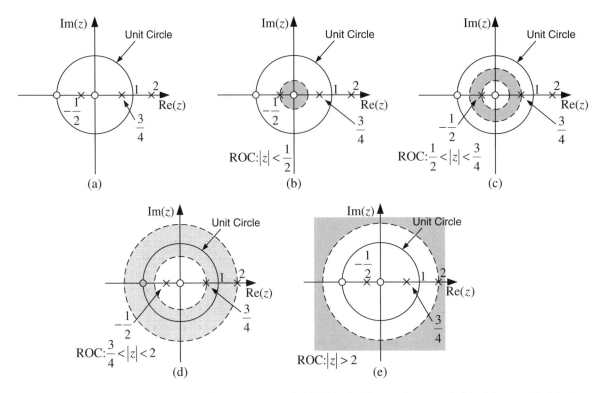

Fig. 9.7.2 (a) Pole–zero plot, (b) ROC : $|z| < .5$, (c) ROC : $(1/2) < |z| < (3/4)$, and (d) ROC : $(3/4) < |z| < 2$, ROC : $|z| > 2$

3. ROC : $(3/4) < |z| < 2$: ROC is a ring and $x[n]$ sequence is two sided:

$$x[n] = \frac{4}{25} \cdot \left(-\frac{1}{2}\right)^n u[n] + \frac{24}{25} \cdot (2)^n$$

$$u[n] + \frac{28}{25} \cdot \left(\frac{3}{4}\right)^n u[-n-1].$$

4. ROC : $|z| > 2$: $x[n]$ is a right-sided sequence:

$$x[n] = \frac{4}{25} \cdot \left(-\frac{1}{2}\right)^n u[n] + \frac{24}{25} \cdot (2)^n$$

$$u[n] - \frac{28}{25} \cdot \left(\frac{3}{4}\right)^n u[n]. \qquad \blacksquare$$

9.7.3 Inverse z-Transforms by Power Series Expansion

From the definition of the z-transform of a sequence $x[n]$, we can write

$$X(z) = \sum_{n=-\infty}^{\infty} x[n]z^{-n} = \cdots + x[-2]z^2 + x[-1]z^1$$

$$+ x[0] + x[1]z^{-1} + x[2]z^{-2} + \cdots. \qquad (9.7.22)$$

If $X(z)$ can be expanded in a power series, $x[n]$ can be determined for positive n(negative n) by identifying the coefficients for the negative powers of z (positive powers of z).

Example 9.7.5 Find the inverse z-transform using power series for the function in (9.7.10) assuming two cases of ROC identified by $a. \ |z| > 1, b. \ |z| < (1/2)$.

Solution: *a.* Since the ROC is $|z| > 1$, the sequence is a right-side sequence and therefore the power series of the function contain *negative* powers of z. Divide the numerator by the denominator in (9.6.23) by division. It can be written by

$$X(z) = 3 + \frac{(7/2)z^2 - (15/4)z + (3/4)}{z^3 - 2z^2 + (5/4)z - (1/4)}$$

$$= 3 + \frac{7}{2}z^{-1} + \frac{(13/4)z - (29/8) + (7/8)z^{-1}}{z^3 - 2z^2 + (5/4)z - (1/4)}$$

$$\Rightarrow x[0] = 3, x[1] = 7/2, x[2] = 13/4, \ldots$$

$$(9.7.23)$$

b. Since the ROC is inside the circle of radius $(1/2)$, the sequence is a left-side sequence. It can be obtained by expanding the function in terms of positive powers. This is achieved by first writing the polynomials in the numerator and the denominator in *reverse* order and then making use of long division to obtain series in terms of positive powers of z:

$$X(z) = \frac{-(5/2)z^2 + 3z^3}{-(1/4) + (5/4)z - 2z^2 + z^3}$$

$$= 10z^2 + \frac{-(38/4)z^3 + 20z^4 - 10z^3}{-(1/4) + (5/4)z - 2z^2 + z^3}$$

$$\Rightarrow x[-1] = 0, x[-2] = 10, x[-3] = 38, \ldots.$$

$$(9.7.24)$$

∎

Notes: If the ROC is in an annular region, we can separate the z-transform corresponding to the right-side sequence and the left-side sequence and follow the above procedure.   ∎

Example 9.7.6 Find the inverse z-transform of the following function:

$$X(z) = \log\left[\frac{1}{1 - az^{-1}}\right], |z| > |a|. \qquad (9.7.25)$$

Solution: Using the power series expansion Spiegel (1968), we have

$$X(z) = -\log(1 - az^{-1}) = \sum_{n=1}^{\infty} \frac{1}{n}(az^{-1})^n$$

$$\Rightarrow x[n] = \begin{cases} (1/n)a^n, & n \geq 1 \\ 0, & n \leq 0 \end{cases}. \qquad (9.7.26)$$

The ROC is outside the circle of radius $|a|$ and $x[n]$ is a right-side sequence.   ∎

9.8 The Unilateral or the One-Sided z-Transform

The unilateral transform is useful since most of our sequences are right sided. The causal part of an arbitrary sequence $y[n]$ is $y[n]u[n]$. The unilateral or one-sided transform is

$$X_I(z) = Z_I\{x[n]\} = \sum_{n=0}^{\infty} x[n]z^{-n}. \qquad (9.8.1)$$

An important property of this transform is its ROC and is outside of a circle in the z-plane.

9.8.1 Time-Shifting Property

Consider the transform pair $x[n] \xleftrightarrow{z} X_I(z)$. The transforms of the delayed and advanced sequences are given below and can be shown by starting with the definition of the unilateral transform of these sequences and reducing them into the appropriate forms:

a. $x[n - m] \xleftrightarrow{z} z^{-m}X_I(z) + z^{-m+1}x[-1]$

$$\qquad + z^{-m+2}x[-2] + \cdots + x[-m], \ m > 0, \qquad (9.8.2)$$

b. $x[n + m] \xleftrightarrow{z} z^m X_I(z) - z^m x[0]$

$$\qquad - z^{m-1}x[1] - \cdots - zx[m - 1], \ m > 0. \qquad (9.8.3)$$

a. First consider (9.8.2). The one-sided transform of the delayed sequence is given by

$$Z_I\{x[n - m]\} = \sum_{n=0}^{\infty} x[n - m]z^{-n} = \sum_{k=-m}^{\infty} x[k]z^{-(m+k)}. \qquad (9.8.4)$$

Separating (9.8.4) into two parts $x[n], n < 0$ and $x[n]$ for $n \geq 0$, we have

$$\sum_{k=-m}^{\infty} x[k]z^{-(m+k)} = \{z^{-m+1}x[-1] + z^{-m+2}x[-2]$$

$$+ \cdots + x[-m]\} + z^{-m}\sum_{k=0}^{\infty} x[k]z^{-k}$$

$$= z^{-m}\{x[-1]z + x[-2]z^2$$

$$+ \ldots + x[-m]z^m\}$$

$$+ z^{-m}X_I(z), m > 0. \qquad (9.8.5)$$

b. Now consider the one-sided transform of the advanced sequence

$$Z_I\{x[n+m]\} = \sum_{n=0}^{\infty} x[n+m]z^{-n}$$

$$= \sum_{k=0}^{\infty} x[k]z^{-(k-m)} - \sum_{k=0}^{m-1} x[k]z^{-(k-m)}$$

$$(9.8.6a)$$

$$= -\{z^m x[0] + x[1]z^{m-1} + \cdots$$
$$+ x[m-1]z^1\} + z^m X_I[z].$$
$$(9.8.6b)$$

The relations in (9.8.2) and (9.8.3) provide a method to solve constant coefficient difference equations. This generally involves two discrete functions, an output $y[n]$ and an input $x[n]$. The procedure parallels that of solving constant coefficient differential equations and the Laplace transform. In the following it is assumed that one-sided transforms are in use and the subscript (I) will not be shown on X explicitly. The procedure involves first finding the z-transform of the difference equation in terms of the two transforms $Y(z)$ and $X(z)$. In determining $Y(z)$, the initial conditions on $y[n]$ need to be known. The input $x[n]$ and therefore $X(z)$ is assumed to be known. Solve for $Y(z)$ and then take the inverse transform of this function resulting in $y[n]$. Two simple examples are considered below, one with zero input, but with initial conditions, and the second one has both input and initial conditions.

Example 9.8.1 Consider the second-order difference equation given by

$$y[n] = y[n-1] + y[n-2], y[-2] = -1, y[-1] = 1.$$
$$(9.8.7)$$

Determine $y[n]$ for several values of n by a. using the equation in (9.8.7) directly and then b. verify this result using the one-sided z-transform. Cadzow (1973) uses (9.8.7) to generate a model for rabbit population.

Solution: a. By the direct method, we have

$$y[0] = 0, y[1] = 1, y[2] = 1,$$
$$y[3] = 2, y[4] = 3, y[5] = 5,$$
$$y[6] = 8, y[7] = 13, \ldots.$$

Note $y[n]$ at location n is the sum of the values at the two previous locations. The sequence results in Fibonacci numbers, see Hershey and Yarlagadda (1986).

b. Taking the one-sided z-transform of the equation in (9.8.7) results in

$$Y_I[z] = Z\{y[n-1]\} + Z\{y[n-2]\}$$
$$= z^{-1}Y_I(z) + y[-1] + z^{-2}Y_I(z)$$
$$+ z^{-1}y[-1] + y[-2]$$
$$= z^{-1}Y_I(z) + 1 + z^{-2}Y_I(z) + z^{-1} - 1 \quad (9.8.8)$$

$$\Rightarrow Y_I(z) = \frac{z^{-1}}{1 - z^{-1} - z^{-2}} = \frac{z}{z^2 - z - 1},$$
$$\text{Poles}: p_1 = (1/2) + (\sqrt{5}/2), p_2 = (1/2) - (\sqrt{5}/2).$$
$$(9.8.9)$$

The region of convergence is $|z| > \frac{1}{2}|1 - j\sqrt{5}|$. The partial fraction expansion is

$$\frac{Y_I(z)}{z} = \frac{1}{z^2 - z - 1} = \frac{A}{z - z_1} + \frac{B}{z - z_2}$$
$$= \frac{A}{z - ((1 - \sqrt{5})/2)} + \frac{B}{z - ((1 + \sqrt{5})/2)},$$
$$A = \frac{1}{\sqrt{5}}, B = -\frac{1}{\sqrt{5}}$$

$$\Rightarrow Y_I(z) = \frac{(1/\sqrt{5})z}{z - ((1 + \sqrt{5})/2)} + \frac{-(1/\sqrt{5})z}{z - ((1 - \sqrt{5})/2)}$$

$$\Rightarrow y[n] = \left\{ \frac{1}{\sqrt{5}} \left[\frac{1 + \sqrt{5}}{2} \right]^n - \frac{1}{\sqrt{5}} \left[\frac{1 - \sqrt{5}}{2} \right]^n \right\} u[n],$$
$$(9.8.10)$$

$$y[n] \approx .4472(1.618)^n, n \gg 1, \quad (9.8.11)$$

$$y[0] = \frac{1}{\sqrt{5}}[1 - 1] = 0, y[1]$$

$$= \frac{1}{\sqrt{5}} \left[\left(\frac{1}{2} + \frac{\sqrt{5}}{2} \right) - \left(\frac{1}{2} - \frac{\sqrt{5}}{2} \right) \right] = 1. \quad \blacksquare$$

Example 9.8.2 Determine $y[n] = x[n] * h[n]$, $x[n] = \{\underset{\uparrow}{1}, 1, 1\}$, and $h[n] = \{\underset{\uparrow}{1}, -1, 1\}$.

Solution: The z-transforms of the two finite length sequences $x[n]$ and $h[n]$ are

$$X(z) = 1 + z^{-1} + z^2, H(z) = 1 - z^{-1} + z^{-2}, \quad (9.8.12a)$$

$$Y(z) = H(z)X(z) = 1 + z^{-2} + z^{-4}$$
$$\Rightarrow y[n] = \{1, 0, 1, 0, 1\}. \qquad (9.8.12b) \quad \blacksquare$$

In the convolution of two sequences $x[n]$ and $h[n]$ are known and $y[n] = x[n] * h[n]$ needs to be found. In Section 9.3.4, the deconvolution problem was identified and three methods were discussed. In the deconvolution problem, $y[n]$ and $x[n]$ are assumed to be known and $h[n]$ is to be determined. Such a problem has practical importance, as it is a system identification problem. That is, determine the unit sample response of a system $h[n]$. In the method of *deconvolution using polynomial long division*, $H(z)$ is obtained $Y(z)/X(z)$. This is illustrated in the following example.

Example 9.8.3 Determine $h[n]$ using (9.8.12b).

Solution:

$$H(z) = \frac{Y(z)}{X(z)} = \frac{1 + z^{-2} + z^{-4}}{1 - z^{-1} + z^{-2}} = 1 + \frac{-z^{-1} + z^{-4}}{1 + z^{-1} + z^{-2}}$$

$$= 1 - z^{-1} + \frac{z^{-2} + z^{-3} + z^{-4}}{1 + z^{-1} + z^{-2}} = 1 - z^{-1} + z^{-2}$$

$$\Rightarrow h[n] = \{1, -1, 1\} \qquad \blacksquare$$

9.9 Discrete-Data Systems

In this section, basic concepts associated with discrete-time systems will be discussed. Presentation will be very similar to the continuous-time systems studied in Chapter 6. Our discussion will be brief. A discrete-time system is represented by a block diagram shown in Fig. 9.9.1 mapping $x[n]$ into $y[n]$. The T inside the block diagram is some transformation that converts the input data into output data. We can characterize a discrete-time data system by putting constraints on the transformation T:

$$y[n] = T\{x[n]\}. \qquad (9.9.1)$$

$$x[n] \rightarrow \boxed{T\{.\}} \rightarrow y[n].$$

Fig. 9.9.1 A discrete-data system

Principles of additivity and proportionality: A system is said to be *additive* if $T\{x_1[n] + x_2[n]\} = T\{x_1[n]\} + T\{x_2[n]\}$. This is sometimes referred to as the superposition property. A system is homogeneous if it satisfies the principle of proportionality, $y[n] = T\{\alpha x[n]\} = \alpha\{x[n]\}$ for a constant α and for any input sequence.

Linear systems: A system that is both additive and homogeneous is called a linear system. A system is linear if for any inputs $x_i[n]$ and for any constants $\alpha_i, i = 1, 2$,

$$T\{\alpha_1 x_1[n] + \alpha_2 x_2[n]\}$$
$$= \alpha_1 T\{x_1[n]\} + \alpha_2 T\{x_2[n]\}. \qquad (9.9.2)$$

Example 9.9.1 Consider the systems described by the following transformations. In each case, determine whether the corresponding system is linear or nonlinear:

a. $y_1[n] = Ax[n] + B, B \neq 0,$

b. $y_2[n] = x^2[n],$

c. $y_3[n] = nx[n].$

Solution: Using (9.9.2), it follows that

$$\alpha_1 T\{x_1[n]\} = \alpha_1 Ax_1[n] + \alpha_1 B,$$
$$\alpha_2 T\{x_2[n]\} = \alpha_2 Ax_2[n] + \alpha_2 B,$$

$$T\{\alpha_1 x_1[n] + \alpha_2 x_2[n]\} = \alpha_1 Ax_1[n] + \alpha_2 Ax_2[n] + B$$
$$\neq \alpha_1 T\{x_1[n]\} + \alpha_2 T\{x_2[n]\}.$$

a. This indicates that the system is nonlinear. It is a linear system if $B = 0$.

b. It is easy to see that the system is nonlinear. Any time, if the transformation has a power of the input other than one, the system is nonlinear.

c. The outputs corresponding to the inputs $x_1(t)$, $x_2(t)$ and $\alpha_1 x_1(t) + \alpha_2 x_2(t)$ are

$$\alpha_i T\{x_i[n]\} = nx_i[n], i = 1, 2 \text{ and } T\{\alpha_1 x_1(n) + \alpha_2 x_2(n)\} = n\{\alpha_1 x_1[n] + \alpha_2 x_2[n]\} = \alpha_1 T\{x_1[n]\} + \alpha_2 T\{x_2[n]\}$$

$\Rightarrow$ System is linear. $\qquad \blacksquare$

Shift (or time) invariance: A discrete-time system is *shift invariant* if and only if $y[n] = T\{x[n]\}$ implies that for every input signal $x[n]$ and every time shift k,

$$T\{x[n-k]\} = y[n-k] \qquad (9.9.3)$$

Example 9.9.2 Are the following systems shift invariant?

a. $y[n] = x^2[n]$,

b. $y[n] = nx[n]$,

c. $y[n] = x[k_0 n]$, k_0 – a positive integer.

Solution:

a. The response of this system corresponding to the input $x[n - n_0]$ is $y_1[n] = T\{x[n - n_0]\} = x^2[n - n_0] = y[n - n_0]$. Therefore, the system is shift invariant.

b. The response of this system corresponding to the input $x[n - n_0]$ is $y_2[n] = nx[n - n_0] \neq (n - n_0)y[n - n_0]$. Therefore, the system is shift variant.

c. The response of this system corresponding to the input $x[n - n_0]$ is $y_3[n] = x[k_0 n - n_0] \neq x[k_0(n - n_0)]$. Therefore, the system is a shift-variant system and is a compressor. If $k_0 = 1$ then the system is shift invariant. ∎

Linear shift-invariant systems: A linear shift-invariant system (LSI) is *linear* and also *shift invariant*. From Chapter 8, we can write that a discrete-time signal $x[n]$ is

$$x[n] = \sum_{k=-\infty}^{\infty} x[k]\delta[n-k]. \qquad (9.9.4)$$

Modeling a discrete-data system is an important topic of study. In the case of analog systems we have defined the impulse response and we called the transform of the impulse response as the transfer function of the system. In a similar manner we can define the *unit sample (discrete impulse) response or simply impulse response*. To stick with the analog impulse response notation, many authors use impulse response rather than a unit sample response. The unit sample response or the impulse response provides a complete description of a linear shift-invariant

system. We will use the impulse response here. If $h[n]$ is the response of a linear time-invariant system to the input unit sample $\delta[n]$, then the response to the input $\delta[n - k]$ is $h[n - k]$. Using the linearity property, we can write the response of a linear system to an input $x[n]$ given in (9.9.4). The output is

$$y[n] = T\{x[n]\} = T\left[\sum_{k=-\infty}^{\infty} x[k]\delta[n-k]\right]$$

$$= \sum_{k=-\infty}^{\infty} x[k]T\{\delta[n-k]\} = \sum_{k=-\infty}^{\infty} x[k]h[n-k]$$

$$= x[n] * h[k]. \qquad (9.9.5)$$

The output of an LSI system is the convolution of the input sequence with the unit impulse response $h[n]$, which gives a complete characterization of the LSI system.

Causality: A causal signal $x[n]$ is zero for $n < 0$. A system is causal if, for any time n, the response of the system $y[n]$ depends only on the present and the past inputs $x[n], x[n-1], ...$, and does *not* depend on the future inputs $x[n+1], x[n+2], ...$ The response of a causal system satisfies the following, where $f\{\cdot\}$ is an arbitrary function:

$$y[n] = f\{x[n], x[n-1], ...\}. \qquad (9.9.6)$$

If a system does not satisfy this constraint, then the system is non-causal. In real-time processing applications, we cannot predict the future values and therefore non-causal systems are not physically realizable. In the case of off-line processing, i.e., if we have all the values of the signal, then it is possible to design a non-causal system to process the data. Such a situation is common in data processing.

Example 9.9.3 Classify each of the following systems are causal or not;

a. $y[n] = x[n] - x[n-1] + x[n-2]$,

b. $y[n] = x[-n] + x[n+1]$,

c. $y[n] = x[2n]$.

Solution: The system in part *a* is causal, whereas the systems in parts *b* and *c* are non-causal as they require the knowledge of future values. ∎

The systems described by constant coefficient difference equations given below are *linear shift-invariant systems*:

$$y[n] = -\sum_{k=1}^{N} a_k y[n-k] + \sum_{k=0}^{L} b_k x[n-k]. \quad (9.9.7)$$

A system described by (9.9.7) is a *recursive linear system* if at least one of the coefficients a_k is not zero and the output depends on previous values of the output as well as the input. For a *non-recursive linear system*, $a[k] = 0$ and is described by the difference equation,

$$y[n] = \sum_{k=0}^{L} b_k x[n-k]. \quad (9.9.8)$$

Stability: The stability of a discrete system can be defined in terms of the input–output behavior as in the analog case. A system is called *BIBO stable* if every *bounded input* produces a *bounded output*. For linear systems BIBO stability requires that the sample response $h[n]$ must be absolutely summable. We can show this by starting with a bounded input such that $|x[n]| \leq M$ for all n and the convolution sum in (9.9.5). That is,

$$|y[n]| \leq \left| \sum_{k=-\infty}^{\infty} h[k] x[n-k] \right| \leq M \sum_{k=-\infty}^{\infty} |h[k]|. \quad (9.9.9a)$$

For a system to be BIBO stable, i.e., to have a bounded output $|y[n]| < \infty$ to a bounded input, the unit sample response of a shift-invariant system must be absolutely summable:

$$\sum_{k=-\infty}^{\infty} |h[k]| < \infty. \quad (9.9.9b)$$

Example 9.9.4 Show the system described by $y[n] = n|x[n]|, x[n] = Au[n], A > 0$ and finite is not BIBO stable.

Solution: The output $y[n] \to \infty$ as $n \to \infty$ and the system is not BIBO stable. ∎

Notes: The response of a discrete-time linear time-invariant system $y[n]$ consists of two parts, one due to the natural response (due to initial conditions) and the second due to the source. A system is said to

be asymptotically stable, if and only if the natural response goes to zero as $n \to \infty$, and is unstable if it grows without bound. We can see this very clearly by making use of the transforms and by expanding the z-domain function into its partial fraction expansion and then taking the inverse terms. The natural response depends upon the characteristic modes of the system. The roots of the characteristic polynomial of the system are the modes. Let a typical root be given by $z = \lambda = |\lambda| e^{j\beta}$. Noting that $\lambda^n = |\lambda|^n e^{j\beta n}$, we can summarize the results using the three cases:

1. $|\lambda| < 1, \lim_{n \to \infty} \lambda^n \to 0,$
2. $|\lambda| > 1, \lim_{n \to \infty} \lambda^n \to \infty,$
3. $|\lambda| = 1, |\lambda|^n \to 1$ for all n. \quad (9.9.10)

The linear discrete-time system is *asymptotically stable* if and only if the characteristic roots, i.e., the poles of the transfer function of the system, are inside the unit circle. It is *unstable* if there is at least one root outside the unit circle and/or if there are multiple roots on the unit circle. It is *marginally stable* if and only if there are no roots outside the unit circle and only simple roots on the unit circle. A marginally stable system is not BIBO stable. Discussion on general stability analysis is beyond the scope here.

Example 9.9.5 Show that the system described by the following equation is stable:

$$y[n+2] + y[n+1] + 2y[n] = x[n+1] + x[n].$$

Solution: The characteristic polynomial can be obtained as follows:

$$z\{y[n+2] + y[n+1] + 2y[n]\} = z^2 + z + 2;$$
$$= (z + z_1)(z + z_2) = 0.$$

$\Rightarrow z_1 = -(1/2) + j\sqrt{7}/2, z_2 = z_1^*, z_1 z_2 = 2$. Since $|z_i| > 1$ the system is unstable. ∎

Classification of LSI systems based on the duration of the impulse response: We can classify the linear shift-invariant (LSI) systems based upon the duration of their (discrete impulse or simply impulse) responses. Without losing any generality, we will consider causal systems. The systems that have *finite-duration impulse response* (FIR) are called FIR systems. On the other hand, the systems with *infinite-duration impulse response* are called *IIR* systems. The system described by (9.9.7) is an IIR system if at least one $a_k \neq 0$, whereas the system described by (9.9.8) is an FIRsystem. As in the analog systems the transform analysis is basic to filter designs. Since the systems need to work with any set of initial conditions, the designs are based on zero initial conditions.

9.9.1 Discrete-Time Transfer Functions

Consider the difference equation in (9.9.7). Assuming the initial conditions are all equal to zero and taking the z-transform of this equation result in

$$Y(z) = -\sum_{k=1}^{N} a_k z^{-k} Y(z) + \sum_{k=0}^{L} b_k z^{-k} X(z), \quad (9.9.11a)$$

$$\Rightarrow Y(z) = \frac{\sum_{k=0}^{L} b_k z^{-k}}{1 + \sum_{k=1}^{N} a_k z^{-k}} X(z) = H(z)X(z),$$

$$H(z) = \frac{Y(z)}{X(z)} = \frac{\sum_{k=0}^{L} b_k z^{-k}}{1 + \sum_{k=1}^{N} a_k z^{-k}}. \quad (9.9.11b)$$

As in the analog case $H(z)$ is called the transfer function and $h[n]$ is the impulse response or the unit sample response. A discrete-time linear time-invariant system can be described by its difference equation, its transfer function, or by its poles and zeros. From our earlier discussion on the z-transforms, we can write the expressions for the system input–output relations in the time domain or in terms of the transform domain. That is,

$$y[n] = h[n] * x[n] \overset{z}{\leftrightarrow} Y(z) = H(z)X(z),$$
$$H(z) = z\{h[n]\}, h[n] = z^{-1}\{H(z)\}. \quad (9.9.11c)$$

Linear time-invariant discrete-time systems are described by either constant coefficient difference equations relating the output to the input or the

impulse response $h[n]$ or the discrete-time transfer function $H(z) \overset{z}{\longleftrightarrow} h[n]$. Since the impulse response and the corresponding transfer function are related, we can discuss the stability of a discrete linear time-invariant system in terms of the poles of the transfer function. The transfer function in (9.9.11b) is

$$H(z) = \frac{\sum_{k=0}^{L} b_k z^{-k}}{1 + \sum_{k=1}^{N} a_k z^{-k}} \quad (9.9.12a)$$

$$= K \frac{(z - z_1)(z - z_2) \cdots (z - z_L)}{(z - p_1)(z - p_2) \cdots (z - p_N)}.$$

Notes: he poles of the transfer function are called the *natural frequencies* or *natural modes* and they determine the time domain behavior of the system response. For example, if we have poles outside the unit circle, the response grows exponentially and the system is unstable. If we have multiple poles on the unit circle, then the response has polynomial growth. If a system has a simple pole on the unit circle, then the system is referred to as marginally stable. If we require that the function is a minimum phase function, then all the zeros and poles must be inside the unit circle. ∎

Special cases of the general model are useful in defining digital filters. These are 1. the *autoregressive moving average filter* (ARMA); *moving average filter* (MA); and the *autoregressive filter* (AR). These are explicitly expressed by

$$H_{ARMA}(z) = \frac{\sum_{k=0}^{L} b_k z^{-k}}{1 + \sum_{k=1}^{N} a_k z^{-k}},$$

$$Y(z) = H_{ARMA}(z)X(z),$$

$$y[n] = -\sum_{k=1}^{N} a_k y[n-k] + \sum_{k=0}^{L} b_k x[n-k],$$

$$(9.9.12b)$$

$$H_{MA}(z) = \sum_{k=0}^{L} b_k z^{-k},$$

$$Y(z) = H_{MA}(z)X(z),$$

$$y[n] = \sum_{k=0}^{L} b_k x[n-k] \quad (9.9.12c)$$

$$H_{AR}(z) = \cfrac{1}{1 + \sum\limits_{k=1}^{N} a_k z^{-k}}, Y(z) = H_{AR}(z)X(z),$$

$$y[n] = x[n] - \sum_{k=0}^{N} a_k y[n-k]. \qquad (9.9.12d)$$

All three models are used in different applications. They are related, at least in the limit, see Marple (1987). The AR models are used extensively in spectral estimation. AR and ARMA models are used in digital filter designs discussed in a later section.

9.9.2 Schur–Cohn Stability Test

If we know the pole locations of a transfer function, the stability of that system can be determined. Factoring an nth order polynomial is not always possible. A stability test that does not require factoring a polynomial is the Schur–Cohn test (Proakis and Manolakis, 1988) and it starts with the denominator polynomial of the transfer function

$$A(z) = 1 + \sum_{k=1}^{N} a_k z^{-k}. \qquad (9.9.13)$$

The procedure involves deriving a set of polynomials (proof is beyond the scope here)

$$A_m(z) = \sum_{k=0}^{m} \alpha_m[k] z^{-k}, \alpha_m[0] = 1 \text{ and } B_m(z) = z^{-m} A_m(z^{-1})$$

$$= \sum_{k=0}^{m} \alpha_m[m-k] z^{-k}. \qquad (9.9.14)$$

The stability test computes a set of coefficients, called *reflection coefficients*, $K_i, i = 1, 2, \ldots, N$ from $A(z)$. The recursive computation uses the following steps:

1. $A_N(z) = A(z)$ and $K_N = \alpha_N(N) = a_N$,   (9.9.15)

2. Compute $A_{m-1}(z) = \frac{A_m(z) - K_m B_m(z)}{(1 - K_m^2)}$, $K_m = \alpha_m(m)$.
   (9.9.16)

The polynomial $A(z)$ given in (9.9.13) has all its roots inside the unit circle if and only if the coefficients $|K_m| < 1, m = 1, 2, \ldots, N$.

Example 9.9.6 Use the Schur–Cohn test to show that all the roots of the polynomial $A(z)$ given below in the transfer function lie inside the unit circle:

$$H(z) = \frac{z}{z^2 - z - (1/2)} = \frac{z^{-1}}{[1 - z^{-1} - (1/2)z^{-2}]} = \frac{z^{-1}}{A(z)}.$$

Solution: The following steps are used to determine the reflection coefficients.

$$A_2(z) = A(z) = 1 - z^{-1} - .5z^{-2} = \sum_{k=0}^{2} \alpha_2[k] z^{-k} \to K_2$$

$$= \alpha_2[2] = -.5, \ B_2(z) = -.5 - z^{-1} + z^{-2},$$

$$\begin{aligned} A_1(z) &= \frac{A_2(z) - K_2 B_2(z)}{1 - K_2^2} \\ &= \frac{(1 - z^{-1} - 0.5z^{-2}) + 0.5(-0.5 - z^{-1} + z^{-2})}{1 - (0.5)(0.5)} \\ &= \frac{.75 - 1.5z^{-1}}{.75} = 1 - 2z^{-1}, K_1 = -2. \end{aligned}$$

Since $|K_1| > 1$, the Schur–Cohn test indicates the system is unstable. Also, the roots of the polynomial are $z = (1/2) \pm ((\sqrt{3})/2)$ indicating one of the roots is outside the unit circle and the system is unstable. ∎

The above recursive algorithm leads to a lattice filter realization. Such implementation has important applications in speech processing, see Rabiner and Schafer (1978).

9.9.3 Bilinear Transformations

Consider the mapping (*bilinear transformation*) of the s-plane to the z-plane by

$$s = c \left[\frac{z-1}{z+1} \right], \quad z = \frac{c+s}{c-s}, \quad c \text{ is a constant.}$$

$$(9.9.17)$$

Some authors use $c = 2/t_s$ and tie it to the sampling interval. Note that $(s/c)|_{s=j\omega_a} = j(\omega_a/c)$, where c is a normalizing factor in the frequency domain and the subscript a on ω is introduced to distinguish from the digital frequency to be discussed later. In filter designs, it is one of the intermediate steps. Using $c = 1$ will not affect the final results in the design. To investigate the effects of this transformation, let

$$s = \sigma + j\omega_a \text{ and } z = re^{j\,\text{r}\Omega}. \quad (9.9.18)$$

If we let $\sigma = 0$, we have the complex variable z in the form

$$z = e^{j\Omega} = \frac{1 + j\omega_a}{1 - j\omega_a} = e^{j2\tan^{-1}(\omega_a)}. \quad (9.9.19)$$

The analog and digital frequencies are related by

$$\Omega = 2\tan^{-1}(\omega_a) \text{ or } \omega_a = \tan(.5\Omega). \quad (9.9.20)$$

This transformation is sketched in Fig. 9.9.1 for *one period*. It is a nonlinear one-to-one transformation that compresses the analog frequency range $-\infty < \omega_a < \infty$ to the digital frequency range $-\pi < \Omega < \pi$. It avoids aliasing at the expense of *distorting* or *warping* the analog frequencies.

For $s = \alpha + j\beta$,

$$|z| = \left|\frac{1 + \alpha + j\beta}{1 - \alpha - j\beta}\right| = \sqrt{\frac{(1 + \alpha)^2 + \beta}{(1 - \alpha)^2 + \beta^2}}. \quad (9.9.21)$$

This implies the following (see Fig. 9.9.2):

$$(\text{Left half of the } s - \text{plane})\alpha < 0 \Leftrightarrow |z| < 1$$
$$(\text{Inside the unit circle in the } z - \text{plane}), \quad (9.9.22a)$$

$$(\text{Imaginary axis on the } s - \text{plane})\alpha = 0 \Leftrightarrow |z| = 11$$
$$(\text{On the unit circle in the } z - \text{plane}), \quad (9.9.22b)$$

$$(\text{Right half of the } s - \text{plane})\alpha > 0 \Leftrightarrow |z| > 1$$
$$(\text{Outside the unit circle in the } z - \text{plane}). \quad (9.9.22c)$$

If we have a polynomial in the z-domain, bilinear transformation converts it into the s-plane and we can use any stability tests that are available for s-domain polynomials.

Example 9.9.7 Use the bilinear transformation and the Routh table to show that the polynomial $D(z) = z^2 - z - 0.5$ represents an unstable system.

Solution: Using (9.9.17) with $c = 1$, we have

$$D(z)\big|_{z=(1+s)/(1-s)} = \frac{s^2 + 2s - (1/3)}{(1-s)^2} = 0$$
$$\Rightarrow s^2 + 2s - (1/3) = 0. \quad (9.9.23a)$$

Routh table:

s^2	1	$-(1/3)$
s	2	
s^0	$-(1/3)$	

$$(9.9.23b)$$

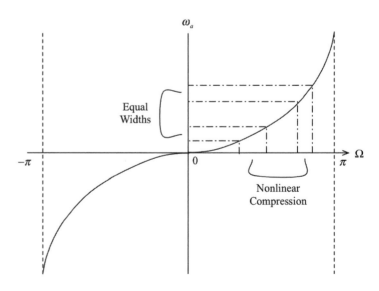

Fig. 9.9.1 Warping effect of the bilinear transformation

Fig. 9.9.2 Regions of stability in the z- and s-planes

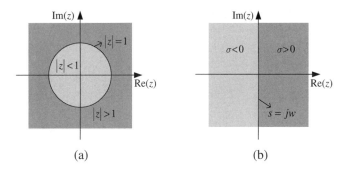

(a)   (b)

There is one sign change in the first column of the Routh table indicating that the s-domain polynomial has one root in the right half of the s-plane implying $D(z)$ has one root outside the unit circle and the system is unstable.   ∎

Notes: Equation (7.12.10), an example by Dahlquist and Bjorck (1974), was given illustrating the effect of a slight change in the coefficient of a polynomial and the corresponding significant change in the roots of the polynomial. It is best to implement filters using second-order sections with a possible one first-order section (if the characteristic polynomial is odd in the case). Second-order systems are important.   ∎

Second-order systems and the stability triangle: The transfer function, the corresponding characteristic polynomial, and its roots are

$$H(z) = K\frac{1 + b_1 z^{-1} + b_2 z^{-2}}{1 + a_1 z^{-1} + a_2 z^{-2}}, \qquad (9.9.24)$$

$$D(z) = z^2 + a_1 z + a_2 = (z + p_1)(z + p_2) = 0, \quad (9.9.25)$$

$$p_1, p_2 = (-a_1/2) \pm \left[\sqrt{(a_1^2 - 4a_2)/2}\right], \qquad (9.9.26)$$

$$a_1 = -(p_1 + p_2), a_2 = p_1 p_2.$$

The system is stable if all the poles lie inside the unit circle in the z-plane.

Stability implies

$$|a_2| = |p_1 p_2| = |p_1||p_2| < 1,$$
$$|a_1| = -(p_1 + p_2) < 1 + a_2. \qquad (9.9.27)$$

These conditions can also be derived from the Schur–Cohn stability test. From the recursive

equations in (9.9.14) and (9.9.15), we can also write (left as an exercise).

$$K_1 = a_1/(a_1 + a_2), \quad K_2 = a_2. \qquad (9.9.28)$$

Since the Schur–Cohn stability criterion requires that the coefficients $|K_1|, |K_2| < 1$ for the system to be stable, the constraints can be expressed by

$$|K_2| = |a_2| < 1 \text{ and } |K_1| = \left|\frac{a_1}{1 + a_2}\right| < 1 \qquad (9.9.29)$$

$$\Rightarrow -1 < a_2 < 1, -1 - a_2 < a_1 < 1 + a_2.$$

The stability conditions can be expressed in (a_1, a_2) coefficient plane shown in Fig. 9.9.3. The second-order system is stable if and only if the point (a_1, a_2) lies inside the triangle and the triangle is called the *stability triangle* (Proakis and Manolakis, 1988).

9.10 Designs by the Time and Frequency Domain Criteria

In Section 6.5.1, we have seen that for a continuous-time linear time-invariant system with a transfer function $H_a(s)$, the system response to the input e^{st} can be expressed by

$$y(t) = H_a(s)e^{st}, \qquad (9.10.1a)$$

where $H_a(s)$ is the system transfer function. To differentiate the s-domain functions from the z-domain functions the subscript is used in (9.10.1a) to denote that it is an analog function. Interestingly, z^n plays a similar role in discrete-time systems as e^{st} played in the continuous-time systems. That is, if $x[n] = z^n$

Fig. 9.9.3 Stability triangle in the (a_1, a_2) coefficient plane

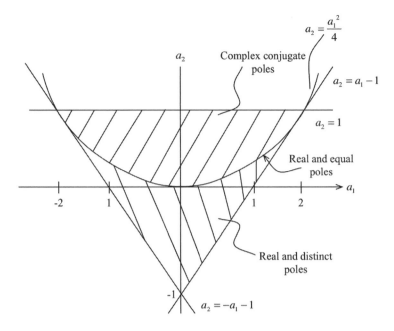

then the response of the discrete-time linear shift-invariant system is given by the convolution

$$y[n] = h[n]z^n = \sum_{k=-\infty}^{\infty} h[k]z^{n-k}$$

$$= \left(\sum_{k=-\infty}^{\infty} h[k]z^{-k}\right)z^n = \left[\sum_{k=-\infty}^{\infty} h[k]z^{-k}\right]z^n = H(z)z^n.$$

(9.10.1b)

The above result is *valid* only for the values of z for which $H(z)$, the transfer function of the discrete-time linear shift-invariant system, exists. Now consider replacing an analog system by an *A/D converter–digital filter–D/A converter* combination as shown in Fig. 9.10.1, which is discrete representation continuous-time filter. If the continuous signal is sampled every $t_s S$ interval, then

$$x[n] = e^{s(nt_s)}\big|_{z=e^{st_s}} = z^n, \quad y[n] = H(e^{st_s})e^{snt_s}. \quad (9.10.2)$$

Considering the block diagrams in Fig. 9.10.1, the nth sample of the output $y(t)$ in (9.10.1a) can now be expressed by

$$y(nt_s) = H_a(s)e^{snt_s}. \quad (9.10.3)$$

If the analog and the discrete system responses are to be equal at the sample values, i.e., the discrete-

data system is designed to mimic the analog system in the sense that the responses are the same at the sampling instants $y(nt_s) = y(t)|_{t=nt_s}$, then

$$H(e^{st_s})\big|_{s=j\omega} = H(e^{j\omega t_s}) \approx H_a(j\omega),$$
$$-\omega_s/2 < \omega < \omega_s/2. \quad (9.10.4)$$

It implies that

$$\sum_{n=-\infty}^{\infty} h[n]e^{-nst_s} = \int_{-\infty}^{\infty} h_a(t)e^{-st}dt \approx \sum_{n=-\infty}^{\infty} [t_s h_a(nt_s)]e^{-snt_s}.$$

Note that $z = e^{st_s}$ in (9.10.2) is not a good transformation since it does not convert a rational transfer function $H_a(s)$ into a rational function in the z-domain. This gives us an impetus to study designs that make use of the impulse responses of the continuous systems at the sample points and other transformations that transform $H_a(s)$ to a rational

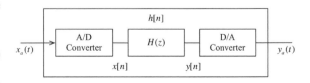

Fig. 9.10.1 A/D converter–digital filter–D/A converter

functions in the z-domain, such as the bilinear transformations.

9.10.1 Impulse Invariance Method by Using the Time Domain Criterion

In this method the discrete system model imitates the continuous-time system by matching the impulse response of a continuous system. We should note that if $t_s \to 0$, the two responses match at every point, which is impractical in reality. In addition, if the analog function is not frequency limited, then aliasing problems will arise in the frequency domain, as the sampling rate may not be sufficient. In the bilinear transformation method, discussed later, aliasing problems will not arise.

The criterion for impulse invariance method is the equivalence of the analog and discrete systems, both having the same impulse response at the sample points, see Fig. 9.10.1. We can write the outputs in the continuous time and discrete time. First,

$$y_a(t) = h_a(t) * x_a(t) = \int_{-\infty}^{\infty} x_a(\alpha)h_a(t-\alpha)d\alpha,$$

$$h_a(t) = L^{-1}\{H_a(s)\}, \qquad (9.10.5)$$

$$\Rightarrow ya(t)|_{t=nt_s} \approx t_s \sum_{k=-\infty}^{\infty} x_a(kt_s)h_a(nt_s - kt_s)$$

$$= \sum_{k=-\infty}^{\infty} x_a[n]\{t_s h_a[n-k]\}. \qquad (9.10.6)$$

Consider the simple transfer function $H_a(s)$ and the corresponding time response $h_a(t)$ by

$$H_a(s) = \frac{A_i}{(s+s_i)} \overset{z}{\leftrightarrow} A_i e^{-s_i t}.$$

At $t = nt_s$, we have the time response and its transform given by

$$h(nt_s) = A_i e^{-s_i nt_s}, n \geq 0, h[n] = h_a(nt_s),$$

$$H(z) = A_i \sum_{n=0}^{\infty} e^{-s_i nt_s} z^{-n}$$

$$= \frac{A_i}{1 - e^{-s_i t_s}z^{-1}}, \left| e^{-s_i t_s}z^{-1} \right| < 1. \quad (9.10.7)$$

This can be generalized for the multiple pole case.

$$H_a(s) = \frac{A_i}{(s+s_i)^m} \overset{LT}{\longleftrightarrow} \frac{A_i}{(m-1)!}(t)^{m-1}e^{-s_i t} = h(t)$$

$$\Rightarrow h_a(nt_s) = \left\{ \begin{array}{c} \frac{A_i}{(m-1)!}(nt_s)^{m-1}e^{-s_i nt_s}, \ n \geq 0 \\ 0, \ n < 0 \end{array} \right\}.$$

$$(9.10.8a)$$

For example, in the case of $m = 2$ in (9.10.8a), we have, by using (9.5.9)

$$h_a(nt_s) = A_i t_s n(e^{-s_i t_s})^n u[n] = A_i t_s a^n u[n], \ a = e^{-s_i t_s},$$

$$A_i t_s n a^n u[n] \overset{z}{\longleftrightarrow} A_i t_s \frac{az}{(z-a)^2}, |z| > |a|. \quad (9.10.8b)$$

Then, the corresponding s-domain and the corresponding z-domain functions are

$$\frac{A_i}{(s+s_i)^2} \Rightarrow A_i t_s \frac{e^{-s_i t_s}z}{(z - e^{-s_i t_s})^2}. \qquad (9.10.8c)$$

Since all the classical filter transfer s-domain functions have simple poles, we need to consider only simple poles. Complex pole cases can be handled by combining the terms corresponding to the complex pole and its conjugate. The two important complex pole cases in the s-domain and the corresponding z-domain functions are as follows:

$$\frac{s+a}{(s+a)^2 + b^2} \Rightarrow \frac{1 - e^{-at_s}\cos(bt_s)z^{-1}}{1 - 2e^{-at_s}\cos(bt_s)z^{-1} + e^{-2at_s}z^{-2}}$$

$$(9.10.9a)$$

$$\frac{b}{(s+a)^2 + b^2} \Rightarrow \frac{e^{-at_s}\sin(bt_s)z^{-1}}{1 - 2e^{-at_s}\cos(bt_s)z^{-1} + e^{-2at_s}z^{-2}}.$$

$$(9.10.9b)$$

Note if $a = 0$, the L-transforms correspond to the cosine and sine functions and the digital resonator functions are obtained from analog functions. Following gives a summary of the procedure to determine the z-domain transfer function from an s-domain function:

Impulse invariance method:

$$H_a(s) = \sum_{i=1}^{N} H_{ai}(s) \Rightarrow H(z) = \sum_{i=1}^{N} H_{ai}(z).$$

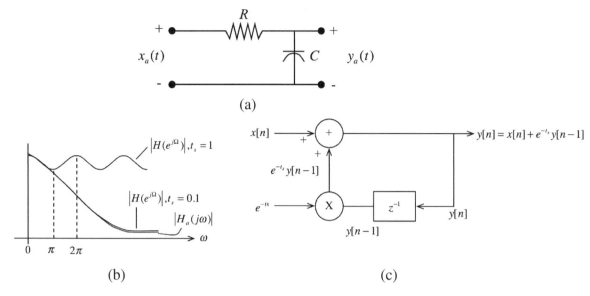

Fig. 9.10.2 (a) RC circuit, (b) normalized sketches of $H(e^{j\Omega}) = H(e^{j\omega t_s})|_{t_s=0.1, 1}$, $H_a(j\omega)$, and (c) simulation diagram

Example 9.10.1 Consider the RC circuit shown in Fig. 9.10.2a with RC = 1

a. Give the expressions for the impulse response and sketch the magnitude response $|H_a(j\omega)|$ and the magnitude responses of the discrete-time filter using the impulse invariance method assuming $t_s = 1$ and $t_s = .1$.

b. Give the difference equation and the corresponding flow diagram that simulates the response of the RC circuit using the symbols in Fig. 9.2.1.

Solution: The transfer function of the RC circuit, its impulse response, the corresponding impulse response $h[n]$, and its z-transform along with its amplitude response are as follows:

$$a.\ H_a(s) = \frac{1}{RCs + 1}\bigg|_{RC=1} = \frac{1}{s+1} \overset{L}{\leftrightarrow} e^{-t}u(t)$$

$$= h_a(t), |H_a(j\omega)| = \frac{1}{\sqrt{\omega^2 + 1}}, \quad (9.10.10a)$$

$$h[n] = (e^{-t_s})^n u[n] \overset{z}{\leftrightarrow} \frac{1}{1 - e^{-t_s}z^{-1}} = H(z), \quad (9.10.10b)$$

$$|H(e^{j\Omega})| = |H(z)|_{z=e^{j\Omega}} = \left|\frac{1}{1 - e^{-t_s}e^{-j\Omega}}\right|$$

$$= \frac{1}{\sqrt{[1 - e^{-2t_s} - 2e^{-t_s}\cos(\Omega)]}}, \Omega = \omega t_s.$$

$$(9.10.11)$$

Note the frequency variables in the discrete and the continuous domains are related by $\Omega = \omega t_s$. For the sampling intervals $t_s = 1$ and $t_s = .1$,

$$|H(e^{j\Omega})|_{t_s=1} = \frac{1}{\sqrt{[1 - e^{-2} - 2e^{-1}\cos(\Omega)]}},$$
$$(9.10.12a)$$

$$|H(e^{j0})|_{\Omega=0, t_s=1} = \frac{1}{\sqrt{[1 - e^{-2} - 2e^{-1}]}},$$

$$|H(e^{j\Omega})|_{t_s=.1} = \frac{1}{\sqrt{[1 - e^{-.2} - 2e^{-.1}\cos(\Omega)]}},$$

$$|H(e^{j0})|_{\Omega=0, t_s=.1} = \frac{1}{\sqrt{[1 - e^{-.2} - 2e^{-.1}]}}. \quad (9.10.12b)$$

The amplitude responses given in (9.10.12a) and (9.10.12b) are *periodic* with period 2π (in the Ω domain) or $2\pi/t_s$ (in the ω domain). Amplitude plots are shown in Fig. 9.10.2b. For comparison purposes, the sketches

$$|H_a(j\omega)|, \ |H(e^{j\Omega})|_{t_s=1}, \ \text{and} \ |H(e^{j\Omega})|_{t_s=.1}$$

are *normalized* so that the peak values at $\omega = 0$ coincide. The shape of the digital filter amplitude response for the sampling interval $t_s = .1$ with period equal to $\omega_s = 20\pi$ in the ω domain is very close to the analog response in the frequency interval $(0, \omega_s/2) = (0, 10\pi)$. On the other hand, the digital filter magnitude frequency response for the sampling interval $t_s = 1$ with period $\omega_s = 2\pi$ shows the periodicity and the aliasing problems created by a low sampling rate.

From the input–output transform relations of the discrete-time system, we have

$$Y(z) = H(z)X(z), \ H(z) = \frac{1}{1 - e^{-t_s}z^{-1}},$$

$$\left[1 - e^{-t_s}z^{-1}\right] Y(z) = X(z), \qquad (9.10.13a)$$

$$y[n] - e^{-t_s}y[n-1] = x[n]. \qquad (9.10.13b)$$

Equation (9.10.13b) can be simulated using three components, namely *delay*, *multiplier*, and *summer* shown in Fig. 9.10.2c. The *simulation* (or a *symbolic*) diagram of the equation in (9.10.13b) is shown in Fig. 9.10.2c. It illustrates the responses of an RC filter and the digital filter for the simple RC circuit.

Notes: In the impulse invariance method, the analog and digital filters have the same values at the sample points of the impulse response. In practice, it is customary to scale $H(z)$ to $KH(z)$ at a convenient frequency location so that the s-domain function and the z-domain functions match at a convenient location, usually at the frequency corresponding to the highest amplitude. It depends upon the type of the filter. In the case of a low-pass and band-elimination filters, the frequency locations are at dc, in the high-pass filter case, it is at $\omega_s/2$ (or $\Omega = \pi$) and in the band-pass case it is at the center frequency. Let the s-domain and its impulse-invariant z-domain functions are

$$H_a(s) = 1/(s + s_k) \Rightarrow H(z) = z/(z - e^{-s_k t_s}), z = e^{j\Omega}.$$

The dc gain of the analog transfer function $H_a(s)$ is $(1/s_k)$ and the dc gain of the digital transfer function is $1/(1 - e^{-s_k t_s})$ obtained by evaluating the z-domain function at $z = 1$ or $\Omega = 0$. In the above discussion we considered the design by matching the time domain analog and digital responses for an impulse input. Obviously filters can be designed to match the time domain analog and digital responses for any input, such as step, ramp; see Strum and Kirk (1988).

The impulse invariance method results in a recursive filter based on the time response of the continuous system, i.e., the *design in time domain*. The drawback is the aliasing error near half of the

sampling frequency. This can only be minimized by increasing the sampling rate, i.e., by reducing the sampling interval making the implementation expensive. Since there is maximum aliasing at $\omega_s/2$, this method cannot be used for high-pass and band-elimination filters and is restricted to low-pass and band-pass filters. ∎

9.10.2 Bilinear Transformation Method by Using the Frequency Domain Criterion

In the ideal analog filter design specifications we generally identify ranges of frequencies for passband, stopband, and transition bands, where, in each frequency range, the amplitude gains of the frequency responses are assumed to be constant. That is, they are assumed to have *piecewise constant amplitude responses over certain bands*. We have considered in Chapter 6 the low-pass, high-pass, band-pass, and band-elimination analog filter ideal and non-ideal filter characteristics. The edges of these bands are identified as the *critical frequencies* of the filter. Before proceeding we note that the digital filter frequency response in terms of the variable ω is periodic with period 2π. The digital frequencies are the normalized frequencies by the sampling frequency. That is, the digital frequency is given by $\Omega = 2\pi f_d/f_s = (\omega_d t_s)$, where f_s is the sampling frequency and the corresponding sampling interval is $t_s = 1/f_s$. To avoid any confusion, we have added a subscript on ω to identify that $\omega_d = \Omega/t_s$ is a digital frequency variable and ω_a is the analog variable in the bilinear transformation (see (9.9.20)):

$$\omega_a = \tan(\omega_d t_s/2). \qquad (9.10.14)$$

Note the subscripts on ω s. The bilinear transformation method of designing a digital filter is based upon the frequency amplitude response values at the frequency edges of the passband and stopband and other important frequencies. The procedure is given below.

1. Identify the critical frequencies and ranges, i.e., identify the passband, stopband, transition band, the maximum attenuation points, etc., of the desired digital filter. Let these frequencies be

identified by Ω_i. Compute a new set of frequencies

$$\omega_a i = \tan(\Omega_i/2) = \tan(\omega_{d_i} t_s/2). \qquad (9.10.15)$$

2. Find a transfer function $H_a(s)$ that satisfies the given specifications of the digital filter at the new frequencies obtained from (9.10.15).
3. Substitute s by $(z-1)/(z+1)$ and obtain $H(z)$ as a ratio of two polynomials in the variable z yielding a digital filter that satisfies the given digital filter specifications.

The digital filter specifications include the required attenuation characteristics of the filter *plus* the desired sampling frequency or the highest frequency in the input signal. In most applications, the highest frequency of a signal is known or assumed. If the highest frequency in the signal to be processed is given by ω_h, then the minimum sampling frequency must be larger than twice the highest frequency in the signal, i.e., $f_s \geq 2\omega_h/2\pi = 2f_h, \omega_h = 2\pi f_h$, or the sampling interval must satisfy $t_s \leq \pi/\omega_h = 1/2f_h$. Following examples illustrate the design of digital filters using bilinear transformation.

Example 9.10.2 Repeat the implementation of a digital filter starting with the transfer function of the RC circuit discussed in Example 9.10.1 using the bilinear transformation method with the 3 dB bandwidth equal to $\Omega_{3dB} = \pi/4$.

Solution: The transfer function corresponding to the simple RC circuit was given in (9.10.9). The first step is to find the critical frequencies.

1. The 3 dB frequency: $\omega_a = \tan(\omega_d t_s/2) = \tan(\pi/8) = .414$.
2. The transfer function corresponding to this specification is

$$H_a(s) = \frac{1}{(s/\omega_a)+1} = \frac{.414}{s+.414}.$$

3. Replace s by $(z-1)/(z+1)$ and simplify

$$H(z) = \frac{.414}{[(z-1)/(z+1)]+.414} = \frac{.414(z+1)}{1.414z-.586}$$

$$= \frac{.293(z+1)}{z-.414} = \frac{.293(1+z^{-1})}{1-.414z^{-1}}$$

The corresponding frequency response of the discrete-time filter (or digital filter) is

$$H(e^{j\Omega}) = H(z)|_{z=e^{j\Omega}} = \frac{.293(1+e^{-j\Omega})}{1-.414e^{-j\Omega}}, \qquad (9.10.16)$$

$$Y(z) = H(z)X(z).$$

It is always a good idea to check the values at important frequencies

$$\Omega = 0 : H_d(e^{j0}) = 1, \Omega_{3dB} = \pi/4 : H_d(e^{j\pi/4}) = .707,$$

$$\Omega = \pi : H_d(e^{j\pi}) = .293(1+e^{-j\pi})(1-.414e^{-j\pi}) = 0.$$

It may be easier to compute these values by evaluating the digital filter function $H(z)$ at $z = 1, z = e^{j\pi/4}$ and at $z = e^{j\pi} = -1$, respectively.

The simulation diagram can be obtained by writing the output transform in terms of three equations given below using an intermediate variable $P(z)$:

$$Y(z) = [1+z^{-1}]P(z), P(z) = \left[\frac{1}{1-.414z^{-1}}\right]X_1(z),$$

or $P(z) = .414z^{-1}P(z) + X_1(z), \qquad (9.10.17a)$

$$X_1(z) = .293X(z). \qquad (9.10.17b)$$

The corresponding difference equations can be obtained by noting that $x[n] \overset{z}{\longleftrightarrow} X(z)$, $x_1[n] \overset{z}{\longleftrightarrow} X_1(z), p[n] \overset{z}{\longleftrightarrow} P(z)$, and $p[n-1] \overset{z}{\longleftrightarrow} z^{-1}P(z)$,

$$x_1[n] = 0.293x[n],$$

$$p[n] = x_1[n] + .414p[n-1], \qquad (9.10.17c)$$

$$y[n] = p[n] + p[n-1].$$

The filter can be implemented by the simulation diagram shown in Fig. 9.10.3. ∎

Example 9.10.3 Design a digital filter, which has a monotonic amplitude frequency response to be within 3 dB in the pass band of 0–1000 Hz and the response goes down to 10 dB at frequencies beyond 2 kHz. Assume the sampling rate is 10 kHz.

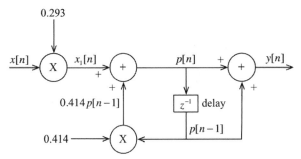

Fig. 9.10.3 Simulation diagram (or realization of (9.10.17c))

Solution: The critical frequencies are

$$\omega_{d_1} t_s = 2\pi(1000)(1/10000)$$
$$= .2\pi \text{ and } \omega_{d_2} t_s = 2\pi(2/10) = .4\pi. \tag{9.10.18}$$

1. Compute $\omega_{ai}, i = 1, 2$ by

$$\omega_{a1} = \omega_c = \tan(0.4\pi/2) = \tan(0.2\pi/2) = 0.3249,$$

$$\omega_{a2} = \omega_r = \tan(0.2\pi) = 0.7265.$$

2. Since a monotonic amplitude response is required, we will use Butterworth filter with $\omega_c = .3249$ and $\omega_r = .7265$. The analog low-pass amplitude characteristic is shown in Fig. 9.10.4. To find the order of the filter, find the value of n such that

$$20 \log|H(j\omega_a)|\big|_{\omega_a=\omega_r} = 10 \log\frac{1}{1 + (\omega_r/\omega_c)^{2n}}$$

$$\leq -10 \text{ dB} \rightarrow [1 + (2.236)^{2n}] \geq 10.$$

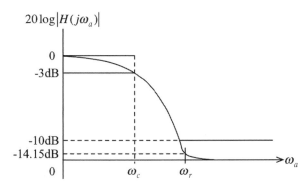

Fig. 9.10.4 Low-pass filter specifications in the analog domain and the resultant response

It shows that $n = 2$ would satisfy the design specifications and the corresponding amplitude at ω_r is -14.15 dB. Figure 9.10.5 gives the sketch for the analog amplitude frequency response. The second-order analog Butterworth filter with a cutoff frequency of $\omega_c = 0.3249$ has poles at $s = s_1, s_2 = 0.3249(-0.707 \pm j0.707) = -0.23 \pm j0.23$. The analog function is

$$H_a(s) = \frac{s_1 s_2}{(s - s_1)(s - s_2)} = \frac{0.1058}{s^2 + 0.46s + 0.1058}.$$

3. Determine the digital filter function by replacing s by $(z - 1)/(z + 1)$ in the analog function $H_a(s)$. This results in the transfer function

$$H(z) = \frac{0.1058}{[(z-1)/(z+1)]^2 + 0.46[(z-1)/(z+1)] + 0.1058}$$

$$= \frac{0.1058(z^2 + 2z + 1)}{1.5658z^2 - 1.7884z + 0.6458}, Y(z) = H(z)X(z).$$

The frequency response of this filter function is

$$H(e^{j\Omega}) = H(z)\big|_{z=e^{j\Omega}}$$
$$= \frac{0.1058(z^2 + 2z + 1)}{1.5658z^2 - 1.7884z + 0.6458}\bigg|_{z=e^{j\Omega}}$$
$$= \frac{0.1058(e^{j2\Omega} + 2e^{j\Omega} + 1)}{1.5658e^{j2\Omega} - 1.7844e^{j\Omega} + 0.6458}.$$

The amplitudes can be obtained at the important frequencies using $H(z)$ and

$$\Omega = 0, z = 1 : H(z)|_{z=1} \cong 1,$$

$$\Omega = \pi, z = -1 : H(z)|_{z=-1} = 0.$$

Evaluation at the two frequencies $\Omega = .2\pi$ and $\Omega = .4\pi$ is left as an exercise. ∎

All-pass filters: A digital all-pass function satisfies

$$\left|H_{ap}(e^{j\Omega})\right| = 1 \quad \text{for all } \Omega. \tag{9.10.19}$$

Example 9.10.4 Show that the functions below satisfy (9.10.19). For Part a, show that $\left|H_a(e^{j\Omega})\right| = 1$. For Part b, show first that $\left|H_b(z)H_b(z^{-1})\right| = 1$ and then $\left|H(e^{j\Omega})\right|^2 = 1$:

a. $H_a(z) = \dfrac{\beta + z^{-1}}{1 + \beta z^{-1}},$ b. $H_b(z) = \dfrac{\alpha + \beta z^{-1} + z^{-2}}{1 + \beta z^{-1} + \alpha z^{-2}}.$

Solution:

a. $\left|H_a(e^{j\Omega})\right| = \left|H_a(z)\right|\Big|_{z=e^{j\Omega}} = \left|\dfrac{\beta + e^{-j\Omega}}{1 + \beta e^{-j\Omega}}\right|^2$

$= \dfrac{[\beta + \cos(\Omega)]^2 + \sin^2(\Omega)}{[1 + \beta\cos(\Omega)]^2 + (\beta\sin(\Omega))}$

$= \dfrac{1 + \beta_k^2 + 2\beta_k\cos(\Omega)}{1 + \beta_k^2 + 2\beta_k\cos(\Omega)} = 1.$

$$(9.10.20a)$$

b. $H_b(z)H_b(z^{-1}) = \dfrac{\alpha + \beta z^{-1} + z^{-2}}{1 + \beta z^{-1} + \alpha z^{-2}} \dfrac{\alpha + \beta z + z^2}{z^2 + \beta z + \alpha z^2}$

$= 1 \Rightarrow \left|H(e^{j\Omega})H(e^{-j\Omega})\right| = \left|H(e^{j\Omega})\right|^2 = 1.$

$$(9.10.20b)$$

∎

Notes: Bilinear transformation method is one of the best ways of designing IIR digital filters and is restricted to the design of approximations to filters with piecewise constant frequency amplitude characteristics, such as low-pass, high-pass; see Rabiner and Gold (1975). It has several advantages over the impulse invariance method. In the impulse invariance method, there is the problem of aliasing. The bilinear transformation method does not create any aliasing problems, as there is one-to-one mapping from the s-domain to the z-domain. It requires a lower sampling rate compared to the impulse invariance method and produces filters that have much sharper roll-off rate in the stopband. The designs result in a rational function with poles and zeros. These have impulse responses with infinite time duration. These are IIR or recursive filters. Finite impulse response (FIR) filters have finite duration. IIR filters have non-linear phase responses, whereas the FIR filters can be designed with linear phase and therefore a constant delay.

The IIR filters are sensitive to the filter coefficient accuracy. Since the implementation of digital filters depend upon the word lengths of the filter coefficients, any inaccuracies in the filter implementation can change the filter behavior. For stability, the roots of the characteristic polynomial of the IIR filters must be inside the unit circle. For example, the ideal digital generators imply $a = 0$ in (9.10.9a and b). This implies the poles of the digital resonator function are located on the unit circle. The

recursive filter designs are well established only for amplitude responses that are piecewise constant, such as low-pass, high-pass. Non-recursive filters can be designed for an arbitrary amplitude response with N coefficients.

To approximate a recursive filter by a nonrecursive filter requires N to be large. The nonrecursive and the recursive filters are equal only in the limit, i.e., $N \to \infty$. Nonrecursive filters do not have any stability problems. Recursive filters take less processing delay and therefore are faster compared to the non-recursive filters. If the processing delay is not a critical factor in an application and the phase response of the filter needs to have a linear phase, then the non-recursive filter is the choice. ∎

9.11 Finite Impulse Response (FIR) Filter Design

The F-series expansion of a periodic function $x_T(t) = x_T(t + T), \omega_0 = 2\pi/T$ is

$$x_T(t) = \sum_{k=-\infty}^{\infty} X_s[k]e^{jk\omega_0 t},$$

$$X_s[k] = \frac{1}{T}\int_{-T/2}^{T/2} x_T(t)e^{-jk\omega_0 t}dt, x_T(t) \overset{\text{FS}}{\longleftrightarrow} X_s[k]. \quad (9.11.1)$$

The time function is continuous and the F-series coefficients are discrete. In Chapter 8, it was seen that if the time function $h[n]$ is discrete, then its DTFT $H(e^{j\Omega})$ is periodic and continuous in the frequency domain. Although FIR filters can be designed with arbitrary periodic functions in the frequency domain in terms of their Fourier series, the interest here is to design nonrecursive digital filters with *linear phase*.

The discrete-time Fourier transform (DTFT) of the sequence $h[n]$ is

$$H(e^{j\Omega}) = \sum_{n=-\infty}^{\infty} h[n]e^{-jn\Omega} \equiv H_{2\pi}(e^{j\Omega}). \quad (9.11.2)$$

For simplicity, the subscript 2π is not shown on H on the left in (9.11.2). The series expansions in

(9.11.1) and (9.11.2) have the same general form, except one is F-series in continuous time and the other one is in continuous frequency. The procedures discussed in Chapter 3 can be used to derive

$$H(e^{j\Omega}) = \sum_{n=-\infty}^{\infty} h[n]e^{-jn\Omega} \xrightarrow{\text{FS}} \frac{1}{2\pi}\int_{-\pi}^{\pi} H(e^{j\Omega})e^{j\Omega n}d\Omega = h[n].$$

(9.11.3)

In addition, note the difference in the signs of the exponentials in (9.11.1) and (9.11.2). The expression for $H(e^{j\Omega})$ is the *synthesis* equation and the integral expression for the filter coefficients $h[n]$ is the *analysis* equation.

For the digital filter design, the design steps include the specifications on the filter transfer function $H(e^{j\Omega})$, derivation of its F-series, and truncating the number of coefficients N to be of reasonable size for practical reasons. As in the time domain, Fourier series approximation of a periodic function with discontinuities results in overshoots and undershoots before and after the discontinuity, i.e., Gibbs phenomenon; see Section 3.7. The same is true in the Fourier series of a periodic frequency function. Noting the sampling interval is t_s s, the Fourier series pair in (9.11.3) can be written in terms of the sampling frequency $\omega_s = 2\pi f_s = 2\pi/t_s$ as follows:

$$H(e^{j\omega t_s}) = \sum_{n=-\infty}^{\infty} h[n]e^{-jn\omega t_s} \xrightarrow{\text{FS}} \frac{1}{2\pi}$$
$$\int_{-\omega_s/2}^{\omega_s/2} H(e^{j\omega t_s})e^{jn\omega t_s}d(\omega t_s)$$
$$= h[n],$$

(9.11.4a)

$$\frac{1}{2\pi}\int_{-\omega_s/2}^{\omega_s/2} H(e^{j\omega t_s})e^{jn\omega t_s}d(\omega t_s)$$
$$= \frac{1}{\omega_s}\int_{-\omega_s/2}^{\omega_s/2} H(e^{j\omega t_s})e^{jn\omega t_s}d\omega = h(nt_s).$$ (9.11.4b)

Consider the four basic filters, low-pass, high-pass, band-pass, and band-elimination FIR filters with linear phase. In Section 8.5.2 the discrete-time Fourier transforms of real signals with symmetries were considered. The transfer function of a causal linear phase filter can be written with filter coefficients with even or odd symmetry by the following equations:

$$H(e^{j\Omega}) = |H(e^{j\Omega})|e^{j\alpha\Omega} : \text{Filter coefficients with even symmetry,}$$

(9.11.5a)

$$H(e^{j\Omega}) = j|H(e^{j\Omega})|e^{j\alpha\Omega} : \text{Filter coefficients with odd symmetry.}$$

(9.11.5b)

9.11.1 Low-Pass FIR Filter Design

Starting with the ideal filter periodic function with period equal to 2π, we have

$$H_{Lp}(e^{j\Omega}) = |H_{Lp}(e^{j\Omega})|e^{j\angle H_{Lp}(e^{j\Omega})}.$$ (9.11.6)

If an arbitrary transfer function $H(e^{j\Omega})$ has a piecewise linear phase characteristics, then

$$\frac{d\angle H_{Lp}(e^{j\Omega})}{d\Omega} = \text{constant.}$$ (9.11.7)

If we have a sequence, say $x[n]$ and is delayed by l samples resulting in $x[n-l]$, then

$$\text{DTFT}\{h[n-l]\} = e^{-j\Omega l}H(e^{j\Omega}).$$ (9.11.8)

The amplitude and phase responses of the ideal digital low-pass filter are shown in Fig. 9.11.1a,b for one period. Our goal is to determine the impulse response $h_{Lp}[n]$ from $H_{Lp}(e^{j\Omega})$.

To simplify the analysis, first neglect the phase part of the transfer function in determining the filter

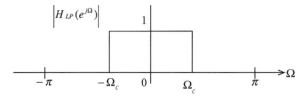

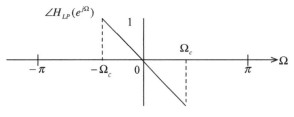

Fig. 9.11.1 Amplitude and phase responses of an ideal digital low-pass filter

Table 9.11.1 Ideal low-pass filter FIR coefficients with $\Omega_c = \pi/4$

n	0	± 1	± 2	± 3	± 4	± 5	± 6	± 7	± 8	± 9	± 10
$\hat{h}_{Lp}[n]$	0.25	0.225	0.159	0.075	0	-0.045	-0.053	-0.032	0	0.025	0.032

coefficients and later consider the delay using (9.11.8). The amplitude and phase responses of the low-pass filter are

$$|H_{Lp}(e^{j\Omega})| = \begin{cases} 1, 0 \le |\Omega| \le \Omega_c, \\ 0, \Omega_c < |\Omega| < \pi \end{cases},$$

$$\angle H_{Lp}(e^{j\Omega}) = \begin{cases} \alpha\Omega, \alpha - \text{constant}, 0 \le |\Omega| \le \Omega_c \\ 0, \Omega_c < |\Omega| < \pi \end{cases}.$$

The complex F-series coefficients of the amplitude response and the corresponding transform are

$$\hat{h}_{Lp}[n] = \frac{1}{2\pi} \int_{-\pi}^{\pi} |H_{LP}(e^{j\Omega})| e^{jn\Omega} d\Omega$$

$$= \frac{1}{2\pi} \int_{-\Omega_c}^{\Omega_c} e^{jn\Omega} d\Omega = \frac{1}{n\pi} \left[\frac{e^{jn\Omega_c} - e^{-jn\Omega_c}}{2j} \right]$$

$$\text{(9.11.9a)}$$

$$= \frac{1}{n\pi} \sin(n\Omega_c), \quad n = 0, \pm 1, \pm 2, ..., \quad \text{(9.11.9b)}$$

$$\hat{h}_{Lp}[n] \overset{\text{DTFT}}{\longleftrightarrow} \hat{H}_{Lp}(e^{j\Omega}) = \sum_{n=-\infty}^{\infty} \hat{h}_{Lp}[n] e^{-jn\Omega}. \quad \text{(9.11.9c)}$$

Note the infinite number of coefficients in (9.11.9c) in the F-series approximation of the ideal low-pass filter approximation, which is impractical for implementation and

$$\hat{h}_{Lp}[-n] = \hat{h}_{Lp}[n], \quad \text{(9.11.10)}$$

$$\hat{h}_{Lp}[0] = \lim_{n \to 0} \left[\frac{\sin(n\Omega_c)}{n\pi} \right] = \lim_{n \to 0} \left[\frac{\Omega_c \cos(n\Omega_c)}{\pi} \right]_{\Omega_c = \pi/4} = \frac{1}{4}. \quad \text{(9.11.11)}$$

The coefficients $\hat{h}_{Lp}$ in (9.11.9c) is real and can be *positive* and *negative* even though the function $|H_{Lp}(e^{j\Omega})|$ is real and nonnegative. They are equal in the sense that the integral squared error between $|H_{Lp}(e^{j\Omega})|$ and $\hat{H}_{Lp}(e^{j\Omega})$ goes to zero. There will be overshoots and undershoots before and after the discontinuity in the F-series approximation. The filter coefficients in (9.11.9) are not only a function of n but also a function of the cutoff frequency Ω_c,

which needs to be specified. As an illustration, let $\Omega_c = \pi/4$. The coefficients $\hat{h}_{Lp}[n]$ are listed in Table 9.11.1 for $n = 0, \pm 1, \pm 2, \ldots, \pm 10$.

From (9.11.9b), for large N, the coefficients are proportional to n^{-1}. The sample response is non-causal, as $\hat{h}_{Lp}[n]$ in (9.11.9b) is nonzero for $n < 0$. Now shifting all the coefficients to the right by 10 results in a causal sequence and

$$h_{Lp}[n] = \hat{h}_{Lp}[n - 10] = \frac{K}{(n-10)\pi} \sin([n-10]\Omega_c),$$

$$n = 0, 1, 2, \ldots, 20. \quad \text{(9.11.12a)}$$

The frequency response of this filter is

$$H_{Lp}(e^{j\Omega}) = \sum_{n=0}^{20} h_{Lp}[n] e^{-jn\Omega}$$

$$= h_{Lp}[0] + h_{Lp}[1] e^{-j\Omega} + \cdots + h_{Lp}[20] e^{-20\Omega}$$

$$= e^{-j10\Omega} \{ h_{Lp}[0] e^{j10\Omega} + h_{Lp}[1] e^{j9\Omega}$$

$$+ \cdots + h_{Lp}[10] + \cdots + h_{Lp}[19] e^{-j9\Omega}$$

$$+ h_{Lp}[20] e^{-10\Omega} \}.$$

$$\text{(9.11.12b)}$$

Consider the sum

$$h_{Lp}[n] e^{-jn\Omega} + h_{Lp}[20-n] e^{-j(20-n)\Omega}$$

$$= 2h_{Lp}[n] \left[\frac{e^{j[10-n]\Omega} + e^{-j[10-n]\Omega}}{2} \right] e^{-j10\Omega}$$

$$= 2h_{Lp}[n] \cos[(10-n)\Omega] e^{-j10\Omega}, n=1,2,\ldots,9 \text{ (9.11.12c)}$$

$$\Rightarrow H_{Lp}(e^{j\Omega})$$

$$= \left\{ h_{Lp}[10] + 2\sum_{n=0}^{9} h_{Lp}[n] \cos[(10-n)\Omega] \right\} e^{-j10\Omega} e^{j\pi\beta(\Omega)}$$

$$\text{(9.11.13a)}$$

$$= \left[A(e^{j\Omega}) e^{j\pi\beta(\Omega)} \right] e^{-j10\Omega} \quad \text{(9.11.13b)}$$

$$= |H_{Lp}(e^{j\Omega})|$$

$$= \left| \left\{ h_{Lp}[10] + 2\sum_{n=0}^{9} h_{Lp}[n] \cos[(10-n)\Omega] \right\} \right|$$

$$\text{(9.11.13c)}$$

$$\angle H_{Lp}(e^{j\Omega}) = -10\Omega + \pi\beta(\Omega) . \quad \text{(9.11.13d)}$$

In (9.11.13b), $\beta(\Omega)$ is 0 or 1, depending on whether $A(e^{j\Omega})$ is positive or negative and the function (-10Ω) is linear with respect to Ω. For practical reasons, the phase plots need to be in the interval $-\pi \le P = [-L\Omega + \pi\beta] \le \pi$. Otherwise, the phase increases without limit and showing the sketches is not possible. The function $-L\Omega + \pi\beta$ is a piecewise linear function of Ω with $180°$ jumps at those values of Ω where $[A(e^{j\Omega})e^{j\pi\beta(\Omega)}]$ changes sign. The term $\pi\beta(\Omega)$ is usually not shown in (9.11.13d). The amplitude response is sketched in Fig. 9.11.2 assuming 11 coefficients $h_{Lp}[0], h_{Lp}[1], \ldots, h_{Lp}[10]$. From (9.11.13a),

$$H_{Lp}(e^{j0}) = h_{Lp}[10] + 2\sum_{n=0}^{9} h_{Lp}[n],$$

$$H_{Lp}(e^{j\pi}) = h_{Lp}[10] + 2\sum_{n=0}^{9} h_{Lp}[n](-1)^n.$$

Increasing the number of coefficients increases the number of ripples in the passband and the stopband due to the discontinuity in the approximating function; see Gibbs phenomenon in Section 3.7.2. At the discontinuity, the F-series converges to the *half-value* and here, it is

$$[|H_{Lp}(e^{j\pi/4})|^- + |H_{Lp}(e^{j\pi/4})|^+]/2 = 0.5.$$

9.11.2 High-Pass, Band-Pass, and Band-Elimination FIR Filter Designs

High-pass filter design: Consider the amplitude responses of the ideal digital low-pass and Hp

filters shown in Fig. 9.11.3a,b for one period. In Fig. 9.11.3b, it is shown as Ω_c. The ideal Hp filter amplitude response function and its F-series are

$$|H_{Hp}(e^{j\Omega})| = \begin{cases} 0, & 0 < |\Omega| < \pi - \Omega_a \\ 1, & \pi - \Omega_a < |\Omega| < \pi \end{cases}, \quad H_{HP}(e^{j\Omega})$$
$$= H_{HP}(e^{j(\Omega+2\pi)}), \quad (9.11.14)$$

$$|H_{Hp}(e^{j\Omega})| \approx \sum_{n=-\infty}^{\infty} \hat{h}_{Hp}[n]e^{-jn\Omega} = \hat{H}_{Hp}(e^{j\Omega}).$$

The coefficients $\hat{h}_{Hp}[n]$ are related to the coefficients in the low-pass case $\hat{h}_{Lp}[n]$. Starting with an ideal

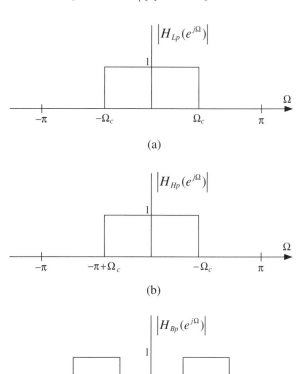

(a)

(b)

(c)

(d)

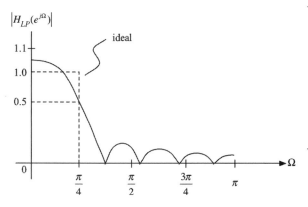

Fig. 9.11.2 Low-pass filter frequency response $|H_{LP}(e^{j\Omega})|$ with 11 coefficients

Fig. 9.11.3 Ideal digital filter amplitude responses: (a) low-pass, (b) high-pass, (c) band-pass, and (d) band elimination

low-pass amplitude response with a cutoff frequency Ω_c and shifting the low-pass frequency response function by π result in the high-pass filter function

$$H_{Hp}(e^{j\Omega}) = H_{Lp}(e^{j(\Omega-\pi)}). \qquad (9.11.15)$$

Using the results in (9.11.9a) and comparing the coefficients results in the following:

$$\left| H_{Hp}(e^{j\Omega}) \right| \approx \sum_{n=-\infty}^{\infty} \hat{h}_{Hp}[n]e^{-jn\Omega} = \hat{H}_{Hp}(e^{j\Omega})$$

$$= \hat{H}_{Lp}(e^{j(\Omega-\pi)}) = \sum_{n=-\infty}^{\infty} \hat{h}_{Lp}[n]e^{-jn\Omega}|_{\Omega\to\Omega-\pi}$$

$$= \sum_{n=-\infty}^{\infty} \hat{h}_{Lp}[n]e^{-jn(\Omega-\pi)}$$

$$= \sum_{n=-\infty}^{\infty} \left\{ \hat{h}_{Lp}[n]e^{jn\pi} \right\} e^{-jn\Omega} \Rightarrow \hat{h}_{Hp}[n]$$

$$= (-1)^n \hat{h}_{Lp}[n].$$

$$(9.11.16a)$$

In approximating the Hp function, the number of coefficients is truncated to $(2L+1)$ resulting in $\hat{h}_{Hp}[n], 0 \leq |n| \leq L$ and the other coefficients are assumed to be zero. The causality problem can be handled by shifting all the coefficients to the right by L. That is,

$$h_{Hp}[n] = \hat{h}_{Hp}[n-L], \quad H_{Hp}(e^{j\Omega}) = \sum_{n=0}^{2L+1} h_{Hp}[n]e^{-j\Omega}.$$

$$(9.11.16b)$$

These results could have been obtained directly using the F-series approximation of the high-pass function instead of deriving Hp filter coefficients from the Lp filter coefficients.

Example 9.11.1 Consider an ideal HP filter with a cutoff frequency of 1 kHz and the sampling frequency is 5 kHz. Using the F-series, determine the coefficients of the filter.

Solution: The digital cutoff frequency is

$$\Omega = \omega t_s|_{f_c=1\text{ kHz}} = \omega/f_s|_{f_c=1\text{ kHz}}$$
$$= 2\pi(10^3)(1/5(10^3)) = .4\pi \text{ rad/s}.$$

Now set the cutoff digital frequency as $\pi - \Omega_c = 0.4\pi \Rightarrow \Omega_c = 0.6\pi$. The cutoff frequency of the

digital Lp is $\Omega_c = 0.6\pi$. Now design a low-pass filter with this cutoff frequency. The filter coefficients are

$$\hat{h}_{Lp}[n] = (1/n\pi)\,\sin(0.6n\pi),\ n = 0, \pm 1, \pm 2, \ldots.$$

Assuming the number of filter coefficients is equal to $2L+1 = 21$, determine the Hp filter coefficients and then shift the coefficients to the right by L samples. These result in

$$h_{Hp}[n] = (-1)^{(n-L)}[1/\pi(n-L)]\,\sin(0.6\pi(n-L)),$$
$$n = 0, 1, 2, \ldots, 2L. \qquad \blacksquare$$

Digital band-pass filter design using the low-pass filter design: Consider the ideal amplitude res-ponses of Bp and Lp functions in Fig. 9.11.3c,a with frequencies as identified. Considering one period, the band-pass function can be obtained by shifting the low-pass function to the right and to the left by Ω_0. That is $\Omega \to (\Omega - \Omega_0)$ and $\Omega \to (\Omega + \Omega_0)$. The band-pass function in terms of the low-pass function is

$$H_{Bp}(e^{j\Omega}) = H_{Lp}(e^{j(\Omega-\Omega_0)}) + H_{Lp}(e^{j(\Omega+\Omega_0)}). \qquad (9.11.17)$$

The upper and lower cutoff frequencies of the BP filter are Ω_u and Ω_l and the center frequency is $\Omega_0 = [\Omega_u + \Omega_l]/2$. In terms of the Bp frequencies, the ideal Lp filter response is centered at $\Omega = 0$ with a cutoff frequency $\Omega_c = [\Omega_u - \Omega_l]/2$. The Bp filter response is

$$\left| H_{Bp}(e^{j\Omega}) \right| \approx \sum_{n=-\infty}^{\infty} \hat{h}_{Bp}[n]e^{-jn\Omega}$$

$$= \sum_{n=-\infty}^{\infty} \hat{h}_{Lp}[n]e^{-jn\Omega}|_{\Omega\to\Omega-\Omega_0}$$

$$+ \sum_{n=-\infty}^{\infty} \hat{h}_{Lp}[n]e^{-jn\Omega}|_{\Omega\to\Omega+\Omega_0}$$

$$= \hat{H}_{Bp}(e^{j\Omega}) = \sum_{n=-\infty}^{\infty} \hat{h}_{Lp}[n][n]e^{-jn(\Omega-\Omega_0)}$$

$$+ \sum_{n=-\infty}^{\infty} \hat{h}_{Lp}[n][n]e^{-jn(\Omega+\Omega_0)}$$

$$= \sum_{n=-\infty}^{\infty} (\hat{h}_{Lp}[n]e^{jn\Omega_0})e^{-jn\Omega}$$

$$+ \sum_{n=-\infty}^{\infty} (\hat{h}_{Lp}[n]e^{-jn\Omega_0})e^{-jn\Omega}.$$

Using Euler's theorem and combining two typical terms in the last equation results in

$$\left|H_{Bp}(e^{j\Omega})\right| \approx \sum_{n=-\infty}^{\infty} \{2\hat{h}_{Lp}[n] \cos(n\Omega_0)\}e^{-jn\Omega_0}$$

$$= \hat{H}_{Bp}(e^{j\Omega}). \tag{9.11.18a}$$

The Bp filter coefficients are related to the LP filter coefficients by

$$\hat{h}_{Bp}[n] = 2\hat{h}_{Lp}[n] \cos(n\Omega_0). \tag{9.11.18b}$$

Example 9.11.2 Consider an ideal Bp filter with a pass band from 1 to 1.5 kHz with a sampling frequency of 5 kHz. Determine the coefficients for the Bp filter.

Solution: Just like in the last example, using the sampling frequency $f_s = 5$ kHz, the upper, the lower, and the center frequencies of the band-pass spectrum are

$$\Omega_u = \omega_u t_s = \omega_u/f_s = 2\pi(1.5)(10^3)/5(10^3) = 0.6\pi,$$

$$\Omega_l = \omega_l t_s = \omega_l/f_s = 2\pi(10^3)/5(10^3) = 0.4\pi,$$

$$\Omega_0 = (\Omega_u + \Omega_l)/2 = 0.5\pi. \tag{9.11.19a}$$

The cutoff frequency of the Lp filter is $\Omega_c = (\Omega_u - \Omega_l)/2 = 0.1\pi$. The coefficients in the low-pass design in Example 9.10.3 and the corresponding band-pass filter coefficients for the filter centered at $\Omega_0 = .5\pi$ are, respectively, given by

$$\hat{h}_{Lp}[n] = (1/n\pi) \sin(n\Omega_c) = (1/n\pi) \sin(0.1n\pi),$$

$$\tag{9.11.19b}$$

$$\hat{h}_{Bp}[n] = 2 \cos(0.5n\pi)\hat{h}_{Lp}[n], \quad n = 0, \pm1, \pm2, \ldots.$$

$$\tag{9.11.19c}$$

Truncating the number of coefficients to $(2L+1)$ coefficients and shifting the coefficients to the right by L result in the coefficients for a causal band-pass filter by

$$h_{Bp}[n] = \hat{h}_{Bp}[n - L]$$

$$= 2 \cos(0.5(n - L)\pi)\frac{\sin(0.1\pi(n - L))}{\pi(n - L)},$$

$$n = 0, 1, 2, \ldots, 2L. \tag{9.11.19d}$$

∎

Digital band-elimination filter design using the band-pass filter design: The derivation for the filter coefficients for the band-elimination filter follows that of the band-pass filter. The proof is left as an exercise. As before, we have

$$\Omega_0 = (\Omega_u + \Omega_L)/2, \ \Omega_c = (\Omega_u - \Omega_l)/2, \tag{9.11.20a}$$

$$\left|H_{Be}(e^{j\Omega})\right| = 1 - \left|H_{Bp}(e^{j\Omega})\right| \approx 1 - \sum_{n=-\infty}^{\infty} \hat{h}_{Bp}[n]e^{-jn\Omega}$$

$$= (1 - \hat{h}_{Bp}[0]) - \sum_{\substack{n=-\infty \\ n\neq 0}}^{\infty} \hat{h}_{Bp}[n]e^{-jn\Omega}$$

$$= \sum_{n=-\infty}^{\infty} \hat{h}_{Be}[n]e^{-jn\Omega} = \hat{H}_{Be}(e^{j\Omega})\hat{h}_{Be}[0]$$

$$= (1 - \hat{h}_{Bp}[0]), \ \hat{h}_{Be}[n] = -\hat{h}_{Bp}[n],$$

$$n = \pm1, \pm2, \ldots, \pm L. \tag{9.11.20b}$$

Table 9.11.2 FIR Filter Coefficients for the Four Basic Filters

Low pass:	$\hat{h}_{lp}[n] = \frac{1}{n\pi}\sin(n\Omega_c), n = 0, \pm1, \pm2, \ldots, \pm L, \left|H_{Lp}(e^{j\Omega})\right| = \begin{cases} 1, |\Omega| < \Omega_c \\ 0, \Omega_c < |\Omega| < \pi \end{cases},$	(9.11.21)
High pass:	$\hat{h}_{Hp}[n] = (-1)^n\hat{h}_{Lp}[n], n = 0, \pm1, \pm2, \ldots, \pm L, \left|H_{Hp}(e^{j\Omega})\right| = \begin{cases} 1, \pi - \Omega_c < |\Omega| < \pi \\ 0, 0 < |\Omega| < \pi - \Omega_c \end{cases},$	(9.11.22)
Band pass:	$\begin{cases} \hat{h}_{Bp}[n] = 2\cos(n\Omega_0)\hat{h}_{Lp}[n], n = 0, \pm1, \pm2, \ldots, \pm L \\ \left|H_{Bp}(e^{j\Omega})\right| = \begin{cases} 1, \Omega_l < |\Omega| < \Omega_u \\ 0, 0 < |\Omega| < \Omega_l \text{ and } \Omega_u < |\Omega| < \pi \end{cases} \end{cases},$	(9.11.23)
Band elimination:	$\begin{cases} \hat{h}_{Be}[0] = 1 - \hat{h}_{Bp}[0], \hat{h}_{Be}[n] = -\hat{h}_{Bp}[n], n = \pm1, \pm2, \ldots, \pm L \\ \left|H_{Be}(e^{j\Omega})\right| = \begin{cases} 0, \Omega_l < |\Omega| < \Omega_u \\ 1, 0 < |\Omega| < \Omega_l \text{ and } \Omega_u < |\Omega| < \pi \end{cases} \end{cases}.$	(9.11.24)

To find the causal filter coefficients, the coefficients need to be shifted by L. The filter coefficients for the non-causal low-pass, high-pass, band-pass, and band-elimination filters are given in Table 9.11.2. To have the causal filter the coefficients need to be shifted to the right by L resulting in $(2L+1)$ coefficients.

9.11.3 Windows in Fourier Design

The F-series approximation of ideal filter functions results in ripples before and after the discontinuity and the series converges to the half-value of the function at the discontinuities. Since the HP, BP, and BE filters are designed using LP filter amplitude responses it is instructive to study the Gibbs phenomenon of the amplitude response of a noncausal low-pass filter. Fourier series is *optimum* only in the

sense of minimizing the *error energy*. Truncating the number of coefficients to a length of $(2L+1)$ results in

$$\hat{h}[n]w_R[n] = \begin{cases} \hat{h}[n], |n| \le L \\ 0, \quad |n| > L \end{cases}, \ w_R[n] = \begin{cases} 1, |n| \le L \\ 0, |n| > L \end{cases}.$$
(9.11.25)

$$\hat{H}_R(e^{j\Omega}) = \sum_{n=-L}^{L} \hat{h}[n]e^{-jn\Omega} = \sum_{n=-\infty}^{\infty} \hat{h}[n]w_R[n]e^{-jn\Omega}.$$
(9.11.26)

The subscript R indicates the coefficients are for the ideal filter function with the rectangular window resulting in ripples in the frequency amplitude response before and after the discontinuity at the cutoff frequency. A *tapered window,* instead of a rectangular window, would reduce the ripples; see Harris (1978) and Ambardar (2007).

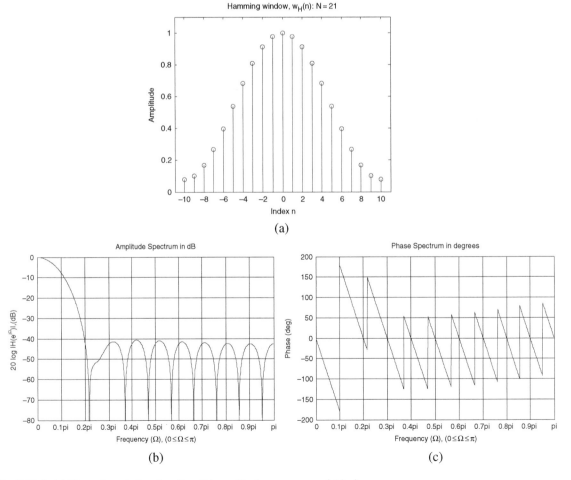

Fig. 9.11.4 (a) Hamming window function, (b) amplitude response, and (c) phase response

Hamming window: The symmetric Hamming window is a popular window defined by

$$w_H[n] = \left\{ \begin{array}{l} 0.54 + 0.46\cos(n\pi/L), |n| \leq L \\ 0, \text{ otherwise} \end{array} \right\} \overset{\text{DTFT}}{\longleftrightarrow}$$

$$0.54 W_R(e^{j\Omega}) + 0.23 W_R(e^{j(\Omega+2\pi/N)})$$
$$+ 0.23 W_R(e^{j(\Omega-2\pi/N)}) = W_H(e^{j\Omega}), N = 2L+1.$$
$$(9.11.27)$$

Figure 9.11.4 gives an example with the number of coefficients equal to 21 illustrating the time domain sequence and the corresponding amplitude and phase responses of a Hamming window. The peak sidelobe attenuation is approximately -42.7 dB compared to -13.3 dB in the rectangular window. Window minimized the sidelobe level at the expense of the high-frequency decay rate. Hamming window is a practical window that has essentially fixed sidelobe levels. Another window that is popular is the *Kaiser window* that has a variable parameter that controls the peak sidelobe level. It is considered to be nearly optimal in the sense that of having the most energy in its main lobe for a given sidelobe amplitude, see Oppenheim and Schafer (1999).

   Illustration of the effect of a window on the filter design: Fourier series expansion of the filter functions that have discontinuities result in ripples before and after the discontinuity. The ripple reduction is illustrated using the simple *cosine window function*. The window and its transform are

$$w_C[n] = (e^{j\pi/N} + e^{-j\pi/N})/2 = \cos(\pi n/N), -L < n < L,$$
$$N = 2L + 1,$$
$$(9.11.28a)$$

$$W_C(e^{j\Omega}) = \pi\delta(\Omega - \Omega_0) + \pi\delta(\Omega - \Omega_0), \Omega_0 = \pi/N.$$
$$(9.11.28b)$$

With these, we have

$$\hat{H}_C(e^{j\Omega}) = \sum_{n=-L}^{L} w_C[n]\hat{h}[n]e^{-jn\Omega},$$

$$w_C[n]\hat{h}[n] \overset{\text{DTFT}}{\longleftrightarrow} W_C(e^{j\Omega}) * \hat{H}(e^{j\Omega}), \quad (9.11.29a)$$

$$\hat{H}(e^{j\Omega}) = \sum_{n=-\infty}^{\infty} \hat{h}[n]e^{-jn\Omega}, \ W_C(e^{j\Omega}) = \sum_{n=-L}^{L} w_C[n]e^{-jn\Omega}.$$
$$(9.11.29b)$$

Noting the time domain multiplication corresponds to frequency domain convolution, we have

$$\hat{H}_{LP,FIR}(e^{j\Omega}) * W_c(e^{j\Omega})$$

$$= \frac{1}{2\pi}\int_{-\pi}^{\pi} \hat{H}_{Lp,FIR}(e^{j\theta}) W_c(e^{j(\theta-\Omega)})d\theta$$

$$= \frac{1}{2\pi}\int_{-\pi}^{\pi} \hat{H}_{Lp,FIR}(e^{j\theta})\pi[\delta(\Omega-\Omega_0-\theta)+\delta(\Omega+\Omega_0-\theta)]d\theta$$

$$= \frac{1}{2}\hat{H}_{Lp,FIR}(e^{j(\Omega+\frac{\pi}{N})})+\frac{1}{2}\hat{H}_{Lp,FIR}(e^{j(\Omega-\frac{\pi}{N})}). \quad (9.11.30)$$

Equation (9.11.30) illustrates that the convolution of the simple cosine window function with the Fourier series expansion of the filter function resulted in the addition of two shifted functions separated by $2\pi/N$. Figure 9.11.5a,b illustrates the effects of superimposing two shifted versions of the frequency responses obtained by using the Fourier series expansion of a low-pass filter to reduce the ripples

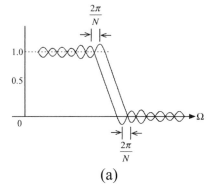
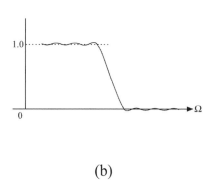

Fig. 9.11.5 (a) Two shifted versions of low-pass filter responses and (b) reducing the Gibbs phenomenon by averaging the two responses

(a)   (b)

around a discontinuity. Noting that the *ripples* *alternate* in sign, adding the responses reduce the overshoots to about 2% at the *expense* of increasing the *transition width* of the filter. This can be extended for the Hamming window in (9.11.27). It has a constant and a cosine term. Now there are three copies and using the above analysis we can see how the Hamming window results in further reduction of the ripple size at the expense of the filter transition width. The equations are

$$\hat{H}_{H,FIR}(e^{j\Omega}) = \sum_{n=-L}^{L} w_H[n]\hat{h}[n]e^{-jn\Omega},$$

$$w_H[n]\hat{h}[n] \overset{\text{DTFT}}{\longleftrightarrow} W_H(e^{j\Omega}) * \hat{H}(e^{j\Omega}), \quad (9.11.31a)$$

$$\hat{H}(e^{j\Omega}) = \sum_{n=-\infty}^{\infty} \hat{h}[n]e^{-jn\Omega},$$

$$W_H(e^{j\Omega}) = \sum_{n=-L}^{L} w_H[n]e^{-jn\Omega}. \quad (9.11.31b)$$

An important measure to compare windows is the peak sidelobe gain. For the rectangular and the Hamming widows, they are 13.3 and 42.7 dB. Another way to evaluate them is by the Poisson summation formula, see Yarlagadda and Allen (1982).

Notes: Fourier series design is a practical method for linear phase filter designs. There are *no* simple answers in selecting a particular window and the number of samples in it. Ludeman (1986) gives a systematic procedure in the design of FIR *low-pass* filters.

1. The stop-band filter amplitude response is relatively insensitive to the number of filter coefficients. Select a window that satisfies the minimum stop-band attenuation. For most applications Hamming would be adequate. The minimum stop-band filter approximate attenuations for the rectangular and the Hamming windows are -21 and -53 dB.

2. Let the edges of the passband and stopband be Ω_1 and Ω_2. Main lobe width of the window is an estimate of the filter transition width. If the window width is N, then the transition widths of the rectangular and Hamming windows are $(4\pi/N)$ and $(8\pi/N)$. As a starting point, select

$$N \geq k\frac{2\pi}{\Omega_2 - \Omega_1}(k = 2, \text{rectangular},$$

$$k = 4 \text{ Hamming window}). \quad (9.11.32a)$$

Oppenheim and Schafer (1989) suggest a good estimate of the filter transition width as $1.81\pi/(N-1)$ for a rectangular window and $6.27\pi/(N-1)$ for the Hamming window. Estimate the number of window samples by

$$N = \frac{\text{Attenuation in the stop band} - 8}{2.285(\text{frequency transition width})}. \quad (9.11.32b)$$

3. Most designs require an integer delay. Select so that N(an odd number)$\geq (2N+1)$.
4. Select the cutoff frequency for the response as $\Omega_c = \Omega_1$ and the delay $\alpha = (N-1)/2$.
5. A trial response is

$$h[n] = \frac{\sin[\Omega_c(n-(N-1)/2)]}{\pi[n-(N-1)/2]}w[n]. \quad (9.11.33)$$

The type of the window selected is based on the stop-band attenuation requirements.

6. Plot $\hat{H}(e^{j\Omega})$ and check if the design specifications are satisfied. If the attenuation requirement at $\Omega = \Omega_1$ is not satisfied, adjust Ω_c. Normally, we increase the value on the first iteration. Go back to the last step and check to see if the attenuation specifications are satisfied. If they are satisfied, check to see if the number of coefficients can be reduced. If so, reduce N. The designs require computer facilities, such as MATLAB, as the procedure is iterative to satisfy the amplitude specifications. For MATLAB, see Ingle and Proakis (2007). Ludeman (1986) gives a systematic procedure to determine the digital filter coefficients in an A/D–digital filter–D/A structure. This is given below.

Example 9.11.3 Consider the low-pass filter design in an A/D–digital filter–D/A with the specifications shown in Fig. 9.11.6a. The filter is required to have a linear phase. Assume the sampling rate is equal to be $f_s = 100$ or the sampling interval is $t_s = (1/100)$. The transition band of the analog filter is assumed to be in the range $30\pi < \omega < 45\pi$. The pass-band and stop-band attenuations must be no more than 3 dB and at least 50 dB.

Fig. 9.11.6 (a) Analog and (b) digital frequency response specifications

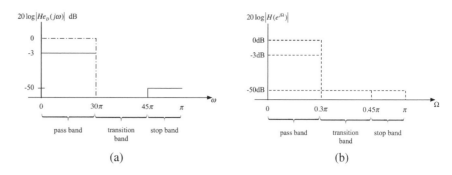

(a)

(b)

The digital frequencies are $\Omega_i = \omega_i t_s$ and the edges of the passband and stopband of the digital filter are $\Omega_1 = 30\pi/100 = .3\pi$, $\Omega_2 = 45\pi/100 = .45\pi$. The digital filter specifications are shown in Fig. 9.11.6b. A Hamming window is used to have the stop-band attenuation over 50 dB.

$$N(\text{Initial sample size}) \geq k \frac{2\pi}{(\Omega_2 - \Omega_1)},$$

$k = 4$ for the Hamming window.

This gives a stop-band attenuation of 53.3 dB. To have an integer delay, select $N = 55$. Set $\Omega_c = \Omega_1 = 0.3\pi$. The integer delay is $\alpha = (N-1)/2 = 27$. The trial impulse response from (9.11.33) is

$$h[n] = \frac{\sin[\Omega_c(n - ((N-1)/2))]}{\pi(n - ((N-1)/2))}[0.54 - 0.46 \cos(2\pi n/54)],$$

$$0 \leq n \leq 54.$$

$w_H[n]$ is a causal sequence and the sign is negative before 0.46 in the above equation, which is different to the one in (9.11.2). Figure 9.11.7a gives the Hamming window sequence and Fig. 9.11.7b gives the corresponding filter response. The corresponding amplitude and phase responses are plotted using MATLAB given in Fig. 9.11.7c,d. The sketches are plotted using the MATLAB (see Section B.11). The attenuation at frequency Ω_1 is 6 dB, which is more than the specified at frequency Ω_1. So increase the cutoff frequency to say $\Omega_c = .33$. Check the frequency response to see if all the specifications are satisfied. If satisfied, then see by trial and error, if the value of N can be reduced and still satisfy the requirements. By trial and error, $N = 29$ and

$$h[n] = \frac{\sin[\Omega_c(n - 29)]}{\pi(n - 29)}[.54 - .46 \cos(2\pi n/29)],$$

$$0 \leq n \leq 28.$$

The final amplitude and phase filter responses are shown in Fig. 9.11.8a,b. Note that if (9.11.32b) had been used to find the initial N, it comes out to be 39, which is closer to the final N. The input–output equation of the FIR filter corresponding to the value N is given below requiring N multiplications and $N - 1$ additions.

$$y[n] = \sum_{k=0}^{N-1} h[k]x[n-k] = h[0]x[n] + h[1]x[n-1]$$

$$+ \cdots + h[N-1]x[n-(N-1)]. \qquad (9.11.34)$$

∎

9.12 Digital Filter Realizations

Consider the realization of IIR filter using summers, multipliers, and delay components given in Fig. 9.2.1, assuming the design specifications resulted in a transfer function (or the difference equation) given below:

$$Y[z] = H(z)X(z), H(z) = \frac{\sum_{k=0}^{M} b_k z^{-k}}{1 + \sum_{k=1}^{N} a_k z^{-k}}, M \leq N,$$

$$(9.12.1a)$$

$$y[n] = -\sum_{k=1}^{N} a_k y[n-k] + \sum_{k=0}^{M} b_k x[n-k]. \quad (9.12.1b)$$

The output transform $Y(z)$ in (9.12.1a) can be written in terms of two parts, an all pole model and an all zero model given below. The output transform

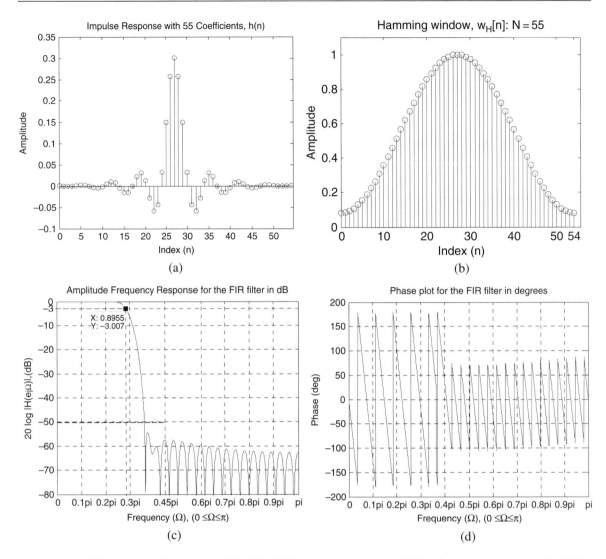

Fig. 9.11.7 (a) Hamming window function ($N = 55$), (b) filter impulse response, (c) filter frequency response, and (d) filter phase response

can be written in terms of an intermediate function $P(z)$ as follows:

$$Y(z) = \left(\sum_{k=0}^{M} b_k z^{-k} \right) P(z), \qquad (9.12.2a)$$

$$P(z) = \frac{1}{1 + \sum_{k=1}^{N-1} a_k z^{-k}} X(z) \text{ or}$$

$$\qquad (9.12.2b)$$

$$P(z) + \sum_{k=1}^{N-1} a_k [z^{-k} P(z)] = X(z).$$

Noting $p[n-k] \overset{z}{\longleftrightarrow} z^{-k} P(z)$ and $y[n-k] \overset{z}{\longleftrightarrow} z^{-k} Y(z)$, the corresponding difference equations are

$$p[n] = x[n] - \sum_{k=1}^{N} a_k p[n - k],$$

$$y[n] = \sum_{k=0}^{M} b_k p[n - k] \text{ (Form 1).} \qquad (9.12.3)$$

Figure 9.12.1a uses (9.12.3) to realize a second-order ($N = 2$) system, which is a *direct form*. Similar procedure can be used to determine a structure for any N. Orders higher than 2 are highly sensitive to small changes in the coefficients. The standard

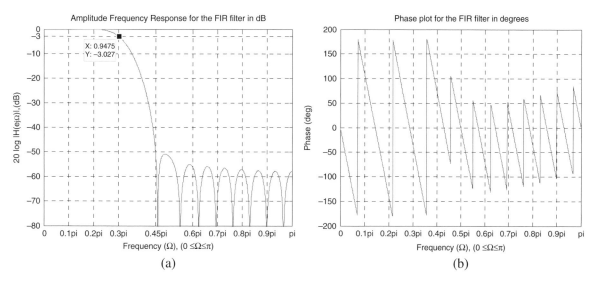

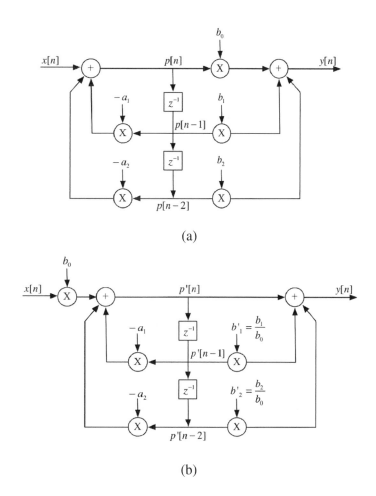

Fig. 9.11.8 (a) Final amplitude and (b) final phase responses ($N = 29$)

Fig. 9.12.1 (a) Direct and (b) modified direct form realizations of a second-order system

approach is to use second-order realizations. The equations in (9.12.2a and b) can be modified by redefining the functions as follows. The realization uses the difference equation in (9.12.5) and is shown in Fig. 9.12.1b, which is referred to as a *modified direct form*:

$$Y(z) = \left(\sum_{k=0}^{M} (b_k/b_0)z^{-k} \right), \quad (9.12.4a)$$

$$P'(z) = \frac{b_0 X(z)}{1 + \sum_{k=1}^{N-1} a_k z^{-k}}$$

$$\text{or } P'(z) = b_0 X(z) - \sum_{k=1}^{N-1} a_k z^{-k} \quad (9.12.4b)$$

$$\Rightarrow p'[n] = b_0 x[n] - \sum_{k=1}^{N} a_k p'[n-k];$$

$$y[n] = \sum_{k=0}^{M} \frac{b_k}{b_0} p'[n-k]. \quad (9.12.5)$$

9.12.1 Cascade Form of Realization

The transfer function is written in terms of a product of second-order functions. That is,

$$H(z) = H_k(z)H_{k-1}(z)...H_1(z). \quad (9.12.6)$$

The second-order terms or the *biquadratic terms* are

$$H_k(z) = \frac{b_{0k} + b_{1k}z^{-1} + b_{2k}}{1 + a_{1k}z^{-1} + a_{2k}z^{-2}}$$
$$= b_{0k} \frac{1 + b'_{1k}z^{-1} + b'_{2k}z^{-2}}{1 + a_{1k}z^{-1} + a_{2k}z^{-2}},$$
$$b_{ik} = b'_{ik}b_{0k}, \ i = 1, 2. \quad (9.12.7)$$

The transfer function $H(z)$ can be realized by the cascade form in Fig. 9.12.2. If N happens to be odd, then one of the above functions, say $H_K(z)$ is a first-order system and $b'_{2K} = a_{2K} = 0$. Furthermore,

since the coefficient b_{0k} is separated from the other part, we can combine all the multipliers into a single multiplier $b'_0 = b_{01}b_{02}...b_{0k}$.

Notes: In a digital filter realization, it is necessary to avoid overflows. This is achieved by using *scaling*. The topic is beyond the scope here, see Ifeachor and Jervis (1993).

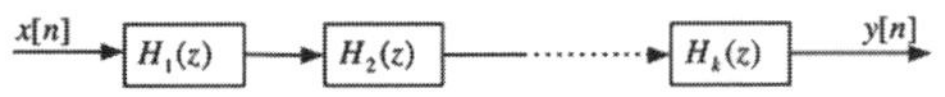

9.12.2 Parallel Form of Realization

In this form the transfer function is expressed as a sum of second-order terms with real coefficients in (9.12.8). Its realization in terms of $H_i(z)$ is shown in Fig. 9.12.3:

$$H(z) = \sum_{i=1}^{k-1} H_i(z),$$

$$H_i(z) = \frac{b_{0i} + b_{1i}z^{-1}}{1 + a_{1i}z^{-1} + a_{2i}z^{-2}}, \quad Y(z) = \sum_{i=1}^{k-1} [H_i(z)X(z)].$$

$$(9.12.8)$$

Special cases:

$$b_{0i} = c \text{ and } b_{1i} = a_{1i} = a_{2i} = 0 \Rightarrow H_i(z) = c, \quad (9.12.9a)$$

$$b_{1i} = a_{2i} = 0 \Rightarrow H_i(z) = b_{0i}/[1 + a_{1i}z^{-1}], \quad (9.12.9b)$$

$$b_{0i} = a_{1i} = a_{2i} = 0 \Rightarrow H_i(z) = b_{1i}z^{-1}. \quad (9.12.9c)$$

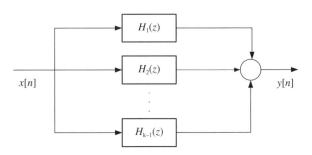

Fig. 9.12.3 Parallel realization in terms of biquadratic sections

$x[n] \rightarrow \boxed{H_1(z)} \rightarrow \boxed{H_2(z)} \rightarrow \cdots \rightarrow \boxed{H_k(z)} \rightarrow y[n]$

Fig. 9.12.2 Cascade form of realization using possibly first- and second-order systems

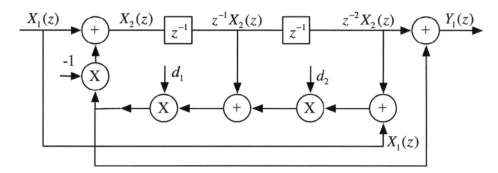

Fig. 9.12.4 Realization of a second-order all-pass function

9.12.3 All-Pass Filter Realization

Example 9.12.1 Show the transfer function of the system in Fig. 9.12.4 is

$$H_{ap}(z) = \frac{d_1 d_2 + d_1 z^{-1} + z^{-2}}{1 + d_1 z^{-1} + d_1 d_2 z^{-2}}, \ Y_1(z) = H_{ap}(z) X_1(z).$$

$$(9.12.10a)$$

Solution: From Fig. 9.12.4, we have

$$X_2(z) = X_1(z) - d_1 z^{-1} X_2(z) - d_1 d_2 z^{-2} X_2(z)$$
$$- d_1 d_2 X_1(z) \Rightarrow X_2(z)$$
$$= X_1(z)[1 - d_1 d_2]/[1 + d_1 z^{-1} + d_1 d_2 z^{-2}]$$
$$Y_1(z) = z^{-2} X_2(z) + d_1 z^{-1} X_2(z)$$
$$+ d_1 d_2 [z^{-2} X_2(z) + X_1(z)] \quad (9.12.10b) \ \blacksquare$$

simplifying these results in (9.12.10a).

9.12.4 Digital Filter Transposed Structures

Additional implementations can be generated using the *transposition (TP) theorem*, see Oppenheim et al. (1999). Each realization contains delay components, branches going from one node to another involving multipliers, summers, or summing nodes and branch points. The TP theorem states that the transfer function is unchanged if the following sequences of network operations are conducted on a digital structure:

1. Reverse the direction of all branches.

2. Change branch points into summing nodes and vice versa.

3. Interchange the input and output.

Figure 9.12.5a,b gives two second-order structures, a modified direct form and a transposed direct form. These forms are equivalent by solving for $y[n]$ in terms of $x[n]$

Direct form:

$$p[n] = -a_1 p[n-1] - a_2 p[n-2] + x[n],$$
$$y[n] = b_0 p[n] + b_1 p[n-1] + b_2 p[n-2].$$

Transposed direct form:

$$q[n] = b_0 x[n] + q_1[n-1], \quad y[n] = q[n],$$
$$q_1[n] = -a_1 y[n] + b_1 x[n] + q_2[n-1],$$
$$q_2[n] = -a_2 y[n] + b_2 x[n].$$

9.12.5 FIR Filter Realizations

Consider the output of the FIR filter

$$y[n] = \sum_{k=0}^{M-1} h[k] x[n-k], \ Y(z) = H(z) X(z), \ H(z)$$
$$= \sum_{k=0}^{M-1} h[k] z^{-k}. \quad (9.12.11)$$

A simple *transversal or a tapped-delay line filter* is shown in Fig. 9.12.6. A transposed structure corresponding to this filter can be used as well. Noting

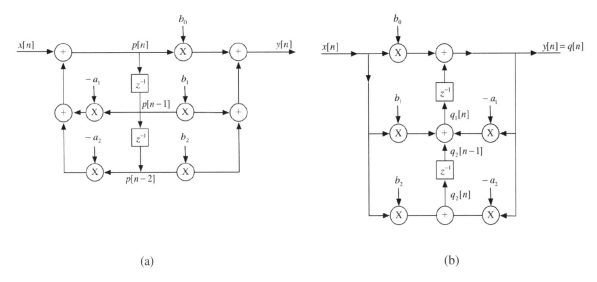

(a)   (b)

Fig. 9.12.5 (a) Direct form and (b) transposed direct form

Fig. 9.12.6 Direct-form realization of the FIR filter

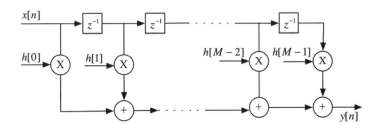

that we are interested in linear phase FIR filters, we can reduce the number of multipliers by noting that these filters satisfy either the symmetry or asymmetry condition

$$h[n] = \pm h[M - 1 - n]. \qquad (9.12.12)$$

We can reduce the number of multiplications to $(M - 1)/2$ for M odd and $M/2$ for M even by using (9.12.12). With even symmetry, the transfer function in (9.12.11) is

$$H(z) = \begin{cases} \displaystyle\sum_{k=0}^{M/2-1} h[k][z^{-k} + z^{-(M-1-k)}], & M \text{ even} \\ \displaystyle\sum_{k=0}^{(M-3)/2} h[k][z^{-k} + z^{-(M-1-k)}] \\ \quad + h[(M-1)/2]z^{-(M-1)/2}, & M \text{ odd} \end{cases}$$
$$(9.12.13)$$

9.13 Summary

This chapter dealt with fast Fourier transforms, convolution, correlation, discrete-time systems analysis, z-transforms, filtering algorithms, and realizations of the discrete-time filters. Following gives a list of some of the specific topics in this chapter.

- Review of the discrete Fourier transforms (DFTs)
- Computation of the DFT using fast Fourier transforms algorithms and its efficiency
- Computation of convolution and correlation of time-limited sequences
- Use of overdetermined system of equations in the deconvolution problem
- One- and two-sided z-transforms
- Solutions of difference equations using one-sided z-transforms
- BIBO stability and the stability of discrete-time linear time-invariant systems

- Basic infinite impulse response (IIR) filter designs
- Digital filter designs by matching the impulse response of an analog system
- The design of finite impulse response (FIR) filters based on window design
- Digital filter simulation diagrams

Problems

9.2.1 Use the flow graph in Fig. 9.2.3 corresponding to the 4-point discrete Fourier transform implementation to write the relations in matrix form in the form $\mathbf{X} = \mathbf{A}_{\text{DFT}}\mathbf{x}$. Show that $\mathbf{A}_{\text{DFT}}$ is the 4×4 DFT matrix.

9.2.2 Make a table that gives the number of complex multiplications required by the direct method of computing the DFT ($N_{\text{Direct}} = N^2$) versus computing the DFT using FFT $N_{\text{FFT}} = (N/2)\log_2(N)$ assuming that $N = 2^\nu$ for $\nu = 2, 3, 4, 5, 6, 7, 8, 9, 10$ and the ratios $N_{\text{DFT}}/N_{\text{FFT}}$ for each of these cases.

9.2.3 Show that DFT can be used to compute the inverse DFT. That is, show $x[n] = (\text{DFT}(X^*[k]))^*/N$.

9.3.1 Consider the data sequence $\{x[0], x[1], x[2], x[3]\}$ and the corresponding DFT sequence $\{X[0], X[1], X[2], X[3]\}$. Now introducing a set of $M = 3$ zeros in the middle of the DFT sequence results in seven DFT values. Let this new sequence be $X_{\text{New}}[k]$. What can you say about its conjugate symmetry of this new sequence? The time sequence obtained from the interpolated sequence will not be real. Now consider the DFT sequence $\{X[0], X[1], (1/2)X[2], 0, 0, 0, (1/2)X[2], X[3]\}$. What can you say about the conjugate symmetry of this sequence and its IDFT? Would this be a proper way of interpolating the time sequence in the case of a time sequence with even number of samples?

9.3.2 Show that the power spectral density can be computed by taking the DFT of the autocorrelation sequence or by using the DFT coefficients directly.

9.3.3 a. Determine the convolution of the sequences $x[n] = \{1, -1, 1\}$ and $h[n] = \{1, 0, -1\}$
  $\qquad\qquad\quad\uparrow\qquad\qquad\qquad\qquad\uparrow$
b. Use the overlap-add method to determine the

convolution of the following sequences: $x[n] = \{1, 2, 3, 4\}$ and $h[n] = \{1, -1\}$.
$\quad\uparrow\qquad\qquad\qquad\qquad\quad\uparrow$

9.3.4 Determine the autocorrelation of the sequence $h[n]$ given in Part b in the last problem directly and by DFT.

9.3.5 Determine the number of multiplications required to compute the convolution of two sequences, each of length 32 and compare this result using FFT.

9.3.6 Consider the following overdetermined system of equations. Solve these equations by minimizing the least-squared error:

$$\underline{y} = \begin{bmatrix} y_1 \\ y_2 \\ y_3 \end{bmatrix} = \begin{bmatrix} 1 \\ 2 \\ 1 \end{bmatrix} = \begin{bmatrix} a_{11} \\ a_{21} \\ a_{31} \end{bmatrix} h = \begin{bmatrix} 1 \\ 1 \\ 1 \end{bmatrix} h = \underline{A}h.$$

Hint: Define $E = (y_1 - a_{11}h)^2 + (y_2 - a_{21}h)^2 + (y_3 - a_{31}h)^2$. Use $\partial E/\partial h = 0$ to solve for h. Use (9.3.28b) and show the results are the same by the two methods.

9.4.1 Find the z-transforms of the following sequences. Give the ROC in each case:

a. $x[n] = (n - 2) \cos(\Omega_0(n - 2))u[n - 2]$

b. $y[n] = nx[n], x[n] \overset{z}{\leftrightarrow} X(z)$ (Use $x[n]$ in Part a to determine the transform)

c. $x[n] = a^n u[n], |a| < 1$,

d. $x[n] = a^n u[-n], |a| > 1$,

e. $x[n] = \delta[n - 3] + 2\delta[n - 1] + \delta[n]$,

f. $x[n] = a^n \sin(\Omega_0 n)$.

9.5.1 Derive the expression given in (9.5.19b).

9.5.2 Determine the final value of the sequence that has the z-transform:

$$X(z) = \frac{z}{z - \alpha}, |z| > |\alpha|.$$

Identify the necessary condition(s) so that the final value theorem is applicable. Give the results for two cases: $a.\ \alpha = 1,\ b.\ |\alpha| < 1$.

9.5.3 Find the initial value $x[0]$ of the sequence $x[n] \overset{z}{\longleftrightarrow} X(z)$ from

$$X(z) = \frac{z}{(1 - .5z^{-1})(1 - .4z^{-2})}, \text{ROC} : |z| > .5.$$

Hint: In your solution, first express $X(z)$ in terms of a ratio of polynomials in z. Check to see its behavior as $z \to \infty$. What can you say about the causality of the sequence? You need to separate the causal part by using the long division and then use the theorem.

9.5.4 Find the z-transform of the following sequences:

a. $x[n] = (n-1)a^{n-1}\cos(\Omega_o(n-1))u[n-1]$,
b. $x[n] = e^{-n}u[n-1]$.

9.5.5 Find the inverse z-transforms of the following functions using the partial fractions and then by using the long division:

a. $X(z) = \dfrac{z(z-1)}{(z-1)^2(z-.5)}$, ROC : $|z| > 1$,

b. $X(z) = \dfrac{z}{(z-1)(z-.5)^2}$, ROC : $|z| > 1$.

9.5.6 Is the system described by $y[n] + .5y[n-1] = x[n]$ stable?

9.5.7 Find the response of the following system assuming the two indicated cases:

$$y[n] - y[n-1] = (.5)^n u[n],$$
$$a.\ y[-1] = 0,\ b.\ y[-1] = 1.$$

9.5.8 Given $x[n]$ below determine the pole–zero plot of $X(z)$. What can you say about the uniqueness of a function $X(z)$ if it is determined from the pole–zero plots?

$$x[n] = \begin{cases} a^n, 0 \le n \le M - 1 \\ 0, \text{otherwise} \end{cases}.$$

9.5.9 Use the z-transforms to compute the convolution of the following sequences:

a. $x[n] = \delta[n] - \delta[n-1], h[n] = \delta[n-2] + 2\delta[n-3]$,
b. $x[n] = (1/2)^n u[n], h[n] = \{1, -1\}$.
$\uparrow$

9.5.10 Determine the autocorrelations of the following sequences using z-transforms:

a. $x[n] = \delta[n] + \delta[n-1]$, b. $h[n] = (1/2)^n u[n]$.

9.5.11 Use the sequences given in the last problem and the z-transforms to determine the following

cross-correlations. Verify the result that $r_{xh}[k] = r_{hx}[-k]$:

a. $r_{xh}[k] = \displaystyle\sum_{n=-\infty}^{\infty} x[n]h[n+k] = x[-k] * h[k]$,

b. $r_{hx}[k] = \displaystyle\sum_{n=-\infty}^{\infty} h[n]x[n+k] = h[-k] * x[k]$.

9.6.1 Some times it is easier to compute the z-transform of a function by considering differentiation with respect to a second variable. Consider the following two cases:

a. Is it possible to find the z-transform in (9.6.17a) from (9.6.17b) byassuming Ω_0 is a continuous variable and by noting that that sine and cosine functions are related through a derivative? If so, derive the corresponding expression. If not, explain.
b. Derive the z-transform of the function $x[n] = na^n u[n]$ using the result in (9.6.15a) and then using a as the second variable.

9.7.1 Consider the function $H(z) = (-z + 1)/(z^2 - z + 1)$, $Y(z) = H(z)X(z)$. a. By using the long division, determine $h[n]$, $n \ge 0$. b. By using long division, determine $h[n]$, $n \le 0$. c. Write the difference equation in terms of $y[n]$ and $x[n]$. The impulse response is $y[n] = h[n]$. Assuming $h[-1] = h[-2] = 0$, determine $h[n]$, $n = 0, 1, 2, 3$. d. Repeat the last part to determine $h[0], h[-1]$ and $h[-2]$ assuming $h[1] = h[2] = 0$.

9.7.2 Find $x[n]$ for the following ROCs: a. $|z| < 1$, b. $|z| > 3$, c. $2 < |z| < 3$ for the function $X(z) = z/[(z-3)(z-2)(z-1)]$ using the partial fraction expansion.

9.7.3 Find the inverse transforms of the following functions:

a. $X(z) = (z+2)^2/z$, ROC : $0 < |z| < \infty$,
b. $X_2(z) = z^2/(z-a)$, ROC : $|a| < |z| < \infty$
c. $X_3(z) = \dfrac{1}{1-az}$, ROC : $|z| < (1/|a|)$.

9.7.4 Find the causal response corresponding to the system function $H(z) = 1/(z - .5)^2$.

9.8.1 Determine the values of a and b so that the linear shift-invariant system described by the following unit sample response is stable:

$$h[n] = \begin{cases} a^n, n \geq 0 \\ b^n, n < 0 \end{cases} \left(\text{Use:} \sum_{n=0}^{\infty} |a|^n = (1/(1-|a|)), |a| < 1 \right)$$

9.8.2 Consider the system described by the difference equation $y[n] = x[n] - 5x[n-1]$. Is this system stable? Noting that this system is an example of an FIR filter, what can you say about the stability of FIR filters in general?

9.8.3 Is the system described by $y[n] - .4y[n-1] = x[n]$ a time-invariant or a time-varying system? Is it a causal system? Determine the stability of the system.

9.8.4 Find the step response of the system described by $y[n] = y[n-1] + x[n]$. What can you say about the stability of this system? Answer this by using the BIBO stability test assuming $x[n] = u[n]$ and then by using the z-transforms.

9.8.5 Consider the transfer function given by $H(z) = 1/[1 - 1.5z^{-1} - 0.5z^{-2}]$. *a.* Determine the stability of this system using the roots of the characteristic polynomial. *b.* Use the Schur–Cohn stability test to verify the result. *c.* Convert the problem to the analog case using the bilinear transformation to determine the stability of the system.

9.8.6 Consider the second-order transfer function $H(z) = 1/[1 + a_1 z^{-1} + a_2 z^{-2}]$. Determine the impulse responses of this system assuming that the poles satisfy the following:

a. real and distinct poles $(z_i = .6, .4)$,

b. real and equal poles $(z_i = 0.5, 0.5)$,

c. complex conjugate poles $(z_i = re^{\pm j\Omega_0} = 0.5e^{\pm j\pi/4})$.

9.9.1 Using the impulse invariance method and convert the following transfer function into a discrete domain transfer function $H(z)$. Assume $f_s = 2\,\text{Hz}$ for the two cases as indicated:

$$H_a(s) = (s+1)/(s^2 + \alpha s + 2), \quad a.\ \alpha = 4, \ b.\ \alpha = 1.$$

9.9.2 Use the step invariance method with $t_s = 1$ to convert the transfer function $H_a(s) = 1/(s+1)$ into a discrete domain transfer function $H(z)$. Use the following steps. Determine $Y_a(s) = H_a(s)X_a(s), X_a(s) = (1/s)$. Find the output response $y(t)$. Determine the sampled response $y(nt_s)$ and the corresponding z-domain function from $y[n] = y(nt_s)$. The transfer function of the digital filter that matches the step response is

$$H_s(z) = Y(z)/X(z), X(z) = Z\{u[n]\} = z/(z-1).$$

9.10.1 Show the transfer function derived in 9.10.18b satisfies all the required specifications.

9.10.2 Show that the all-pass functions of the form given below satisfy $|H(e^{j\Omega})| = 1$:

$$H(z) = \prod_{k=1}^{P} \frac{z^{-1} - a_k^*}{1 - a_k z^{-1}} H.$$

9.11.1 Show that the following result is true:

$$H(e^{j\Omega}) = \sum_{n=-\infty}^{\infty} h[n]e^{-jn\Omega}$$

$$\xrightarrow{FS} \frac{1}{2\pi} \int_{-\pi}^{\pi} H(e^{j\Omega})e^{j\Omega n} d\Omega = h[n].$$

9.11.2 In Section 9.11.2, the Fourier series expansions of the high-pass, band-pass, and band-elimination filters in terms of the low-pass filter Fourier series coefficients were derived. *a.* Derive these expressions by using the integrals directly. *b.* Derive these by using impulse functions.

9.11.3 Consider a *comb filter* defined by the impulse response $h[n] = \delta[n] - \delta[n-k]$. Determine the transfer function in the z-domain first and then determine the frequency amplitude and phase responses for two cases $k = 5, 6$. The amplitude response of this filter looks like the rounded teeth of a comb and therefore is referred to as a comb filter.

9.11.4 Find the Fourier series expansion of a differentiation function $H_a(j\omega) = j\omega, |\omega| \leq \pi/T$. Use a 11-point window to determine the coefficients of the series expansion using *a.*rectangular and *b.*Hamming windows. Sketch the amplitude responses:

$$H_a(j\omega) = j\omega, \ |\omega| \leq \pi/T.$$

9.12.1 Show the two realizations in Fig. 9.12.5a,b result in the same transfer function.

9.12.2 Find a realization of the all-pass function given below using two multipliers plus a multiplication by (-1) and four delay components. Hint: Use the form $Y(z) = \alpha[X(z) - z^{-2}Y(z)] + \beta[z^{-1}X(z) - z^{-1}Y(z)] + z^{-2}X(z)$ in your solution:

$$H_{ap}(z) = \frac{\alpha + \beta z^{-1} + z^{-2}}{1 + \beta z^{-1} + \alpha z^{-2}}, \quad Y(z) = H_{ap}(z)X(z).$$

Chapter 10
Analog Modulation

10.1 Introduction

In this chapter we will consider some of the fundamental concepts associated with analog modulation. Communication of analog signals from one location to another is accomplished by using either a wire channel or a radio channel. The source signals, such as voice, pictures, and in general baseband signals are not always suitable for direct transmission over a given channel. These signals are first converted by an input transducer into an electrical waveform referred to as the *baseband signal* or *the message signal*. The spectral contents of the baseband signals are located in the low-frequency region. The wire channels have a low-pass transfer function and can be used for transmitting signals that have a bandwidth less than the channel bandwidth. The radio channels have a band-pass characteristic. Low-pass signals can be transmitted through radio channels by using a modulator. *Modulation* is the process of alteration of a carrier wave in accordance with the message (modulating signal). The signal obtained through modulation is called the *modulated signal*. The term baseband is used to denote the band of frequencies of the signal delivered by the source or the input transducer. As an example, the audio voice band, i.e., the range of frequencies, 0 to 3.5 kHz, is a baseband. There are several reasons for modulation. Some of these are discussed below.

The audio band is about the same for all humans. Speech production, in simple terms, can be visualized as a filtering operation in which a sound source excites a vocal tract filter O'Shaughnessy (1987). In a room, if every one is talking at the same time, we cannot understand what each one is saying. If one person is talking loud enough, we may be able to understand him or her. Similarly if every radio station transmits at the same frequency, then the received signal, at the front of the radio receiver, will have a linear combination of all the signals that are transmitted. There is no way we can understand one signal from another. One way to avoid this problem is shift each radio station band of frequencies corresponding to a different location, i.e., use modulation and then transmit. This allows for simultaneous transmission of several stations. A second example is, if telephone communications need to be established between two cities, we need to have thousands of telephone lines. Can you imagine how much copper we need for such construction? Instead, we could modulate each signal and put several telephone conversations on the same telephone line. Such a process is called frequency-division multiplexing. The dual to *frequency-division multiplexing (FDM)* is the *time-division multiplexing (TDM)*. The signals are sampled and allocated at recurring time slots for transmission.

We use antennas to transmit and receive signals. A rule of thumb for the length of xz an antenna is at least quarter wavelength. The wavelength is given by the parameter $\lambda = c/f_c$, where c is the speed of light and $(1/f_c)$ gives a typical period. (1/4) comes from the fact that if we know (1/4) of a period of a cosine or a sine function, we know the entire period of these functions. To put this in perspective, consider a speech signal. The antenna length required to recover the speech signal (note the speed of light is $3(10^8)$ m/s) is

$$\frac{1}{4}\left[\frac{3(10^8)}{f}\right]\frac{\text{m/s}}{\text{cycles/s}}. \qquad (10.1.1)$$

R.K.R. Yarlagadda, *Analog and Digital Signals and Systems*, DOI 10.1007/978-1-4419-0034-0_10,
© Springer Science+Business Media, LLC 2010

The antenna length for a 1 kHz in the voice band requires the antenna length by

$$\frac{1}{4}\frac{3(10^8)}{f}\Big|_{f=1\,\text{kHz}} \approx 47\text{miles}. \qquad (10.1.2)$$

Obviously, such an antenna is impractical. Noting that the antenna length is inversely proportional to the frequency, we can use reasonable sizes of antennas by shifting or translating the frequencies to a higher level by modulation. Other reasons include, for example, that if we want to transmit signals that can penetrate through different media, we use different frequency bands. Again, modulation is necessary.

In summary, modulation is used to translate the spectral contents of signals so that they can be efficiently transmitted and the spectral content of the modulated message signal lies in the operating frequency band of a communication channel. Second, the modulation puts the information content of a signal in a form that is less vulnerable to noise and/or interference. Third, modulation allows for simultaneous transmission of several signals that occupy essentially the same frequency ranges over a channel to the respective destinations. Finally, modulation is necessary to reduce noise and/or interference. Theoretically, any type of modulation, amplitude, frequency modulation, etc., could be used at any transmission frequency. However, to have some semblance and have high efficiencies, regulations specify the modulation type, bandwidth, and type of information that can be transmitted over the designated frequency bands.

Divisions of frequency spectrum Poularikas and Seely (1991):

Telephony, navigation, industrial communication	3–300 kHz
AM broadcasting, military communication and amateur and citizen band radio	.3–30 MHz
FM broadcasting, TV broadcasting, and land transportation	30–300 MHz
UHF TV, radar, and military applications	.3–3 GHz
Satellite and space communications, microwave, and radar	3–30 GHz
Research and radio astronomy	Above 30 GHz

There are also assigned amateur radio bands above 30 kHz. Analog modulation schemes can be divided into two schemes, namely continuous-wave modulation and pulse modulation, which allow for propagation of a low-frequency message signal using a high-frequency carrier. The carrier is generally assumed to be a sinusoid of the form

$$x_c(t) = A(t)\cos(\omega_c t + \phi(t)). \qquad (10.1.3)$$

Usually, for analog signals, one of the functions $A(t)$ or $\phi(t)$ is a function of the message $m(t)$ and the other is a constant. The frequency $\omega_c[\text{or } f_c = \omega_c/2\pi]$ is called the carrier frequency. Although in (10.1.3), we have used a cosine function, a sine function can also be used, as the sine and the cosine functions are the same except one is a phase-shifted version of the other. The functions $A(t)$ and $\phi(t)$ are called the *instantaneous amplitude* and *phase angle of the carrier*. We will see later that sinusoidal carrier is not necessary. The *continuous-wave (CW) modulation* uses two types of modulation. When $A(t)$ is linearly related to the message signal and $\phi(t)$ is a constant, then the result is a *linear modulation*. When $\phi(t)$ or its derivative is linearly related to the message signal, then we have the cases of *phase* or *frequency modulation*. Frequency and phase modulations are referred as angle modulation schemes, as the phase angle of the carrier has the message information. *Phase and frequency modulations* are *non-linear* modulations. In the analog pulse modulation, the message is sampled at discrete-time intervals and the amplitude, width, or position of a pulse has the information about the message signal. In the strict sense, *pulse modulation* is a message-processing technique. We will be interested in *pulse amplitude modulation (PAM), pulse width (or duration) modulation (PWM),* and *pulse position modulation (PPM).* If the value of each sample is quantized, i.e., only a finite number of levels, say a power of 2 are kept. Then each sample is represented by n bits. Such a process is called *pulse code modulation.* There are other modulation schemes Ziemer and Tranter (2002).

After the signal is received at the destination we need to undo the modulation, which is referred to as the *demodulation*. In the case of the multiplexed signal, it needs to be demultiplexed and demodulated before it can be used. Most of this chapter deals with the mathematical development that essentially involves the evaluation and the description of the spectra of the signals at various locations of the signal as it goes through modulation, transmission, demodulation, and the recovery of the signal. The performance of various modulation schemes are judged on the basis of how well a particular scheme performs in the presence of noise. This requires knowledge of random process, which is beyond the scope of our study. Therefore, our analysis will be based on simple concepts. Before we study various types of modulation schemes, we like to introduce few components that are non-linear in nature and are useful in building communication systems. For references on analog communications, see Ziemer and Tranter (2002), Simpson and Houts (1971), Carlson (1975), Lathi (1983), and others.

10.2 Limiters and Mixers

The input–ouput relations of a non-linear system were discussed in Section 6.9. A limiter is a non-linear circuit that has output saturation. A *hard (ideal) limiter* takes the input signal and the output is a constant v_L if the input is positive and is $-V_L$ if

the input is negative. Limiter–band-pass filter combination is a device that takes the input waveform, limits the waveform between two limits $\pm V_L$, and is then passed through a band-pass filter. As an example consider the input, a band-pass signal given by

$$v_{in}(t) = R(t)\cos(\omega_c t + \phi(t)),$$
$$R(t) \geq 0 \text{ for all } t. \qquad (10.2.1)$$

$R(t)$ is the *envelope* and $\phi(t)$ is the phase angle of the signal. Figure 10.2.1 gives the block diagram of a limiter cascaded by a band-pass filter. Passing $v_{in}(t)$ through a limiter, results in a clipped signal. Hard limiters introduce dc as well as high frequencies. Passing the clipped signal through an ideal BP filter centered on the frequency ω_c with a BW greater than the message BW results in

$$v_o(t) = KV_L \cos[\omega_c t + \phi(t)]. \qquad (10.2.2)$$

Figure 10.2.2a gives a simple limiter and Fig. 10.2.2b gives sketches of an input and the corresponding output of the limiter. Passing the output $v_{out}(t)$ through a band-pass filter makes it a sinusoidal signal.

Fig. 10.2.1 Block diagram of a hard limiter with a band-pass filter

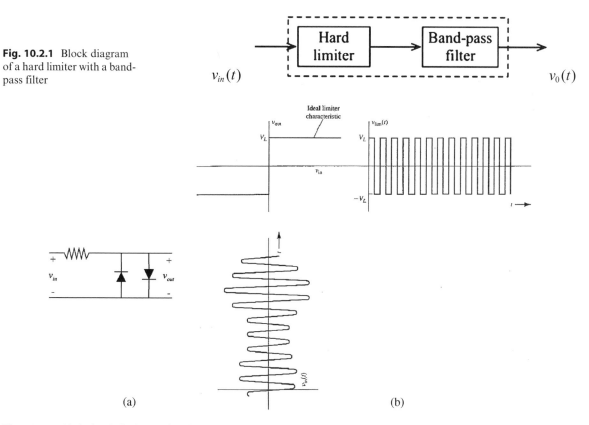

(a)   (b)

Fig. 10.2.2 (a) A simple limiter and (b) ideal limiter characteristics with an input and the corresponding output wave forms (printed with permission from the book Couch (1997))

10.2.1 Mixers

An *ideal mixer* is an electronic circuit that implements a mathematical multiplier operation of the message signal $v_{in}(t)$ and a sinusoidal signal generated by a local oscillator $v_{Lo}(t) = A_c \cos(\omega_c t)$. Consider the block diagram in Fig. 10.2.3. The output of the non-linear device is described below where the higher-order terms are neglected.

$$v_1(t) = K(v_{in}(t) + v_{Lo}(t))^2 + \text{other terms}$$
$$\approx K[v_{in}^2(t) + 2v_{in}(t)v_{Lo}(t) + v_{Lo}^2(t)]. \quad (10.2.3)$$

Assuming the BP filter removes all the terms except the cross-product term, $v_{in}(t)v_{Lo}(t)$ we have

$$v_0(t) = K_1 v_{in}(t) \cos(\omega_c t), \quad (10.2.4)$$

where K_1 is a constant that includes the filter gain. In obtaining the expression in (10.2.4), we assumed $v_{in}(t)$ and $v_{in}^2(t)$ are low-frequency functions. We will see that the function given in (10.2.4) will be called later as a *double-sideband (DSB) modulated signal*. Furthermore, $v_{Lo}^2(t) = A_c^2 \cos^2(\omega_c t) = A_c^2[(1/2) + (1/2)\cos(2\omega_c t)]$ contains the dc component and the frequency component at ($(2\omega_c)$ or $(2f_c)$ in Hertz) and the other terms in (10.2.3) are outside the frequency band of the band-pass filter. There are other methods that can be used in achieving this goal. These include the use of a continuously variable transconductance device, such as a dual-gate FET and using a square wave oscillator. See Couch (2001).

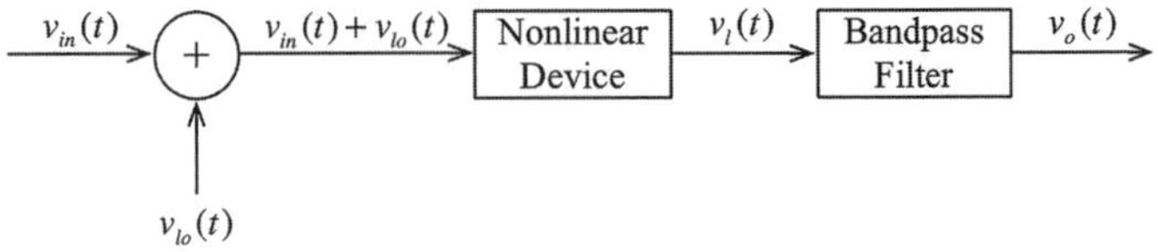

10.3 Linear Modulation

When the carrier amplitude is linearly related to the message signal, it is called as linear modulation

Fig. 10.2.3 Generation of $v_0(t) = K_1 v_{in}(t) \cos(\omega_c t)$

described by (10.3.1). The phase angle $\phi(t) = \phi_0$ is assumed to be a constant. Without losing any generality for our analysis here we assume $\phi_0 = 0$.

$$x_c(t) = A(t)\cos(\omega_c t + \phi_0) \equiv A(t)\cos(\omega_c t). \quad (10.3.1)$$

There are many modulation schemes. A few are listed below:

1. Double-sideband (DSB) modulation.
2. Amplitude modulation (AM).
3. Single-sideband (SSB) modulation.
4. Vestigial sideband (VSB) modulation.

10.3.1 Double-Sideband (DSB) Modulation

The DSB signal is

$$x_{\text{DSB}}(t) = y(t) = A_c m(t)\cos(\omega_c t), \quad \omega_c = 2\pi f_c. \quad (10.3.2)$$

$A_c \cos(\omega_c t)$ is the *carrier signal*, f_c (or ω_c) is the *carrier frequency*, and $m(t)$ is the *message or the information signal*. Frequencies will be referred in terms of either $\omega = 2\pi f$ in radians per second or f in Hertz. The amplitude of the time function multiplying $\cos(\omega_c t)$ in (10.3.2) given by $a(t) = |A_c m(t)|$ is called the *envelope* of the DSB signal. The message signal is generally a low-pass signal and is assumed to be band limited to $B = f_m$ Hz. That is,

$$F[m(t)] = M(j\omega) = 0, \omega > 2\pi f_m.$$

The transform of the modulated signal is (see the modulation theorem in Chapter 4)

$$X_{\text{DSB}}(j\omega) = \frac{1}{2}M(j(\omega - \omega_c)) + \frac{1}{2}M(j(\omega + \omega_c)). \quad (10.3.3)$$

The DSB signal can be generated using the block diagram shown in Fig. 10.3.1. Figure 10.3.2 illustrates the time and frequency waveforms for a simple case.

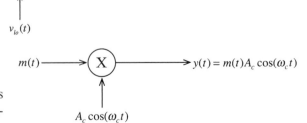

Fig. 10.3.1 Generation of a double-sideband modulated signal

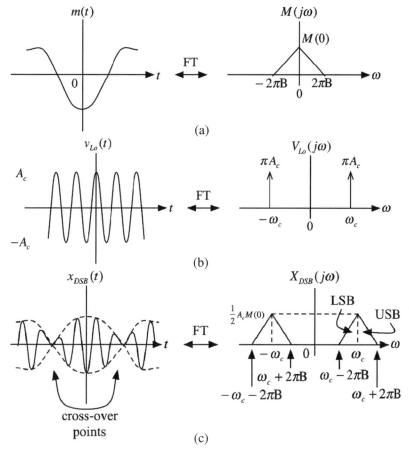

Fig. 10.3.2 Signals and their spectra associated with double-sideband modulation: (a)$M(j\omega) = F[m(t)]$, (b)$V_{Lo}(j\omega) = F[v_{Lo}(t)]$, and (c)$X_{DSB}(j\omega) = F[x_{DSB}(t)]$

Figure 10.3.2a gives an arbitrary message signal with its F-transform. For simplicity we assumed the transform is real, even, and positive. The message signal $m(t)$ is assumed to cross the time axis for the reasons and these are referred to as crossover points to be discussed later. Figure 10.3.2b shows the carrier and its transform. Figure 10.3.2c shows the DSB signal, where the dotted line is the carrier envelope of the modulated signal. The spectrum of the modulated signal is obtained by translating and scaling the message spectrum $M(j\omega)$ to $(1/2)M(j(\omega\pm\omega_c))$. It has two parts: upper and the lower sidebands. The spectra above the carrier frequency, f_c i.e., the portion of the spectrum between f_c and $f_c + B$ (and also before $-f_c$), i.e., the spectrum between $-(f_c + B)$ and $-f_c$ is the *Upper SideBand (USB)*. The spectrum below the carrier frequency, f_c i.e., the portion of the spectra between $(f_c - B)$ and f_c (and also between $-f_c$ and $-(f_c - B)$) is the *Lower*

Sideband (LSB). The bandwidth of the message is B Hz and the bandwidth of the DSB signal is $(2B)$ Hz. The carrier frequency is assumed to be large enough that the spectra around f_c and $(-f_c)$ do not overlap, i.e., $(f_c - B) > 0$. In the broadcast applications, a radiating antenna can radiate only a narrowband of frequencies without distortion. To avoid this distortion, we assume that $f_c \gg B$, which is a reasonable assumption. Otherwise, we will not be able to recover the message signal.

10.3.2 Demodulation of DSB Signals

Recovery of the message signal from the modulated signal is *demodulation or detection*. First, let us consider the *coherent (or synchronous) detection* scheme shown in Fig. 10.3.3. The received DSB

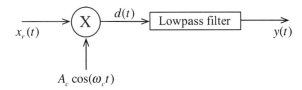

Fig. 10.3.3 Synchronous demodulation

signal is $x_r(t) = y(t) + n(t)$. $n(t)$ is the noise signal, and it can only be described statistically, which is beyond our scope here. For simplicity, we will assume that $x_r(t) \approx y(t)$ here. In the case of the synchronous or coherent modulation scheme the received signal is multiplied by the local carrier signal $\cos(\omega_c t)$. The output of the multiplier is

$$d(t) = x_r(t) \cos(\omega_c t) \approx x_{DSB}(t) \cos(\omega_c t)$$
$$= A_c m(t) \cos^2(\omega_c t)$$
$$= \frac{1}{2} A_c m(t) + \frac{1}{2} A_c m(t) \cos(2\omega_c t). \quad (10.3.4)$$

It has two parts, the message part $(1/2)A_c m(t)$, a low-frequency signal, occupying the frequency range $(0, B)$ Hz. The second part $(1/2)A_c m(t) \cos(2\omega_c t)$ is a band-pass signal with its positive spectrum between $(2f_c - B)$ and $(2f_c + B)$ and a negative spectrum between $(-2f_c - B)$ and $(-2f_c + B)$. Since $f_c \gg B$, a low-pass filter with a BW greater than B Hz with a gain constant H_0 will result in the output

$$y(t) \approx (H_0/2)A_c m(t). \quad (10.3.5)$$

All systems have an inherent delay. Since an ideal delay in the received signal does not matter, as far as the understanding the modulation scheme and the delay will *not be explicitly shown*. Also, we did not consider any noise in the received signal.

Coherent demodulation requires that the oscillators at the transmitter and the receiver must have the same phase as well as the frequency. Any difference in the frequency and/or phase between the oscillators at the transmitter and the receiver require additional discussion. Let the expressions for the received signal and the local oscillator be given by $x_r(t) = A_c m(t) \cos(\omega_c t)$ and $\cos((\omega_c + \Delta\omega)t + \delta)$, respectively. We should note that we are assuming for simplicity δ as a constant, rather than a function of time. The frequency and the phase errors between the

transmitter and the receiver oscillators are, respectively, given by $\Delta\omega$ and δ. By using trigonometric formulas, we have

$$d(t) = A_c m(t) \cos(\omega_c t) \cos((\omega_c + \Delta\omega)t + \delta),$$
$$= \frac{A_c}{2} m(t) \cos((\Delta\omega)t + \delta)$$
$$+ \frac{A_c}{2} m(t) \cos((2\omega_c + \Delta\omega)t + \delta)\}. \quad (10.3.6)$$

Note the second term in (10.3.6) is centered at $(2\omega_c + \Delta\omega)$, which is much larger than the BW of the low-frequency signal $m(t)$. Passing $d(t)$ through a LP results in the output

$$y(t) = \frac{1}{2} A_c m(t) \cos((\Delta\omega)t + \delta). \quad (10.3.7)$$

In the case of coherence, i.e., when $\Delta\omega = 0$ and $\delta = 0$, the output

$$y(t) = (1/2)A_c m(t)$$

is a scaled version of the message signal. We now consider the cases with respect to the difference in the oscillator frequency and the phase shift at the transmitter and the receiver separately. That is, if $\Delta\omega = 0$, then (10.3.7) reduces to $y(t) = A_c m(t) \cos(\delta)$. The output is proportional to the input if δ is a constant and *not* equal to $\pm(\pi/2)$, resulting in only attenuation and no distortion. If $\delta = \pm(\pi/2)$ then the output is zero. If the phase error δ varies randomly, say due to variations in signal propagation paths, $\cos(\delta)$ varies randomly, which is undesirable. If $\delta = 0$ and $\Delta\omega \neq 0$, we have

$$y(t) = (A_c/2)m(t) \cos((\Delta\omega)t). \quad (10.3.8)$$

The output is the desired message multiplied by a low-frequency sinusoid $\cos((\Delta\omega)t)$. This is referred to as the *beating effect* and the distortion can be serious. Since the bandwidth of the signal $B \ll f_c = (\omega_c/2\pi)$, $\Delta\omega/(2\pi B)$ can be appreciable. Human year can tolerate a drift in frequency of a few Hertz. To avoid this problem, one can send a carrier signal, usually referred to as a pilot at a reduced level. At the receiver the pilot is separated by using a narrowband bandpass filter centered at the pilot frequency.

Another approach to generate a phase synchronized receiver signal uses a non-linear device

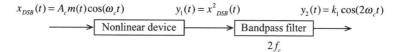

$$x_{DSB}(t) = A_c m(t)\cos(\omega_c t) \quad\quad y_1(t) = x^2_{DSB}(t) \quad\quad y_2(t) = k_1\cos(2\omega_c t)$$

Nonlinear device → Bandpass filter
$2f_c$

Fig. 10.3.4 Generation of $y_2(t)$

described by the block diagram in Fig. 10.3.4. The output of the squarer is

$$y_1(t) = (A_c m(t)\cos(\omega_c t))^2$$
$$= (A_c^2/2)m^2(t) + (A_c^2/2)m^2(t)\cos(2\omega_c t). \quad (10.3.9)$$

First,

$$(A_c^2/2)F[m^2(t)] = (A_c^2/2)[M(j\omega) * M(j\omega)]$$
$$\equiv (A_c^2/2)M_c(j\omega),$$
$$\Rightarrow Y_1(j\omega) = \frac{A_c^2}{2}M_c(j\omega) + \frac{A_c^2}{4}[M_c(j(\omega + 2\omega_c))$$
$$+ M_c(j(\omega - 2\omega_c))]. \quad (10.3.10)$$

Since the message signal is a low-pass signal with a bandwidth of B Hz, $m^2(t)$ is also low-pass signal with a bandwidth of $2B$ Hz, whereas the signal $(1/4)[m^2(t)\cos(2\omega_c t)]$ is a band-pass signal centered at $(2f_c)$ with a bandwidth of $4B$. Passing the signal $y_1(t)$ through a narrowband band-pass filter with a bandwidth $\Delta f \ll 4B$, the transform of the output of the band-pass filter consists of two narrow pulses centered at $f = \pm 2f_c$. Since these are assumed to be very narrow pulses, we can approximate them by impulses and the output transform and its inverse are as follows, where Ks are constants.

$$Y_2(j\omega) \simeq K[\delta(\omega + 2\omega_c) + \delta(\omega - 2\omega_c)] \xrightarrow{FT} y_2(t)$$
$$= K_1\cos(2\omega_c t). \quad (10.3.11)$$

This signal has the same phase as the input carrier, or, in other words, this signal is in phase synchronism with the input carrier. The frequency of the input

carrier is f_c, whereas the frequency of the sinusoid $y_2(t)$ is $2f_c$ and we will use this concept below.

10.4 Frequency Multipliers and Dividers

A frequency multiplier has the input and output sinusoids $x(t) = A\cos(\omega_c t)$ and $y(t) = C\cos(N\omega_c t)$, respectively, where N is an integer; A and C are some constants. Since the output frequency is an integer multiple of the input frequency, an N-law non-linear device can be used to generate the harmonics of the original signal. The output of this device is fed into a narrowband band-pass filter centered at the frequency Nf_c resulting in $y(t)$. Figure 10.3.4 works as a *frequency multiplier*.

Now consider a simple *frequency divider* with the system in Fig. 10.4.1. Let the input be a sinusoid $x(t) = A\cos(2\pi f_{in}t)$. The output of the narrowband band-pass filter is assumed to be $c\cos(2\pi f_{out}t)$, where c is a constant. The output of the frequency multiplier is assumed to be $r(t) = \cos(2\pi(M-1)f_{out}t)$. Now

$$e(t) = A\cos(2\pi f_{in}t)\cos(2\pi f_{out}(M-1)t)$$
$$= \frac{A}{2}\cos(2\pi((M-1)f_{out} - f_{in})t)$$
$$+ \frac{A}{2}\cos(2\pi((M-1)f_{out} + f_{in})t).$$

Frequencies in $e(t)$:

$$f_1 = [f_{in} + (M-1)f_{out}] \text{ and } f_2 = [f_{in} - (M-1)f_{out}].$$
$$(10.4.1)$$

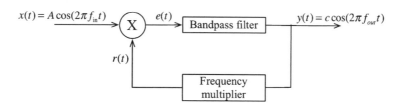

$$x(t) = A\cos(2\pi f_{in}t) \xrightarrow{\quad} X \xrightarrow{e(t)} \text{Bandpass filter} \xrightarrow{\quad} y(t) = c\cos(2\pi f_{out}t)$$

$r(t)$

Frequency multiplier

Fig. 10.4.1 Frequency divider

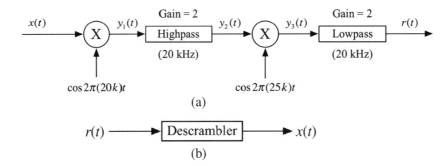

Fig. 10.4.2 (a) Speech scrambler and (b) descrambler

A narrowband BP filter with a center frequency of f_2 can eliminate the frequency f_1, resulting in the output frequency $f_2 = f_{in} - (M-1)f_{out}$. Since $f_2 = f_{out}$, we have

$$f_{out} = f_{in}/M \text{ and } y(t) = c\cos(2\pi f_{out}t)$$

$$= c\cos(2\pi(f_{in}/M)t). \qquad (10.4.2)$$

The feedback circuit in Fig. 10.4.1 works as a frequency divider.

Example 10.4.1 The input $x(t)$ in Fig. 10.4.2a is assumed to be real signal band limited to 5 kHz. It is modulated by a carrier with frequency $f_c = 20$ kHz. Show that the system can be used as a speech scrambler by identifying the spectra at various locations by assuming all the filters are ideal with the cut-off frequencies as identified. See Carlson (1975).

Solution: From the block diagram, $y_1(t) = x(t)\cos 2\pi(20\,k)t$, $y_2(t) =$ HP filtered version of $y_1(t)$, $y_3(t) = y_2(t)\cos 2\pi(25\,k)t$, and $r(t) =$ LP-filtered version of $y_3(t)$. Now

$$Y_1(j\omega) = 0.5[X(j\omega - 2\pi(20\,k))$$
$$+ X(j\omega + 2\pi(20\,k))],$$
$$Y_2(j\omega) = H_{\text{Hp}}(j\omega)Y_1(j\omega),$$

$$Y_3(j\omega) = 0.5[Y_2(j(\omega + 2\pi(25\,k)))$$
$$+ Y_2(j(\omega - 2\pi(25\,k)))]$$
$$= Y_3(j\omega), \quad R(j\omega) = H_{Lp}(j\omega)Y_3(j\omega).$$

Noting that the filters are assumed to be *ideal* and ignoring all the delays involved in the system for simplicity, the spectra at various locations on the block diagram are sketched in Fig. 10.4.3. From the input and the output spectra, we see that we have

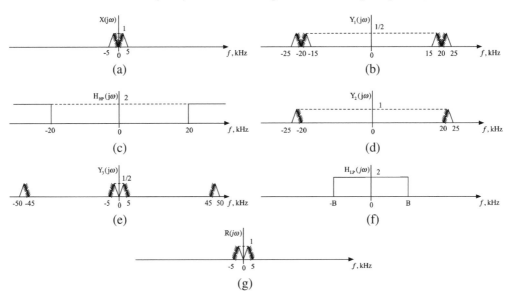

Fig. 10.4.3 Spectra of the functions at locations identified in Fig. 10.4.2

used lower sideband and the output spectrum is the inverted frequency spectrum of the input. Considering only the positive frequencies, we have

$$R(j\omega) = X(j(2\pi(5k) - \omega)), 0 \le \omega \le 2\pi(5k). \quad (10.4.3)$$

The system is a *speech scrambler* based on frequency inversion. Use of a *descrambler*, same as the scrambler, recovers the original signal. Higher level of security can be achieved by dividing the signal spectrum into bands and encrypting each band. ∎

Notes: There are several products available in the market that performs the analog frequency inversion. In Example 8.6.11, *frequency inversion* is implemented in the *digital domain* by replacing a discrete signal $x_1[n]$ by $(-1)^n x_1[n]$. ∎

10.5 Amplitude Modulation (AM)

We note that $x_{\text{DSB}}(t_i) = 0$ when $m(t_i) = 0$, where $t_i s$ are *zero crossover points*. See Fig. 10.3.2c. If $m(t) \gg 0$ then the envelope of the DSB signal is proportional to the message signal. In this case, we can recover the message signal by detecting the envelope of the DSB signal, which is simpler to do than the coherent detection scheme. In coherent detection, we need the same carrier frequency and the phase at the input and the outputs.

Message signals can take both negative and positive values. To avoid zero crossovers, add a constant A to the message signal $m(t)$, so that

$$(A + m(t)) > 0 \text{ or } A \ge |\min\{m(t)\}| \text{ for all } t. \quad (10.5.1)$$

The *AM modulation index* or the *sensitivity* of the AM signal is defined by

$$\mu = |\min m(t)|/A. \quad (10.5.2)$$

With μ, $(A + m(t))$ can be written in terms of a scaled version, $m_n(t)$ by

$$(A + m(t)) = A \left[1 + \frac{m(t)}{|\min m(t)|} \frac{|\min m(t)|}{A} \right]$$
$$= A[1 + \mu m_n(t)],$$
$$m_n(t) = m(t)/|\min m(t)|. \quad (10.5.3)$$

If $0 \le \mu \le 1$, we have $(1 + \mu m_n(t)) \ge 0$. The amplitude modulated signal is

$$\begin{aligned} y(t) &= [A + m(t)]A_c \cos(\omega_c t) \\ &= AA_c[1 + (m(t)/A)] \cos(\omega_c t) \\ &\equiv A_c'[1 + (m(t)/A)] \cos(\omega_c t) \\ &= A_c'[1 + \mu m_n(t)] \cos(\omega_c t), \ A_c' = AA_c. \end{aligned}$$
$$(10.5.4)$$

Noting $A_c'(1 + \mu m_n(t)) > 0$, the envelope of the signal $A_c'[1 + \mu m_n(t)] \cos(\omega_c t)$ is

$$a(t) = \left| A_c'(1 + \mu m_n(t)) \right| = A_c'(1 + \mu m_n(t)). \quad (10.5.5)$$

This points out that an envelope detector will recover the function $A_c'(1 + \mu m_n(t))$.

Notes: Computation of the minimum value of the function $m(t)$ is not always easy to find. Most signals approximately have symmetrical maximum and minimum values. That is, $\max m(t) \approx -\min m(t)$ and

$$m_n(t) \approx m(t)/|m(t)|_{\max}. \quad (10.5.6)$$

Assuming the minimum value of $m_n(t)$ is -1, the AM signal is

$$x_{\text{AM}}(t) = A_c'(1 + \mu m_n(t)) \cos(\omega_c t), A_c' = A_c A. \quad (10.5.7) \ \blacksquare$$

Example 10.5.1 Give the expressions for the AM signal for the message signals

a. $m_a(t) = 4\cos(\omega_0 t) + 3\cos(3\omega_0)t$, $\omega_0 = 2\pi(100)$
$$(10.5.8)$$

b. $m_b(t) = 4\cos(\omega_0 t) - 3\cos(2\omega_0 t) + 2\cos(4\omega_0 t)$,
$$\omega_0 = 2\pi(100). \quad (10.5.9)$$

Solution: a. Cosine function is bounded by 1 and $\max|m_a(t)| \le 7$. At $t = k/(2f_0)$, for $k-$ odd we have $\min m_a(t) = -7$. The magnitudes of the minimum and the maximum are the same. Using (10.5.7), the AM signal is

$$x_{AM}(t) = A_c'[1 + \mu m_n(t)] \cos(\omega_c t),$$
$$m_n(t) = \frac{1}{7}[4\cos(\omega_0 t) + 3\cos(3\omega_0)t]$$

$$\Rightarrow x_{AM}(t) = A'_c(1 + \mu\frac{1}{7}[4\cos(2\pi f_0 t)$$
$$+ 3\cos(2\pi(3f_0)t])\cos(\omega_c t). \quad (10.5.10)$$

b. We can find the minimum and maximum values by using MATLAB code given below.

```
t = 0 : .0001 : .01;
y = 4 * cos(2 * pi * 100 * t) − 3 * cos(2 * pi * 300 * t)
+ 2 * cos(2 * pi * 400 * t)
[Y1, I] = max(y), [Y2, J] = min(y).
```

The maximum, the minimum values of $m_b(t)$, and the corresponding AM signal are

$$\max(m_b(t)) = 5.1647, \ \min(m_b(t))$$
$$= -5.9643, \quad (10.5.11)$$
$$x_{AM}(t) = A(1 + \mu(1/5.9643)[4\cos(\omega_0 t)$$
$$- 3\cos(2\omega_0)t + 2\cos(4\omega_0 t)])\cos(\omega_c t).$$
$$(10.5.12) \quad \blacksquare$$

10.5.1 Percentage Modulation

The modulation index μ is usually expressed as a percentage. For an envelope demodulation, we require $0 \le \mu \le 1$. When $\mu = 1$, we have *100%* modulation. If $\mu > 1$, we have *over-modulation* and phase cross overs result in the AM signal, which result in envelope distortion. This is undesirable since our goal is to detect the envelope of the AM signal. There is *no modulation* if $m(t) = 0$. Let

$$A_{\max} = \max[A'_c(1 + \mu m_n(t))],$$
$$A_{\min} = \min[A'_c(1 + \mu m_n(t))]. \quad (10.5.13)$$

The modulation percentages of an AM signal are defined by

$$\% \text{ Positive modulation} = \frac{A_{\max} - A'_c}{A'_c} \times 100$$
$$= \max[\mu m_n(t)] \times 100,$$
$$(10.5.14a)$$
$$\% \text{ Negative modulation} = \frac{A'_c - A_{\min}}{A'_c} \times 100$$
$$= -\min[\mu m_n(t)] \times 100,$$
$$(10.5.14b)$$

$$\% \text{ Modulation} = \frac{A_{\max} - A_{\min}}{2A'_c} \times 100$$
$$= \frac{\max[\mu m_n(t)] - \min[\mu m_n(t)]}{2} \times 100.$$
$$(10.5.14c)$$

Example 10.5.2 Illustrate the percentage modulations for the normalized signal given by $m_n(t) = \cos(\omega_m t)$ and $\mu = .5$. Use the expression $x_{AM}(t) = A'_c[1 + .5\cos(\omega_m t)]\cos(\omega_c t)$.

Solution: This is *tone modulation*, as the message signal is a sinusoid (or a tone). The maximum and minimum values of $m_n(t) = \cos(\omega_m t)$ are $+1$ and -1, respectively and

$$x_{AM}(t) = A'_c[1 + .5\cos(\omega_m t)]\cos(\omega_c t),$$
$$\% \text{ Positive modulation} = \frac{A'_c(1 + .5) - A'_c}{A'_c} = 50\%,$$
$$\% \text{ Negative modulation} = \frac{A'_c - .5A'_c}{A'_c} = 50\%,$$
$$\% \text{ Percentage modulation} = \frac{A'_c(1 + .5) - A'_c(.5)}{2A'_c} \quad \blacksquare$$
$$= 50\%.$$

10.5.2 Bandwidth Requirements

Assuming the message signal is $m(t) \overset{\text{FT}}{\longleftrightarrow} M(j\omega)$, the AM signal and its transform are

$$x_{AM}(t) = A_c[A + m(t)]cos(\omega_c t) \overset{\text{FT}}{\longleftrightarrow} X_{AM}(j\omega),$$

$$X_{AM}(j\omega) = \pi A A_c[\delta(\omega - \omega_c) + \delta(\omega + \omega_c)]$$
$$+ \frac{A_c}{2}[M(j(\omega - \omega_c)) + M(j(\omega + \omega_c))].$$
$$(10.5.15)$$

Figure 10.5.1 shows an assumed spectrum of the message signal $m(t)$ and the corresponding spectrum of the AM signal. If the message signal bandwidth is B Hz, then the bandwidth of the AM signal is $2B$ Hz, i.e., two times the bandwidth of the message signal. If $A = 0$, then

$$x_{AM}(t)|_{A=0} = A_c(A + m(t))\cos\omega_c t)|_{A=0}$$
$$= A_c m(t)\cos\omega_c t) = x_{DSB}(t).$$

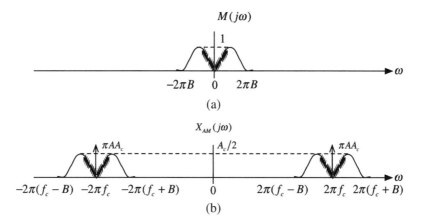

Fig. 10.5.1 (a) Spectrum of a message signal and (b) spectrum of the AM signal

Since there is no carrier term in the DSB signal, it is called *amplitude modulated signal with a suppressed carrier*. The transmission bandwidths of AM and DSB are the same.

10.5.3 Power and Efficiency of an Amplitude Modulated Signal

Modulation processes are compared on the basis of *modulation efficiency η*. It is the ratio of the percentage of the power in the sidebands conveying the message to the total power in the modulated-carrier signal. The time average of $x_{AM}^2(t)$ is defined over the period of the carrier $T_c = 2\pi/\omega_c$ using the notaion $\langle . \rangle$ by

$$\langle x_{AM}^2(t)\rangle = \frac{1}{T_c}\int_{T_c} x_{AM}^2(t)dt$$

$$= \left\langle \{A_c(A + m(t))\cos(\omega_c t)\}^2 \right\rangle$$

$$= \frac{A_c^2}{2}\left\langle A^2 + m^2(t) + 2Am(t) \right\rangle$$

$$+ \frac{A_c^2}{2}\left\langle [A^2 + m^2(t) + 2Am(t)]\cos(2\omega_c t)\right\rangle.$$

$$(10.5.16)$$

The time average of the message signal is assumed to be zero, i.e., it does not have a DC component and $\langle m(t)\rangle = 0$, a standard assumption for most signals. Second, $m(t)$ is a low-frequency signal and the highest frequency in the message is much smaller than the carrier frequency. It varies slowly

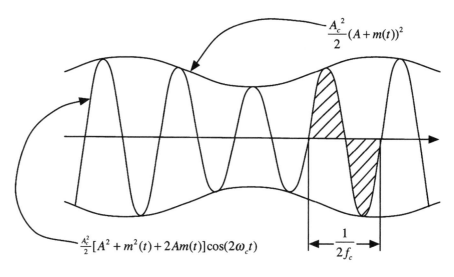

Fig. 10.5.2 AM signal

compared to the carrier signal. This implies that the time average over one period of the second term in (10.5.16) is (period of $\cos(2\omega_c t)$ is $T_c/2$.)

$$\int_{T_c/2} \frac{A_c^2}{2}[A^2 + m^2(t) + 2Am(t)]\cos(2\omega_c t)dt \approx 0. \quad (10.5.17)$$

See Fig. 10.5.2. We are assuming $m(t)$ is approximately *flat* for one period of the carrier. The time average of a sum is equal to the sum of the time averages. With $A_c' = AA_c$, and assuming $\langle m(t)\cos(2\omega_c t)\rangle \simeq 0$, we have

$$\begin{aligned}\langle x_{AM}^2(t)\rangle &\approx \frac{A_c^2 A^2}{2} + \frac{A_c^2}{2}\langle m^2(t)\rangle \\ &= \frac{A_c^2}{2}[A^2 + \langle m^2(t)\rangle] \\ &= \frac{(A_c')^2}{2}[1 + \mu^2\langle m_n^2(t)\rangle]. \quad (10.5.18)\end{aligned}$$

First term does not have $m(t)$ and the second term contains $m(t)$. The average power contained in the message is $P_{inf} = \langle m^2(t)\rangle$. The *modulation efficiency* is defined by the ratio

$$\eta = \frac{\langle m^2(t)\rangle}{A^2 + \langle m^2(t)\rangle}. \quad (10.5.19)$$

The AM signal (see 10.5.8.) and the *efficiency of the AM signal* in terms of $m_n(t)$ are

$$\begin{aligned}x_{AM}(t) &= A_c(A + m(t))\cos(\omega_c t) \\ &= A_c'(1 + \mu m_n(t))\cos(\omega_c t).\end{aligned}$$

$$\eta = \frac{\mu^2\langle m_n^2(t)\rangle}{1 + \mu^2\langle m_n^2(t)\rangle}. \quad (10.5.20)$$

In addition to the average power, another measure is the *peak envelope power (PEP)*. It is the average power if $m(t)$ is a constant at its maximum value. The PEP of the AM signal is

$$P_{PEP} = ((A_cA)^2/2)[1 + \max(m(t))]^2. \quad (10.5.21)$$

Example 10.5.3 Compute the efficiencies of the following assuming $m_{bn}(t) = m_{bn}(t + T)$.

$$a.\ m_{an}(t) = \cos(\omega_c t),$$

$$b.\ m_{bn}(t) = \begin{cases} 1, & 0 < t < (T/2) \\ -1, & (T/2) < t < T \end{cases}. \quad (10.5.22)$$

Solution: *a.* The average power in $m_{an}(t)$ is $\langle\cos^2(\omega_m t)\rangle = .5 + .5\langle\cos(2\omega_m t)\rangle = .5$. The efficiency of a tone modulated signal is

$$\eta = \text{Efficiency} = (1/2)\mu^2/[1 + (1/2)\mu^2]. \quad (10.5.23)$$

With $\mu \leq 1$, the upper limit on η for a single sine or a cosine wave is $(1/3)$ or 33.33%. *b.* For the square wave, $\langle m_{bn}^2(t)\rangle = 1$ and the efficiency is $\eta = \mu^2/[1 + \mu^2]$ and its upper limit on η is $\eta = 1/2 = 50\%$. For a single tone modulation, only 33% of total power is useful. For a square wave, only 50% of the total power is useful. The efficiency of an AM signal of a periodic function can be determined using the Parseval's theorem. ∎

10.5.4 Average Power Contained in an AM Signal

The Federal Communication Commission (FCC) rates AM broadcast band transmitters by the average power in the carrier $A_c\cos(\omega_c t)$ (see Couch (1997)), i.e., $P = (A_c)^2/2$.

Example 10.5.4 Find the PEP for a 5000 W AM transmitter connected to a 50 Ω load assuming $m(t) = \cos(2\pi(1000)t)$.

Solution: Since $P = 5000 = .5(A_c)^2/50$, the peak voltage is $A_c = 707$ V across the load during the times of no message. The total average power contained in the AM signal $x_{AM}(t) = A_c(A + m(t))\cos(\omega_c t) = A_cA(1 + \mu m_n(t))\cos(\omega_c t)$ is

$$\begin{aligned}P &= \frac{1}{2}\left(\frac{(A_c')^2}{50}\right) + \frac{\mu^2}{2}\left(\frac{(A_c')^2}{50}\right)\langle m_n^2(t)\rangle \\ &= \frac{(A_c')^2}{100}\left(1 + \frac{1}{2}\right) = 1.5(A_c^2)/100 = 7500\ W.\end{aligned}$$

Note that we assumed $A = 1$, $A_c' = AA_c = A_c$ and $\mu = 1$ in computing the above power. The peak voltage across the 50 Ω resistor and the PEP are, respectively, given by

$$A'_c(1 + \max(\mu\cos(\omega_m t)) = 2A'_c = 1414\ V \Rightarrow P_{\text{PEP}}$$

$$= 4\left[\frac{1}{2}\frac{(A'_c)^2}{50}\right] = 20,000\ W.$$

The average and peak powers are 7500 and 20,000 W, respectively. $\blacksquare$

10.6 Generation of AM Signals

Generation of an AM signal is considered next. See Lathi (1983) and Couch (1993).

10.6.1 Square-Law Modulators

A diode can be used as a non-linear device in a square-law modulator. See Fig. 10.6.1. The voltage across $e(t)$ and the current through the diode $i(t)$ can be approximated by

$$i \approx ae + be^2. \tag{10.6.1}$$

Assuming the diode and the resistor R as a composite non-linear component, we can write

$$e(t) = k + m(t) + A\cos(\omega_c t), \tag{10.6.2}$$

$$\Rightarrow i(t) \approx \left[ak + bk^2 + \frac{bA^2}{2} + (a + 2bk)m(t) + bm^2(t)\right]$$
$$+ A[a + 2bk + 2bm(t)]\cos(\omega_c t)$$
$$+ \left\{\frac{bA^2}{2}\cos(2\omega_c t)\right\}. \tag{10.6.3}$$

The input voltage to the BP filter is $Ri(t)$. The first term in (10.6.3) is a low-frequency signal, as it is a function a dc component and the message signal. Noting that the second and third terms are $(\cos(\omega_c t))$ and in $(\cos(2\omega_c t))$, respectively, we can see that the third term has a much higher-frequency component compared to the second term. Passing the signal $Ri(t)$ through a band-pass filter centered at the carrier frequency $\omega_c = 2\pi f_c$ results in the output signal

$$x_{AM}(t) = RA[a + 2bk + 2bm(t)]\cos(\omega_c t)$$
$$= RA(a + 2bk)[1 + \frac{2b}{a + 2bk}m(t)]\cos(\omega_c t)$$
$$= A'_c\left[1 + \mu\frac{m(t)}{|m(t)|_{\max}}\right]\cos(\omega_c t),$$

$$A'_c = RA(a + 2bk), \quad \mu = \frac{2b|m(t)|_{\max}}{(a + 2bk)}. \tag{10.6.4}$$

The constants a, b, and k are selected so that the modulation index is less than 1.

10.6.2 Switching Modulators

AM signal can be generated using a switching circuit shown in Fig. 10.6.2. The first part of the circuit with the diode and the resistor R acts as a switching circuit assuming $c \gg \max(|m(t)|)$. The voltage $v_{bb'}$ can be expressed in terms of Fourier series with harmonic terms $\cos(n\omega_c t)$. A band-pass filter centered at f_c with twice the bandwidth of the message signal recovers the frequencies around the carrier frequency resulting in an AM signal.

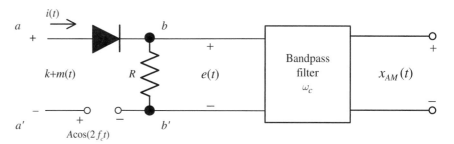

Fig. 10.6.1 Square-law AM modulator

Fig. 10.6.2 Switching modulator

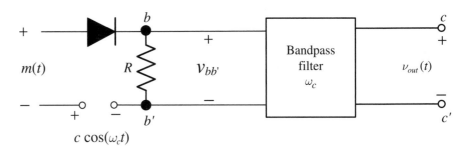

$c \cos(\omega_c t)$

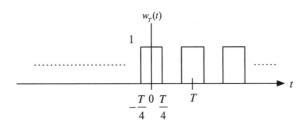

Fig. 10.6.3 Square wave $w_T(t)$

The voltage $v_{bb'}$ is

$$v_{bb'} = [c\cos(\omega_c t) + m(t)]w_T(t). \quad (10.6.5)$$

$w_T(t)$ is a periodic square wave switching function shown in Fig. 10.6.3. Its F-series is

$$w_T(t) = \frac{1}{2} + \frac{2}{\pi}[\cos(\omega_c t) - \frac{1}{3}\cos(3\omega_c t)$$
$$+ \frac{1}{5}\cos(5\omega_c t) - + \ldots], \quad (10.6.6)$$

$$\Rightarrow v_{bb'}(t) = \frac{1}{2}[c\cos(\omega_c t) + m(t)] + \frac{2}{\pi}[c\cos(\omega_c t)$$

$$+ m(t)]\cos(\omega_c t) - \frac{2}{3\pi}[c\cos(\omega_c t) + m(t)]\cos(3\omega_c t)\ldots$$

$$= \left[\frac{c}{2} + \frac{2}{\pi}m(t)\right]\cos(\omega_c t) + \text{other terms.} \quad (10.6.7)$$

The "other terms" identified in (10.6.7) correspond to the frequencies outside of the passband of the BP filter. When this signal is passed through a BP filter, centered at the carrier frequency ω_c with twice the bandwidth of $m(t)$, the output of the filter results in the AM signal given below, where constants are selected so that μ is less than 1.

$$v_{out}(t) \approx \left[\frac{c}{2} + \frac{2}{\pi}m(t)\right]\cos(\omega_c t). \ (c \gg \max|m(t)|).$$
$$(10.6.8)$$

$$\Rightarrow v_{out}(t) = \frac{c}{2}[1 + \frac{4}{\pi c}m(t)]\cos(\omega_c t)$$
$$= \frac{c}{2}[1 + \mu m_n(t)]\cos(\omega_c t). \quad (10.6.9)$$

10.6.3 Balanced Modulators

A DSB modulated signal can be generated using two AM signal generators. See Fig. 10.6.4. Note

$$y_1(t) = m(t)\cos(\omega_c t) + A\cos(\omega_c t),$$
$$y_2(t) = -m(t)\cos(\omega_c t) + A\cos(\omega_c t) . \quad (10.6.10)$$

$$y(t) = y_1(t) - y_2(t) = 2m(t)\cos(\omega_c t). \quad (10.6.11)$$

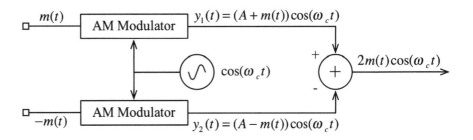

Fig. 10.6.4 Balanced modulator

The carrier term cancels out in (10.6.11) provided the modultors match.

10.7 Demodulation of AM Signals

Now consider recovering the message signal (i.e., demodulation) from an AM signal.

10.7.1 Rectifier Detector

Consider the circuit in Fig. 10.7.1. The input to the demodulator is assumed to be

$$x_{AM}(t) = [A + m(t)]\cos(\omega_c t). \qquad (10.7.1)$$

Noting that the diode is short when its voltage across is positive and open when it is negative. The rectified signal can be written as

$$v_{rec}(t) = \{[A + m(t)]\cos(\omega_c t)\}w_T(t). \qquad (10.7.2)$$

The switching function $w_T(t)$ is the same as before and its F-series in (10.6.6). The rectified signal is

$$v_{rec}(t) = \{[A + m(t)]\cos(\omega_c t)\}\left\{\frac{1}{2} + \frac{2}{\pi}[\cos(\omega_c t)\right.$$

$$\left. -\frac{1}{3}\cos(3\omega_c t) + \frac{1}{5}\cos(5\omega_c t) - + \ldots]\right\}$$

$$= \frac{1}{2}\{[A + m(t)]\cos(\omega_c t)\} + \frac{2}{\pi}[A + m(t)]\frac{1}{2}$$

$$[1 + \cos(2\omega_c t)] + \text{ higher order terms.}$$

$$(10.7.3)$$

Passing this signal through a LP filter with the message signal BW $B(B \ll f_c)$ results in

$$v_0(t) = H_0(A + m(t)). \qquad (10.7.4)$$

The dc component $(H_0 A)$ can be removed using a *high-pass filter* (or a *dc blocking capacitor*) resulting in the output $H_0 m(t)$.

10.7.2 Coherent or a Synchronous Detector

If a copy of the carrier is available at the receiver, coherent, or synchronous detection shown in Fig. 10.7.2 can be used. The output of the multiplier is

$$d(t) = (A + m(t))\cos^2(\omega_c t)$$

$$= \frac{1}{2}(A + m(t)) + \frac{1}{2}(A + m(t))\cos(2\omega_c t). \qquad (10.7.5)$$

Passing this through a LP filter results in

$$y(t) = \frac{1}{2}(A + m(t)). \qquad (10.7.6)$$

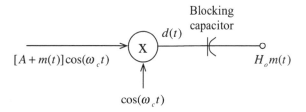

Fig. 10.7.2 Coherent AM demodulation

A blocking capacitor can be used to suppress the dc term $(A/2)$. The rectifier detection circuits operate without the carrier at the receiver. Coherent detector can demodulate *any AM signal* regardless of the value of A. Synchronous detectors are also referred as *product detectors*.

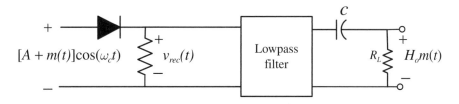

Fig. 10.7.1 Rectifier AM detector

10.7.3 Square-Law Detector

This detector uses a simple non-linear device, such as a squarer. The output of the square-law device in Fig. 10.7.3 ($A_c = 1$ is assumed for simplicity) is

$$x^2_{AM}(t) = \{(A + m(t)) \cos(\omega_c t)\}^2$$
$$= \frac{1}{2}[A^2 + m^2(t) + 2Am(t)][1 + \cos(2\omega_c t)].$$
$$(10.7.7)$$

Assuming $A \gg \max(|m(t)|)$ and $(m(t)/A)^2 \ll (m(t)/A)$, the LP filter filters out all the terms that involve $\cos(2\omega_c t)$ resulting in

$$y_0(t) = \frac{A^2}{2}\left[1 + \frac{2m(t)}{A} + \left(\frac{m(t)}{A}\right)^2\right] \approx (.5A^2 + Am(t)).$$
$$(10.7.8)$$

Blocking capacitor suppresses the dc term $A^2/2$ and the output is $Am(t)$. Coherent detector can demodulate any AM signal regardless of any A.

10.7.4 Envelope Detector

If the message signal takes positive and negative values, the DSB modulated signal has crossover points. See Fig. 10.3.2. Detecting the envelope of the DSB signal will not recover the message signal. AM signal with less than 100% modulation does not have crossover points and the envelope detector recovers the message. Figure 10.7.4 gives a simple envelope

capacitor voltage cannot follow the input and the charge across the capacitor is discharged through the resistor R until the input voltage becomes greater than the capacitor voltage. At that time, the diode turns on again and the output voltage follows to the input peak voltage and the process is repeated. During the time the diode is off, the charge across the capacitor will discharge through the resistor. The rate of discharge during this portion of the cycle depends upon the RC time constant and the peak voltage across the capacitor v_p. The capacitor voltage follows the equation

$$v_0(t) = v_p e^{-(1/RC)t} u(t).$$
$$(10.7.9)$$

The time interval for the positive half cycle is $(1/2f_c)$. Figure 10.7.5a,b shows two extreme cases. In the first case, the time constant $\tau = RC$ is *too small*, i.e., $(1/RC)$ is large, and the capacitor discharges too fast and the capacitor voltage may reach zero value before the input is in the next positive half cycle (see Fig. 10.7.5a.). In the second case, the time constant τ is large, i.e., $(1/RC)$ is small and the capacitor cannot discharge fast enough and the output may miss some pulses (see Fig. 10.7.5b.). These point out that the time constant must be much smaller than the inverse of the bandwidth of the message signal, $(1/B)$, and much larger than the period of the carrier $(1/f_c)$. That is,

$$(1/f_c) \ll RC \ll (1/B).$$
$$(10.7.10)$$

The output is

$$y(t) \approx (A + m(t)).$$
$$(10.7.11)$$

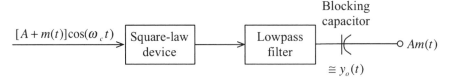

Fig. 10.7.3 Square-law AM detector

detector. During the positive half cycle of the received signal, i.e., when $x_c(t) = A_c[A + m(t)]\cos(\omega_c t) > 0$, the diode is forward biased and it acts like a short circuit and the capacitor C charges up to the peak value of the received signal. As the input signal falls below its peak value, the diode turns off as the

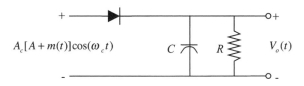

Fig. 10.7.4 AM envelope detector

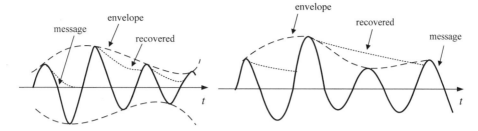

Fig. 10.7.5 Response of the envelope detector for two cases: (a) RC time constant too small and (b) RC time constant too large

The dc part can be removed using a bias removal capacitor. The envelope detector has a *poor response* at *low frequencies*.

Notes: AM radios can be used as *storm detectors*. When there is lightning, the input to the envelope detector is the received signal plus noise. We can assume the noise is a large spike of voltage and it discharges according to (10.7.9). See Fig. 10.7.6. The demodulated signal is simply the response of the *RC* circuit, not the message signal.

Example 10.7.1 Determine the upper bound on the *RC* time constant to ensure the capacitor voltage of the detector follows the envelope with $m(t) = \cos(\omega_m t)$.

Solution: Without loosing any generality, we can assume that the AM wave form and the capacitor voltage as in Fig. 10.7.7.

$$x_{AM}(t) = A'_c[1 + \mu \cos(\omega_m t)] \cos(\omega_c t),\ 0 < \mu < 1,$$
$$f_c = \omega_c/2\pi \gg \omega_m/2\pi = f_m. \qquad (10.7.12)$$

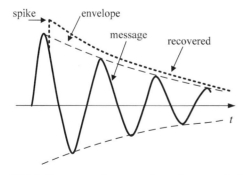

Fig. 10.7.6 Response of an envelope detector during a thunderstorm

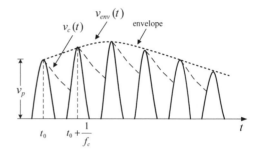

Fig. 10.7.7 Approximation of the capacitor voltage

During the negative half cycle, the charge across the capacitor is discharged through the resistor and is described earlier by (10.7.9). It can be approximated by using the first two terms of the power series expansion of the exponential decaying function (see (10.7.13)) in the time interval $0 \leq t \leq \pi/\omega_m$. Envelope of the tone modulated signal and its derivative are (see Fig. 10.7.7)

$$v_c(t) = v_p e^{-t/RC} \approx v_p(1 - (t/RC)),$$
$$v_{env}(t) = A'_c[1 + \mu \cos(\omega_m t)]. \qquad (10.7.13)$$

Slope of the envelope : $\dfrac{dv_{env}(t)}{dt}$

$$= -\mu A'_c \omega_m \sin(\omega_m t). \qquad (10.7.14)$$

To avoid missing the next peak, the magnitude of the slope of the voltage response of the *RC* circuit must be greater than the slope of the envelope. That is,

$$\left|\frac{dv_c}{dt}\right| \approx \left|\frac{dv_p(1 - t/RC)}{dt}\right| = \frac{v_p}{RC} \geq \left|\frac{dv_{env}}{dt}\right|.$$

Using (10.7.12) and (10.7.13), we have $[A'_c(1 + \mu\cos(\omega_m t))/RC] \geq \mu A'_c\omega_m|\sin(\omega_m t)|$. Since $|\sin(\omega_m t)| \geq \sin(\omega_m t)$, we can drop the absolute value sign and

$$\frac{A'_c(1 + \mu\cos(\omega_m t))}{RC} \geq \mu A'_c\omega_m \sin(\omega_m t) \text{ or}$$

$$RC \leq \frac{(1 + \mu\cos(\omega_m t))}{\mu\omega_m\sin(\omega_m t)}. \quad (10.7.15)$$

The upper bound in (10.7.15) can be determined by first taking its partial derivative with respect to t and equating it to zero. That is,

$$\frac{\sin(\omega_m t)(-\mu^2\omega_m)\sin(\omega_m t) - [1 + \mu\cos(\omega_m t)](\mu\omega_m)\cos(\omega_m t)}{\mu\sin^2(\omega_m t)} = 0.$$

$$\Rightarrow (-\mu\omega_m)[\sin^2(\omega_m t) + \cos^2(\omega_m t)] - \omega_m\cos(\omega_m t) = 0 \Rightarrow \cos(\omega_m t) = -\mu, \sin(\omega_m t) = \sqrt{1 - \mu^2}.$$

$$RC \leq \frac{(1 + \mu\cos(\omega_m t))}{\mu\omega_m\sin(\omega_m t)} = \frac{1}{\omega_m}\left[\frac{\sqrt{1 - \mu^2}}{\mu}\right]. \quad (10.7.16) \quad \blacksquare$$

10.8 Asymmetric Sideband Signals

The amplitude spectrum of a real signal is even and the phase spectrum is odd. Knowing the spectral information of a signal for $\omega \geq 0$ can be used to determine its spectral information for $\omega < 0$. If we know the spectral information on one side of a double-sideband modulated signal, i.e., the spectrum above (or below) the carrier frequency then we can determine the complete spectrum. We need only one of the sideband.

10.8.1 Single-Sideband Signals

An *upper single-sideband (USSB) signal* has a zero-valued spectrum for $|f| < f_c$, where f_c is the carrier

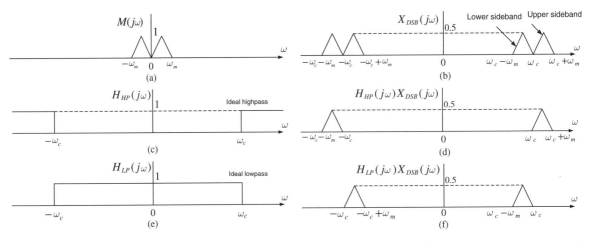

Fig. 10.8.1 (a) Message spectrum, (b) DSB spectrum, (c) ideal high-pass filter spectrum, (d) spectrum of the upper sideband signal, (e) ideal low-pass filter spectrum, and (f) spectrum of the lower-sideband signal

frequency. Similarly a *lower single-sideband signal (LSSB)* signal has zero-valued spectrum for $|f| > f_c$. An obvious way to have a single-sideband signal is by using an ideal low-pass (or an ideal high-pass filter) to obtain an LSSB (USSB) signal. Figure 10.8.1a gives an example of a double-sideband signal spectrum and Fig. 10.8.1d,f shows the spectra of the LSSB and USSB signals using ideal filters with the responses shown in Fig. 10.8.1c,e. Since the ideal filters are not physically realizable, LSSB and USSB signals can only be *approximated*.

Example 10.8.1 Using $m(t) = \cos(\omega_m t)$, derive the expressions for the USSB and the LSSB signal using the Hilbert transforms (see Section 5.10).

Solution: The Hilbert transform of $m(t)$, DSB, USSB, and LSSB signals (see Fig. 10.8.2 for the spectra) are

$$\hat{m}(t) = H\{m(t)\},$$

$$m(t) = \cos(\omega_m t) \overset{\text{HT}}{\longleftrightarrow} \hat{m}(t) = \sin(\omega_m t), \quad (10.8.1)$$

$$x_{\text{DSB}}(t) = A_c m(t) \cos(\omega_c t)$$

$$= \frac{A_c}{2}\cos(\omega_c - \omega_m)t + \frac{A_c}{2}\cos(\omega_c + \omega_m)t, \quad (10.8.2)$$

$$x_{\text{USSB}}(t) = \frac{A_c}{2}\cos(\omega_c + \omega_m)t$$

$$= \frac{A_c}{2}[\cos(\omega_m t)\cos(\omega_c t) - \sin(\omega_m t)\sin(\omega_c t)], \quad (10.8.3)$$

$$x_{\text{LSSB}}(t) = \frac{A_c}{2}\cos(\omega_c - \omega_m)t$$

$$= \frac{A_c}{2}[\cos(\omega_m t)\cos(\omega_c t) + \sin(\omega_m t)\sin(\omega_c t)]. \quad (10.8.4)$$

The sideband signals can be expressed for the tone (and for any message $m(t)$) signal by

$$x_{\text{USSB}}(t) = .5A_c[m(t)\cos(\omega_c t) - \hat{m}(t)\sin(\omega_c t)],$$

$$x_{\text{LSSB}}(t) = \frac{A_c}{2}m(t)\cos(\omega_c t) + \frac{A_c}{2}\hat{m}(t)\sin(\omega_c t). \quad (10.8.5) \quad \blacksquare$$

10.8.2 Vestigial Sideband Modulated Signals

Ideal filters are not physically realizable and the use of non-ideal filters results in generating an SSB signal is not exactly an SSB signal. The resultant sideband signal will have *most* of one of the sidebands and a *vestige* (a trace) of the other sideband. Such a signal is referred to as a *vestigial sideband modulated signal (VSB)*. The following example illustrates a simple case of a VSB signal spectrum (Ziemer and Tranter, (2002)).

Example 10.8.2 Assuming $m(t) = a_1\cos(\omega_1 t) + a_2\cos(\omega_2 t)$, illustrate the VSB spectra.

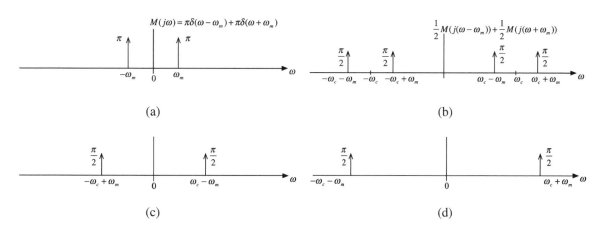

Fig. 10.8.2 (a) $M(j\omega)$, (b) DSB signal spectrum, (c) LSB signal spectrum, and (d) USSB signal spectrum

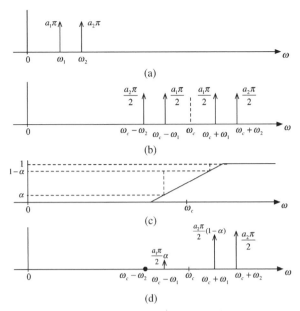

Fig. 10.8.3 (a) Message spectrum, (b) DSB spectrum, (c) non-ideal HP filter spectrum, and (d) vestigial upper single-sideband spectrum

Solution:

$$x_{\text{DSB}}(t) = m(t)\cos(\omega_c t) = (1/2)a_1\cos((\omega_c - \omega_1)t)$$

$$+ (1/2)a_2\cos((\omega_c - \omega_2)t)$$

$$+ (1/2)a_1\cos((\omega_c + \omega_1)t)$$

$$+ (1/2)a_2\cos((\omega_c + \omega_2)t). \qquad (10.8.6)$$

The spectra of $m(t)$ and the DSB signal are shown for $\omega > 0$ in Fig. 10.8.3a,b. Passing the DSB signal through a high-pass filter with the characteristic shown in Fig. 10.8.3c, the output of the VSB filter is

$$x_{\text{VSB}}(t) = \frac{1}{2}a_1\alpha\cos((\omega_c - \omega_1)t)$$

$$+ \frac{1}{2}a_1(1 - \alpha)\cos((\omega_c + \omega_1)t)$$

$$+ \frac{1}{2}a_2\cos((\omega_c + \omega_2)t),\ 0 < \alpha < 1. \quad (10.8.7)$$

The VSB signal given above contains most of the upper sideband and a small portion or a trace of

the lower sideband as can be seen from Fig. 10.8.3. The size of the trace depends upon the parameter α. In this simple example, the term corresponding to the frequency component $(f_c - f_2)$ is eliminated, whereas the amplitude of the frequency component $(f_c - f_1)$ term is significantly reduced. Note the multiplication factor α. ∎

10.8.3 Demodulation of SSB and VSB Signals

The SSB and the VSB signals can be recovered using coherent demodulation in Fig. 10.8.4. Following gives the analysis of the demodulation scheme for the SSB signal. The analysis for VSB is left as an exercise. By using (10.8.4) and (10.8.5)

$$d(t) = x_{\text{SSB}}(t)\cos(\omega_c t) = \frac{A_c}{2}m(t)\cos^2(\omega_c t)$$

$$\mp \frac{A_c}{2}\hat{m}(t)\cos(\omega_c t)\sin(\omega_c t)$$

$$= \frac{A_c}{4}m(t) + \frac{A_c}{4}m(t)\cos((2\omega_c)t)$$

$$\mp \frac{A_c}{4}\hat{m}(t)\sin((2\omega_c)t). \qquad (10.8.8)$$

Note $|F[\hat{m}(t)]| = |-j\text{sgn}(\omega)M(j\omega)| = |M(j\omega)|$; $m(t)$ and $\hat{m}(t)$ are both LP signals with a bandwidth of f_m. Since $f_c \gg f_m$, passing $d(t)$ through an ideal LP filter results in

$$y(t) \approx km(t). \qquad (10.8.9)$$

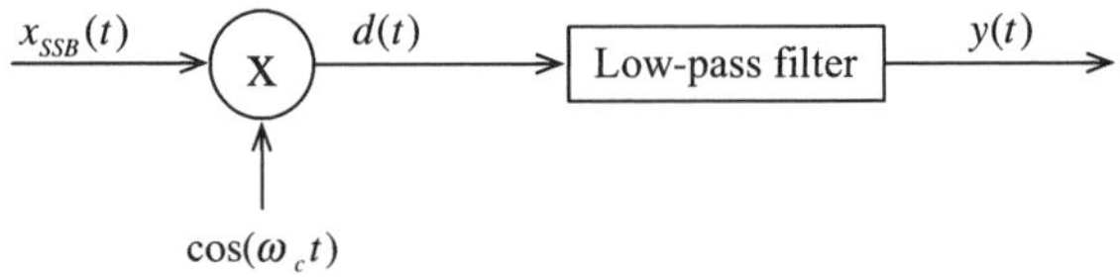

$x_{SSB}(t)$ → ⊗ → $d(t)$ → [Low-pass filter] → $y(t)$

$\cos(\omega_c t)$

Fig. 10.8.4 Synchronous demodulation of SSB

10.8.4 Non-coherent Demodulation of SSB

The effect of a non-coherent demodulation scheme in SSB is similar to the one on the DSB case. Let the local carrier at the receiver be $\cos((\omega_c + \Delta\omega)t + \delta)$. Note the phase difference δ is assumed to be a constant. Using the product demodulation, using $(-)$ sign for the USB, the $(+)$ sign for the LSB signal, we have

$$d(t) = \frac{A_c}{2}[m(t)\cos(\omega_c t)$$
$$\mp \hat{m}(t)\sin(\omega_c t)]\cos((\omega_c + \Delta\omega)t) + \delta)$$
$$= \frac{A_c}{4}[m(t)\cos(\Delta\omega t + \delta) + m(t)\cos((2\omega_c + \Delta\omega)t + \delta)]$$
$$\pm\frac{A_c}{2}\hat{m}(t)[\sin(\Delta\omega t + \delta) - \sin((\omega_c + \Delta\omega)t) + \delta)].$$
$$(10.8.10)$$

Passing $d(t)$ through a low-pass filter results in

$$y(t) \approx \frac{A_c}{4}[m(t)\cos(\Delta\omega + \delta) \pm \hat{m}(t)\sin(\Delta\omega + \delta)].$$
$$(10.8.11)$$

If $\Delta\omega$ and δ are both zero, then the output is $y(t) \approx (A_c/4)m(t)$. The recovered signal will be distorted when they are not equal to zero. Specifically, when $\Delta\omega = 0$, the output is

$$y(t) \approx \frac{A_c}{4}[m(t)\cos(\delta) \pm \hat{m}(t)\sin(\delta)]. \quad (10.8.12)$$

Since $F[\hat{x}(t)] = -j\,\text{sgn}(\omega)X(j\omega)$, it follows that $|F[\hat{x}(t)]| = |X(j\omega)|$. The phase error in the local carrier results in a phase distortion in the detector output. When $\delta = 0$, the effect of the frequency error is

equivalent to generating another SSB signal with a new carrier frequency $\Delta\omega$. Voice signals sound slightly different and a frequency shift of $\pm 20\,\text{Hz}$ are tolerable for these signals. For voice signals, $\delta = 0$ is usually not required, as the detector output is a linear combination of $m(t)$ and $\hat{m}(t)$ and the distortion is tolerable. In video and data transmission, phase distortion can be critical. Ideal filters are physically unrealizable and good sideband conventional filters are difficult to design. Good sideband suppression is possible with crystal filters (Couch (2001)).

Notes: A compromise between DSB and SSB is vestigial sideband modulation (VSB). VSB signals are easy to generate using realizable filters. The increase in bandwidth of VSB from SSB is approximately 25%. Partial transmission of the two bands allows for exact recovery of the baseband signal using a synchronous detector. ∎

10.8.5 Phase-Shift Modulators and Demodulators

Another way to generate SSB signals is by the use of phase-shift modulators. See Fig. 10.8.5. The design of these is not simple and the imperfections usually result in distortion in the low-frequency components. The basic unit in these modulators is a $-90°$ phase-shift network, a Hilbert transformer with the transfer function (see 5.11.2)

$$H(j\omega) = -j\,\text{sgn}(\omega). \quad (10.8.13)$$

Hilbert transformer is a *quadrature filter*, as its output is a sine function if its input is a cosine function. The signals identified at different locations on the

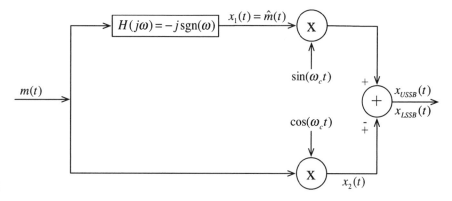

Fig. 10.8.5 Phase-shift modulator ((−) upper SSB signal, (+) lower SSB signal)

block diagram in Fig. 10.8.5 are given by $x_1(t) = \hat{m}(t)$, $x_2(t) = m(t)\cos(\omega_c t)$ and $x_3(t) = \hat{m}(t)\sin(\omega_m t)$. The output of the modulator for the upper and the lower sideband modulations are given by

$$y(t) = m(t)\cos(\omega_c t) \pm \hat{m}(t)\sin(\omega_c t). \quad (10.8.14)$$

[$(+)$, lower sideband, $(-)$ upper sidebands.]

Example 10.8.3 Using the single-tone modulating signal $m(t) = \cos(\omega_m t)$, determine the output of the modulator shown in Fig. 10.8.5.

Solution: Hilbert transform of the message signal is $\hat{m}(t) = H[\cos(\omega_m t)] = \sin(\omega_m t)$. Then

$$y(t) = \cos(2\pi f_m t)\cos(2\pi f_c t)$$
$$\mp \sin(2\pi f_m t)\sin(2\pi f_c t) = \cos(2\pi(f_c \pm f_m)t),$$

$$\Rightarrow \text{USSB signal} = \cos(2\pi(f_c + f_m)t),$$
$$\text{LSSB signal} = \cos(2\pi(f_c - f_m)t)). \quad (10.8.15)$$

Phase-shift demodulator: Figure 10.8.6 gives the phase-shift SSB demodulator. Now we will show that $x_{\text{USSB}}(t) = m(t)\cos(\omega_c t) - \hat{m}(t)\sin(\omega_c t)$. The corresponding output is

$$d(t) = [m(t)\cos(\omega_c t) - \hat{m}(t)\sin(\omega_c t)]\cos(\omega_c t)$$
$$+ H\{[m(t)\cos(\omega_c t) - \hat{m}(t)\sin(\omega_c t)]\sin(\omega_c t)\}$$
$$= \frac{m(t)}{2} + \frac{m(t)}{2}\cos(2\omega_c t) - \frac{\hat{m}(t)}{2}\sin(2\omega_c t)$$
$$+ H\left\{\left[\frac{m(t)}{2}\sin(2\omega_c t)\right]\right\}$$
$$- \frac{1}{2}\hat{m}(t) + H\left\{\left[\frac{\hat{m}(t)}{2}\cos(2\omega_c t)\right]\right\}. \quad (10.8.16)$$

Hilbert transform of a sum is the sum of the Hilbert transforms of the individual terms. If $x(t)$ is a Lp function and $y(t)$ is a non-overlapping Hp function, then (see (5.10.23))

$$H\{x(t)y(t)\} = x(t)H\{y(t)\}. \quad (10.8.17)$$

In addition, we note that $m(t)$ and $\hat{m}(t)$ are low-pass functions with a bandwidth of B Hz. The spectra of $m(t)$ (and $\hat{m}(t)$) do not overlap the spectra of $\cos(2\omega_c t)$ and $\sin(2\omega_c t)$, assuming $f_c \gg B$, which is a valid assumption. Using this and (10.8.16), we have

$$\hat{\hat{m}}(t) = -m(t), H\{\cos(2\omega_c t)\}$$
$$= \sin(2\omega_c t), H\{\sin(2\omega_c t)\} = -\cos(2\omega_c t)$$

$$H\{(1/2)\hat{m}(t)\cos(2\omega_c t)\} = (1/2)\hat{m}(t)\sin(2\omega_c t),$$
$$H\{(1/2)m(t)\sin(2\omega_c t)\} = -(1/2)m(t)\cos(2\omega_c t).$$
$$\Rightarrow d(t) = m(t)$$

In the phase-shift modulators and demodulators, it is common to generate $\sin(\omega_c t)$ from $\cos(\omega_c t)$ using a $90°$ phase shifter. The transmitted SSB signal contains only one sideband, and the envelope of the SSB signal does *not* correspond to the message signal.

Bandwidth comparisons:

$$\text{BW}_{\text{SSB}} < \text{BW}_{\text{VSB}} < \text{BW}_{\text{DSB}} = \text{BW}_{\text{AM}}. \quad (10.8.18)$$

10.9 Frequency Translation and Mixing

So far we considered linear frequency translation in the modulation schemes. We assumed a single carrier frequency and used *mixing or heterodyning*.

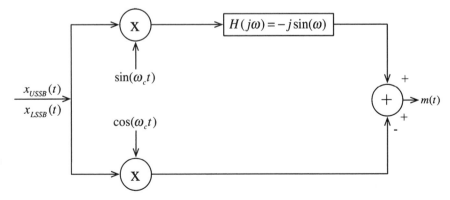

Fig. 10.8.6 Phase-shift demodulator: $(+)$ for upper single-sideband input and $(-)$ for lower single-sideband input

A radio should be able to receive all signals and recover any of these. Frequency location of a modulated signal depends upon the carrier frequency ω_c and it is captured using a band-pass filter centered at this frequency with a bandwidth twice that of the message signal. First, consider the basic associated with mixers and *up/down converters*.

The inputs to the mixer are (see Fig. 10.9.1) $m(t) \overset{FT}{\longleftrightarrow} M(j\omega)$ and $v_{Lo}(t) = 2\cos((\omega_1 + \omega_2)t)$. The input and the output spectra are given below.

$$v_m(t) = m(t)\cos(\omega_1 t) \overset{FT}{\longleftrightarrow} (1/2)M(j(\omega + \omega_1)) + (1/2)M(j(\omega - \omega_1)) = V_m(j\omega), \quad (10.9.1)$$

$$e_m(t) = 2m(t)\cos(\omega_1 t)\cos(\omega_1 + \omega_2)t = m(t)\cos(\omega_2 t) + m(t)\cos((2\omega_1 + \omega_2)t), \quad (10.9.2a)$$

$$\begin{aligned} E_m(j\omega) = F[e_m(t)] &= (1/2)M(j(\omega + \omega_2)) \\ &+ (1/2)M(j(\omega - \omega_2)) \\ &+ (1/2)M(j(\omega + 2\omega_1 + \omega_2)) \\ &+ (1/2)M(j(\omega - 2\omega_1 - \omega_2)). \quad (10.9.2b) \end{aligned}$$

The spectra at various locations are shown in Fig. 10.9.2. The subscript *mon e(t)* in Fig. 10.9.1 indicates that it corresponds to the input $m(t)$. Signal spectrum of $m(t), M(j\omega)$ is shown in Fig. 10.9.2a. For simplicity, it is assumed to be real. The spectra $V_m(j\omega)$ and $E_m(j\omega)$ are shown in Fig. 10.9.2c,d. By using a BP filter with a center frequency of ω_2 and with a bandwidth (BW) larger than twice the bandwidth of the message signal $m(t)$, we obtain the output

$$\begin{aligned} y_m(t) = m(t)\cos(\omega_2 t) &\overset{FT}{\longleftrightarrow} (1/2)M(j(\omega + \omega_2)) \\ &+ (1/2)M(j(\omega - \omega_2)). \quad (10.9.3) \end{aligned}$$

Its spectrum is shown in Fig. 10.9.2e. The block diagram in Fig. 10.9.1 shifts the spectrum of the signal $m(t)\cos(\omega_1 t)$ centered at the frequency ω_1 to the spectrum of the signal $m(t)\cos(\omega_2 t)$ centered at the frequency ω_2. That is, it translates a signal spectrum from ω_1 to ω_2. It is possible that a signal located at a frequency other than ω_1 may be translated to the same location as ω_2 by the above system. Consider the signal $k(t)$, whose spectrum $K(j\omega)$ is shown in Fig. 10.9.2b, which is assumed to be different than $M(j\omega)$. Now let $v_k(t) = k(t)\cos((\omega_1 + 2\omega_2)t)$ is the input to the system in Fig. 10.9.1. The spectrum of $v_k(t)$ is shown in Fig. 10.9.2e and is given by

$$\begin{aligned} V_k(j\omega) = (1/2)K(j(\omega - \omega_1 - 2\omega_2)) \\ + (1/2)K(j(\omega + \omega_1 + 2\omega_2)). \end{aligned}$$

The signal $e_k(t)$ and its transform $E_k(j\omega)$ (see Fig. 10.9.2 f.) are as follows:

$$\begin{aligned} e_k(t) &= 2k(t)\cos((\omega_1 + 2\omega_2)t)\cos((\omega_1 + \omega_2)t) \\ &= k(t)\cos(\omega_2 t) + k(t)\cos((2\omega_1 + 3\omega_2)t). \end{aligned}$$

$$\begin{aligned} E_k(j\omega) = F[e_k(t)] &= (1/2)K(j(\omega + \omega_2)) \\ &+ (1/2)M(j(\omega - \omega_2)) \\ &+ (1/2)M(j(\omega + (2\omega_1 + 3\omega_2))) \\ &+ (1/2)M(j(\omega - (2\omega_1 + 3\omega_2))). \end{aligned}$$
$$(10.9.4)$$

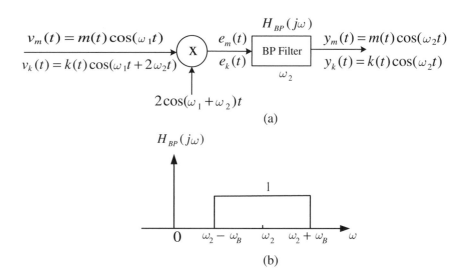

$$v_m(t) = m(t)\cos(\omega_1 t)$$
$$v_k(t) = k(t)\cos(\omega_1 t + 2\omega_2 t)$$

$$2\cos(\omega_1 + \omega_2)t$$

$$e_m(t)$$
$$e_k(t)$$

BP Filter ω_2

$$H_{BP}(j\omega)$$

$$y_m(t) = m(t)\cos(\omega_2 t)$$
$$y_k(t) = k(t)\cos(\omega_2 t)$$

(a)

$$H_{BP}(j\omega)$$

1

0 $\omega_2 - \omega_B$ ω_2 $\omega_2 + \omega_B$ ω

(b)

Fig. 10.9.1 (a) Mixer and (b) $H_{Bp}(j\omega)$

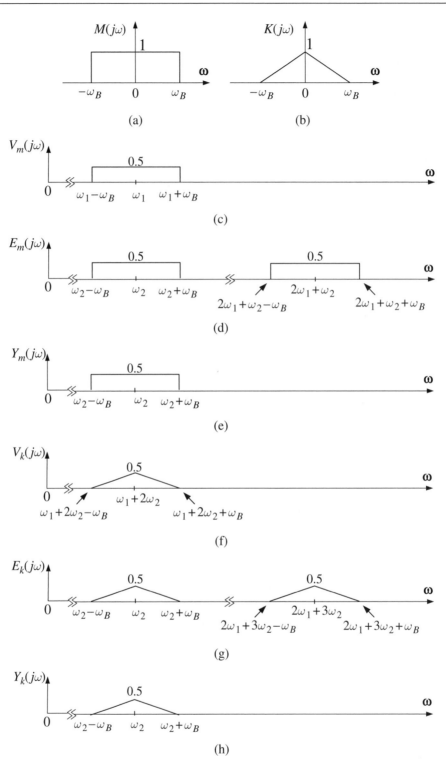

Fig. 10.9.2a (a) $M(j\omega)$, (b) $K(j\omega)$, (c) $V_m(j\omega)$, (d) $E_m(j\omega)$, (e) $E_k(j\omega)$, (f) $V_k(j\omega)$, (g) $E_k(j\omega)$, (h) $Y_k(j\omega)$, $\omega \geq 0$

Table 10.9.1 Inputs and outputs of the system in Fig. 10.9.1

Input	Output
$v_m(t) = m(t)\cos(\omega_1 t)$	$y_m(t) = m(t)\cos(\omega_2 t)$
$v_k(t) = k(t)\cos((\omega_1 + 2\omega_2)t)$	$y_k(t) = k(t)\cos(\omega_2 t)$

It is shown in Fig. 10.9.2 g. Passing $e_k(t)$ through a Bp filter with the center frequency ω_2 with a bandwidth twice that of results in (see Fig. 10.9.2 h for the spectra of $y_k(t)$.)

$$y_k(t) = k(t)\cos(\omega_2 t) \xleftrightarrow{\text{FT}} (1/2)K(j(\omega + \omega_2))$$
$$+ (1/2)K(j(\omega - \omega_2)). \tag{10.9.5}$$

See Table 10.9.1 for the input and the output functions for the two cases. Both the output spectra are centered at the frequency ω_2. The input frequency $\omega_1 + 2\omega_2$ is the *image frequency* of the desired frequency ω_1. If the input to the system in Fig. 10.9.1 is $v_m(t) + v_k(t)$, neither $y_m(t)$ nor $y_k(t)$ can be recovered. Such a problem can appear in radio receivers. *Superheterodyne receiver provides* a way to avoid this problem.

10.10 Superheterodyne AM Receiver

Block diagram of the superheterodyne AM receiver is shown in Fig. 10.10.1. The receiver consists of a *radio-frequency (RF) section* that has an *amplifier and a band-pass filter*; a *mixer*; an *intermediate-frequency (IF) section* that has an *amplifier and a fixed filter*; an *envelope detector*; and an *audio section* that has an *amplifier and a speaker*. Although we use

$\omega (= 2\pi f)'s$, the frequencies $f's$ are used as well in Hz. The front end of the AM radio receiver has all the AM signals in the form of a signal:

$$\text{Received radio signal} = \sum v_i(t) + \text{Noise.}$$

$(v_i)(t)$ is one of the radio signals).

We are interested in the AM signal $v_L(t) = g(t)\cos(\omega_c t)$, where $g(t) = A[1 + \mu m(t)]$; A, a constant; $m(t) \xleftrightarrow{\text{FT}} M(j\omega)$ is the message signal; and μ, the modulation index. The bandwidth of $m(t)$ is assumed to be B Hz.

The RF section consists of a tunable band-pass filter-amplifier. Turning the knob of the oscillator, identified by $A_c \cos(\omega_c + \omega_{IF})t$, automatically adjusts the center frequency of the filter to f_c (note the two arrows in Fig. 10.10.1) resulting in an amplified signal at point L by

$$v_L(t) = g(t)\cos(\omega_c t) + \text{noise} \approx g(t)\cos(\omega_c t), \omega_c$$
$$= 2\pi f_c. \tag{10.10.1}$$

Shortly we will consider the image frequencies. For simplicity, the noise is assumed to be negligible. $v_L(t)$ is the input to the mixer. The carrier frequency is assumed to be $\omega_c + \omega_{IF}; f_{IF} = \omega_{IF}/2\pi$ is the *intermediate frequency (IF)*. Then, at location M, on the block diagram in (10.10.1),

$$v_M(t) = A_c g(t)\cos(\omega_c t)\cos((\omega_c + \omega_{IF})t)$$
$$= (1/2)A_c g(t)\cos(\omega_{IF}t)$$
$$+ (1/2)A_c g(t)\cos((2\omega_c + \omega_{IF})t). \tag{10.10.2}$$

The spectrum of the first term on the right in (10.10.2) is centered at the intermediate frequency (IF) ω_{IF}. The spectrum of the second term is centered at

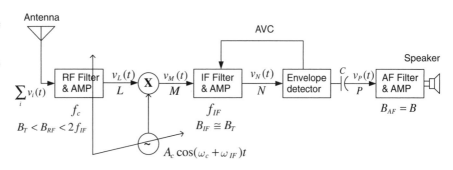

Fig. 10.10.1 Superheterodyne AM receiver, B = message BW, $B_T = 2B$ = AM transmission BW, RF = radio frequency, IF = intermediate frequency, AF = audio frequency, $B_{RF} \approx 2B$ = RF filter bandwidth, $B_{IF} = 2B$ = IF filter bandwidth, $B_{IF} = B$ = Audio filter bandwidth

$(2\omega_c + \omega_{IF})$, which is removed by the BP IF filter centered at ω_{IF} with a bandwidth of $2B$ kHz. A standard IF frequency for an AM radio is 455 kHz. Then, at point N, the output of the IF filter and amplifier is

$$v_N(t) = k_2 g(t) \cos(\omega_{IF} t), k_2 - \text{ a constant. } (10.10.3)$$

This signal is demodulated using an envelope detector and a bias removal capacitor. The signal at point P is the recovered message signal. Assuming that the AF filter (with a bandwidth of BHz) rejected most of the noise present in the signal, the signal v_P at P is

$$v_P(t) \approx k_3 m(t), k_3 - \text{ a constant.}$$

Now briefly consider image frequencies. If we are trying to recover a signal with a carrier frequency ω_c, we also receive a signal having a carrier frequency at $\omega_c + 2\omega_{IF}$. The image frequency is separated from the desired frequency by $2\,\omega_{IF}$ or . The image frequency signal is eliminated by the RF band-pass filter amplifier. Since the image frequency is separated from the desired signal frequency by 900 kHz, i.e., close to 1 MHz and the RF filter need not be a narrowband filter, see Ziemer and Tranter (2002).

The AM broadcast band occupies the frequency range from 540 kHz to 1.6 MHz. Noting the IF is 455 kHz, the tunable range of frequencies is from

$$(540\,\text{kHz} + 455\,\text{kHz}) = 995\,\text{kHz to } (1600\,\text{kHz} + 455\,\text{kHz}) = 2055\,\text{kHz}.$$

Example 10.10.1 Illustrate the superheterodyne receiver assuming the desired station is centered at $f_c = 550$ kHz and the message BW bandwidth B Hz. Discuss the image frequency signal and the bandwidth (BW) requirements on the RF filter.

Solution: First consider the RF filter BW so that the image frequency signal is eliminated by it. The signal generated at point M is located at the center frequency $f_c + f_{IF} = (550 + 455) = 1005$ kHz. The corresponding image frequency signal (say $k(t)$) is located at the center frequency $f_c + 2f_{IF} = (550 + 2(455)) = 1460$ kHz, which is removed by the RF filter with a BW larger than two times the

BW of the message signal, i.e., the transmission BW. The output of the mixer for the input $g(t)$ at point L is

$$lv_L(t) = A_c g(t) \cos(2\pi(550\,k)t) \cos((2\pi(1005\,k)t)$$
$$= (A_c/2)g(t)[\cos(2\pi(1555\,k)t)$$
$$+ \cos(2\pi(455\,k)t)].$$

This is filtered by a good IF filter centered at 455 kHz with bandwidth $2B$. The signal at point N is

$$v_N(t) \approx \beta g(t) \cos(2\pi(455\,k)t). \, (\beta \text{ is a constant}).$$

This is fed into a proper demodulator, such as an envelope detector followed by a bias removal circuit. Assuming that the AF filter–amplifier (with a bandwidth of BHz) amplifier rejected most of the noise present in the signal, the signal at point P is

$$v_P(t) \approx \alpha m(t), \alpha - \text{ a constant.} \quad (10.10.4)$$

The output of the AF filter–amplifier drives a speaker resulting in the desired audio signal. ∎

Notes: There are three main frequencies: *audio (A), intermediate (I), and the radio (R) frequencies* of interest in a superheterodyne receiver. The audio frequency (AF) band is assumed to be between 0 and 5 kHz; the radio frequency (RF) band corresponding to the carrier frequency f_c is in the frequency range $f_c - 5\,kHz < f < f_c + 5\,kHz$. The IF band range is $450\,kHz < f < 460\,kHz$. The RF filter amplifier in the superheterodyne receiver is a variable filter centered at the desired frequency $f_c = \omega_c/2\pi$ and the BW of the band-pass RF filter $BW_{RF} > 2B = 10\,kHz$. The image frequency is separated from the desired signal frequency by approximately 1 MHz. The IF filter is a *fixed filter* centered at $f_{IF} = 455$ kHz with the bandwidth of 10 kHz and is designed to do most of the filtering. The superheterodyne receiver can separate closely spaced signals.

In Fig. 10.10.1 there is a feedback loop taking the information from the envelope detector to the IF filter amplifier providing the *automatic gain or volume (AGC or AVC)* control if there is fading in the signal. The AGC detects the signal level and either increases or decreases its amplitude before it passes the signal to the next stage. It is achieved by rectifying the receiver's audio signal and finding its average value. It is used to increase (or decrease) the IF stage's gain. ∎

AM broadcast range standards (see Couch (2001) and Roden (1996))

Assigned frequency, f_c	10 kHz increments from 540 to 1700 kHz
Channel bandwidth	10 kHz
Carrier frequency stability	±20 Hz of the assigned frequency
Maximum power licensed	50 kW

AM broadcast stations in the United States are licensed by the *Federal Communications Commission (FCC)*. Since the spacing between the carrier's frequencies of AM radio stations must be at least 10 kHz, the entire frequency band can support only a few over 100 stations. There are several factors FCC uses in the decision process. The *location and the height of the antenna* are important as a low-power station in a rural area. It is less critical compared to a high-power station in a city. For stations that are nearby, the spacing between the carrier frequencies must be separated by 30 kHz or more. To avoid interference between stations, directional antennas are used. The *radiated power* affects the range of transmission. FCC takes this into consideration when assigning carrier frequencies. The *antenna pattern* controls the range as a function bearing from the antenna. Radio stations use directional antennas to provide service to their customers. The *time of broadcast* affects the range of transmission. The transmission characteristics at medium frequencies depend upon temperature, humidity, and the time of operation, day, or night and others. Clear channel stations operate full time (day and night) and have maximum licensed power of 50 kW and they have assigned special carrier frequencies (see Couch, 2001). These stations cover large areas. Nonclear-channel stations are assigned special frequencies. Some of them may have the same as the clear-channel frequencies if they can be operated without interference of the clear stations. These secondary stations operate using directional antennas with low power for a local area with a power of 1 kW. If they are allowed to operate at night, they use a night time power of 250 W or less. Night time sky wave interference is large at the frequencies these stations operate. International AM broadcast stations operate in the shortwave band (3–30 MHz) at 500 kW to 1 MW range.

Distortions: In Example 6.5.8, the response of a simple RC network for a square pulse resulted in a rounded and spread out pulse causing *linear distortion*. These distortions can be partially corrected using a deconvolution filter. Systems that cause *non-linear distortion* were studied briefly in Section 6.9.

Example 10.10.2 Consider that an AM signal is transmitted through a second-order non-linear channel and the received signal is given below. Identify the terms in $y(t)$.

$$y(t) = a_1 x_{\mathrm{AM}}(t) + a_2 x_{\mathrm{AM}}^2(t) + a_3 x_{\mathrm{AM}}^3(t), x_{\mathrm{AM}}(t)$$
$$= g(t)\cos(\omega_c t), g(t) = A_c(A + m(t)).$$

Solution:
$$y(t) = a_1 g(t)\cos(\omega_c t) + a_2 g^2(t)\cos^2(\omega_c t)$$
$$+ a_3 g^3(t)\cos^3(\omega_c t)$$

$$= (a_2/2)g^2(t) + [a_1 g(t) + (3/4)a_3 g^3(t)]\cos(\omega_c t)$$
$$+ (a_2/2)g^2(t)\cos(2\omega_c t) + (1/4)a_3 g^3(t)\cos(3\omega_c t).$$
$$(10.10.5)$$

Equation (10.10.5) indicates that the channel non-linearity affected the output. The term that multiplies $\cos(\omega_c t)$ is $[a_1 g(t) + (3/4)a_3 g^3(t)]$, where the non-linear distortion term is $(3/4)a_3 g^3(t)$. The BW of $g^3(t)$ is three times that of $g(t)$. Channel non-linearities increase the BW of the signal and interfere with other signals on the channel. Meteorological conditions change with the weather conditions. These change the effective channel transfer function, causing random attenuation of the signal referred to as *fading*. One way to reduce the slow variations due to fading is use *automatic gain control (AGC)*, see Lathi (1983). ∎

10.11 Angle Modulation

AM tends to be noisy during thunderstorms. *Angle modulation* provides much better quality of the signal at the receiver at the expense of higher transmission bandwidth. The angle modulated signal is

$$x_c(t) = A_c \cos(\omega_c t + \phi(t)). \qquad (10.11.1)$$

The constant A_c is the carrier constant and the instantaneous phase angle and the instantaneous frequency of the carrier are, respectively, given by

$$\theta_i(t) = \omega_c t + \phi(t), \quad \omega_i(t) = \frac{d\theta_i(t)}{dt} = \omega_c + \frac{d\phi}{dt}.$$
(10.11.2a)

The term $d\phi/dt$ is the frequency deviation in radians per second. The maximum or peak frequency deviation in Hertz is defined by

$$\Delta f = |\omega_i - \omega_c|/2\pi, \ \max \Delta f$$
$$= \max|\omega_i - \omega_c|/2\pi.$$
(10.11.2b)

Special cases of the angle modulated schemes:

1. *Phase modulation (PM)*.
2. *Frequency modulation (FM)*.

In the phase modulation, the phase $\phi(t)$ is varied according to the message signal $m(t)$. In the frequency modulation, the frequency is varied according to the message signal and a transfer function $h(t)$ to be defined shortly. These can be expressed by

$$\phi(t) = \int_{-\infty}^{t} m(\alpha)h(t-\alpha)d\alpha.$$
(10.11.3)

Using this in (10.11.1) results in the angle modulated signal

$$x_c(t) = A_c \cos[\omega_c t + \int_{-\infty}^{t} m(\alpha)h(t-\alpha)d\alpha].$$
(10.11.4)

If $h(t) = k_p\delta(t)$, then

$$x_{PM}(t) = A \cos[\omega_c t + \int_{-\infty}^{t} m(\alpha)\delta(t-\alpha)d\alpha]$$
$$= A \cos[\omega_c t + k_p m(t)].$$
(10.11.5)

k_p is the *phase deviation constant* in radians per unit of $m(t)$. The phase angle $\phi(t)$ is varied linearly with the message signal $m(t)$. If $h(t) = k_f u(t)$, then

$$x_{FM}(t) = A \cos[\omega_c t + k_f \int_{-\infty}^{t} m(\alpha)u(t-\alpha)d\alpha$$

$$= A_c \cos[\omega_c t + k_f \int_{-\infty}^{t} m(\alpha)d\alpha]$$

$$= A_c \cos[\omega_c t + k_f \int_{}^{t} m(\alpha)d\alpha].$$
(10.11.6)

This is a *frequency modulated (FM) signal*. It is common to leave the lower limit out as the starting time may not be known and the message signal $m(t)$ is assumed to have *no dc component*. Otherwise, the integral in (10.11.6) would diverge as $t \to \infty$. The *instaneous frequency* and the *frequency deviation* of the carrier are defined as follows.

Instantaneous frequency:

$$\omega_i(t) = d\theta_i/dt = \omega_c + d\phi/dt.$$
(10.11.7)

Frequency deviation of the carrier:

$$= d\phi/dt = k_f m(t) \propto m(t).$$
(10.11.8)

Integrating this, the phase deviation of the frequency modulated signal is

$$\phi(t) = k_f \int_{t_0}^{t} m(\alpha)d\alpha + \phi_0, \ k_f = 2\pi f_d.$$
(10.11.9)

k_f is the *frequency-deviation constant* with units, radians per second per unit $m(t)$. It is generally written in terms of the frequency deviation constant f_d, and the units are Hertz per unit of $m(t)$. In addition, ϕ_0 is the phase deviation at $t = t_0$.

Notes: A PM signal corresponding to $m(t)$ is also an FM signal corresponding to the modulating signal $[dm(t)/dt]$. With FM, the frequency deviation is proportional to the message $m(t)$. Whereas, with PM, the frequency deviation is proportional to $dm(t)/dt$. FM and PM are similar and the following gives a summary.

Replacing $k_p m(t)$ by $k_f \int^t m(\alpha)d\alpha$ changes the PM signal to an FM signal

Repalcing $k_f \int m(\alpha)d\alpha$ by $k_p m(t)$ changes FM signal to a PM signal ∎

Example 10.11.1 Give the expression for the FM signal with $m(t) = A\cos(\omega_m t)$.

Solution: From (10.11.6), the FM signal is

$$x_{\text{FM}}(t) = A_c \cos\left(\omega_c t + A\frac{2\pi f_d}{2\pi f_m}\sin(\omega_m t)\right)$$

$$= A_c \cos[\omega_c t + \beta\sin(\omega_m t)], \quad \beta = A\frac{f_d}{f_m}.$$

$$\tag{10.11.10}$$

The term β is called the *FM modulation index.* ∎

Example 10.11.2 Give the expression for the PM signal assuming $m(t) = A\sin(\omega_m t)$.

Solution: From (10.11.5), the PM signal is

$$x_{\text{PM}}(t) = A_c \cos[\omega_c t + k_p A \sin(\omega_m t)]$$

$$= A_c \cos[\omega_c t + \beta\sin(\omega_m t)], \quad \beta = k_p A.$$

$$\tag{10.11.11}$$

Interestingly the phase and frequency modulated signals have the same general form given in (10.11.10) and (10.11.11) corresponding to the sinusoidal message signals, $A\cos(\omega_m t)$ for the FM and $A\sin(\omega_m t)$ for the PM signal. The modulation index is

$$\beta = \begin{cases} A\frac{f_d}{f_m}, & \text{FM} \\ Ak_p, & \text{PM} \end{cases}. \tag{10.11.12}$$

Since the difference between a sine and a cosine function is a 90° phase shift and the frequency is the same in both cases, the spectral analysis for both PM and FM signals can be discussed for a tone signal modulation at the *same time.* ∎

Example 10.11.3 Show that angle modulation does not satisfy the linearity property and therefore is a *non-linear modulation* scheme.

Solution: Consider PM and FM of the signal $m_1(t) + m_2(t)$ by

$$x_{\text{PM}}(t) = A_c \cos[\omega_c t + k_p(m_1(t) + m_2(t))]$$

$$\neq A_c \cos[\omega_c t + k_p m_1(t)]$$

$$+ A_c \cos[\omega_c t + k_p m_2(t)],$$

$$x_{\text{FM}}(t) = A_c \cos[\omega_c t + (k_f\int^t m(\alpha)d\alpha + k_f\int^t m_2(\alpha)d\alpha],$$

$$\neq A_c \cos[\omega_c t + k_f\int^t m_1(\alpha)d\alpha]$$

$$+ A_c \cos[\omega_c t + k_f\int^t m_2(\alpha)d\alpha]. \quad ∎$$

The angle modulated signals (see (10.11.5) and (10.11.6).) can be written in the form

$$x_c(t) = A_c \cos(\omega_c t + \phi(t))$$

$$= A_c \text{Re}[e^{j[\omega_c t + \phi(t)]}] \tag{10.11.13}$$

$$= A_c \text{Re}[e^{j\omega_c t}e^{j\phi(t)}],$$

$$x_{\text{PM}}(t) = A_c \text{Re}[e^{j(\omega_c t + k_p m(t))}]$$

$$x_{\text{FM}}(t) = A_c \text{Re}[e^{j(\omega_c t + \phi(t))}],$$

$$\phi(t) = 2\pi f_d \int^t m(\alpha)d\alpha. \tag{10.11.14}$$

Re [.] corresponds to the real part of the function inside the brackets. Angle modulation is also called *exponential modulation.* Figure 10.11.1 illustrates the waveforms of a sinusoidal message signal

Modulating signal

AM

FM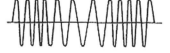

PM

Fig. 10.11.1 A sinusoidal modulating signal and the corresponding AM, FM, and PM signals

along with the corresponding AM, FM, and PM signals. The AM signal is assumed to be $A_c(A + m(t))\cos(\omega_c t)$ with $(A + m(t)) > 0$. We will consider the spectra of the sinusoidally modulated AM signal and the angle modulated signals shortly. Note $|x_c(t)| = |A_c \cos(\omega_c t + \phi(t))| = |A_c|$.

The frequency content of an angle modulated signal changes according to the message signal and exact spectral description of all angle modulated signals is not possible. The frequency deviation of an angle modulated signal from the carrier frequency is given in (10.11.2b). The frequency varies in the range

$$\left[\omega_c - \max\left|\frac{d\phi}{dt}\right|\right] \leq \text{Frequency deviation}$$
$$\leq \left[\omega_c + \max\left|\frac{d\phi}{dt}\right|\right]. \qquad (10.11.15)$$

FM signal has a maximum (minimum) frequency deviation wherever the message signal is negative maximum (negative minimum). The phase modulated signal has maximum (minimum) frequency deviation wherever the message signal has positive maximum (negative minimum) slope. Later we will consider these in analytical terms. In the case of the sinusoidal modulation, the modulated signals are very similar, whereas for square wave modulation the FM and PM waves are distinctly different. Earlier we pointed out that PM and FM modulations are intimately related. By just looking at an angle modulated signal, we cannot tell whether it is a PM signal or an FM signal.

10.11.1 Narrowband (NB) Angle Modulation

By approximating the cosine and the sine functions using power series, $\cos(\phi(t)) \approx 1$ and $\sin(\phi(t)) \approx \phi(t)$ for $|\phi(t)| \ll 1$, the angle modulated signal reduces to a NB angle modulation scheme given by

$$x_c(t) = A_c \cos[\omega_c t + \phi(t)]$$
$$= A_c[\cos(\omega_c t)\cos(\phi(t)) - \sin(\omega_c t)\sin(\phi(t))], \qquad (10.11.16)$$
$$\approx A_c \cos(\omega_c t) - A_c \phi(t)\sin(\omega_c t). \qquad (10.11.17)$$

This approximate representation is very similar to the amplitude modulated signal.

$$x_{\text{AM}}(t) = A_c(A + m(t))\cos(\omega_c t)$$
$$= A_c A \cos\omega_c t + A_c m(t)\cos(\omega_c t). \quad (10.11.18)$$

Both the NB angle modulated signal and the AM signal have the carrier component. In the second part of the AM signal, we have the DSB part $[A_c m(t)\cos(\omega_c t)]$; whereas, in the second part of the NB angle modulated signal we have the term $[-\phi(t)\sin(\omega_c t)]$ and in AM we have the term $A_c m(t)\cos(\omega_c t)$. The NBFM and NBPM signals are given by

$$x_{\text{NBPM}}(t) = A_c \cos(\omega_c t)$$
$$- A_c k_p m(t)\sin(\omega_c t), \qquad (10.11.19)$$

$$x_{\text{NBFM}}(t) = A_c \cos(\omega_c t) - A_c k_f \left[\int^t m(\alpha)d\alpha\right]\sin(\omega_c t).$$
$$(10.11.20)$$

Example 10.11.4 Illustrate the differences between the AM and the NB angle-modulated signals using tone modulation, see (10.11.10) and (10.11.11).

Solution: The narrowband FM (NBFM) and narrowband PM (NBPM) modulated signals can be approximated by

$$x_{\text{FM}}(t) = A_c \cos\left[\omega_c t + A\frac{f_d}{f_m}\sin(\omega_m t)\right] \approx A_c \cos(\omega_c t)$$
$$- A_c\left(A\frac{f_d}{f_m}\right)\sin(\omega_m t)\sin(\omega_c t) = x_{\text{NBFM}}(t),$$
$$(10.11.21)$$

$$x_{\text{PM}}(t) = A_c \cos[\omega_c t + A k_p \sin(\omega_m t)]$$
$$\approx A_c \cos(\omega_c t) - A_c(A k_p)\sin(\omega_m t)\sin(\omega_c t)$$
$$= x_{\text{NBPM}}(t). \qquad (10.11.22)$$

They both have the same general form. Using the trigonometric formulas, we have

$$x_{\text{NBFM}}(t) = A_c \cos(\omega_c t) + \frac{1}{2}A_c\left(A\frac{f_d}{f_m}\right)[\cos(\omega_c + \omega_m)t - \cos(\omega_c - \omega_m)t], \qquad (10.11.23)$$

$$x_{\text{NBPM}}(t) = A_c \cos(\omega_c t) + \frac{1}{2}A_c(A k_p)[\cos(\omega_c + \omega_m)t - \cos(\omega_c - \omega_m)t], \qquad (10.11.24)$$

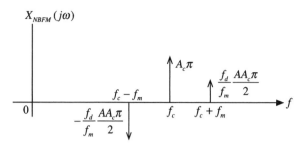

Fig. 10.11.2 Spectrum of a narrowband FM signal

$$\Rightarrow X_{\text{NBFM}}(j\omega) = A_c\pi[\delta(\omega - \omega_c) + \delta(\omega + \omega_c)]$$
$$+ \frac{AA_c\pi f_d}{2 \quad f_m}[\delta(\omega - (\omega_c + \omega_m))$$
$$- \delta(\omega - (\omega_c - \omega_m))$$
$$+ \delta(\omega + (\omega_c + \omega_m))$$
$$- \delta(\omega + (\omega_c - \omega_m))]. \quad (10.11.25)$$

The NBFM spectrum is sketched in Fig. 10.11.2 for positive frequencies. Note the horizontal axis is given in terms of frequency $f = (\omega/2\pi)$ in Hertz rather than in terms of radians per second. The derivation and sketch for the spectrum of the NBPM is left as an exercise. Comparing this with the AM signal we see that both the NBFM and the AM signals require the same bandwidth, i.e., $2f_m$. NB angle modulation schemes are not that much of use as they are, except that they can be used as a first

step in generating wideband modulation discussed next. Note the bandwidth of the NBFM is $2\,\omega_m$.

10.11.2 Generation of Angle Modulated Signals

Implementations of the NBFM and NBPM schemes are illustrated Fig. 10.11.3a,b based on (10.11.21) and (10.11.22). Since the sine is a phase-shifted cosine function, it can be generate from the cosine function using a 90° phase shifter, see Fig. 10.11.3c.

Narrowband-to-wideband conversion: There are two standard methods referred to as the direct and indirect methods to generate a wideband angle modulated signal.

Indirect method: A simple way to generate a wideband signal is first generate a NB signal and use a frequency multiplier and a band-pass filter that filters out the undesired frequencies. Let the input to the square-law device is a NB angle modulated signal. The NB signal and the output of the square-law device are ($\phi(t)$ is a function of $m(t)$.)

$$x_1(t) = A_1\cos(\omega_1 t + \phi(t)), |\phi(t)| \ll 1,$$
$$y_1(t) = x_1^2(t) = A_1^2\cos^2(\omega_1 t + \phi(t))$$
$$= \frac{1}{2}A_1^2 + \frac{1}{2}A_1^2\cos(2\omega_1 t + 2\phi(t)). \quad (10.11.26)$$

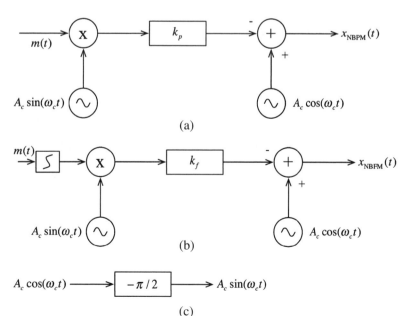

Fig. 10.11.3 Generation of narrowband angle modulated signal (a) NBFM, (b) NBFM, and (c) phase shifter

Passing $y_1(t)$ through a BP filter with the center frequency of $(2f_1)$ with sufficient bandwidth (discussed in the next section) results in

$$y_0(t) = A_1^2 \cos(2\omega_1 t + 2\phi_1(t)). \qquad (10.11.27)$$

One can use an n-law device and obtain the output $A_1^n \cos^n(n\omega_1 t + n\phi_1(t))$. Using trigonometric identities (10.11.27) can be written in terms of a sum of harmonic terms. Using a NBBP filter with the center frequency of $f_c = nf_1$ with sufficient BW, the output is

$$y(t) = A_c \cos(\omega_c t + \phi(t)), \ \omega_c = n\omega_1, \ \phi(t) = n\phi_1(t). \qquad (10.11.28)$$

Selecting a large n will result in a wide-band modulated signal. Note the increase in the carrier frequency from f_1 to nf_1.

Direct method: In this method FM is generated using a *voltage-controlled oscillator* (VCO), whose frequency is controlled by the modulating signal. One way to generate an FM directly is by varying a capacitance or an inductance of a tuned oscillator. Direct method provides large frequency deviations. One of the major disadvantages is that carrier frequency tends to drift and frequency stabilization is needed, see Couch (2001).

10.12 Spectrum of an Angle Modulated Signal

To find the frequency content of a signal, such as an angle modulated signal, all we need to do is take the Fourier transform of this signal. Unfortunately, closed form solutions for the transforms of the angle modulated signals do not exist for all *most all* message signals. To start with we will consider tone angle modulated signals. Recall that the angle-modulated signal can be expressed as (see (10.11.10) and (10.11.11))

$$x_c(t) = A_c \cos[\omega_c t + \beta \sin(\omega_m t)],$$

$$\beta = \begin{cases} (Af_d/f_m), & \text{FM} \\ Ak_p, & \text{PM} \end{cases}. \qquad (10.12.1)$$

Recall that in deriving this expression we have used two different message signals, a cosine function for the FM case and a sine function for the PM case. Since the sine and the cosine functions of the same frequency are the same except for the phase shift, spectral analysis of the signal in (10.12.1) gives us the necessary information, and therefore, we can spectral analyze the signals for both FM and PM at the same time.

In Chapter 1, Section 1.6.3 functions of periodic functions were considered. With $\phi(t) = \beta \sin(\omega_m t)$ in (10.11.13), we have

$$x_c(t) = A_c \text{Re}\left[e^{j\omega_c t} e^{j\beta \sin(\omega_m t)} \right], \omega_m = 2\pi f_m. \quad (10.12.2)$$

Since $\beta \sin(\omega_m t)$ is periodic with period $T = (1/f_m)$, so is $e^{j\beta \sin(\omega_m t)}$. It can be seen from (10.12.3) below. Since it is periodic, it can be expressed in terms of F-series in (10.12.4).

$$e^{j\beta[\sin(\omega_m(t+T))]} = e^{j\beta \sin(\omega_m t)} = y_T(t), \omega_m T = 2\pi, \qquad (10.12.3)$$

$$e^{j\beta \sin(\omega_m t)} = \sum_{k=-\infty}^{\infty} X_s[k] e^{j2\pi k f_m t},$$

$$X_s[k] = \frac{1}{(1/f_m)} \int_{-(1/2f_m)}^{(1/2f_m)} e^{j\beta \sin(\omega_m t)} e^{-j2\pi k f_m t} dt,$$

$$\omega_m = 2\pi f_m. \qquad (10.12.4)$$

Using $\lambda = 2\pi f_m t$ in the above integral and $t = \pm(1/2f_m) \Rightarrow \lambda = \pm\pi$ and $d\lambda = 2\pi f_m dt$, we have the F-series coefficients

$$X_s[k] = \frac{1}{2\pi} \int_{-\pi}^{\pi} e^{-j(k\lambda - \beta \sin(\lambda))} d\lambda \equiv J_k(\beta). \qquad (10.12.5)$$

This integral cannot be determined analytically and numerical methods are needed. $J_k(\beta)$ is the Bessel function of the first kind of order k with respect to real β. See Poularikas (1996) for a detailed discussion on Bessel functions. The Bessel functions $J_k(\beta)$ can be expanded using power series for $\beta \neq 0$ and $k \neq 0$. We will discuss this special case shortly.

$$J_k(\beta) = \sum_{n=0}^{\infty} \frac{(\beta/2)^{k+2n}(-1)^n}{(n+k)!(n)!}, \quad -\infty < \beta < \infty. \qquad (10.12.6)$$

10.12.1 Properties of Bessel Functions

Some properties of Bessel functions can be derived by using either (10.12.5) or (10.12.6) assuming β is real, see Whittaker and Watson (1927). These are given below.

1. $J_{-k}(\beta) = (-1)^k J_k(\beta), \; k = 0, 1, 2, \ldots$ (10.12.7)

By replacing k by $-k$ in (10.12.6) and noting that $1/[(n-k)!] = 0$ for $n < k$ and $0! = 1$,

$$J_{-k}(\beta) = \sum_{n=0}^{\infty} \frac{(-1)^n (\beta/2)^{2n-k}}{n!(n-k)!} = \sum_{n=k}^{\infty} \frac{(-1)^n (\beta/2)^{2n-k}}{n!(n-k)!}$$

$$= \sum_{m=0}^{\infty} \frac{(-1)^{m+k}(\beta/2)^{2m+k}}{m!(m+k)!}. \qquad (10.12.8)$$

The last equality in (10.12.8) is obtained by seting $n = m + k$. From this equation, (10.12.7) follows. Furthermore, from the integral in (10.12.5), we have the special case

2. $J_0(0) = \dfrac{1}{2\pi} \displaystyle\int_{-\pi}^{\pi} e^{-jk\lambda} dy = 1,$

$$J_k(0) = \frac{1}{2\pi} \int_{-\pi}^{\pi} e^{-jk\lambda} d\lambda = \frac{e^{-jk\lambda}}{2\pi(-jk)}\Big|_{\lambda=-\pi}^{\lambda=\pi} = 0, k \neq 0.$$

$$(10.12.9)$$

For small values of β, by keeping only the first term in the series in (10.12.6), we have

$$J_k(\beta) \approx \beta^k/2^k k!, \qquad (10.12.10a)$$

For $\beta \leq .3$, $J_0(\beta) \approx 1, J_1(\beta) \approx \beta/2, J_k(\beta)$
$$\approx 0, \; k > 1. \qquad (10.12.10b)$$

For large values of β, $J_k(\beta)$ can be approximated by

$$J_k(\beta) \approx \sqrt{\frac{2}{\pi\beta}} \cos\left(\beta - \frac{\pi}{4} - \frac{n\pi}{2}\right). \qquad (10.12.10c)$$

That is, $J_k(\beta)$ behaves like a sinusoidal function with decreasing amplitude and

$$\lim_{k\to\infty} J_k(\beta) = 0 \text{ for a fixed } \beta. \qquad (10.12.10d)$$

3. For $k = 0, 1$ and $\beta \geq 4$, good approximations for $J_0(\beta)$ and $J_1(\beta)$ are

$$J_0(\beta) \approx \frac{\cos(\beta - \frac{\pi}{4})}{\sqrt{(\pi/2)\beta}},$$

$$J_1(\beta) \approx \frac{\sin(\beta - \frac{\pi}{4})}{\sqrt{(\pi/2)\beta}}. \qquad (10.12.10e)$$

4. Bessel functions can be derived using a difference equation (see Poularikas (1996))

$$(2k/\beta)J_k(\beta) = J_{k-1}(\beta) + J_{k+1}(\beta). \qquad (10.12.11)$$

5. $$\sum_{k=-\infty}^{\infty} J_k^2(\beta) = 1 \text{ for all } \beta. \qquad (10.12.12)$$

This is shown below.

$$\sum_{k=-\infty}^{\infty} J_k^2(\beta) = \frac{1}{(2\pi)^2} \sum_{k=-\infty}^{\infty} \int_{-\pi}^{\pi} \int_{-\pi}^{\pi} e^{(j\beta\sin(\theta)-jk\theta-j\beta\sin(\phi)+jk\phi)} d\theta d\phi. \qquad (10.12.13)$$

Interchanging the order of summation and double integration, we have

$$\sum_{k=-\infty}^{\infty} J_k^2(\beta) = \frac{1}{(2\pi)^2} \int_{-\pi}^{\pi} \int_{-\pi}^{\pi} e^{j\beta[\sin(\theta)-\sin(\phi)]} \sum_{k=-\infty}^{\infty} e^{jk(\phi-\theta)} d\theta d\phi.$$

The summation can be simplified by using the relationship derived in Example 3.4.4 and

$$\sum_{n=-\infty}^{\infty} \delta(t-nT) = \frac{1}{T}\sum_{k=-\infty}^{\infty} e^{jk\omega_0 t}, \; \omega_0 = \frac{2\pi}{T}. \quad (10.12.15)$$

Noting the limits on the integral, we are only interested in the limited range of $-\pi \le (\phi - \theta) \le \pi$

and $T = 2\pi$ and the summation can be expressed by

$$\frac{1}{2\pi}\sum_{k=-\infty}^{\infty} e^{jk\omega_0 t} = \delta(\phi - \theta). \quad (10.12.16)$$

The summation in (10.12.15) can be simplified and

$$\sum_{k=-\infty}^{\infty} J_k^2(\beta) = \frac{1}{(2\pi)^2}\int_{-\pi}^{\pi}\int_{-\pi}^{\pi} e^{j\beta[\sin(\theta)-\sin(\phi)]}(2\pi)\delta(\phi - \theta)d\phi d\theta = \frac{1}{2\pi}\int_{-\pi}^{\pi} d\theta = 1.$$

Line spectra of tone modulated signals: From (10.12.2), the angle modulated tone signal can be expressed using Bessel functions.

$$x_c(t) = A_c \text{Re}\left[e^{j\omega_c t}e^{j\beta\sin(\omega_m t)}\right]$$

$$= A_c \text{Re}\left[e^{j\omega_c t}\sum_{k=-\infty}^{\infty} J_k(\beta)e^{j2\pi k f_m t}\right]$$

$$= A_c \text{Re}\left[\sum_{k=\infty}^{\infty} J_k(\beta)e^{j(\omega_c + k\omega_m)t}\right]$$

$$= A_c \sum_{k=-\infty}^{\infty} J_k(\beta)\cos((\omega_c + k\omega_m)t). \quad (10.12.17a)$$

$$= A_c\{J_0(\beta)\cos(\omega_c t) + J_1(\beta)[\cos(\omega_c + \omega_m)t$$
$$- \cos(\omega_c - \omega_m)t] + J_2(\beta)[\cos(\omega_c + 2\omega_m)t$$
$$+ \cos(\omega_c - 2\omega_m)t] + \dots \quad (10.12.17b)$$

Figure 10.12.1 shows the *one-sided line spectra* of $x_c(t)$. From (10.12.17b), the *spectrum* of a *tone angle modulated signal* consists of a carrier-frequency f_c plus an *infinite number* of

sideband frequency frequencies $(f_c \pm k f_m)$, $k = 1, 2, 3, \dots$ The spectral amplitudes depend on the value of $J_k(\beta)$ and the amplitudes become small for large values of k. How many terms we need to keep in (10.12.17b) so that it is a good approximation of the modulated signal? This depends upon the modulation index β. If $\beta \ll 1$, only J_0 and J_1 are significant and the other coefficients are not. The spectrum can be approximated by the carrier and the two sidebands for small β.

$$x_c(t) \approx A_c \cos(\omega_c t) + A_c J_1(\beta)\cos(\omega_c + \omega_m)t$$
$$- A_c J_1(\beta)\cos(\omega_c - \omega_m)t, \beta \ll 1. \quad (10.12.18)$$

For *tone modulation, narrowband angle modulation is a low-deviation* modulation. For $\beta \gg 1$, several terms in (10.12.17b) are needed. Large β implies wider bandwidth. From the power series expansion of the Bessel functions in (10.12.6),

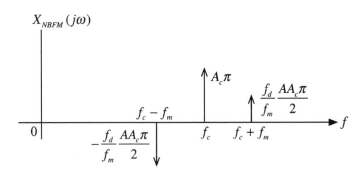

Fig. 10.12.1 One-sided line spectra

$|J_k(\beta)| \ll 1$ if $|k/\beta| \gg 1$. Selected values of $J_k(\beta)$ are given in Table 10.12.1. Blanks in the table correspond to $|J_k(\beta)| < .01$. See Gradshteyn and Ryzhik (1980) for an extensive table of values of Bessel functions. The number of terms in approximating the angle modulated signal depends on what percentage of power need to be kept in the approximated signal. The average power can be computed using the Parseval's theorem.

We can see from Table 10.12.1 that $|J_k(\beta)|$ falls off rather rapidly for $|k/\beta| > 1$.

10.12.2 Power Content in an Angle Modulated Signal

The average power in $x_c(t)$ is given by $P = \langle x_c^2(t) \rangle$ and is

$$P = \lim_{T \to \infty} \frac{1}{T} \int_{-T/2}^{T/2} x_c^2(t)dt = \langle A_c^2 \cos^2(\omega_c t + \phi(t)) \rangle$$

$$= \frac{1}{2}A_c^2 + \frac{1}{2}A_c^2 \langle \cos(2\omega_c t + 2\phi(t)) \rangle. \quad (10.12.19)$$

Since the carrier frequency, f_c is very large compared to the BW of the message signal W, $x_c(t)$ has *negligible frequency* content in the *dc region*. The envelope of the angle modulated signal is approximately flat over one period $(1/2f_c)$. The time average over this period is zero and

$\langle \cos(2(\omega_c t + \phi(t)) \rangle \approx 0$. The average power in an angle modulated signal is *independent* of the message signal and the carrier frequency and is

$$P = \frac{1}{2}A_c^2. \quad (10.12.20)$$

Returning to the sinusoidal modulated signal, let the *power* contained in the *carrier* and k *upper lower sidebands* be identified by P_r^K. The power contained in a sinusoid, $J_k(\beta) \cos((\omega_c \pm k\omega_m)t)$ is equal to $(J_k^2(\beta)/2)$. Therefore

$$P_r^K = \frac{1}{2}A_c^2 \sum_{k=-K}^{K} J_k^2(\beta), \quad (10.12.21)$$

$$\lim_{K \to \infty} P_r^K = \frac{1}{2}A_c^2 \sum_{k=-\infty}^{\infty} J_k^2(\beta) = \frac{1}{2}A_c^2$$

$$\text{(see (10.12.13))}. \quad (10.12.22)$$

To keep 98% or more of the power in angle modulated signal for a single message tone, K frequencies need to be kept above and below the carrier frequency such that

$$P_r = P_r^K/P \geq .98. \quad (10.12.23)$$

The corresponding BW of the angle modulated signal is $BW \approx 2 Kf_m$; f_m is the frequency of the single tone signal. It turns out that K is equal to the integer part of $(\beta + 1)$ for $P_r \geq .98$ and the bandwidth is approximated by

$$BW \approx 2(\beta + 1)f_m. \quad (10.12.24)$$

Table 10.12.1 Bessel function values

$k/J_k(\beta)$	$J_k(.1)$	$J_k(.2)$	$J_k(.5)$	$J_k(1)$	$J_k(2)$	$J_k(5)$	$J_k(8)$	$J_k(10)$
0	0.997	0.990	0.938	0.765	0.224	−0.178	0.172	−0.246
1	0.050	0.100	0.242	0.440	0.577	−0.328	0.235	0.043
2	0.001	0.005	0.031	0.115	0.353	0.047	−0.113	0.255
3		0.003	0.003	0.020	0.129	0.365	−0.291	0.058
4				0.020	0.034	0.391	−0.105	−0.220
5				0.002	0.007	0.261	0.286	−0.234
6					0.001	0.131	0.338	−0.014
7						0.053	0.321	0.217
8						0.018	0.224	0.318
9						0.006	0.126	0.292
10						0.001	0.061	0.208
11							0.026	0.123
12							0.010	0.063
13							0.003	0.029
14							0.001	0.012

Example 10.12.1 Consider the phase modulated signal with $A_c = 10$, $Ak_p = 5$ and $m(t) = A\sin(\omega_m t)$. Find the average power, bandwidths contained in the PM signal assuming three message signal frequencies $f_m = 1, 2, 4\,\text{kHz}$.

Solution: The PM signal is (see (10.12.1))

$$x_{\text{PM}}(t) = A_c \cos[\omega_c t + k_p m(t)]$$
$$= 10\cos(\omega_c t + 5\sin(\omega_m t)). \quad (10.12.25)$$

The average power contained in the signal is $P = A_c^2/2 = 50\,\text{W}$. From (10.12.25), $\beta = Ak_p = 5$ which is *independent* of the message signal frequency and the BWs are

$$BW = 2(\beta+1)f_m = \begin{cases} 2(5+1)1 = 12\,\text{kHz}, fm = 1\,\text{kHz} \\ 2(5+1)2 = 24\,\text{kHz}, fm = 2\,\text{kHz}\,. \\ 2(5+1)4 = 48\,\text{kHz}, fm = 4\,\text{kHz} \end{cases} \quad ∎$$

Example 10.12.2 Assuming the transmitted signal is an FM signal with $A_c = 10$ and $m(t) = A\cos(\omega_m t)$

and $Af_d = 5k$, give an approximate FM transmission BW.

Solution: The average power is $P = 50\,\text{W}$. The FM signal is

$$x_{FM}(t) = A_c \cos(\omega_c t + \phi(t))$$

$$= A_c \cos\left(\omega_c t + k_f \int^t m(\alpha)d\alpha\right)$$

$$= A_c \cos\left[\left\{\omega_c t + \frac{Ak_f}{\omega_m}\sin(\omega_m t)\right\}\right],$$

$$= A_c \cos[(\omega_c t + \beta\sin(\omega_m t))],$$

$$\beta = A\frac{k_f}{\omega_m} = \frac{A(2\pi f_d)}{2\pi f_m} = Af_d/f_m.$$

$$(10.12.26)$$

Message frequencies and the BWS:

$$\beta = \begin{cases} 5k/1k = 5, f_m = 1k \\ 5k/2k = 2.5, f_m = 2k \\ 5k/4k = 1.25, f_m = 4k \end{cases}, BW = 2(\beta+1)f_m = \begin{cases} 2(5+1)1 = 12\,\text{kHz}, f_m = 1\,\text{kHz} \\ 2(2.5+1)2 = 14\,\text{kHz}, f_m = 2\,\text{kHz} \\ 2(1.25+1)4 = 18\,\text{kHz}, f_m = 4\,\text{kHz} \end{cases}. \quad ∎$$

The expression for BW of an angle modulated can be generalized from tone signals to arbitrary signals. First, let

$D = \text{Deviation constant}$

$$= \frac{\text{peak frequency deviation}}{\text{bandwidth of the message signal } m(t)}. \quad (10.12.27)$$

For an arbitrary signal we replaced β by the message signal signal BW in for D. The *transmission BW* of an angle modulated signal can be estimated by the following rules:

$$B_T = 2(D+1)W \text{(Carson's rule)}, \quad (10.12.28)$$

$$B_T \approx 2(D+2)W \text{(modifiedCarson's srule)}. \quad (10.12.29)$$

Carson's rule underestimates the bandwidth of FM signals in the cases of $2 < D < 10$.

Notes: If $D \ll 1$ then we have the narrowband (NB) angle modulated signal and its transmission BW is 2 W. If $D \gg 1$ then the BW is approximately equal to 2 DW and is a wideband signal. The spectrum of an angle modulated signal is centered at the carrier frequency with a BW B_T and is independent of the carrier frequency. ∎

Example 10.12.3 The peak frequency deviation of a commercial FM signal is limited to 75 kHz. Let the message signal BW is 15 kHz. Determine the FM transmission BW.

Solution: The deviation ratio and the transmission bandwidth from (10.12.28) are

$$D = (75/15) = 5 \Rightarrow B_T = 2(5+1)15 = 180\text{kHz}$$
$$(10.12.30) \quad \blacksquare$$

FCC standards suggest the *commercial FM transmission bandwidth* to be *200* kHz. $\quad \blacksquare$

Notes: BW of an angle modulated signal is significantly larger than the AM BW. It is more immune to noise allowing for a smaller transmission power independent of the message. FM trades signal power to BW. Constant amplitude of an angle modulated signal makes it less susceptible to non-linear distortion. $\quad \blacksquare$

10.13 Demodulation of Angle Modulated Signals

The simplest method converts an angle modulated signal to an AM signal and the message is recovered using envelope detection. For other methods, see Couch (2001).

10.13.1 Frequency Discriminators

A block diagram of the steps involved in demodulating an FM signal is shown in Fig. 10.13.1. The instantaneous frequency of an FM signal is proportional to the message signal and an ideal FM detector (a demodulator) is a device that produces an output that is proportional to the instantaneous input frequency. Neglecting the delay and noise, the received signal $x_r(t)$ is approximately the same as the transmitted angle modulated signal $x_c(t)$, i.e.,

$$x_r(t) = x_c(t).$$
$$x_c(t) = A_c \cos[\omega_c t + \phi(t)]$$
$$= \begin{cases} A_c \cos[\omega_c t + k_f \int^t m(\alpha)d\alpha], \ k_f = 2\pi f_d, FM \\ A_c \cos(\omega_c t + k_p m(t)), PM \end{cases}.$$
$$(10.13.1)$$

Note that transmission delays are ignored here, as they do not affect the message recovery

process. The amplitude variations of an angle modulated signal can be eliminated by a limiter–band-pass system in Fig. 10.2.1. BP filter is centered at ω_c with sufficient bandwidth. The input to the differentiator is assumed to be $x_c(t)$ in (10.13.1).

Simple frequency discriminators – basic concepts: A simple frequency discriminator is an ideal differentiator followed by an envelope detector, see Fig.10.13.1. Assuming the input to the discriminator is $x_c(t) = A_c \cos(\omega_c t + \phi(t))$, the output of the differentiator is

$$x_c'(t) = \frac{dx_c(t)}{dt} = -A_c \left[\omega_c + \frac{d\phi}{dt}\right] \sin(\omega_c t + \phi(t)).$$
$$(10.13.2)$$

$x_c'(t)$ is amplitude and angle modulated. The coefficient term in (10.13.2)

$$y(t) = A_c \left[\omega_c + \frac{d\phi}{dt}\right] > 0 \text{ if } \omega_c > -\frac{d\phi}{dt} \text{ for all } t.$$
$$(10.13.3)$$

Note the carrier frequency $f_c \gg$ Message signal bandwidth. Therefore the second inequality in (10.13.3) is valid. Then we can say that the envelope of $x_c'(t)$ is $y(t)$.

$$y(t) = A_c(\omega_c + d[2\pi f_d \int^t m(\alpha)d\alpha]/dt])$$
$$= A_c(\omega_c + 2\pi f_d)m(t). \quad (10.13.4)$$

Removal of dc term $A\omega_c$ results in the output $A_c(2\pi f_d)m(t)$. Note the filter gains are not included in (10.13.4). Including all the gains and all the constants as discriminator constant K_D, the output is $K_D m(t)$. The frequency discriminator can also be used to demodulate PM signals when $\phi(t) = k_p m(t)$, see (10.13.1). Correspondingly, $y_D(t)$ in (10.13.4) for this case is

Fig. 10.13.1 FM discriminator with a band-pass limiter

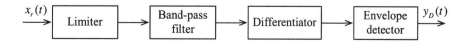

$$x_r(t) \rightarrow \boxed{\text{Limiter}} \rightarrow \boxed{\begin{array}{c}\text{Band-pass}\\\text{filter}\end{array}} \rightarrow \boxed{\text{Differentiator}} \rightarrow \boxed{\begin{array}{c}\text{Envelope}\\\text{detector}\end{array}} \xrightarrow{y_D(t)}$$

Fig. 10.13.2 (a)
Differentiation – a block
diagram, (b) a simple low-
pass circuit used as a
differentiator, (c) a simple
low-pass circuit used as a
differentiator, (d) a simple
band-pass circuit used for a
differentiator, (e) low-pass
amplitude transfer
characteristic, and (f) band-
pass amplitude transfer
characteristic

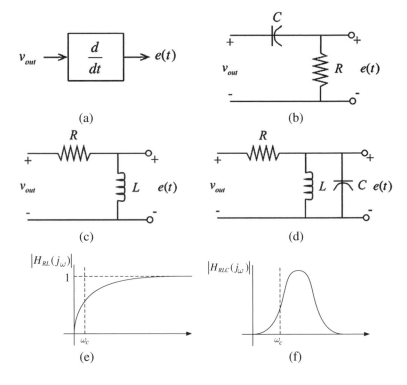

$$y_D(t) = A_c k_p \frac{dm(t)}{dt}. \qquad (10.13.5)$$

Integration of this function results in the message
signal proportional to $m(t)$. Including all the gain
constants in the detection process, we can write the
output as $K_{D1} m(t)$.

Simple discriminator circuits: Figure 10.13.2a
gives the block diagram of the derivative operation.
Figure 10.13.2b,c,d gives three simple differentiator
circuits. The transfer functions and the amplitude
responses of the RC and RL circuits are

$$H_{RC}(j\omega) = \frac{R}{R + (1/j\omega C)} = \frac{j\omega}{j\omega + (1/RC)}, |H_{RC}(j\omega)|$$

$$= \frac{|\omega|}{\sqrt{(\omega)^2 + (1/RC)^2}}. \qquad (10.13.6a)$$

$$H_{RL}(j\omega) = \frac{j\omega L}{R + j\omega L}, |H_{RL}(j\omega)|$$

$$= \frac{|\omega L|}{\sqrt{R^2 + (\omega L)^2}} = \frac{|\omega|}{\sqrt{\omega^2 + (R/L)^2}}. \qquad (10.13.6b)$$

See Fig. 10.13.2e for their amplitude responses. For
low frequencies,

$$|H_{RC}(j\omega)| \approx RC|\omega|, \; |f| \ll (1/2\pi RC),$$

$$|H_{RL}(j\omega)| \approx (L/R)|\omega|, \; |f| \ll (1/2\pi (L/R)).$$

$$(10.13.7a)$$

These circuits act as differentiators in the low-frequency range (see $(j\omega)$ below)

$$H_{RC}(j\omega) \approx j\omega RC \text{ and } H_{RL}(j\omega) = j\omega L/R. \quad (10.13.7b)$$

RC circuit output:

$$v(t) = RC \frac{dx_{FM}(t)}{dt}$$

$$= -RCA_c \left[\omega_c + k_f m(t) \right] \sin(\omega_c t + k_f \int^t m(\alpha)d\alpha).$$

$$(10.13.8)$$

Passing $v(t)$ through an envelope detector and a
blocking capacitor gives the output proportional
to $m(t)$. The 3 dB frequencies of the RC and the
RL circuits are $(1/2\pi RC)$ and $(R/2\pi L)$ and must
exceed the carrier frequency (see Fig. 10.13.2e.).

Discriminator constants: $K_d = A_c(2\pi)RC$ and K_d
$$= A_c(2\pi)L/R.$$

The RC and RL circuit amplitude responses point out that they have a small linear region of operation. A tuned circuit in Fig. 10.13.2d increases the linear region. Now

$$H_{RLC}(j\omega) = \frac{j\omega L}{R(1 - \omega^2 LC) + j\omega L}, \; |H_{RLC}(j\omega)|$$
$$= \frac{|\omega L|}{\sqrt{R^2(1 - \omega^2 LC)^2 + (\omega L)^2}}.$$

See Fig.10.13.2 f. The linearity of the RLC circuit can be improved by using two balanced circuits shown in Fig. 10.13.3a, each tuned to slightly different frequency. The amplitude characteristics of the top and the bottom RLC circuits in Fig. 10.13.3b with the subscipts a and b are $|H_A(j\omega)|$ and $|H_B(j\omega)|$.). The center frequencies of the Bp filters are $f_{ia} = 1/2\pi\sqrt{(L_{ia}C_{ia})}$, $f_{ib} = 1/2\pi\sqrt{(L_{ib}C_{ib})}$, with $f_{1a} > f_c$ and $f_{1b} < f_c$. The overall amplitude characteristic is approximately equal to (see Ziemer and Tranter (2002).)

$$|H(j\omega)| \approx |H_A(j\omega)| - |H_B(j\omega)|. \quad (10.13.9)$$

The two tuned circuits are used to balance out the dc when the input has a carrier frequency equal to f_c. That is $|H(j\omega_c)| = 0$. The second part of the circuit in Fig. 10.13.3a is a simple *balanced envelope detector*. The input to the cascaded tuned circuit with the balaced envelope detector is supplied by a center-tapped transformer with the input $x_{FM}(t)$. Since $|H(j\omega_c)| = 0$, we do not need the DC block after the envelope detectors and therefore the low-frequency response is significantly improved.

10.13.2 Delay Lines as Differentiators

Delay lines can be used since

$$y(t) = x_c'(t) = \frac{dx_c(t)}{dt} \simeq \frac{x_c(t) - x_c(t - \tau)}{\tau}, \quad (10.13.10)$$

$$\Rightarrow Y(j\omega) = \frac{1}{\tau}[X_c(j\omega) - e^{-j\omega\tau}X_c(j\omega)]$$
$$= \frac{1}{\tau}X_c(j\omega)[1 - e^{-j\omega\tau}]. \quad (10.13.11)$$

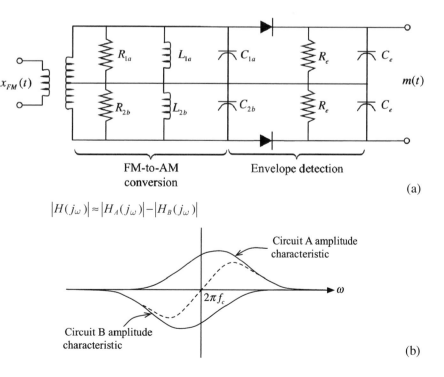

FM-to-AM conversion

Envelope detection

(a)

$$|H(j_\omega)| \approx |H_A(j_\omega)| - |H_B(j_\omega)|$$

Circuit A amplitude characteristic

Circuit B amplitude characteristic

(b)

Fig. 10.13.3 (a) FM-to-AM conversion discriminator and (b) amplitude characteristic

If $|\omega\tau| \ll 1$, then $1 - e^{-j2\pi f\tau} \approx j2\pi f\tau$. This can be seen by expanding the exponential using power series and keeping only the first two terms. It follows that

$$Y(j\omega) \approx j\omega K_1 X_c(j\omega), K_1 - \text{a constant.} \quad (10.13.12)$$

The system shown in Fig. 10.13.4 approximates a differentiator provided

$$\tau \ll 1/(f_c + \text{Maximum frequency deviation}).$$
$$(10.13.13)$$

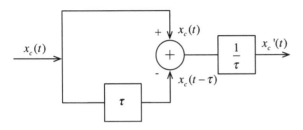

Fig. 10.13.4 Differential or Approximation

See Couch (2001) and Ziemer and Tranter (2001) for other popular FM detectors.

10.14 FM Receivers

Most FM receivers are of the superheterodyne type shown in Fig. 10.14.1. The basic operations in AM and FM receivers are the same except that we have an AM detector after the IF amplifier and filter in the AM case, whereas we have an FM detector or discriminator in the FM case. The commercial FM radios have a tuning range from 88 to 108 MHz compared to 540 to 1600 kHz for AM. The intermediate frequency is 10.7 MHz in FM compared to 455 kHz in the AM case. The FM transmission bandwidth is approximately 200 kHz compared to 10 kHz for AM.

10.14.1 Distortions

The angle-modulated signal has the property that $|x_c(t)| = A_c$ providing less susceptibilty to *non-linear distortion*. Now consider the effects of a non-linear channel on an angle modulated signal Lathi, (1983). Consider that the angle modulated signal $x_c(t) = A_c \cos(\omega_c t + \phi(t))$ is transmitted over a non-linear channel with the output

$$y(t) = a_1 x_c(t) + a_2 x_c^2(t) + a_3 x_c^3(t),$$

$$= A_c a_1 \cos(\omega_c t + \phi(t)) + a_2 A^2 \cos^2(\omega_c t + \phi(t))$$
$$+ a_3 A^3 \cos^3(\omega_c t + \phi(t)),$$

$$= a_2(A^2/2) + [a_1 A + (3/4)a_3 A^3] \cos(\omega_c t + \phi(t))$$
$$+ (a_2/2)A^2 \cos(2\omega_c t + 2\phi(t))$$
$$+ (a_3/4)A^3 \cos(3\omega_c t + 3\phi(t)). \quad (10.14.1)$$

It has four terms corresponding to a third-order non-linearity. The second term is the angle modulated signal. Its spectrum is centered at f_c. The spectra of the remaining terms are centered at dc, $2f_c$, and $3f_c$. The linear term is undistorted and can be recovered from the received signal by passing $y(t)$ through a band-pass filter centered at f_c. See Example 10.10.2

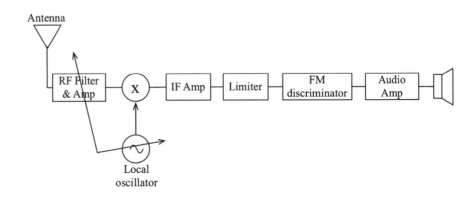

Fig. 10.14.1 FM superheterodyne receiver

for the effects of channel non-linearities of an AM signal. Channel non-linearities cause unwanted modulation with carrier frequencies $2\omega_c$ and $3\omega_c$. They also produced distortion of the desired signal. Angle modulated signals are more *robust to noise corruption* compared to AM. The constant amplitude of angle modulated signals gives immunity against fading. Angle modulated signals use larger bandwidth compared to AM. They exchange SNR for the transmission bandwidth.

Relations between PM and FM signals:

FM signal of the message $m(t)$ is a PM signal of the message signal $\int^t m(\alpha)d\alpha$.

PM signal of the message $m(t)$ is an FM signal of the message $[dm(t)/dt]$

Lathi (1983) states that "if $m(t)$ has a large peak amplitude, and its derivative has a relatively small amplitude, PM tends to be superior to FM. For opposite conditions, FM tends to be superior to PM". Comparing PM and FM with respect to SNRs, PM is superior if the message signal is concentrated at lower end of the frequency baseband and FM is superior if it is concentrated at the high end of the baseband.

10.14.2 Pre-emphasis and De-emphasis

Commercial broadcasting FM uses a pre-emphasis filter that emphasizes the high frequencies in the baseband at the transmitter, see Lathi (1983), Couch (2001), and others. In simple terms, it does the derivative operation boosting high-frequency components of the baseband signal at the transmitter. At the receiver the high-frequency components de-emphasize are de-emphasized by attenuating the high frequencies emphasized at the transmitter. Since most noise injected into the signal in the channel during transmission is at the high end, attenuation of the high-frequency components reduces the noise part significantly in the baseband at the receiver. Figure 10.14.2 gives the block diagram identifying the pre-emphasis circuit–FM modulator–FM demodulator–de-emphasis circuit. The circuits for the pre-emphasis and the de-emphasis and their corresponding Bode plot frequency responses are shown Fig. 10.14.3a,b.

Transfer functions:

$$\frac{V_{1o}}{V_{1i}} = H_{PE}(j\omega) = K\frac{1+j\omega\tau_1}{1+j\omega\tau_2}, K=\frac{R_2}{R_1+R_2}, \tau_1$$

$$= RC, \tau_2 = \frac{R_1 R_2 C}{R_1+R_2}, \tag{10.14.2a}$$

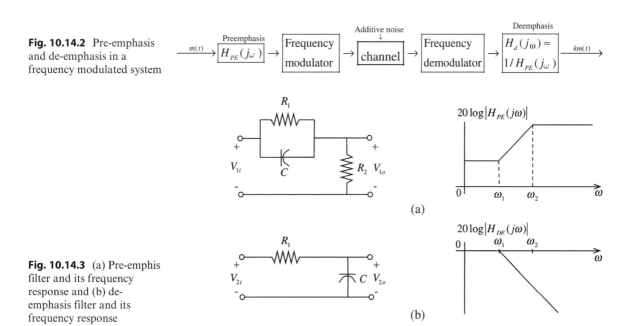

Fig. 10.14.2 Pre-emphasis and de-emphasis in a frequency modulated system

Fig. 10.14.3 (a) Pre-emphis filter and its frequency response and (b) de-emphasis filter and its frequency response

$$\frac{V_{2o}}{V_{2i}} = H_{DE}(j\omega) = \frac{1/j\omega C}{R_1 + (1/j\omega C)} = \frac{1}{1 + j\omega\tau_1}, \tau_1 = R_1 C.$$
(10.14.2b)

Standard values are $\tau_1 = R_1 C = 75\,\mu s$ and $\tau_2 = R_2 C \ll R_1 C$ with $\omega_1 = 1/\tau_1$ and $\omega_2 = 1/\tau_2$. We can show that

$$H_{PE}(j\omega)H_{DE}(j\omega) \approx K, \text{ a constant for } \omega \ll \omega_2.$$
(10.14.2c)

10.14.3 Distortions Caused by Multipath Effect

When a transmitted signal $x(t)$ arrives at the receiver by two or more paths with *different delays*, the received signal is a sum of these signals and its effect on the signal to be received is referred to as *multipath effect*. A simple model for a multipath communication channel, assuming two paths, is Hsu (1995)

$$y(t) = x(t) + \alpha x(t-\tau), \alpha < 1 \text{ and } \tau > 0. \quad (10.14.3a)$$

$$\Rightarrow Y(j\omega) = X(j\omega) + \alpha e^{-j\omega\tau} X(j\omega) = [1 + \alpha e^{-j\omega\tau}]X(j\omega)$$

$$= H(j\omega)X(j\omega)H(j\omega) = [1 + \alpha e^{-j\omega\tau}]$$

$$= 1 + \alpha\cos(\omega\tau) - j\alpha\sin(\omega\tau)$$

$$= |H(j\omega)|e^{-(j\omega\tau + \tan^{-1}(\alpha\sin(\omega\tau)/(1+\alpha\cos(\omega\tau))))},$$

$$|H(j\omega)| = [1 + \alpha^2 + 2\alpha\cos(\omega\tau)]^{1/2}. \quad (10.14.3b)$$

The amplitude response $|H(j\omega)|$ is *periodic* with period $T = 2\pi/\tau$ (see Fig. 10.14.4a.). For $\alpha = 1$, $|H(j0)| = 2$ and $|H(j(\pi/\tau))| = (1-\alpha)$. To compensate for the multipath channel induced distortion, an equalization filter with the following transfer function is needed.

$$H_{eq}(j\omega) \approx [1/H(j\omega)] = 1/[1 + \alpha e^{-j\omega\tau}]$$

$$= 1 - \alpha e^{-j\omega\tau} + \alpha^2 e^{-j2\omega\tau} - \dots. \quad (10.14.4)$$

Since $e^{-j\omega k\tau}$ is a periodic function, a *tapped delay line filter* can be used as shown in Fig. 10.14.4b to approximate the equalization. The output is

$$z(t) = \sum_{k=1}^{N} a_k y[t - (k-1)\tau] \xleftrightarrow{FT} \left(\sum_{k=1}^{N} a_k e^{-j\omega(k-1)\tau}\right) Y(j\omega)$$

$$= H_{eq}(j\omega)Y(j\omega), \quad (10.14.5)$$

$$H_{eq}(j\omega) = \sum_{k=1}^{N} a_k e^{-j\omega(k-1)\tau} = 1 - \alpha e^{-j\omega\tau} + \alpha^2 e^{-j2\omega\tau} - \dots,$$
(10.14.6)

$$\rightarrow a_1 = 1, a_2 = -\alpha, a_3 = \alpha^2, \dots, a_N = (-\alpha)^{N-1}.$$
(10.14.7)

Noting that $|\alpha| < 1$, the needed number of coefficients depends upon the size of α.

Notes: Echo suppression is important in telephone communication. $|H(j\omega)|$ has dips or nulls at $\omega = \pm(2k-1)\pi/\tau, k = 1, 2, \dots$. Sound echoes

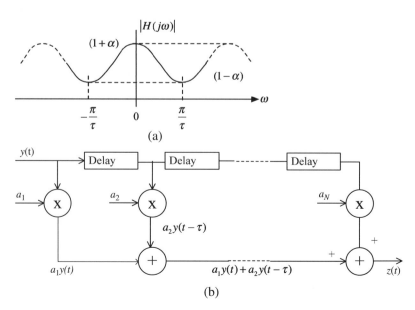

Fig. 10.14.4 (a) $|H(j\omega)|$ and (b) tapped delay line

with a delay of 5 ms correspond to a transmission path of distance of a few feet. For delays longer than 50 msec, humans hear the delayed speech signal as a distant echo, see McClellen et al. (2003).

10.15 Frequency-Division Multiplexing (FDM)

In telephone communications, for example, we may need several hundred channels, from one point to another. One way to do is string 100 different lines, i.e., one conversation per line, which is highly inefficient, see Carlson (1975). By transmitting a single telephone conversation, we are using only a small frequency band and not using the remaining channel frequency band. At the front end of a receiver we have all the stations, i.e. it is a frequency-division multiplexed signal. FDM is a technique, wherein several message signals are first modulated with different center frequencies and with sufficient bandwidth so that they do not interfere with each other and transmitted simultaneously on one channel.

FDM scheme is illustrated in Fig. 10.15.1 using single-sideband modulation. We assumed three different signals with real spectra of different shapes in Fig. 10.15.2. Shapes of the spectra are not critical, whereas the location of the spectra and the corresponding bandwidths of the signals are important. We need to first band limit the signals and adjacent frequency bands

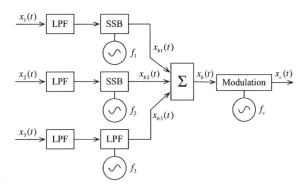

Fig. 10.15.1 Frequency-division multiplexing

need to be placed apart to minimize any interference. These are then modulated using the carrier frequencies $f_i = \omega_{bi}/2\pi, i = 1, 2, 3$ with $\omega_{b1} + \omega_1 < \omega_{b2} + \omega_2 < \omega_{b3}$, see Fig. 10.15.2. It illustrates the spectras using the SSB modulation. The signals are then summed resulting in the composite signal referred to as the baseband signal with a bandwidth of $\omega_{b3} + \omega_3$ r/s. Here upper sidebands. One can use either upper or low sidebands, not both, for the signals to be multiplexed. In addition to band limiting the signals, to minimize any overlaps of the spectra in the adjacent bands in the multiplexed signal, *guard bands* are included.

The received RF signal $x_c(t)$ is demodulated to recover the baseband signal $x_b(t)$, which is band-pass filtered to obtain the three baseband signals $x_{b1}(t), x_{b2}(t),$ and $x_{b3}(t)$. These are then

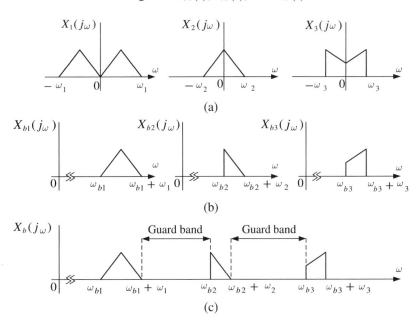

Fig. 10.15.2 (a) Spectra of the message signals, (b) spectra of the SSB signals, $\omega \geq 0$, and (c) spectrum of the baseband signal, $\omega \geq 0$

Fig. 10.15.3 Frequency-division demultiplexing

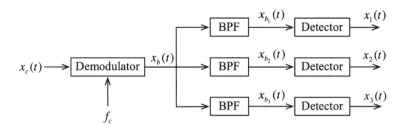

demodulated to recover the message signals. Figure 10.15.3 gives a block diagram that illustrates the demultiplexing scheme and the demodulators recover the signals in the demultipexing scheme. We will now illustrate FDM using AM and FM stereo multiplexing. AM stereo systems were not implemented until the 1980s. Five different types of AM stereo systems were proposed, see Mennie (1978). Receiver manufacturers did not want to build all five decoding circuits to recover the transmitted stereo AM signal. FCC decided to let the market place decide on the best system.

10.15.1 Quadrature Amplitude Modulation (QAM) or Quadrature Multiplexing (QM)

This scheme is shown in Fig. 10.15.4. $m_1(t)$ and $m_2(t)$ are *different* messages. The DSB signals are $m_1(t)\cos(\omega_c t)$ and $m_2(t)\sin(\omega_c t)$ that occupy the *same frequency space*.

The carrier signals $A_c\cos(\omega_c t)$ and $A_c\sin(\omega_c t)$ are usually generated by one oscillator. For simplicity, A_c is assumed to be 1. The *QAM signal* is

$$x_{\text{QAM}}(t) = m_1(t)\cos(\omega_c t) + m_2(t)\sin(\omega_c t), \quad (10.15.1)$$

$$\Rightarrow X_{\text{QAM}}(j\omega) = \frac{1}{2}[M_1(j(\omega - \omega_c)) + M_1(j(\omega + \omega_c)) + e^{-j(\pi/2)}M_2(j(\omega - \omega_c)) + e^{j(\pi/2)}M_2(j(\omega + \omega_c))]. \quad (10.15.2)$$

Figure 10.15.4 gives the coherent demodulation scheme to recover $m_1(t)$ and $m_2(t)$. Note

$$x_r(t)\cos(\omega_c t + \theta) \equiv r_U(t) = 2m_1(t)\cos(\omega_c t)$$
$$\times \cos(\omega_c t + \theta) + 2m_2(t)\sin(\omega_c t)$$
$$\times \cos(\omega_c t + \theta), \quad (10.15.3a)$$
$$x_r(t)\sin(\omega_c t + \theta) \equiv r_L(t) = 2m_1(t)\cos(\omega_c t)$$
$$\times \sin(\omega_c t + \theta) + 2m_2(t)\sin(\omega_c t)$$
$$\times \sin(\omega_c t + \theta). \quad (10.15.3b)$$

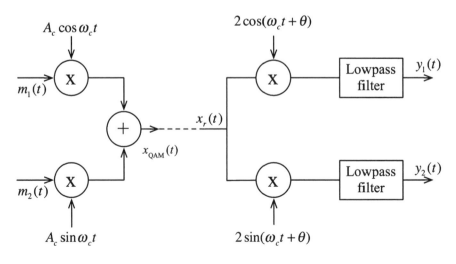

Fig. 10.15.4 Quadrature multiplexing and demultiplexing

Multiplexer   Demultiplexer

Using the trigonometric identities, the outputs of the LP filters can be derived as follows.

$$2\sin(\alpha)\cos(\beta) = \sin(\alpha - \beta) + \sin(\alpha + \beta),$$
$$2\cos(\alpha)\cos(\beta) = \cos(\alpha - \beta) + \cos(\alpha + \beta)$$
$$2\sin(\alpha)\sin(\beta) = \cos(\alpha - \beta) - \cos(\alpha + \beta)$$

$$r_U(t) = m_1(t)\cos(\theta) - m_2(t)\sin(\theta)$$
$$+ m_1(t)\cos(2\omega_c t + \theta) + m_2(t)\sin(2\omega_c t + \theta),$$
$$(10.15.4a)$$

$$r_L(t) = m_1(t)\sin(\theta) + m_2(t)\cos(\theta)$$
$$+ m_1(t)\sin(2\omega_c t + \theta)$$
$$- m_2(t)\cos(2\omega_c t + \theta). \qquad (10.15.4b)$$

The terms $m_1(t)\sin(\theta) \pm m_2(t)\cos(\theta)$ are low-frequency terms. The remaining terms are centered at the frequency $(2f_c)$. LP filtering $r_U(t)$ and $r_L(t)$ results in

$$y_1(t) = m_1(t)\cos(\theta) - m_2(t)\sin(\theta),$$
$$y_2(t) = m_2(t)\cos(\theta) + m_1(t)\sin(\theta). \quad (10.15.5)$$

If $\theta = 0$, i.e., the input and the output oscillators are phase synchronized, then $y_i(t) = m_i(t), i = 1, 2$ and we have recovered the input signals. If $\theta \neq 0$, then we have *co-channel interference* and the output is a mixture of the scaled versions of the inputs. Clearly, the phase difference between the transmitter and the receiver oscillators identified by the phase angle θ is critical. If θ is small, we can approximate the outputs to the input message signals. The phase error results in attenuation, which can be time-varying and crosstalk results. The interference is measured by the amplitude ratio $\sin(\theta)/\cos(\theta) = \tan(\theta)$. For the interference to be 20 dB or below, the ratio must be less than or equal to $(.1)$ and the mismatch must be less than $\theta = \tan^{-1}(.1) \approx 5.7^0$.

Quadrature multiplexing is an efficient way of transmitting two signals that are located within the same frequency band. If the input signals $m_i(t)$, $i = 1, 2$ have bandwidths B_i, then the transmission bandwidth of the QAM signal is given by $2B_{max}$, $B_{max} = \max(B_i)$. We are transmitting two signals in the same band and therefore we have saving in the transmission bandwidth. The quadrature multiplexing can be used for both DSB and SSB siganls. Non-coherent demodulation of QM is not possible since the envelope of the modulated signal is $A_c\sqrt{[m_1^2(t) + m_2^2(t)]}$. Quadrature multiplexing is very useful and has significant number of applications. One, in particular, is the use of QM for transmission of color signals in commercial television broadcasting, see Lathi (1983). QAM is referred to as *frequency-domain multiplexing* rather than frequency-division multiplexing as the modulated spectra of the signals in QM overlap rather than apart.

AM stereo systems: There are several AM stereo systems that are available on the market. Most widely used system is the Motorola compatible quadrature amplitude modulation or C-QUAM, a trade mark of Motorola. See Couch (2001) and Stremler, (1992) for a good discussion on this topic. We should keep in mind that all AM radios do not have the stereo capabilities. In its simplest form of the AM stereo signal generation, the inputs consist of a dc offset (V_0), left-channel, and the right-channel audio signals $m_L(t)$ and $m_R(t)$. The signal $m_1(t) = V_0 + m_L(t) + m_R(t)$ is used to modulate a cosine carrier. The quadrature carrier is modulated by the difference signal $m_2(t) = m_L(t) - m_R(t)$. QAM signal allows for mono reception using $m_1(t)$. Stereo AM receivers use an envelope detector to recover the sum signal $[m_L(t) + m_R(t)]$ and a quadrature *product detector* to recover the difference signal $[m_L(t) - m_R(t)]$. Sum and difference networks can be used to recover the left-channel and the right-channel audio signals.

10.15.2 FM Stereo Multiplexing and the FM Radio

Consider the block diagram shown in Fig. 10.15.5a, where the inputs are the left (L) and the right (R) signals $m_L(t)$ and $m_R(t)$. These two signals are used to generate $m_L(t) + m_R(t)$ and $m_L(t) - m_R(t)$ using the sum and the difference networks. The message signals are assumed to be band limited to the frequency band 30 Hz to 15 kHz. Three baseband signals, namely $m_L(t) - m_R(t)$, $m_L(t) + m_R(t)$ and a *"pilot signal"*$\cos(2\pi(19k)t)$ are generated as illustrated in the block diagram in Fig. 10.15.5a. The

Fig. 10.15.5 (a) FM stereo multiplexing block diagram and (b) baseband spectrum

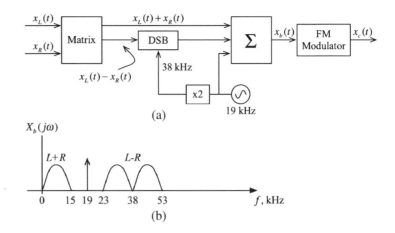

pilot signal is used to generate a 38 kHz signal using a frequency doubler and then the left minus right signal is double-side modulated by the carrier. The output is

$$x_b(t) = (m_L(t) + m_R(t)) + A_c[m_L(t) - m_R(t)]$$
$$\times \cos(2\pi(38\,k)t) + A_p\cos(2\pi(19\,k)t).$$
$$(10.15.6)$$

The pilot signal is part of the baseband signal and allows for *synchronization* at the receiver. The baseband signal $x_b(t)$ is then modulated using a frequency modulator. Figure 10.15.5b gives the baseband spectrum $X_b(j\omega)$, where the spectral shapes of the signals are assumed. The message signal is assumed to have a 15 kHz bandwidth. The bandwidth of the baseband signal is 53 kHz. At the receiver (see Fig. 10.15.6), FM signal is demodulated using an FM detector. Baseband signal is passed through three filters, a LP filter

with a bandwidth of 15 kHz, a BP filter centered at 38 kHz with bandwidth equal to 30 kHz, and a pilot filter, a narrowband BP filter centered at 19 kHz. Using the pilot signal and a frequency doubler, the carrier signal is generated with the carrier frequency of 38 kHz. Then the DSB signal is demodulated using this carrier, which is then passed through a low-pass filter with a bandwidth equal to 15 kHz. We can recover $m_L(t)$ and $m_R(t)$ from $m_L(t) + m_R(t)$ and $m_L(t) - m_R(t)$.

10.16 Pulse Modulations

If a signal $x(t)$ is adequately described by its sample values $x(nt_s)$, it can be transmitted using analog pulses. In these schemes, the sampled values are represented by a set of pulses with some characteristic of the input analog waveform. We will consider the pulse amplitude modulation and later very

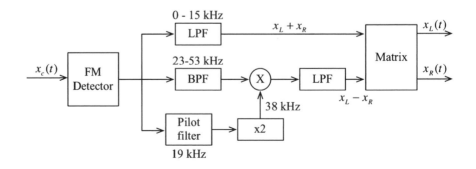

Fig. 10.15.6 Demultiplexing of FM stereosignal

briefly on pulse duration and pulse position modulations.

10.16.1 Pulse Amplitude Modulation (PAM)

The signal $x(t)$ is ideally sampled at a sampling rate equal to $f_s > 2B$, and the sampling interval is $t_s = (1/f_s)$. From (8.2.2), the ideally sampled signal is

$$x_s(t) = \sum_{n=-\infty}^{\infty} x(nt_s)\delta(t - nt_s) = x(t) * \delta_{t_s}(t),$$

$$\delta_{t_s}(t) = \sum_{n=-\infty}^{\infty} \delta(t - nt_s). \qquad (10.16.1)$$

Pulse amplitude modulated signal $x_{PAM}(t)$ results when $x_s(t)$ is passed through a filter with the impulse response $h(t)$ given below. Now

$$h(t) = \Pi\left[\frac{t - (\tau/2)}{\tau}\right], \qquad (10.16.2a)$$

$$x_{PAM}(t) = x_s(t) * h(t)$$
$$= \sum_{n=-\infty}^{\infty} x(nt_s)\Pi\left[\frac{t - (\tau/2) - nt_s}{\tau}\right]. \qquad (10.16.2b)$$

Figure 10.16.1 illustrates a smooth function, $x(t)$, and the pulse amplitude modulated signal obtained by passing the sampled signal $x_s(t)$ through a filter with the impulse response given in (10.16.2b) resulting in $x_{PAM}(t)$.

Let us briefly review the frequency analysis of the PAM signal (see Chapter 8). Let $F[x(t)] = X(j\omega)$ and $F[x_s(t)] = X_s(j\omega)$. The transforms of the ideally sampled signal and the transform of the PAM signal are

$$X_s(j\omega) = f_s \sum_{n=-\infty}^{\infty} X(j(\omega - n\omega_s)), \qquad (10.16.3)$$

$$x_{PAM}(t) \overset{FT}{\longleftrightarrow} X_{PAM}(j\omega) = X_s(j\omega)H(j\omega),$$

$$h(t) \overset{FT}{\longleftrightarrow} H(j\omega), \qquad (10.16.4)$$

$$H(j\omega) = F[h(t)] = \tau\,\mathrm{sinc}(\omega\tau/2)e^{-j\omega\tau/2},$$

$$\mathrm{sinc}(\omega\tau/2) = \frac{\sin(\omega\tau/2)}{(\omega\tau/2)}, \qquad (10.16.5a)$$

$$X_{PAM}(j\omega) = f_s\tau\,\mathrm{sinc}(\omega\tau/2)e^{-j\pi f\tau}$$

$$\sum_{n=-\infty}^{\infty} X(j(\omega - n\omega_s)), \quad \omega = 2\pi f. \qquad (10.16.5b)$$

The high-frequency roll-off characteristic of $|H(j\omega)|$ acts as a low-pass filter and the filter attenuates the upper portion of the message spectrum. This loss of high-frequency content is referred to as the *aperture effect*. This depends on the size of the pulse width τ. The ratio (τ/t_s) is a measure of the flatness of the function $|H(j\omega)|$. Passing the signal $x_{PAM}(t)$ through an ideal low-pass filter gives the signal that approximates the original signal. If τ is small, i.e., $(1/\tau) \gg (f_s/2)$ then the magnitude spectrum $|H(j\omega)|$ is fairly constant in the region of interest and the recovered signal approximates the original signal. Aperture effect can be neglected if $\tau/t_s \le (.1)$. If not, an equalization filter is needed with the transfer function $H_e(j\omega) \approx (1/H(j\omega))$ to reduce distortion.

Among many analog pulse modulation schemes, *pulse width (duration) modulation (PWM or PDM)* and *pulse position modulation (PPM)* have been popular. A PWM signal consists of a sequence of pulses, in which each pulse has a width proportional to the value of the signal at the sampling instant. A PPM signal consists of pulses, in which the pulse displacement from a specified time reference is proportional to the sample values of the message signal. For a discussion on these, see Ziemer and Tranter (2002).

10.16.2 Problems with Pulse Modulations

PAM, PDM, and PPM are vulnerable to noise. When a pulse of width τ passes through a channel, the output will be significantly distorted. See Example 6.6.3 for an illustration in transmitting a finite width pulse

$x(t)$ and $x_{PAM}(t)$

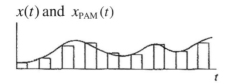

Fig. 10.16.1 $x(t)$(smooth function) and $x_{PAM}(t)$(pulsed wave form)

through a simple RC circuit. The output of the RC circuit is distorted with infinite time width (or not time limited). This is a simple illustration showing the problems associated with a band-limited channel. Other problems may include signal interference from other sources, and noise, in general. Analog pulse modulation schemes have problems as the variables, amplitudes for PAM, durations for PDM, and positions for PPM can vary over a wide continuous range. Digital schemes use a set of bits to represent a number and each bit can be transmitted individually. As an example, we can assume that a pulse of amplitude A corresponds to a one and a pulse with zero amplitude (i.e., no pulse) corresponds to a zero. At the receiver all we need to do is make a decision if the pulse exists or not at the recever, making decoding simpler and robust, leading to pulse code modulation (PCM), see Ziemer and Tranter (2002).

There is no exact formula to determine the output pulse width of an amplitude modulated signal. A rough rule of thumb that relates the output pulse duration and channel bandwidth is Carlson (1975)

$$\tau_{min} \geq 1/2B. \qquad (10.16.6)$$

Maximum number of resolved output pulses per unit time is about

$$2B = 1/\tau_{min}. \qquad (10.16.7)$$

Intersymbol interference: If the channels are linear and distortionless, the transmitted signal can be decoded without any problem. In reality, channels are band limited and, as a result, the output pulses spread during transmission. See the discussion on transmitting a rectangular pulse through a simple RC circuit in Example 6.6.3. Since several of pulses in sequence are transmitted at a time, spreading of these pulses cause overlap of pulses into adjacent and nearby time slots due to the limitations on the channel bandwidth. This is *intersymbol interference (ISI)*. In Chapter 8 sinc pulses of the form $p(t) = \text{sinc}(2\pi Bt)$ were used to represent a signal with a bandwidth of B Hz (see (8.2.17))

$$y(t) = \sum_{n=-\infty}^{\infty} x(nt_s)\text{sinc}(\pi(f_s t - n)), \ f_s = 2B. \quad (10.16.8)$$

Noting that $y(nt_s) = x(nt_s)$ in (10.16.8), $f_s = 2B$ samples per second can be transmitted, assuming there is no interference from one pulse to the next. Obviously if the channel is not distortionless, then the pulses read at the receiver will not be free of interference from the other samples. There are several problems associated with the sinc pulses. First, the sinc pulse is physically unrealizable, as the pulse exists for all times. If the pulse is truncated, the bandwidth of such a pulse will be larger than B. Second, the sinc pulses decay slowly at a rate proportional to $(1/t)$ from the peak. This is a serious practical problem. For example, if the value of B in $\text{sinc}(2\pi Bt)$ deviates from its nominal value, then the pulse amplitudes will not be zero at multiples of $(1/2B)$ causing interference with all the pulses. The interference at the center of a pulse caused by other pulses has the form $\sum(1/n)$ adding up to large amplitude. Same thing would happen if the sampling of the pulses is not measured exactly at $t = nt_s$. *Time jitter,* caused by the variations of the pulse positions at the sampling instants, is a problem. In Chapter 8, various interpolation methods were discussed. One of them is the raised cosine pulse that decays at a rate proportional to $(1/t^3)$ from the peak.

Notes: The raised cosine function and its transform were given in (4.11.9 and b). See Fig. 4.11.1 for their sketches. Tapering of the frequency responses is controlled by the *roll-off factor* β with $0 \leq \beta \leq 1$. If $\beta = 0$, the spectrum of the raised cosine function reduces to the ideal LP filter spectrum with a minimum bandwidth of $1/t_s$. The roll-off factor β represents excess bandwidth as a fraction of the minimum bandwidth $(1/2t_s)$. The bandwidth of a raised cosine pulse varies from a minimum of $f_B = 1/(2t_s)$Hz (for $\beta = 0$) to a maximum of $f_B = 1/(t_s)$Hz (for $\beta = 1$). With $\beta = 1$, the pulse is a raised cosine pulse with 100% excess bandwidth. If the desired pulse transmission rate is $(1/t_s)$ pulses/s, then the BW is

$$f_B = (1 + \beta)/2t_s \text{ Hz}. \qquad (10.16.9a)$$

Given f_B, then the allowable pulse rate is

$$1/t_s = 2f_B/(1 + \beta). \qquad (10.16.9b) \quad \blacksquare$$

10.16.3 Time-Division Multiplexing (TDM)

Frequency-division multiplexing (FDM) assigns a frequency band for each signal. In the TDM, a time slot is allotted for each signal. TDM is used to transmit a composite signal consisting of several different signals over a single channel. It is like a student taking several courses during a semester and each course is offered at different times. TDM is easier to visualize compared to FDM. TDM is a *serial process*, whereas FDM is a *parallel process*. They are duals, one is in time and the other is in frequency. Figure 10.16.2a illustrate the conceptual scheme of TDM. Each of the input signals is band limited by a low-pass filter, which removes frequencies that are not essential to an adequate representation of the signal under consideration. The outputs of the low-pass filters are the inputs to a commutator. Some older systems use a rotary mechanical system and in the newer systems, this process is implemented by electronic switching. Each signal is sampled at a rate higher than (about 1.1 times) the Nyquist sampling rate to avoid aliasing problems. The samples are then *interleaved*. This interleaving process is illustrated assuming two signals, for simplicity, in Fig. 10.16.2b. The composite signal, a frame, consists of the interleaved pulses of the input signal is transmitted over the channel. If all the signals have the same bandwidth, one frame of the composite signal consists of the samples $s_1 s_2 \ldots s_N$ (i.e., N input signals). If the input signals have unequal bandwidths, then the signals with larger bandwidths must transmit more samplesin each frame. The implementation is simpler if the bandwidths of the signals are harmonically related. As an example, consider that there are four sources$m_1(t), m_2(t), m_3(t)$, and $m_4(t)$ with bandwidths $3B, B, B$, and B Hz, respectively. The Nyquist rates for these signals are $6B, 2B, 2B$, and $2B$, respectively. If the commutator rotates at $2B$ rotations per second, then in one rotation we have three samples from the first signal and one sample from each of the three remaining signals. That is, there are six samples in each frame. The commutator must have poles connected to the signals shown in Fig. 10.16.2a at the transmitter and the receiver and they must be synchronized.

TDM bandwidth: Let t_{TDM} be the time spacing between adjacent samples in the time-multiplexed signal with N input signals. Assuming that all the input signals have the *same bandwidth* B Hz and the sampling interval is the same for each signal, then $t_{\text{TDM}} = t_s/N$ and the sampling rate for each signal is $t_s = 1/f_s, f_s \geq 2B$. Assuming the TDM signal is a LP signal with bandwidth B_{TDM}, then the required

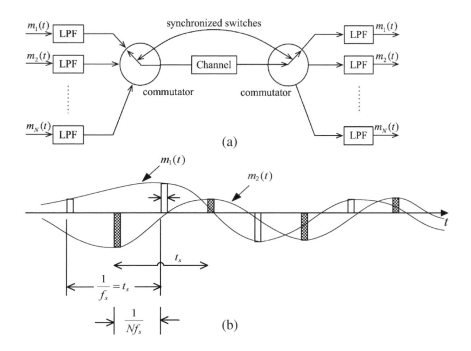

Fig. 10.16.2 (a) TDM switching (several inputs) and (b) TDM waveforms ($N = 2$)

minimum sampling rate of the TDM signal is $2B_{\text{TDM}}$ and therefore

$$B_{\text{TDM}} = \frac{1}{2t_{\text{TDM}}} = \frac{N}{2t_s} = \frac{1}{2}Nf_s \geq NB\,\text{Hz}. \quad (10.16.10a)$$

If we have N signals with bandwidths equal to B_i, $i = 1, 2, \ldots, N$, then

$$B_{\text{TDM}} \geq \left(\sum_{k=1}^{n} B_i\right). \quad (10.16.10b)$$

Interestingly, the required bandwidths for TDM and FDM are the same.

TDM receiver: At the receiving end, the composite signal is demultiplexed using a commutator, see Fig.10.16.2a. The outputs are then demodulated using low-pass filters. The input and the output commutators have to be *synchronized* to recover the signals. Without precise synchronization, TDM is useless. A popular method used to maintain synchronization is to insert a relatively high dc component in one of the channels allowing the receiver commutator in sync with the transmitter commutator.

Comparison between FDM and TDM: In time-division multiplexing, each signal is assigned a time slot, whereas a frequency slot is allocated for each of the signals. In TDM, the signals are separate in time and the corresponding frequencies are jumbled. In FDM, the signals are separated in the frequency domain and are jumbled in the time domain. TDM is implemented in a series operation and is simpler to visualize than FDM. FDM requires sub-carrier modulators, band-pass filters, and demodulators for each message signal. TDM requires commutators and the demodulators, which are low-pass filters. Since FDM is in frequency, high-quality channel is desired, as harmonic distortion caused by non-linearities can result in intermodulation distortion. Such problems do not appear in TDM, as each channel is assigned a separate recurring time interval. There is no cross talk in TDM if the pulses are separated and non-overlapping. To reduce cross talk, guard times are provided between pulses. This is

analogous to the FDM guard bands. Carlson (1975) gives a nice presentation quantifying these.

10.17 Pulse Code Modulation (PCM)

In Section 10.16.2 some of the problems associated with analog pulse modulations were discussed. If digital schemes are used with ones and zeros, for example, then when the pulses are received, only the existence or non-existence of these pulses needed to be decided. Digital modulation is more robust.

PCM is a three-step process consisting of *sampling*, *quantizing*, and *encoding* as shown in Fig. 10.17.1. A continuous-time signal $x(t)$ is sampled by measuring the amplitudes of the continuous signal at times kt_s seconds, where the sampling interval t_s is determined by sampling theorem, see Section 8.2. The sample values are $x(kt_s) = x[k]$.

These values are represented by a finite set of levels using quantization methods and are then coded, see Haykin (2001) and Couch (1995).

10.17.1 Quantization Process

Here, only *uniform quantization* is discussed. The sample values are divided into regions of uniform width and an integer code is assigned to each region.

Midrise quantization: Figure 10.17.2 gives an eight $8 = 2^3$-level, a 3-bit *mid-riser quantizer*. The mid-riser is so-called because the origin is in the middle of a rising part of the staircase-like function. It has the same number of positive and negative levels, symmetrically positioned about the origin. On top of the characteristics, a 3-bit normal code $(b_2b_1b_0)$ is identified from the lowest value of 000 (level 0) to the highest value 111 (level 7). The normal binary representation of an integer between 0 and 2^{n-1} is

Fig. 10.17.1 Pulse code modulation

$\xrightarrow{x(t)}$ **Sampler** $\xrightarrow{x(kt_s)=x[k]}$ **Quantizer** $\xrightarrow{\hat{x}[k]}$ **Encoder** $\rightarrow$ PCM code

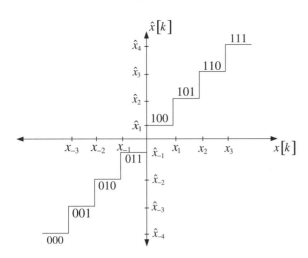

Fig. 10.17.2 Uniform symmetric mid-riser quantizer input–output characteristic

Table 10.17.1 Quantization values and codes corresponding to Fig. 10.17.2

Range of $x[n]$	Quantized values	A simple binary code, $b_2 b_1 b_0$
$-A < x[k] < x_{-3}$	$\hat{x}_{-4}$	000
$x_{-3} < x[k] < x_{-2}$	$\hat{x}_{-3}$	001
$x_{-2} \le x[k] < x_{-1}$	$\hat{x}_{-2}$	010
$x_{-1} \le x[k] < x_0$	$\hat{x}_{-1}$	011
$x_{-1} \le x[k] < x_1$	$\hat{x}_1$	100
$x_1 \le x[k] < x_2$	$\hat{x}_2$	101
$x_2 \le x[k] < x_3$	$\hat{x}_3$	110
$x_3 \le x[k] < A$	$\hat{x}_4$	111

$$L = b_{n-1} 2^{n-1} + b_{n-2} 2^{n-2} + \ldots + b_0 2^0,$$
$$b_i = 0 \text{ or } 1, \ (L)_2 = b_{n-1} b_{n-2} \ldots b_0. \quad (10.17.1)$$

The sampled values are assumed to be bounded and $-A < x[k] < A$. This range is divided into $L = 8$ equal intervals in Fig. 10.17.2. Each interval is mapped onto a quantized value. For example, the output of the quantizer is assigned by the quantized value

$$Q[x_1 \le x[k] < x_2] = \hat{x}_2[k].$$

See Table 10.17.1 for the others. The quantization step size is $x_i - x_{i-1} = \Delta$ and $\hat{x}_i - \hat{x}_{i-1} = \Delta$. Note that a simple 3-bit binary code is assigned for each range corresponding to eight possible levels.

Now consider how to decide on the number of levels needed in the quantization process. The input sample amplitude can take any value in the range $|x[k]| < A$, where A is chosen using the statistics of the signal, see Rabiner and Schafer (1978). If the input goes outside this range, then there is overload. The range is then divided into L intervals. Assuming equal width (i.e, uniform quantization), in each interval, the *step size* is

$$\Delta = (2A/L). \quad (10.17.2a)$$

Encoder assigns a unique code word to each of the L quantization levels to each of the sampled values of $x[n]$ using (10.17.1). Let $q_e[n]$ be the *quantization*

error generated by the quantization process expressed by

$$q_e[n] = x[n] - \hat{x}[n] \text{ and } -(\Delta/2) \le q_e[n] \le (\Delta/2). \quad (10.17.2b)$$

Since the error is random, it can only be described in statistical terms. When the quantization process is fine enough, the distortion produced by the quantization can be viewed as additive and independent noise source with zero mean and the variance determined by the quantization step size Δ. The number of levels is usually assumed to be greater than or equal to 64. In Example 1.7.1, the uniform density function was discussed. Using $b = \Delta/2$ and $a = -\Delta/2$ in (1.7:8) and (1.7.10) results in

$$\text{Mean value} = 0 \text{ and Variance} = \sigma_e^2 = \Delta^2/12. \quad (10.17.3)$$

Signal distortion due to quantization can be controlled by choosing a small enough step size Δ related to the number of levels used in the quantization so that $L = 2^n$. Figure 10.17.3 gives an example of a continuous waveform $x(t)$, its sample values, quantized sample values, code numbers, and the *pulse code modulated (PCM)* sequence, see Taub and Schilling (1971). Signal varies between -4 and 4 V. The quantization step size between levels is set at 1 V resulting in $L = 8$ quantization levels located at $-3.5, -2.5, -1.5, -.5, +.5, +1.5, +2.5, +3.5$ V. Table 10.17.2 gives the nearest quantization levels, code numbers, and their binary representations. Each sample value $x(kt_s)$ is approximated by the quantized value $q[x(kt_s)]$, one of the $L = 2^n$ levels closest to $x(kt_s)$ and a binary code is used representing this value.

Fig. 10.17.3 Natural sample values, quantized samples, code numbers, and the binary sequence representing the code numbers (reprinted with permission from Taub and Schilling (1971))

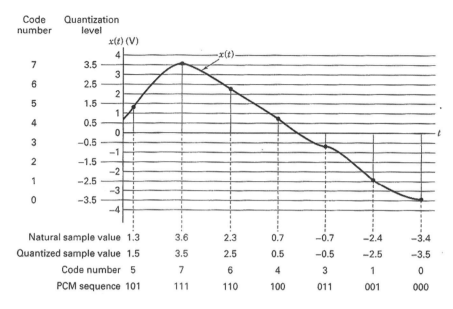

Natural sample value	1.3	3.6	2.3	0.7	−0.7	−2.4	−3.4
Quantized sample value	1.5	3.5	2.5	0.5	−0.5	−2.5	−3.5
Code number	5	7	6	4	3	1	0
PCM sequence	101	111	110	100	011	001	000

Table 10.17.2 Binary representation of quantized values

Nearest quantized value	−3.5	−2.5	−1.5	−0.5	0.5	1.5	2.5	3.5
Code number	0	1	2	3	4	5	6	7
PCM sequence	000	001	010	011	100	101	110	111

For example, the sample value of 1.3 (see Fig. 10.17.3.) is assigned the quantized value 1.5 with the code number 5, $((5)_2 = (101))$. A simple signaling format uses a positive pulse for a 1 and a negative pulse for a 0. Such a format is preferable compared to a pulse and no pulse, since the average value of a positive negative pulse is zero. To transmit 101, three pulses, a positive, a negative, and a positive pulse, are used.

Signal-to-noise ratios and the number of levels in uniform quantization: The transmitted waveform is not known. For simplicity, to determine the signal-to-noise ratios, the input signal is assumed to be a sinusoid $x_T(t) = A\cos(\omega_0 t)$. Its peak-to-peak excursion is $2A$. The average power of the sinusoid is $P = A^2/2$. Assuming the number of levels in the quantization is L, the quantizer step size is then $\Delta = 2A/L$, see (10.17.1). Assuming the noise is uniform, the variance given in (10.17.3) is used as a measure of the noise power. The signal-to-noise ratios (SNRs) are

$$\text{SNR} = \frac{P}{\sigma_e^2} = \frac{A^2/2}{\Delta^2/12} = \frac{\Delta^2 L^2/(4)(2)}{\Delta^2/12} = \frac{3}{2}L^2. \quad (10.17.4)$$

$$\begin{aligned}(\text{SNR})_{\text{dB}} &= 1.76 + 20\log L = 1.76 + 20\log(2^n)\\ &= 1.76 + 6.02n\,\text{dB}. \quad (10.17.5)\end{aligned}$$

A 6 dB/bit, some times referred to as the *6 dB rule*, requires one bit for every 6 dB signal-to-noise ratio applicable to the uniform quantization. This is used to determine the required number of bits for a given signal-to-noise ratio. For example, CD recording systems use $n = 16$ bits A/D converter to convert stereo music from an analog signal to a digital signal. This gives a signal-to-noise ratio of approximately 96 dB.

Quantization is an irreversible process and the noise produced by it is part of the transmitted signal and the original signal is not recoverable. It is serious when the signal level is small and covers only a few of the lower quantizing levels. In speech, for example, low amplitudes are more probable than larger amplitudes. Hence, smaller step sizes at low amplitudes and larger step sizes at larger, less likely, amplitudes. This allows for adaptive quantization based on statistical measures, see Rabiner and Schafer (1978).

10.17.2 More on Coding

In the normal binary code presented above, each code word consists of n bits representing 2^n distinct

Table 10.17.3 Normal binary and Gray code representations for $N=8$.

k	Normal Binary Code $b_2b_1b_0$	Normal Gray Code $g_2g_1g_0$
0	000	000
1	001	001
2	010	011
3	011	010
4	100	110
5	101	111
6	110	101
7	111	100

levels, see (10.17.1). Table 10.17.3 gives the ordinal number between $0 \leq k \leq L = 2^n - 1, n = 3$, the *normal binary code* and the corresponding *normal (or reflected) Gray code*. In Fig. 10.17.3, eight levels, $0-7$ were considered and the binary codes by their equivalents $(0)_2 = 000$ to $(7)_2 = 111$. One of the problems with the normal binary code is that *all the bits* associated with a function at adjacent times can change from one level to the next level. Note $(3)_2 = 011$ and $(4)_2 = 100$. For example, going from level 3 to level 4, each bit has to be changed from 0 to 1 and 1 to 0. Gray code alleviates this problem. There are many possible Gray codes and one of them is the reflected Gray code. In this code, two successive code words of weight α (i.e., the number of ones in the code is α), differ in only and exactly two positions. A procedure to convert a normal binary code to reflected Gray code and vice versa is given below ($\oplus$ represents modulo 2 addition).

Normal binary code word:

$$b_{n-1}b_{n-2}\ldots b_1b_0 \qquad (10.17.6)$$

Normal Gray code word:

$$g_{n-1}g_{n-2}\cdots g_1g_0. \qquad (10.17.7)$$

Normal binary code to normal Gray code:

$$g_i = \begin{cases} b_i, & i = n-1 \\ b_i \oplus b_{i+1}, & i \neq n-1. \end{cases} \qquad (10.17.8)$$

Normal Gray code to normal binary code:

$$b_i = \begin{cases} g_i, & i = n-1 \\ g_i \oplus b_{i+1}, & i \neq n-1 \end{cases} \qquad (10.17.9)$$

For example, the normal Gray corresponding to the normal binary code $b_3b_2b_1b_0 = 1000$ is

$$g_3 = b_3 = 1, g_2 = b_2 \oplus b_3 = 1,$$
$$g_1 = b_1 \oplus b_2 = 0, g_0 = b_0 \oplus b_1 = 0. \qquad (10.17.10)$$

Table 10.17.2 gives the normal binary code and the corresponding normal Gray code for $N = 8$.

Notes: The normal Gray code is a reflected Gray code as its generation involves iterated reflections, see Problem 17.10.4. The weight of a code is the number of 1s in it. For example, the code word 110 has a weight of 2. A relevant property of the normal Gray code is that *successive n-bit code words of weight k differ in only and exactly *two positions*, see Hershey and Yarlagadda (1986). ∎

10.17.3 Tradeoffs Between Channel Bandwidth and Signal-to-Quantization Noise Ratio

From the low-pass Nyquist sampling theorem, a signal $x(t)$ with a bandwidth of B Hz requires the minimum sampling rate of $2B$ samples per second. Each sample is coded into n binary pulses and the pulse rate is $2nB$ pulses per second.

PCM transmission bandwidth: For a binary PCM, L quantizing levels are used with $L = 2^n$ ($n = \log_2(L)$) and n binary pulse needs to be transmitted for each sample of the message signal. Let the bandwidth of the message be W Hz. Using the low-pass sampling theorem, the minimum sampling rate is $f_s(\geq 2W)$ and nf_s binary pulses per second needs to be transmitted. Assuming bipolar signaling, i.e., using positive and negative pulses of equal amplitude corresponding to 1 and 0, the transmission bandwidth is

$$B = 2\,W\log_2(L). \qquad (10.17.11a)$$

For the minimum sampling rate and the minimum value for bandwidth in transmitting pulses, (10.17.12a) gives a lower bound, as the bandwidth of a pulse is inversely proportional to the pulse width, see Section 4.7. A better expression for the bandwidth is

$$B = 2\,W\alpha\log_2(L), \alpha \geq 1. \qquad (10.17.11b)$$

If the PCM requires a smaller quantization error, then a larger value of L and a larger transmission bandwidth is required. Error can be exchanged for bandwidth.

Example 10.17.1 Bandwidths of telephone lines vary, as they are operated by different private companies. Assuming the bandwidth of the telephone line is 3.5 kHz, determine the pulse rate in bits/second to transmit binary data on a telephone line using the raised cosine pulses with $\beta = 0.5$.

Solution: Using (10.16.10c) results in

$$\text{Data rate} = \frac{1}{t_s} = \frac{2f_B}{1+\beta} = \frac{2(3.5)10^3}{1+.5} = 4667 \text{ bits/s}.$$

$$(10.17.12) \quad \blacksquare$$

10.17.4 Digital Carrier Modulation

Pulse signals cannot be transmitted as they are, as they have significant power at low frequencies. They are suitable only for transmission over a pair of wires and cannot be transmitted over a radio link, as this would require very large size of antennas,

see Section 10.1. To alleviate these problems, modulation is used. PCM waveforms have both low and high frequencies. The pulses representing the sample values can be transmitted on an RF carrier by using analog pulse modulation schemes. Digital modulation process corresponds to *switching or keying* the *amplitude, phase, or frequency* of the continuous-wave carrier between either of the two values corresponding to 0 and 1. The three types of binary digital modulation schemes are *amplitude-shift keying (ASK)*(or *on–off keying [OOK]*), *frequency-shift keying (FSK)*, and *phase-shift keying (PSK) (or phase reversal keying [PRK])*.

Amplitude-shift keying (ASK): In binary ASK, the modulated signal is

$$x_{\text{ASK}}(t) = .5A_c[1 + d(t)]\cos(\omega_c t). \quad (10.17.13)$$

The function $d(t)$ is the NRZ data waveform and it takes on values ± 1. See Fig. 10.17.4a for an example. The corresponding coherent ASK signal is shown in Fig. 10.17.4b. Every time the signal is turned on by the presence of 1 in the data, the phase is the same as the continuation of the previous pulse shown by the

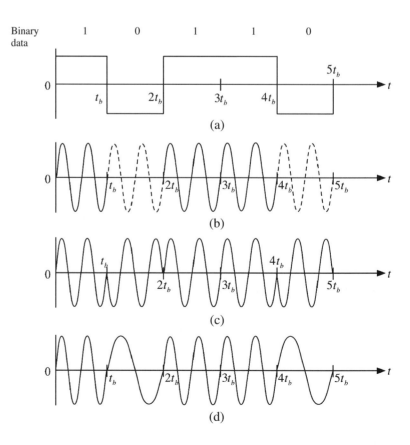

Fig. 10.17.4 Data and modulated signals: (a) nonzero binary data, (b) ASK signal, (c) PSK signal, and (d) FSK signal waveform

dotted line. Furthermore, the expression in (10.16.14) contains the unmodulated carrier component. The amplitude-shift keying signal has the same form as the amplitude modulated signal.

Phase-shift keying (PSK) or phase reversal keying (PRK): In PSK, the modulated signal is expressed by

$$x_{PSK}(t) = A_c d(t) \cos(\omega_c t). \qquad (10.17.14)$$

Using the waveform in Fig. 10.17.4a and noting that the signal for 0 is the negative of the signal for a 1, PSK produces the waveform shown in Fig. 10.17.4c. Note the phase reversals in the modulated waveform corresponding to the locations at the discontinuities in $d(t)$. For example, there is a discontinuity at $t = t_b$ resulting in the phase reversal in the modulated signal at this location.

Frequency-shift keying (FSK): In binary FSK, the modulated signal is expressed by

$$x_{FSK}(t) = \begin{cases} A_c \cos(\omega_1 t), & \text{symbol } 0 \\ A_c \cos(\omega_2 t), & \text{symbol } 1 \end{cases} \qquad (10.17.15)$$

An example of this function is shown in Fig. 10.17.4d corresponding to the waveform in Fig. 10.17.4a. The binary FSK waveform can be viewed as the superposition of two binary ASK waveforms with two different carrier frequencies ω_1 and ω_2. There are several options for selecting the carrier frequencies ω_1 and ω_2. For a detailed discussion on this, see Roden (1991).

Generation and detection of binary modulated waves: The equations given above can be used to generate ASK, PSK, and FSK waveforms, see Fig.10.17.5a,b,c. For ASK and PSK product modulators are used. Figure 10.17.5c illustrates a simple modulator that consists of two oscillators and a switch for FSK.

Power spectral densities (PSDs): The *data are random* and *random spectral analysis* is needed to determine the PSDs of the three modulation schemes, which is beyond the scope here. Interested reader should consult the book by Lathi (1983).

Detection: Simple schemes use multipliers (i.e., product modulator [s]) with a locally generated sinusoidal carrier(s) to demodulate the three modulated waveforms. For the ASK and PSK, the scheme in Fig. 10.17.5a can be used. An integrator that operates on the multiplier output for successive bit intervals acts as a low-pass filter. By using a decision device that compares the output of the integrator with a preset threshold, a decision is made that the output is 1 if the threshold exceeds the preset value and 0 otherwise. In the absence of noise, the output of the product multiplier of a PSK is

$$2A_c d(t)\cos(\omega_c t)\cos(\omega_c t) = A_c d(t) + A_c d(t)\cos(2\omega_c t). \qquad (10.17.16)$$

Noting that $\omega_c \gg$ data rate, the integrator eliminates the second term resulting in NRZ data wave form. The difference in the demodulation of a binary ASK and a binary PSK is the choice of the threshold level. See Fig. 10.17.5c for FSK.

For the demodulation of a binary FSK wave, two multipliers and two integrators are used as in Fig. 10.17.6b. It results in two outputs and the comparator compares them. If the output l_{top} produced in the upper path associated with the frequency ω_1 is greater than the output l_{bottom} produced in the bottom path associated with the frequency ω_2, the output is 0. Otherwise, the ouput is 1. Products of the two multipliers are given below for the input frequencies ω_i, $i = 0, 1$ on the top and the bottom paths

$$A_c \cos(\omega_i t)\cos(\omega_0 t) = .5A_c \cos((\omega_i - \omega_0)t)$$
$$+ .5A_c \cos((\omega_1 + \omega_0)t). \qquad (10.17.17a)$$

Fig. 10.17.5 Generation of (a) ASK, (b) PSK, and (c) FSK

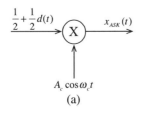

(a)

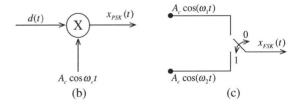

(b)　　　　　　　(c)

Fig. 10.17.6 Coherent detectors: (a) ASK and PSK and (b) FSK

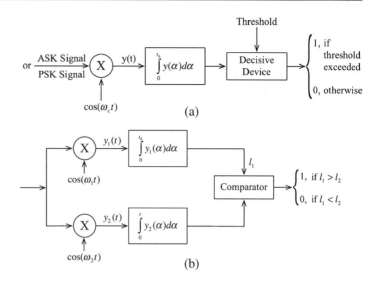

$$A_c \cos(\omega_i t) \cos(\omega_1 t) = .5A_c \cos((\omega_i - \omega_1)t)$$
$$+ .5A_c \cos((\omega_i + \omega_1)t)$$
$$(10.17.17\text{b})$$

If the input bit corresponds to 0 and $\omega_i = \omega_1$, then the output of the integrator on the top gives $l_{\text{top}} = .5A_c$ and the integrator at the bottom gives $l_{\text{bottom}} = 0$. The comparator decides that the input bit is 0. Similarly if the input bit is 1 and $\omega_i = \omega_2$, then the output of the integrator on the top gives $l_{\text{top}} = 0$ and the output at the bottom gives $l_{\text{bottom}} = .5A_c$ resulting in the decision the input bit is 1.

10.18 Summary

This chapter presented on some of the basics on analog modulation, demodulation, time-division, and frequency-division multiplexing, and a brief presentation on binary digital communications. Specific topics that were discussed in this chapter are:

- Linear modulation: double-sideband modulation, amplitude modulation, single-sideband modulation, and vestigial sideband modulation.
- Modulation and demodulation of DSB, AM, SSB, and VSB.
- Power, efficiencies, and bandwidths of the modulation schemes.
- Frequency translation and mixing along with superheterodyne receivers.
- Distortions caused by non-linear channels.
- Frequency and phase modulations.
- Generation of angle modulated signals.
- Spectral analysis of angle modulated signals.
- Bessel functions and bandwidth requirements for angle modulated signals.
- Demodulation of angle modulated signals along with superheterodyne receivers.
- Distortions caused by non-linear channels and multipath are briefly discussed.
- Frequency-division and time-division multiplexing.
- Analog pulse modulation schemes.
- Pulse code modulation.
- A brief presentation of binary digital communciations.

Problems

10.2.1 Consider that we have a non-linear device, whose output is $y(t) = \alpha(A_1 \cos(\omega_0 t + \theta_1))^2$. Show that the non-linear device in cascade with a BP filter can be used as a frequency doubler and give the center frequency of the BP filter.

10.3.1 Show that coherent demodulation can be used to demodulate an AM signal.

10.4.1 Consider the block diagram shown in Fig. P10.4.1 with the input signal as given. Sketch the spectrum of the output signal, which is the spectrum of a "scrambled" version of the input signal. Then, show

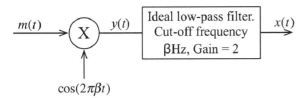

Fig. P10.4.1 A simple Scrambler

that an identical system can be used to "unscramble" the scrambled signal. That is, recover $m(t)$ from $x(t)$.

10.5.1 Find the average power in $x_{AM}(t) = A_c[1 + \mu m(t)]\cos(\omega_c t)$,

$$m(t) = \sum_{i=0}^{N-1} a^i \cos(\omega_i t + \theta_i), \ |a| \neq 1.$$

10.5.2 Sketch the function $x_{AM}(t) = (1 + \mu m_n(t))\cos(\omega_c t)$ assuming the following for the modulation index: $\mu < 1, \mu = 1,$ and $\mu > 1$.

a. $m_n(t) = \cos(\omega_m t)$,

b. $m_n(t) = \begin{cases} 1, \ 0 < t < T/2 \\ -1, \ -T/2 < t < 0 \end{cases}, \ m_n(t) = m_n(t + T).$

10.5.3 Give the expression for the AM signal assuming $m(t) = 4\cos(\omega_0 t) - 3\cos(3\omega_0 t)$

10.5.4 What is the largest a so that an envelope detector detects the modulating signal $m(t) = [\sin(\omega_m t) + a\cos(\omega_m t)]$ from the AM signal $x_{AM}(t) = A_c[1 + \mu m(t)]\cos(\omega_c t)$?

10.6.1 Let the input–output characteristic of a system be $y(t) = ax(t) + bx^2(t), x(t) = [m(t) + \cos(\omega_c t)]$. a. Find $Y(j\omega)$.

$$m(t) \xleftarrow{\text{FT}} m(j\omega) = \wedge\left[\frac{\omega}{2\pi\beta}\right]$$

Use a filter that can recover the message. Identify the center frequency, constraints on the constants, filter type, and its bandwidth. b. Assuming $m(t) = \cos(\omega_m t)$ show that the results derived in Part a. are valid.

10.7.1 The AM and DSB signals can be demodulated using a signal with its F-series

$$p(t) = X_s[0] + \sum_{k=1}^{\infty} a[k]\cos(k\omega_c t).$$

Give the expressions for the outputs of the demodulator for the DSB and AM cases. What is the difference in the two demodulators in these cases? Assume the bandwidth of the message signal ω_B is much, much smaller than the frequency, i.e., $\omega_B \ll \omega_c$.

10.7.2 A method of detecting a message signal from an AM signal is by using a *rectifier detector*. First, assume the rectified AM signal is $y(t) = (A + m(t))|\cos(\omega_c t)|$.

Now express $|\cos(\omega_c t)|$ in terms of its trigonometric F-series and then use a LP filter to filter out all the high-frequency components. Use a bias removal capacitor to obtain the message signal. Go through the mathematical details of these steps.

10.7.3 An AM signal is to be demodulated by an envelope detector. The message is band limited to 5 kHz and the carrier frequency is 100 kHz. Give the limits on the detector time constant to avoid significant distortion.

10.8.1 Figure P10.8.1 gives a block diagram to generate an SSB signal, see Carlson (1975). Assuming the input is a band-limited voice signal to a bandwidth of 3500 Hz, sketch the spectra at each point identified in the block diagram.

10.8.2 Use Example 10.8.2 to show that a synchronous detector recovers the message.

10.8.3 Consider the SSB modulated signal $x_{SSB}(t) = [A + m(t)]\cos(\omega_c t) - \hat{m}(t)\sin(\omega_c t)$.

Express this function in the form $x_{SSB}(t) = A(t)\cos(\omega_c t + \theta(t))$ using tables. Note that the expression has the time-varying amplitude $A(t)$ and the time-varying phase $\theta(t)$. a.Use an envelope detector, a low-pass filter, and a bias removal capacitor to obtain $m(t)$. Give the constraints on $[A + m(t)]$ and $\hat{m}(t)$.

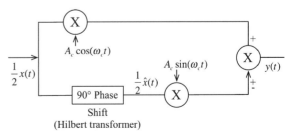

Fig. P10.8.1 An SSB signal generator

b. Show that a square-law detector does not work to demodulate an SSB signal.

10.8.4 Consider the SSB signal $x_{SSB}(t) = m(t)\cos(\omega_c t) + \hat{m}(t)\sin(\omega_c t)$. At the receiver, it is demodulated using the local carrier $\cos[(\omega_c + \Delta\omega)t + \delta]$ and passed through a low-pass filter. *a.* Give the output of the demodulator. Show that if the local oscillator is coherent with the transmitter, then the output is a scaled version of the input message.

   b. Give the expression for the output assuming $\Delta\omega = 0$ in terms of $m(t), \hat{m}(t)$ and δ.

   c. What is the transform of the output in Part *b.*? Show that the phase error in the local oscillator gives rise to phase distortion. Comment on the distortions in the amplitude?

10.9.1 Find the image frequency for the following receiver. The RF frequency we are tuning is assumed to be 10 MHz. The local oscillator frequency is assumed to be equal to 11 MHz. These give us the intermediate frequency of 1 MHz. Give the image frequencies.

10.10.1 A superheterodyne receiver uses an intermediate frequency of 455 kHz. The receiver is tuned to a transmitter having the carrier frequency $f_c = 780$ kHz. Find some permissible frequencies of the local oscillator and the corresponding image frequencies. Assuming the received signal is given by $x_{AM}(t) = A_c[A + m(t)]\cos(\omega_c t), \omega_c = 2\pi f_c$ express the functions at each location of the superheterodyne receiver.

10.11.1 Determine the instantaneous frequency of the angle modulated signal $x_a(t) = A_c\cos(\omega_c t + \phi(t))$ by assuming $\omega_c = 2\pi(10^6)$ and $\phi(t) = 20t^2$.

10.11.2 Derive the expression for the narrowband (NB) phase modulated signal and sketch the spectrum. Give the bandwidth of the corresponding NBPM signal.

10.11.3 Consider the message waveform shown in Fig. P10.11.3. Sketch the plots showing how the instaneous frequency changes with time of the FM and PM signals.

10.11.4 Show that an FM demodulator–integrator can be used as a PM demodulator. Show that an integrator–PM modulator can be used as an FM modulator.

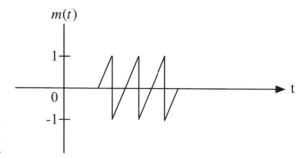

Fig. P10.11.3 A message signal

10.11.5. Find *a.* the average power and *b.* the maximum frequency and phase deviations of $x_c(t)$ given below and its bandwidth.

$$x_c(t) = 10\cos[2\pi(10^8)t + (.1)\sin(2\pi(2000)t)]$$

10.11.6 Consider the angle modulated signal given in Problem 10.11.5 with $k_p = 10$. *a.* Assuming the modulation is PM, determine the message signal $m(t)$. Estimate the bandwidth of the modulated signal. *b.* Assuming the modulation is frequency modulation with $k_f = 10\pi$. Find the message signal $m(t)$ and estimate the bandwidth.

10.12.1 Consider the frequency modulated signal with a carrier 100 MHz and is modulated with a 100 kHz sinusoid. The frequency deviation assumed to be 1 MHz. Determine the signal bandwidth by keeping eight sidebands. If the modulated signal is passed through an ideal band-pass filter with a center frequency 100 MHz with a bandwidth equal to the bandwidth calculated earlier, what percentage of the signal power will be passed? You need to make use of the Bessel function table given in Table 10.12.1.

10.14.1 Pre-emphasis and de-emphasis filters were given in Fig. 10.14.3a,b. Derive their transfer functions. Show that $H_{PE}(j\omega)H_{DE}(j\omega) \approx K$, a constant for $\omega \ll \omega_2$. Hint: For frequencies in the range $\omega \ll \omega_2$, approximate $H_{PE}(j\omega)$ by its numerator term. See Lathi, (1983) on the use of these and the improvements on the signal-to-noise ratios.

10.14.2 In Section 10.14.3 a tapped delay line was considered to reduce the effect of multipath transmission errors by making use of an inverse filter function $H_{eq}(j\omega)$ given below. Make use of the identity below in obtaining the impulse response of such a filter.

$$H_{eq}(j\omega) = \frac{1}{1 + \alpha e^{-j\omega\tau}}, \quad \sum_{k=0}^{\infty} \beta^k = \frac{1}{1 - \beta}, |\beta| < 1$$

10.15.1 Quadrature multiplexing is used in AM stereo broadcasting. Harris Corporation proposes a modulated signal $x_c(t) = [A\cos(\omega_c t) + x_L(t) \cos(\omega_c t - \phi_0) + x_R(t)\cos(\omega_c t + \phi_0)]$, $(\phi_0 = 15^0)$. *a.* Use $x_c(t) = m_1(t)\cos(\omega_c t) + m_2(t)\sin(\omega_c t)$ and identify $m_1(t)$ and $m_2(t)$. *b.* Show that sum signal can be recovered using an envelope detector and the difference signal can be detectd using a product detector.

10.16.1 In the demodulation of a pulse amplitude modulation scheme, a simple method is simply hold the level of a given pulse until next pulse appears. The holding circuit can be considered as a linear system with a rectangular unit-impulse response. We studied zero-order hold devices before. Give a simple block diagram that has a transfer function, which results in the approximation of the impulse response $h(t) = \Pi[(t - t_s/2)/t_s]$.

10.16.2 Let $H(j\omega) = F[h(t)]$ with the identity given below. Show $h(nt_s)$ as indicated.

$$\sum_{n=-\infty}^{\infty} H(j(\omega + (2\pi k/t_s))) = 1, |\omega| \le (\pi/t_s),$$

$$h(nt_s) = F^{-1}[H(j\omega)] = \begin{cases} 1/t_s, n = 0 \\ 0, \quad n = 0 \end{cases}$$

This is the *Nyquist's pulse-shaping criterion*. It is basic to the baseband signaling systems through a band-limited channel, see Ziemer and Tranter (2002).

10.17.1 A speech signal with a bandwidth of 4 kHz is sampled, quantized, and transmitted using a PCM system. It is desired that the data sample at the receiving end must be known within $\pm 5\%$ of the peak-to-peak full-sacle value of the amplitude $\pm A$. Determine the number of samples must each digital word contain. Estimate the bandwidth of the PCM signal assuming the pulse is $\Pi[(t - t_0)/\tau]$ by assuming $k = 1$ and 2 in (10.17.11b).

10.17.2 A standard assumption for the the frequency band of an audio signal is in the frequency range 300 Hz to 3.2 kHz. A PCM signal is generated with a sampling rate of 8000 samples per second with signal-to-quantizing ratio of 44 dB. *a.* Determine the number of bits needed to satisfy the signal-to-noise ratio constraint. *b.* Determine the minimum required PCM bandwidth.

10.17.3 Start with a binary channel with a bit rate of 64000 bits/s and we like to transmit a speech signal with a bandwidth of 4 kHz. Find the sampling rate, the number of quantization levels, and the number digits per sqample to transmit using this channel.

10.17.4 The two bit normal Gray code is generated below.

		0		0	0
	0	1		0	1
$n = 1:$	$\longrightarrow$		Prefix the entries above the line with a zero $\longrightarrow$ and prefix those below with alone	$-\quad-$	$: n = 2$
	1	1		1	1
		0		1	0

In the above, we have written a zero and one for the normal Gray code for $n = 1$. In the second step, we draw a line of reflection and use the bits generated earlier on top of the line. Below the line we write the reflected bits as shown. To obtain the Gray code for $n = 2$, we prefix the entries above the line with a zero and with a one below the line. Continue this procedure to determine the 3 bit normal Gray code.

10.17.5 Sketch the ASK, PSK, and FSK signals assuming the input bits are 1011001.

Appendix A

Matrix Algebra

Analog and discrete-time signal analysis and design involves significant involvement of matrices. They provide a compact way of representing circuit and system equations. Obtaining a solution of a set of equations is one of the important steps in the circuit and system analysis. Cramer's rule allows us to solve a system of equations. Finding the discrete Fourier transform of a set of data essentially corresponds to the multiplication of a matrix times a vector. The presentation of fast Fourier transforms in terms of matrices shows the beauty and the simplicity associated with the fast algorithms. Matrix theory is an important topic and every undergraduate student in electrical engineering should have a basic knowledge. In this short appendix we present the basic properties and some useful results, including determinants, inverses, generalized inverses, vector norms, and solutions of over- and underdetermined system of equations, eigenvalue–eigen vector decompositions, singular value decompositions, and others. At the end of this appendix, two numerical-based interpolation methods are presented to complement the spectrally based interpolation methods discussed in Section 8.2. Matrix theory is an established area and there are good texts available. See, for example, the books by Hohn (1958), Perlis (1958), and many others. Some of the examples are worked out using MATLAB. See Appendix B for a discussion on MATLAB.

A.1 Matrix Notations

In this section we will consider matrix notations, representation of a set of equations in terms of matrices and their solutions.

A matrix is a rectangular array of numbers or transforms of dimension $m \times n$ (read "m by n" matrix, generally written as $(m \times n)$, where m denotes the number of rows and n is the number of columns. In the following, we have a matrix $\mathbf{A}$ of dimension $m \times n$, a row vector $\mathbf{y}$ of dimension $(1 \times n)$, or simply an n-dimensional row vector and a column vector $\mathbf{x}$ of dimension $(m \times 1)$.

$$\mathbf{A} = \begin{bmatrix} a_{11} & a_{12} & \cdots & a_{1n} \\ a_{21} & a_{22} & \cdots & a_{2n} \\ \vdots & \vdots & \vdots & \vdots \\ a_{m1} & a_{m2} & \cdots & a_{mn} \end{bmatrix},$$

$$\mathbf{y} = \begin{bmatrix} y_1 & y_2 & \cdots & y_n \end{bmatrix}, \mathbf{x} = \begin{bmatrix} x_1 \\ x_2 \\ \vdots \\ x_m \end{bmatrix}. \quad (A.1.1)$$

We will use *boldfaced capital letters*, such as $\mathbf{A}$, for *matrices* and *boldface lower case letters*, such as $\mathbf{x}$ for *vectors*. A typical entry or element in $\mathbf{A}$ is given by a_{ij} and is located in the ith row and the jth column and use lowercase letters with subscripts for the matrix entries. A square matrix is an array of $m \times m$ entries or elements. A diagonal matrix $\mathbf{D}$ is a square matrix with each of the off-diagonal elements is zero and the main diagonal elements are given by d_{ii} and is written sometimes as $\mathbf{D} = \text{dia}(d_{11}, d_{22}, ..., d_{mm})$ for simplicity. An identity matrix of dimension $(n \times n)$ is a diagonal matrix identified by $\mathbf{I}_n$ with each diagonal entry equal to 1 and zeros everywhere else. The subscript on $\mathbf{I}$ is usually not included for simplicity. A null matrix of dimensions $m \times n$ is identified by the null matrix $\mathbf{0}$

R.K.R. Yarlagadda, *Analog and Digital Signals and Systems*, DOI 10.1007/978-1-4419-0034-0,
© Springer Science+Business Media, LLC 2010

and it consists of all zeros. The sum of the diagonal elements of a square matrix $\mathbf{A}$ is the *trace* and is

$$tr(\mathbf{A}) = \sum_{i=1}^{m} a_{ii}. \qquad (A.1.2)$$

$$\mathbf{D} = \begin{bmatrix} 1 & 0 & 0 \\ 0 & 3 & 0 \\ 0 & 0 & 5 \end{bmatrix}, \quad \mathbf{T} = \begin{bmatrix} 2 & 2 & 3 \\ 0 & -1 & 2 \\ 0 & 0 & 1 \end{bmatrix},$$

$$\mathbf{N} = \begin{bmatrix} 1 & 0 & 0 \\ 2 & 1 & 0 \\ 3 & 4 & 9 \end{bmatrix}, \quad \mathbf{X} = \frac{1}{\sqrt{2}} \begin{bmatrix} 1 & 1 \\ 1 & -1 \end{bmatrix}, \qquad (A.2.1b)$$

$$\mathbf{U} = \frac{1}{2} \begin{bmatrix} 1 & 1 & 1 & 1 \\ 1 & -j & -1 & j \\ 1 & -1 & 1 & -1 \\ 1 & j & -1 & j \end{bmatrix}. \qquad (A.2.1c) \quad \blacksquare$$

A.2 Elements of Matrix Algebra

Equality of two matrices of the same dimensions $\mathbf{A} = \mathbf{B}$ means that the corresponding elements of $\mathbf{A}$ and $\mathbf{B}$ are equal, i.e., $a_{ij} = b_{ij}$. The addition of two matrices $\mathbf{C} = \mathbf{A} + \mathbf{B}$ means that a typical entry in the matrix $\mathbf{C}$ is given by $c_{ij} = a_{ij} + b_{ij}$ for all i,j and two matrices must be of the same dimensions. The negative of a matrix $\mathbf{B} = -\mathbf{A}$ means that $b_{ij} = -a_{ij}$. The transpose of a matrix $\mathbf{F} = \mathbf{A}^T$ means that $f_{ij} = a_{ji}$. In the case of a matrix with complex entries, the conjugate of a matrix is given by $\mathbf{G} = \bar{\mathbf{A}}$. Note the bar above the matrix and the entries in the matrix $\mathbf{G}$ are given by $g_{ij} = \bar{a}_{ij}$. That is, $a'_{ij}s$ are replaced by the conjugates of a_{ij}. The matrix $\mathbf{A}^*$ denotes that $\mathbf{A}^* = (\bar{\mathbf{A}})^T$. That is, the superscript (*) denotes the complex-conjugate transpose of the matrix. $\mathbf{A}$ is a symmetric matrix if $\mathbf{A} = \mathbf{A}^T$. That is $a_{ij} = a_{ji}$. The matrix $\mathbf{A}$ is a skew symmetric matrix if $\mathbf{A} = -\mathbf{A}^T$, i.e., $a_{ij} = -a_{ji}$. The matrix $\mathbf{A}$ is called *Hermitian* if $\mathbf{A} = \mathbf{A}^*$ and *skew Hermitian* if $\mathbf{A} = -\mathbf{A}^*$. A matrix $\mathbf{B}$ is called *orthogonal* if $\mathbf{BB}^T = \mathbf{B}^T\mathbf{B} = \mathbf{I}$, an identity matrix. A complex matrix $\mathbf{C}$ is called a *unitary matrix* if $\mathbf{CC}^* = \mathbf{C}^*\mathbf{C} = \mathbf{I}$.

Example A.2.1 Matrix examples of some of these are as follows. $\mathbf{A}$ is symmetric, $\mathbf{B}$ is skew symmetric, $\mathbf{C}$ is a Hermitian, $\mathbf{D}$ is diagonal, $\mathbf{T}$ is an upper triangular, $\mathbf{N}$ is a lower upper triangular, $\mathbf{X}$ is an orthogonal, and $\mathbf{U}$ is a unitary matrix.

$$\mathbf{A} = \begin{bmatrix} 2 & 1 & 0 \\ 1 & -1 & 1 \\ 0 & 1 & 5 \end{bmatrix}, \quad \mathbf{B} = \begin{bmatrix} 0 & 1 & 2 \\ -1 & 0 & 3 \\ -2 & -3 & 0 \end{bmatrix},$$

$$\mathbf{C} = \begin{bmatrix} 1 & 1+j & 2-j3 \\ 1-j & 2 & 3 \\ 2+j3 & 3 & 0 \end{bmatrix}, \qquad (A.2.1a)$$

Matrix operations: Let α be a scalar then the entries in the matrix $\mathbf{B} = \alpha\mathbf{A}$ are $b_{ij} = \alpha a_{ij}$ for all i,j. Let $\mathbf{A}$ and $\mathbf{B}$ be two matrices of dimensions $(m \times n)$ and $(p \times q)$, respectively. Then the matrix product $\mathbf{AB}$ is defined only for $n = p$. That is, the matrix product $\mathbf{AB}$ is defined only when the number of columns in $\mathbf{A}$ is equal to the number of rows in $\mathbf{B}$. The matrix $\mathbf{A}$ is *conformable* to the matrix $\mathbf{B}$ for multiplication $\mathbf{AB}$. Note that the matrix $\mathbf{B}$ may not be conformable to the matrix $\mathbf{A}$ for the matrix multiplication $\mathbf{BA}$. The dimension of the matrix product $\mathbf{AB}$ is $(m \times q)$. The ik th entry in $\mathbf{C} = \mathbf{AB}$, is as follows:

$$c_{ik} = \sum_{j=1}^{n} a_{ij}b_{jk}, \quad i = 1, 2, ..., m, k = 1, 2, ..., q \quad (A.2.2)$$

We obtain the ikth element in the matrix product $\mathbf{AB}$ by multiplying the elements in the ith row of $\mathbf{A}$ by the corresponding elements of the kth column in $\mathbf{B}$ and add the terms. Note the outer subscript j on a and the outer subscript j on b should be the same. The transpose of the sum of two matrices is the sum of their transposes. That is $(\mathbf{A} + \mathbf{B})^T = \mathbf{A}^T + \mathbf{B}^T$. The transpose of the product of two matrices $\mathbf{A}$ and $\mathbf{B}$ is

$$(\mathbf{AB})^T = \mathbf{B}^T\mathbf{A}^T. \qquad (A.2.3)$$

Order of the two matrices should be kept in tact in dealing with transposes of matrices. Interestingly if $\mathbf{AB} \neq \mathbf{BA}$ and $\mathbf{AB} = \mathbf{0}$, it does not necessarily imply $\mathbf{A} = \mathbf{0}$ or $\mathbf{B} = \mathbf{0}$. Also, $\mathbf{AB} = \mathbf{AC}$ does not imply $\mathbf{B} = \mathbf{C}$. Let $\mathbf{A}$ and $\mathbf{B}$ are square matrices of the same dimensions with $\mathbf{AB} = \mathbf{BA}$, then we say that the two matrices are *commutative*. Given the two matrices **A** and **B** of the same dimensions $m \times n$ and $n \times m$, respectively, then

$$tr(\mathbf{AB}) = \sum_{i=1}^{m} \left[\sum_{j=1}^{n} a_{ij} b_{ji} \right], \ \ tr(\mathbf{AB}) = \sum_{j=1}^{n} \left[\sum_{i=1}^{m} b_{ji} a_{ij} \right],$$

$$tr(\mathbf{AB}) = tr(\mathbf{BA}). \tag{A.2.4}$$

Matrices are generally not commutative. They satisfy the associative and distributive properties, provided they are conformable for addition and multiplication. Multiplication by the identity matrix is commutative. These are stated as follows:

$$(\mathbf{AB})\mathbf{C} = (\mathbf{AB})\mathbf{C} \text{ and } \mathbf{A}(\mathbf{B} + \mathbf{C}) = \mathbf{AB} + \mathbf{AC}, \tag{A.2.5a}$$

$$\mathbf{AI} = \mathbf{IA} = \mathbf{A}. \tag{A.2.5b}$$

Example A.2.2 Given the two matrices **A** and **B** as below, find the matrix products **AB** and **BA** and the traces of the two matrices.

$$\mathbf{A} = \begin{bmatrix} 1 & 2 & 1 \\ 0 & 1 & 2 \end{bmatrix}, \quad \mathbf{B} = \begin{bmatrix} 1 & 0 \\ 2 & 1 \\ 1 & 2 \end{bmatrix},$$

$$\mathbf{AB} = \begin{bmatrix} 6 & 4 \\ 4 & 5 \end{bmatrix}, \quad \mathbf{BA} = \begin{bmatrix} 1 & 2 & 1 \\ 2 & 5 & 4 \\ 1 & 4 & 5 \end{bmatrix}.$$

The matrix product **AB** is a 2 by 2 matrix and the product **BA** is a 3 by 3 matrix. Noting that $\mathbf{B} = \mathbf{A}^T$, the matrix products are both symmetric. The traces are given by

$$tr(\mathbf{AB}) = 6 + 5 = 11, \quad tr(\mathbf{BA}) = 1 + 5 + 5 = 11. \ \blacksquare$$

Example A.2.3 Consider the following DFT matrices and identify their properties.

$$\mathbf{A}_1 = \frac{1}{\sqrt{2}} \begin{bmatrix} 1 & 1 \\ 1 & -1 \end{bmatrix}, \ \mathbf{A}_2 = \frac{1}{\sqrt{3}} \begin{bmatrix} 1 & 1 & 1 \\ 1 & e^{-j2\pi/3} & e^{-j4\pi/3} \\ 1 & e^{-j4\pi/3} & e^{-j2\pi/3} \end{bmatrix}.$$

$$\tag{A.2.6}$$

Solution: $\mathbf{A}_1$ is a real symmetric matrix and $\mathbf{A}_2$ is a symmetric complex matrix. Furthermore, since $\mathbf{A}_1 \mathbf{A}_1^T = \mathbf{A}_1^T \mathbf{A}_1 = \mathbf{I}_2, \mathbf{A}_2 \mathbf{A}_2^* = \mathbf{A}_2^* \mathbf{A}_2 = \mathbf{I}_3$, it follows that $\mathbf{A}_1$ is an orthogonal matrix and $\mathbf{A}_2$ is a unitary matrix. $\blacksquare$

A.2.1 Vector Norms

Length of a vector is an important measure of the vector. To generalize this concept, associate a vector with a nonnegative scalar that gives in some sense a measure of its magnitude. This will be important when the errors between the desired and computed values are evaluated. The L_p norm is defined for a vector **x** of dimension n by

$$\|\mathbf{x}\|_p = (|x_1|^p + |x_2|^p + \cdots + |x_n|^p)^{1/p}, \ 1 \le p < \infty. \tag{A.2.7}$$

If the subscript p is not explicitly identified, then the properties given below apply for all p. The following properties of the norms are useful:

$$\|\mathbf{x}\| > 0 \text{ if } \mathbf{x} \ne 0, \ \|\mathbf{x}\| = 0 \ \text{ if } \mathbf{x} = 0, \tag{A.2.8a}$$

$$\|\alpha \mathbf{x}\| = |\alpha| \|\mathbf{x}\|, \alpha \text{ a scalar}, \tag{A.2.8b}$$

$$\|\mathbf{x} + \mathbf{h}\| \le \|\mathbf{x}\| + \|\mathbf{h}\|. \tag{A.2.8c}$$

In particular, L_2 norm of a vector **x** is of interest and is defined by

$$\|\mathbf{x}\|_2 = (|x_1|^2 + |x_2|^2 + \cdots + |x_n|^2)^{1/2}. \tag{A.2.8d}$$

For simplicity, the subscript 2 in (A.2.8) will be omitted in the following. The inequality in (A.2.8c) is an important property. Specifically,

$$\|\mathbf{x} + \mathbf{h}\| = [(\mathbf{x} + \mathbf{h})^* (\mathbf{x} + \mathbf{h})]^{1/2}$$
$$= [\mathbf{x}^* \mathbf{x} + \mathbf{h}^* \mathbf{h} + \mathbf{x}^* \mathbf{h} + \mathbf{h}^* \mathbf{x}]^{1/2} \le [\mathbf{x}^* \mathbf{x} + \mathbf{h}^* \mathbf{h}]^{1/2}.$$

An important special case is when $\mathbf{x}^* \mathbf{h} = 0$ resulting in

$$\|\mathbf{x} + \mathbf{h}\| = \|\mathbf{x}\| + \|\mathbf{h}\|. \tag{A.2.8e}$$

For a detailed discussion on the vector and matrix norms, see Wilkinson (1965).

The norm $\|\mathbf{x}\|_\infty$ is interpreted as $\max |x_i|$. The simplest of all the distributions is Gaussian, which is of interest here and the L_2 norm is applicable. Interestingly, if $\mathbf{U}$ is a unitary matrix, then

$$\|\mathbf{U}\mathbf{x}\|_2 = \|\mathbf{x}\|_2. \qquad (A.2.9)$$

Notes: L_p norms give some information about the vectors. For example, L_1 norm relates to the absolute sum of the entries in the vector, L_2 norm relates to the power in the vector, and $L\infty$ norm gives the peak absolute deviation from zero. In Section 3.3 the selection of p in the L_p norm based on the type of distribution the data came from was considered. In this appendix, L_2 norm is of interest. This measure allows for simple solutions for over- and underdetermined system of equations. ∎

Example A.2.4 Compute the L_p norms of the following vector for $p = 1, 2, \infty$:

$$\mathbf{x} = [1 \quad 1+j \quad 1-j].$$

Solution:

$$|\mathbf{x}| = [1 \quad \sqrt{2} \quad \sqrt{2}], \|\mathbf{x}\|_1 = \sum_{n=1}^{3} |x[n]| = 1 + 2\sqrt{2}, \|\mathbf{x}\|_2$$

$$= \sqrt{\sum_{n=1}^{3} |x[n]|^2} = \sqrt{5}, \|\mathbf{x}\|_\infty = \sqrt{2}.$$

MATLAB codes for finding these are

```
>> x1= norm (x,1), x2
   = norm (x,2), x3= norm (x,inf)
```

The answers are

$$\|\mathbf{x}\|_1 = 3.8284, \ \|\mathbf{x}\|_2 = 2.2361, \ \|\mathbf{x}\|_\infty = 1.4142.$$

A.3 Solutions of Matrix Equations

The solution of a set of linear equations is basic in matrix analysis. Considering a $m \times n$ matrix $\mathbf{A}$ and an n-dimensional vector $\mathbf{x}$ given in (A.1.1), we define the matrix equation

$$\mathbf{y} = \begin{bmatrix} y_1 \\ y_2 \\ \vdots \\ y_m \end{bmatrix} = \mathbf{A}\mathbf{x} = \begin{bmatrix} a_{11} & a_{12} & \cdots & a_{1n} \\ a_{21} & a_{22} & \cdots & a_{2n} \\ \vdots & \vdots & \vdots & \vdots \\ a_{m1} & a_{m2} & \cdots & a_{mn} \end{bmatrix} \begin{bmatrix} x_1 \\ x_2 \\ \vdots \\ x_n \end{bmatrix}$$

$$= \begin{bmatrix} a_{11}x_1 + a_{12}x_2 + \cdots + a_{1n}x_n \\ a_{21}x_1 + a_{22}x_2 + \cdots + a_{2n}x_n \\ \vdots \\ a_{m1}x_1 + a_{m2}x_2 + \cdots + a_{mn}x_n \end{bmatrix}. \qquad (A.3.1)$$

The matrix $\mathbf{A}$ transforms a vector $\mathbf{x}$ into another vector $\mathbf{y}$. In most cases the dimensions of $\mathbf{x}$ and $\mathbf{y}$ are the same. For example, the discrete Fourier transform (DFT) vector of a set of data $\mathbf{x}$ (see (8.9.16a)) is expressed by a transformation and the entries in the corresponding matrix that executes this transformation are

$$\mathbf{y} = \mathbf{A}_{\text{DFT}}\mathbf{x}, \ (\mathbf{A}_{\text{DFT}})_{ik} = e^{-j2\pi(i-1)(k-1)/N}. \qquad (A.3.2a)$$

The 3-point DFT of the data vector, written in terms of a column vector $\mathbf{x}$ is

$$\mathbf{y} = \begin{bmatrix} y_1 \\ y_2 \\ y_3 \end{bmatrix} = \mathbf{A}_{\text{DFT}}\mathbf{x} = \begin{bmatrix} 1 & 1 & 1 \\ 1 & e^{-j2\pi/3} & e^{-j4\pi/3} \\ 1 & e^{-j4\pi/3} & e^{-j2\pi/3} \end{bmatrix} \begin{bmatrix} x_1 \\ x_2 \\ x_3 \end{bmatrix}.$$

$$\qquad (A.3.2b)$$

Since $(1/\sqrt{3})\mathbf{A}_{DFT}$ is a unitary matrix, it follows that $\|(1\sqrt{3})\mathbf{A}_{DFT}\mathbf{x}\|_2 = \|\mathbf{y}\|_2$. Note that length of the vector $\mathbf{y}$ is equal to the length of the vector $\mathbf{x}$.

A.3.1 Determinants

The determinant of a matrix $\mathbf{A}$, a scalar quantity, is defined for a square matrix

$$\Delta = |\mathbf{A}| = |\mathbf{A}^T| = \det(\mathbf{A}). \qquad (A.3.3a)$$

It may be positive or negative despite the bars in the notation in (A.3.3a). The determinant of a 2×2 matrix is

$$|\mathbf{A}| = \begin{vmatrix} a_{11} & a_{12} \\ a_{21} & a_{22} \end{vmatrix} = a_{11}a_{22} - a_{12}a_{21}. \qquad (A.3.3b)$$

The determinants of larger matrices are evaluated by expansions involving *reduced* arrays. For each element a_{ij} in the square matrix $\mathbf{A}$, a *cofactor* is defined by

$$\Delta_{ij} = (-1)^{i+j}\left|\mathbf{M}_{ij}\right|. \qquad (A.3.4)$$

The submatrix $\mathbf{M}_{ij}$ is the matrix obtained from the coefficient matrix $\mathbf{A}$ by deleting the ith row and the jth column and $\left|\mathbf{M}_{ij}\right|$ is the determinant of the matrix $\mathbf{M}_{ij}$. Since $(-1)^{i+j} = \pm 1$, it follows that the signs depend on whether $(i+j)$ is even or odd. The determinant of a $n \times n$ matrix $\mathbf{A}$ is by using the i th row of the matrix $\mathbf{A}$ by *Laplace's expansion*

$$\Delta = |\mathbf{A}| = a_{i1}\Delta_{i1} + a_{i2}\Delta_{i2} + \cdots + a_{in}\Delta_{in}. \quad (A.3.5a)$$

The ith row can be any one of the rows and the expansion is simpler if there are many zeros in that row. Determinant is a scalar quantity and it can be determined using either a row or a column. By using the i th column of the matrix $\mathbf{A}$ results in

$$\Delta = |\mathbf{A}| = a_{1j}\Delta_{1j} + a_{2j}\Delta_{2j} + ... + a_{nj}\Delta_{nj}. \quad (A.3.5b)$$

The determinant of a 3×3 matrix reduces to a sum of three terms in (A.3.4), involving 2×2 determinants, which can be computed using (A.3.3). The determinant of a triangular matrix is the product of diagonal entries. The determinants of orthogonal and unitary matrices equal to 1.

Example A.3.1 Expand the determinant given below. Give the trace of the matrix $\mathbf{A}$.

Solution: By using the first row, we have

$$|\mathbf{A}| = \Delta = \begin{vmatrix} a_{11} & a_{12} & a_{13} \\ a_{21} & a_{22} & a_{23} \\ a_{31} & a_{32} & a_{33} \end{vmatrix} = a_{11}(a_{22}a_{33} - a_{32}a_{23})$$

$$- a_{12}(a_{21}a_{33} - a_{31}a_{23}) + a_{13}(a_{21}a_{32} - a_{31}a_{22}).$$

$$= a_{11}a_{22}a_{33} - a_{11}a_{32}a_{23} - a_{12}a_{21}a_{33} + a_{12}a_{31}a_{23}$$

$$+ a_{13}a_{21}a_{32} - a_{13}a_{31}a_{22}. \qquad (A.3.6)$$

The trace of the matrix A is

$$tr(\mathbf{A}) = a_{11} + a_{22} + a_{33}. \qquad (A.3.7) \quad \blacksquare$$

A.3.2 Cramer's Rule

Given a set of equations in (A.3.8) with the coefficient matrix $\mathbf{A}$ being square, i.e., $m = n$, we would like to find its solution, that is, solve for $\mathbf{x}$. The set is written symbolically as

$$\mathbf{Ax} = \mathbf{y}. \qquad (A.3.8)$$

There are several ways to obtain the solution. Two ways of solution are considered below, one by Cramer's rule and the other by finding the inverse of the coefficient matrix. A square matrix $\mathbf{A}$ has an inverse $\mathbf{A}^{-1}$ if and only if $|\mathbf{A}| \neq 0$. In that case, it is called a *nonsingular* matrix, whereas if $|\mathbf{A}| = 0$, it is a *singular* matrix. *Cramer's rule* makes use of the determinant of the matrix $\mathbf{A}$ and the cofactors obtained from this matrix. The ith unknown is

$$x_i = \frac{1}{\Delta}(\Delta_{1i}y_1 + \Delta_{2i}y_2 + \cdots + \Delta_{ni}y_n),$$
$$\Delta = |\mathbf{A}|, i = 1, 2, \ldots, n. \qquad (A.3.9)$$

Noting that Δ is in the denominator in (A.3.9), the necessary condition for the solution of the set of equations with $n = m$ in (A.3.8) is that $\Delta = |\mathbf{A}| \neq 0$. An interesting way of visualizing the Cramer's rule is as follows. Replace the jth column of the matrix $\mathbf{A}$ by the column vector $\mathbf{y}$ and calculate the determinant of this matrix, which is defined by Δ_j. The solutions are given by

$$x_j = \frac{\Delta_j}{\Delta}, j = 1, 2, ..., n; \quad \Delta_j = \begin{vmatrix} a_{11} & a_{12} & ..y_1.. & a_{1n} \\ a_{21} & a_{22} & ..y_2.. & a_{2n} \\ . & . & ...... & . \\ . & . & ..... & . \\ . & . & ..... & . \\ a_{n1} & a_{n2} & ...y_n.. & a_{nn} \end{vmatrix},$$

$$\Delta_j = y_1\Delta_{1j} + x_2\Delta_{2j} + ... + y_n\Delta_{nj}. \qquad (A.3.10)$$

Computations become laborious for a set of equations with the order of the coefficient matrix n greater than 3. MATLAB provides good numerical solutions.

Example A.3.2 Write the following equations in matrix form, see (A.3.1). Using (A.3.10), solve them by Cramer's rule

$$x_1 + 2x_2 + 3x_3 = 9$$
$$-x_1 + x_2 + x_3 = 2 . \qquad \text{(A.3.11)}$$
$$x_1 + 2x_2 + x_3 = 5$$

Solution: In matrix form, we have

$$\begin{bmatrix} y_1 \\ y_2 \\ y_3 \end{bmatrix} = \begin{bmatrix} 9 \\ 2 \\ 5 \end{bmatrix} = \begin{bmatrix} 1 & 2 & 3 \\ -1 & 1 & 1 \\ 1 & 2 & 1 \end{bmatrix} \begin{bmatrix} x_1 \\ x_2 \\ x_3 \end{bmatrix}, \quad \mathbf{y} = \mathbf{Ax}.$$

$$\text{(A.3.12)}$$

The determinants are as follows:

$$|\mathbf{A}| = \begin{vmatrix} 1 & 2 & 3 \\ -1 & 1 & 1 \\ 1 & 2 & 1 \end{vmatrix} = 1(1 - 2) - (-1)(2 - 6)$$
$$+ 1(2 - 3) = -6, \qquad \text{(A.3.13)}$$

$$\Delta_1 = \begin{vmatrix} 9 & 2 & 3 \\ 2 & 1 & 1 \\ 5 & 2 & 1 \end{vmatrix} = 9(1 - 2) - 2(2 - 6)$$
$$+ 5(2 - 3) = -6, \qquad \text{(A.3.14a)}$$

$$\Delta_2 = \begin{vmatrix} 1 & 9 & 3 \\ -1 & 2 & 1 \\ 1 & 5 & 1 \end{vmatrix} = -9(-1 - 1) + 2(1 - 3)$$
$$- 5(1 - 3) = -6, \qquad \text{(A.3.14b)}$$

$$\Delta_3 = \begin{vmatrix} 1 & 2 & 9 \\ -1 & 1 & 2 \\ 1 & 2 & 5 \end{vmatrix} = 9(-2 - 1) - 2(2 - 2)$$
$$+ 5(1 + 3) = -12. \qquad \text{(A.3.14c)}$$

Solution:

$$x_1 = \Delta_1/\Delta = 1, x_2 = \Delta_2/\Delta = 1, x_3 = \Delta_3/\Delta = 2.$$
$$\text{(A.3.15)} \quad \blacksquare$$

A.3.3 Rank of a Matrix

Rank of a matrix is defined as the largest number of independent columns (or rows) in the matrix. When a square matrix of order n is nonsingular, i.e., $|\mathbf{A}| \neq 0$, its rank is n. This implies that if there are

n columns of A given by $\mathbf{a}_1, \mathbf{a}_2, ..., \mathbf{a}_n$, then the only solution of the equation

$$\beta_1 \mathbf{a}_1 + \beta_2 \mathbf{a}_2 + \cdots + \beta_n \mathbf{a}_n = 0 \qquad \text{(A.3.16)}$$

is $\beta_i = 0, i = 1, 2, ..., n$. That is, the vectors $\mathbf{a}_i$, $i = 1, 2, ..., n$ are independent. If they are *dependent*, then there is a set of β_i, not all equal to zero, such that (A.3.16) is satisfied. If a matrix is singular, then the rank of that matrix is less than n.

A rectangular matrix $\mathbf{A}$ is of *rank r* if and only if it has at least one determinant of a submatrix of order r, that is, not zero, but has no submatrix of order more than r that its determinant is not zero. *Elementary operations* given below will not alter the rank of a matrix.

1. Interchange of two rows (or columns).
2. Multiplication of a row (or a column) by a nonzero.
3. Addition to one row (column) by a scalar multiple of a different row (column).

Example A.3.3 Find the ranks of the following two matrices using elementary operations:

$$\mathbf{A} = \begin{bmatrix} 1 & 1 & 1 \\ 2 & 2 & b \\ 1 & a & 1 \end{bmatrix} \text{ (}a \text{ and } b \text{ can take any value.)},$$

$$\mathbf{B} = \begin{bmatrix} 2 & 1 & 1 \\ 1 & 0 & 2 \end{bmatrix}$$

Solution: The matrices can be reduced to a triangular matrix by the following operations:

(1) Multiply the first row by 2 and subtract from the second row 2 and replace it by the second row.
(2) Multiply the first row by –1 and add to the third row and replace the third row by this.
(3) Interchange the second and the third rows. All these operations are indicated by the following. Note the symbol ($\sim$) indicates the rank of the matrix does not change.

$$A = \begin{bmatrix} 1 & 1 & 1 \\ 2 & 2 & b \\ 1 & a & 1 \end{bmatrix} \sim \begin{bmatrix} 1 & 1 & 1 \\ 0 & 0 & b - 2 \\ 1 & a & 1 \end{bmatrix}$$
$$\sim \begin{bmatrix} 1 & 1 & 1 \\ 0 & 0 & b - 2 \\ 0 & a - 1 & 0 \end{bmatrix} \sim \begin{bmatrix} 1 & 1 & 1 \\ 0 & a - 1 & 0 \\ 0 & 0 & b - 2 \end{bmatrix}.$$

If $a = 1$ and $b \neq 2$ or $a \neq 1$ and $b = 2$ then the rank of the matrix **A** is 2. If $a = 1$ and $b = 2$ then the rank of the matrix **A** is 1. If $a \neq 1$ and $b \neq 2$, the rank of the matrix **A** is 3.

In the case of the matrix **B,** the rank can be at most 2, as it has 2 rows and 3 columns. Note that there is one 2×2 triangular matrix given below and its determinant is

$$\det \begin{bmatrix} 1 & 1 \\ 0 & 2 \end{bmatrix} \neq 0 \Rightarrow \mathrm{Rank}(\mathbf{B}) = 2. \qquad \blacksquare$$

A.4 Inverses of Matrices and Their Use in Determining the Solutions of a Set of Equations

Inverse of the square matrix **A**, denoted by $\mathbf{A}^{-1}$ exists only if $|\mathbf{A}| \neq 0$. It satisfies

$$\mathbf{A}\mathbf{A}^{-1} = \mathbf{A}^{-1}\mathbf{A} = \mathbf{I}, \text{ an identity matrix.} \quad (A.4.1)$$

Given a system of matrix equations $\mathbf{y} = \mathbf{A}\mathbf{x}$ with $|\mathbf{A}| \neq 0$, then the solution of this system is

$$\mathbf{x} = \mathbf{A}^{-1}\mathbf{y}. \qquad (A.4.2)$$

A typical entry in the vector **x**, x_i is given in terms of y_i are

$$x_i = \alpha_{i1}y_1 + \alpha_{i2} + \ldots + \alpha_{in}y_n \qquad (A.4.3)$$

Comparing this with the Cramer's rule, it follows from (A.3.9) that $\alpha_{ij} = \Delta_{ji}/\Delta$. The following matrix with entries Δ_{ji} is called the *adjoint* matrix

$$Adj[\mathbf{A}] = \begin{bmatrix} \Delta_{11} & \Delta_{21} & \ldots & \Delta_{n1} \\ \Delta_{11} & \Delta_{22} & \ldots & \Delta_{n2} \\ \vdots & \vdots & \ldots & \vdots \\ \Delta_{1n} & \Delta_{2n} & \ldots & \Delta_{nm} \end{bmatrix}. \qquad (A.4.4)$$

Note the subscripts of the entries in $adj[\mathbf{A}]$. The inverse of **A** in terms of $adj[\mathbf{A}]$ is

$$\mathbf{A}^{-1} = \frac{1}{\Delta} adj[\mathbf{A}]. \qquad (A.4.5)$$

Interestingly, the inverse of the product of two non-singular matrices **A** and **B** is

$$(\mathbf{A}\mathbf{B})^{-1} = \mathbf{B}^{-1}\mathbf{A}^{-1}. \qquad (A.4.6)$$

Example A.4.1 Solve the matrix equation given in (A.3.12) by first finding the inverse of the matrix and then verify the solution obtained in that example.

Solution: The entries in the adjoint matrix are given by

$$\Delta_{11} = \begin{vmatrix} 1 & 1 \\ 2 & 1 \end{vmatrix} = -1, \Delta_{12} = -\begin{vmatrix} -1 & 1 \\ 1 & 1 \end{vmatrix} = 2,$$

$$\Delta_{13} = \begin{vmatrix} -1 & 1 \\ 1 & 2 \end{vmatrix} = -3, \Delta_{21} = -\begin{vmatrix} 2 & 3 \\ 2 & 1 \end{vmatrix} = 4,$$

$$\Delta_{22} = \begin{vmatrix} 1 & 3 \\ 1 & 1 \end{vmatrix} = -2,$$

$$\Delta_{23} = -\begin{vmatrix} 1 & 2 \\ 1 & 2 \end{vmatrix} = 0, \Delta_{31} = \begin{vmatrix} 2 & 3 \\ 1 & 1 \end{vmatrix} = -1,$$

$$\Delta_{32} = -\begin{vmatrix} 1 & 3 \\ -1 & 1 \end{vmatrix} = -4, \Delta_{33} = \begin{vmatrix} 1 & 2 \\ -1 & 1 \end{vmatrix} = 3.$$

The determinant of **A** was computed earlier in (A.3.13) and is equal to (-6). The adjoint matrix and the inverse of the matrix **A** are given by

$$adj[\mathbf{A}] = \begin{bmatrix} -1 & 4 & -1 \\ 2 & -2 & -4 \\ -3 & 0 & 3 \end{bmatrix},$$

$$\mathbf{A}^{-1} = \frac{adj[\mathbf{A}]}{|\mathbf{A}|} = \frac{1}{(-6)} \begin{bmatrix} -1 & 4 & -1 \\ 2 & -2 & -4 \\ -3 & 0 & 3 \end{bmatrix}.$$

The solution is the same as before.

$$\mathbf{x} = \mathbf{A}^{-1}\mathbf{y} = \frac{1}{6} \begin{bmatrix} 1 & -4 & 1 \\ -2 & 2 & 4 \\ 3 & 0 & -3 \end{bmatrix} \begin{bmatrix} 9 \\ 2 \\ 5 \end{bmatrix}$$

$$= \frac{1}{6} \begin{bmatrix} 9 - 8 + 5 \\ -18 + 4 + 20 \\ 27 - 15 \end{bmatrix} = \begin{bmatrix} 1 \\ 1 \\ 2 \end{bmatrix}. \qquad \blacksquare$$

Notes: If **A** is an $n \times n$ square nonsingular matrix, then the solution of the system of equations $\mathbf{A}\mathbf{x} = \mathbf{y}$ is unique and is $\mathbf{x} = \mathbf{A}^{-1}\mathbf{y}$. Discrete Fourier transform matrices are very useful and are used in Chapters 8 and 9. Since the matrix $(1/\sqrt{N})\mathbf{A}_{\text{DFT}}$ is a unitary matrix, i.e., $[(1/\sqrt{N})\mathbf{A}_{\text{DFT}}][(1/\sqrt{N})\mathbf{A}_{\text{DFT}}^*] = \mathbf{I}$, it follows that the inverse discrete transform (IDFT) can be used to obtain the data vector from the DFT vector. That is, if $\mathbf{y} = \mathbf{A}_{\text{DFT}}\mathbf{x}$, the DFT vector, then the data vector $\mathbf{x}$ can be obtained from $\mathbf{y}$ from $\mathbf{x} = (1/N)\mathbf{A}_{\text{DFT}}^*\mathbf{y}$.

A.5 Eigenvalues and Eigenvectors

The German word "eigen" can be translated as "characteristic." Every square matrix $\mathbf{A}$ is associated with an eigenvalue λ and an eigenvector $\mathbf{x}$ with

$$\mathbf{A}\mathbf{x} = \lambda\mathbf{x} \text{ or } (\mathbf{A} - \lambda\mathbf{I})\mathbf{x} = 0. \quad (A.5.1a)$$

The constant λ is the *characteristic* root of the *characteristic polynomial* of A given by

$$d(\lambda) = |\lambda\mathbf{I} - \mathbf{A}| = 0. \quad (A.5.1b)$$

The vector $\mathbf{x}$ is the *characteristic vector*. The system in (A.5.1a) is a *homogeneous system of equations* and has a solution $\mathbf{x} \neq 0$ only if $|\lambda\mathbf{I} - \mathbf{A}| = 0$. The roots of this equation are $\lambda_i, i = 1, 2, ..., n$ and some of these values may be equal.

Example A.5.1 Find the eigenvalues and the eigenvectors of the matrix

$$\mathbf{A} = \begin{bmatrix} 0 & 1 \\ -2 & -3 \end{bmatrix}. \quad (A.5.2)$$

Solution: First,

$$|\lambda\mathbf{I} - \mathbf{A}| = \begin{vmatrix} \lambda & -1 \\ 2 & \lambda+3 \end{vmatrix} = \lambda^2 + 3\lambda + 2 = (\lambda+2)(\lambda+1)$$

$$= 0 \Rightarrow \lambda_1 = -2, \lambda_2 = -1. \quad (A.5.3)$$

The eigenvalues are distinct and the corresponding eigenvectors can be computed from

$$\mathbf{A}\mathbf{x}_1 = \lambda_1\mathbf{x}_1 \text{ or } (\lambda_1\mathbf{I} - \mathbf{A})\mathbf{x}_1$$

$$= 0 \Rightarrow \begin{bmatrix} -2 & -1 \\ 2 & -2+3 \end{bmatrix} \begin{bmatrix} x_{11} \\ x_{21} \end{bmatrix} = \mathbf{0}. \quad (A.5.4a)$$

Note the rank of the coefficient matrix $(\lambda_1\mathbf{I} - \mathbf{A})$ is 1, which indicates that there is a single vector that satisfies this equation. The corresponding equation and a solution is given by

$$x_{12} = -2x_{11}, x_{11} = 1, \ x_{12} = -2 \Rightarrow \mathbf{x}_1 = \begin{bmatrix} 1 \\ -2 \end{bmatrix}. \quad (A.5.4b)$$

Note that $k\mathbf{x}_1, \ k \neq 0$ satisfies (A.5.4a). For simplicity k is taken as 1. Similarly,

$$\mathbf{A}\mathbf{x}_2 = \lambda_2\mathbf{x}_2 \text{ or } (\lambda_2\mathbf{I} - \mathbf{A})\mathbf{x}_2$$

$$= 0 \Rightarrow \begin{bmatrix} -1 & -1 \\ 2 & -1+3 \end{bmatrix} \begin{bmatrix} x_{12} \\ x_{22} \end{bmatrix} = \mathbf{0} \quad (A.5.5)$$

$$\Rightarrow x_{12} = 1, x_{22} = -1, \mathbf{x}_2 = \begin{bmatrix} x_{12} \\ x_{22} \end{bmatrix} = \begin{bmatrix} 1 \\ -1 \end{bmatrix}.$$

With $\mathbf{A}\mathbf{x}_i = \mathbf{x}_i\lambda_i, \ i = 1, 2$ results in

$$\mathbf{A}[\mathbf{x}_1 \ \ \mathbf{x}_2] = [\mathbf{x}_1\lambda_1 \ \ \mathbf{x}_2\lambda_2] = [\mathbf{x}_1 \ \ \mathbf{x}_2] \begin{bmatrix} -2 & 0 \\ 0 & -1 \end{bmatrix} \quad (A.5.6)$$

$$\mathbf{X} = [\mathbf{x}_1 \ \ \mathbf{x}_2] = \begin{bmatrix} 1 & 1 \\ -2 & -1 \end{bmatrix}, \quad \mathbf{X}^{-1} = \begin{bmatrix} -1 & 2 \\ 2 & -2 \end{bmatrix}.$$

(A.5.6) can be written in the form

$$\mathbf{A}\mathbf{X} = \mathbf{X}\Lambda \Rightarrow \mathbf{A} = \mathbf{X}\Lambda\mathbf{X}^{-1}. \quad (A.5.7)$$

The order of the eigenvalues can be arranged in the increasing or decreasing or any desired order. The vectors can be normalized so that the length of each of the vectors is 1. That is, for example, the length of $\mathbf{x}_2$ is $\sqrt{2}$. The corresponding normalized vector is $(\mathbf{x}_2/\sqrt{2})$. Arranging the eigenvalues in increasing values in terms of their magnitudes, with the normalized, the solution is

$$\mathbf{A} = \begin{bmatrix} 0.7071 & -0.4472 \\ -0.7071 & 0.8944 \end{bmatrix} \begin{bmatrix} -1 & 0 \\ 0 & -2 \end{bmatrix}$$

$$\times \begin{bmatrix} 2.8284 & 1.4142 \\ 2.2361 & 2.2361 \end{bmatrix} = \mathbf{X}\mathbf{D}\mathbf{X}^{-1}. \quad (A.5.8)$$

MATLAB code to determine the (A.5.8) is

$$[\mathbf{X}, \mathbf{D}] = eig(\mathbf{A}). \quad (A.5.9) \quad \blacksquare$$

In the above example, the matrix **A** has distinct roots. If a matrix has multiple roots, then these results need to be generalized. Every matrix **A** can be expressed in the *Jordan form* identified by

$$\mathbf{A} = \mathbf{XJX}^{-1}. \qquad (A.5.10)$$

The matrix **J** is *block diagonal*, where each block is a triangular matrix and is given by

$$
\begin{bmatrix}
\mathbf{J}_1 & & & & \\
& \mathbf{J}_2 & & & \\
& & \ddots & & \\
& & & \ddots & \\
& & & & \ddots & \\
& & & & & \mathbf{J}_K
\end{bmatrix},
$$

$$
\mathbf{J}_i =
\begin{bmatrix}
\lambda_i & 1 & & & \\
& \lambda_i & 1 & & \\
& & \ddots & \ddots & \\
& & & \ddots & \ddots \\
& & & & \lambda_i & 1 \\
& & & & & \lambda_i
\end{bmatrix}. \qquad (A.5.11)
$$

The matrix **J** is the *Jordan matrix*. If the eigenvalues are distinct, then **J** is a diagonal matrix consisting of the eigenvalues $\lambda_i, i = 1, 2, ..., n$ as diagonal entries. The complete analysis of this is beyond the scope here. In the following a simple example is given to illustrate the multiple root case. See Pearl (1973).

Example A.5.2 Express the following matrix in the form $\mathbf{A} = \mathbf{XJX}^{-1}$.

$$
\mathbf{A} =
\begin{bmatrix}
3 & 1 & -2 \\
-1 & -1 & 4 \\
0 & -1 & 3
\end{bmatrix}. \qquad (A.5.12)
$$

Solution: The characteristic polynomial is

$$
|\lambda \mathbf{I} - \mathbf{A}| =
\begin{vmatrix}
\lambda - 3 & -1 & 2 \\
-1 & \lambda + 1 & -4 \\
0 & 1 & \lambda - 3
\end{vmatrix}
$$

$$
= \lambda^3 - 5\lambda^2 + 8\lambda - 4 = (\lambda - 2)^2 (\lambda - 1). \qquad (A.5.13)
$$

It has a double root $\lambda = 2$ and a simple root at $\lambda = 1$. The eigenvector corresponding to the eigenvalue $\lambda = 2$ is

$$
(\mathbf{A} - \lambda_1 \mathbf{I})\mathbf{x}_1 |_{\lambda_1 = 2} =
\begin{bmatrix}
2 & 1 & -2 \\
-1 & -2 & 4 \\
0 & -1 & 2
\end{bmatrix}
\begin{bmatrix}
x_{11} \\
x_{21} \\
x_{31}
\end{bmatrix}
$$

$$
= 0 \Rightarrow \mathbf{x}_1 =
\begin{bmatrix}
1 \\
1 \\
1
\end{bmatrix}.
$$

Rank of $(\mathbf{A} - 2\mathbf{I})$ is 2 and there is only *one* characteristic vector of **A** (up to a multiplication within a constant) corresponding to the characteristic root 2 that satisfies the equation $(\mathbf{A} - 2\mathbf{I})\mathbf{x} = 0$. The eigenvector corresponding to the eigenvalue $\lambda = 1$ is

$$
(\mathbf{A} - \lambda_3 \mathbf{I})\mathbf{x}_3 |_{\lambda_3 = 1} =
\begin{bmatrix}
2 & 1 & -2 \\
-1 & -2 & 4 \\
0 & -1 & 2
\end{bmatrix}
\begin{bmatrix}
x_{13} \\
x_{23} \\
x_{33}
\end{bmatrix}
$$

$$
= 0 \Rightarrow \mathbf{x}_3 =
\begin{bmatrix}
0 \\
2 \\
1
\end{bmatrix}.
$$

There are only two characteristic vectors. The matrix **A** is not similar to a diagonal matrix and is similar to a Jordan matrix. A third-order transformation matrix is needed that transforms **A** to the Jordan form. Two vectors $\mathbf{x}_1$ and $\mathbf{x}_3$ are already determined. A third vector $\mathbf{x}_2$ is needed. Consider the following matrix equation:

$$\mathbf{A}[\mathbf{x}_1 \ \mathbf{x}_2 \ \mathbf{x}_3] = [\mathbf{x}_1 \ \mathbf{x}_2 \ \mathbf{x}_3]\mathbf{J}$$

$$
\begin{bmatrix}
3 & 1 & -2 \\
-1 & -1 & 4 \\
0 & -1 & 3
\end{bmatrix}
[\mathbf{x}_1 \ \mathbf{x}_2 \ \mathbf{x}_3] = [\mathbf{x}_1 \ \mathbf{x}_2 \ \mathbf{x}_3]
\begin{bmatrix}
2 & 1 & 0 \\
0 & 2 & 0 \\
0 & 0 & 1
\end{bmatrix}.
$$

The vector $\mathbf{x}_2$ can be obtained by equating the second column on each side. That is,

$$\begin{bmatrix} 3 & 1 & -2 \\ -1 & -1 & 4 \\ 0 & -1 & 3 \end{bmatrix} \mathbf{x}_2 = (1)\mathbf{x}_1 + (2)\mathbf{x}_2, \ \mathbf{x}_2 = \begin{bmatrix} x_{21} \\ x_{22} \\ x_{23} \end{bmatrix}$$

$$\Rightarrow \begin{bmatrix} 1 & 1 & -2 \\ -1 & -3 & 4 \\ 0 & -1 & 1 \end{bmatrix} \begin{bmatrix} x_{12} \\ x_{22} \\ x_{32} \end{bmatrix} = \begin{bmatrix} x_{11} \\ x_{21} \\ x_{31} \end{bmatrix} = \begin{bmatrix} 1 \\ 1 \\ 1 \end{bmatrix}.$$

A solution is

$$\mathbf{x}_2 = \begin{bmatrix} 1 \\ -2 \\ -1 \end{bmatrix}.$$

The transformation matrix $\mathbf{X}$ and the matrix decomposition $\mathbf{A} = \mathbf{XJX}^{-1}$ are, respectively, given by

$$\mathbf{X} = [\mathbf{x}_1 \ \ \mathbf{x}_2 \ \ \mathbf{x}_3] = \begin{bmatrix} 1 & 1 & 0 \\ 1 & -2 & 2 \\ 1 & -1 & 1 \end{bmatrix},$$

$$\mathbf{X}^{-1} = \begin{bmatrix} 0 & -1 & 2 \\ 1 & 1 & -2 \\ 1 & 2 & -3 \end{bmatrix},$$

$$\mathbf{A} = \mathbf{XJX}^{-1} = \begin{bmatrix} 1 & 1 & 0 \\ 1 & -2 & 2 \\ 1 & -1 & 1 \end{bmatrix} \begin{bmatrix} 2 & 1 & 0 \\ 0 & 2 & 0 \\ 0 & 0 & 1 \end{bmatrix} \begin{bmatrix} 0 & -1 & 2 \\ 1 & 1 & -2 \\ 1 & 2 & -3 \end{bmatrix}.$$
(A.5.14)

Notes: This example illustrated the computation of the transformation matrix $\mathbf{X}$ that transforms the given matrix $\mathbf{A}$ into a Jordan form in the form $\mathbf{J} = \mathbf{X}^{-1}\mathbf{AX}$. This is a *similarity transformation*. Finding the structure of the Jordan matrix is complicated, especially if the multiplicity of an eigenvalue is large. The transformation matrix $\mathbf{X}$ is the *right modal matrix* or simply the modal matrix of $\mathbf{A}$. Its inverse $\mathbf{X}^{-1}$ is the *left modal matrix* of $\mathbf{A}$. For a nice compact presentation on this topic, see Frame (1964). MATLAB code to determine the modal matrix and the Jordan matrix is

$$[\mathbf{X}, \mathbf{J}] = \text{jordan}(\mathbf{A}). \quad (A.5.15)$$

The output gives the modal matrix $\mathbf{X}$ and the Jordan matrix $\mathbf{J}$ given the matrix is $\mathbf{A}$.

Symmetric matrices are an important class of matrices in signal and system analysis. A real symmetric matrix $\mathbf{S} = \mathbf{S}^T$ is similar to a diagonal matrix. That is, there exists an orthogonal matrix $\mathbf{P}$ with $\mathbf{P}^{-1} = \mathbf{P}^T$, such that

$$\mathbf{S} = \mathbf{PDP}^T. \quad (A.5.16)$$

Ayres (1962) gives an example with simple numbers that illustrates the case of multiple eigenvalues, which is given below.

Example A.5.3 Find the eigenvalue–eigenvector decomposition of the symmetric matrix $\mathbf{S} = \mathbf{PDP}^T$.

$$\mathbf{S} = \begin{bmatrix} 7 & -2 & 1 \\ -2 & 10 & -2 \\ 1 & -2 & 7 \end{bmatrix}. \quad (A.5.17)$$

Solution: The eigenvalues of the matrix are

$$|\lambda \mathbf{I} - \mathbf{S}| = \lambda^3 - 24\lambda^2 + 180\lambda - 432$$
$$= (\lambda - \lambda_1)(\lambda - \lambda_2)(\lambda - \lambda_3)$$
$$= (\lambda - 6)^2(\lambda - 12).$$

Since $\lambda_1 = 6, \lambda_2 = 6, \lambda_3 = 12$, an extra step is needed to find the modal matrix, as there is a double root. For the eigenvalue $\lambda = 6$, the corresponding eigenvector is

$$(\mathbf{A} - \lambda_1)\mathbf{p}_1 = \begin{bmatrix} -1 & 2 & -1 \\ 2 & -4 & 2 \\ -1 & 2 & -1 \end{bmatrix} \begin{bmatrix} p_{11} \\ p_{12} \\ p_{13} \end{bmatrix} = 0$$

$$\Rightarrow \mathbf{p}_1 = \begin{bmatrix} p_{11} \\ p_{12} \\ p_{13} \end{bmatrix} = \begin{bmatrix} 1 \\ 0 \\ -1 \end{bmatrix}.$$

For the second eigenvalue, $\lambda_2 = 6$ a vector $\mathbf{p}_2$ needs to be determined that is orthogonal to the vector $\mathbf{p}_1$. That is $\mathbf{p}_1^T \mathbf{p}_2 = 0$. A solution is

$$\mathbf{p}_2 = \begin{bmatrix} p_{12} \\ p_{22} \\ p_{32} \end{bmatrix} = \begin{bmatrix} 1 \\ \alpha \\ 1 \end{bmatrix} = \begin{bmatrix} 1 \\ 1 \\ 1 \end{bmatrix}.$$

Note α is arbitrary in $\mathbf{p}_2$ and is selected as 1. The eigenvector corresponding to the eigenvalue 12 is

$$(\mathbf{A} - \lambda_3\mathbf{I})\mathbf{p}_3 = \begin{bmatrix} 5 & 2 & -1 \\ 2 & 2 & 2 \\ -1 & 2 & 5 \end{bmatrix} \begin{bmatrix} p_{31} \\ p_{32} \\ p_{33} \end{bmatrix} = 0$$

$$\Rightarrow \mathbf{p}_3 = \begin{bmatrix} 1 \\ -2 \\ 1 \end{bmatrix}.$$

The lengths of the three vectors defined earlier as L_2 norms are

$$\|\mathbf{p}_1\| = \sqrt{1^2 + 0^2 + 1^2} = \sqrt{2}, \|\mathbf{p}_2\| = \sqrt{1^2 + 1^2 + 1^2}$$
$$= \sqrt{3}, \|\mathbf{p}_3\| = \sqrt{1^2 + 2^2 + 1^2} = \sqrt{6}.$$

Putting these eigenvectors and normalizing the vectors by their lengths, the corresponding orthogonal transformation matrix is

$$\mathbf{P} = \begin{bmatrix} \mathbf{p}_1/\sqrt{2} & \mathbf{p}_2/\sqrt{3} & \mathbf{p}_3/\sqrt{6} \end{bmatrix}.$$

The eigenvalue–eigenvector decomposition of the matrix $\mathbf{S}$ is given by

$$\mathbf{S} = \mathbf{PDP}^T = \begin{bmatrix} 1/\sqrt{2} & 1/\sqrt{3} & 1/\sqrt{6} \\ 0 & 1/\sqrt{3} & -2/\sqrt{6} \\ -1/\sqrt{2} & 1/\sqrt{3} & 1/\sqrt{6} \end{bmatrix} \begin{bmatrix} 6 & 0 & 0 \\ 0 & 6 & 0 \\ 0 & 0 & 12 \end{bmatrix}$$

$$\times \begin{bmatrix} 1/\sqrt{2} & 0 & -1/\sqrt{2} \\ 1/\sqrt{3} & 1/\sqrt{3} & 1/\sqrt{3} \\ 1/\sqrt{6} & -2/\sqrt{6} & 1/\sqrt{6} \end{bmatrix}. \quad (A.5.18)$$

The decomposition is not unique since not all the diagonal entries in $\mathbf{D}$ are different. ∎

Example A.5.4 Determine the eigenvalue–eigenvector decompositions of the following matrices, where the matrix $\mathbf{A}$ is given in A.5.2.

$$a. \ \mathbf{B}_1 = \mathbf{AA}^T, \quad b. \ \mathbf{B}_2 = \mathbf{A}^T\mathbf{A}. \quad (A.5.19a)$$

Solution: The eigenvalue–eigenvector decompositions of such matrices are useful in the singular value decomposition in the next section. The computation is left as an exercise.

$$\mathbf{B}_1 = \begin{bmatrix} 1 & -3 \\ -3 & 13 \end{bmatrix}, \ \mathbf{B}_2 = \begin{bmatrix} 4 & 6 \\ 6 & 10 \end{bmatrix}. \quad (A.5.19b)$$

$$\mathbf{B}_1 = \mathbf{P}_1\Lambda_1\mathbf{P}_1^T = \begin{bmatrix} -0.9732 & -0.2298 \\ -0.2298 & 0.9732 \end{bmatrix}$$

$$\times \begin{bmatrix} 0.2918 & 0 \\ 0 & 13.7082 \end{bmatrix} \begin{bmatrix} -0.9732 & -0.2298 \\ -0.2298 & 0.9732 \end{bmatrix}. \quad (A.5.20a)$$

$$\mathbf{B}_2 = \mathbf{P}_2\Lambda_2\mathbf{P}_2^T = \begin{bmatrix} -0.8507 & 0.5257 \\ 0.5257 & 0.8507 \end{bmatrix}$$

$$\times \begin{bmatrix} 0.2918 & 0 \\ 0 & 13.7082 \end{bmatrix} \begin{bmatrix} -0.8507 & 0.5257 \\ 0.5257 & 0.8507 \end{bmatrix}. \quad (A.5.20b) \ \blacksquare$$

Cayley–Hamilton theorem Horn and Johnson (1990) is stated below without proof.

Cayley–Hamilton theorem: Consider an $n \times n$ matrix $\mathbf{A}$ and its characteristic polynomial $d(\lambda) = |\lambda\mathbf{I} - \mathbf{A}| = 0$. Every square matrix satisfies its characteristic polynomial. That is,

$$d(\mathbf{A}) = 0. \quad (A.5.21)$$

Example A.5.5 *a*. Show that the matrix given in (A.5.2) satisfies its characteristic polynomial $d(\lambda) = \lambda^2 + 3\lambda + 2 = 0$. *b*. Find the inverse of the matrix $\mathbf{A}$ using (A.5.21).

Solution: *a*. $d(\mathbf{A}) = \mathbf{A}^2 + 3\mathbf{A} + 2\mathbf{I} = \begin{bmatrix} -2 & -3 \\ 6 & 7 \end{bmatrix}$

$$+ \begin{bmatrix} 0 & 3 \\ -6 & -9 \end{bmatrix} + \begin{bmatrix} 2 & 0 \\ 0 & 2 \end{bmatrix} = \begin{bmatrix} 0 & 0 \\ 0 & 0 \end{bmatrix}. \quad (A.5.22)$$

b. The polynomial matrix $d(\mathbf{A})$ can be written in the form

$$\left[-\frac{1}{2}\mathbf{A}^2 - \frac{3}{2}\mathbf{A} \right] = \mathbf{I} \text{ or } \mathbf{A}\left[-\frac{1}{2}\mathbf{A} - \frac{3}{2}\mathbf{I} \right] = \mathbf{I} \Rightarrow \mathbf{A}^{-1}$$

$$= \left[-\frac{1}{2}\mathbf{A} - \frac{3}{2}\mathbf{I} \right].$$

$$\mathbf{A}^{-1} = -\begin{bmatrix} 0 & 1/2 \\ -1 & -3/2 \end{bmatrix} - \begin{bmatrix} 3/2 & 0 \\ 0 & 3/2 \end{bmatrix}$$

$$= \begin{bmatrix} -3/2 & -1/2 \\ 1 & 0 \end{bmatrix}. \quad (A.5.23)$$

Inverse exists here since $|\mathbf{A}| \neq 0$. This theorem can be used to find integer powers of $\mathbf{A}$. ∎

Notes: Section A.2.1 considered the L_p norms of vectors. An important matrix norm is the L_2 norm. The L_2 norm of an $n \times n$ matrix $\mathbf{A}$ is

$\|\mathbf{A}\|_2 = [\text{maximum characteristic root of } \mathbf{A}^*\mathbf{A}]^{1/2}.$

$$(A.5.24)$$

This is also called the *spectral norm*. Note the determinant of the matrix $\mathbf{A}$ is the product of its characteristic roots.   ∎

A.6 Singular Value Decomposition (SVD)

Singular value decomposition (SVD) is an important tool in signal processing, as it provides robust solutions in spectral analysis, filter design, system identification, estimation theory, applications in statistics, and many others. See Scharf (1991) for applications.

Given a $m \times n$ matrix $\mathbf{A}$ of rank r, there exists three matrices identified as follows: an $n \times n$ unitary matrix $\mathbf{V}$; $m \times m$ unitary matrix $\mathbf{U}$; and a $r \times r$ diagonal matrix $\mathbf{D}$ with the diagonal entries that are *strictly positive*; and the SVD is given by

$$\mathbf{A} = \mathbf{U}\Sigma\mathbf{V}^*, \quad \Sigma = \begin{bmatrix} \mathbf{D} & \mathbf{0} \\ \mathbf{0} & \mathbf{0} \end{bmatrix}. \quad (A.6.1)$$

If the rank of the matrix $\mathbf{A}$ is $r = m = n$ then the matrix Σ reduces to the simple case $\Sigma = \mathbf{D}$ and the diagonal elements in $\mathbf{D}$ are strictly positive; that is, $d_{ii} > 0$. The decomposition is valid for both real and complex matrices. For a proof and a general discussion, see Pearl (1973). The SVD decomposition can be determined as follows. First, find the eigenvalue–eigenvector decomposition of the two matrices

$$(\mathbf{A}^*\mathbf{A}) = \mathbf{V}(\Sigma^*\Sigma)\mathbf{V}^* \text{ and } (\mathbf{A}\mathbf{A}^*) = \mathbf{U}(\Sigma\Sigma^*)\mathbf{U}^*.$$

$$(A.6.2)$$

Note $(\mathbf{A}\mathbf{A}^*)$ and $(\mathbf{A}^*\mathbf{A})$ have the *same nonzero eigenvalues*. See Example A.5.4 for an illustration. Equation (A.6.2) indicates that the unitary matrices $\mathbf{U}$ and $\mathbf{V}$ are modal matrices of $(\mathbf{A}\mathbf{A}^*)$ and $(\mathbf{A}^*\mathbf{A})$, respectively, thus providing a way to determine these matrices. Let the nonzero eigenvalues of $(\mathbf{A}^*\mathbf{A})$ be $\sigma_1^2, \sigma_2^2, ..., \sigma_r^2$. The diagonal entries of Σ are called the *singular values* of the matrix $\mathbf{A}$ and are ordered as

$$\sigma_1 \geq \sigma_2 \geq ... \geq \sigma_r > 0. \quad (A.6.3)$$

They are determined from the diagonal matrix $(\Sigma\Sigma^*)$. Singular values are fairly insensitive to perturbations in the matrix $\mathbf{A}$ compared to its eigenvalues. Noting that $\mathbf{A} = \mathbf{U}\Sigma\mathbf{V}^*$ and assuming Σ is a nonsingular matrix, we have $\Sigma^{-1} = \mathbf{D}^{-1}$. Then

$$\mathbf{U} = \mathbf{A}\mathbf{V}\Sigma^{-1} = \mathbf{A}\mathbf{V}\mathbf{D}^{-1}. \quad (A.6.4)$$

The SVD can be determined by first finding $\mathbf{V}$, Σ^{-1} and then $\mathbf{U}$ by (A.6.4). If the matrix Σ is singular, then its inverse will be its generalized inverse discussed below.

Example A.6.1 Determine the SVD for the real matrix $\mathbf{A}$ given in (A.5.2).

Solution: The matrices $(\mathbf{A}\mathbf{A}^T)$ and $(\mathbf{A}^T\mathbf{A})$ are, respectively, given by

$$(\mathbf{A}\mathbf{A}^T) = \begin{bmatrix} 1 & -3 \\ -3 & 13 \end{bmatrix}, (\mathbf{A}^T\mathbf{A}) = \begin{bmatrix} 4 & 6 \\ 6 & 10 \end{bmatrix}. \quad (A.6.5a)$$

Eigenvalue–eigenvector decomposition of $\mathbf{A}^T\mathbf{A}$:

$$\mathbf{A}^T\mathbf{A} = \begin{bmatrix} 4 & 6 \\ 6 & 10 \end{bmatrix} = \mathbf{V}\mathbf{D}^T\mathbf{D}\mathbf{V}^T = \begin{bmatrix} -0.8507 & 0.5257 \\ 0.5257 & 0.8507 \end{bmatrix}$$

$$\times \begin{bmatrix} 0.2918 & 0 \\ 0 & 13.7082 \end{bmatrix} \begin{bmatrix} -0.8507 & 0.5257 \\ 0.5257 & 0.8507 \end{bmatrix}$$

$$= \begin{bmatrix} 0.5257 & -0.8507 \\ 0.8507 & 0.5257 \end{bmatrix} \begin{bmatrix} 13.7082 & 0 \\ 0 & 0.2918 \end{bmatrix}$$

$$\times \begin{bmatrix} 0.5257 & 0.8507 \\ -0.8507 & 0.5257 \end{bmatrix}.$$

Using the positive square roots of the diagonal entries of $\mathbf{D}\mathbf{D}^T$ and computing $\mathbf{D}^{-1}$ results in

$$\mathbf{U} = \mathbf{A}\mathbf{V}\mathbf{D}^{-1} =$$

$$\begin{bmatrix} 0 & 1 \\ -2 & -3 \end{bmatrix} \begin{bmatrix} -0.8507 & 0.5257 \\ 0.5257 & 0.8507 \end{bmatrix} \begin{bmatrix} 0.2701 & 0 \\ 0 & 1.8512 \end{bmatrix}$$

$$= \begin{bmatrix} -0.2298 & 0.9732 \\ 0.9732 & 0.2298 \end{bmatrix}.$$

Arranging the singular values in decreasing order in the SVD of the nonsingular matrix $\mathbf{A}$, the SVD of $\mathbf{A}$ is

$$\mathbf{A} = \mathbf{U}\mathbf{D}\mathbf{V}^T = \begin{bmatrix} -.2298 & .9732 \\ .9732 & .2298 \end{bmatrix} \begin{bmatrix} 3.7025 & 0 \\ 0 & .5402 \end{bmatrix}$$

$$\times \begin{bmatrix} -.5257 & -.8507 \\ -.8507 & .5257 \end{bmatrix}. \quad (A.6.5b)$$

The singular values are 3.7025 and 0.5402. MATLAB command for the SVD decomposition is

$$[\mathbf{U}, \Sigma, \mathbf{V}] = \text{svd}(\mathbf{A}). \qquad (A.6.6) \quad \blacksquare$$

When $\mathbf{A}$ is not a square matrix, the SVD requires some knowledge of *generalized* or *conditional inverses*. See Rao and Mitra (1971) and Graybill (1983).

A.7 Generalized Inverses of Matrices

Given a matrix $\mathbf{A}$ of dimension $m \times n$ with rank r, we like to find a *generalized inverse matrix* $\mathbf{A}^-$ which has some of the properties of $\mathbf{A}^{-1}$. The matrix $\mathbf{A}^-$ is also called as *pseudo inverse* or *Moore–Penrose inverse*. The conditions for existence are

1. $\mathbf{AA}^-$ is symmetric $\qquad\qquad$ (A.7.1a)

2. $\mathbf{A}^-\mathbf{A}$ is symmetric $\qquad\qquad$ (A.7.1b)

3. $\mathbf{AA}^-\mathbf{A} = \mathbf{A}$ $\qquad\qquad\qquad$ (A.7.1c)

4. $\mathbf{A}^-\mathbf{AA}^- = \mathbf{A}^-$. $\qquad\qquad$ (A.7.1d)

If a generalized inverse of a $m \times n$ matrix $\mathbf{A}$ exists, then its inverse $\mathbf{A}^-$ has the dimension $n \times m$ and $(\mathbf{A}^-)^- = \mathbf{A}$. Each matrix has a unique generalized inverse. The generalized inverse of a null matrix is also a null matrix. The discussion here will be concentrated on matrices that are of full rank. The generalized inverses are as follows:

Rank of $\mathbf{A}$ is m, then $\mathbf{A}^- = \mathbf{A}^T(\mathbf{AA}^T)^{-1}$ and $\mathbf{AA}^- = \mathbf{I}$

$$(A.7.2a)$$

Rank of $\mathbf{A}$ is n, then $\mathbf{A}^- = (\mathbf{A}^T\mathbf{A})^{-1}\mathbf{A}$ and $\mathbf{A}^-\mathbf{A} = \mathbf{I}$

$$(A.7.2b)$$

In the case of a nonsingular square matrix $\mathbf{A}$, it follows that

$$\mathbf{A}^- = \mathbf{A}^T(\mathbf{AA}^T)^{-1} = \mathbf{A}^T(\mathbf{A}^T)^{-1}\mathbf{A}^{-1} = \mathbf{A}^{-1}. \quad (A.7.2c)$$

MATLAB command for the generalized inverse of $\mathbf{A}$ is

$$\mathbf{A}^- = \text{pinv}(\mathbf{A}). \qquad (A.7.2d)$$

Example A.7.1 Find the generalized inverses for the following matrices:

$$a.\ \mathbf{A} = \begin{bmatrix} 1 & 0 \\ 2 & 1 \\ 0 & 2 \end{bmatrix},$$

$$b.\ \mathbf{B} = \begin{bmatrix} 1 & 2 & 0 \\ 0 & 1 & 2 \end{bmatrix},$$

$$c.\ \Sigma_1 = \begin{bmatrix} 1 & 0 \\ 0 & 2 \\ 0 & 0 \end{bmatrix},$$

$$d.\ \Sigma_2 = \begin{bmatrix} 1 & 0 & 0 \\ 0 & 2 & 0 \end{bmatrix},$$

$$e.\ \Sigma = \begin{bmatrix} \mathbf{D} & \mathbf{0} \\ \mathbf{0} & \mathbf{0} \end{bmatrix}, \qquad (A.7.3a)$$

$\mathbf{D}$ is a diagonal matrix.

Solution: *a*. The generalized inverse of the matrix $\mathbf{A}$ is

$$\begin{aligned} \mathbf{A}^- = (\mathbf{A}^T\mathbf{A})^{-1}\mathbf{A} &= \begin{bmatrix} 5 & 2 \\ 2 & 5 \end{bmatrix}^{-1} \begin{bmatrix} 1 & 2 & 0 \\ 0 & 1 & 2 \end{bmatrix} \\ &= \frac{1}{21} \begin{bmatrix} 5 & 8 & -4 \\ -2 & 1 & 10 \end{bmatrix} \\ &= \begin{bmatrix} 0.2381 & 0.3810 & -0.1905 \\ -0.0952 & 0.0476 & 0.4762 \end{bmatrix}. \quad (A.7.3b) \end{aligned}$$

b. Noting that $\mathbf{B} = \mathbf{A}^T$, it follows:

$$\mathbf{B}^- = \begin{bmatrix} 0.2381 & -0.0952 \\ 0.3810 & 0.0476 \\ -0.1905 & 0.4762 \end{bmatrix}. \quad (A.7.3c)$$

Similarly,

$$\Sigma_1^{-1} = \begin{bmatrix} 1 & 0 \\ 0 & 1/2 \\ 0 & 0 \end{bmatrix}, \Sigma_2^{-1} = \begin{bmatrix} 1 & 0 & 0 \\ 0 & 1/2 & 0 \end{bmatrix},$$

$$\Sigma^{-1} = \begin{bmatrix} \mathbf{D}^{-1} & \mathbf{0} \\ \mathbf{0} & \mathbf{0} \end{bmatrix}. \qquad (A.7.3d) \quad \blacksquare$$

Notes: Singular value matrix decomposition is often used in computing the generalized inverses and ranks of *defective matrices*. See Stewart (1973).

A.8 Over- and Underdetermined System of Equations

Consider the system of equations with real entries

$$\mathbf{y} = \mathbf{A}\hat{\mathbf{x}} + \mathbf{e}. \qquad (A.8.1)$$

In (A.8.1) $\mathbf{y}$ is a known m-dimensional known vector, $\hat{\mathbf{x}}$ is an unknown n-dimensional vector, $\mathbf{A}$ is a known $m \times n$-dimensional matrix, and $\mathbf{e}$ is an m-dimensional unknown error vector. Given an estimated vector $\mathbf{y}$, $\hat{\mathbf{y}} + \mathbf{e}$, we are interested in finding the vector $\hat{\mathbf{x}}$ from the set of equations

$$\hat{\mathbf{y}} = \mathbf{y} - \mathbf{e} = \mathbf{A}\hat{\mathbf{x}}. \qquad (A.8.2)$$

by minimizing the error vector for the overdetermined and underdetermined cases

$$a.\, m > n, \quad b.\, m \le n. \qquad (A.8.3)$$

A. 8.1 Least-Squares Solutions of Overdetermined System of Equations ($m > n$)

In the case of overdetermined system of equations, there are more equations than unknowns. Except in trivial cases there is no solution for such a system. If no solution exists, then a solution is desired that minimizes the error using some error measures. The error vector is

$$\mathbf{e} = \mathbf{y} - \hat{\mathbf{y}} = \mathbf{y} - \mathbf{A}\hat{\mathbf{x}} = \begin{bmatrix} y_1 - \hat{y}_1 \\ y_2 - \hat{y}_2 \\ . \\ . \\ . \\ y_m - \hat{y}_m \end{bmatrix}. \qquad (A.8.4)$$

The errors can only be considered in statistical terms. L_p, $1 \le p \le \infty$ errors were considered in Section 9.3 and are

$$E_p = \sum_{i=1}^{m} |(y_i - \hat{y}_i|^p. \qquad (A.8.5)$$

The simplest of these is the L_2, i.e., the *least-squares error measure*, which assumes the error density function is Gaussian. It can be expressed in the form

$$E_2 = (\mathbf{y} - \mathbf{A}\hat{\mathbf{x}})^T (\mathbf{y} - \mathbf{A}\hat{\mathbf{x}}). \qquad (A.8.6)$$

Example A.8.2 Consider the system of equations given below. Find the error measure in (A.8.6) and find the solution by minimizing the least-squares error:

$$\mathbf{y} = \begin{bmatrix} 9 \\ 2 \\ 5 \end{bmatrix} = \begin{bmatrix} 1 \\ -1 \\ 1 \end{bmatrix} \hat{x} = \mathbf{A}\hat{\mathbf{x}}. \qquad (A.8.7)$$

Solution: There are three equations in one unknown and there is no solution of $\mathbf{x}$ that satisfies all the equations. The error vector and its L_2 measure are

$$\mathbf{e} = (\mathbf{y} - \mathbf{A}\hat{\mathbf{x}}) = \begin{bmatrix} 9 - \hat{x} \\ 2 + \hat{x} \\ 5 - x \end{bmatrix}, \qquad (A.8.8)$$

$$E_2 = \mathbf{e}^T \mathbf{e} = (9 - \hat{x})^2 + (2 + \hat{x})^2 + (5 - \hat{x})^2. \quad (A.8.9)$$

The *least-squares* error is minimized by taking the partial of E_2 with respect to the unknown vector (in this case, a scalar x) and solve for the vector $\mathbf{x} = x$. Taking the partial with respect to x and solving for it results in

$$\frac{\partial E_2}{\partial \hat{x}} = -2(9 - \hat{x}) + 2(2 + \hat{x}) - 2(5 - \hat{x})$$
$$= 0 \Rightarrow x = \hat{x} = 4. \qquad (A.8.10)$$

Note that *least-squares solution* does not satisfy any one of the equations in (A.8.7). The least-squares solution is machine like. Suppose there is a fork in the road and like to get an answer to decide which road to take. Least-squares solution suggests going between the two roads, i.e., go into bushes! For this reason L_1 measure is considered superior in some cases. Moon and Stirling (2000) suggest the selection of p on the basis of error density function. Least-squares solution gives good results in most applications. It can be used to obtain other L_p solutions using iterative algorithms, such as *Iteratively*

Reweighted Least Squares (IRLS) algorithm. See Byrd and Payne (1979).   ∎

Least-squares solution in general terms: consider

$$\sum_{i=1}^{n} a_{ji}\hat{x}_i - y_j = e_j, \; j = 1, 2, ..., m. \qquad (A.8.11)$$

Our goal is to choose $\hat{x}_i's$ so that

$$E_2 = \sum_{j=1}^{m} e_j^2 = \text{minimum}. \qquad (A.8.12)$$

Taking the partial derivatives with respect to $\hat{x}_i$ and equating them to zero results in

$$\frac{\partial}{\partial x_k} \sum_{j=1}^{m} e_j^2 = 0 \rightarrow \sum_{j=1}^{m} e_j \frac{\partial e_j}{\partial x_k} = 0, \; k = 1, 2, ..., n.$$
$$(A.8.13)$$

From (A.8.11), it follows that

$$\frac{\partial e_j}{\partial x_k} = a_{kj}. \qquad (A.8.14a)$$

Using these in (A.8.13) results in

$$\sum_{j=1}^{m} \left[\sum_{i=1}^{n} \hat{x}_i a_{ji} - y_j \right] a_{jk} = 0, \, k = 1, 2, ..., m. \quad (A.8.14b)$$

This can be written in the form

$$\sum_{i=1}^{n} \hat{x}_i \left[\sum_{j=1}^{m} a_{ji} a_{jk} \right] = \sum_{j=1}^{m} y_j a_{jk}, \; k = 1, 2, ..., n.$$
$$(A.8.14c)$$

This is a system of n linear equations in n unknowns $\hat{x}_1, \hat{x}_2, ..., \hat{x}_n$ and can use any method to compute the solution. Considering the double summations in the above equation, it is hard to visualize the above analysis and the following example illustrates the ideas.

Example A.8.2 Illustrate the above procedure using the following system of equations:

$$\begin{bmatrix} a_{11} & a_{12} \\ a_{21} & a_{22} \\ a_{31} & a_{32} \end{bmatrix} \begin{bmatrix} \hat{x}_1 \\ \hat{x}_2 \end{bmatrix} = \begin{bmatrix} y_1 \\ y_2 \\ y_3 \end{bmatrix} \Rightarrow \mathbf{y} = \mathbf{A}\hat{\mathbf{x}}, \; m = 3, n = 2.$$
$$(A.8.15)$$

Find the best least-squares solution for this over-determined system of equations.

Solution: First

$$a_{11}\hat{x}_1 + a_{12}\hat{x}_2 - y_1 = e_1$$
$$a_{21}\hat{x}_1 + a_{22}\hat{x}_2 - y_2 = e_2 \, .$$
$$a_{31}\hat{x}_1 + a_{32}\hat{x}_2 - y_3 = e_3$$

$$e_1^2 + e_2^2 + e_3^2 = \text{minimum}$$

Using (A.8.14),

$$k = 1: \; \left(\sum_{j=1}^{m=3} a_{j1}a_{j1} \right) \hat{x}_1 + \left(\sum_{j=1}^{m=3} a_{j1}a_{j2} \right) \hat{x}_2 = \sum_{j=1}^{m=3} a_{j1}y_j$$

$$k = 2: \; \left(\sum_{j=1}^{m=3} a_{j2}a_{j1} \right) \hat{x}_1 + \left(\sum_{j=1}^{m=3} a_{j2}a_{j2} \right) \hat{x}_2 = \sum_{j=1}^{m=3} a_{2j}y_j.$$

These equations can be written in the following forms:

$$\begin{bmatrix} a_{11}a_{11} + a_{21}a_{21} + a_{31}a_{31} & a_{11}a_{12} + a_{21}a_{22} + a_{31}a_{32} \\ a_{12}a_{11} + a_{22}a_{21} + a_{32}a_{31} & a_{12}a_{12} + a_{22}a_{22} + a_{32}a_{32} \end{bmatrix} \begin{bmatrix} \hat{x}_1 \\ \hat{x}_2 \end{bmatrix} = \begin{bmatrix} a_{11}y_1 + a_1 y_2 + a_{31}y_3 \\ a_{12}y_1 + a_{22}y_2 + a_{32}y_3 \end{bmatrix}.$$

$$\Rightarrow \begin{bmatrix} a_{11} & a_{21} & a_{31} \\ a_{12} & a_{22} & a_{32} \end{bmatrix} \begin{bmatrix} a_{11} & a_{12} \\ a_{21} & a_{22} \\ a_{31} & a_{32} \end{bmatrix} \begin{bmatrix} \hat{x}_1 \\ \hat{x}_2 \end{bmatrix} = \begin{bmatrix} a_{11} & a_{21} & a_{31} \\ a_{12} & a_{22} & a_{32} \end{bmatrix} \begin{bmatrix} y_1 \\ y_2 \\ y_3 \end{bmatrix} \Rightarrow \mathbf{A}^T\mathbf{A}\hat{\mathbf{x}} = \mathbf{A}^T\mathbf{y}. \qquad (A.8.16a)$$

This set of equations is referred to as the *normal equations*. Interestingly, in obtaining the last equation, first, multiply the equation in (A.8.15) by $\mathbf{A}^T$, find the inverse of the matrix $(\mathbf{A}^T\mathbf{A})$ and then determine $\mathbf{x}$ by

$$\mathbf{x} = (\mathbf{A}^T\mathbf{A})^{-1}\mathbf{A}^T\mathbf{y}. \qquad (A.8.16b)$$

The matrix $\mathbf{A}^- = (\mathbf{A}^T\mathbf{A})^{-1}\mathbf{A}$ is the *generalized inverse* discussed earlier. Inverse of the matrix $\mathbf{A}\mathbf{A}^T$ may *not always* exist. This happens in real data and, in these cases, $(\mathbf{A}^T\mathbf{A})$ is replaced by $(\mathbf{A}^T\mathbf{A}) + \delta\mathbf{I}$, where δ is some small positive number. This is referred to as *diagonal loading*.

Example A.8.3 Use the equation in (A.8.16b) to verify the solution given in (A.8.10).

Solution:

$$(\mathbf{A}^T\mathbf{A}) = [1 \quad -1 \quad 1]\begin{bmatrix} 1 \\ -1 \\ 1 \end{bmatrix} = 3,$$

$$\hat{\mathbf{x}} = (\mathbf{A}^T\mathbf{A})^{-1}\mathbf{A}^T\mathbf{y} = \frac{1}{3}[1 \quad -1 \quad 1]\begin{bmatrix} 9 \\ 2 \\ 5 \end{bmatrix} = 4. \quad \blacksquare$$

Solutions of overdetermined system of equations are often used to reduce the effects of random noise in deconvolution (see Section 9.3.4) and in curve fitting. See Moon and Sterling (2000), and others.

A.8.2 Least-Squares Solution of Underdetermined System of Equations ($m \le n$)

Here, the number of equations m is assumed to be equal or fewer than the number of unknown's n. The following assumes the rank of the matrix $\mathbf{A}$ is m. Consider the matrix equation in (A.8.2)

$$\mathbf{y} = \mathbf{A}\hat{\mathbf{x}}. \qquad (A.8.17)$$

If $m = n$ and the determinant of $\mathbf{A} \ne 0$, the solution is $\hat{\mathbf{x}} = \mathbf{A}^{-1}\mathbf{y}$. In the case of $m < n$, there are an infinite number of solutions and like to find a vector $\hat{\mathbf{x}}$ with

$$E_2 = \hat{\mathbf{x}}^T\hat{\mathbf{x}}, \qquad (A.8.18)$$

being minimal. This is the *minimum energy condition* and the corresponding solution is the least-squares solution. There are two sets of equations to consider, one is the given underdetermined system of equations and the other is the minimum energy condition. The problem is to find $\hat{\mathbf{x}}$ in (A.8.17) using the minimum

error condition with E_2 being minimum. In the first step, the set of equations is written in the form

$$\mathbf{y} = \mathbf{A}\hat{\mathbf{x}} = [\mathbf{A}_{11} \quad \mathbf{A}_{12}]\begin{bmatrix} \hat{\mathbf{x}}_1 \\ \hat{\mathbf{x}}_2 \end{bmatrix}. \qquad (A.8.19)$$

Note $\mathbf{A}_{11}$ is a $m \times m$ nonsingular matrix. This involves rearranging the columns in the matrix $\mathbf{A}$ and the corresponding entries in the vector $\hat{\mathbf{x}}$. A *particular solution* is

$$\hat{\mathbf{x}}_1 = \mathbf{A}_{11}^{-1}\mathbf{y} - \mathbf{A}_{11}^{-1}\mathbf{A}_{12}\hat{\mathbf{x}}_2, \ \hat{\mathbf{x}}_2 = \mathbf{0}.$$

These two equations can be written in the matrix form

$$\hat{\mathbf{x}} = \begin{bmatrix} \hat{\mathbf{x}}_1 \\ \hat{\mathbf{x}}_2 \end{bmatrix} = \begin{bmatrix} \mathbf{A}_{11}^{-1}\mathbf{y} - \mathbf{A}_{11}^{-1}\mathbf{A}_{12}\hat{\mathbf{x}}_2 \\ \mathbf{0} \end{bmatrix}. \qquad (A.8.20)$$

This can be treated as an overdetermined system of equations. Minimization of the scalar function $E_2 = \hat{\mathbf{x}}_1^T\hat{\mathbf{x}}_1 + \hat{\mathbf{x}}_2^T\hat{\mathbf{x}}_2$ involves the solution of the over-determined system of equations given by

$$\begin{bmatrix} \mathbf{A}_{11}^{-1}\mathbf{A}_{12} \\ \mathbf{I} \end{bmatrix}\hat{\mathbf{x}}_2 = \begin{bmatrix} \mathbf{A}_{11}^{-1}\mathbf{y} \\ \mathbf{0} \end{bmatrix}. \qquad (A.8.21)$$

First, solve for $\hat{\mathbf{x}}_2$ using the least-squares solution of the overdetermined system of equations in (A.8.21) and then use $\hat{\mathbf{x}}_1 = \mathbf{A}_{11}^{-1}\mathbf{y} - \mathbf{A}_{11}^{-1}\mathbf{A}_{12}\hat{\mathbf{x}}_2$ to determine $\hat{\mathbf{x}}_1$.

Least-squares solution using generalized inverses: As mentioned earlier, there are many solutions for underdetermined system of equations. It can be seen the solution $\hat{\mathbf{x}} = {}^-\mathbf{y}$ satisfies the matrix equation $\mathbf{y} = \mathbf{A}\hat{\mathbf{x}}$. That is,

$$\mathbf{y} = \mathbf{A}\hat{\mathbf{x}} = \mathbf{A}[\mathbf{A}^-\mathbf{y}] = \mathbf{A}\mathbf{A}^T(\mathbf{A}\mathbf{A}^T)^{-1}\mathbf{y} = \mathbf{y}. \quad (A.8.22)$$

Example A.8.4 Find the least-squares solution of the system of equations given below using (A.8.20) and then verify the results using the generalized inverse of the matrix $\mathbf{A}$.

$$\mathbf{y} = \begin{bmatrix} 1 \\ 1 \end{bmatrix} = \begin{bmatrix} 1 & 0 & 2 \\ 0 & 1 & -1 \end{bmatrix}\begin{bmatrix} \hat{x}_1 \\ \hat{x}_2 \\ \hat{x}_3 \end{bmatrix} = \mathbf{A}\hat{\mathbf{x}}. \quad (A.8.23)$$

Solution: First

$$\mathbf{y} = \begin{bmatrix} 1 \\ 1 \end{bmatrix} = \begin{bmatrix} 1 & 0 & 2 \\ 0 & 1 & -1 \end{bmatrix} \begin{bmatrix} \hat{x}_1 \\ \hat{x}_2 \\ \hat{x}_3 \end{bmatrix} = [\mathbf{A}_{11}|\mathbf{A}_{12}] \begin{bmatrix} \hat{x}_1 \\ \hat{x}_2 \\ \hat{x}_3 \end{bmatrix},$$

$$\mathbf{A}_{11} = \mathbf{I}, \mathbf{A}_{12} = \begin{bmatrix} 2 \\ -1 \end{bmatrix}, \hat{\mathbf{x}}_1 = \begin{bmatrix} x_1 \\ x_2 \end{bmatrix}, \hat{\mathbf{x}}_2 = \hat{x}_3$$

Using (A.8.21) results in

$$\mathbf{A}_{11}^{-1}\mathbf{A}_{12} = \begin{bmatrix} 2 \\ -1 \end{bmatrix}, \begin{bmatrix} 2 \\ -1 \\ 1 \end{bmatrix}\hat{x}_3 = \begin{bmatrix} 1 \\ 1 \\ 0 \end{bmatrix}. \quad (A.8.24)$$

Now solve the above set of overdetermined system of equations on the right for $\hat{x}_3$. It is

$$\hat{x}_3 = \frac{1}{6}\begin{bmatrix} 2 & -1 & -1 \end{bmatrix}\begin{bmatrix} 1 \\ 1 \\ 0 \end{bmatrix}\frac{1}{6}.$$

Using $\hat{\mathbf{x}}_1 = \mathbf{A}_{11}^{-1}\mathbf{y} - \mathbf{A}_{11}^{-1}\mathbf{A}_{12}\hat{\mathbf{x}}_2$ (see the first set of equations in (A.8.20).) results in

$$\begin{bmatrix} \hat{x}_1 \\ \hat{x}_2 \\ \hat{x}_3 \end{bmatrix} = \begin{bmatrix} -2 \\ 1 \\ 1 \end{bmatrix}(1/6) + \begin{bmatrix} 1 \\ 1 \\ 0 \end{bmatrix} = \begin{bmatrix} 2/3 \\ 7/6 \\ 1/6 \end{bmatrix} \quad (A.8.25)$$

The solution can be verified using the generalized inverse $\mathbf{A}^-$. That is,

$$\mathbf{A} = \begin{bmatrix} 1 & 0 & 2 \\ 0 & 1 & -1 \end{bmatrix}, (\mathbf{A}\mathbf{A}^T)^{-1} = \begin{bmatrix} 5 & -2 \\ -2 & 2 \end{bmatrix}^{-1} = \frac{1}{6}\begin{bmatrix} 2 & 2 \\ 2 & 5 \end{bmatrix},$$

$$\mathbf{A}^- = \mathbf{A}^T(\mathbf{A}\mathbf{A}^T)^{-1} = \begin{bmatrix} 1 & 0 \\ 0 & 1 \\ 2 & -1 \end{bmatrix}\frac{1}{6}\begin{bmatrix} 2 & 2 \\ 2 & 5 \end{bmatrix} = \begin{bmatrix} 1/3 & 1/3 \\ 1/3 & 5/6 \\ 1/3 & -1/6 \end{bmatrix},$$

$$\hat{\mathbf{x}} = \mathbf{A}^T(\mathbf{A}\mathbf{A}^T)^{-1}\mathbf{y} = \begin{bmatrix} 1 & 0 \\ 0 & 1 \\ 2 & -1 \end{bmatrix}\frac{1}{6}\begin{bmatrix} 2 & 2 \\ 2 & 5 \end{bmatrix}\begin{bmatrix} 1 \\ 1 \end{bmatrix}$$

$$= \begin{bmatrix} 1/3 & 1/3 \\ 1/3 & 5/6 \\ 1/3 & -1/6 \end{bmatrix}\begin{bmatrix} 1 \\ 1 \end{bmatrix} = \begin{bmatrix} 2/3 \\ 7/6 \\ 1/6 \end{bmatrix}.$$

$$\Rightarrow \hat{x}_1 = 2/3, \hat{x}_2 = 7/6, \hat{x}_3 = 1/6. \quad (A.8.26) \quad \blacksquare$$

A.9 Numerical-Based Interpolations: Polynomial and Lagrange Interpolations

In Section 8.2 spectrally based interpolation methods were considered. The methods presented here do not use the spectrum of the signal. Taylor's series expansion of a continuous function $x(t)$ with continuous derivatives was considered earlier (see (3.1.2)). It uses its value at $t = a$ and the derivatives of $x(t)$ at this location. In many cases they may not be known or may not even exist. Following methods use the values of the function at discrete locations and the function is approximated by a polynomial that matches exactly at the discrete locations. Discussion on this topic is minimal and is included here to complement spectral-based interpolations.

A.9.1 Polynomial Approximations

A function $x(t)$ is assumed to be known at discrete locations $t_0 < t_1 < \cdots < t_{N-1}$ by $x(t_0), x(t_1), \ldots, x(t_{N-1})$. Consider the approximation function $x_a(t)$

$$x_a(t) = \sum_{k=0}^{N-1} a_k t^k. \quad (A.9.1)$$

At $t = t_i$ (A.9.1) can be written in algebraic, matrix, and symbolic forms as follows:

$$x(t_i) = x_a(t)|_{t=t_i} = \sum_{k=0}^{N-1} a_k t_i^k, \quad i = 0, 1, 2, .., N-1. \quad (A.9.2)$$

$$\begin{bmatrix} x_a(t_0) \\ x_a(t_1) \\ \cdot \\ \cdot \\ \cdot \\ x_a(t_{N-1}) \end{bmatrix} = \begin{bmatrix} 1 & t_0 & t_0^2 & \cdots & t_0^{N-1} \\ 1 & t_1 & t_1^2 & \cdots & t_1^{N-1} \\ \cdot & \cdot & \cdot & & \cdot \\ \cdot & \cdot & \cdot & & \cdot \\ \cdot & \cdot & \cdot & & \cdot \\ 1 & t_{N-1} & t_{N-1}^2 & \cdots & t_{N-1}^{N-1} \end{bmatrix}\begin{bmatrix} a_0 \\ a_1 \\ a_2 \\ \cdot \\ \cdot \\ a_{N-1} \end{bmatrix},$$

$$\mathbf{x} = \mathbf{T}\mathbf{a}, \quad (A.9.3)$$

x and **a** are n-dimensional vectors and **T** is an $N \times N$ nonsingular *Vandermonde* matrix, see Hohn (1958). Its determinant is not zero provided $t_i's$ are distinct as

$$\det(\mathbf{T}) = |\mathbf{T}| = \prod_{i > j} (t_i - t_j) = [(t_1 - t_0)][(t_2 - t_1)(t_2 - t_0)] \ldots [(t_{N-1} - t_{N-2}) \ldots (t_{N-1} - t_0)].$$

Example A.9.1 Approximate the function $x(t) = \cos((\pi/2)t)$ over the interval $[0, 2]$, with $t_0 = 0$, $t_1 = 1$, and $t_2 = 2$ using (A.9.3).

Solution: Noting that $x(t_0) = 1$, $x(t_1) = 0$ and $x(t_2) = -1$ results in

$$\begin{bmatrix} 1 \\ 0 \\ -1 \end{bmatrix} = \begin{bmatrix} 1 & 0 & 0 \\ 1 & 1 & 1 \\ 1 & 2 & 4 \end{bmatrix} \begin{bmatrix} a_0 \\ a_1 \\ a_2 \end{bmatrix} \Rightarrow a_0 = 1,$$

$$a_1 = -1, \text{ and } a_2 = 0. \qquad \text{(A.9.4)}$$

The cosine function is approximated by a straight line (see Fig. A.9.1).

$$x_a(t) = a_0 + a_1 t + a_2 t^2 = 1 - t. \qquad \text{(A.9.5)} \quad \blacksquare$$

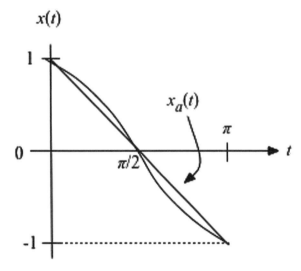

Fig. A.9.1 Approximation of $x(t)$, $x_a(t)$

A.9.2 Lagrange Interpolation Formula

The polynomial approximation discussed above requires the solution of a set of equations. The Lagrange interpolation formula avoids solving a system of equations. $x(t)$ is known at $t_0, t_1, t_2, \ldots, t_{N-1}$. The *interpolation formula* is

$$\begin{aligned} x_a(t) = &\, x(t_0) \frac{(t - t_1)(t - t_2) \ldots (t - t_{N-1})}{(t_0 - t_1)(t_0 - t_2) \ldots (t_0 - t_{N-1})} \\ &+ x(t_1) \frac{(t - t_0)(t - t_2) \ldots (t - t_{N-1})}{(t_1 - t_0)(t_1 - t_2) \ldots (t_1 - t_{N-1})} \\ &+ x(t_{N-1}) \frac{(t - t_0)(t - t_1) \ldots (t - t_{N-2})}{(t_{N-1} - t_0)(t_{N-1} - t_2) \ldots (t_{N-1} - t_{N-2})}. \end{aligned}$$
$$\text{(A.9.6)}$$

From this, it follows that $x_a(t_i) = x(t_i), i = 0, 1, \ldots, N - 1$. For other values of t, i.e., for $t \neq t_i$, $x_a(t)$ interpolates the values using the interpolation formula.

Example A.9.2 Find the Lagrange interpolation formula using the information

$$t_0 = 0, \ t_1 = 1, \ t_2 = 2; \quad x(0) = 1, \ x(1) = 1, \ x(2) = 2.$$

Solution: The Lagrange interpolation formula is

$$\begin{aligned} x_a(t) = &\, (1) \frac{(t - 1)(t - 2)}{(0 - 1)(0 - 2)} + (1) \frac{(t - 0)(t - 2)}{(1 - 0)(1 - 2)} \\ &+ 2 \frac{(t - 0)(t - 1)}{(2 - 0)(2 - 1)} = \frac{1}{2}(t^2 - t + 2). \end{aligned}$$
$$\text{(A.9.7)} \quad \blacksquare$$

Problems

A.2.1 Show that a real matrix **A** is a sum of a symmetric and a skew symmetric matrix.

A.2.2 Show the equality in (A.2.4).

A.2.3 Show that the L_2 norms of the vectors **x** and (**Ux**) are the same.

$$\mathbf{x} = \begin{bmatrix} 1 \\ 1 \\ 1 \end{bmatrix}, \mathbf{U} = \frac{1}{\sqrt{3}} \begin{bmatrix} 1 & 1 & 1 \\ 1 & e^{-j2\pi/3} & e^{-j2\pi(2)/3} \\ 1 & e^{-j2\pi(2)/3} & e^{-j2\pi(4)/3} \end{bmatrix}.$$

A.3.1 Show that the determinant of a skew symmetric matrix is zero.

A.3.2 *a.* Write the following equations in a matrix form. *b.* Use Cramer's rule to find $I_1(s)$ and $I_2(s)$.

$$V_i(s) = R_1 I_1(s) - (1/C_1 s)I_2(s)$$
$$0 = -(1/C_1 s)I_1(s) + [L_1 s + (1/C_1 s)$$
$$+ (R_2/(R_2 C_2 s + 1))]I_2(s).$$

A.3.3 Find the ranks of the following matrices:

$$a.\ \mathbf{A} = \begin{bmatrix} 3 & 2 & 1 \\ 2 & 3 & 2 \\ 1 & 1 & 3 \end{bmatrix},\ b.\ \mathbf{B} = \begin{bmatrix} 2 & 2 & 2 \\ 1 & 1 & 0 \end{bmatrix}.$$

A. 4.1 Find the inverses of the following matrices:

$$a.\ \mathbf{A} = \begin{bmatrix} 0 & 1 & 0 \\ 0 & 0 & 1 \\ -3 & -2 & -1 \end{bmatrix},\ b.\ \mathbf{B} = \begin{bmatrix} 1 & 2 & 3 \\ 0 & 1 & 2 \\ 0 & 0 & 1 \end{bmatrix}.$$

A.5.1 *a.* Show that $D(\lambda) = |\lambda I - \mathbf{A}| = \lambda^3 + a_1\lambda^2 + a_2\lambda + a_3$ for the matrix given below.

$$\mathbf{A} = \begin{bmatrix} 0 & 1 & 0 \\ 0 & 0 & 1 \\ -a_3 & -a_2 & -a_1 \end{bmatrix}.$$

The matrix $\mathbf{A}$ is referred to as the companion matrix of the characteristic polynomial, as the coefficients in the polynomial are explicitly included in the companion matrix. *b.* Assuming the values, $a_1 = 5$, $a_2 = -8, a_3 = -4$, determine the decomposition in the form

$$\mathbf{A} = \mathbf{X}^{-1}\mathbf{J}\mathbf{X}, \mathbf{J} = \begin{bmatrix} 2 & 1 & 0 \\ 0 & 2 & 0 \\ 0 & 0 & 1 \end{bmatrix}.$$

A.5.2 Consider the matrix given below with the following three cases. Express it, in each case, in the form $\mathbf{A} = \mathbf{X}\mathbf{J}\mathbf{X}^{-1}$: *a.* $\varepsilon = 0$ and $\alpha = 0$, *b.* $\varepsilon \neq 0$ and $\alpha = 2$, *c.* $\varepsilon = 0$ and $\alpha = 2$.

$$\mathbf{A} = \begin{bmatrix} 1 & \alpha \\ 0 & 1+\varepsilon \end{bmatrix}.$$

A.5.3 Discrete convolutions and correlations can be expressed in terms of *circulant matrices* Moon and Stirling (2000). One of the classic books in the general area of Toeplitz matrices is by Grenander and Szego (1958). A circulant matrix C is defined by

$$\mathbf{C} = \begin{bmatrix} c_0 & c_1 & \cdot & \cdot & \cdot & c_{m-1} \\ c_{m-1} & c_0 & \cdot & \cdot & \cdot & c_{m-2} \\ c_{m-2} & c_{m-1} & \cdot & \cdot & \cdot & c_{m-3} \\ \cdot & & \cdot & \cdot & \cdot & \cdot \\ \cdot & & & \cdot & \cdot & \cdot \\ \cdot & & & & \cdot & \cdot \\ c_1 & c_2 & \cdot & \cdot & \cdot & c_0 \end{bmatrix},$$

$$\mathbf{C} = \sum_{i=0}^{m-1} c_i \mathbf{P}^i, \mathbf{P} = \begin{bmatrix} 0 & 1 & & & 0 & 0 \\ 0 & 0 & & & 0 & 0 \\ \cdot & & & & & \cdot \\ \cdot & & & & & \cdot \\ \cdot & & & & & \cdot \\ 0 & 0 & 0 & \cdot\cdot\cdot & 0 & 1 \\ 1 & 0 & 0 & \cdot\cdot\cdot & 0 & 0 \end{bmatrix}.$$

The eigenvalues and the corresponding eigenvectors of a circulant matrix are given by

$$\lambda_i = \sum_{k=0}^{m-1} c_k e^{-j2\pi(ik)/m},\ \mathbf{x}_i = \begin{bmatrix} 1 \\ e^{-j2\pi i/m} \\ e^{-j2\pi(2i)/m} \\ \cdot \\ \cdot \\ \cdot \\ e^{-j2\pi(m-1)i/m} \end{bmatrix},$$

$$i = 1, 2, ..., m.$$

Using these results, determine $\mathbf{X}$ such that $\mathbf{C} = (1/m)\mathbf{X}\mathbf{D}\mathbf{X}^*$.

$$\mathbf{C} = \begin{bmatrix} 0 & 1 & 0 \\ 0 & 0 & 1 \\ 1 & 0 & 0 \end{bmatrix}.$$

Notes: Eigenvalues of circulant matrices can be obtained using the DFT of the sequence $\{c_0, c_1, ..., c_{m-1}\}$. The eigenvectors of every $m \times m$ circulant matrix depend only on $c_i, i = 0, 1, ...,$ $m - 1$, thus allowing for the ease in computing the sums, products of these same size matrices by using their eigenvalue–eigenvector decompositions.

A.6.1 Determine the SVD of

$$\mathbf{A} = \begin{bmatrix} 2 & 1 \\ 0 & 2 \end{bmatrix}.$$

A.7.1 Find the generalized inverse of the matrix **A** given below using the following:

$$\mathbf{A} = \begin{bmatrix} 1 & 0 \\ 2 & 1 \\ 0 & 2 \end{bmatrix}, \mathbf{A} = \mathbf{U}\Sigma\mathbf{V}^*,$$

$$\mathbf{U} = \begin{bmatrix} -0.2673 & 0.4082 & 0.8729 \\ -0.8018 & 0.4082 & -0.4364 \\ -0.5345 & -0.8165 & 0.2182 \end{bmatrix},$$

$$\Sigma = \begin{bmatrix} 2.6458 & 0 \\ 0 & 1.7321 \\ 0 & 0 \end{bmatrix}, \mathbf{V} = \begin{bmatrix} -.7071 & .7071 \\ -.7071 & -.7071 \end{bmatrix}.$$

A.8.1 Determine the least-squares solution of the following system of equations:

$$\mathbf{y} = \begin{bmatrix} 1 \\ 2 \\ 1 \end{bmatrix} = \mathbf{Ax} = \begin{bmatrix} 1 & 0 \\ 1 & 1 \\ 0 & 1 \end{bmatrix} \begin{bmatrix} x_1 \\ x_2 \end{bmatrix}.$$

A.8.2 Show that the vector $\mathbf{x} = \mathbf{A}^-\mathbf{y}$ is a least-squares solution to the matrix equation $\mathbf{Ax} = \mathbf{y}$ among all vectors such that the L_2 norm of the vector $\mathbf{x}$ is minimized.

A.9.1 Approximate the function $x(t) = \cos((\pi/2)t)$ in the interval [0, 2] using the polynomial approximation with four equally spaced points. Sketch the given function and the polynomial approximation function.

A.9.2 Use the function $x(t) = \cos((\pi/2)t)$ using four equally spaced points and their values of the function in the interval [0, 2] using Lagrange interpolation formula. Sketch the function $x(t)$ and the function obtained by the formula.

Appendix B

MATLAB® for Digital Signal Processing

B.1 Introduction

MATLAB is short for MATrix LABoratory software developed by *The Math Works, Inc*. It is a premier scientific software package for numeric computation, data analysis, graphics, and design is extensively used by students and researchers in the scientific community. It was first designed to do math operations on matrices. It has quickly grown to include an array of special toolboxes and functions to facilitate solving most technical problems, especially those related to signal processing. One of the strongest assets of MATLAB is that there are over 1000 built-in functions, with the toolboxes providing additional functions that are useful in almost every aspect of science. In addition, MATLAB is user friendly and allows the user to write custom functions. It also supports a wide array of graphic plots and visualization to efficiently represent the data. In the following it will be assumed that the student is familiar with basics of MATLAB. This appendix will focus on some common uses of MATLAB for digital signal processing.

Most electrical engineering departments require a basic MATLAB course before the signal analysis course. There are several texts that provide the basics of MATLAB programming. See Etter (1993), Palm (2001), Chapman (2005), Etter, Kuncicky with Hull (2002). In addition to the general references given earlier, there are several books that use MATLAB for digital signal processing. See, for example, Ingle and Proakis (2007), Mitra (2006), and others. There are several links to online MATLAB tutorials available on the web:

ftp://ftp.eng.auburn.edu/pub/sjreeves/
  matlab_primer_40.pdf
http://www.mathworks.com/academia/
  student_center/tutorials/launchpad.html
http://users.ece.gatech.edu/~bonnie/book/
  book.html
http://www.ee.ucr.edu/EESystems/docs/
  Matlab/

B.2 Signal Representation

In any computer-based application it is important to realize that all signal are processed as digital signals. Continuous-time signals are approximated by their sampled values for processing. In MATLAB signals are represented as vectors or arrays. For example, consider the sinc function (see Section 1.2.8) defined as

$$y(x) = \mathrm{sinc}(x) = \frac{\sin(\pi x)}{\pi x} \qquad (B.2.1)$$

The sinc function is plotted in Fig. B.2.1a. Even though the curve is continuous in nature, it is an interpolated version of the discrete signal in Fig. B.2.1b. These discrete values are stored in the computer and are used in any computation involving the signal. The MATLAB code used to generate these plots and its explanation is given in the code block below. It provides good online help for every keyword. More information about the built-in command may be obtained by typing "help keyword" or "doc keyword" at the MATLAB prompt. For example, help sinc, doc sinc, help title, help plot, etc.

Matlab Code for representing a sinc function

```
%Clear all pre-existing variables from memory
clear all

% Create a vector or array with values -3pi/2 to 3pi/2 with increments of 0.1
x = -3*pi/2:0.1:3*pi/2;

% Evaluate (B.2.1) for these values using the function sinc
y = sinc(x);

% Make x,y continuous plot and enable grid
plot(x,y); grid on;

% Set and label tick marks for the x-axis
set(gca,'XTick,-3*pi/2:pi/2:3*pi/2);
set(gca,'XTickLabel,{'-3*pi/2,'-pi,'pi/2,'0,'pi/2,'pi,'3*pi/2'});

% Label x & y axes
xlabel('x'); ylabel('y = sinc(x)');

% Title figure
title('Sinc Function');

% Set x and y axes range
axis([-3*pi/2 3*pi/2 -0.3 1.1]);

% Make discrete stem x,y plot and enable grid
stem(x,y); grid on;

% Adjust and set axes
set(gca, 'XTick, -3*pi/2:pi/2:3*pi/2);
set(gca, 'XTickLabel, {'-3*pi/2, '-pi, 'pi/2, '0,'pi/2,'pi,'3*pi/2'});
xlabel('x'); ylabel('y = sinc(x)');
title('Sinc Function');
axis([-3*pi/2 3*pi/2 -0.3 1.1]);
```

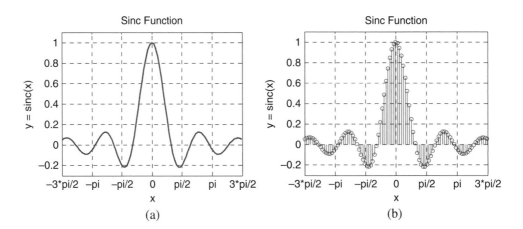

(a)   (b)

Fig. B.2.1 (a) Continuous-time plot of sinc function and (b) underlying discrete signal

B.3 Signal Integration

Many of the DSP algorithms in signal processing require the approximation of integrals. These include Fourier transforms, analog convolution, analog correlation, and others. A simple way to implement integration is to use the rectangular integration formula, see (1.3.3a).)

$$A = \int_a^b x(t)dt = \sum_{n=0}^{N-1} x(a+n\Delta t)\Delta t. \qquad (B.3.1)$$

Example B.3.1 Using rectangular integration, illustrate the evaluation of $x(t) = \sqrt{t}$ the integral above assuming $a = 0$, $b = 1$ and $\Delta t = 0.1$. Give the MATLAB script and verify the result in (B.3.2).

$$A = \int_a^b \sqrt{t}dt = t^{3/2}(2/3)\big|_a^b = \frac{2}{3}(b^{3/2} - a^{3/2})$$

$$= \frac{2}{3} = 0.6667. \qquad (B.3.2)$$

Solution: Using the rectangular integration formula with $x(t) = \sqrt{t}$, we have

$$A \approx 0.1 \sum_{n=0}^{10} x(0 + .1n),$$

$$\text{where } x(0 + .1n) = \sqrt{(0 + (.1)n)}. \qquad (B.3.3)$$

First construct the following vector and use it in B.3.3 to approximate the integral:

$$\mathbf{x} = [x(0) \quad x(0.1) \quad x(0.2) \quad x(0.3) \quad x(0.4) \quad x(0.5) \quad x(0.6) \quad x(0.7) \quad x(0.8) \quad x(0.9) \quad x(1)].$$

The MATLAB code to implement the integral approximation is given below.

MATLAB Code for example B.3.1

```
%Clear all pre-existing variables from memory
clear all

%Set the variable delta_t that controls number of samples
delta_t = 0.1;

%Create a vector with values from 0 to 1 with increments of delta_t=1
t = 0:delta_t:1;

%Create vector x as described above
x = sqrt(t);

%Make stem plot and set axes
stem(t,x); grid on;
xlabel('t'); ylabel('x(t)');
axis([0 1.1 0 1.1]);

%Approximate and display integral value using B.3.3
A=sum(x)*delta_t;
sprintf('Integral value A with delta_t=%g is %g',delta_t,A);
```

With delta_t = 0.1, the integral is approximated as $A = 0.710509$. Whereas, with delta_t = 0.01, the integral is approximated as $A = 0.671463$, which is much closer to the theoretical value of 0.6667. The signals corresponding to the different delta_t are shown in Fig. B.3.1a,b. Observe that a larger number of samples result in an approximation much closer to the theoretical value of the integral.

B.4 Fast Fourier Transforms (FFTs)

In Chapters 8 and 9 the discrete Fourier transform (DFT) and its fast implementation (FFT) algorithms were considered. The MATLAB function "**fft**" can be used to compute the frequency response. The command $Y = \text{fft}(X)$ returns the discrete Fourier transform (DFT) of X, computed

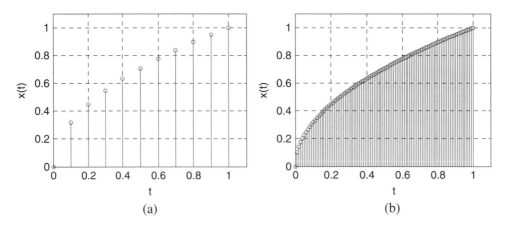

Fig. B.3.1 (a) $x(t)$ with delta_t = 0.1, (b) $x(t)$ with delta_t = 0.01.

using a fast Fourier transform (FFT) algorithm. Y = fft(X,n) returns the n-point DFT of X. If the length of X is less than n, X is padded with trailing zeros up to length n. If the length of X is greater than n, the sequence is truncated. For example, if the data are given by $x = [1, 2, 3]$, then fft(x), fft($x,4$), and fft($x,2$) are given by

$$x = [1, 2, 3] \xrightarrow{\textit{fft}(x)} [6.0000, -1.5000 + j0.8660,$$
$$- 1.5000 - j0.8660] = \mathbf{X},$$

$$X = [1, 2, 3] \xrightarrow{\textit{fft}(x,4)} [6, -2 - j2, 2 - 2 + j2] = X,$$

$$x = [1, 2, 3] \xrightarrow{\textit{fft}(x,2)} [3, -1] = X.$$

In the following, a simple cosine function is used to illustrate the use of fft to find the frequency spectrum of a given signal. We generate a 64-point

cosine signal with 10 samples per period. Then the fft function is used to find the frequency spectrum, see Fig. B.4.1a,b.

The fft function produces complex spectral values. The complex values are difficult to visualize and therefore we will plot the magnitude of the complex value. Notice how the peak of the frequency spectrum occurs at 0.1*fs and −0.1*fs. This is reasonable as the cosine function has 10 samples per period. Ideally we would expect to see two impulses in the frequency spectrum; however, that is not the case because the 64-point cosine signal has one incomplete period in the end. If the MATLAB code were to be modified such that t = 0:59 and N = 60, we get the plots in Fig. B.4.2. Notice that having complete periods in the original signal results in a more ideal frequency spectrum.

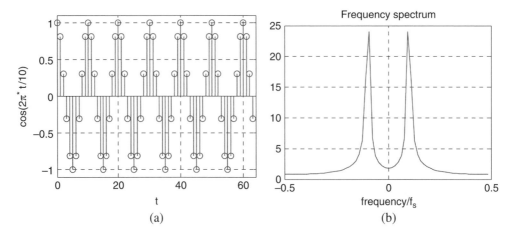

Fig. B.4.1 (a) 64-point cosine signal and (b) 64-point FFT magnitude spectrum

MATLAB Code for Fig. B.4.1

```
clear all

%set signal length
t=0:63;
%set FFT length
N=64;

%Construct and plot the cosine signal
x = cos (2*pi*t/10);
stem(t,x); grid on;
xlabel(' t');
ylabel(' cos( 2\pi * t/10 )');
axis([ 0 64 -1.1 1.1]);

%Compute N point FFT of the signal and take absolute its value to find
%magnitude of the frequency response
X=abs(fft(x,N));

%Shift 0 frequency to the center
X=fftshift(X);

f_ratio =[ -N/2:N/2-1] /N;

%Plot frequency response
figure,plot(f_ratio,X); grid on;
xlabel(' frequency / f_s');
title(' Frequency spectrum');
```

Fig. B.4.2 (a) 60-point cosine signal and (b) 60-point FFT magnitude spectrum

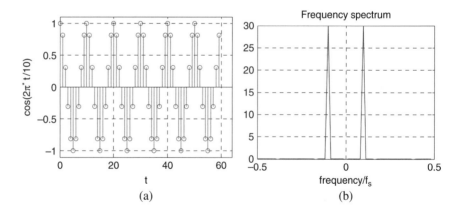

(a)   (b)

B.5 Convolution of Signals

MATLAB convolution function can be used to multiply two polynomials.

Example B.5.1 Use the MATLAB function, *conv*, to determine the multiplication $A(x)B(x) = C(x)$ below.

$$A(x) = 2x^2 + x + 1, \quad B(x) = 3x^3 - 2x^2 + 1,$$
$$C(x) = A(x)B(x) \tag{B.5.1}$$

Solution: First write the coefficients of the polynomials in the form of vectors as a = [2,1,1], b = [3, −2, 0, 1]. Now compute the convolution using the command below and the corresponding polynomial $C(x)$.

$$c = \text{conv}\,(a,b) \Rightarrow c = [6 \quad -1 \quad 1 \quad 0 \quad 1 \quad 1], \tag{B.5.2}$$

$$\Rightarrow C(x) = 6x^5 - x^4 + x^3 + (0)x^2 + x + 1. \tag{B.5.3}$$

MATLAB Code for Example B.5.1

```
clear all

%Input the signals x1[ n]  and x2[ n]
x1=[0 0 1 2 3 4 3 2 1 0 0];
x2=[0 0 0 0 1 1 1 0 0 0 0];

%Compute convolution result
y=conv(x1,x2);

%Make plots using subplot to have multiple graphs in one figure.
figure,subplot(3,1,1)
stem(x1); grid on;
xlabel('n');
ylabel('x_1[ n]');
axis([0 11 0 5]);

subplot(3,1,2)
stem(x2); grid on;
xlabel('n');
ylabel('x_2[ n]');
axis([0 11 0 2]);

subplot(3,1,3)
stem(y); grid on;
xlabel('n');
```

Example B.5.2 Use the MATLAB function, *conv*, to determine the convolution of signals $x_1[n]$ and $x_2[n]$. See the functions inside the MATLAB code block for the values. See the sketches for $x_1[n]$ and $x_2[n]$ in Fig B.5.1.

Solution: The MATLAB code for the convolution is given above. Note the length of the convolution sequence is longer than either of the input signals. See the width property of discrete convolution in Section 8.3.2.

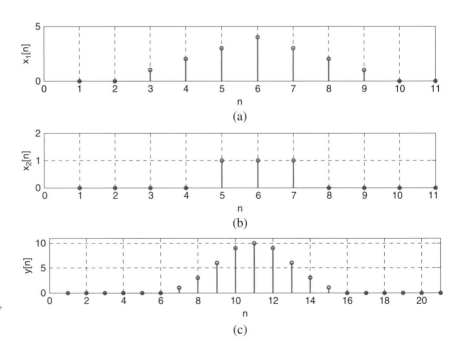

Fig. B.5.1 (a) Signals $x_1[n]$, (b) $x_2[n]$ and (c) $y[n] = x_1[n] * x_2[n]$.

B.6 Differentiation Using Numerical Methods

Numerical differentiation methods estimate the derivative of a function, $g(x)$ at $x = x_n$, i.e., $g'(x_n)$ using the *backward difference, forward difference*, and the *central difference* functions, see (1.3.1). These approximate the slope of the function at x_n.

MATLAB Code for Example B.6.1

```
clear all

x = 0:0.1:4;
%compute function f(x)
f = x.^3 - 6*x.^2 + 11*x -6;

%plot function and label axes
subplot(2,1,1);
plot(x,f); grid on;
xlabel('x'); ylabel('f(x)');
title('Third order polynomial');

%compute backward difference using function diff, then take ratio to obtain
%derivative
df = diff(f)./diff(x);

%plot the derivative and label axes
subplot(2,1,2);
plot(x(2:length(x)),df); grid on;

title('Derivative of third order polynomial');
xlabel('x'); ylabel('df');
```

The polynomial $f(x)$ has three roots at $x = 1, 2, 3$. The zeros of the derivative polynomial correspond to the local minima or local maxima of the function $f(x)$. It does not have a global

Example B.6.1 Consider the polynomial

$$f(x) = x^3 - 6x^2 + 11x - 6.$$

Use MATLAB to sketch the functions $f(x)$ and $f'(x)$ using a backward difference equation.

Solution: The following MATLAB code gives the sketches for $f(x)$ and $f'(x)$ given in Fig. B.6.1. ∎

minimum or a global maximum, as the range of the function is $-\infty < x < \infty$. Critical points can be determined using the following statements and are 1.4 and 2.6.

```
Product = (df(1:length(df)-1).*df(2:length(df)));
Critical = x(find(Product<0)+1)
```

B.7 Fourier Series Computation

Exponential F-series coefficients (see Chapter 3) can be determined using the discrete Fourier transforms discussed in Chapter 9. See for a brief discussion in approximating integrals in Section B.3. The period is T and is divided into N intervals of t_s seconds each.

F-series coefficents: $X_s[k] = \dfrac{1}{T} \displaystyle\int_T x_T(t)e^{-jk\omega_0 t}dt\big|_{\omega_0=2\pi/T}$

$$= \frac{1}{Nt_s}\sum_{n=0}^{N-1}(t_s x(nt_s))e^{-j(2\pi/Nt_s)nt_s k} = \frac{1}{N}\sum_{n=0}^{N-1}x[n]e^{-j(2\pi/N)kn} \ . \qquad \text{(B.7.1)}$$

Fig. B.6.1 Plot of the third-order polynomial and its derivative over the range $0 \le x \le 4$

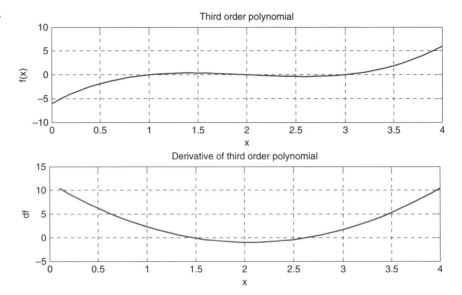

The spectral coefficients repeat periodically with period N. That is $X_s[k] = X_s[k + N]$. We assumed that the first sample starts at $t = 0$ and the last sample at $(T - t_s)$ *not at* $t = T$. As we have seen in Chapter 3 that the Fourier series coefficients of most functions decay as a function of k. There is spectral overlap due to the periodic nature of the spectra of the sampled signal. Select N large enough so as to minimize the overlap.

Example B.6.1 Compute the F-series coefficients using MATLAB of (see Lathi (1998)

$$x(t) = e^{-t/2}, \ 0 \le t < \pi, \ x(t) = x(t + T), \ T = \pi.$$
$$\text{(B.7.2)}$$

Solution: The samples start at $t = 0$, which is the same as the sample at $t = T$ and is not equal to the sample value at $t = T - t_s$. We know that the F series converge to the average of the two values on two sides if there is a discontinuity. Therefore the value of the function at $t = 0$ is not equal to 1 but it is taken as $(e^{-\pi/2} + 1)/2 = 0.604$. From our discussion in Chapter 3, the signal has a jump discontinuity and therefore we can predict that the Fourier series coefficients decay slowly as $(1/n)$. We further assume that the number of Fourier series coefficients $X_s[n]$ to be determined is based on the constraint that the amplitudes of the coefficients $X_s[n]$ are negligible for $n \ge N/2$. It is safe to select a value of n such that $|X_s[n]| < 0.01$. This implies that the

100th harmonic is about 1% of the fundamental. $N = 200$ satisfies the requirement. Use N, a power of 2, say $N = 256$ and then use the fft algorithm.

Extracting the first ten of these gives the first ten coefficients of the complex F series coefficients in terms of their magnitudes and phase angles and are given in Table B.7.1.

Table B.7.1 Amplitudes and phase angles of the harmonic Fourier series coefficients (Example B.7.1)

$$x_T(t) = X_s(0) + \sum_{k=1}^{\infty} d[k]\cos(k\omega_0 t + \theta[k]) \ ;$$
$$\text{where } T = \pi; \omega_0 = 2.$$

Amplitudes		Phase angle (in degrees)	
$|X_s(0)|$	.5058	$\angle X_s(0)$	0
d[1]	.2454	$\theta[1]$	−75.9622
d[2]	.1255	$\theta[2]$	−82.8719
d[3]	.0840	$\theta[3]$	−85.2317
d[4]	.0631	$\theta[4]$	−86.4175
d[5]	.0506	$\theta[5]$	−87.1299
d[6]	.0422	$\theta[6]$	−87.6048
d[7]	.0362	$\theta[7]$	−87.9437
d[8]	.0316	$\theta[8]$	−88.1977
d[9]	.0281	$\theta[9]$	−88.3949

These approximate values can be verified by using the compact harmonic Fourier series coefficients.

$$X_s[0] = 0.504; \ k > 1, \ d[k] = 0.504\left[\frac{2}{\sqrt{1 + 16\,k^2}}\right],$$

$$\theta[k] = -\tan^{-1}(4\,k) \qquad\qquad \text{(B.7.3)}$$

For $k = 9$, the amplitude and the phase angle of the coefficient are, respectively, given by $0.504*(2/\text{sqrt}(1+16*81)) = 0.0280$, $-\text{atan}(36)*$ $180/\text{pi} = -88.4074°$.   ∎

MATLAB Code for Example B.7.1

```
clear all
%Select the value of M, i.e., the number coefficients equal to 10
T = pi; N = 256; M=10 ;

%Generate N values of time starting at t = 0 at time intervals of
ts = T/N;
t=0:ts:ts*(N-1);

%Convert the above as a column vector
t = t';

%Define the function x = exp(-t/2) , we define x(0)=0.604
%Note that MATLAB sequence starts at index n = 1
xk = exp(-t/2);
x(1)=.604;

%Use fft to determine the spectrum and divide it by the number
%of sample points N
Xn = fft(xk)/N;

%Find the amplitudes and the phase angles of the complex FS coefficients
Xamp = abs(Xn);
Xphase = angle(Xn)*180/pi;

%Note that amplitudes will be twice the values computed above for k=2,...,M.
X0=Xamp(1);
Xk=2*Xamp(2:M);

%Display the coefficients and phase angle
[ X0;Xk]
Xphase(1:10)
```

B.8 Roots of Polynomials, Partial Fraction Expansions, Pole−Zero Functions

Given a polynomial $D(s)$, we like to find the roots of this polynomial using MATLAB.

$$D(s) = 3s^3 - 2s^2 + 1. \tag{B.8.1}$$

MATLAB function roots can provide the required solution:

r = roots([3, −2, 0, 1]) ⇒ $0.5974 + j0.5236$,
$0.5974 - j0.5236, -0.5282$. \qquad (B.8.2)

MATLAB gives the results of complex numbers in the form $0.5974 + .5236i$ rather than $0.5974 + j0.5236$. If the roots are given, then we

can obtain the polynomial using the MATLAB script and the results as given below.

poly(r) ⇒ 1.0000 0−0.6667   0.0000
$0.3333 \Rightarrow s^3 - 0.6667s + 0.3333.$ \qquad (B.8.3)

We can use MATLAB to convert a transfer function given in terms of a ratio of two polynomials into a transfer function expressed in terms of its zeros, poles, and the gain constant, and vice versa. The corresponding commands are

Zp2tf converts zero−pole−gain
    to transfer function, \qquad (B.8.4)

Tf2zp converts transfer function to
    zero−pole−gain. \qquad (B.8.5)

B.8.1 Partial Fraction Expansions

A rational transfer function can be written as

$$H(s) = \frac{B(s)}{A(s)} = \sum_{n=0}^{M-N} k_n s^n + \frac{N(s)}{A(s)}. \qquad \text{(B.8.6)}$$

Example B.8.1 Use the MATLAB residue function to find the partial fraction expansions of the following functions:

a. $H_a(s) = \dfrac{2s^3 + 2s^2 + 6s + 7}{s^2 + s + 5}$

$= 2s + \dfrac{-2 - j2.0647}{s + 0.5 - j2.1794} + \dfrac{-2 + j2.0647}{s + 0.5 + j2.1794}$,

b. $H_b(s) = \dfrac{1}{s^3 + 4s^2 + 5s + 2}$

$= \dfrac{1}{(s+1)^2} - \dfrac{1}{(s+1)} + \dfrac{1}{(s+2)}$.

Solution: *a.* The vectors and the MATLAB statement for the partial fraction expansion are as follows:

$$\mathbf{b} = [2 \quad 2 \quad 6 \quad 7], \ \mathbf{a} = [1 \quad 1 \quad 5],$$
$$[\mathrm{r,p,k}] = \text{residue (b, a)} \qquad \text{(B.8.7)}$$

Outputs:

$r = -2.0000 - 2.0647\mathrm{i}, -2.0000 + 2.0647\mathrm{i}$,

$p = -0.5000 + 2.1794\mathrm{i}, -0.5000 - 2.1794\mathrm{i}$,

$k = [2 \quad 0]$.

MATLAB Code for example B.9.1

The residues are in the vector r, the poles are in vector p, and the entries in the vector k are the coefficients identified in (B.8.6).

b. Vectors in the MATLAB statement: $\mathbf{b} = [1]$, $\mathbf{a} = [1 \quad 4 \quad 5 \quad 2]$.

Outputs:

$r = 1.0000, \ -1.0000, \ 1.0000,$

$p = -2.0000, \ -1.0000, \ -1.0000,$

$k = [\ \ ](k \text{ is identically zero})$

B.9 Bode Plots, Impulse and Step Responses

There are several MATLAB functions available for analysis and designs of linear systems. These functions include bode, nyquist, rlocus, step, and others.

B.9.1 Bode Plots

Example B.9.1 Use MATLAB to sketch the magnitude in dB and the corresponding phase angle plots for the fourth-order Butterworth function

$$H(s) = \frac{1}{[s^4 + 2.6131s^3 + 3.4142s^2 + 2.6131s + 1]}.$$

Solution: Following MATLAB statements would produce the Bode magnitude in dB and the phase angle in degrees plots shown in Fig. B.9.1.

```
clear all

%define numerator and denominator
num=1; den=[ 1 2.6131 3.4142 2.6131 1];

%Use function bode to make plot
bode(num,den); grid on;
title(' Bode plot for a fourth order Butterworth function');
```

B.9.2 Impulse and Step Responses

MATLAB has built in routines for impulse and step responses for a given transfer function. The coefficients of the numerator (num) and the denominator (den) polynomials are written in decreasing powers.

Example B.9.2 Use MATLAB to sketch the impulse and step responses of the function

$$H(s) = 1/[s^2 + 1.7967s + 2.114].$$

Solution: The required MATLAB code and plots is given below. See Fig B.9.2 for plots

MATLAB Code for Example B.9.2

```
clear all

%define numerator and denominator
num=1; den=[ 1 1.7967 2.1140];

%Use function tf to construct system with transfer function with given
%numerator and denominator
sys1= tf(num,den);

%use function impulse to make plot of impulse response
subplot(121);
impulse(sys1); grid on;

%use function step to make plot of impulse response
subplot(122);
step(sys1); grid on;
```

Fig. B.9.1 Bode frequency plot for the fourth-order filter defined in Example B.9.1

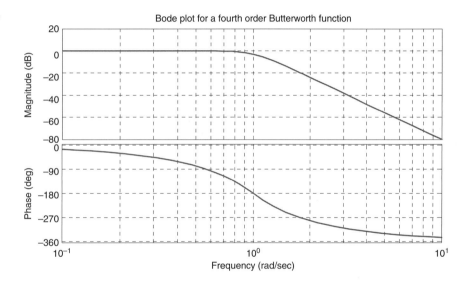

Fig. B.9.2 Impulse and step responses for exapmle B.9.2

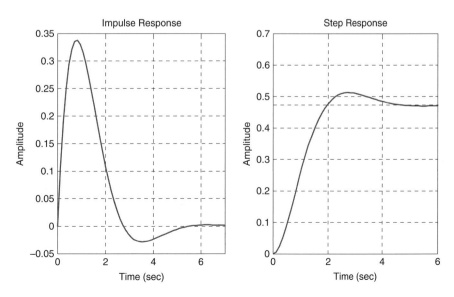

B.10 Frequency Responses of Digital Filter Transfer Functions

Consider the function

$$H(z) = \frac{B(z)}{D(z)} = \frac{b_0 + b_1 z^{-1} + b_2 z^{-2} + \cdots + b_n z^{-n}}{d_0 + d_1 z^{-1} + d_2 z^{-2} + \cdots + d_n z^{-n}}$$

(B.10.1)

The MATLAB function *freqz* function uses three inputs B, A, and N given below. The argument N specifies the number of normalized frequency values over the interval $[0, \pi]$. The MATLAB program that determines the plots of the magnitudes uses the following statement:

MATLAB Code for Example B.10.1

$$B = [b_0, \ b_1, \ldots, b_n]; \ A = [d_0, d_1, \ldots, d_n];$$
$$[Hz, wT] = \text{freqz}(B, A, N).$$

(B.10.2)

Example B.10.1 Use the MATLAB *freqz* function to sketch the frequency amplitude and phase responses of

$$H(z) = 0.1432 \frac{1 + 3z^{-1} + 3z^{-2} + z^{-3}}{1 - 0.1801 z^{-1} + 0.3419 z^{-2} - 0.0165 z^{-3}}.$$

(B.10.3)

Solution: MATLAB code is given below and the plots are shown in Fig. B.10.1.

```
clear all

num = 0.1432.*[ 1  3  3  1];
den =[ 1 -.1801 .3419 -.0165];

%Use freqz function to determine frequency response
[ H,w] = freqz(num,den);

%Find magnitude and phase of complex values stored in H
magH=abs(H); phaH=angle(H);

%Normalize frequency to be between 0 and 1 for use in plot
w_norm=w/pi;

%Make plots of magnitude and phase
figure,subplot(2,1,1),plot(w_norm,magH); grid on;
title(' Magnitude Response' )
xlabel(' Normalized frequency (x \pi rad/sample)' );
ylabel(' Magnitude (dB)' );

subplot(2,1,2); plot(w_norm,phaH/pi); grid on;
title(' Phase response' );
xlabel(' Normalized frequency (x \pi rad/sample)' );
ylabel(' Phase in pi units' );
```

B.11 Introduction to the Construction of Simple MATLAB Functions

Instead of typing commands directly, we can type functions (subroutines). These are called script files or *M-files*. They use the variables that are local to themselves and do not appear in the main

workspace. The following illustrates a *simple function* dB conversion. M-file begins with a word function, followed by the output argument(s), an equal sign, and the name of the function, such as the simple function given below that converts magnitude to dB with an illustration of how to construct this function, see Etter (1993).

```
function [ dB] = dbconversion(mag)
```

Fig. B.10.1 Plots of amplitude and phase responses for example B.10.1

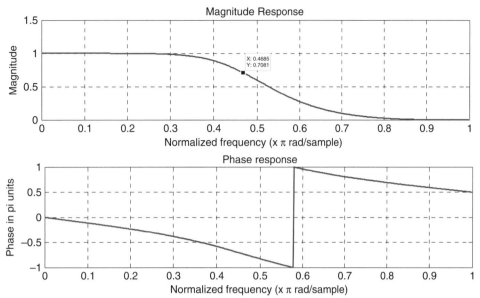

1. Open a New M-File (corner of the menu bar). A window editor will pop up.
2. Type the following script and save it with file name dbconversion.
3. To use this function, type dB = dB conversion (67).

MATLAB Script for Fig. 9.11.6

4. Press "enter" and the MATLAB will respond the dB value 36.5215.

B.12 Additional MATLAB Code

```
clear all

%Compute a Hamming Window
N=55;                                     %Set number of points
for n=0:N-1,
    w(n+1)=(0.54-(0.46*cos((2*pi*n)/(N-1))));
end

figure(1)                                 %Corresponds to Figure 9.11-6a.
stem(0:N-1,w)                             %Plot Hamming window, w(n)
xlabel('Index (n)'); ylabel('Amplitude')
title('Hamming window, w_H[n]: N=55')
axis([0 55.5 0 1.1])

%Compute the impulse response, h(n) for a Hamming window
for n=0:N-1,
    if n==(N-1)/2,
        h(n+1)=0.3;                       %Passband
    else
        h(n+1)=(sin(0.3*pi*(n-((N-1)/2)))/(pi*(n-((N-1)/2))))*w(n+1);
    end
end
figure(2)                                 %Corresponds to Figure 9.11-6b.
stem(0:N-1,h)                             %Plot the impulse response
xlabel('Index (n)'); ylabel('Amplitude')
title('Impulse Response with 55 Coefficients, h(n)')
axis([0 55.5 -0.1 0.35])
```

```
%Compute the Frequency Response for the FIR filter in dB
Xw=fft(h,3024);                                 %Zero pad input and return
                                                %3024 point DFT
XwdB=db(abs(Xw));                               %Compute magnitude (in dB)
NXwdB=XwdB-max(XwdB);                           %Normalized magnitude spectrum
ws=[ 0:1/length(XwdB):0.5-0.5/length(XwdB)];    %Normalized Frequency
                                                %(from 0 to 0.5*ws)


figure(3)                                       %Corresponds to Figure 9.11-6c.
plot(ws*2*pi,NXwdB(1:length(XwdB)/2));          %Plot the Frequency response
set(gca,'XTick',0:0.1*pi:pi)
set(gca,'XTickLabel,{' 0,' 0.1pi,' 0.2pi,' 0.3pi,' 0.4pi,' 0.5pi,' 0.6pi,' 0.
7pi,' 0.8pi,' 0.9pi' pi'})
xlabel(' Frequency (\Omega), (0 \leq \Omega \leq \pi)')
ylabel(' 20 log |H(e^j^\Omega)|,(dB)')
title(' Amplitude Frequency Response for the FIR filter in dB')
axis([ 0 pi -80 0])
grid

%Compute the Phase plot for the FIR filter in degrees
Xphase = angle(Xw)*180/pi;
figure(4)                                       %Corresponds to Figure 9.11-6d.
plot(ws*2*pi,Xphase(1:length(XwdB)/2))          %Plot the filter Phase response
set(gca,'XTick',0:0.1*pi:pi)
set(gca,'XTickLabel,{' 0,' 0.1pi, ' 0.2pi, ' 0.3pi, ' 0.4pi, ' 0.5pi,' 0.6pi, ' 0.
7pi, ' 0.8pi, ' 0.9pi,' pi'})
xlabel(' Frequency (\Omega), (0 \leq \Omega \leq \pi)')
ylabel(' Phase (deg)')
title(' Phase plot for the FIR filter in degrees ')
axis ([ 0 pi -200 200])
grid
```

Appendix C

Mathematical Relations

Following is a summary of mathematical relations that are useful. For an extensive list of the formulas, see Abramowitz and Stegun, Editors [1964], Spiegel [1966], and others.

$$\cos(\theta) = \left(e^{j\theta} + e^{-j\theta}\right)/2 \qquad \text{(C.1.14)}$$

$$\sin(\theta) = \left(e^{j\theta} - e^{-j\theta}\right)/2j \qquad \text{(C.1.15)}$$

C.1 Trigonometric Identities

$$\sin(x \pm y) = \sin(x)\cos(y) \pm \cos(x)\sin(y) \qquad \text{(C.1.1)}$$

$$\cos(x \pm y) = \cos(x)\cos(y) \mp \sin(x)\sin(y) \qquad \text{(C.1.2)}$$

$$\tan(x \pm y) = \frac{\tan(x) \pm \tan(y)}{1 \mp \tan(x)\tan(y)} \qquad \text{(C.1.3)}$$

$$\cos(x \pm (\pi/2)) = \mp \sin(x) \qquad \text{(C.1.4)}$$

$$\sin(x \pm (\pi/2)) = \pm \cos(x) \qquad \text{(C.1.5)}$$

$$\cos(2x) = \cos^2(x) - \sin^2(x) \qquad \text{(C.1.6)}$$

$$\sin(2x) = 2\sin(x)\cos(x) \qquad \text{(C.1.7)}$$

$$\cos(x)\cos(y) = [\cos(x-y) + \cos(x+y)]/2 \qquad \text{(C.1.8)}$$

$$\sin(x)\sin(y) = [\cos(x-y) - \cos(x+y)]/2 \qquad \text{(C.1.9)}$$

$$\sin(x)\cos(y) = [\sin(x-y) + \sin(x+y)]/2 \qquad \text{(C.1.10)}$$

$$\cos^2(x) = [1 + \cos(2x)]/2 \qquad \text{(C.1.11)}$$

$$\sin^2(x) = [1 - \cos(2x)]/2 \qquad \text{(C.1.12)}$$

$$A\cos(x) + B\sin(x) = C\cos(x + \theta),$$
$$C = \sqrt{A^2 + B^2}, \theta = -\tan^{-1}(B/A)$$
$$A = R\cos(\theta), B = \sin(\theta) \qquad \text{(C.1.13)}$$

C.2 Logarithms, Exponents and Complex Numbers

$$\log_a(AB) = \log_a(A) + \log_a(B) \qquad \text{(C.2.1)}$$

$$\log_a(A/B) = \log_a(A) - \log_a(B) \qquad \text{(C.2.2)}$$

$$\log_a(A^p) = p\log_a(A) \qquad \text{(C.2.3)}$$

$$\log_a(A) = \frac{\log_b(A)}{\log_b(a)} \qquad \text{(C.2.4)}$$

$$\log_e(A) = \ln(A) = (2.30258..)\log_{10}(A) \qquad \text{(C.2.5)}$$

$$e^A e^B = e^{(A+B)} \ , \ e^A/e^B = e^{(A-B)} \qquad \text{(C.2.6)}$$

Complex Numbers:

$$A = \text{Re}(A) + j\text{Im}(A) = |A|e^{j\theta_A},$$
$$|A| = \sqrt{[\text{Re}(A)]^2 + [\text{Im}(A)]^2}, \qquad \text{(C.2.7a)}$$

$$\theta_A = \begin{cases} \tan^{-1}[\text{Im}(A)/\text{Re}(A)], & \text{Re}(A) > 0 \\ \pm 180^0 - \tan^{-1}[\text{Im}(A)/(-\text{Re}(A))], & \text{Re}(A) < 0 \end{cases}$$
$$\text{(C.2.7b)}$$

$$\text{Re}(A) = |A|\cos(\theta_A), \text{Im}(A) = |A|\sin(\theta_A), \qquad \text{(C.2.7c)}$$

$$A^* = \text{Re}(A) - j\text{Im}(A), \qquad \text{(C.2.7d)}$$

$$j^2 = -1, 1/j = -j, \qquad 6 \qquad \text{(C.2.8)}$$

$$A^p = r^p e^{jp\theta_A}, \qquad \text{(C.2.9)}$$

$$(A)^{1/n} = [re^{j(\theta_A + 2k\pi)}]^{1/n} = r^{1/n} e^{j(\theta_A + 2k\pi)/n}, \quad \text{(C.2.10)}$$

$$\ln(re^{j\theta_A}) = \ln(r) + j(\theta_A + 2k\pi), k = \text{integer}. \qquad \text{(C.2.11)}$$

C.3 Derivatives

$$\frac{d(cx^n)}{dx} = ncx^{n-1} , \qquad \text{(C.3.1)}$$

$$\frac{de^{sx}}{dx} = se^{sx}, \qquad \text{(C.3.2)}$$

$$\frac{d\cos(u)}{dx} = -\sin(u)\frac{du}{dx}, \qquad \text{(C.3.3)}$$

$$\frac{d\sin(u)}{dx} = \cos(u)\frac{du}{dx}, \qquad \text{(C.3.4)}$$

$$\frac{d\tan(u)}{dx} = \sec^2(u)\frac{du}{dx}, \qquad \text{(C.3.5)}$$

$$\frac{d(uv)}{dx} = u\frac{dv}{dx} + v\frac{du}{dx}, \qquad \text{(C.3.6)}$$

$$\frac{d(u/v)}{dx} = \frac{1}{v^2}\left[v\frac{du}{dx} - u\frac{dv}{dx}\right], \qquad \text{(C.3.7)}$$

$$\frac{d(u^n)}{dx} = nu^{n-1}\frac{du}{dx}, \qquad \text{(C.3.8)}$$

$$\frac{dy}{dx} = \frac{dy}{du}\frac{du}{dx}, \qquad \text{(C.3.9)}$$

$$\frac{d}{dx}\ln(u) = \frac{1}{u}\frac{du}{dx}, \qquad \text{(C.3.10)}$$

$$\frac{d\log_a(u)}{dx} = \frac{\log_a(e)}{u}\frac{du}{dx}, \ a \neq 0, 1, \qquad \text{(C.3.11)}$$

$$\frac{d(e^u)}{dx} = e^u\frac{du}{dx}, \qquad \text{(C.3.12)}$$

$$\frac{du^v}{dx} = vu^{v-1}\frac{du}{dx} + u^v\ln(u)\frac{dv}{dx}, \qquad \text{(C.3.13)}$$

$$\frac{d\sinh(u)}{dx} = \cosh(u)\frac{du}{dx}, \qquad \text{(C.3.14)}$$

$$\frac{d\cosh(u)}{dx} = \sinh(u)\frac{du}{dx}, \qquad \text{(C.3.15)}$$

$$\frac{d\tanh(u)}{dx} = \text{sech}^2(u)\frac{du}{dx}, \qquad \text{(C.3.16)}$$

$$\frac{d\sinh^{-1}(u)}{dx} = \sqrt{\frac{1}{u^2 + 1}}\frac{du}{dx}, \qquad \text{(C.3.17)}$$

$$\frac{d\cosh^{-1}(u)}{dx} = \frac{\pm 1}{\sqrt{u^2 - 1}}\frac{du}{dx},$$
$$\left\{\begin{array}{l} +\text{if}\cosh^{-1}(u) > 0, u > 1 \\ -\text{if}\cosh^{-1}(u) < 0, u > 1 \end{array}\right\}, \qquad \text{(C.3.18)}$$

$$\frac{d\tanh^{-1}(u)}{dx} = \frac{1}{1 - u^2}\frac{du}{dx}, \quad -1 < u < 1, \quad \text{(C.3.19)}$$

C.4 Indefinite Integrals

$$\int u\,dv = uv - \int v\,du, \qquad \text{(C.4.1)}$$

$$\int f^{(n)}g\,dx = f^{(n-1)}g - f^{(n-2)}g' + -\ldots(-1)^n\int fg^{(n)}dx,$$
$$f^{(n)} = \frac{d^n f(x)}{dx^n}, g^{(n)} = \frac{d^n g}{dx^n}, \qquad \text{(C.4.2)}$$

$$\int (a + bx)^n dx = \frac{(a + bx)^{n+1}}{b(n + 1)}, \ 0 < n, \qquad \text{(C.4.3)}$$

$$\int \frac{dx}{a + bx} = \frac{1}{b}\ln|a + bx|, \qquad \text{(C.4.4)}$$

$$\int \frac{dx}{(a + bx)^n} = \frac{-1}{(n - 1)(a + bx)^{n-1}}, \ 1 < n, \quad \text{(C.4.5)}$$

$$\int \frac{dx}{x^2 + a^2} = \frac{1}{a}\tan^{-1}(x/a), \qquad \text{(C.4.6)}$$

$$\int \frac{dx}{a^2 - x^2} = \frac{1}{2a}\ln\left[\frac{a + x}{a - x}\right], \qquad \text{(C.4.7)}$$

$$\int \cos(ax)dx = (1/a)\sin(ax), \qquad \text{(C.4.8)}$$

$$\int x\cos(ax)dx = (1/a^2)\cos(ax) + (1/a)x\sin(ax), \tag{C.4.9}$$

$$\int x^2\cos(ax)dx = (2x/a^2)\cos(ax) + [(x^2/a) - (2/a^3)]\sin(ax), \tag{C.4.10}$$

$$\int \sin(ax)dx = -(1/a)\cos(ax), \tag{C.4.11}$$

$$\int x\sin(ax)dx = (1/a^2)\sin(ax) - (x/a)\cos(ax), \tag{C.4.12}$$

$$\int x^2\sin(ax)dx = (2x/a^2)\sin(ax) + [(2/a^3) - (x^2/a)]\cos(ax), \tag{C.4.13}$$

$$\int a^x dx = \frac{a^x}{\ln(a)}, a > 0, a \neq 1, \tag{C.4.14}$$

$$\int xe^{ax}dx = e^{ax}\left[\frac{x}{a} - \frac{1}{a^2}\right], \tag{C.4.15}$$

$$\int x^2 e^{ax}dx = \frac{e^{ax}}{a}\left[x^2 - \frac{2x}{a} + \frac{2}{a^2}\right], \tag{C.4.16}$$

$$\int e^{ax}\cos(bx)dx = \frac{e^{ax}[a\cos(bx) + b\sin(bx)]}{a^2 + b^2}, \tag{C.4.17}$$

$$\int e^{ax}\sin(bx)dx = \frac{e^{ax}[a\sin(bx) - b\cos(bx)]}{a^2 + b^2}, \tag{C.4.18}$$

$$\int \cos(bx)e^{ax}dx = \frac{e^{ax}}{a^2 + b^2}[a\cos(bx) + b\sin(bx)], \tag{C.4.19}$$

$$\int \sin(bx)e^{ax}dx = \frac{e^{ax}}{a^2 + b^2}[a\sin(bx) - b\sin(bx)], \tag{C.4.20}$$

$$\int \cos(ax)\cos(bx)dx = \frac{\sin(a-b)x}{2(a-b)} + \frac{\sin(a+b)x}{2(a+b)}, \\ a^2 \neq b^2, \tag{C.4.21}$$

$$\int \sin(ax)\sin(bx)dx = \frac{\sin(a-b)x}{2(a-b)} - \frac{\sin(a+b)x}{2(a+b)}, \\ a^2 \neq b^2, \tag{C.4.22}$$

$$\int \sin(ax)\cos(bx)dx = -\frac{\cos[(a-b)x]}{2(a-b)} - \frac{\cos[(a+b)x]}{2(a+b)}, \\ a^2 \neq b^2, \tag{C.4.23}$$

C.5 Definite Integrals and Useful Identities

$$\int_a^b f(x)dx = -\int_b^a f(x)dx, \tag{C.5.1}$$

$$\int_a^b f(x)dx = \int_a^c f(x)dx + \int_c^b f(x)dx, \tag{C.5.2}$$

$$\int_0^\infty x^n e^{-ax}dx = \frac{n!}{a^{n+1}}, a > 0, \tag{C.5.3}$$

$$\int_0^\infty e^{-r^2 x^2}dx = \frac{\sqrt{\pi}}{2r}, r > 0, \tag{C.5.4}$$

$$\int_0^\infty \frac{\sin(px)}{x}dx = \begin{cases} \pi/2, & p > 0 \\ 0, & p = 0, \\ -\pi/2, & p < 0 \end{cases} \tag{C.5.5}$$

$$\int_0^\infty \frac{dx}{x^2 + a^2} = \frac{\pi}{2a}. \tag{C.5.6}$$

C.6 Summation Formulae

$$\sum_{k=0}^n a^k = \frac{1 - a^{n+1}}{1 - a}, a \neq 1, \tag{C.6.1a}$$

$$\sum_{i=0}^{N-1} e^{j(m-k)(2\pi/N)n} = \begin{cases} 0, & m \neq k \\ N, & m = k \end{cases}, \tag{C.6.1b}$$

$$\sum_{k=0}^{\infty} a^k = \frac{1}{(1-a)}, |a| < 1, \qquad \text{(C.6.2)}$$

$$\sum_{0}^{\infty} ka^k = \frac{a}{(1-a)^2}, |a| < 1, \qquad \text{(C.6.3)}$$

$$\sum_{k=0}^{n} k = \frac{n(n+1)}{2}, \qquad \text{(C.6.4)}$$

$$\sum_{k=0}^{n} k^2 = \frac{n(n+1)(2n+1)}{6}. \qquad \text{(C.6.5)}$$

$$e^x = 1 + x + \frac{x^2}{2!} + \frac{x^3}{3!} + \dots, -\infty < x < \infty, \quad \text{(C.7.2)}$$

$$\ln(1+x) = x - \frac{x^2}{2} + \frac{x^3}{3} -, \dots, -1 < x \le 1, \text{(C.7.3)}$$

$$\sin(x) = x - \frac{x^3}{3!} + \frac{x^5}{5!} -, \dots, -\infty < x < \infty, \quad \text{(C.7.4)}$$

$$\cos(x) = 1 - \frac{x^2}{2!} + \frac{x^4}{4!} - \frac{x^6}{6!} +, \dots, -\infty < x < \infty, \quad \text{(C.7.5)}$$

$$\tan(x) = x + \frac{x^3}{3} + \frac{2x^5}{15} + \dots, |x| < \frac{\pi}{2}, \qquad \text{(C.7.6)}$$

C.7 Series Expansions

$$(1+x)^n = 1 + nx + \frac{n(n-1)}{2!}x^2 + \dots, \qquad \text{(C.7.1)}$$

$$\sin^{-1}(x) = x + \frac{x^3}{2(3)} + \frac{1(3)}{2(4)}\frac{x^5}{5} + \frac{1(3)(5)}{2(4)(6)}\frac{x^7}{7}$$
$$+, \dots, |x| < 1, \qquad \text{(C.7.7)}$$

$$\cos^{-1}(x) = \frac{\pi}{2} - \sin^{-1}(x), |x| < 1, \qquad \text{(C.7.8)}$$

$$\tan^{-1}(x) = \begin{cases} x - (x^3/3) + (x^5/5) - (x^7/7) + \dots, |x| < 1 \\ \pm(\pi/2) - (1/x) + (1/3x^3) - (1/5x^5) + \dots, [+\text{if } x \ge 1, - \text{ if } x \le -1] \end{cases}. \qquad \text{(C.7.9)}$$

C.8 Special Constants and Factorials

$$\pi = 3.14159\dots, \qquad \text{(C.8.1)}$$

$$e = 2.71828\dots, \qquad \text{(C.8.2)}$$

$$1 \text{ rad } = 180°/\pi, \qquad \text{(C.8.3)}$$

$$n! = 1.2.3, \dots, n, ! = 1. \qquad \text{(C.8.4)}$$

Bibliography

Abramowitz, M. and I. A. Stegun (eds.) (1964) *Handbook of Mathematical Functions with Formulas, Graphs, and Mathematical Tables,* National Bureau of Standards Applied Mathematics Series 55, U.S. Government Printing Office, Washington, D.C.

Ahmad N. and K. R. Rao (1975) *Orthogonal Transforms for Digital Signal Processing,* Springer-Verlag, New York.

Allen, J. B. (1977) Short-time Spectral Analysis, Synthesis, and Modification of Discrete Fourier Transform, *IEEE Trans. ASSP,* ASSP-25, pp. 235–38.

Ayres, F. (1962) *Theory and Problems of Matrices,* Shaum Publishing, New York.

Ambardar, A. (1995) *Analog and Digital Signal Processing,* PWS Foundations in Engineering Series, Boston.

Ambardar, A. (2007) *Digital Signal Processing,* Thompson, Toronto, Canada.

Baher, H. (1990) *Analog & Digital Signal Processing,* Wiley, New York.

Bainter, J.R. (1975) Active Filter Has Stable Notch, and Response Can be Regulated, *Electronics,* pp. 115–17.

Balbanian, N., T. A. Bickart, and S. Seshu (1969) Electrical Network Theory, Wiley, New York.

Barna, A. (1971) *Operational Amplifiers,* Wiley, New York.

Beckmann, P. (1971) *A History of π,* St. Martin's Press.

Blinchikoff, H. J. and A. I. Zverev (1976), *Filtering in the Time and Frequency Domain,* John Wiley, New York.

Bode, H. (1945) *Network Analysis and Feedback Amplifier Design,* Van Nostrand Reinhold, New York.

Bracewell, R. N. (1986), *The Hartley Transform,* Oxford University Press, New York.

Brogan, W. L. (1991) *Modern Control Theory,* Prentice-Hall, Englewood Cliffs, NJ.

Budak, A. (1974) *Passive and Active Network Analysis and Synthesis,* Houghton Mifflin, Boston, MA.

Burrus, C. S, J. H. McClellan, A. V. Openheim, T. W. Parks, R. W. Schafer, and H. W. Schuessler, (1994) *Computer Based Exercises for Signal Processing Using MATLAB,* Prentice-Hall, Englewood Cliffs, NJ.

Bird, G. (1974) *Radar Precision and Resolution,* Wiley, New York.

Bird, R. H. and D. A. Payne (1979) Convergence of the Iteratively Reweighted Least Squares Algorithm for Robust Regression, The John Hopkins University, Baltimore, MD, Technical Report 313.

Cadzow, J. A. (1973) *Discrete-time Systems,* Prentice-Hall, Englewood Cliffs, NJ.

Cadzow, J. A. (1987) *Foundations of Digital Signal Processing and Data Analysis,* Macmillan, New York.

Carlson, A. B. (1975) *Communication Systems,* McGraw-Hill, New York.

Carlson, A. B. (2000) *Circuits,* Brooks/Cole, Pacific Grove, CA.

Carlson, A. B. (1998) *Signal and Linear System Analysis,* John Wiley, New York.

Carlson, A. B., P. B. Crilly and J. C. Rutledge, (2002) *Communication Systems,* McGraw-Hill, New York.

Carslaw, H. S. (1950) *An Introduction to the Theory of Fourier's series and Integrals,* Dover, New York.

Cartinhour, J. (2000) *Digital Signal Processing,* Prentice-Hall, Englewood Cliffs, NJ.

Cauer, W. (1958) *Synthesis of Linear Communication Networks,* McGraw-Hill, New York, New York.

Chapman, S. J. (2005) *Essentials of MATLAB® Programming,* Thompson, Toronto, Canada.

Childers, D. and A. Durling (1975) *Digital Filtering and Signal Processing,* West Publishing, New York.

Christian, E. and E. Eisenmann (1977) *Filter Design Tables and Graphs,* Transmission Networks International, Inc. Knightdale, N.C.

Chuanyi, Z. (2003) *Almost Periodic Type Functions and Ergodicity,* Science press/Kluwer Academic Publishers, Boston, MA.

Churchill, R. V. (1958) *Operational Mathematics,* McGraw-Hill, New York.

Churchill, R. V. (1948) *Introduction to Complex Variables and Applications,* McGraw-Hill, New York.

Cioffi, J. M. and Y. S. Byun (1993) Adaptive filtering in *Handbook for Digital Signal Processing* by S. K. Mitra and J. F. Kaiser, Wiley, New York.

Close, C. M. (1966) *The Analysis of Linear Circuits,* Harcourt, Brace & World, New York.

Cooley, J. W. and J.W Tukey (1965) An Algorithm for the Machine Calculation of Complex Fourier Series, *Math. Computation,* Vol. 19, pp. 297–301.

Cooper, G. R. and C. D. McGillem (1999) *Probabilistic Methods of Signal and System Analysis,* Oxford University, New York.

Couch II, L. W. (1997) *Digital and Analog Communication Systems,* Prentice-Hall, Englewood Cliffs, NJ.

Dahlquist, A and A. Bjorck (translated by N. Anderson), (1974) *Numerical Methods,* Prentice-Hall, Englewood Cliffs, NJ.

Daniels, R. W. (1974) *Approximation Methods for Electronic Filter Design,* McGraw-Hill, New York.

Daryanani, G. (1976) *Active and Passive Network Synthesis*, Wiley, New York.

DeFatta, D. J., J. G. Lucas, and W. S. Hodgkiss (1988) *Digital Signal Processing, a System Design Approach*, Wiley, New York.

Deliyannis, T, Y. Sun and J. K. Fidler (1999) *Continuous-time Active Filter Design*, CRC Press, Boca Raton, Florida.

Deliyannis, T. (1969) *RC* Active All-pass Sections, *Electronics Letters*, Vol. 5, pp. 59–60.

DiStefano, III, J, A. R. Stubberud, and I. J. Williams (1990) Schaum's Outline Series, *Feedback and Control* Systems, McGraw-Hill, New York.

D'Souza, A. F. (1988) *Design of Control Systems*, Prentice-Hall, Englewood Cliffs, NJ.

Etter, D. M. (1993) *Engineering Problem Solving with MATLAB*, Prentice Hall, Englewood Cliffs, NJ.

Etter, D. M., D. C. Kuncicky with D. Hull (2002) Introduction to *MATLAB 6*, Prentice Hall, Englewood Cliffs, NJ.

Fourier, J. B. J. (1955) *The Analytical Theory of Heat*, A. Freeman, translation, Dover, New York.

Frame, J. S. (1964) Matrix Functions and Applications, *IEEE Spectrum*, Part I, March, pp. 208–20, Part II, April, pp. 102–108; Part III, May, pp. 100–109; Part IV, June, pp. 123–31, Part V, July, pp. 103–109.

Gantmacher, F. R. (1960) *The Theory of Matrices*, Volume 1, Chelsea Publishing, New York.

Gardner, F. M. (1966) *Phase Lock Techniques*, Wiley, New York.

Geher, K. (1971) *Theory of Network Tolerances*, Academiai Kiado, Budapest, Hungary.

Gibson, J. T. (1993) *Principles of Digital and Analog Communications*, Macmillan, New York.

Gibson, J. T. (1996) *The Communications Handbook*, CRC Press, Boca Raton, Florida.

Gilbert G. and R. Hatcher (2000) *Mathematics beyond the Numbers*, Wiley, New York.

Gold, B. and C. M. Rader with A. V. Oppenheim and T. G. Stockham, Jr. (1969) *Digital Processing of Signals*, McGraw-Hill, New York.

Gradshteyn, I. S. and I. M. Ryzhik (1980) *Table of Integrals, Series, and Products*, Academic Press, New York.

Grenander, U. and G. Szego (1958) *Toeplitz Forms and Their Applications*, University of California Press, Berkeley. CA.

Graybill, F. A. (1983) *Matrices with Applications in Statistics*, Wadsworth International Group, Belmont, CA.

Grupa, D. (1976) *Identification of systems*, Krieger Publishing, Huntington, New York.

Hahn, S. L. (1996) *Hilbert Transforms* in *The Transforms and Applications Handbook*, CRC Press, Boca Raton, FL.

Halfin, S. (1970) An Optimization Method for Cascade Filters, *Bell System Tech. J.* Vol. 44, pp. 185–90.

Hamming, R. W. (1988) *Digital Filters*, Prentice-Hall, Englewood Cliffs, NJ.

Harris, F. J. (1978) On the Use of Windows for Harmonic Analysis with the Discrete Fourier Transform, *Proceedings of the IEEE*, Vol. 66, pp. 51–3.

Hassan, A. A., J. Hershey, J. Schroeder, Guy R. L. Sohie, and R. K. Rao Yarlagadda (1994) *The Elements of System Design*, Academic Press, San Diego, CA.

Hawking, S. (2005) *God created the Integers, Running Press,* Philadelphia.

Haykin, S. (1989) *An Introduction to Analog and Digital Communications*, Wiley, New York.

Haykin, S. and B. Van Veen (1999) *Signals and Systems,* Wiley, New York.

Haykin, S. (2001) *Communication Systems*, Wiley, New York.

Haykin, S. and B. Van Veen (2002) *Signals and Systems*, John Wiley, New York.

Hershey, J. E. and R. K. Yarlagadda (1986) *Data Transportation and Protection,* Plenum, New York.

Hilberman, D. (1973) An Approach to the Sensitivity and Statistical variability of Biquadratic Filters, *IEEE Trans. Circuit Theory*, CT-20, No. 4, pp. 382–90.

Hohn, F. E. (1958) *Elementary Matrix Algebra*, Macmillan, New York.

Horn, R. A. and C. R. Johnson (1990) *Matrix Analysis*, Cambridge University Press, Cambridge.

Houts, R. C. and O. Alkin (1999) *Signal Analysis in Linear Systems*, Sanders, Philadelphia, PA.

Hsu, H. P. (1967) *Fourier Analysis*, Simon and Schuster Tech Outlines.

Hsu, H. P. (1995) *Signals and Systems*, Schaum's Outline Series, McGraw-Hill, New York.

International Telecommunication Union (1984) CCITT *Red Book*, Volume VI. Fascicle VI.1.

Ifeachor, E. C. and B. W. Jervis (1993) *Digital Signal Processing*: A Practical Approach, Addison-Wesley.

Ingle, V. K. and J. G. Proakis (2007) *Digital Signal Processing using MATLAB*, Thomson, Toronto, Canada.

Jackson, L. B. (1970) On the Interaction of Round-off Noise and Dynamic Range in Digital Filters, *Bell system Technical Journal*, Vol. 49, pp. 159–84.

Jansson, P. S. (1984) *Deconvolution*, Academic Press, Orland, Florida.

Jeffrey, R. L. (1956) *Trigonometric series,* University of Toronto Press, Toronto.

Johnson, D. E. (1976) *Introduction to Filter Theory*, Prentice-Hall, Englewood Cliffs, NJ.

Kaiser, J. F. (1966) *Digital Filters*, F. K. Kuo and J. F. Kaiser (eds.), *System Analysis by Digital Computer*, Wiley, New York.

Karni, S. (1966) *Network Theory: Analysis and Synthesis*, Allyn and Bacon, Boston, MA.

Korn, G.A. and T. M. Korn (1961) *Mathematical Handbook for Scientists and Engineers*, McGraw-Hill, New York.

Kuh, E. S. and R. A. Rohrer (1967) *Theory of Linear Active Networks*, Holden-Day, San Francisco.

Kuo, B. C. (1987) *Automatic Control Systems*, Prentice-Hall, Englewood Cliffs, New Jersey.

Lathi, B. P. (1983) *Modern Digital and Analog Communication Systems,* Oxford University Press.

Lathi, B. P. (1998) *Signal Processing & Linear Systems*, Berkeley Cambridge Press, Carmichael, CA.

Lee, T. H. (2004) *The Design of CMOS Radio-Frequency Integrated Circuits*, Cambridge University Press, New York.

Lighthill M. J. (1958) *An Introduction to Fourier Analysis and Generalized Functions,* Cambridge University Press, Cambridge, England.

Ludeman, L. C. (1986) *Fundamentals of Digital Signal Processing*, Harper & Row, New York.

Lueder, E. (1970) A Decomposition of a Transfer Function Minimization Distortion and in Band Losses, *Bell System Tech. J.*, Vol. 49, pp. 455–569.

Lyons, R. G. (2004) *Understanding Digital Signal Processing*, Prentice Hall, Englewood Cliffs, New Jersey.

Lyons, R. G. and A. Bell (2004) The Swiss Army Knife of Digital Networks, *IEEE Signal Processing Magazine*, Vol. 21(3), pp 90–97.

Marple, S. L. (1987) *Digital Spectral Analysis*, Prentice-Hall, Englewood Cliffs, New Jersey.

McCollum, P. A. and B. F. Brown (1965) *Laplace Transform Tables and Theorems*, Holt, Rhinehart and Winston, New York.

McClellan, R. W. Schafer and M. A. Yoder (2003) *Signal Processing First*, Pearson prentice-Hall, Upper Saddle River, New Jersey.

McGillem, C. D. and G. R. Cooper (1991) *Continuous & Discrete Signal and System Analysis*, Saunders College Publishing, Philadelphia, PA.

Melsa, J. L. and D. G. Schultz (1969) *Linear Control Systems*, McGraw-Hill, New York.

Mennie, D. (1978) AM Stereo: Five Competing Options, *IEEE Spectrum*, Vol. 18, pp. 56–58.

Mitra, Sanjit K. (1969) *Analysis and Synthesis of Linear Active Networks*, Wiley, New York.

Mitra, Sanjit K. (2006) *Digital Signal Processing*, McGraw-Hill, New York.

Money, A. H. (1982) The Linear Regression Model: L_p Normed Estimation and the Choice of p, *Commun. Statist. Simulation Computation*, Vol. 11, pp. 89–109.

Morrison, N (1994) *Introduction to Fourier Analysis*, John Wiley, New York.

Moon, T. K. and W. C. Stirling (2000) *Mathematical Methods and Algorithms for Signal Processing*, Prentice Hall, Englewood Cliffs, New Jersey.

Moschytz, G. S. and P. Horn (1981) *Active Filter Design Handbook*, Wiley, New York.

Nilsson, J. W. and S. A. Riedel (1966) *Electric Circuits*, Addison-Wesley, Reading, MA.

Nise, N. S. (1992) *Control Systems Engineering*, Benjamin/ Cummings Publishing, Redwood City, CA.

Ogata, K. (1961) *State Space Analysis of Control Systems*, Prentice-Hall, Englewood Cliffs, NJ.

Ogata, K. (2004) *System Dynamics*, Pearson/Prentice Hall, Upper Saddle River, NJ

Olejniczak, K. J. (1996) *The Hartley Transform*, in *The Transforms and Applications Handbook*, Edited by A. D. Poularikas, CRC Press, Boca Raton, FL.

Oppenheim, A. V. and R. W. Schafer (with J. R. Buck) (1999) *Discrete-Time Signal processing*, Prentice-Hall, Englewood Cliffs, New Jersey.

Oppenheim, A.V. and A. S. Willsky, with S. H. Nawab (1997) *Signals & Systems*, Prentice-Hall, Englewood Cliffs, New Jersey.

O'Shaughnessy, D. (1987) *Speech Communication*, Addison-Wesley, Reading, MA.

Otnes, R. K. and L. Enochson (1972) *Digital Time Series Analysis*, Wiley, New York.

Palm III, W. J. (2001) *Introduction to MATLAB 6 FOR ENGINEERS*, McGraw-Hill, New York.

Papoulis, A. (1955) Displacement of the Zeros of the Impedance $Z(p)$ due to an Incremental Variation in the Network Elements, *Proceedings IRE*, 43, pp 79–82.

Papoulis, A. (1962) *The Fourier Integral and its Applications*, McGraw-Hill, New York.

Papoulis, A. (1977) *Signal Analysis*, McGraw-Hill, New York.

Pearl, M. (1973) *Matrix Theory and Finite Mathematics*, McGraw-Hill, New York.

Peebles, P. Z. (2001) *Probability Random Variables, and Random Signal Principles*, McGraw-Hill, New York.

Perlis, S. (1958) *Theory of Matrices*, Addison-Wesley, Reading, MA.

Peterson, G. and H. Barney (1952) Control Methods Used in Study of Vowels, *J. Acoust. Soc. Am.* Vol. 24, pp. 175–84.

Poularikas, A. D. (1996) *The Transforms and Applications Handbook*, CRC Press, Boca Raton, FL.

Poularikas, A. D. and S. Seely (1990) *Signals and Systems*, PWS-Kent, Boston, MA.

Pozar, D. M. (1998) *The Microwave Engineering*, John Wiley, New York.

Press, W. H., B. P. Flannery, S. A. Teukolsky and W. T. Vetterling (1990) *Numerical Recipes*, Cambridge University press, Cambridge.

Proakis, J. G. and D. G. Manolakis (1988) *Introduction to Digital Signal processing*, MacMillan, New York.

Rabiner L. R. and B. Gold (1975) *Theory and Application of Digital Signal Processing*, Prentice-Hall, Englewood Cliffs, NJ.

Rabiner L. R. and R. W. Schafer (1978) *Digital Processing of Speech Signals*, Prentice-Hall, Englewood Cliffs, NJ.

Ramirez, R. W. (1975) The FFT: Fundamentals and Concepts, Tektronix, Inc., Beaverton, OR.

Rao, C. R. and Sujit K. Mitra (1971) *Generalized Inverse of Matrices and its Applications*, John Wiley, New York.

Roberts, M. J. (2007) *Fundamentals of Signals and Systems*, McGraw-Hill, New York.

Robinson, E. A. (1967) Multi channel Time Series Analysis with Digital Computer Programs, Holden-Day, San Francisco.

Robinson, E. A. and S. Treitel (1980) *Geophysical Signal Analysis*, Prentice-Hall, Englewood Cliffs, NJ.

Roden, M. A. (1991) *Analog and Digital Communications Systems*, Prentice-Hall, Englewood Cliffs, NJ.

Rosen K. H. (1984) *Elementary Number Theory and its Applications*, Addison-Wesley, Reading, MA.

Sallen R. P. and E. L. Key (1955) A Practical Method of Designing *RC* Active Filters, *IRE Trans. on Circuit Theory*, *CT-2*, pp. 74–85.

Scharf, L. (1991) *Statistical Signal Processing Detection, Estimation and Time Series Analysis, Addison-Wesley, Reading Massachusetts.*

Scheid, F. (1968) *Theory and Problems of Numerical analysis*, Schaum's Outline Series, McGraw-Hill, New York.

Shenoi, K. (1995) *Digital Signal Processing in Telecommunications*, Prentice Hall, Englewood Cliffs, New Jersey.

Schroeder, J. and R. Yarlagadda, Linear Predictive Spectral Estimation via the L_1 norm. *Signal Processing*, Vol. 17, May 1989, pp. 19–29.

Scott, N. R. (1960) *Analog and Digital Computer Technology*, McGraw-Hill, New-York.

Sedra, A. S. and P. O. Bracket (1978) *Filter Theory and Design*: *Active and Passive*, Pitman, London.

Semmelman, C. L., E. D. Walsh, and G. Daryanani (1971) Linear Circuits and Statistical Design, *Bell System Tech. J.*, Vol. 50, pp. 1149–171.

Seshu, S. and N. Balabanian (1959) *Linear Network Analysis*, John Wiley & Sons, New York.

Simpson, R. S. and R. C. Houts (1971) *Fundamentals of Analog and Digital Communication Systems*, Allyn and Bacon, Boston.

Sipress, J. M. (1960) Synthesis of Active *RC* Networks, *Doctoral Dissertation*, Polytechnic Institute of Brooklyn, New York.

Sklar, B. (2001) *Digital Communications*, Prentice-Hall, Englewood Cliffs, NJ.

Smith III, J. O. (2007) *Mathematics of the Discrete Fourier Transform (DFT)*, W(3)K Publishing, CA.

Smith, S. W. (2002) *Digital Signal Processing: A Practical Guide for Engineers*, California Technical Publishing, San Diego, CA.

Spiegel, M. R. (1965) *Theory and Problems of Laplace Transforms*, Schaum's outline series, McGraw-Hill, New York.

Spiegel, M. R. (1968) *Mathematical Handbook*, Schaum's outline series, McGraw-Hill, New York.

Spilker, J. J. (1977) *Digital Communications by Satellite*, Prentice-Hall, Englewood Cliffs, NJ.

Stanley, W. D., G. R. Dougherty and R. Dougherty (1984) *Digital Signal Processing*, Reston, Reston, Virginia.

Stern H. P. E. and S. A. Mahmoud (2004) *Communication Systems*, Prentice-Hall, Englewood Cliffs, NJ.

Stimpson, G. W. (1983) *Introduction to Airborne Radar*, Hughes Aircraft Company.

Stine, J. E. (2003) *Digital Computer Arithmetic Datapath Design using Verilog HDL*, Kluwer Academic Press, Boston, MA.

Stone, M. L., P. R. Armstrong, X. Zhang, G. H. Brusewitz, and D. D. Chen, (1996) Watermelon Maturity Determination in Field Using Acoustic Impulse Response, *American Society of Agricultural Engineers*, Vol. 39(6), pp. 2325–330.

Storch, L. (1954) Synthesis of Constant-Time-Delay Ladder Networks Using Bessel Polynomials. *Proceedings IRE*, Vol. 42, no.11, pp. 1666–675.

Storer, J. E. (1957) *Passive Network Synthesis*, McGraw-Hill, New York.

Stremler, F. G. (1990) *Introduction to Communication Systems*, Addison-Wesley, Reading, MA.

Strum, R. D. and D.E. Kirk (1988) *First Principles of Discrete Systems and Digital Signal Processing*, Addison-Wesley, Reading, Massachusetts.

Swartzlander, Jr., E. E., (2001) *Computer Arithmetic, in Computer Engineering Handbook*, Edited by V. Oklobdzija, CRC Press, Boca Raton, Florida.

Tarantola, A. (1987) *Inverse Problem Theory*, Elsevier, Amsterdam, Netherlands.

Taub, H. and D. L. Schilling (1971) *Principles of Communications Systems*, McGraw-Hill, New York.

Tellegen, B. D. H. (1948) The Gyrator: A New Network Element, *Phillips Res. Rept.*, 3, pp. 81–101.

Temes, G. C. and S. K. Mitra (1973) *Modern Filter Theory and Design*, John Wiley & Sons, New York.

Thaler, G. H. and R. G. Brown (1962) *Analysis and Design of Feedback Control Systems*, McGraw-Hill, New York.

Thomas, L.C. (1971) The Biquad: Part I-some Practical Design considerations, *IEEE Trans. on Circuit Theory* CT-18, pp. 350–357.

Thomas, L. C. (1971) The Biquad: Part II-a Multi Purpose Active Filtering System, *IEEE Trans. on Circuit Theory* CT-18, pp. 358–61.

Tolstov, G. P. (Translated by R. A. Silverman) (1962) *Fourier series*, Dover, New York.

Tomovic, R. and M. Vukobratovic (1972) *General Sensitivity Theory*, Elsevier, New York.

Tou J. and R. C. Gonzalez (1974) Pattern *Recognition Principles*, Addison-Wesley Publishing Company, Reading, MA.

Tow, J. (1968) Active RC filters-A State Space Realization, *Proceedings IEEE* Vol. 56, pp. 1137–139.

Tribolet, J. M. (1979) *Seismic Applications of Homomorphic Signal Processing*, Prentice-Hall, Englewood Cliffs, NJ.

Truxal, J. G. (1955) *Automatic Feedback Control System Synthesis*, McGraw-Hill, New York.

Van Valkenburg, M. E. (1982) *Analog Filter Design*, Holt, Rinehart and Winston, New York.

Van Valkenburg, M. E. (1960) *Introduction to Modern Network Synthesis*, Wiley, New York.

Wagner, J. N. (1983) Inspecting the "impossible," *Food Engineering*, Vol. 55, No. 6.

Weinberg, L. (1962) *Network Analysis and Synthesis*, McGraw-Hill, New York.

Whittaker, E. T. and G.N. Watson (1927) *A Course of Modern Analysis*, Cambridge University Press, London, England.

Wilf, H. S. (1986) *Algorithms and Complexity*, Prentice-Hall, Inc., Englewood Cliffs, NJ.

Wilkinson, J. H. (1965) *The Algebraic Eigenvalue Problem*, Clarendon Press, Oxford.

Yarlagadda, R. and J. Allen (1982) Aliasing Errors in Short-Time Analysis, *Signal Processing* 4, pp. 79–84.

Yarlagadda, R., J. B. Bednar, and T. Watt (1985) Fast Algorithms for L_p, deconvolution, *IEEE Transactions on Acoustics, Speech and Signal Processing*, Vol. 33, pp. 174–82.

Yarlagadda, R. and J. Hershey (1985) A Naturalness-Preserving Transform for Image Coding and Reconstruction, *IEEE Trans. on Acoustics, Speech and Signal Processing*, Vol.33, pp. 1005–1012.

Yarlagadda, R. and J. Hershey (1997) *Hadamard Matrix Analysis and Synthesis with Applications to Communications*, Klewer Academic Press, New York.

Yates, R. D. and D. J. Goodman (1999) *Probability and Statistic Processes*, John Wiley, New York.

Yip, P. (1996) *Sine and Cosine Transforms*, in *Transforms and Applications Handbook*, Edited by A. D. Poularikis, CRC Press, Boca Raton, Florida.

Zaguskin, V. L. (1961) *Handbook of Numerical Methods for Solution of Equations*, Macmillan, New York.

Ziemer, R. E. and W. H. Tranter (2002) *Principles of Communications*, Wiley, New York.

Ziemer, R. E. and R. L. Peterson (2002) *Introduction to Digital communication*, Prentice-Hall, Englewood Cliffs, NJ.

Zverev, A. I. (1967) *Handbook of Filter Synthesis*, Wiley, New York.

Zygmund, A. (1955) *Trigonometrical Series*, Dover, New York.

Author Index

Subject Index

Printed in the United States of America